COUNTY COLLEGE OF MORRIS

EXPOSURE ANALYSIS

Edited by

Wayne R. Ott
Anne C. Steinemann
Lance A. Wallace

Taylor & Francis
Taylor & Francis Group
Boca Raton London New York

CRC is an imprint of the Taylor & Francis Group,
an informa business

CRC Press
Taylor & Francis Group
6000 Broken Sound Parkway NW, Suite 300
Boca Raton, FL 33487-2742

© 2007 by Taylor & Francis Group, LLC
CRC Press is an imprint of Taylor & Francis Group, an Informa business

No claim to original U.S. Government works
Printed in the United States of America on acid-free paper
10 9 8 7 6 5 4 3 2 1

International Standard Book Number-10: 1-56670-663-7 (Hardcover)
International Standard Book Number-13: 978-1-56670-663-6 (Hardcover)

This book contains information obtained from authentic and highly regarded sources. Reprinted material is quoted with permission, and sources are indicated. A wide variety of references are listed. Reasonable efforts have been made to publish reliable data and information, but the author and the publisher cannot assume responsibility for the validity of all materials or for the consequences of their use.

No part of this book may be reprinted, reproduced, transmitted, or utilized in any form by any electronic, mechanical, or other means, now known or hereafter invented, including photocopying, microfilming, and recording, or in any information storage or retrieval system, without written permission from the publishers.

For permission to photocopy or use material electronically from this work, please access www.copyright.com (http://www.copyright.com/) or contact the Copyright Clearance Center, Inc. (CCC) 222 Rosewood Drive, Danvers, MA 01923, 978-750-8400. CCC is a not-for-profit organization that provides licenses and registration for a variety of users. For organizations that have been granted a photocopy license by the CCC, a separate system of payment has been arranged.

Trademark Notice: Product or corporate names may be trademarks or registered trademarks, and are used only for identification and explanation without intent to infringe.

Library of Congress Cataloging-in-Publication Data

Exposure analysis / edited by Wayne R. Ott, Anne C. Steinemann, Lance A. Wallace.
    p. cm.
  Includes bibliographical references and index.
  ISBN 1-56670-663-7 (alk. paper)
  1. Biological monitoring. 2. Environmental monitoring. I. Ott, Wayne. II. Steinemann, Anne C. III. Wallace, Lance A.

RA1223.B54E97 2006
615.9'02--dc22
                                                  2006043890

Visit the Taylor & Francis Web site at
http://www.taylorandfrancis.com

and the CRC Press Web site at
http://www.crcpress.com

RA
1223
B54
E97
2007

Foreword

Our greatest progress in improving public health has occurred in those environmental regulatory programs most closely based on accurate measurements of human exposure. Probably no single topic in the environmental sciences is more important than exposure analysis. Yet exposure science has not always received the attention warranted. Knowing quantitatively what people are exposed to, by how much, when, and where it comes from is the centerpiece of exposure analysis. Only by accurately determining the sources of exposure can we protect public health by reducing exposure. As measurements of population body burden continue to show elevated concentrations of many toxic pollutants in humans, reasonable questions will be asked about the sources of these pollutants. Exposure analysis provides the basic tools and methods needed to identify the sources, understand the causes, alter the exposures, and track the changes over time. This book is dedicated to the development of exposure analysis as a new scientific field in its own right. That is, exposure analysis is not just a collection of related interdisciplinary approaches; it is a field unto itself. This book takes an important first step by identifying the building blocks of this new field.

Preface

Exposure analysis is at the heart of methods to protect public health from the harmful effects of pollution. Though not well developed when early environmental laws that affect exposure were written, exposure analysis has become a rapidly growing science. Perhaps the most important contribution of programs dealing with environmental pollution has been the initiation and development of this new field of exposure analysis as a formal science.

This book introduces the reader to the science of exposure analysis. It is assumed that the reader has at least a college-level understanding of mathematics and chemistry but may not be familiar with the methods of exposure analysis. Due to the breadth of exposure science as a field, this book is intended for a wide audience, including environmental engineers, public health officials, environmental scientists, risk assessors, indoor air quality specialists, government regulators, and members of the public interested in environmental science. The book is also relevant to disciplines concerned with occupational health, industrial hygiene, toxicology, epidemiology, statistics and biostatistics, and environmental engineering.

When Stanford University began teaching its Introduction to Human Exposure Analysis course in the Department of Civil and Environmental Engineering, no classroom textbook was available. To help fill this need, a course reader on exposure analysis was developed. This new course was open to graduate and undergraduate students of all disciplines, and it included a laboratory component in which students form small groups and carry out experimental laboratory studies using state-of-the-art exposure monitors. After a decade of teaching this course, the authors of many of the scientific papers used in the course were invited to contribute chapters to this book. Every year the students provided written comments on the chapters, and the student feedback was used to revise the chapters and improve their clarity.

Planning a textbook requires decisions about its level and content — whether to organize the book by pollutant groups, such as dioxin and pesticides, or by general methods, such as measurement techniques. This book combines the two approaches, building most of the chapters upon the individual pollutants and pollutant classes, which are grouped by air, dermal, ingestion, and multimedia routes of exposure. The book's structure is outlined briefly below.

Chapter 1, Exposure Analysis: A Receptor-Oriented Science, introduces an important principle of exposure analysis: it often begins by measuring exposure — what people actually breathe, eat, drink, or have skin contact with — and then moves backward to find the true sources of that exposure. Thus, it follows a receptor-to-source (receptor-oriented) approach as well as the more traditional source-to-receptor (source-oriented) approach. The receptor-oriented approach has demonstrated its effectiveness by discovering many new sources of exposure that previously were overlooked but are important for protecting public health. The receptor-oriented approach is not unique to exposure analysis; it also is a fundamental part of the forensic process of criminal investigations, as illustrated by Agatha Christie's Monsieur Hercules Poirot, who discovers a body and then works backward to find the causes (persons responsible) for the death.

Chapter 2, Basic Concepts and Definitions of Exposure and Dose, summarizes the existing definitions of exposure, dose, and related concepts. It also provides a mathematical framework that is at the heart of these conceptual definitions. The glossary included at the end of Chapter 2 presents definitions of common terms used in the field, which are consistent with the definitions adopted by the International Society of Exposure Analysis (ISEA) and other international groups.

Chapter 3, Probability-Based Sampling for Human Exposure Assessment Studies, discusses the statistical methods used in human exposure measurement field studies. The science of exposure analysis has made a major contribution to the older science of environmental monitoring by conducting large-scale population exposure field studies that have clearly stated objectives, good statistical protocols, and well-planned probability sampling designs.

Chapter 4, Inhalation Exposure, Uptake, and Dose, introduces a collection of chapters on human exposure through the air carrier medium (Chapter 4 through Chapter 10). Chapter 4 reviews the physical and physiological processes affecting the movement of pollutants in air and their transfer to the human respiratory system.

Chapter 5, Personal Monitors, describes small, portable devices that people can wear or carry with them to measure their exposure to air pollutants. Many personal monitors are *real-time* instruments that can measure and record data on a minute-by-minute basis. Other types (active samplers) use a small pump that the person wears along with a collection filter, while still others (passive samplers) have no pump but rely on diffusion to collect a sample.

Chapter 6, Exposure to Carbon Monoxide (CO), shows that ambient concentrations and personal exposures to CO have greatly decreased in the United States over several decades due to effective regulatory programs. Exposure to this pollutant is still widespread in other parts of the world, however. One important component of this successful environmental program has been the availability of excellent CO measurement methods for ambient, indoor, and personal exposures. The full story of exposure to CO and the scientific history of its decrease in the United States provide an important lesson in exposure analysis, because CO has been explored so thoroughly from an exposure standpoint.

Measurement studies using personal monitors have shown that Exposure to Volatile Organic Compounds (VOCs) (Chapter 7) is both significant and widespread. This class of thousands of pollutants is present in countless consumer products and building materials, providing typically two to five times as much exposure as the major outdoor sources of "air toxics" that are subject to ambient regulation. The more important indoor sources are largely unregulated. Exposure studies were able to pinpoint many of these sources, providing findings that may allow persons to reduce their exposure by removing sources from the home, changing their buying habits, or carrying out other lifestyle changes. However, it is important to convey the findings from these studies to the general public.

Chapter 8, Exposure to Particles, points out that the World Health Organization estimates that 1.6 million deaths a year are due to indoor air pollution, mostly in developing countries, from particles created from combustion during cooking and heating. An unanswered question is the relative influence of indoor and outdoor sources on exposure, and new ways to separate the contributions of indoor and outdoor sources to particle exposure are explained.

Chapter 9, Exposure to Secondhand Smoke, illustrates the effectiveness of exposure analysis studies in changing human behavior and improving public health. Although many cities, counties, states, and nations have acted to ban smoking in public places, these bans would not have occurred without the scientific measurements of concentrations in the locations where smokers and other people congregate.

Chapter 10, Intake Fraction, compares the relationship of sources to exposures when people are in indoor vs. outdoor locations. It shows that a pollutant released in a residence has odds of being inhaled that are typically 1000 times the odds of inhaling the same pollutant if released outdoors. The high efficiency of small indoor sources literally under our noses helps explain why studies find indoor pollutant concentrations are often very high.

Chapter 11, Dermal Exposure, Uptake, and Dose, introduces the reader to the emerging science of the dermal route of exposure and dose to toxic agents. It describes the factors affecting absorption by the human skin, the mechanisms of exposure, techniques for directly measuring dermal exposure, and techniques for measuring absorption. It includes a glossary of terms on dermal exposure that may be unfamiliar to some readers, especially biological terms about the human skin.

Chapter 12, Dermal Exposure to VOCs while Bathing, Showering, or Swimming, deals with an important aspect of dermal exposure — the absorption of volatile organic compounds while showering or bathing. It discusses the first set of experiments that were able to measure directly the dermal uptake of chloroform (a by-product of water treatment) while bathing.

As noted in Chapter 13, Ingestion Exposure, the gut (gastrointestinal system) is one of our three major barriers to absorbing environmental pollutants (the others being the lung and the skin). Calculating intake requires knowledge of the amount and kinds of food eaten or liquids drunk, gained from activity pattern studies, Market Basket surveys, USDA statistics, and similar sources. Also required is knowledge of the absorption, metabolism, and excretion of organic and inorganic substances.

Chapters 14 through 17 deal with exposure through multiple carrier media. Many pollutants, such as lead and polycyclic aromatic hydrocarbons, reach people through more than one carrier medium, such as air breathed and food eaten, and therefore they come under the category of *multimedia* exposure.

Chapter 14, Exposure to Pollutants from House Dust, summarizes studies on the concentrations of pollutants measured in the house dust of American homes. Adults and children can be exposed to pollutants in house dust by resuspension in the air, hand-to-mouth contact, ingestion, and skin contact. Pollutants in house dust often exceed the levels that would trigger a risk assessment at a Superfund site, and the causes of these elevated concentrations usually are common everyday sources, such as house paint in older homes. Effective vacuuming is one way to reduce the exposure of adults and children in homes, although efficient vacuuming requires an agitator brush and an embedded dirt finder.

Chapter 15 provides an overview of exposure to pesticides, including the use of personal monitors in pesticides exposure monitoring field studies and the findings from these studies. Although many pesticides reach people through the food they eat, air is the dominant carrier medium for exposure to many pesticides, especially indoor air.

Chapter 16 describes exposure to dioxin and dioxin-like compounds, which are multimedia pollutants that reach humans primarily through the food carrier medium. Although widespread in the environment, human exposure to dioxins in the United States has exhibited a decreasing trend over time.

Biomarkers of Exposure (Chapter 17) deals with measurements of pollutants in the body, particularly in blood, urine, and exhaled breath. A great advantage of biomarkers is that they can show immediately which pollutants are most important in terms of entering the human body. For example, two nationwide studies, one using biomarkers in exhaled breath and the other in blood, agreed in identifying a small number of about a dozen volatile organic chemicals (out of the thousands present in the environment) as the most prevalent in our bodies. Comparing body burden to measured levels of exposure can identify previously unsuspected routes of exposure—an example described in the chapter is the discovery that smokers receive close to 90% of their benzene exposures from mainstream cigarette smoke.

In addition to progress in measuring the chemical and physical properties of pollutants reaching people, there has been progress in the development of mathematical models to estimate and predict exposures. Because of the importance of indoor air quality in estimating human exposure to air pollutants, Chapter 18, Mathematical Modeling of Indoor Air Quality, describes quantitative methods that have been validated for predicting the concentrations in enclosed everyday locations, such as automobiles and rooms of the home. Modeling Human Exposure to Air Pollution (Chapter 19) describes quantitative techniques for predicting personal exposure to air pollution, and Models of Exposure to Pesticides (Chapter 20) reviews an emerging area of exposure modeling.

Finally, because of the policy implications of many of the scientific exposure studies, Chapter 21, Environmental Laws and Exposure Analysis, reviews how federal laws in the United States deal with human exposure to environmental pollution. The laws generally focus on large outdoor sources while underemphasizing smaller indoor sources of the same pollutants that cause greater exposure. Human exposure studies reveal that most of our exposure to toxic pollutants comes from common everyday sources that are close to us and are often overlooked, such as consumer products and building materials.

Accurate measurements of exposure are some of the most promising ways to help guide our environmental programs to protect and improve pubic health. These measurements allow us to focus more precisely than ever before on traditional sources as well as on important sources found close at hand in our homes and that are often overlooked, such as cleaning supplies, air fresheners, fragrances, paints, adhesives, chlorinated water, interior furnishings, attached garages, particle board, house dust, molds, pesticides, secondhand smoke, cooking activity, fireplaces, candles, incense, and the multitude of products we use daily in our lives. Great strides have been made in improving the science of exposure analysis, and its relevance for protecting health is so far reaching that it is hoped that exposure science will receive greater emphasis in future policies and laws.

Many of the authors of this book have made fundamental pioneering contributions to the field of exposure analysis. As editors and readers, we are fortunate to have their contributions to this textbook. We believe the chapters in this book are authoritative reviews and will be helpful to readers for a considerable length of time.

About the Editors

Wayne R. Ott has worked on developing new methods to measure, quantify, and predict human exposure to environmental pollutants for 4 decades. His Ph.D. research at Stanford University included a large-scale personal exposure monitoring field survey. He was an environmental engineer in the Commissioned Corps of the U.S. Public Health Service assigned to the USEPA's Office of Research and Development for 30 years. As team leader of USEPA's Air, Toxics, and Radiation Monitoring Research Staff, he coordinated scientific research on measurement methods, nationwide monitoring networks, exposure modeling, and human exposure field surveys. He received the 1995 Jerome J. Weselowski award for his contributions to the knowledge and practice of exposure analysis and the Public Health Service Commendation Medal for developing a nationally uniform air pollution index. Since coming to Stanford from USEPA in 1996, he has continued working on methods to measure and predict human exposure for a variety of common everyday sources, such as cigarettes, cigars, pipes, cooking, candles, incense, motor vehicles, and wood smoke. He conducted field measurements of exposure to secondhand smoke in California taverns, restaurants, gaming casinos, homes, and automobiles and has authored or co-authored over 100 journal articles, conference presentations, and research reports, including the CRC Press book, *Environmental Statistics and Data Analysis*. He is consulting professor, civil and environmental engineering, Stanford University, and visiting scholar at the department of statistics.

Anne C. Steinemann is Professor of Civil and Environmental Engineering, Professor of Public Affairs, and Director of The Water Center at the University of Washington. She received her Ph.D. in Civil and Environmental Engineering from Stanford University. Dr. Steinemann specializes in environmental impact assessment and regulatory policy, water resources management, hazard prediction and mitigation, and health effects of pollutants, combining expertise in engineering, economics, policy, and public health. She received the National Science Foundation CAREER Award in addition to university and national teaching awards. Dr. Steinemann has investigated more than 100 sick buildings to identify pollutant sources, reduce exposures, and improve occupants' health. She conducted the first national epidemiological study of chemical sensitivity, its causes and symptoms related to exposures, and its overlaps with asthma. Dr. Steinemann has directed more than $5 million of funded research, and serves as adviser to agencies and industries on environmental issues. Among her recent publications is the textbook, *Microeconomics for Public Decisions* (South-Western, 2005).

Lance Wallace has concentrated throughout his career on developing methods such as personal air quality monitors and breath analysis to measure human exposure to air pollution. Dr. Wallace, whose Ph.D. is in physics, conceived and implemented the USEPA TEAM studies of human exposure. These very large-scale studies used personal monitors carried by more than 2000 persons (selected to represent more than 2 million residents of a number of U.S. cities) to measure their exposure to volatile organic compounds, pesticides, carbon monoxide, and respirable particles. Many new unexpected sources of these pollutants were found, leading to an increased emphasis on personal behavior, consumer products, and building materials as sources of exposure. Dr. Wallace received a number of awards from the USEPA for these and other studies carried out during his 30-year career with the agency. He spent 2 years as a visiting scholar at Harvard University and has served as an advisor to a number of doctoral students at Harvard, Johns Hopkins, and the University of Washington. He has published 80 articles in peer-reviewed journals, one book and

15 book chapters, and more than 100 research reports and presentations at conferences. Dr. Wallace received the Jerome J. Weselowski award for lifetime contributions to human exposure and the Constance Mehlman award for scientific contributions affecting environmental policy from the International Society for Exposure Analysis (ISEA).

Acknowledgments

Grateful appreciation is given to the members of the Advisory Board of this book for providing thoughtful suggestions in the initial planning stages. The Advisory Board members were Prof. Richard L. Corsi (Department of Civil Engineering, University of Texas at Austin), Prof. Naihua Duan (University of California at Los Angeles), Prof. Peter G. Flachsbart (Department of Urban and Regional Planning, University of Hawaii at Manoa), Prof. Petros Koutrakis (Harvard School of Public Health, Harvard University), Profs. James Leckie and Lynn Hildemann (Department of Civil and Environmental Engineering, Stanford University), Prof. P. Barry Ryan (Rollins School of Public Health, and Department of Chemistry, Emory University), and Prof. Kirk R. Smith (University of California at Berkeley).

Grateful appreciation also is given to Rashida Basrai, Sabiha Basrai, Bayard Colyear, Karen Johnson, and Willa AuYeung for their contributions to the graphics and artwork of this book, to Dan Ribeiro and Deborah Livingstone for their editing of the chapters, and to Paloma Beamer for providing thoughtful comments and student input on the chapters.

Special appreciation is given to Prof. James Leckie for his contributions to dermal exposure science and for developing a pioneering course on the Introduction to Human Exposure Analysis at Stanford's Department of Civil and Environmental Engineering and to Valerie Zartarian, Robert Canales, Alesia Ferguson, Kelly Naylor, Andrea Ferro, Paloma Beamer, and Timothy Julian for teaching this new course.

Contributors

Harold A. Ball
Environmental Engineer
Superfund Technical Support
U.S. Environmental Protection Agency
  Region 9
San Francisco, California

Robert A. Canales
Yerby Postdoctoral Fellow
School of Public Health
Harvard University
Boston, Massachusetts

Naihua Duan
Professor
Psychiatry and Biostatistics
University of California Center for Community
  Health/Neuropsychiatric Institute
Los Angeles, California

Alesia C. Ferguson
Assistant Professor
College of Public Health
University of Arkansas for Medical Sciences
Little Rock, Arkansas

Andrea R. Ferro
Assistant Professor
Department of Civil and Environmental
  Engineering
Clarkson University
Potsdam, New York

Peter G. Flachsbart
Associate Professor
Department of Urban and Regional Planning
University of Hawai'i at Manoa
Honolulu, Hawai'i

Sydney M. Gordon
Exposure Research Leader
Battelle Memorial Institute
Columbus, Ohio

Lynn M. Hildemann
Associate Professor
Department of Civil and Environmental
  Engineering
Stanford University
Stanford, California

Neil E. Klepeis
Human Exposure Researcher
Department of Statistics
Stanford University
Stanford, California

James O. Leckie
Professor
Department of Civil and Environmental
  Engineering
Stanford University
Stanford, California

Robert G. Lewis
Environmental Research Chemist
U.S. Environmental Protection Agency (ret.)
Research Triangle Park, North Carolina

Julian D. Marshall
Postdoctoral Research Fellow
School of Occupational and Environmental
  Hygiene
University of British Columbia
Vancouver, British Columbia, Canada

William W. Nazaroff
Professor
Department of Civil and Environmental
  Engineering
University of California
Berkeley, California

Wayne R. Ott
Exposure Research Scientist
Department of Civil and Environmental
  Engineering and Department of Statistics
Stanford University
Stanford, California

James L. Repace
Visiting Assistant Clinical Professor
Tufts University School of Medicine and
  Repace Associates, Inc.
Secondhand Smoke Consultants
Bowie, Maryland

John W. Roberts
Exposure Research Engineer
Engineering-Plus, Inc.
Sammamish, Washington

P. Barry Ryan
Professor
Department of Environmental and
  Occupational Health
Rollins School of Public Health and
  Department of Chemistry
Emory University
Atlanta, Georgia

Kirk R. Smith
Professor of Environmental Health Sciences
Brian and Jennifer Maxwell Endowed Chair in
  Public Health
School of Public Health
University of California
Berkeley, California

Anne C. Steinemann
Professor
Civil and Environmental Engineering;
  Professor, Public Affairs; Director,
  The Water Center
University of Washington
Seattle, Washington

Daniel J. Stralka
Regional Toxicologist
Superfund Technical Support
U.S. Environmental Protection Agency
  Region 9
San Francisco, California

Lance A. Wallace
Environmental Research Scientist
U.S. Environmental Protection Agency (ret.)
Reston, Virginia

Nancy J. Walsh
Adjunct Professor
School of Law
Emory University
Atlanta, Georgia

Roy Whitmore
Manager
Environmental Epidemiology and Statistics
  Research Program
RTI International
Research Triangle Park, North Carolina

Valerie G. Zartarian
Research Environmental Engineer
U.S. Environmental Protection Agency Office
  of Research and Development
National Exposure Research Laboratory
Boston, Massachusetts

Table of Contents

PART I   *Exposure Concepts*

Chapter 1
Exposure Analysis: A Receptor-Oriented Science 3
Wayne R. Ott

Chapter 2
Basic Concepts and Definitions of Exposure and Dose 33
Valerie G. Zartarian, Wayne R. Ott, and Naihua Duan

Chapter 3
Probability-Based Sampling for Human Exposure Assessment Studies 65
Roy Whitmore

PART II   *Inhalation*

Chapter 4
Inhalation Exposure, Uptake, and Dose 81
Andrea R. Ferro and Lynn M. Hildemann

Chapter 5
Personal Monitors 99
Lance A. Wallace

Chapter 6
Exposure to Carbon Monoxide 113
Peter G. Flachsbart

Chapter 7
Exposure to Volatile Organic Compounds 147
Lance A. Wallace and Sydney M. Gordon

Chapter 8
Exposure to Particles 181
Lance A. Wallace and Kirk R. Smith

Chapter 9
Exposure to Secondhand Smoke 201
James L. Repace

Chapter 10
Intake Fraction 237
Julian D. Marshall and William W. Nazaroff

PART III   Dermal Exposure

Chapter 11
Dermal Exposure, Uptake, and Dose 255
Alesia C. Ferguson, Robert A. Canales, and James O. Leckie

Chapter 12
Dermal Exposure to VOCs while Bathing, Showering, or Swimming 285
Lance A. Wallace and Sydney M. Gordon

PART IV   Ingestion

Chapter 13
Ingestion Exposure...303
P. Barry Ryan

PART V   Multimedia Exposure

Chapter 14
Exposure to Pollutants from House Dust..319
John W. Roberts and Wayne R. Ott

Chapter 15
Exposure to Pesticides 347
Robert G. Lewis

Chapter 16
Exposure to Dioxin and Dioxin-Like Compounds 379
Daniel J. Stralka and Harold A. Ball

Chapter 17
Biomarkers of Exposure 395
Lance A. Wallace

PART VI   Models

Chapter 18
Mathematical Modeling of Indoor Air Quality 411
Wayne R. Ott

Chapter 19
Modeling Human Exposure to Air Pollution 445
Neil E. Klepeis

Chapter 20
Models of Exposure to Pesticides 471
Robert A. Canales and James O. Leckie

PART VII   Policy

Chapter 21
Environmental Laws and Exposure Analysis 487
Anne C. Steinemann and Nancy J. Walsh

Index 515

Part I

Exposure Concepts

1 Exposure Analysis: A Receptor-Oriented Science

Wayne R. Ott
Stanford University

CONTENTS

1.1 Synopsis...3
1.2 Introduction...4
1.3 Full Risk Model..4
1.4 Total Human Exposure Concept 6
    1.4.1 Indirect Approach 10
1.5 Receptor-Oriented Approach..12
1.6 Indoor Air Quality 14
1.7 Activities That Increase Exposures..17
    1.7.1 *p*-DCB in Residential Microenvironments..19
1.8 Source Apportionment of Exposure..21
1.9 Public Education..24
1.10 Exposurist as a Profession 26
1.11 Conclusions...27
1.12 Questions for Review 28
References 28

1.1 SYNOPSIS

Exposure science differs from past approaches in the environmental sciences, because, instead of beginning with an assumed source of pollution and then trying to find out who or what may be affected, exposure science works from both directions — not just from the source but also backward from the receptor. In the past, environmental policy analysts have assumed some obvious source, such as a smokestack or a leaky drum, may be important and then tried to trace where the pollutant goes, often losing its complicated trail long before the much-diluted pollutant ever reaches any real people, plants, or animals. By following this older *source-oriented* approach, it has been difficult to show in epidemiological studies that traditional sources actually affect anyone's health or that these assumed sources cause adverse exposure of humans to harmful concentrations. Exposure science, on the other hand, begins with the target itself, measuring the pollutant concentrations that actually reach people by using, for example, personal monitors worn by individuals. Then the analyst works backward from the personal exposure to find the actual source. By following this *receptor-oriented* approach and measuring pollutant concentrations at the contact boundary of the person, new sources have been discovered, and some sources originally thought to be important were found to make negligible contributions to exposure. Using direct measurements of exposure, many significant new indoor sources — consumer products, building materials, and personal activities — emerge as the main contributors to human exposure and thus are more likely to be the causes of environment-related illness than traditional outdoor sources. Activity pattern surveys show that people spend on average more than 90% of their time indoors or in a vehicle, and efforts to protect public health from environmental pollution and to reduce personal exposure must consider both indoor air quality and outdoor air quality as well as other routes of exposure, such as dermal contact, drinking water, nondietary contact, and food.

1.2   INTRODUCTION

Historically, environmental risk problems often were discussed without identifying the target of the risk. As noted in definitions of the concepts of exposure and related terms (see Chapter 2), risk assessors must always answer a critically important question, "Whose risk is to be reduced?" If the goal is to protect public health, then the target of the risk becomes the human being. If the goal is to protect an ecological system, then the target may become some organism within the system — for example, an animal or a plant. Failure to identify the target of the risk in environmental problems causes much confusion.

In this book, we are concerned primarily with the risks of pollutants to public health, so the target of concern is the human being. In particular, we are more interested in the risks caused by environmental pollution to the *general public* than in the risks faced by workers in certain occupational settings that are protected by occupational health and safety laws. Environmental protection laws, unlike occupational health laws, are intended to protect the general public. The science of exposure analysis has provided important insight by focusing more clearly than ever before on individual members of the general public and the manner in which pollutants reach and adversely affect them. Included, of course, are those persons in special high-risk occupations, but worker exposure constitutes a special subset of the general population.

1.3   FULL RISK MODEL

To reduce the risks of environmental pollutants to human beings, a relationship between the sources of pollution and their effects must be found. If the risk is to be assessed accurately, then all the sources of the pollutant must be included. The sources considered cannot be limited to just the traditional or more obvious ones (smokestacks, sewage outfalls, hazardous waste sites, etc.) but need to include the nontraditional sources as well (for example, building materials in indoor settings, consumer products, food, drinking water, and house dust).

Establishing the links between a particular target and a particular agent, or pollutant, requires a knowledge of five fundamental components that may be viewed as links in a chain — from source to effect — comprising the full risk model (Figure 1.1). Such a link sometimes is called a *route of exposure*. In this full risk model, each of the five components is linked to the others and is dependent on the one before it. Thus, the pollutant output from one component is the input into the next component. Therefore, if information on one component of the model is lacking, then it is not possible to fully characterize the relationship between the sources of pollutants and the resulting effects. With a missing component, the effect of controlling a source on the exposure received by the target cannot be determined. The sources in this conceptual model are *all* sources of a particular pollutant that may cause risk, regardless of whether the sources are found indoors, outdoors, or in-transit, or whether they are carried to the target by air, water, food, or dermal contact.

Despite the importance of each of the five components in Figure 1.1 for determining the public health risk associated with environmental pollution, historically our scientific understanding of all five components has not been equal. Usually, environmental pollution has come to the attention of public officials because traditional pollutant sources, such as smokestack plumes or leaking drums, have caused alarm because they were so obvious. The obviousness of certain traditional sources has caused an overemphasis on this source category in the complete risk model. Consequently, a

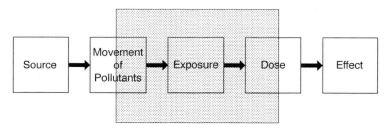

FIGURE 1.1 Five components of the conceptual full risk model. To determine how much a change in a source causes a corresponding change in an effect, it is necessary to know: (a) all five components and (b) the linkage between each pair of components. The shaded zone indicates the primary emphasis of the science of exposure analysis. (From Ott 1985. With permission.)

great body of knowledge exists today about source abatement and control of traditional sources, and many of the environmental laws deal with direct regulation of traditional sources, regardless of whether their contribution to health risks is large or small. By comparison, nontraditional sources, which release pollutants that reach people by nontraditional routes of exposure (for example, consumer products emitting pollutants in the home), have received relatively little attention either from regulators or the public.

Once a traditional source of pollution is known and identified, environmental analysts usually focus on the manner in which it moves from its source — sometimes called its fate and transport — until it ultimately is converted into other harmless chemicals or reaches humans. Like the source component (first box in Figure 1.1), the pollutant movement component (second box) has received considerable research attention. Meteorologists have developed atmospheric dispersion models, and water quality and groundwater specialists have developed geophysical models for the movement of pollutants in streams, groundwater, surface water, soil, and the food chain. Research on the movement, or fate and transport, of pollutants has dealt primarily with traditional routes of exposure, tracing the movement of pollutants through geophysical carrier media on a large physical scale (1–100 km), while nontraditional routes of exposure (for example, microenvironments with sources within 30 m of the person) usually have been overlooked.

As with the first two components, the fifth component in Figure 1.1 — the effects of a pollutant on the target — also has received considerable past research attention. Numerous studies have looked at the effects on animals and humans of specific concentration levels, often without knowing whether suspected sources can create these same high exposure levels. Because research on the health effects of each pollutant is difficult, much uncertainty still remains despite years of studies of the effect of a given exposure level on rats and other animals in the laboratory, humans in clinical experimental settings, and community epidemiology studies.

In contrast with the source, movement, and effect components of the full risk model, very little knowledge has been available up until the last 20 years for the remaining two components — exposure and dose. As the science advances, exposure and dose, because they have been neglected, are proving to be the two components of the full risk model for which major contributions to knowledge are possible. Without accurate knowledge of human exposure (the middle box), it is often impossible to determine which sources should be controlled and by how much, or the likely effect on public health of controlling a source. The shaded zone in Figure 1.1 shows the main areas of emphasis in exposure analysis.

Filling the critical gaps in the full risk model is necessary for implementing a risk-based approach to environmental management. Completing the full risk model allows an analyst to determine if traditional source control efforts are actually reducing the risks to public health in the manner expected and to the degree needed. If we do not know whether the right sources (or other contributors to exposure and risk) are being controlled, or whether they are being controlled by the correct amount, then our environmental programs will become ineffective and inefficient in reducing public health risks.

Fortunately, exposure research completed over two decades has demonstrated for a handful of pollutants that the all-important missing exposure and dose data can be obtained and that the full risk model can be completed, thus making possible a true risk-based approach to environmental management. This exposure research also has identified a variety of new nontraditional pollutant sources and other important contributors to exposure that environmental programs currently are not addressing adequately, and the data show that many of these newly identified sources contribute more to public health risk than many traditional sources now subject to environmental laws and regulations. As a consequence of these new findings, some observers believe that our environmental priorities and policies need to be reshaped and our laws revised.

Although it is important to link sources to exposures to effects in the full risk model, even linking sources to exposures (but not necessarily to dose or effects) provides a great new body of practical knowledge that is important to regulators, decision makers, public health officials, and the general public. If an accurate source-exposure relationship can be established for a particular environmental pollutant, then it is possible to discover the most economical and efficient way to reduce risk by reducing exposures, with a consequent reduction in potential risk. This benefit is possible because it is reasonable to assume that a monotonic relationship exists between exposure and risk — that is, decreases in exposure lead to corresponding decreases in health risk, even though the exact form of the relationship may not be known. Indeed, by completing even the partial risk model between sources (first box) and exposure (middle box) — that is, finding the relationships for just the first three boxes — we may discover that our mitigation activities are placing too much emphasis on the wrong sources and that other overlooked nonregulatory approaches may be more effective in reducing exposures than the ones we have chosen. Indeed, the evidence from some exposure research findings points toward many simple steps not usually associated with environmental laws that can be taken by individuals to reduce their own risks.

1.4 TOTAL HUMAN EXPOSURE CONCEPT

The total human exposure concept seeks to provide the missing component in the full risk model: estimates of the total exposure of the population to an environmental pollutant with known accuracy and precision, usually averaged over a time period of interest, such as 24 hours. Ideally, the exposure information needed is not for just one member of the population, nor is it the average exposure of the entire population. Rather, what usually is sought is a frequency distribution of all members of the population, giving not only the population mean exposure but also the highest 1%, 5%, and 10% of the exposures of the population.[1] This measurement methodology has been completely developed and demonstrated successfully for one major pollutant, carbon monoxide (CO; see Chapter 6), and the total human exposure methodology also has been applied effectively to a number of other toxic environmental pollutants.

The total human exposure approach begins first by applying the conceptual definitions of exposure that can be described quantitatively (Chapter 2). A target must be identified, and a contact boundary also must be identified, even though it might not be formally stated in a real-life exposure analysis. Any pollutant in any carrier medium that comes into contact with this conceptual contact boundary — either through the air, food, water, or dermal route — is considered to be an exposure to that pollutant at that instant of time (Figure 1.2). The *instantaneous exposure* usually is expressed as a pollutant concentration (for example, mass/volume) in a particular carrier medium at a particular instant of time. Some pollutants, such as CO, can reach humans through only one carrier medium,

---

[1] These percentages are based on the quantiles of the distribution of population exposures. For example, if we rank all the population exposures we have from highest to lowest, and 99% of the exposures are found to be below 65 $\mu g/m^3$ of some pollutant, then can we say that 1% of the population of our sample have exposures at or above 65 $\mu g/m^3$.

CONCEPTUAL MODEL OF TOTAL HUMAN EXPOSURE

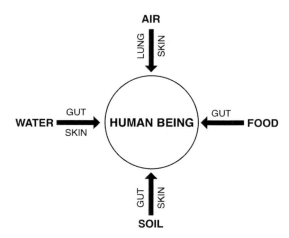

FIGURE 1.2 Exposure occurs when any pollutant makes direct contact with the human being through one of four possible carrier media.

the air inhaled. Other pollutants, such as lead and chloroform, can reach humans through two or more routes of exposure in multiple carrier media (for example, air, food, dermal, and drinking water). If multiple routes of exposure are involved, then the total human exposure approach seeks to determine a person's exposure (concentration in each carrier medium at a particular instant of time) through all routes of exposure from all the sources of that pollutant (Figure 1.3).

Once applied, the total human exposure methodology seeks to provide information, with known precision and accuracy, about the exposures of the general public through all environmental media, regardless of whether the pathways of exposure are air, drinking water, food, or dermal contact. It seeks to provide reliable, quantitative data on the number of people exposed and their levels of

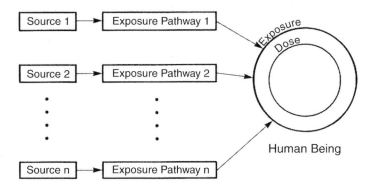

FIGURE 1.3 Full risk model when there is more than one source of the same agent. Here the movement of the pollutant from the source to the target (a human being) is shown as an "exposure pathway" for each source. The entire journey of a pollutant from a source to a target is a "route of exposure."

exposure, as well as the sources and other contributing factors responsible for these exposures, including the reasons for the individual person-to-person differences in exposure. Ideally, these exposure data are presented in an exposure distribution — a frequency distribution showing the proportion of the population exposed to different concentration level intervals over a specific time period.

Placing the human being at the center of attention makes the total human exposure approach different from the older and more traditional environmental analysis approaches — total human exposure is a receptor-oriented approach, because it begins with the receptor, the human being. The total human exposure approach first considers all routes of exposure by which a pollutant may reach a target. Then, it focuses on those particular routes that are relevant for the pollutant of concern, developing measurements of the concentrations reaching the target. Activity pattern information from diaries maintained by respondents can help identify those activities and microenvironments that are of greatest importance, which, in many cases, helps to identify the contributing sources. *Body burden*, the amount of pollutant present in the person's body — for example, blood concentrations of lead (Pb) or breath concentrations of volatile organic compounds (VOCs) — often is included in exposure studies as an indication of previous exposure, thereby helping to confirm and strengthen our understanding of the exposure measurements.

The *direct approach* consists of direct measurements of exposures of the general population to the pollutants of concern, such as the U.S. Environmental Protection Agency's (USEPA) Total Exposure Assessment Methodology (TEAM) studies. In a TEAM study, a representative random sample of the population is selected based on a carefully planned statistical design using the probability sampling methods outlined in Chapter 3. Then, for the particular pollutant (or class of pollutants) under study, the pollutant concentrations reaching the respondents selected according to this statistical design are measured from all sources and for all relevant carrier media. A sufficient number of people are sampled using statistical sampling techniques — sometimes called a multi-stage probability design — to permit inferences to be drawn, with known precision, about the exposures of the larger population from which the sample is drawn. Often a stratified random sample is used to select a greater number of those persons who are of special interest from an exposure standpoint (for example, long-distance commuters if a vehicular air pollutant is under investigation), and then sample weights are used in the subsequent data analysis to adjust for the over-sampling. From statistical analyses of the diaries (activities and locations visited), it often is possible to identify the likely sources, microenvironments, and human activities that contribute most to pollutant exposures, including both traditional and nontraditional sources. Many of these large-scale exposure field studies, such as the TEAM studies (Table 1.1), have been made possible because of new compact personal exposure monitors (PEMs) capable of measuring exposure to some air pollutants with high precision (see Chapters 5–8 and Chapter 15). An exposure field study of this kind typically samples 25 to 800 respondents and includes the following four basic elements:

1. Use of a representative probability sample of the population
2. Direct measurement of the pollutant concentrations reaching these people through all carrier media (air, food, water, and dermal contact)
3. Direct measurement of body burden to infer dosage
4. Direct recording of each person's daily activities through diaries

For volatile pollutants that are readily absorbed and given up by the blood, it is possible to determine with reasonable precision the concentration of that pollutant present in the blood by asking respondents to exhale their deep-lung breath into a sampling bag and then to analyze the contents of the bag. This is basically the same method that is used by police officers who stop drivers on the highway to measure the concentration of their blood alcohol using a breath analyzer test. Breath measurement works satisfactorily for a number of other pollutants of interest besides alcohol (ethanol) that are absorbed by the blood (for example, CO and benzene). The resulting

TABLE 1.1
Locations of Total Exposure Assessment Methodology (TEAM) Studies

City	Pollutants	Date	No. of Households
Chapel Hill, NC	VOCs	1980	6
Beaumont, TX	VOCs	1980	11
Elizabeth–Bayonne, NJ (3 seasons)	VOCs	1980	9
Research Triangle Park, NC (3 seasons)	VOCs	1980	3
Los Angeles, CA	CO	1981	9
Elizabeth–Bayonne, NJ (3 seasons)	VOCs	1981–83	355
Greensboro, NC	VOCs	1982	25
Devils Lake, ND	VOCs	1982	25
Denver, CO	CO	1982-83	450
Washington, DC	CO	1982-83	712
Los Angeles, CA (3 seasons)	VOCs	1984	120
Antioch–Pittsburg, CA (3 seasons)	VOCs	1984	75
Jacksonville, FL	Pesticides	1985	9
Jacksonville, FL (3 seasons)	Pesticides, House dust	1986–87	200
Springfield, MA (3 seasons)	Pesticides	1987	100
Baltimore, MD	VOCs	1987	150
Los Angeles, CA	VOCs	1987	50
Bayonne–Elizabeth, NJ	VOCs	1987	11
Azusa, CA	Particles, metals	1989	9
Riverside, CA	Particles, metals	1991	178
Valdez, AK[1]	VOCs, benzene	1990–91	58
Toronto, Ontario[2]	Particles, metals	1995–96	750
Indianapolis, IN[2]	Particles, metals	1996	240

[1] Study with personal monitors by Alyeska Pipeline Service Co. through research contractors.

[2] Approach similar to TEAM by Ethyl Corp. under a contract with Research Triangle Institute, NC.

concentrations of the pollutant in the breath can be used to assess the concentration of the pollutant in the blood and is a useful measure of body burden at that time. Measurements of body burden in some bodily fluids, such as breath or urine, have the advantage that they provide a useful indicator of previous exposure and are noninvasive — they do not require a sample of blood or tissue from the person. In addition, pharmacokinetic models have been developed that express the mathematical relationship between prior exposure and predicted blood and breath concentrations.

Although the final sample sizes for some of the TEAM studies in Table 1.1 were not extremely large, the representative probability designs of the larger studies enable the analyst to make inferences about the exposures of the much larger populations from which these samples were drawn, such as the population of an entire community. The TEAM studies were the first to show for large populations that indoor sources of toxic chemicals not only greatly outnumber outdoor sources but also cause the bulk of human exposures, often two to five times the amount caused by outdoor sources (Figure 1.4). For example, tetrachloroethylene (sometimes called perchloroethylene, or PERC) comes from clothes recently brought home from the dry cleaners; chloroform is emitted by water boiled during cooking and released when showering; bathroom and kitchen deodorizers and mothballs are sources of *para*-dichlorobenzene (*p*-DCB); solvents stored in the home and cleaning fluids are sources of 1,1,1-trichloroethane and tetrachloroethylene; automobiles and gasoline stored in attached garages emit benzene; aerosol sprays release a variety of organic compounds, including *p*-DCB; other organic pollutants arise from paints, glues, varnishes, turpentine, and adhesives, as well as furniture polish, carpets, and building materials. Finally, pesticides and herbicides stored in homes or applied indoors can be important sources of personal exposure, in addition to house dust tracked in from outdoors. There is a greater likelihood of receiving a high exposure from sources that are nearby (usually indoors), while the lower probability of receiving an elevated exposure from a traditional source suggests that the outdoor sources depicted in Figure 1.4 should be shown in the far distance and should be smaller than they are now depicted.

Indoor exposures are significant for a number of reasons. A great variety and number of chemicals are present indoors close to people, and the small physical scale (under 30 m) limits dilution, causing relatively high concentrations. Also, the infiltration of outdoor air into indoor settings is relatively slow, causing a low air change rate and a long *residence time* — the time required for the volume of air equal to the volume of the indoor setting to be fully replaced. Air change rates in homes typically range from 0.2 to 1 air change per hour, giving residence times (the reciprocal of the air change rate) of 1–5 hours. These factors — intensity and variety of sources combined with low dilution and low air change rate — make indoor air pollutants especially important contributors to the total exposure of occupants. Finally, *activity pattern*, or *time budget*, studies based on large representative probability samples in which participants completed 24-hour diaries show that people spend most of their time indoors (Klepeis et al. 2001), as discussed briefly in the following sections. All these factors combine to make indoor air pollution an important contributor to total exposure and to total risk.

1.4.1 INDIRECT APPROACH

The direct approach described above is valuable for *measuring* directly the distribution of exposures across a population by using personal exposure monitors to measure air exposure and other methods to measure dermal, food, and drinking water exposure, making it possible to draw inferences from a representative sample to a much larger population. By contrast, the *indirect approach* seeks to measure and understand the basic relationships between causative factors and the resulting exposures in the small settings called *microenvironments*, such as homes or motor vehicles. By combining microenvironmental air pollutant concentrations with diary-based activity patterns in an exposure model, it is possible to predict population exposure distributions. Similar approaches are applicable to dermal exposure.

Thus, the indirect approach is basically a modeling approach combining measurements from small-scale microenvironmental experiments with large-scale activity pattern data in order to predict population exposure distributions. The resulting exposure model is intended to complement the direct approach field studies by helping to translate their findings to other locales and to other situations. These models also are designed to allow the analyst to consider "what if" questions and to predict how exposures might change in response to different policies and regulatory (and nonregulatory) actions. Exposure models are not the same as the traditional outdoor pollutant transport models for predicting outdoor concentrations in ambient air, surface water, or groundwater. Rather, they are designed to predict human exposure of a rather mobile human being (see Chapter 19 and Chapter 20). Thus, they require information on typical personal activities, locations visited during the day, and time budgets of people, as well as information on the likely concentration

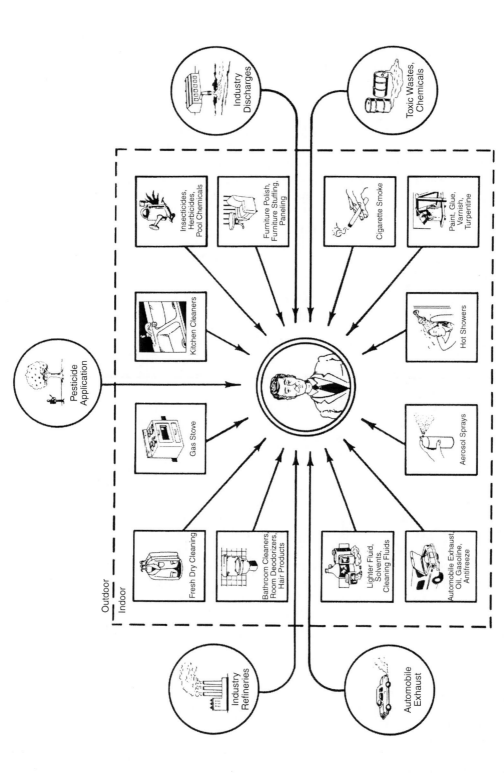

FIGURE 1.4 Examples of traditional (outdoor) and nontraditional (indoor) sources of exposure, based on findings from the total exposure measurement studies. (From Ott 1990. With permission.)

distributions in the places the people spend their time (ordinary microenvironments such as homes, motor vehicles, stores, restaurants, schools, etc.).

An example of an early human activity pattern-exposure model was the Simulation of Human Activities and Exposures (SHAPE) model, which was designed to predict the exposures of the population to CO in urban areas (Ott 1984). The SHAPE model used the CO concentrations measured in microenvironments in Denver and Washington, DC, to determine the contributions from commuting, cooking, cigarette smoke, and other factors. Once an exposure model is validated by demonstrating that it accurately predicts exposure distributions measured in a total exposure field study, then, in theory, it should be possible in a new city to make a reasonably accurate prediction of that population's exposures using the new city's data on human activities, travel habits, and outdoor concentrations. USEPA has developed an exposure model for particulate matter that is similar to the SHAPE model, the Stochastic Human Exposure and Dose Simulation (SHEDS-PM) computer model (Burke, Zufall, and Özkaynak 2001). A SHEDS model also has been developed for predicting the exposure of adults and children to pesticides. Another modeling approach, called the Random Component Superposition (RCS) model, uses empirical exposure measurements to predict population exposure distributions in cities (Ott, Wallace, and Mage 2000). The RCS model is a statistical approach that describes the difference between indoor, outdoor, and personal exposures. Other statistical approaches treat short-term and long-term exposure distributions (Wallace, Duan, and Ziegenfus 1994, Wallace and Williams 2005).

1.5 RECEPTOR-ORIENTED APPROACH

There is a fundamentally different conceptual approach between the new science of exposure analysis and traditional approaches used historically in the environmental fields. To illustrate this difference intuitively, consider an Agatha Christie murder mystery story. At some point in the tale — usually fairly early in the plot — we usually encounter a body, the victim of murder. The principal investigator — Monsieur Hercules Poirot — follows various clues and traces events surrounding the murder to try to discover the cause of the crime and to identify the murderer.

Like Sherlock Holmes stories and like real crimes, each Agatha Christie story begins with the victim — a dead person — and then works backward to find the cause of the crime (that is, the source of the mortal wound). If the death occurred by gunshot, M. Poirot may analyze bullet fragments to find out more details about the gun. Samples may be collected from the victim's body and at the crime scene to try to find the possible killer, the murder weapon, and any other factors that help explain the crime. All these investigative efforts seek to find the facts surrounding the firing of the gun, which is analogous to the emission of a pollutant from its source.

The same approach that is used to investigate a crime — working backward from the victim — applies to the exposure sciences. By making appropriate measurements very close to the person, the exposure analyst tries to determine if the person is being exposed to a particular pollutant. By making measurements of body burden (breath, blood, etc.), the exposure analyst tries to determine if a person has been exposed to a particular pollutant. Similarly, Agatha Christie's Monsieur Poirot first determines if the person has been exposed to a bullet and then seeks to find the source of the bullet. Both fields — exposure analysis and criminology — use the same conceptual approach: they work backward from the point of contact to identify the source. Thus, both techniques are receptor-oriented.

Many of the environmental regulatory approaches, by contrast, proceed in the opposite direction. They use a source-oriented rather than receptor-oriented approach. Most environmental laws are concerned with emissions or with effluents. They do not, for example, begin with the people and attempt to see whether these pollutants actually are reaching any members of the general public. One probable reason is that the framers of these laws were overly impressed by the appearance of leaking storage drums, smokestack emissions, effluents discharged into waterways, and other traditional images. However, as progress has been made in reducing source emissions, effluents, leaking underground storage tanks, etc., it becomes extremely important to determine whether pollutants actually still are reaching the people, by how much, and through what routes. That is, it becomes extremely important to measure the actual exposures that the population receives very close to the body and to answer the four basic questions mentioned above concerning: (1) the number of people exposed, (2) the level of their exposure, (3) the causes of their exposure, and (4) the manner in which exposure can be reduced most efficiently.

A source-oriented crime approach — unlike the Agatha Christie and Sherlock Holmes stories — would not originate with a dead body. Rather, it would originate by discovering a gun and its spent cartridges. All we would know is that this gun had been fired, but we would not know where the bullets went or whether they actually hit anyone. Even though there was no body — or even any evidence that a bullet had injured anyone — the source-oriented approach would try to trace the path of the bullets forward to find out if anyone was hit by a bullet. Thus, we would investigate even if we did not know if a crime had been committed. Conducting source-oriented criminology would be expensive to society, because the police would have to investigate carefully the firing of every gun, even if no person were known to be hit. As readers of fiction, we probably all would find novels based on a source-oriented crime approach to be rather boring — much ado about nothing. Yet, it appears that much of the environmental regulatory community has embarked on a source-oriented approach. Usually, the process of tracing the pollutant to its receptor in a source-oriented approach is so difficult that the investigator runs out of resources before discovering who is exposed, if anyone.

The public health community, like criminology and exposure analysis, also follows a receptor-oriented approach. Many harmful viruses exist in the environment. If a harmful virus has been identified, public health officials usually immediately ask, "Has anyone been exposed (i.e., come into contact) with the virus?" Simply knowing a virus is present somewhere in the environment is not assumed automatically to cause a threat to public health. Someone must be exposed first. Once it is known that an exposure has occurred, public health officials usually next ask, "How many people were exposed?" "What was the cause of their exposure?" "How can exposures be reduced?" Only by knowing the extent of the exposure and the causes of the exposures can public health and environmental officials intervene to reduce exposures in the most effective manner possible and thereby reduce public health risks.

The importance of the receptor-oriented approach in all these fields is obvious. It is not possible for one to get sick from a virus *unless an exposure has occurred*. Similarly, it is not possible to die from a bullet unless the bullet makes contact with a person (that is, the victim must be exposed to the bullet by being shot). Similarly, without exposure to an environmental pollutant, one cannot experience adverse health effects. More importantly, changing one's exposure often may be achieved more effectively through nonregulatory approaches than by reducing some sources directly through regulations. Often those sources that can be changed by making regulations contribute the least to one's personal exposure. Both regulatory and nonregulatory approaches should be considered for reducing human exposure to pollutants. The public health practitioner uses public information, as in the war against AIDS, to advise people to modify their activities and thereby to reduce their exposures. Once the pollutant enters the body, however, it becomes a *dose*, and little can be done to reduce its effects. For example, one cannot usually take a pill to reduce the effects of one's exposure. The best approach is to reduce one's exposure first. Of course, like Agatha Christie solving a crime, it is not possible to reduce exposures unless we know they have occurred, so determining population exposures by direct measurement and attempting to answer the four questions mentioned above are essential steps for protecting public health.

An important scientific contribution to the field of exposure analysis was the development and successful implementation of the TEAM studies (Wallace 1987; see Table 1.1), followed a decade later by the National Human Exposure Assessment Survey (NHEXAS) studies (Whitmore et al. 1999; Clayton et al. 1999; Robertson et al. 1999). These total human exposure field studies combine probability sampling with direct measurement of exposure. Conceptually, exposure is measured in the food eaten, water drunk, skin surface contacted, and air breathed using small, portable personal exposure monitors (Chapter 5), and body burden is measured in breath and urine. For the pollutants that are not present in food or water or soil, air usually is the main route of exposure. By using a good statistical design (Chapter 3), inferences can be made from a carefully designed sample of respondents to larger populations of the community. By using direct measurement of exposure, important discoveries are possible that cannot be made with predictive models alone, at least not with the present state of the art of modeling.

1.6 INDOOR AIR QUALITY

Why is indoor air quality important in the exposure sciences? A receptor-oriented approach focuses on humans and on measuring pollutants very close to the body. The monitors must be located very close to the person — preferably carried by people. Large-scale field studies in which people recall their locations and activities through diaries show that people usually are indoors — either at home, at the office, or in a motor vehicle — with very little time spent outdoors. An activity pattern database of 1,579 California adults (Jenkins et al. 1992) — a representative sample of the state's population — was analyzed using custom-designed computer programs to find where people are located over 24 hours (Figure 1.5). Examining the proportion of the California population located in each of six microenvironments by time of day (beginning and ending at midnight with time on the chart moving from top to bottom) shows that the most frequently occupied location for Californians is indoors at home, accounting for approximately 15 hours on the average. The large proportion of adults found inside a vehicle by time of day is striking when it is compared with the proportion of adults found outdoors. For all but 2 hours of the day (10:00 A.M. and 11:00 A.M.), the proportion of California adults inside a vehicle is greater than the proportion of adults outdoors. This proportion varies considerably with the time of day. At 5:00 P.M., the maximum difference between these two locations occurs: 18% of the adults are in a vehicle but only 8% are outdoors.

Comparing the *time budgets* for these three locations — indoors, outdoors, and inside a vehicle — shows that outdoors accounts for the smallest amount of total time (Table 1.2) and is even smaller than the average time spent in a motor vehicle. Despite the widely held view that Californians are outdoors much more than people living elsewhere in the United States, the differences between California and the United States are extremely small (Jenkins et al. 1992; Klepeis et al. 2001). These diary studies show that Californians aged 12 and older were outdoors 6% of the day (86 minutes), while the U.S. population was outdoors 7.6% of the day (109 minutes). On average, both the U.S. and California population was indoors 87% of the day, and members of the U.S. population were indoors or in a vehicle more than 92% of the day.

The graph showing the locations of Californians by time of day was plotted so that time over the 24-hour period from midnight to midnight flows from top to bottom (Figure 1.5). An alternative approach is to plot time over 24 hours from left to right (Figure 1.6), and fortunately diary-based activity pattern data similar to that in California are available for the entire nation. The National Human Activity Pattern Survey (NHAPS) study was a probability study of a sample of 9,386 persons selected to represent the entire U.S. population (Klepeis et al. 2001). As with the California activity pattern study, more people were found to be indoors at home (top, lightly shaded labeled "Residence–Indoors" in Figure 1.6) than in any of nine other locations throughout the day. Figure 1.5 and Figure 1.6 are essentially the same type of 24-hour diurnal graphs with the axes interchanged, except that Figure 1.5 is for California adults and Figure 1.6 is for the entire U.S. population.

In addition to spending a large amount of time indoors, especially in their homes, people breathe air pollutant concentrations in indoor settings that generally are higher than those measured outdoors. The results from the TEAM studies of CO (Akland et al. 1985; Hartwell et al. 1984; Johnson 1983; Ott et al. 1988; Ott, Mage, and Thomas 1992; Wallace et al. 1988b; Wallace and Ott 1982; see Chapter 6), volatile organic compounds (VOCs) (Pellizzari et al. 1987a,b; Wallace and O'Neill

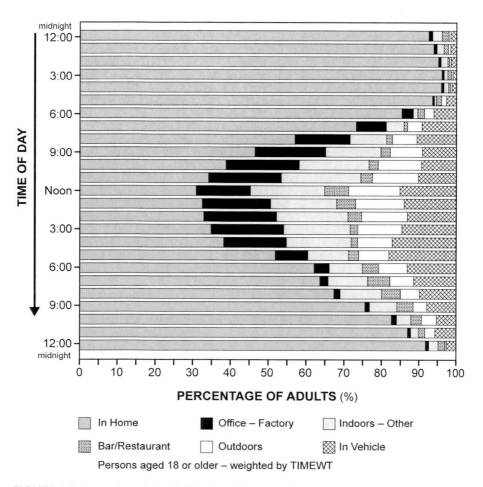

FIGURE 1.5 Proportion of the California adult population present in each of six locations by time of day, based on a computer analysis of the diaries from a stratified representative random samples of 1579 persons aged 18 or older throughout the state. (From Ott 1995a. With permission.)

TABLE 1.2
Time Budgets for Persons Aged 12 and Older

Location	California[a]		U.S.[b]	
	Minutes per Day	Percent of Day	Minutes per Day	Percent of Day
Indoors	1256	87%	1252	87%
In-vehicle	98	7%	79	5.5%
Outdoors	86	6%	109	7.6%
Total	1440	100%	1440	100%[c]

[a] Jenkins et al. (1992).
[b] Klepeis et al. (2001).
[c] Does not add to 100.0% due to rounding.

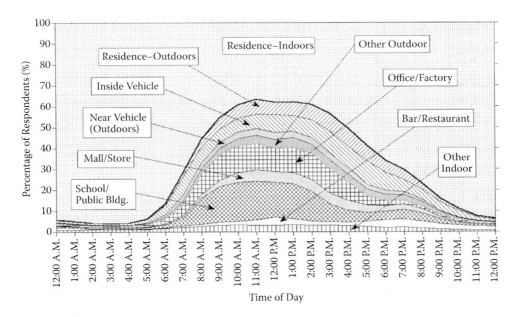

FIGURE 1.6 Stacked plot showing the weighted percentage of respondents in the U.S. in each of 10 different locations by time of day. The original minute-by-minute diary data have been smoothed for clarification. As with the California activity pattern data, the percentage of people outdoors (either at "Residence–Outdoors" or "Near Vehicle (Outdoors)" or in "Other Outdoor" locations) is relatively small, though it varies in a diurnal pattern. When people were outdoors, they were most likely to be outdoors at their residence ("Residence–Outdoors"), while the most likely location for finding anyone indoors was at their residence ("Residence–Indoors"), which varied from 96% of the time to about 36% over the 24-hour period. (From Klepeis et al. 2001. With permission.)

1987; Wallace 1986, 1991a,b, 1993a,b,c, 1997, 2001; Wallace et al. 1982, 1984, 1985, 1986, 1987a,b,c, 1988a, 1989; Zweidinger et al. 1982; see Chapter 7), particles (Clayton et al. 1993; Özkaynak et al. 1996; Pellizzari et al. 1993a,b; Thomas et al. 1993b; Wallace 1996; Yakovleva, Hopke, and Wallace 1999; see Chapter 8), and pesticides (Whitmore et al. 1994; see Chapter 15), as well as various indoor air quality studies (Thomas et al. 1993a) show that indoor concentrations are generally higher than outdoor concentrations for a large number of pollutants (Table 1.3). This table lists all those pollutants for which the average concentration indoors has been found to be higher than the corresponding average concentration outdoors. The main reason for the higher indoor concentration is the presence of sources indoors. A number of these same pollutants are found on California's list of toxic air pollutants under Proposition 65, a referendum passed by the voters that requires products containing toxic pollutants to be labeled. To place these findings in perspective, the pollutants in Table 1.3 are grouped into several different categories: criteria air pollutants, pesticides, toxic air pollutants, and others.

The 3-year Valdez Health Study was interesting because it focused on a single large point source, the Alyeska Marine Terminal (Yocum, Murray, and Mikkelsen 1991; Goldstein et al. 1992). This large petroleum storage and loading terminal was approximately 3 miles from the 3600 residents living in Valdez, AK. As with the TEAM VOC studies, personal monitoring of the residents showed that the contribution of indoor sources and personal activities to benzene exposures was much higher than the contribution of the marine terminal, accounting for 89% of the total exposure to benzene in this community. The Marine Terminal accounted for only 11% of the population exposures, showing the importance of everyday sources that are close to people vs. distant point sources, even if their total emissions may be large.

TABLE 1.3
Examples of Air Pollutants Found Indoors at
Higher Concentrations than Outdoors

Criteria Air Pollutants:	**Toxic Air Pollutants:**
Carbon monoxide*	Chloroform**
Nitrogen dioxide	1,1,1-Trichloroethane
Particles ($PM_{2.5}$, PM_{10})	Benzene**
Lead*	Carbon tetrachloride**
	Trichloroethylene**
Pesticides:	Tetrachloroethylene**
Dichlorovos*	Styrene
Chlorothalonil**	*meta, para*-Dichlorobenzene**
Hexachlorobenzene**	Ethylbenzene
Heptachlor**	*ortho*-Xylene
Chlorpyrifos*	*meta, para*-Xylene
Aldrin**	Formaldehyde**
Oxychlordane	Methylene chloride
Captan**	Polychlorinated biphenyls (PCBs)
Dieldrin**	
Chlordane**	**Other Categories:**
4,4'-DDT**	Radon**
4,4'-DDE**	Nicotine**
ortho-Phenylphenol	Asbestos**
Propoxur	

* Toxic chemical listed by California under Proposition 65 as known to cause reproductive toxicity.
**Toxic chemical listed by California under Proposition 65 as known to cause cancer.

Source: Updated from Ott 1995a.

1.7 ACTIVITIES THAT INCREASE EXPOSURES

An important finding of the exposure field studies is that a person's activities greatly affect their exposures. In one microenvironmental experiment, seven persons volunteered to perform 25 common activities to see how their activities might change their personal exposures to VOCs during a 3-day monitoring activity (Wallace et al. 1989). Exposures to most chemicals in the absence of the specified activities (that is, during sleep) ranged between 1 and 10 µg/m³. About 20 of the 26 activities resulted in increased exposures during the 5- to 11-hour monitoring period (Table 1.4). At times, the increases were dramatic, reaching levels 100 times the background exposures over the same averaging times. Those activities with the largest associated increases in personal exposure to the target chemicals included visiting a dry cleaners, cleaning a car engine with solvents, painting and using paint remover, and using air fresheners and toilet deodorants.

Breath concentrations, a measure of body burden in the blood, sometimes showed very large increases following certain activities. For example, decane levels increased by a factor of 100 (from 2.9–290 µg/m³) and xylenes by a factor of 50 (from 4–200 µg/m³) in one subject's breath following painting and use of solvents. Other increases by factors of 10 were not uncommon.

Persons who lived in the homes with continuously elevated levels of certain chemicals showed continuously elevated breath concentrations. For example, the median level of 1,1,1-trichloroethane

TABLE 1.4
Activities Associated with Increased Personal Exposures or Indoor Air Concentrations ($\mu g/m^3$)

	Personal Exposure			Indoor Air Concentration		
Chemical/Activity	N	Median	Maximum	N	Median	Maximum
para-Dichlorobenzene						
Toilet deodorizer (solid)	12	330	500	6	340	630
Spray deodorizer	16	33	84	9	37	59
Liquid deodorizer	5	12	28	3	25	30
No deodorizer use	10	2.4	6.6	6	2.6	5.2
meta-Dichlorobenzene						
Liquid deodorizer	5	10	23	3	23	25
Unknown activity	4	36	57	2	48	59
No deodorizer use	13	ND	2.4	10	ND	QL
1,1,1-Trichloroethane						
Visit dry cleaners	2	675	1000	--	--	--
Work in chem lab	2	86	100	--	--	--
Work as lab technician	3	24	63	--	--	--
Household cleaning/whitener	1	260	260	1	380	380
Use pesticide	3	64	110	3	82	94
Use paint	6	44	110	6	40	380
No activities	8	QL	16	6	4.8	14
Chloroform						
Wash clothes	4	17	24	4	20	44
Wash dishes	14	17	36	14	24	44
Work as lab technician	3	93	95	--	--	--
Work in chem lab	2	60	110	--	--	--
No activities	3	5.5	12	3	QL	5
Trichloroethylene						
Unknown indoor source	15	20	69	12	24	43
Work at chem lab	2	75	110	--	--	--
Work apt. maintenance	1	220	220	--	--	--
No activities	8	ND	ND	6	ND	ND
Tetrachloroethylene						
Cleaning carburetor	1	220	220	--	--	--
Visiting dry cleaners	2	52	52	--	--	--
No activities	14	QL	11	11	ND	5.8
Carbon tetrachloride						
Using paint remover	1	26	26	--	--	--
Unknown activity	1	12	12	--	--	--
No activities	13	ND	QL	11	ND	6.3
Benzene						
Smoking cigarettes	18	12	23	9	5.6	12
Driving	25	7.9	23	--	--	--
Lawn mower exhaust	1	32	32	--	--	--
No activities	14	QL	11	11	QL	9.9
Styrene						
Cleaning carburetor	1	26	26	--	--	--
No activities	14	ND	QL	11	ND	13
o-Xylene						
Cleaning engine	1	410	410	--	--	--
Painting	6	16	420	6	10	130

TABLE 1.4 (CONTINUED)
Activities Associated with Increased Personal Exposures or Indoor Air Concentrations (μg/m³)

Chemical/Activity	Personal Exposure			Indoor Air Concentration		
	N	Median	Maximum	N	Median	Maximum
Operating combustion source	1	54	54	--	--	--
Driving	23	6.6	24	--	--	--
No activities	14	QL	12	11	QL	11
Ethylbenzene						
Cleaning engine	1	800	800	--	--	--
Painting	6	18	470	6	8.6	120
Operating combustion source	1	50	50	--	--	--
Driving	23	4.5	17	--	--	--
No activities	14	QL	10	11	QL	6.9
m + p-Xylene						
Cleaning engine	1	970	970	--	--	--
Painting	6	44	1200	6	29	300
Operating combustion source	1	150	150	--	--	--
Driving	23	15	75	--	--	--
No activities	10	10	26	8	11	22
Decane						
Unknown indoor source	17	25	150	9	44	160
Painting	6	48	350	2	102	200
Driving	18	10	53	--	--	--
No activities	10	3.3	7.4	8	5.7	16
Undecane						
Painting	6	30	280	2	32	60
Unknown indoor source	17	34	280	9	70	290
Driving	18	7.8	43	--	--	--
No activities	10	QL	12	8	3.6	16

N = Number of monitoring periods when activity occurred.
ND = Not detected.
QL = Below quantifiable limit.
"--" = Not applicable.

Source: Wallace et al. (1989). With permission.

in the breath of one subject, whose home had the highest indoor levels, was 30 μg/m³, compared to levels of 7 and 4 μg/m³ in the breath of subjects in another home. Similarly, the median levels of p-DCB in the breath of two subjects in another home (which had an unknown source of p-DCB) were 40 and 47 μg/m³ — far above the median level of 1.5 μg/m³ in the breath of the person living in the other home with the high 1,1,1-trichloroethane. In all, about 120 cases of sharp increases in breath levels (more than doubling previous breath levels) were noted; of these, 100 reflected elevated personal exposure due to one or more of the planned activities.

1.7.1 *P*-DCB in Residential Microenvironments

It is of interest to present an example of one of these compounds, p-DCB, for the seven subjects in the residential microenvironment study discussed above (Wallace et al. 1989). This compound is used in moth crystals, toilet deodorants, and room air fresheners. It often shows up in public

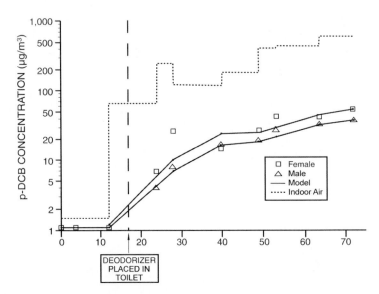

FIGURE 1.7 Increase in breath concentrations of *p*-dichlorobenzene following placement of a toilet bowl deodorant in the home. Indoor air concentrations and personal exposures of male and female residents increased from less than 5 µg/m³ to 500 µg/m³ during the 48 hours following introduction of the deodorant. Breath concentrations increased from 1 µg/m³ to 40–50 µg/m³ in the same time. Residence times of 28 and 30 hours were calculated using a least-squares fit to a one-compartment pharmacokinetic mode. The curves predicted by the model are fitted with the observed breath values. (From Wallace et al. 1989. With permission.)

restrooms as a small white cake with a pungent odor. In this experiment, toilet bowl deodorant containing *p*-DCB was purchased and placed in a toilet in one of the houses following two consecutive monitoring period measurements that established a baseline indoor air concentration of approximately 1 µg/m³. After the deodorant was placed, the indoor air value during the next 48 hours increased by two orders of magnitude, ranging between 100 and 600 µg/m³ (Figure 1.7). Breath values of both inhabitants of the home (Persons No. 3 and 4 in this study) increased from 1–40 µg/m³ during the 48-hour exposure (Table 1.5). For two residents, a least-squares fit to their breath concentrations using a simple pharmacokinetic model (Wallace et al. 1986) resulted in estimated biological residence times of 28 and 30 hours. Using the same pharmacokinetic model and assuming a constant exposure of 500 µg/m³, the breath levels of the male and female subjects were estimated to approach asymptotic values of 90 and 135 µg/m³, respectively.

In a second home, one drop of liquid deodorizer was placed in four locations throughout the house during monitoring period 4. Both of the isomers, *p*-DCB and *meta*-dichlorobenzene, increased immediately from low or undetectable concentrations to 23–28 µg/m³. These levels from the single drop of deodorizer were maintained for two monitoring periods (about 15 hours) and then declined over the remainder of the study.

In a third home, the residents reported using a spray deodorant for a 2-minute period during the first and third monitoring periods. However, the consistently elevated concentrations of *p*-DCB in this home indicate that an additional source of *p*-DCB had been present in the home before the start of the monitoring period. Indoor levels were elevated (37–59 µg/m³) at night, dropping to 13–15 µg/m³ during the day (when windows were opened), through the first seven monitoring periods. During the last two monitoring periods, indoor levels dropped sharply (to less than 4 µg/m³), suggesting that the unknown source had been removed. Personal exposures for both persons were stable at 32–42 µg/m³ and 39–54 µg/m³. The stability of the breath levels following several large fluctuations of personal exposures is additional indication of a relatively long biological half-life of *p*-DCB in humans.

TABLE 1.5
Effect of Toilet Bowl Deodorizer on Indoor Air Concentrations, Personal
Exposures, and Breath Levels (μg/m³) of *para*-Dichlorobenzene

Time Period	Averaging Time (hr)	Indoor Air Concentration	Personal Exposure		Breath Concentration	
			Person 3	Person 4	Person 3	Person 4
1	3.7	1	1	1	1	1
2	8.2	2	2	3	2	2
3[a]	11.6	65	8	43	4	7
4	4.0	250	240	250	8 (7)	27(10)
5	12.0	120	300	330	17 (17)	15 (25)
6	8.6	190	220	---[b]	20 (19)	28 (26)
7	3.5	430	425	340	28 (23)	44 (30)
8	11.5	450	470	410	34 (34)	44 (45)
9	8.5	630	450	500	39 (39)	54 (55)

Note: ( ) Predicted value using the model described in Wallace et al. (1986), $C_{ALV} = C_{ALV}(0) \, e^{-t/\tau} + fA \, (1 - e^{-t/\tau})$, where C_{ALV} is alveolar breath value; t is averaging time; f is fraction exhaled at equilibrium; A is average exposure; and τ is residence time. Values of f (0.18 and 0.27) and (28 and 30 hours) were determined for each subject by nonlinear least squares.

[a] Deodorant placed in toilet bowl during this period.
[b] Sample lost.

Source: Wallace et al. (1989). With permission.

1.8   SOURCE APPORTIONMENT OF EXPOSURE

Duan and Ott (1992) proposed a mathematical model for determining the optimal mix of source reductions that is least costly for reducing one's total exposure, including both indoor and outdoor sources. Other investigators (Sexton and Hayward 1987; Wallace 1989, 1990, 1991a; Smith 1988a,b, 1993) have suggested a similar approach. The basic idea is to spend the *greatest effort* controlling those sources that contribute the *most* to one's exposure — and therefore to one's risks — while spending the *least effort* controlling those sources that contribute the *least* to one's exposures and to one's risks. This concept was believed to be important enough to be given a name: Relative Source Apportionment of Exposure (RSAE), a fundamental part of the receptor-oriented approach. Here relative means that one examines the contribution to exposures of each and every source of the same pollutant relative to all the others. The RSAE approach is intended to help determine the least costly combination of steps that society can take to protect public health. Indeed, often significant reductions in exposure can be achieved by very modest changes in activities and lifestyles. An example is the exposure of children and toddlers to lead, pesticides, and toxic substances in house dust. It has been estimated that the concentration of lead in rugs can be reduced by a factor of 13 by removal of shoes, by 6 by use of large walk-in door mats, and by 3 by a vacuum with an agitator brush (Roberts et al. 1992; see topics on house dust in Chapter 14).

From the standpoint of number of tons emitted of particles, Smith (1993) estimate that power plants are 25 times more polluting than cigarettes, but that environmental tobacco smoke (ETS) contributes 50 times as much exposure to particles. Thus, he concludes that "...a 2-percent decrease in ETS would be equivalent to eliminating all the coal-fired power plants in the country." A similar conclusion occurs for most other pollutants in Table 1.3 as well, because most of our exposure to toxic pollutants comes from "...relatively small localized sources that are, literally, right under our noses: cigarettes, spray cans, and dry-cleaned clothes, for example. More often than not, these exposures occur indoors" (Smith 1993). The reader wishing to explore the topic of indoor pollutant concentrations in greater detail is referred to various literature surveys (Behar et al. 1979; Lioy 1990; Ott 1982, 1985, 1990, 1995a; Ott et al. 1986; Ott and Roberts 1998; Spengler and Sexton 1983; Spengler and Soczek 1984; Smith, 1988a,b, 1993; Wallace 1993a, 1995, 1996, 2000; Wallace and Ott 1982; Weselowski, 1984).

For an everyday example of an RSAE approach, consider benzene exposure to nonsmokers in a local community, the San Francisco Bay Area. (Benzene exposure to smokers is contributed almost wholly by their cigarettes, which account for about 90% of their total exposure to benzene.) The Bay Area Air Quality Management District (BAAQMD) considered benzene as contributing the largest risk of all the ambient toxic air pollutants, but because of the traditional emphasis of this and other similar air pollution control agencies on outdoor ambient air, their calculations did not take into account indoor air quality and assumed that the residents spend 24 hours outdoors each day (Hill 1991). At one point, the Board of the BAAQMD called for a 50% reduction in benzene emissions from industrial sources (Levy 1991). If they had applied the RSAE approach to benzene exposure and health risk in the Bay Area, or more specifically to Contra Costa County, where most of the industrial benzene emissions occur, what would this exposure-based approach tell us about this emission reduction plan?

The California TEAM study has provided data on population exposures in Contra Costa County (Wallace et al. 1988a), a highly industrialized community in the Bay Area with numerous refineries and factories and 91,000 residents, although these exposure data do not appear to be considered in the BAAQMD's official risk assessments. Figure 1.8 shows the frequency distribution of 12-hour personal exposures to benzene in Contra Costa County.[2] The average of the 24-hour exposures to benzene in this populous county was approximately 8 $\mu g/m^3$. We can do a *source apportionment* of these personal exposures to benzene following the approach of Wallace (1989, 1990). The average benzene level measured in outdoor air in Contra Costa County was 2 $\mu g/m^3$, giving 8 $\mu g/m^3$ – 2 $\mu g/m^3$ = 6 $\mu g/m^3$ attributable to indoor and personal sources. Wallace (1990) estimated the national contribution of passive smoking to indoor benzene as 2 $\mu g/m^3$, so we shall assume that 2 $\mu g/m^3$ of the total of indoor concentration of 6 $\mu g/m^3$ in Contra Costa County came from passive smoking. Wallace (1990) estimated that the average benzene concentration inside automobiles in traffic in California at that time was 40 $\mu g/m^3$. Assuming that Contra Costa residents commuted 1 hour per day on average, then approximately 2 $\mu g/m^3$ of their 6 $\mu g/m^3$ average personal exposure was attributable to commuting in traffic. Finally, the remaining 6 $\mu g/m^3$ – 4 $\mu g/m^3$ = 2 $\mu g/m^3$ was attributed to consumer products in the home and the effect of an attached garage, since studies have shown benzene occurs widely in a variety of common household products (Sack et al. 1992; Wallace 1989; Wallace et al. 1987c; Westat, 1987) and benzene emissions from hot engine vapors and stored gasoline can infiltrate into the home (Thomas et al. 1993a).

Although outdoor air contributes 2 $\mu g/m^3$, or 25%, of the total exposure, the greatest share of this outdoor portion comes from automobile emissions. Wallace (1990) estimated that 85% of the outdoor benzene emissions are due to mobile sources (motor vehicles) and 15% were due to stationary sources. Applying these estimates to Contra Costa County, only 0.3 $\mu g/m^3$ of the total average population exposure of 8 $\mu g/m^3$ was attributable to stationary sources. Thus, passive smoking, auto travel, and indoor activities each contributed 2 $\mu g/m^3$ (25%) to the average total personal exposures; outdoor air quality also contributed 2 $\mu g/m^3$ (25%), of which 22% came from vehicular emissions and 3% was due to industry (Table 1.6 and Figure 1.9).

These results suggest that the large reduction of 50% in stationary sources called for by the BAAQMD Board would cause a reduction of 0.15 $\mu g/m^3$ in average benzene exposures, or a reduction of only 1.5% in cancer risk, which should have virtually no observable effect. Wallace

---

[2] If the concentrations are lognormally distributed (that is, the logarithm of the concentration is normally distributed), then they plot as a straight line on this type of graph (Ott 1995b). In that case, the geometric mean will equal the median and the arithmetic mean will lie approximately at the 70–80% quantile.

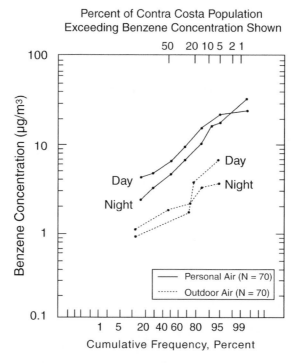

FIGURE 1.8 Example of logarithmic-probability plot for benzene, a known human carcinogen, measured in Contra Costa County in the California TEAM study. The graph shows the frequency distribution of daytime and nighttime personal exposures (solid lines) and daytime and nighttime outdoor concentrations (dotted lines). The data represent 91,000 people living in this industrialized Bay Area community with many factories and refineries. A vertical line drawn from 5% at the top to 95% at the bottom of this figure would intersect 18 $\mu g/m^3$ for the nighttime personal exposure and 3.2 $\mu g/m^3$ for the nighttime outdoor benzene concentration. This means that 5% of the 91,000 people (i.e., 4,550 persons) are estimated to have benzene personal exposures above 18 $\mu g/m^3$, mostly indoors at night. By contrast, 5% of the outdoor benzene concentrations were above 3.2 $\mu g/m^3$. At night, therefore, it was safer for people to be outdoors in their backyards than indoors in their homes. In the daytime, outdoor concentrations also were lower than personal exposures to benzene. (From Ott 1995a. With permission.)

TABLE 1.6
Sources of Benzene Exposures in the San Francisco Bay Area

Source	Concentration ($\mu g/m^3$)	Percentage (%)
Auto travel (1 hr/day)	2	25
Indoor sources/activities	2	25
Passive smoke	2	25
Auto emissions	1.7	22
Industry	0.3	3
Total	8	100

Source: Ott 1995a.

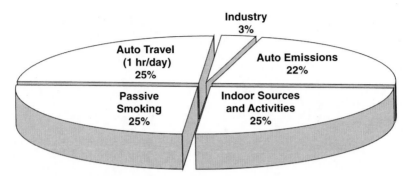

FIGURE 1.9 Sources of exposure to benzene in the San Francisco Bay Area, illustrating the relative source apportionment of exposure (RSAE) using personal exposure data and indoor air quality data from the California TEAM study in one of California's most industrialized counties, Contra Costa County, CA. (From Ott 1995a. With permission.)

(1990) reported an identical percentage of benzene exposure estimates applied on a nationwide basis. Evidently the BAAQMD Board and the local industries affected did not fully consider the implications of the local TEAM study findings: the small contribution that stationary sources make to benzene population exposures in the Bay Area compared with indoor and personal sources. The RSAE approach suggests this regulatory effort will reduce cancer risks in the Bay Area by 1.5% — a negligible amount — while the cost of these reductions may be very high. Although this analysis applies to benzene, the same analysis, showing the same marginal results of controlling outdoor stationary sources, can be applied to many other toxic air pollutants. Because of the large contribution to exposures made by common indoor sources (Wallace 1987, 1993a, 1995; Ott and Roberts 1998), we are likely to find the same relatively small effect of controlling outdoor stationary sources on population exposures.

1.9 PUBLIC EDUCATION

Over the last 15 years, dramatic reductions in the exposures of the population to environmental tobacco smoke (ETS) have occurred in the United States, particularly in California. Actions to reduce or eliminate smoking indoors have occurred on the county, city, and individual level. As a result, there have been enormous reductions in the exposures to the more than 4,000 chemicals in ETS that once were common in restaurants, stores, and bingo parlors (see Chapter 9). Most of these actions were taken by local governments or by individual restaurant owners, without federal laws or regulatory intervention on a national level. These changes suggest that, if the people are properly informed and educated, they will make intelligent choices to reduce their health risks. A critical reason for this change was research measuring the exposure of people to secondhand smoke and public information to educate and inform the citizens.

Consider for comparison the disease of AIDS. Our efforts to combat AIDS follow a receptor-oriented, public health approach that does not depend on regulations and lawyers. The primary approach is through public education. For many toxic environmental pollutants, the greatest reduction in public health risks can be achieved by reducing exposures at the personal level — by changing personal activities, habits, and lifestyles. As with AIDS, government can make major reductions in the health risks of the population from toxic pollutants through public education. It is critical, however, to inform the public about the ubiquitous nature of toxic pollutants in our homes, our carpets, and our consumer products, based on our scientific findings of these studies.

As indicated by Morgan (1993):

If public bodies are to make good decisions about regulating potential hazards, citizens must be well informed. The alternative of entrusting policy to panels of experts working behind closed doors has proved a failure, both because the resulting policy may ignore important social considerations and because it may prove impossible to implement in the face of grass-roots resistance.

Morgan (1993) concludes that:

...the only way to communicate risks reliably is to start by learning what people already know and what they need to know, then develop messages, test them and refine them until surveys demonstrate that the messages have conveyed the intended information.

A pilot focus group study of five groups of ten persons in Los Angeles by the RAND Corporation (Geschwind, Meredith, and Duan 1995) found that "...participants did not appear to know much about their indoor exposure risks, nor did they express a great deal of concern about them." Indoor pollutants and their related health risks did not dominate the initial discussion of risks they faced in their daily lives. Thus, there appears to be a serious lack of knowledge by the public about these important topics.

A well-known example of a source-oriented approach is the *Toxics Release Inventory*, a consequence of The Emergency Planning and Community Right-to-Know Act passed by Congress in 1986 listing substances whose use by industry must be reported. As we now know from exposure measurement field studies, actual exposure has little to do with the tons released. Indeed, as the earlier example on personal benzene exposures indicated, the concept of tons released usually exaggerates the contribution that stationary sources make to human health risks. What some environmental scientists and policy analysts have advocated is a Community Right-to-Know Act about our personal exposures.

Most exposure calculations performed by engineers at Superfund sites are based on generalized "default values" and have little or no significance for the populations living near these sites. Indeed, it often is accepted practice in making these estimates to theorize about hypothetical scenarios in the future — imagining hypothetical future events and artificial exposures — rather than making accurate measurements of exposure. Henry (2000) states, "Exposure is the wasteland of risk assessment because we tend to rely on default assumptions when we don't have methodologies or information." Many existing laws were written before the emergence of the new science of exposure analysis, and few of these laws show an understanding of exposure in their policies or call for accurate scientific information on exposure (see Chapter 21). Thus, a lack of exposure science often influences major environmental decisions about whether to clean up a site, because the policies on which these cleanups are based use default exposure factors instead of solid measurements of personal exposure that would include all the sources affecting residents near these sites. That is, these decisions nearly always follow a source-oriented approach and are estimated rather than measured, overlooking other nontraditional sources that may be more important in the lives of residents living near these sites. As a result, it is difficult to show that any of these costly actions has had a positive effect in reducing exposure or improving public health.

An important challenge facing those who try to educate the public about personal exposure to environmental pollutants is how to convey these ideas in an understandable manner that does not exclude important quantitative information. One graphical approach suggested for public education compares exposure from VOCs using a thermometer-like scale that is logarithmic and shows the exposure levels associated with everyday activities (Figure 1.10). This pollutant "exposimeter" compares the exposures associated with activities such as visiting a dry cleaners, repairing an automobile, pumping gasoline, cleaning a house with solvents, using spray cans, with the typically low levels measured near chemical plants, in urban areas, and near hazardous waste sites. Although the health effects associated with these exposure levels are difficult to convey in absolute terms, the exposure levels themselves are expressed in quantitative ranges. While the pollutants on the chart may differ from each other in toxicity, the logarithmic scale facilitates approximate comparisons of sources based on the levels of exposure. The exposimeter is intended only as a rough

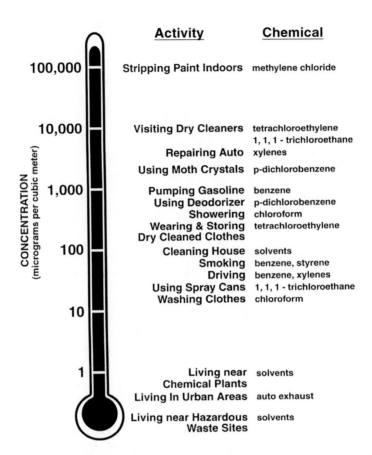

FIGURE 1.10 Toxic pollutant "exposimeter" suggested as a simple graph for communicating to the public about human exposure to VOCs has a single logarithmic scale that compares the exposure levels associated with personal activities and the levels associated with traditional outdoor sources. (Lance Wallace, personal communication, 2005.)

descriptive tool, although it is based on a number of exposure measurements from scientific studies in the United States, Canada, and other countries.

1.10 EXPOSURIST AS A PROFESSION

The environmental community includes a variety of professions, such as the chemist, the meteorologist, the biologist, the epidemiologist, the industrial hygienist, and the toxicologist. These professions are necessary and their members perform an important role. To this list of environmental professionals, perhaps, we must add a new profession — the practitioner of the exposure sciences, namely, the *exposurist*. Just as the toxicologist studies how pollutants affect the body, the exposurist studies how pollutants come into contact with the body. The exposurist's tools would include both measurements and validated models of exposure.

Perhaps the exposurist should evolve as a distinct professional specialty by itself, because none of the existing disciplines adequately accommodates all the tools required by the exposure sciences, but probably the best existing field for preparing the exposurist is the field of environmental engineering. The other fields do not emphasize the quantitative disciplines of making physical and chemical measurements or constructing and evaluating mathematical models on the small scale that is needed. The meteorologist, for example, considers the scale of miles rather than feet or inches from the body. Traditionally, the epidemiologist has limited expertise in making physical or chemical measurements. The exposurist's discipline requires rigorous scientific training — with considerable emphasis on chemistry, physics, engineering, biology, mathematics, and statistics. Early efforts have been made to develop a graduate program in the exposure sciences (Lioy 1991), and Stanford's course on the Introduction to Exposure Analysis, like other similar university courses elsewhere, is an important signal of progress.

Overall, the new field of exposure assessment is a *measurement science*. The most inappropriate way to predict exposures is to use an unvalidated model. Because human exposure — the manner in which pollutants make contact with the body — is extremely complex, unvalidated models nearly always give the wrong answer, or sometimes a ridiculous answer. Discoveries such as the personal cloud (Rodes, Kamens, and Wiener 1991; McBride et al. 1999), which explains why personal exposures contacting the body often differ from surrounding concentrations indoors or the strong effect on indoor air quality of opening doors or windows, show that many important details operate on a very small scale (less than 20 meters) that can exert enormous effects on human exposure. Today's unvalidated models — and certainly today's outdoor meteorological diffusion models — do not take these phenomena into account. Nor should they, because usually the largest share of one's exposure is due to sources that are indoors or in traffic, and these sources are not included in outdoor models.

As a result, a field of quantitative exposure modeling is emerging, the human activity pattern-microenvironmental exposure models (Burke, Zufall, and Özkaynak 2001; Duan 1982; Ott 1984; Ott et al. 1988; Ott, Mage, and Thomas 1992; see also Chapter 19 and Chapter 20). The purpose of these models is to predict the same exposure information obtained in the TEAM studies. But this new exposure modeling science is in its infancy. At this point in exposure science, no existing model has been adequately validated using field study data on exposures, with the possible exception of the early partial validation of the CO exposure model (Ott et al. 1988). While the concepts used in the models are interesting, at present no exposure model is sufficiently validated to be used by policy makers for making real environmental decisions.

1.11 CONCLUSIONS

Human exposure analysis is the science of determining what human beings on this Earth are exposed to. Over the last two decades, great strides have been made in using the direct approach — probability sampling combined with personal monitoring — to measure the exposures of the population of the United States to more than 100 chemicals. Great strides also have been made in understanding human activity patterns, developing personal monitors and indoor measurement methods, and conducting total exposure field studies. Despite the successes of the last 2 decades, more needs to be done to measure, with known accuracy, and to predict the exposures of the population to a variety of chemicals. Exposure analysis is a complex science, and we are only beginning to understand many important topics — the personal cloud, for example, which causes personal exposures to be higher than concentrations measured simultaneously indoors.

Although there have been a great many scientific achievements, exposure scientists seem to be confronted by a regulatory community and a public that has been unaware of this topic or of its importance for protecting their public health. Very often, people can reduce their exposures most by making relatively small changes in lifestyle or activities. An example is the use of doormats to reduce tracked-in dust, which thereby reduces pesticides, lead, and other toxic substances in the home (Chapter 14). Unless the public is aware of these scientific findings, people will not take the action that is necessary to reduce their exposures. An informed public — as shown by the example of reduction in exposure to secondhand smoke indoors (Chapter 9) — is a powerful force in bringing about change.

It has been suggested that more needs to be done to make the results and implications of exposure science more widely known and understood by the public. Most importantly, because of the complexity of this new science, possibly this science must develop a new environmental specialist — the exposurist — a practitioner of the science of exposure assessment. A number of scientific and measurement gaps exist that future research should address. For example, research should develop better methods to measure concentrations in the food eaten, the water drunk, and the skin contacted. Emphasis should be placed on the development, testing, and validation of new measurement methods for personal monitoring and indoor air quality. Additional studies of the exposures of children in daycare centers and schools are needed. Partially explained phenomena, such as the personal cloud, needs greater emphasis. How cooking activities generate indoor pollutants, and the effect of these pollutants on personal exposure and health, needs more study. Studies are needed to quantify the presence of toxic chemicals in consumer products. Follow-up TEAM studies and microenvironmental studies are needed to determine long-term trends in exposures. As the science of exposure assessment develops and matures, it is likely that many of these diverse research needs will be addressed in the future.

1.12 QUESTIONS FOR REVIEW

1. Explain the role of each of the five components of the *full risk model*. How do these five components relate to each other and what might be moving through the arrows connecting each box?
2. Based on the concepts in this chapter, can you give several examples of a *route of exposure*?
3. How would the analysis of health risks and their prevention and control be affected if information on any of the first three components of the full risk model is missing?
4. If information on the last component of the full risk model (health effects) is missing, explain why the other components still can be useful for evaluating alternatives for reducing human exposure by controlling different sources of the same pollutant.
5. Discuss the concept of *body burden*. Can you provide an example of body burden? How does body burden differ from exposure?
6. What is your initial definition of *dose* and *exposure*? How does dose differ from exposure? These concepts and definitions are discussed in greater detail in the next chapter.
7. Explain why the *total human exposure* approach is often described as *receptor-oriented*.
8. Why is the *total human exposure* concept referred to as total?
9. Describe the basic differences between the *direct* and *indirect* approaches of exposure analysis.
10. Discuss the reasons for using a *multistage probability design* in a total human exposure field study of an urban population.
11. What two main factors make an air pollutant emitted *indoors* usually much more serious from a health standpoint than when the same quantity of pollutant is emitted *outdoors*?
12. What findings from human *activity pattern studies* strongly suggest that exposure to pollutants while people are indoors should be considered when evaluating health risks?
13. Explain the rationale for *Relative Source Apportionment of Exposure (RSAE)*.
14. What type of scientific and professional training would be helpful to the profession of the *exposurist*?

REFERENCES

Akland, G.G., Hartwell, T.D., Johnson, T.R., and Whitmore, R.W. (1985) Measuring Human Exposure to Carbon Monoxide in Washington, DC, and Denver, Colorado During the Winter of 1982–83, *Environmental Science & Technology,* **19**: 911–918.

Behar, J.V., Schuck, E.A., Stanley, R.E., and Morgan, G.B. (1979) Integrated Exposure Assessment Monitoring, *Environmental Science and Technology,* **13**(1): 34–39.

Burke, J.M., Zufall, M.J., and Özkaynak, H. (2001) A Population Exposure Model for Particulate Matter: Case Study Results for $PM_{2.5}$ in Philadelphia, PA, *Journal of Exposure Analysis and Environmental Epidemiology*, **11**: 470–489.

Clayton, C.A., Perritt, R.L., Pellizzari, E.D., Thomas, K.W., Whitmore, R.W., Özkaynak, H., Spengler, J.D., and Wallace, L.A. (1993) Particle Total Exposure Assessment Methodology (PTEAM) Study: Distributions of Aerosol and Elemental Concentrations in Personal, Indoor, and Outdoor Air Samples in a Southern California Community, *Journal of Exposure Analysis & Environmental Epidemiology*, **3**: 227–250.

Clayton, E.D., Pellizzari, E.D., Whitmore, R.W., Perritt, R.L., and Quackenboss, J.J. (1999) National Human Exposure Assessment (NHEXAS): Distributions and Associations with Lead, Arsenic, and Volatile Organic Compounds in EPA Region 5, *Journal of Exposure Analysis and Environmental Epidemiology*, **9**(5): 381–392.

Duan, N. (1982) Microenvironment Types: A Model for Human Exposure to Air Pollution, *Environment International*, **8**: 305–309.

Duan, N. and Ott, W. (1992) An Individual Decision Model for Environmental Exposure Reduction, *Journal of Exposure Analysis and Environmental Epidemiology*, **2**(suppl. 2): 155–174.

Geschwind, S.A., Meredith, L.S., and Duan, N. (1995) A Breath of Fresh Air: Rethinking National Environmental Policy from Pollutant Source to Exposure Recipient, RAND Social Policy Development, Santa Monica, CA.

Goldstein, B.D., Tardiff, R.G., Baker, S.R., Hoffnagle, G.F., Murray, D.R., Catizone, P.A., Kester, R.A., and Caniparoli, D.G. (1992) *Valdez Air Health Study: Technical Report & Summary Report*, Alyeska Pipeline Service Co., Environmental Department, MS-538, Anchorage, AK.

Hartwell, T.D., Clayton, C.A., Mitchie, R.W., Whitmore, R.W., Zelon, H.S., Jones, S.M., and Whitehurst, D.A. (1984) *Study of Carbon Monoxide Exposure of Residents of Washington, DC, and Denver, CO*, Report No. EPA-600/54/84-031, NTIS PB-84-183516, U.S. Environmental Protection Agency, Research Triangle Park, NC.

Henry, C. (2000) Human Exposure: The Key to Better Risk Assessment, *Environmental Health Perspectives*, **108**(12): A559–A656.

Hill, S. (1991) Toxic Air Contaminant Reduction Plan, Bay Area Air Quality Management District, San Francisco, CA.

Jenkins, P.L, Phillips, T.J., Mulberg, E.J., and Hui, S.P. (1992) Activity Patterns of Californians: Use of and Proximity to Indoor Pollutant Sources, *Atmospheric Environment*, **26A**(12): 2141–2148.

Johnson, T. (1983) *A Study of Personal Exposure to Carbon Monoxide in Denver, Colorado*, Report No. EPA-600/4-84-014, NTIS PB-84-146125, U.S. Environmental Protection Agency, Research Triangle Park, NC.

Klepeis, N.E., Nelson, W.C., Ott, W.R., Robinson, J.P., Tsang, A.M., Switzer, P., Behar, J.V., Hern, S.C., and Engelmann, W.H. (2001) The National Human Activity Pattern Survey (NHAPS): A Resource for Assessing Exposure to Environmental Pollutants, *Journal of Exposure Analysis and Environmental Epidemiology*, **11**: 231–252.

Levy, D. (1991) New Rules Target Toxics in the Air: 50% Cut in Emissions by Businesses, *San Francisco Chronicle*, 8 August 1991.

Lioy, P.G. (1990) Assessing Total Human Exposure to Contaminants, *Environmental Science & Technology*, **24**: 938–945.

Lioy, P.G. (1991) Human Exposure Assessment: A Graduate Level Course, *Journal of Exposure Analysis & Environmental Epidemiology*, **1**(3): 271–281.

McBride, S.J., Ferro, A.R., Ott, W.R., Switzer, P., and Hildemann, L.M. (1999) Investigations of the Proximity Effect for Air Pollutants in the Indoor Environment, *Journal of Exposure Analysis and Environmental Epidemiology*, **9**: 602–621.

Morgan, M.G. (1993) Risk Analysis and Management, *Scientific American*, July: 32–41.

Ott, W.R. (1982) Concepts of Human Exposure to Air Pollution, *Environment International*, **7**: 179–196.

Ott, W.R. (1984) Exposure Estimates Based on Computer Generated Activity Patterns, *Journal of Toxicology: Clinical Toxicology*, **21**: 97–128.

Ott, W.R. (1985) Total Human Exposure: An Emerging Science Focuses on Humans as Receptors of Environmental Pollution, *Environmental Science & Technology*, **19**: 880–886.

Ott, W.R. (1990) Total Human Exposure: Basic Concepts, EPA Field Studies, and Future Research Needs, *Journal of the Air & Waste Management Association,* **40**(7): 966–975.

Ott, W.R. (1995a) Human Exposure Assessment: The Birth of a New Science, *Journal of Exposure Analysis and Environmental Epidemiology,* **5**(4): 449–472.

Ott, W.R. (1995b) *Environmental Statistics and Data Analysis,* Lewis Publishers, CRC Press, Boca Raton, FL.

Ott, W.R. and Roberts, J. (1998) Everyday Exposure to Toxic Pollutants, *Scientific American,* February **278**(2): 86–91.

Ott, W.R., Mage, D.T., and Thomas, J. (1992) Comparison of Microenvironmental CO Concentrations in Two Cities for Human Exposure Modeling, *Journal of Exposure Analysis and Environmental Epidemiology,* **2**(2): 249–267.

Ott, W.R., Thomas, J., Mage, D., and Wallace, L. (1988) Validation of the Simulation of Human Activity and Pollutant Exposure (SHAPE) Model Using Paired Days from the Denver, CO, Carbon Monoxide Field Study, *Atmospheric Environment,* **22**: 2101–2113.

Ott, W.R., Wallace, L.A., and Mage, D.T. (2000) Predicting Particulate (PM10) Personal Exposure Distributions Using a Random Component Superposition Statistical Model, *Journal of the Air and Waste Management Association,* **50**: 1390–1406.

Ott, W.R., Wallace, L., Mage, D., Akland, G., Lewis, R., Sauls, H., Rodes, C., Kleffman, D., Kuroda, K., and Morehouse, K. (1986) The Environmental Protection Agency's Research Program on Total Human Exposure, *Environment International,* **12**: 475–494.

Özkaynak, H., Xue, J., Spengler, J.D., Wallace, L.A., Pellizzari, E.D., and Jenkins, P. (1996) Personal Exposure to Airborne Particles and Metals: Results from the Particle TEAM Study in Riverside, CA, *Journal of Exposure Analysis and Environmental Epidemiology,* **6**: 57–78.

Pellizzari, E.D., Perritt, R., Hartwell, T.D., Michael, L.C., Sheldon, L.S., Sparacino, C.M., Whitmore, R., Leninger, C., Zelon, H., Handy, R.W., and Wallace, L.A. (1987a) *Total Exposure Assessment Methodology (TEAM) Study: Elizabeth and Bayonne, New Jersey; Devils Lake, North Dakota; and Greensboro, North Carolina, Volume II,* Report No. EPA 600/6-87/002b, NTIS PB 88-100078, U.S. Environmental Protection Agency, Washington, DC.

Pellizzari, E.D., Perritt, K., Hartwell, T.D., Michael, C., Whitmore, R., Handy, R.W., Smith, D., and Zelon, H. (1987b) *Total Exposure Assessment Methodology (TEAM) Study: Selected Communities in Northern and Southern California, Volume III,* Report No. EPA/600/6-87/002c, NTIS PB 88-100086, U.S. Environmental Protection Agency, Washington, DC.

Pellizzari, E.D., Thomas, K.W., Clayton, C.A., Whitmore, R.W., Shores, R.C., Zelon, H.S., and Perritt, R.L. (1993a) *Particle Total Exposure Assessment Methodology (PTEAM): Riverside, California Pilot Study — Volume I,* NTIS PB 93-166 957/AS, U.S. Environmental Protection Agency, Washington, DC.

Pellizzari, E.D., Thomas, K.W., Clayton, C.A., Whitmore, R.W., Shores, R.C., Zelon, H.S., and Perritt, R.L. (1993b) *Project Summary: Particle Total Exposure Assessment Methodology (PTEAM): Riverside, California Pilot Study — Volume I,* Report No. EPA/600/SR-93/050, U.S. Environmental Protection Agency, Research Triangle Park, NC.

Roberts, J.W., Budd, W.T., Ruby, M.G., Camann, D.E., Fortmann, R.C., Lewis, R.G., Wallace, L.A., and Spittler, T.M. (1992) Human Exposure to Pollutants in the Floor Dust of Homes and Offices, *Journal of Exposure Analysis & Environmental Epidemiology* (suppl. 1): 127–146.

Robertson, G.L., Lebowitz, M.D., O'Rourke, M.K., Gordon, S., and Moschandreas, D. (1999) National Human Exposure Assessment Survey (NHEXAS) Study in Arizona — Introduction and Preliminary Results, *Journal of Exposure Analysis and Environmental Epidemiology,* **9**(5): 427–445.

Rodes, C.E., Kamens, R.M., and Wiener, R.W. (1991) The Significance and Characteristics of the Personal Activity Cloud on Exposure Assessment Measurements for Indoor Contaminants, *Indoor Air,* **2**: 123–145.

Sack, T.M., Steele, D.H., Hammerstrom, K., and Remmers, J. (1992) A Survey of Household Products for Volatile Organic Compounds, *Atmospheric Environment,* **26A**(6): 1063–1070.

Sexton, K. and Hayward, S.B. (1987) Source Apportionment of Indoor Air Pollution, *Atmospheric Environment,* **21**(2): 407–418.

Smith, K.R. (1988a) Total Exposure Assessment: Part 1, Implications for the United States, *Environment,* **30**(8): 10–15, 33–38.

Smith, K.R. (1988b) Total Exposure Assessment: Part 2, Implications for Developing Countries, *Environment,* **30**(10): 16–20, 28–35.

Smith, K.R. (1993) Taking the True Measure of Air Pollution: We Have to Look Where the People Are, *EPA Journal,* **19**(4): 6–8.

Smith, K.R. (1994) *Looking for Pollution Where the People Are*, Asia-Pacific Issues, No. 10, East-West Center, Honolulu, HI, 8.

Spengler, J.D. and Sexton, K. (1983) Indoor Air Pollution: A Public Health Perspective, *Science,* **221**(4605): 9–17.

Spengler, J.D. and Soczek, M.L. (1984) Evidence for Improved Ambient Air Quality and the Need for Personal Exposure Research, *Environmental Science and Technology,* **18**: 268–280.

Thomas, K.W., Pellizzari, E.D., Clayton, C.A., Perritt, R.L., Dietz, R.N., Goodrich, R.W., Nelson, W.C., and Wallace, L.A. (1993a) Temporal Variability of Benzene Exposure for Residents in Several New Jersey Homes with Attached Garages or Tobacco Smoke, *Journal of Exposure Analysis and Environmental Epidemiology,* **3**: 49–73.

Thomas, K.W., Pellizzari, E.D., Clayton, C.A., Whitaker, D.A., Shores, R.C., Spengler, J.D., Özkaynak, H., and Wallace, L.A. (1993b) Particle Total Exposure Assessment Methodology (PTEAM) Study: Method Performance and Data Quality for Personal, Indoor, and Outdoor Aerosol Monitoring at 178 Homes in Southern California, *Journal of Exposure Analysis and Environmental Epidemiology,* **3**: 203–226.

Wallace, L.A. (1986) Personal Exposures, Indoor and Outdoor Air Concentrations, and Exhaled Breath Concentrations of Selected Volatile Organic Compounds Measured for 600 Residents of New Jersey, North Dakota, North Carolina, and California, *Toxicology and Environmental Chemistry,* **12**: 215–236.

Wallace, L.A. (1987) *The TEAM Study: Summary and Analysis: Volume I*, Report No. EPA 600/6-87/002a, NTIS PB 88-100060, U.S. Environmental Protection Agency, Washington, DC.

Wallace, L.A. (1989) Major Sources of Benzene Exposure, *Environmental Health Perspectives,* **82**: 165–169.

Wallace, L.A. (1990) Major Sources of Exposure to Benzene and Other Volatile Organic Compounds, *Risk Analysis,* **10**(1): 59–64.

Wallace, L.A. (1991a) Comparison of Risks from Outdoor and Indoor Exposure to Toxic Chemicals, *Environmental Health Perspectives,* **95**: 7–13.

Wallace, L.A. (1991b) Volatile Organic Compounds, in *Indoor Air Pollution: A Health Perspective,* Samet, J.M. and Spengler, J.D., Eds., The Johns Hopkins University Press, Baltimore, MD and London, U.K.

Wallace, L.A. (1993a) A Decade of Studies of Human Exposure: What Have We Learned? *Risk Analysis,* **13**(2): 135–143.

Wallace, L.A. (1993b) Exposure Assessment from Field Studies, in *Environmental Carcinogens — Methods of Analysis and Exposure Measurement, Volume 12: Indoor Air,* Seivert, B., Van de Wiel, H.J., Dodet, B., and O'Neill, I.K., Eds., IARC Scientific Publication No. 109, International Agency for Research on Cancer, Lyon, France, 136–152.

Wallace, L.A. (1993c) VOCs and the Environment and Public Health — Exposure, in *Chemistry and Analysis of Volatile Organic Compounds in the Environment,* Bloemen, H.J.Th. and Burn, J., Eds., Blackie Academic & Professional, Glasgow, Scotland.

Wallace, L.A. (1995) Human Exposure to Environmental Pollutants: A Decade of Experience, *Clinical and Experimental Allergy,* **25**: 4–9.

Wallace, L.A. (1996) Indoor Particles: A Review, *Journal of the Air and Waste Management Association,* **46**: 98–126.

Wallace, L.A. (1997) Human Exposure and Body Burden for Chloroform and Other Trihalomethanes, *Critical Reviews in Environmental Science and Technology,* **27**: 113–194.

Wallace, L.A. (2000) Correlations of Personal Exposure to Particles with Outdoor Air Measurements: A Review of Recent Studies, *Aerosol Science and Technology,* **32**: 15–25.

Wallace, L.A. (2001) Human Exposure to VOCs, in *Indoor Air Quality Handbook,* Spengler, J., Samet, J., and McCarthy, J., Eds., McGraw-Hill, New York, NY, 33.1–33.35.

Wallace, L.A. and O'Neill, I.K. (1987) Personal Air and Biological Monitoring of Individuals for Exposure to Environmental Tobacco Smoke, in *Environmental Carcinogenesis: Selected Methods of Analysis: Volume 9 — Passive Smoking,* International Agency for Research on Cancer (IARC), Lyon, France.

Wallace, L.A. and Ott, W.R. (1982) Personal Monitors: A State-of-the-Art Survey, *Journal of the Air Pollution Control Association,* **32**: 601–610.

Wallace, L.A. and Williams, R. (2005) Validation of a Method for Estimating Long-Term Exposures Based on Short-Term Measurements, *Risk Analysis,* **25**(3): 687–694.

Wallace, L.A., Duan, N., and Ziegenfus, R. (1994) Can Long-Term Exposure Distributions be Predicted from Short-Term Measurements? *Risk Analysis,* **14**: 75–85.

Wallace, L.A., Pellizzari, E.D., Hartwell, T.D., Davis, V., Michael, L.C., and Whitmore, R.W. (1989) The Influence of Personal Activities on Exposure to Volatile Organic Compounds, *Environmental Research,* **50**: 37–55.

Wallace, L., Pellizzari, E., Hartwell, T.D., Perritt, R., and Ziegenfus, R. (1987a) Exposure to Benzene and Other Volatile Compounds from Active and Passive Smoking, *Archives of Environmental Health,* **42**(5): 272–279.

Wallace, L.A., Pellizzari, E., Hartwell, T., Rosenzweig, R., Erickson, M., Sparacino, C., and Zelon, H. (1984) Personal Exposure to Volatile Organic Compounds, I. Direct Measurement in Breathing-Zone Air, Drinking Water, Food, and Exhaled Breath, *Environmental Research,* **35**: 293–319.

Wallace, L.A., Pellizzari, E., Hartwell, T., Sparacino, C., Sheldon, L., and Zelon, H. (1985) Personal Exposures, Indoor-Outdoor Relationships and Breath Levels of Toxic Air Pollutants Measured for 355 Persons in New Jersey, *Atmospheric Environment,* **19**: 1651–1661.

Wallace, L.A., Pellizzari, E.D., Hartwell, T.D., Sparacino, C., Whitmore, R., Sheldon, L., Zelon, H., and Perritt, R. (1987b) The TEAM Study: Personal Exposures to Toxic Substances in Air, Drinking Water, and Breath of 400 Residents of New Jersey, North Carolina, and North Dakota, *Environmental Research,* **43**: 290-307.

Wallace, L.A., Pellizzari, E., Leaderer, B., Hartwell, T., Perritt, K., Zelon, H., and Sheldon, L. (1987c) Emissions of Volatile Organic Compounds from Building Materials and Consumer Products, *Atmospheric Environment,* **21**: 385–393.

Wallace, L.A., Pellizzari, E.D., Hartwell, T.D., Whitmore, R., Sparacino, C., and Zelon, H. (1986) Total Exposure Assessment Methodology (TEAM) Study: Personal Exposures, Indoor-Outdoor Relationships, and Breath Levels of Volatile Organic Compounds in New Jersey, *Environment International,* **12**: 369–387.

Wallace, L.A., Pellizzari, E.D., Hartwell, T.D., Whitmore, R., Zelon, H., Perritt, R., and Sheldon, L. (1988a) The California TEAM Study: Breath Concentrations and Personal Exposures to 26 Volatile Compounds in Air and Drinking Water of 188 Residents of Los Angeles, Antioch, and Pittsburg, CA, *Atmospheric Environment,* **22**(10): 2141–2163.

Wallace, L., Thomas, J., Mage, D., and Ott, W. (1988b) Comparison of Breath CO, CO Exposure, and Coburn Model Predictions in the US EPA Washington-Denver CO Study, *Atmospheric Environment,* **22**(10): 2183–2193.

Wallace, L.A., Zweidinger, R., Erickson, M., Cooper, S., Whitaker, D., and Pellizzari, E. (1982) Monitoring Individual Exposure: Measurement of Volatile Organic Compounds in Breathing-Zone Air, Drinking Water, and Exhaled Breath, *Environment International,* **8**: 269–282.

Weselowski, J.J. (1984) An Overview of Indoor Air Quality, *Journal of Environmental Health,* **46**(6): 311–316.

Westat, Inc. (1987) and Midwest Research Institute, *Household Solvent Products: A 'Shelf' Survey with Laboratory Analysis,* report prepared for the Exposure Evaluation Division, Office of Pesticides and Toxic Substances, U.S. Environmental Protection Agency, Washington, DC.

Whitmore, R.W., Immerman, F.W., Camann, D.E., Bond, A.E., Lewis, R.G., and Schuam, J.L. (1994) Non-occupational Exposures to Pesticides for Residents of Two U.S. Cities, *Archives of Environmental Contamination and Toxicology,* **26**: 47–59.

Whitmore, R.W., Byron, M.Z., Clayton, C.A., Thomas, K.W., Zelon, H.S., Pellizzari, E.D., Lioy, P.J., and Quackenboss, J.J. (1999) Sampling Design, Response Rates, and Analysis Weights for the National Human Exposure Assessment Survey (NHEXAS) in EPA Region 5, *Journal of Exposure Analysis and Environmental Epidemiology,* **9**(5): 369–380.

Yakovleva, E., Hopke, P., and Wallace, L. (1999) Positive Matrix Factorization in Determining Sources of Particles Measured in EPA's Particle TEAM Study, *Environmental Science & Technology,* **33**: 3645–3652.

Yocum, J.E., Murray, D.R., and Mikkelsen, R. (1991) Valdez Air Health Study — Personal, Indoor, Outdoor, and Tracer Monitoring, Paper No. 91-172.12, presented at the 84th Annual Meeting of the Air and Waste Management Association, Vancouver, BC.

Zweidinger, R., Erickson, M., Cooper, S., Whitaker, D., Pellizzari, E., and Wallace, L. (1982) Direct Measurement of Volatile Organic Compounds in Breathing-Zone Air, Drinking Water, Breath, Blood, and Urine, U.S. Environmental Protection Agency, Washington, DC.

2 Basic Concepts and Definitions of Exposure and Dose[1]

Valerie G. Zartarian
U.S. Environmental Protection Agency

Wayne R. Ott
Stanford University

Naihua Duan
University of California

CONTENTS

2.1 Synopsis...34
2.2 Introduction...34
2.3 Criteria for a Framework of Human Exposure Definitions....................................37
2.4 Background..37
2.5 Definitions Related to Exposure and Dose...39
    2.5.1 Agent...39
    2.5.2 Target 39
    2.5.3 Exposure and Related Definitions...39
        2.5.3.1 Exposure...39
        2.5.3.2 Contact Boundary 40
        2.5.3.3 Contact Volume 41
        2.5.3.4 Concentration and Exposure Concentration 42
        2.5.3.5 Spatially Related Exposure Definitions..43
        2.5.3.6 Temporally Related Exposure Definitions.....................................44
    2.5.4 Dose and Related Definitions..45
    2.5.5 Practical Implications of the Theory of Exposure.......................................47
2.6 Examples Illustrating the Definitions..49
    2.6.1 Inhalation Exposure of a Person to Carbon Monoxide..............................49
    2.6.2 Dermal Exposure to DDT 52
    2.6.3 Ingestion Exposure to Manganese in a Vitamin Pill and to Lycopene in Tomatoes..54
2.7 Discussion...55
2.8 Glossary of Exposure and Dose-Related Terms 57
2.9 Questions for Review 59
References 60

---

[1] Adapted from Zartarian, Ott, and Duan (1997) and from WHO (2004). Although the material in this chapter was reviewed by the USEPA and approved for publication, it may not necessarily reflect official Agency policy.

2.1 SYNOPSIS

This chapter presents a quantitative framework for exposure to environmental pollutants and other agents, to help provide a common language for the exposure sciences. It reviews briefly the scientific literature to reveal the diverse and often confusing ways in which the terms "exposure" and "dose" have been used historically. Using six criteria for a new framework, it describes a set of quantitative definitions that encompass and expand upon earlier definitions. After "agent" (e.g., a pollutant) and "target" (e.g., a person) are defined, "exposure" is defined as the contact between an agent and a target. Contact takes place at a contact boundary over an exposure period. An "instantaneous point exposure" is defined as the joint occurrence of two events: (1) point i of a target is located at (x_i, y_i, z_i) at time t, and (2) an agent of concentration C_i is present at location (x_i, y_i, z_i) at time t. The definition of instantaneous point exposure is fundamental in that all other functions of exposure with respect to space or time — such as the average exposure and the integrated exposure — can be derived from it. Because exposure and dose are closely related and often confused, this framework also includes a quantitative definition of dose — the amount of agent that enters a target after crossing a contact boundary. Other commonly used, but often confusing, terms related to exposure and dose are also presented in this chapter. The definitions in this theoretical framework apply readily to human inhalation exposure, dermal exposure, and ingestion exposure to chemicals, as well as to other agents and targets. The glossary of terms and several examples illustrating their usage are based on the previously published framework as well as additional definitions adopted by both the International Programme on Chemical Safety (IPCS) and the International Society of Exposure Analysis (ISEA). Thus, they represent the most current definitions related to exposure and dose as of this book publication.

2.2 INTRODUCTION

A primary goal of environmental regulatory programs is to protect public health from the adverse effects of environmental pollutants. As discussed in Chapter 1, determining the risk to humans posed by environmental chemicals involves a conceptual human health risk model, which is a chain composed of five links: (1) pollutant sources; (2) concentrations of chemicals in environmental media (e.g., air, water, soil); (3) human exposure (i.e., contact) to chemicals; (4) dose (i.e., the amount of agent that enters a human organism); and (5) resulting health effects (Ott 1985). Each link in the chain depends on the previous one: without human contact with chemicals, there can be no exposure; without exposure, there can be no dose or risk. Understanding each of these components and the relationship among them can help determine effective risk reduction strategies (Ott 1990; Ott 1995; Akland 1991; Ott et al. 1986; Lioy 1990; Sexton et al. 1992). Historically, despite its importance, the single component of the risk model that has received the least scientific and regulatory attention is exposure.

The existing environmental health literature contains many different definitions of "exposure," "dose," and related terms. Some of the definitions are narrowly focused; some are vague; some are illogical or inconsistent with the others. Few efforts have been made to place these concepts within a consistent mathematical framework or to develop uniform quantitative definitions. In their paper on a conceptual approach to exposure and dose characterization, Georgopolous and Lioy (1994) refer in their Appendix to the earlier definitions suggested by Duan, Dobbs, and Ott (1989) as "...the most complete and consistent set of [exposure and dose definitions] available in the literature." This chapter summarizes a unified "theoretical framework" originally presented by Zartarian, Ott, and Duan (1997) in the peer-reviewed literature, based on Duan, Dobbs, and Ott (1989). Zartarian, Ott, and Duan (1997) include scientific definitions of exposure, dose, and related concepts developed to facilitate communication and inquiry among the exposure-related sciences. Their theoretical framework was designed to embrace practical measurements collected in the exposure sciences, and it was intended to improve the understanding and precision in thinking about such measurements. The Zartarian, Ott, and Duan (1997) framework was originally proposed as a unifying theoretical system for the exposure sciences, intended to embrace all the definitions likely to be used by exposure assessors, risk assessors, exposure modelers, scientific researchers, and others.

Exposure is defined here as contact between an agent and a target, with contact taking place at a contact boundary (i.e., exposure surface) over an exposure period. The definitions presented in this chapter build on a mathematical framework from the definition of exposure at a single point in space at a single instant in time. Exposure is commonly specified as pollutant concentration integrated over time. In addition to time-integrated exposure, time-averaged exposure can also be important. The definitions allow us to describe mathematically spatially integrated and spatially averaged exposures (i.e., exposure mass and exposure loading, respectively) that are relevant to exposure measurement methods such as wipe samples. A dermal exposure measurement based on a skin wipe sample, expressed as a mass of residue per skin surface area, is an example of a spatially averaged exposure, or exposure loading. The total mass on the wipe sample is an example of spatially integrated exposure, or exposure mass.

With the definition of a contact boundary, the framework inherent in our glossary (presented at the end of the chapter) emphasizes the need for exposure assessors to specify where the contact between an agent and a target occurs, to help facilitate communication and clarify the difference between exposure and dose. We define dose as amount of agent that enters a target in a specified time duration by crossing a contact boundary. If the contact boundary is an absorption barrier (e.g., exposure surface specified as a surface on the skin, lung, gut), the dose is an absorbed dose; otherwise (e.g., exposure surface specified as a conceptual surface over the nostrils and open mouth), it is an intake dose. This concise definition simplifies and is consistent with the numerous dose-related terms used in exposure-related fields. Terms such as internal dose, bioavailable dose, delivered dose, applied dose, active dose, and biologically effective dose that refer to agent crossing an absorption barrier are consistent with our definition of an absorbed dose. Terms such as administered dose and potential dose, which refer to the amount of agent in contact with an exposure surface, are consistent with our definitions of either intake dose or exposure mass depending on where the contact boundary is specified. While it is recognized that the term dose is often used in a way that does not refer to the crossing of a contact boundary (e.g., fields of toxicology, pharmacology), it is being defined this way here to eliminate confusion between exposure mass and dose.

While this framework is intended primarily for human exposure, it was formulated to apply to all carrier media, agents, and targets, building upon some of the concepts presented in the existing literature (e.g., exposure in terms of contact). The definitions in this chapter are worded and described mathematically in such a way that they can apply to both human and nonhuman species — plants and animals — and even to inanimate objects, such as buildings or photographic paper. To clarify the discussion and help develop a common language for the exposure sciences, a glossary of terms and several examples of usage are included at the end of this chapter. These examples, illustrating how the terms apply to the inhalation, dermal, and dietary ingestion routes, are not intended to be comprehensive, but to provide several contextual frameworks that could be applied to other case studies of interest.

To help develop their original theoretical framework, Zartarian, Ott, and Duan (1997) established six criteria that they believe a set of scientific definitions should possess. Following a review of the definitions found in the literature, they then proposed quantitative definitions of human exposure and related terms for use in the environmental sciences. To illustrate that their definitions were both mathematically rigorous and consistent with common sense, they presented several examples included in this chapter, showing how these concepts apply to the inhalation, dermal, and ingestion exposure routes and also to fields beyond the environmental sciences. The concepts presented were designed to help the environmental scientists make progress toward adopting a common language of exposure assessment by providing terminology that is scientifically consistent, concise, and understandable.

The concepts in Zartarian, Ott, and Duan (1997) were adopted by an international committee, the International Programme on Chemical Safety (IPCS, a joint program of the International Labour Organization, United Nations Environment Programme, and World Health Organization [WHO]) exposure terminology workgroup, concerned with "harmonizing" the language used in the field of exposure assessment (Callahan et al. 2001; Hammerstrom et al. 2001, 2002; IPCS, 2002; WHO, 2004). In 2004, the IPCS glossary was adopted as the official glossary of the International Society of Exposure Analysis (ISEA) (Zartarian, McKone and Bahadori 2004). The IPCS glossary was also officially adopted in 2004 by the ISEA to harmonize language used by its members, e.g., at ISEA conferences and in *Journal of Exposure Analysis and Environmental Epidemiology* publications (Zartarian, Bahadori, and McKone 2004).

Although the basic definitions of exposure, dose, and related terms in Zartarian, Ott, and Duan (1997) were adopted for the IPCS/ISEA glossary, some refinements to the original terms were made. In general, Zartarian, Ott, and Duan (1997) were more theoretical and focused on mathematically defining "contact," the fundamental concept behind exposure. Terms such as *contact boundary element, contact volume element, contact volume thickness, exposure point, instantaneous point exposure, intensity,* were included in Zartarian, Ott, and Duan (1997) but not included in the IPCS glossary. Likewise, the IPCS glossary contains a number of additional terms not included in Zartarian, Ott, and Duan (1997): *absorption, absorption barrier, activity pattern data, acute exposure, background level, bioavailability, biomarker, bounding estimate, chronic exposure, exposure assessment, exposure duration, exposure event, exposure frequency, exposure loading, exposure mass, exposure model, exposure pathway, exposure period, exposure route, exposure scenario, intake, microenvironment, pica, source, stressor, subchronic exposure.* The terms in both glossaries can be applied to the primary routes of human exposure (inhalation, dietary, dermal) to chemicals, whereas the Zartarian, Ott, and Duan (1997) definitions were designed to apply more generally to all agents and targets. This chapter presents both the theoretical and more generally applicable framework in Zartarian, Ott, and Duan (1997) as well as the refined and additional terms based on the harmonization efforts of the IPCS terminology workgroup. Several definitions (e.g., biomarker, moving average, total exposure) presented here have been modified slightly or added by the authors of this chapter.

The IPCS workgroup identified four terms that were particularly difficult to define due to their relatively recent emergence. These are aggregate exposure, aggregate dose, cumulative exposure, and cumulative dose. In studying the literature, the terminology workgroup found that "aggregate" and "cumulative" seem to be used interchangeably, suggesting (1) exposures that are from multiple sources, received via multiple exposure pathways, or doses received through multiple routes; (2) exposures or doses which accumulate over time, often over a lifetime; or (3) exposures or doses from more than one chemical or stressor simultaneously or sequentially. The U.S. Environmental Protection Agency (USEPA 2002), in its *Framework for Cumulative Risk Assessment*, uses "aggregate" as a term referring to the risks over time from multiple sources, pathways, and routes for a single chemical or stressor, reserving "cumulative" for assessments where (aggregate exposures or doses for) multiple chemicals or stressors are evaluated together. These definitions are based more on the contextual language of the 1996 *Food Quality Protection Act* than a study of how the terms are being used worldwide, so it remains to be seen whether these particular definitions will come into general usage within the scientific community. At this time, the authors have chosen to postpone inclusion of "aggregate" and "cumulative" in the glossary, awaiting further clarification in the field regarding usage of these terms. We have, however, included a definition of *total exposure*.

TABLE 2.1
Criteria for Definitions of
Exposure and Related
Concepts

1. Build on Previous Definitions
2. Logically Consistent Framework
3. Parsimonious
4. Expressed Mathematically
5. Agree with Common Sense
6. Consistent with Common Usage

2.3 CRITERIA FOR A FRAMEWORK OF HUMAN EXPOSURE DEFINITIONS

To develop their definitions of exposure, Zartarian, Ott, and Duan (1997) established six criteria that definitions of exposure should meet (Table 2.1). First, they proposed that the definitions should build upon previous ones but be more specific and self-consistent than earlier definitions. Second, they argued that the terminology related to exposure should be integrated into a logical theoretical framework such that the definitions are applicable to different types of exposures with respect to time and space and are self-consistent across environmental media (e.g., water, air, soil), agents (e.g., CO, pesticides), and targets (e.g., humans, trees, animals). Third, the definitions should be stated as concisely as possible. Fourth, to achieve precision, the concepts should possess mathematical clarity, which is especially important for practitioners of the field of exposure modeling (Sexton and Ryan 1988). Finally, they proposed that the definitions should agree with common sense and be reasonably consistent with common usage. Thus, the objective of Zartarian, Ott, and Duan (1997) was to propose a theoretical framework that could meet the six criteria in Table 2.1.

2.4 BACKGROUND

A review of the literature in the diverse fields of exposure assessment, environmental policy and management, risk assessment, industrial hygiene, environmental health, toxicology, and epidemiology reveals inconsistent schools of thought about the definition of exposure, with potential for confusion and miscommunication. Several researchers, for example, discuss exposure in terms of ambient environmental pollutant sources (Weinstein 1988; IPCS, 1983; Landers and Yu 1995). In the book *Principles of Exposure Measurement in Epidemiology*, exposure is defined as "...any of a subject's attributes ...that may be relevant to his or her health," implying that a behavior, such as smoking, is an exposure (Armstrong, White, and Saracci 1992). Monson (1980) defines exposure as "a (potential) cause of disease." Lisella (1994) states four different definitions of exposure: (1) "the amount of radiation or pollutant that represents a potential health threat to the living organism in an environment"; (2) "the opportunity of a susceptible host to acquire an infection..."; (3) "the amount of a biological, physical, or chemical agent that reaches a target population"; and (4) "the means by which an organism comes in contact with a toxicant." Another unique definition for exposure exists in the field of radiobiology: "the quotient dQ by dm where dQ is the absolute value of the total charge of the ions of one sign produced in air when all the electrons liberated by photons in a volume of air having mass dm are completely stopped" (ICRU 1979). In addition to the diversity of exposure definitions in the scientific literature, some references use the term "exposure" without defining it at all (IPCS 1994), while others define it in a circular fashion (e.g., "the amount of a factor to which a group or individual was exposed" [Last 1995]).

Despite the variations in definitions, the predominant definition of exposure in the literature involves contact between a target (e.g., human) and a chemical, physical, or biological agent in an environmental medium (Ott 1982; USEPA 1992a; ATSDR 1993; USEPA 1991; Last 1995; Armstrong, White, and Saracci, 1992; NRC 1991; Lioy 1991; Duan and Ott 1992; Duan, Dobbs, and Ott 1990; Calabrese, Gilbert, and Pastides 1989; CRC 1995; Lippmann 1987; Herrick 1992; Lisella 1994; Georgopoulos and Lioy 1994). The quantitative definition of exposure presented in this chapter is based on this concept of contact with an agent. Thus, there is a need to define contact mathematically as part of the quantitative meaning of exposure (Ott 1995). Zartarian, Ott, and Duan (1997) addressed that need by examining exposure at a boundary in terms of concentration at a point.

Zartarian, Ott, and Duan (1997) also recognized that it was important to address the time interval over which contact occurs in an exposure event. Several references, for example, state that exposure can be quantified by multiplying concentration contacted by the duration of contact, so that the unit of exposure is concentration multiplied by time (NRC 1991; USEPA 1992; ATSDR 1993; Georgopolous and Lioy 1994). While such a time-integrated exposure is one possible formulation, there are, in fact, a number of different possible formulations of exposure with respect to time that have units of concentration rather than concentration multiplied by time (Ott 1995). These other time-related formulations of exposure include instantaneous exposure, average exposure, and peak exposure (Ott 1982; Duan and Ott 1992; Duan, Dobbs, and Ott 1990; Duan, Dobbs, and Ott 1989; Tardiff and Goldstein 1991; Georgopolous and Lioy 1994; Armstrong, White, and Saracci 1992). In addition to discussing various temporally related exposures (i.e., at a point or a boundary), the framework discussed in this paper covers various spatially related exposures.

Because the concepts of exposure and dose are closely related, these two terms have often been used interchangeably, causing confusion and miscommunication. For example, Armstrong, White, and Saracci (1992, p. 11) states:

> Dose may be measured either as the total accumulated dose (cumulative exposure), for example, the total number of cigarettes smoked...or as the dose or exposure rate, for example, number of cigarettes smoked per day...Cumulative exposure is calculated as the sum of the products of dose rates by durations of the periods of time for which they apply...Examples of available dose include the concentration of asbestos fibres per mL of ambient air in a small time interval (an exposure rate), or average asbestos fibres per mL of ambient air multiplied by years of exposure (a cumulative exposure)...

Lisella (1994) uses the term "exposure dose" to refer to "a measure of the ... radiation at a certain place, based upon the ability of the radiation to produce ionization." The term "exposure dose" has also been defined as "the total mass of a xenobiotic that is actually inhaled, ingested, or applied on the skin" (Hrudey, Chen, and Rousseaux 1996). Most references, however, define dose as an "amount," "quantity," or "presence" of an agent at a site of toxic effect, resulting from a penetration across a boundary into the target (Atherley 1978; IPCS 1994; USEPA, 1992a; Weinstein, 1988; Armstrong, White, and Saracci 1992; Calabrese, Gilbert, and Pastides 1989; ATSDR 1993; USEPA 1991; Duan, Dobbs, and Ott 1990; Duan, Dobbs, and Ott 1989; NRC 1991; Tardiff and Goldstein 1991). Because of its close relationship to exposure in the human health risk model described above, dose is defined in our framework as the amount of agent that enters a target (e.g., human, lung, stomach) after crossing a contact boundary (e.g., skin, oral passage, nasal passage, gut wall). As shown below, various types of doses discussed in the literature are consistent with this general definition of dose.

Because the variety of exposure definitions across different scientific disciplines has caused confusion for people who read the literature, the following section presents a unified set of definitions, based on the criteria given above, to facilitate future communication in exposure-related fields.

2.5 DEFINITIONS RELATED TO EXPOSURE AND DOSE

2.5.1 AGENT

Despite the numerous exposure definitions in the literature, there appears to be general agreement that an exposure agent is a chemical, biological, or physical entity that may cause deleterious effects in a target after contacting the target (Cohrssen and Covello 1989; USEPA 1992a). Not all agents, however, cause deleterious effects (e.g., vaccines, oxygen). A *stressor* is an entity, stimulus, or condition that can modulate normal functions of a target organism or induce an adverse response (e.g., lack of food, drought, toxic chemical or other agent). Thus, we define an *agent* as a chemical, biological, or physical entity that contacts a target. There are two types of agents: energy-form (e.g., light, heat, sound, radiation, magnetism, electricity) and matter-form (e.g., chemical mass, bacterial count, particle count).

Agents are usually carried in a liquid (e.g., water, beverage), gas (e.g., air, cigarette smoke), or solid (e.g., food, soil) *medium*, defined as material (e.g., air, water, soil, food, consumer products) surrounding or containing an agent. A matter-form agent may be of molecular dimensions (e.g., gas phase pollutants) or larger (e.g., aerosol pollutants). For a matter-form agent, the amount of agent per unit volume is called *concentration* (e.g., μg particulate matter per m^3 air). For an energy-form agent such as light, the amount of agent per unit area is called *intensity*.

2.5.2 TARGET

There also seems to be general agreement over the concept of a *target* of exposure as a physical, biological, or ecological object exposed to an agent. Examples of targets are humans, biological organs, buildings, walls, trees, or photographic paper. Selection of a target depends in part on the agent of interest. For example, one might want to know the exposure of a human to carbon monoxide; of a building to acid rain; of the skin to a pesticide; of paper to light; of an eardrum to sound. To discuss exposure, both the agent and the target should be clearly specified. In considering human exposure, the target could be the entire human body (an external exposure) or a particular organ such as the lung (an internal exposure). The course an agent takes from the source to the target is the *exposure pathway*, and the way an agent enters a target after contact (e.g., by ingestion, inhalation, or dermal absorption) is the *exposure route*. The origin of an agent for the purposes of an exposure assessment is called the *source*.

2.5.3 EXPOSURE AND RELATED DEFINITIONS

2.5.3.1 Exposure

Exposure is defined as contact between an agent and a target. Zartarian, Ott, and Duan (1997) defined *instantaneous point exposure* $\xi(x,y,z,t)$, or ξ, as contact between an agent and a target at a single point in space and at a single instant in time. Such contact can be represented as the joint occurrence of two events, as shown in the following expression (Ott 1995):

$$\left\{ \begin{array}{l} \text{Point } i \text{ of the target is} \\ \text{located at } (x_i, y_i, z_i) \text{ at time } t \end{array} \right\} \cap \left\{ \begin{array}{l} \text{Agent of Concentration } C_i \text{ is present at} \\ \text{location } (x_i, y_i, z_i) \text{ at time } t \end{array} \right\}$$

As we shall see, all spatially and time-related concepts of exposure are built upon this definition of ξ (Duan, Dobbs, and Ott 1989, 1990). Instantaneous point exposure can be measured as concentration at the point of contact, and thus $\xi(x,y,z,t)$ and $C(x,y,z,t)$ are equivalent only under conditions of contact. An *exposure point* is a point on an exposure surface (contact boundary, defined below) at which contact with an agent occurs.

It is clear that there are three components of exposure: a target, an agent, and contact. The following sections introduce three more concepts essential to the framework: (1) the contact boundary, which relates to the target component, (2) the contact volume, which relates to the agent component, and (3) the definition of concentration at a point, which relates to the contact component. With these concepts defined, Zartarian, Ott, and Duan (1997) then built a set of spatially related and temporally related exposure definitions from which all exposure concepts of interest can be discussed.

2.5.3.2 Contact Boundary

A *contact boundary* (also referred to as an *exposure surface*; IPCS 2002; WHO 2004) is defined as a surface on a target where an agent is present, or a surface on a target containing at least one exposure point (based on Duan, Dobbs, and Ott 1989, 1990). Locations of human contact boundaries could include the lining of the stomach wall, the surface of the lung, the exterior of an eyeball, and the surface of the skin. The contact boundary defines exactly what is being exposed and where, and is an important concept because different points on a target can receive different exposures at the same time (i.e., a concentration at one point on the target can differ from the concentration at another point) (Duan, Dobbs, and Ott 1989, 1990). Georgopolous and Lioy (1994) state that it is important to define exposure in such a way that it can apply to points or areas in space in addition to individuals or populations. Whereas Georgopolous and Lioy (1994) focus on probability distributions of exposures for individuals and populations, the Zartarian, Ott, and Duan (1997) framework emphasizes the nature of contact between a point or set of points and an agent.

While specification of the location of the contact boundary is important for discussing or assessing exposure, there are no rules for this specification; selection of the contact boundary depends on the target and application of interest. Consider, for example, a living target such as a leaf (Figure 2.1). Because the leaf has an irregular shape, one might define the contact boundary as the upper surface of the leaf, a locus of points that can be viewed as a curved conceptual film that is shown partially peeled away in this figure. The agent of the exposure of the leaf might be light, but it also might be hydrogen ions in acid rain reaching the leaf. The concentration $C(x_1,y_1,z_1,t)$ shown at the point above the leaf may differ from the concentration $C(x_2,y_2,z_2,t)$ on the contact boundary. The concentration $C(x_2,y_2,z_2,t)$, not $C(x_1,y_1,z_1,t)$, is the concentration to which a point on the leaf is exposed, because Point 2 is coincident with the leaf's contact boundary.

Although the contact boundary is conceptual, specifying its location will facilitate communication among exposure assessors and other scientists, eliminating possible confusion over where contact occurs. One can specify a contact boundary on parts of the body (e.g., eyes, tongue) other than the traditional "exchange boundaries" of the skin, lung, and digestive tract (USEPA 1992).

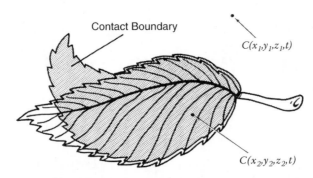

Contact Boundary

$C(x_1,y_1,z_1,t)$

$C(x_2,y_2,z_2,t)$

FIGURE 2.1 Conceptual contact boundary of a leaf. This contact boundary is peeled back for illustration. The leaf is exposed to the concentration (x_2,y_2,z_2) at time t. It is not exposed to the concentration (x_1,y_1,z_1) because point 1 is not located on the leaf's contact boundary. (From Zartarian, Ott, and Duan 1997. With permission.)

Both internal and external human exposures can be considered, depending on how the target and contact boundary are specified. The contact boundary concept allows us to clearly define nonhuman exposure (e.g., exposure of a statue to acid rain) as well as human exposure. Georgopolous and Lioy (1994) point out that definitions of exposure must continue to evolve and highlight the complexity of exposure systems. The theoretical framework discussed in Zartarian, Ott, and Duan (1997) addresses this need by first defining instantaneous contact at a point, and then expanding that idea to other temporal and spatial exposure concepts.

2.5.3.3 Contact Volume

Figure 2.2 illustrates a contact boundary, with the set of points $z = h_1(x,y)$ constituting the lower surface of a volume bounded by surfaces h_1 and h_2. This conceptual volume, denoted as the *contact volume* (IPCS 2002; WHO 2004), is a volume adjoining a contact boundary in which the agent has a high probability of contacting the contact boundary in the time interval of interest.

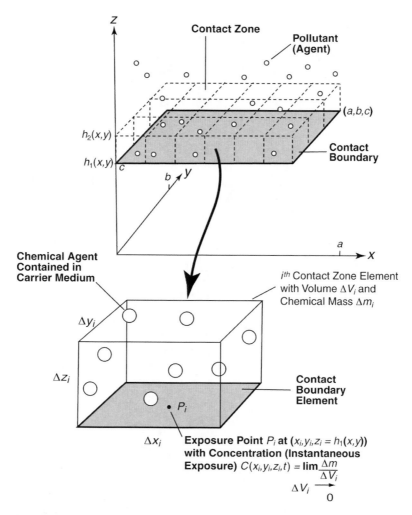

FIGURE 2.2 Illustration of pollutant exposure. (From Zartarian, Ott, and Duan 1997. With permission.)

The thickness of the contact volume, h_2-h_1, can be discussed theoretically as the distance (from the contact boundary) in which a particle of agent has at least a pre-specified probability p of intersecting the contact boundary within a pre-specified time interval t. Note that the thickness is defined as a function of both p and t: for a fixed probability, the contact volume thickness increases with increasing time; for a fixed time, the contact volume thickness decreases with increasing probability. The dermal exposure example in Zartarian, Ott, and Duan (1997) describes how the contact volume thickness may be estimated.

The contact volume shown in Figure 2.2 is divided into n adjacent boxes (*contact volume elements* each with volume ΔV_i whose lower surfaces are *contact boundary elements* each with area ΔA_i) according to the assumption that the ith contact volume element contains only one type of carrier medium (e.g., air, liquid, soil). A point P_i on the contact boundary, located at (x_i, y_i, $h_1(x_i, y_i)$) on the ith contact boundary element, is considered to be an exposure point if the agent is contained within (i.e., present at least at one point) the ith contact volume element.

Defining a contact volume is important because all exposure measurements collected from devices such as personal air monitors or skin wipes contain implicit information about a volume in which the agent is contained. For example, a personal air monitor measures the amount of chemical collected in the volume of air flowing into the monitor. A skin wipe collects mass of chemical in a thin volumetric region just above the skin surface, even though the measurement is reported as a mass of chemical per surface area of skin (our definition of *spatially averaged boundary exposure*, or *exposure loading*, discussed below). Thus, while in practice we may not always be able to explicitly determine the volume or thickness of the contact volume, the contact volume concept is inherent in all exposure measurements. The thickness concept inherent in the contact volume is also relevant when thinking about dermal exposure to air containing a gas phase pollutant or skin immersion to a chemical-containing liquid. The concentration of the air or liquid can be measured as mass per volume, and to convert this to a loading on the skin (mass per area, which can be measured in field studies), one needs to know the thickness of the relevant boundary layer in which mass transfer occurs. The general mass transfer equation is given by Fick's First Law of Diffusion. There are multiple theories for predicting mass transfer coefficients in membranes (film theory, penetration theory, surface renewal theory, boundary layer theory) that incorporate the concept of a film thickness used in calculating mass transfer from a fluid near a surface such as the skin surface (see Zartarian 1996 for a discussion of mass transfer theories and calculation of contact volume thicknesses).

2.5.3.4 Concentration and Exposure Concentration

If the mass of the agent contained in the ith contact volume element is Δm_i, the concentration of the ith contact volume element is $\Delta m_i/\Delta V_i$. The instantaneous point exposure at point P_i is expressed as the limiting value of this ratio as the contact volume element becomes small. Prior to the definition proposed here, a detailed review of the literature indicates that no one has previously proposed a mathematical definition of a concentration at a point. To meet the need for a quantitative definition of "concentration," this Zartarian, Ott and Duan (1997) framework introduced the following definition of concentration at a point as given by Equation 2.1:

$$C(x_i, y_i, z_i) = \lim_{\Delta V_i \to 0} \left(\frac{\Delta m_i}{\Delta V_i} \right) = \frac{dm}{dV} \qquad (2.1)$$

This definition of concentration is analogous to the definition of density at a point found in elementary physics books. To find density of a fluid at any point, we can isolate a small volume element ΔV around the point and measure the mass Δm of the fluid contained within that element. The density at a point is the limit of the ratio $\Delta m/\Delta V$ as the volume element at that point becomes smaller and smaller (Resnick, Halliday, and Krane 1992; Fishbane, Gasiorowicz, and Thornton 1993). In the new exposure theory, the concept of concentration is the same as density, except that density usually refers to the *pure* gas, liquid, or solid, while concentration usually refers to one chemical constituent within the gas or liquid mixture. In practice, one needs to assume that a fluid sample is large compared to atomic dimensions, allowing us to compute concentration as a mass per volume of a sample. *Exposure concentration* is defined as the exposure mass divided by the contact volume or the exposure mass divided by the mass of contact volume, depending on the medium. The exposure at a single point is expressed as the limiting value of this ratio as the contact volume element becomes small.

Instantaneous point exposure has units of amount of agent per volume of medium in the contact volume (e.g., µg chemical/m³ air). Recall the physicist Louis de Broglie's theory that our observable universe is composed entirely of radiation and matter, and that both radiation and matter have a wave-particle nature: frequency and wavelength describe the wave aspect; position and momentum describe the particle aspect (Halliday and Resnick, 1981). Particle- and wave-form agents can, in theory, be discussed in a similar manner so that the definitions of exposure apply consistently to both types of agents.

2.5.3.5 Spatially Related Exposure Definitions

Defining exposure at a point is essential because the point exposures may vary from point to point over a contact boundary. An exposure assessor, however, will often be interested in the average of all point exposures in the contact volume, denoted as the *spatially averaged exposure* ξ_{sa}. It is measured as the amount of agent per unit volume of the contact volume and defined mathematically as shown in Equation 2.2:

$$\xi_{sa}(t) = \frac{\iiint\limits_{V} C(x,y,z,t)\,dxdydz}{\iiint\limits_{V} dxdydz} \tag{2.2}$$

In some cases, spatially averaged exposures expressed as amount of agent per area, rather than per volume, will be of interest (e.g., chemical mass per area of the skin surface; estimation of the total number of particles collected on an area of plate). Thus, *spatially averaged boundary exposure* ξ_{sab}, or the *exposure loading*, is defined mathematically as the spatially integrated exposure divided by the contact boundary area (Equation 2.3), where the *spatially integrated exposure* ξ_{si}, or *exposure mass*, is the amount of agent present in the contact volume (Equation 2.4).

$$\xi_{sab}(t) = \frac{\iiint\limits_{V} C(x,y,z,t)\,dxdydz}{\iint\limits_{A} dxdy} \tag{2.3}$$

$$\xi_{si}(t) = \iiint\limits_{V} C(x,y,z,t)\,dxdydz \tag{2.4}$$

It may be useful to compare ξ_{sab} against dermal exposure measurement devices (e.g., skin patches, hand rinses), which yield mass values that are typically divided by the surface area for which the measurement is applied. In the case of a chemical mass agent and a contact boundary area on the human skin, this definition is consistent with that of "skin loading" by Fenske (1993): "the amount of material reaching the skin and available for absorption...e.g., μg cm^{-2}" and of the USEPA (1992a) definition of applied dose, "the amount of a substance in contact with the...skin...and available for absorption." For an energy-form agent (e.g., light), the spatially averaged boundary exposure would be the average of all point exposures over the contact boundary, where the point exposures are expressed as intensities (e.g., lumens/ft^2) rather than concentrations.

2.5.3.6 Temporally Related Exposure Definitions

The above sections defined the spatially related concepts of exposure (i.e., exposure at a point or set of points) as *instantaneous exposures* (i.e., exposures at an instant in time). *Time-integrated exposure* ξ_{ti} is the integral (or summation for the discrete form) of instantaneous exposures over the exposure duration (Equation 2.5). The *exposure duration* of interest[2] is the length of time over which continuous or intermittent contacts occur between an agent and a target. For example, if an individual is in continuous contact with an agent for 10 minutes a day, for 300 days over a 1-year time period, the exposure duration is 1 year. An *exposure event* is the occurrence of continuous contact between an agent and a target. The *exposure frequency* is the number of exposure events in an exposure duration (300 in this example). An *exposure period* is the time of continuous contact between an agent and a target (10 minutes in this example).

Time-integrated exposure is the type of exposure primarily emphasized by the National Academy of Sciences and in the U.S. Environmental Protection Agency's exposure assessment guidelines, since measurements (e.g., via personal air monitors) usually provide incremental data on exposure (NRC 1991; USEPA 1992a). *Time-averaged exposure* ξ_{ta} is the time-integrated exposure divided by the exposure duration (Equation 2.6) (Duan, Dobbs, and Ott 1989, 1990).

$$\xi_{ti}(x,y,z) = \int_{t_1}^{t_2} C(x,y,z,t)\,dt \tag{2.5}$$

$$\xi_{ta}(x,y,z) = \frac{\int_{t_1}^{t_2} C(x,y,z,t)\,dt}{t_2 - t_1} \tag{2.6}$$

These different time-related exposure terms are illustrated in Figure 2.3. An *exposure time profile* is a continuous record of instantaneous exposures (e.g., ξ_{sa} or ξ) over a time period, i.e., a plot of concentration as a function of time (Duan, Dobbs, and Ott 1989, 1990). The researcher must decide what time period to consider with the exposure time profile (e.g., a biologically relevant time period).

This section and the previous one described exposure first as a function of space, and then as a function of time. With these definitions, we can understand and specify concepts of exposure that are functions of both time and space. Four additional concepts, therefore, are: time-averaged spatially averaged exposure ξ_{tasa}; time-averaged spatially integrated exposure ξ_{tasi}; time-integrated spatially averaged exposure ξ_{tisa}; and time-integrated spatially integrated exposure ξ_{tisi}. Equations describing them mathematically are analogous to Equation 2.2 through Equation 2.6.

---

[2] Although it may be counterintuitive, this definition of exposure duration is based on common usage in environmental risk assessment practice.

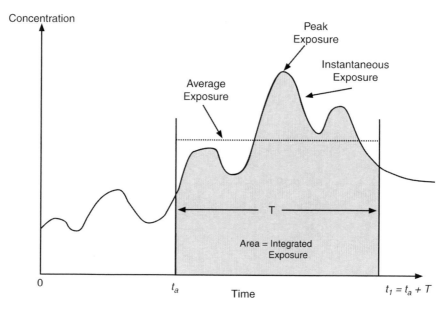

FIGURE 2.3 Hypothetical exposure time profile; pollutant exposure as a function of time illustrating how the average exposure, integrated exposure, and peak exposure relate to the instantaneous exposure. (From Zartarian, Ott, and Duan 1997. With permission.)

2.5.4 Dose and Related Definitions

Because the terms "exposure" and "dose" are closely linked, this chapter specifies the general definition of *dose* as the amount of agent that enters a target in a specified time duration after crossing a contact boundary. While there can be exposure without a corresponding dose, there can be no dose without a corresponding exposure (Figure 2.4). For example, if the skin surface is specified as the contact boundary, exposure of the skin occurs if chemical is placed on the skin surface. If the chemical does not partition into the stratum corneum, then there is no dose.

The *instantaneous dose rate* is the rate at which the agent passes through a unit area of the contact boundary, and it has units of quantity of agent (e.g., mass) per area per unit time. *Time-integrated spatially integrated dose* can be calculated by integrating the instantaneous dose rate over the area of the contact boundary and the exposure time interval. As with exposure, other time-related terms, such as instantaneous, profile, peak, and average dose, can be discussed.

There are two primary classifications of dose: intake dose and absorbed dose (USEPA 1992a). If the agent crosses the contact boundary without subsequently diffusing through a resisting boundary layer, the dose is classified here as an *intake dose. Intake* is the process by which an

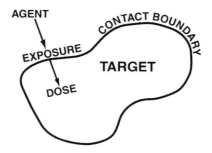

FIGURE 2.4 Conceptual drawing illustrating the difference between exposure and dose.

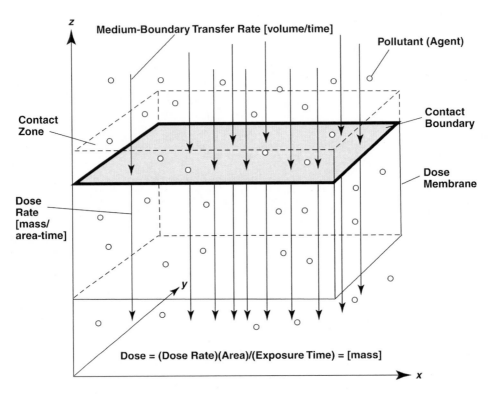

FIGURE 2.5 Illustration of absorbed pollutant dose. (Adapted from Zartarian 1996.)

agent crosses an outer exposure surface of a target without passing an absorption barrier (e.g., through ingestion or inhalation). The *instantaneous spatially integrated intake dose* is the total amount of agent crossing the contact boundary at an instant in time, and it can be computed by multiplying the instantaneous spatially averaged exposure [M/L^3] by a carrier *medium intake rate*, the rate at which the medium crosses the contact boundary [L^3/T].

If the agent diffuses through an *absorption barrier* after crossing the contact boundary, then the dose is classified as an *absorbed dose* (Figure 2.5). *Absorption*, also referred to as *uptake*, is the process by which an agent crosses an absorption barrier. An *absorption barrier* is defined as any contact boundary or zone that may retard the rate of penetration of an agent into a target. Examples of absorption barriers are the skin, respiratory tract lining, and gastrointestinal tract wall. In Figure 2.5, it is a volume of thickness Δd whose upper surface is coincident with the contact boundary. Mathematically, the absorbed dose rate is a vector, a multiple of the gradient of the instantaneous exposure, pointing in the direction of flow across the contact boundary. When the agent crosses the resisting dose membrane, a concentration gradient and flux is created. For an energy-form or matter-form agent, the absorbed dose rate is the flux of agent across the contact boundary (e.g., lumens/cm^2-sec; mg/cm^2-sec).

In keeping with criteria 1 and 6 in Table 2.1, it is useful here to discuss the other dose terms stated in the literature in terms of the definitions presented in this framework. "Internal dose," "absorbed dose," and "bioavailable dose" are defined as the amount of an agent penetrating across the absorption barriers of an organism (USEPA 1992a; Hrudey, Chen, and Rousseaux 1996; NRC 1991). Those definitions are consistent with the "absorbed dose" definition included in this chapter. "Applied dose," defined as the amount of substance presented to an absorption barrier and available for absorption (USEPA 1992a), and "external dose," defined as total mass of an agent at the applied site available for absorption (Hrudey, Chen, and Rousseaux 1996) are equivalent to our definition of a spatially integrated exposure (e.g., mass of chemical in the contact volume), where the contact boundary is the skin, alveoli, or gut wall. The terms "administered dose" and "potential dose," defined as the amount of agent that is actually inhaled, ingested, or applied to the skin (USEPA 1992a; Hrudey, Chen, and Rousseaux 1996), are consistent with our definition of intake dose via the nasal/oral contact boundary and with our definition of spatially integrated exposure where the skin is the contact boundary. The National Research Council (NRC 1991) defines "potential dose" as an exposure value multiplied by a contact rate (e.g., rates of inhalation, ingestion, or skin absorption) assuming total absorption of the agent. This latter definition is consistent with our definition of "intake dose" for inhalation and ingestion, and with our definition of "absorbed dose" subsequent to skin exposure. "Delivered dose," "active dose," and "biologically effective dose," the amount of chemical available for interaction for an organ or cell (Lioy 1990; USEPA 1992a; NRC 1991; Armstrong, White, and Saracci 1992), are additional classifications for our definition of "absorbed dose," based again on specification of the contact boundary where an adverse effect occurs. "Available dose" defined by Armstrong, White, and Saracci (1992) is the same as our definition of exposure (e.g., concentration of asbestos fibers per mL of ambient air). "Eliminated dose" in Duan, Dobbs, and Ott (1990) is similar to our definition of a negative dose, the amount of agent that is eliminated from the target. "Net dose," defined by Duan, Dobbs, and Ott (1989, 1990) is intake or absorbed dose (positive dose) minus the eliminated dose (the negative dose). "Accumulated dose," the amount of agent accumulated inside the target, is equivalent to our definition of time- and spatially averaged dose. The above discussion illustrates how our framework encompasses the multitude of dose definitions found in the literature. The primary difference among the various dose terms is the implied location of the contact boundary.

Several other dose-related terms that tend to cause confusion are bioavailability and biomarker (or biological marker). *Bioavailability* is defined as the rate and extent to which an agent can be absorbed by an organism and is available for metabolism or interaction with biologically significant receptors. Bioavailability involves both release from a medium (if present) and absorption by an organism. A *biomarker* or *biological marker* is defined as an indicator (cellular, biochemical, analytical, or molecular) of a recent or previous exposure in a biological system, intended to measure body burden. The biomarker may be the actual agent itself or a metabolite. For example, measurement of CO in breath is a body burden indicator of CO in the blood (carboxyhemoglobin). We define body burden here as the amount of chemical in the body at a given instant in time.

2.5.5 PRACTICAL IMPLICATIONS OF THE THEORY OF EXPOSURE

Current methods are not always able to measure factors such as exposure concentration, exposure mass, and contact volume with complete accuracy. In the field of inhalation, for example, the exposure concentration is calculated as the amount of agent collected on a filter in a personal air monitor (a surrogate for the exposure mass) divided by the volume of air sampled (a surrogate for the contact volume). In fact, the measured exposure concentration is not identical to the concentration inhaled. Variation in breathing rate throughout the monitoring period will affect the amount inhaled, and the personal air monitor may not retain 100% of the agent that is drawn into the air filter. Likewise for dermal exposure, the exposure mass and exposure loading that actually come into contact with the skin are usually only fractions of the amount removed from the skin by a wipe sample because only a thin layer of agent directly in contact with the skin is capable of being absorbed. These discrepancies reflect limitations in the measurement methods, rather than in the definitions, and should be noted as uncertainties in the exposure assessment. An *exposure assessment* is defined as the process of estimating or measuring the magnitude, frequency and duration of exposure to an agent, along with the number and characteristics of the population exposed. Ideally, it describes the sources, pathways, routes, and the uncertainties in the assessment.

The basic concepts in exposure theory have major practical implications. It is customary, for example, in some risk assessments to use atmospheric dispersion models to predict "community exposure," even though this term has not been clearly defined. With our definitions, the target could be defined as the city and the contact boundary could be the city boundary. The city is exposed to the agent coming from its surrounding; the amount of agent that enters the city is its dose. Although these concepts apply to a gross level and lack the fine precision on an individual (e.g., the city-level dose does not tell us how the dose is distributed across its citizens), these concepts might provide useful input for risk assessments.

Esmen and Marsh (1996), evaluating air dispersion modeling in epidemiology, state that "...exposure is strongly influenced by seasonal or even daily lifestyle preferences, travel and excursion habits, and indoor/outdoor concentration differences, as well as the complexities involved in the estimation of exposures in general." Although new *exposure models* (conceptual or mathematical representations of the exposure process) have been developed to account for human activity patterns (Duan 1982, 1991; Ott 1983–1984; Ott et al. 1988, 1992; Zartarian 1996; Zartarian et al. 2000), few earlier models account for concentrations actually contacted by humans via their daily activities (e.g., indoor concentrations in air breathed or on surfaces touched). As time has passed, there has been increasing agreement about the need to include the presence and activities of people in these models to predict exposure and subsequent risk more accurately. The use of a formal exposure framework can help address this problem by emphasizing contact between an agent and a target.

In addition to improving exposure models, the framework originally proposed by Zartarian, Ott, and Duan (1997) also was intended to enhance the quality and understanding of measurements taken in exposure field studies. For example, the basic design of all the Total Exposure Assessment Methodology (TEAM) large-scale personal exposure monitoring field studies embodied the ideas of the exposure framework by seeking to measure concentrations as close to each person as possible. In the TEAM studies, a representative probability sample of several hundred respondents was equipped with *personal exposure monitors* (PEMs; devices worn on or near the contact boundary that measures concentration) that were carried during their daily activities; if applicable, concentrations also are measured in food and drinking water. These studies have generated a rich information data base about the exposures of the population to carbon monoxide (Akland et al., 1985), volatile organic compounds (Wallace 1987, 1993, 1995; Wallace et al. 1985, 1986, 1987, 1988a, 1989; Zweidinger et al. 1982), pesticides (Whitmore et al. 1994), and particles (Clayton et al. 1993; Özkaynak et al. 1996; Pellizzari et al. 1993a,b; Thomas et al. 1993). The data have provided important findings about the number of people exposed to various chemicals, the levels of their exposures, and the causes of their exposures (Ott 1985, 1990, 1995; Ott and Roberts 1998; Wallace 1987, 1993, 1995).

In the TEAM studies, it was assumed implicitly that the contact boundary for an air pollutant is at the nasal opening. Unfortunately, it was not possible for the hundreds of TEAM respondents to wear inlets attached to their noses. Practical necessity dictated a simpler method: the TEAM particle studies, for example, used a small inlet mounted on the person's lapel (Pellizzari et al., 1993a) and the TEAM CO studies used an electronic monitor carried at waist level with a strap over the respondent's shoulder (Akland et al., 1985). An implicit assumption of these TEAM studies is that concentrations observed at the monitor's inlet do not differ significantly from those reaching the nose. In general, the assumption that a PEM carried by a respondent measures the same concentration that reaches a contact boundary in front of the nasal opening seems reasonable, although there may be a "personal cloud effect" bias due to a person's thermal boundary layer, pollutants emitted by clothing, and other factors (Rodes, Camens, and Wiener 1991; Ferro 2002). Future exposure field studies will need to demonstrate that a personal monitor worn on the body truly reflects the air pollutant concentrations to which the person is exposed (i.e., the concentration at the contact boundary). Thus, it is hoped that the framework in this chapter adds precision in thinking regarding the meaning of exposure measurements.

2.6 EXAMPLES ILLUSTRATING THE DEFINITIONS

The following exposure scenarios are intended to illustrate the application of the definitions presented in this chapter and to show that these definitions are self-consistent across agents, targets, and exposure routes, and over space and time. An *exposure scenario* is a combination of facts, assumptions, and inferences that define a discrete situation where potential exposures may occur. These may include the source, the exposed population, the timeframe of exposure, *microenvironment*(s) (surroundings that can be treated as homogeneous or well characterized in the concentrations of an agent), and *activity pattern data* (information on human activities used in exposure assessments; these may include a description of the activity, frequency of activity, duration spent performing the activity, and the microenvironment in which the activity occurs). Exposure scenarios are often created to aid exposure assessors in estimating exposure. This discussion is intended to clarify important concepts that previously have been treated inconsistently in the literature. The three case studies below focusing on a human target are based on those published in Zartarian, Ott, and Duan (1997); however, they have been modified and expanded to reflect the IPCS glossary definitions in IPCS (2002) and WHO (2004). Glossary terms are italicized when first used.

2.6.1 INHALATION EXPOSURE OF A PERSON TO CARBON MONOXIDE

Because most studies in the exposure assessment field to date have focused on human exposure to air pollutants, this first example looks at carbon monoxide (CO) exposure. In this example, inhalation *exposure* refers to contact between an air pollutant and a human prior to inhalation. The *exposure route* is inhalation; the *agent* of interest is carbon monoxide; the *target* is a person; the *medium* is air; and the *contact boundary* is specified as a locus of points over the entrance to the mouth and nose (shown as S1 in Figure 2.6). Theoretically, the *exposure concentration* is the average of the air concentrations at each point on the exposure surface. The exposure at each point is the limiting value of the concentration in a *contact volume element* containing the point. Practical necessity dictates that in actual field studies, the air in the vicinity of a person's nose is implicitly assumed to be well mixed, and a measured exposure concentration (e.g., 20 ppm) is assumed to

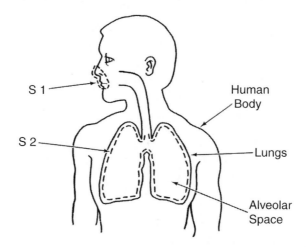

S1: Exposure boundary for whole body exposure
S2: Exposure boundary for lung exposure

FIGURE 2.6 Human inhalation exposure showing different contact boundaries. (Adapted from Duan, Dobbs, and Ott 1990.)

be the exposure at the person's nose (assuming that the measurement was in close proximity to the person). The *contact volume* is the theoretical volume of air available for inhalation in the *exposure period* of interest. The volume of air inhaled during the exposure period is a surrogate for the volume of the contact volume. Often a personal air monitor is used to estimate the exposure concentration of the agent in the contact volume.

Figure 2.7 illustrates an actual diurnal carbon monoxide profile of a 36-year-old female homemaker. This exposure *time profile* was plotted for 1 of the 450 Denver participants in the 1982–1983 Denver–Washington, DC, carbon monoxide personal exposure monitoring field study. Each point on the exposure time profile represents the instantaneous inhalation exposure to carbon monoxide and is measured by a personal exposure monitor. The woman's peak exposure on the study day can be seen as approximately 33 ppm CO. The exposure period that contains the peak exposure appears to be approximately an hour. This high exposure is probably due to her proximity to emissions from motor vehicle tailpipes while shopping. However, the woman could also be exposed to a *background level* of CO (the amount of an agent in a medium that is not attributed to the source(s) under investigation in an exposure assessment). The tailpipes in this *exposure scenario* are the *sources*. The physical trajectory that the CO takes from the tailpipe to the woman's exposure surface is the *exposure pathway*. The time profile in Figure 2.7 depicts the woman's *acute exposure*, since the *exposure duration* is one day. Each spike on the profile represents an *exposure event*, and the *exposure frequency* appears to be ~6 events per day. This exposure profile could have been estimated with an exposure model that combines the woman's *activity pattern data* with measured or predicted concentrations in the *microenvironments* in which she spent time. If longer exposure durations are of interest, such a study could be repeated for several days (*subchronic exposure*) or several years (*chronic exposure*).

The area under the profile shown in Figure 2.7 is the *time-integrated exposure*. The *time-averaged exposure* over the entire day can be found by dividing the area under the curve by the total time the monitor was worn (i.e., a 24-hour exposure duration. Also shown in the figure is the *moving average* 8-hour exposure, which is computed from the exposure profile by taking the average of the measured concentration over the previous 8 hours once every hour. The numbers at the top of the exposure profile are activity codes describing the person's microenvironments in her activity pattern data based on the diary that each person maintained (Ott et al. 1988). Finally, the *biomarker*, blood carboxyhemoglobin (COHb), was computed from the measured CO exposure profile using the Coburn pharmacokinetic model (Coburn, Foster, and Cane 1965; Wallace et al. 1988b) that provides an estimate of the *absorbed dose* of CO, which agrees fairly well with a blood-breath measurement of this respondent later in the day (see small dot marked "Observed COHb" on figure, which was derived from a breath measurement). A *dose rate* could be computed from this profile by computing the dose per unit time.

Inhalation dose refers to the mass of carbon monoxide that crosses the theoretical contact boundary at the entrance to the mouth and nose during the exposure duration, and enters the target. With the way we have defined inhalation exposure, the woman receives a dose as soon as carbon monoxide crosses the oral/nasal region into the body. This is an example of an *intake dose*, because the agent does not cross an *absorption barrier* before entering the target.

The *medium intake rate* is equivalent to the woman's inhalation rate, the volume of air breathed per unit time. An intake dose *time profile* could be obtained by multiplying each point on the exposure profile by the inhalation rate to obtain the various time formulations of inhalation dose (i.e., peak, maximum, time-integrated, time-averaged). Note that the concentration reaching the inhalation contact boundary at any instant of time will always be greater than or equal to zero, but the air flows in and out of the body with each breath, so the airflow carrying the pollutant, and hence the dose, will be positive or negative from one moment to the next.

We could also define an internal exposure surface such as the epithelial lining of the lung (S2 in Figure 2.6) and define the target to be the lung. The dose process in this case would be *uptake* (*absorption*) rather than intake, because the agent would pass through an absorption barrier before

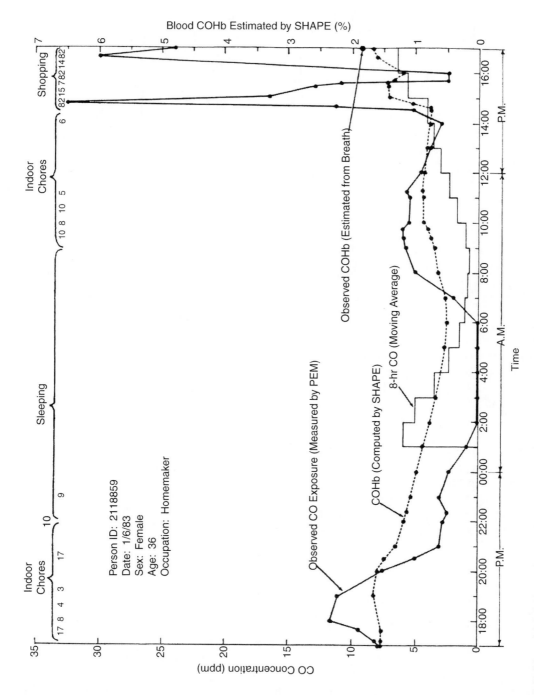

FIGURE 2.7 Diurnal inhalation exposure time profile of a selected participant in the Denver area. (From Ott 1995. With permission.)

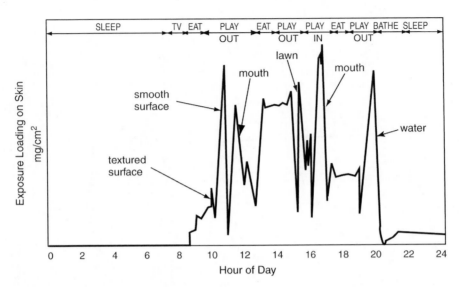

FIGURE 2.8 Hypothetical dermal exposure time profile. (Adapted from IPCS 2002; WHO 2004.)

entering the target. Therefore the specification of the exposure surface depends on the question to be answered.

2.6.2 Dermal Exposure to DDT

This second example focuses on a dermal *exposure scenario*; the *exposure route* is dermal absorption. Figure 2.8 illustrates a dermal exposure *time profile*, with the person's relevant *activity pattern data* indicated (including information about contacts with different media and surfaces). One can compute the *time-integrated exposure* and *time-averaged exposure* using Figure 2.8 in a way similar to that described in the inhalation example. However, it is often helpful in the case illustrated in Figure 2.8 to plot the *exposure loading,* rather than the *exposure concentration,* on the y-axis, since concentrations at different points on the skin surface are for different media and therefore have different units.

Dermal *exposure* is the contact between an *agent* and the external skin surface (the *contact boundary*) of a *target* (e.g., a human) (Figure 2.9). A point on the skin surface is considered to be exposed if chemical mass is present in the *contact volume* containing the point. Dermal exposure can occur via skin contact with a chemical in different *media*. Figure 2.9 illustrates the exposure of an area of a hand, during one *exposure event*, to the pesticide dichlorodiphenyl-trichloroethane (DDT) (the *agent* or *stressor*) carried in air, water, and soil media. Some points are exposed to DDT on aerosols, some to aqueous phase DDT, and some to DDT molecules in a soil matrix. The contact boundary was selected here for the purpose of illustration as a rectangular region on the stratum corneum surface, as shown. *Instantaneous point exposures* on the contact boundary in Figure 2.9 vary spatially because different media are in contact with the skin surface. Current dermal exposure measurement devices, including skin patches, fluorescent tracers, and skin wipes, measure dermal exposure as exposure loading.

The *contact volume* for the dermal route is the volume above the skin surface in which the chemical is considered to be in contact with the skin. The thickness of the contact volume (Δz in Figure 2.9), can be estimated as the height above the skin within which any molecule has a high probability of intersecting the exposure surface during the *exposure period.* This height will vary as a function of the exposure period since the probability of a far-away molecule intersecting the exposure surface will increase with time if diffusion is in the direction toward the skin. Zartarian Ott, and Duan (1997) presents an approach for estimating the thickness of the contact volume using

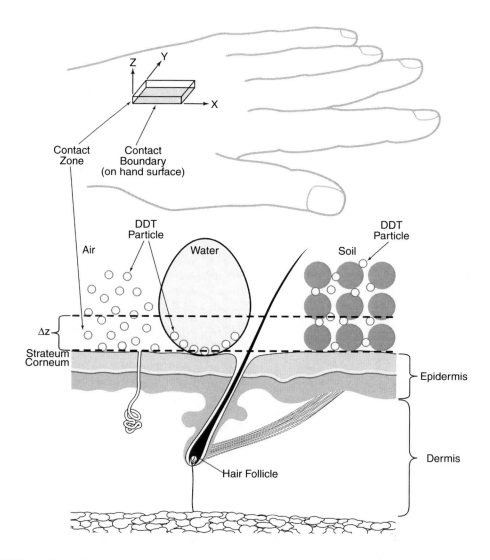

FIGURE 2.9 Dermal exposure to DDT. (From Zartarian, Ott, and Duan 1997. With permission.)

several well-established theories of mass transfer and a range of contact times. The results yield estimates of contact volume thickness in air, water, and soil that agree reasonably well with typical measured film thicknesses. The contact volume concept, based on sound engineering models, allows us to discuss the theory behind what we measure in practice.

When the skin is immersed in a fluid medium, such as water or air containing the agent (Figure 2.10), *uptake (absorption)* is usually estimated as a function of the exposure concentration, the area of the exposure surface (i.e., the immersed skin surface area, shown as S1 and S2 in Figure 2.10), and the exposure period using an empirical, chemical-specific permeability coefficient. The agent in the medium is assumed to be an infinite, well-mixed source. While a contact volume could be defined for this exposure scenario in the same way that it is defined for the dermal residue deposition scenario (i.e., the volume above the skin surface in which any molecule has a high probability of contacting the skin surface), the contact volume is not needed to estimate uptake for this scenario.

Chemicals in media contacting the skin surface partition to the stratum corneum, the outermost layer of the skin, and then diffuse through the stratum corneum into the viable epidermis and dermis, then into general circulation in the body. Because the agent diffuses through an *absorption*

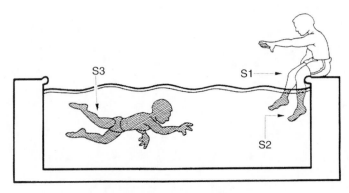

S1: Exposure boundary for residual shower water
S2: Exposure boundary for pool water
S3: Exposure boundary for swimmer

FIGURE 2.10 Dermal exposure of swimmer. (From Duan, Dobbs, and Ott 1990. With permission.)

barrier, the *dose* process is uptake (absorption) and dermal dose is classified as an *absorbed dose*. The stratum corneum provides the major barrier to chemical absorption in the skin, and thus is the dermal absorption barrier. Dermal dose is complex not only because there can be multiple carrier media on a given exposure boundary, but also because the dose membrane is composed of different media. Because chemicals migrate through the stratum corneum via diffusion, the *absorbed dose rate* under steady-state conditions can be calculated using basic principles of diffusion.

2.6.3 INGESTION EXPOSURE TO MANGANESE IN A VITAMIN PILL AND TO LYCOPENE IN TOMATOES

One also can speak of exposure and dose to chemicals consumed in food and drinking water. In these types of exposure scenarios, the *exposure route* is ingestion. Although ingestion dose may be of greater interest than ingestion exposure, we provide the following unusual examples to illustrate that the definition of exposure is consistent across all exposure routes.

Suppose someone were interested in the total amount of manganese (Mn), the *agent*, entering the body when a person, the *target*, takes a vitamin pill containing 5 mg of Mn, a typical formulation for nonprescription multivitamin products. The *contact volume* in this case is the volume of the pill (i.e., 400 mm³). If the analyst selects a *contact boundary* directly in front of the mouth (S1 in Figure 2.6), the same theoretical surface used earlier to illustrate inhalation exposure, the oral *exposure* to Mn will be zero up until the instant that the tablet first touches the exposure surface. Exposure occurs for the second that it takes for the tablet to cross the exposure surface, and then drops to zero again. The *exposure mass* in this example is 5 mg, and the *exposure concentration* is 1.25×10^7 mg/m³ (5 mg divided by the 400 mm³ volume of the tablet). The *exposure period* is the one second that it takes for the pill to cross the exposure surface. The vitamin pill container in this example is the *source*, and the *exposure pathway* is the course the pill takes from the container to the person's mouth. The vitamin pill is the *medium* here, and the *medium intake rate* is 400 mm³ per second. Because the pill crosses an exposure surface that is not an *absorption barrier*, this is an example of *intake*, and the dose is an *intake dose*. The person's oral intake dose from the tablet will be 5 mg, even though other parts of the person's body may receive a different exposure and dose later. Alternatively, an internal exposure surface could be defined as the epithelial lining of the gastrointestinal tract, and the Mn from the vitamin pill that crossed this epithelial *absorption barrier* would be an *uptake dose* for the specific internal target.

A person who has difficulty swallowing solid pills might grind up the tablet and dissolve it in a glass of water. If the liquid in the glass is 200 mL, then the concentration of the tablet when diluted in water will be 5 mg/200 mL = 25,000 mg/m³. If the person drinks the entire contents, the values on the person's *time profile* of exposure concentrations would be zero as the glass moves toward the lips, followed by 25,000 mg/m³ for several seconds (as it crosses the exposure surface), followed again by zero. Regardless of whether the tablet is eaten or dissolved in water and drunk, the same amount of Mn crosses the oral exposure surface, and the dose is 5 mg in both cases.

We could plot an exposure time profile as in the inhalation example. If the person takes a vitamin pill once a day every day for a year, then the *exposure frequency* is one *exposure event* per day, and the *exposure duration* is 1 year. The *time-integrated exposure* would be (25,000 mg/m³) (1 second/event) (365 events) and the *time-averaged* exposure would be (25,000 mg/m³) (1 second/event) (365 events)/365 days. The daily time profile would illustrate the person's *acute exposure*; the 1-year time profile would illustrate the person's *chronic exposure*. The person's behaviors regarding consumption of vitamin pills would be the relevant *activity pattern data* for this example.

Additives, nutrients, and chemical residues in food items can be treated in a similar way. Consider, for example, the ingestion exposure to lycopene[3] from consumption of a tomato. In this case, lycopene is the *agent*. The *contact volume* is the volume of the tomatoes consumed. *Exposure* occurs when the tomatoes cross the *contact boundary* in front of the mouth in the same way as it does in the vitamin pill example. As lycopene would be absorbed from the gastrointestinal (GI) tract, one could define the contact boundary of interest as the epithelium of the GI tract, an *absorption barrier*. The concentration of lycopene at the surface of the GI tract could be considered as a function of time at any given point. This exposure would be similar to that in the example of dermal exposure, rising from zero to a maximum, followed by a decline back to zero as a result of absorption or passage with other materials out of the GI tract via excretion.

It is impractical to measure the concentration of lycopene as it passes through the body and is metabolized or eliminated. Typically, the concentration of lycopene in consumed foods would be measured, and the intake of those foods would be combined with the measured concentrations in each food type to estimate exposure.

There are a number of techniques for estimating exposure to ingredients such as lycopene in a tomato product, additives such as a high intensity sweetener in a beverage, or contaminants such as methylmercury in fish. Market basket studies, and duplicate diet studies provide information concerning the level of the substance in foods. In the duplicate diet approach, for example, a second helping of all the food items a person eats at a given meal is prepared and submitted for laboratory analysis. Then the pollutant concentration in each food item or composites of several food items is measured, and the person's intake dose is estimated by multiplying the pollutant concentration by the quantity of each food item that the person eats. This practical method for estimating exposure and dose via ingestion is useful in many applications, and it is consistent with the Zartarian, Ott, and Duan (1997) conceptual framework and the IPCS/ISEA glossary.

2.7 DISCUSSION

This chapter discusses a theoretical framework of exposure- and dose-related definitions developed in accordance with six criteria (Table 2.1). The three examples given above help the reader understand the wide applicability of the definitions given in this chapter. The framework is relevant to a variety of exposure-related scenarios, situations, or events. For example, one can apply the above exposure and dose definitions quantitatively using the mass balance equation to describe indoor air quality. Suppose the target is the volume of air contained in a room with a single open door through which air enters and a single open window through which air exits, the agent is particulate matter, and the contact boundary is the flat surface area bound by the door frame. One can then discuss the "exposure" of the room in terms of the outdoor air concentration, and the "dose" of the room in terms of the mass of pollutant inside the room. Each term in the mass balance

---

[3] Lycopene is a fat-soluble red plant pigment (carotenoid) known to have antioxidant effects.

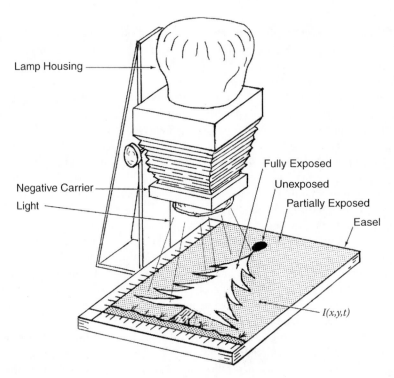

Lamp Housing

Negative Carrier

Light

Fully Exposed

Unexposed

Partially Exposed

Easel

$I(x,y,t)$

FIGURE 2.11 Photographic enlarger projecting the image of a tree, thereby exposing the photographic paper underneath to the light image of the tree, illustrating variation of light intensity across the surface of the print paper.

equation for describing indoor air quality as a function of time can be discussed in terms of our mathematical framework.

Because the concept of exposure is well developed in the field of photography, these same concepts can be applied to light on a photographic print of a negative using an enlarger. Here, the target of the exposure is the photographic print paper, and the agent of the exposure is the light striking the print paper (Figure 2.11). The contact boundary can be represented by a conceptual film of infinitesimal thickness coating the surface of the print. The photography example illustrates the concepts for an energy-form agent, namely light. There are many other situations in which exposure to an energy-form agent may be of interest: exposure of an eardrum to sound; exposure of skin to ultraviolet radiation; exposure of an eye to light; exposure of a dentist's lead shield to X-ray beams. Again, all of these examples can be understood and expressed mathematically using the framework discussed in this chapter.

The older definitions proposed by the USEPA, the National Academy of Sciences, and others actually can be viewed as a subset of the framework presented in this chapter. This framework, however, is intended to be broader, encompassing various time- and space-related formulations of exposure and dose that are commonly used in exposure science. This framework was originally proposed in the published paper by Zartarian, Ott, and Duan (1997) as the unifying theoretical system for the exposure sciences, intended to embrace key definitions likely to be used by exposure assessors, risk assessors, exposure modelers, scientific researchers, and others. The Zartarian, Ott, and Duan (1997) framework was the basis for a glossary developed by an international committee, the International Programme on Chemical Safety (IPCS) exposure terminology workgroup (Hammerstrom et al. 2002; Callahan et al. 2001; IPCS 2002; WHO 2004), concerned with "harmonizing" the language used in the field of exposure assessment. The IPCS glossary has been adopted as the official glossary of the International Society for Exposure Analysis (Zartarian, Bahadori, and

McKone 2004). The glossary included below is a combination of the terms and definitions in Zartarian, Ott, and Duan (1997) and those in the official IPCS/ISEA glossary. Thus, it represents the state-of-the-science for definitions related to exposure and dose as of this book publication.

2.8 GLOSSARY OF EXPOSURE AND DOSE-RELATED TERMS

Absorbed Dose Dose resulting from an agent crossing an absorption barrier.

Absorption (Uptake) The process by which an agent crosses an absorption barrier (see Dose).

Absorption Barrier A contact boundary or zone that may retard the rate of penetration of an agent into a target. Examples of absorption barriers are the skin, respiratory tract lining, and gastrointestinal tract wall.

Activity Pattern Data Information on human activities used in exposure assessments. These may include a description of the activity, frequency of activity, duration spent performing the activity, and the microenvironment in which the activity occurs.

Acute Exposure A contact between an agent and a target occurring over a short time, generally less than a day. (Other terms, such as "short-term exposure" and "single dose," are also used.)

Agent A chemical, biological, or physical entity that contacts a target.

Background Level The amount of an agent in a medium (e.g., water, soil) that is not attributed to the source(s) under investigation in an exposure assessment. Background level(s) can be naturally occurring or the result of human activities. (Note: natural background is the concentration of an agent in a medium that occurs naturally or is not the result of human activities.)

Bioavailability The rate and extent to which an agent can be absorbed by an organism and is available for metabolism or interaction with biologically significant receptors. Bioavailability involves both release from a medium (if present) and absorption by an organism.

Biomarker/biological Marker An indicator (cellular, biochemical, analytical, or molecular) of a recent or previous exposure in a biological system, intended to measure body burden. The biomarker may be the actual agent itself or a metabolite. For example, measurement of CO in breath is a body burden indicator of CO in the blood (carboxyhemoglobin).

Body Burden The amount of chemical in the body at a given instant in time.

Bounding Estimate An estimate of exposure, dose, or risk that is higher than that incurred by the person with the highest exposure, dose, or risk in the population being assessed. Bounding estimates are useful in developing statements that exposures, doses, or risks are "not greater than" the estimated value.

Chronic Exposure A continuous or intermittent long-term contact between an agent and a target. (Other terms, such as "long-term exposure," are also used.)

Concentration The amount of matter–form agent per unit volume.

Contact Boundary (Exposure Surface) A surface on a target where an agent is present (i.e., a surface on a target containing at least one exposure point). Examples of outer human contact boundaries include the exterior of an eyeball, the skin surface, and a conceptual surface over the nose and open mouth. Examples of inner human contact boundaries include the gastrointestinal tract, the respiratory tract and the urinary tract lining. As a contact boundary gets smaller, the limit is an exposure point.

Contact Volume A volume containing the mass of agent that contacts the exposure surface; a volume adjoining a contact boundary in which the agent has a high probability of contacting the contact boundary in the exposure period.

Contact Volume Thickness The distance (from the contact boundary) in which a particle of agent has at least a pre-specified probability p of intersecting the contact boundary within a pre-specified time interval t.

Dose The amount of agent that enters a target in a specified time duration after crossing a contact boundary. If the contact boundary is an absorption barrier, the dose is an absorbed dose/uptake dose (see Uptake); otherwise it is an intake dose (see Intake).

Dose Rate Dose per unit time; the rate at which the agent crosses through a unit area of the contact boundary.

Exposure Contact between an agent and a target. Contact takes place at an exposure surface over an exposure period.

Exposure Assessment The process of estimating or measuring the magnitude, frequency and duration of exposure to an agent, along with the number and characteristics of the population exposed. Ideally, it describes the sources, pathways, routes, and the uncertainties in the assessment.

Exposure Concentration (Spatially Averaged Exposure) The exposure mass divided by the contact volume or the exposure mass divided by the mass in the contact volume, depending on the medium. The exposure at a single point is expressed as the limiting value of this ratio as the contact volume element becomes small.

Exposure Duration The length of time over which continuous or intermittent contacts occur between an agent and a target. For example, if an individual is in contact with an agent for 10 minutes a day, for 300 days over a 1-year time period, the exposure duration is 1 year.

Exposure Event The occurrence of continuous contact between an agent and a target.

Exposure Frequency The number of exposure events in an exposure duration.

Exposure Loading (Spatially Averaged Boundary Exposure) The exposure mass divided by the contact boundary area. For example, a dermal exposure measurement based on a skin wipe sample, expressed as a mass of residue per skin surface area, is an exposure loading.

Exposure Mass (Spatially Integrated Exposure) The amount of agent present in the contact volume. For example, the total mass of residue collected with a skin wipe sample over the entire exposure surface is an exposure mass.

Exposure Model A conceptual or mathematical representation of the exposure process.

Exposure Pathway The course an agent takes from the source to the target.

Exposure Point A point on an exposure surface (contact boundary) at which contact with an agent occurs.

Exposure Period The time of continuous contact between an agent and a target.

Exposure Route The way an agent enters a target after contact (e.g., by ingestion, inhalation, or dermal absorption).

Exposure Scenario A combination of facts, assumptions, and inferences that define a discrete situation where potential exposures may occur. These may include the source, the exposed population, the timeframe of exposure, microenvironment(s), and activities. Scenarios are often created to aid exposure assessors in estimating exposure.

Instantaneous Point Exposure Contact between an agent and a target at a single point on a contact boundary at a single instant in time; the joint occurrence of two events: (1) point i of the target is located at (x_i, y_i, z_i), and (2) an agent of concentration C_i, the exposure concentration, is present at location (x_i, y_i, z_i); measured as a concentration.

Intake The process by which an agent crosses an outer exposure surface of a target without passing an absorption barrier, i.e., through ingestion or inhalation (see Dose).

Intake Dose Dose resulting from the agent crossing the contact boundary without subsequently diffusing through an absorption barrier.

Intensity Amount of energy-form agent per unit area.

Medium Material (e.g., air, water, soil, food, consumer products) surrounding or containing an agent.

Medium Intake Rate The rate at which the medium crosses the contact boundary.

Microenvironment Surroundings that can be treated as homogeneous or well characterized in the concentrations of an agent (e.g., home, office, automobile, kitchen, store). This term is generally used for estimating inhalation exposures.

Moving Average A set of multiple sequential averages using a fixed averaging time, that are computed over a larger time duration.

Negative Dose The amount of agent exiting a target over a specified exposure duration.

Personal Exposure Monitor A device worn on or near the contact boundary that measures concentration.

Pica A behavior characterized by deliberate ingestion of non-nutritive substances, such as soil.

Positive Dose The amount of agent entering a target over a specified exposure duration.

Source The origin of an agent.

Stressor Any entity, stimulus, or condition that can modulate normal functions of the organism or induce an adverse response (e.g., agent, lack of food, drought).

Subchronic Exposure A contact between an agent and a target of intermediate duration between acute and chronic. (Other terms, such as "less-than-lifetime exposure" are also used.)

Target Any physical, biological, or ecological object exposed to an agent.

Time-Averaged Exposure The time-integrated exposure divided by the exposure duration. An example is the daily average exposure of an individual to carbon monoxide. (Also called time-weighted average exposure.)

Time-Integrated Exposure The integral of instantaneous exposures over the exposure duration. An example is the area under a daily time profile of personal air monitor readings, with units of concentration multiplied by time.

Time Profile A continuous record of instantaneous values (e.g., exposure, dose, or medium intake rate) over a time period.

Total Exposure Exposure received by a target (typically a person) to an agent (e.g., an air pollutant) from all sources as the target visits various microenvironments (e.g., indoor, outdoor, and in-transit locations) over a time period of interest. Examples are the continuously measured profile (e.g., minute-by-minute concentration readings) measured over 24 hours using a personal exposure monitor worn by the person, or the average personal exposures to particulate matter collected on a filter worn by the person for 24 hours. Total exposure to some pollutants can occur through multiple environmental carrier media (for example, exposure by air, water, food, or dermal contact).

Uptake (Absorption) The process by which an agent crosses an absorption barrier (see Dose).

2.9 QUESTIONS FOR REVIEW

1. Explain why the definition of exposure presented in this chapter can apply to both human and nonhuman species — plants and animals — and even to inanimate objects, such as buildings. Give several examples of nonhuman targets besides those mentioned in this chapter.
2. In the fewest words possible, what is a salient characteristic of the definition of exposure, as presented in this chapter?
3. Discuss the properties of the *contact boundary*.
4. Discuss the concept of the *instantaneous concentration*.
5. Discuss the mathematical definition of concentration at a point as the first derivative of mass with respect to volume. Discuss the similarity between the concepts of *density* and *concentration*. Consider a three-dimensional graph for the mass of a pollutant m that is uniformly mixed within a volume v. Ignoring the atomic or subatomic structure, make

a sketch and discuss how the spatially integrated concentration changes with distance along the axes.

6. Discuss the concepts of *intake dose* and *absorbed dose*.

7. Discuss the concept of a *dose rate*.

8. How does time enter into the concept of exposure? In a personal exposure time profile, discuss the differences between the *instantaneous exposure*, *integrated exposure*, *average exposure*, *peak exposure*, *minimum exposure*.

9. Given a personal exposure time profile, write equations for the *average exposure* over time *T*, the *variance of exposure* over time *T*, and the *moving average exposure* with time increment t_s.

10. Discuss how the concept of exposure in this chapter could be applied to a person who falls asleep on a sunny beach, receiving a sunburn on the back and exposed arms from ultraviolet solar radiation. Can you apply the concept of exposure to *noise* pollution? Although anthrax may be common in some outdoor soils, explain why people living in these locations might not experience any adverse health effects.

REFERENCES

Akland, G.G. (1991) Organizational Components and Structural Features of EPA's New Human Exposure Research Program, *Journal of Exposure Analysis and Environmental Epidemiology*, **1**(2): 129–141.

Akland, G.G., Hartwell, T.D., Johnson, T.R., and Whitmore, R.W. (1985) Measuring Human Exposure to Carbon Monoxide in Washington, DC, and Denver, Colorado, During the Winter of 1982–1983, *Environmental Science and Technology*, **19**: 911–918.

Armstrong, B.K., White, E., and Saracci, R. (1992) *Principles of Exposure Measurement in Epidemiology*, Oxford University Press, New York, NY, 11.

Atherley, G.R.C. (1978) *Occupational Health and Safety Concepts: Chemical and Processing Hazards*, Applied Science Publishers Ltd., Barking, Essex, England.

ATSDR (1993) *Public Health Assessment Guidance Manual*, Lewis Publishers, Chelsea, MI.

Calabrese, E.J., Gilbert, C.E., and Pastides, H. (1989) *Safe Drinking Water Act: Amendments, Regulations and Standards*, Lewis Publishers, Chelsea, MI.

Callahan, M.A., Jayewardene, R., Norman, C., Zartarian, V., Dinovi, M., Graham, J., Hammerstrom, K., Olin, S., and Sonich-Mullin, C. (2001) International Programme on Chemical Safety (IPCS) Project on the Harmonization of Risk Assessment Approaches: Exposure Assessment Terminology, Paper No. 273 presented at the 11th Annual Meeting of the International Society of Exposure Analysis, Charleston, SC, November 6, 2001.

Clayton, C.A., Perritt, R.R., Pellizzari, E.D., Thomas, K.W., Whitmore, R.W., Özkaynak, H., Spengler, J.D., and Wallace, L.A. (1993) Particle Total Exposure Assessment Methodology (PTEAM) Study: Distributions of Aerosol and Elemental Concentrations in Personal, Indoor, and Outdoor Air Samples in a Southern California Community, *Journal of Exposure Analysis & Environmental Epidemiology*, **3**: 227–250.

Coburn, R.R., Foster, R.E., and Cane, P.B. (1965) Considerations of the Physiologic Variables that Determine the Blood Carboxyhemoglobin Concentrations in Man, *Journal of Clinical Investigation*, **44**: 1899–1910.

Cohrssen, J.J. and Covello, V.T. (1989) *Risk Analysis: A Guide to Principles and Methods for Analyzing Health and Environmental Risks*, Executive Office of the President of the United States, Council on Environmental Quality, Springfield, VA.

Derelanko, M.J., Hollinger, M.A., Eds. (1995) Risk assessment, in *CRC Handbook of Toxicology*, CRC Press, Boca Raton, FL, 591–676.

Duan, N. (1982) Models for Human Exposure to Air Pollution, *Environment International,* **8**: 305.

Duan, N. (1989) Estimation of Microenvironment Concentration Distribution Using Integrated Exposure Measurements, in *Proceedings of the Research Planning Conference on Human Activity Patterns*, Starks, T.H., Ed., Report No. EPA/600/4-89/004, Environmental Monitoring Systems Laboratory, U.S. Environmental Protection Agency, Las Vegas, NV.

Duan, N. (1991) Stochastic Microenvironment Models for Air Pollution Exposure, *Journal of Exposure Analysis and Environmental Epidemiology*, **1**(2): 235–257.

Duan, N. and Ott, W. (1992) An Individual Decision Model for Environmental Exposure Reduction, *Journal of Exposure Analysis and Environmental Epidemiology*, **2**(suppl. 2): 155–174.

Duan, N., Dobbs, A., and Ott, W. (1989) Comprehensive Definitions of Exposure and Dose to Environmental Pollution, in *Proceedings of the EPA/A&WMA Specialty Conference on Total Exposure Assessment Methodology*, November 1989, Las Vegas, NV.

Duan, N., Dobbs, A., and Ott, W. (1990) *Comprehensive Definitions of Exposure and Dose to Environmental Pollution*, SIMS Technical Report No. 159, Department of Statistics, Stanford University, Stanford, CA.

Esmen, N.A. and Marsh, G.M. (1996) Applications and Limitations of Air Dispersion Modeling in Environmental Epidemiology, *Journal of Exposure Analysis and Environmental Epidemiology*, **6**(3): 339–353.

Ferro, A.R. (2002) The Effects of Proximity, Compartments, and Resuspension on Personal Exposure to Indoor Particulate Matter, Ph.D. diss., Department of Civil and Environmental Engineering, Stanford University, Stanford, CA.

Fishbane, P.M., Gasiorowicz, S., and Thornton, S.T. (1993) *Physics for Scientists and Engineers*, Prentice-Hall, Inc., Englewood Cliffs, NJ.

Georgopolous, P.G. and Lioy, P.J. (1994) Conceptual and Theoretical Aspects of Human Exposure and Dose Assessment, *Journal of Exposure Analysis and Environmental Epidemiology*, **4**(3): 253–285.

Halliday, D. and Resnick, R. (1981) *Fundamentals of Physics*, John Wiley & Sons, New York, NY.

Hammerstrom, K., Sonich-Mullin, C., Olin, S., Callahan, M., Dinovi, M., Graham, J., Jayewardene, R., Norman, C., and Zartarian, V. (2002) Glossary of Key Exposure Assessment Terms, Harmonization of Approaches to the Assessment of Risk from Exposure to Chemicals, International Programme on Chemical Safety Harmonization Project, Exposure Assessment Planning Workgroup, Terminology Subcommittee, http://www.who.int/ipcs/publications/methods/harmonization/en/index.html (accessed April 25, 2006).

Hammerstrom, K., Sonich-Mullin, C., Olin, S., Jantunen, M., Jayewardene, R., Norman, C., Beauchamp, R., Callahan, M., Dinovi, M., Graham, J., Guiseppe-Ellie A., Heinemeyer, G., Kohrman, K., Özkaynak, H., Younes, M., and Zartarian, V. (2001) International Programme on Chemical Safety Project on the Harmonization of Risk Assessment Approaches: Exposure Related Activities, 11th Annual Meeting of the International Society of Exposure Analysis, Charleston, SC, November 4–8, 2001.

Herrick, R.F. (1992) Exposure Assessment in Risk Assessment, in *Conference on Chemical Risk Assessment in the DOD: Science, Policy, and Practice*, Clewell, H.J., Ed., American Conference of Governmental Industrial Hygienists, Inc., Cincinnati, OH.

Hrudey, S.E., Chen, W., and Rousseaux, C.G. (1996) *Bioavailability in Environmental Risk Assessment*, Lewis Publishers, Boca Raton, FL.

ICRU (1979) *Quantitative Concepts and Dosimetry in Radiobiology*, International Commission on Radiation Units and Measurements Report 30, Washington, DC.

IPCS (1983) *Guidelines on Studies in Environmental Epidemiology*, International Programme on Chemical Safety Environmental Health Criteria 27, World Health Organization (WHO), Geneva, Switzerland.

IPCS (1994) *Guidelines on Studies in Environmental Epidemiology*, International Programme on Chemical Safety Environmental Health Criteria 27, World Health Organization (WHO), Geneva, Switzerland.

IPCS (2002) *Harmonization of Approaches to the Assessment of Risk from Exposure to Chemicals*, International Programme on Chemical Safety Harmonization Project, Exposure Assessment Planning Workgroup, Terminology Subcommittee, December 2002.

Landers, W.G. and Yu, M. (1995) *Introduction to Environmental Toxicology: Impacts of Chemicals upon Ecological Systems*, CRC Press, Inc, Boca Raton, FL.

Last, J.M. (1995) *A Dictionary of Epidemiology*, Oxford University Press, New York, NY.

Lioy, P.J. (1990) Assessing Total Human Exposure to Contaminants, *Environmental Science and Technology*, **24**(7): 938–945.

Lioy, P.J. (1991) Human Exposure Assessment: A Graduate Level Course, *Journal of Exposure Analysis and Environmental Epidemiology*, **1**(3): 271–281.

Lippman, M. (1987) Toxic Chemical Exposure and Dose to Target Tissues, in *Toxic Chemicals, Health, and the Environment*, Lave, L.B. and Upton, A.C., Eds., The Johns Hopkins University Press, Baltimore, MD.

Lisella, F.S., Ed. (1994) *The VNR Dictionary of Environmental Health and Safety*, Van Nostrand Reinhold, New York, NY.

McKone, T.E. and Bahadori, T. (2003) *Exposure Terminology Workshop: Toward Adopting an Official Glossary for the ISEA Session WS-13*, An Interactive Workshop at the 13th Annual Meeting of the International Society of Exposure Analysis Annual Meeting, September 21–25, 2003, Stresa, Italy, Conference and Session Program available at http://www.ktl.fi/isea2003/programframe.html.

Monson, R.R. (1980) *Occupational Epidemiology*, CRC Press, Boca Raton, FL.

NRC (1991) *Human Exposure Assessment for Airborne Pollutants: Advances and Opportunities*, National Research Council, Committee on Advances in Assessing Human Exposure to Airborne Pollutants, National Academy Press, Washington, DC.

Ott, W.R. (1982) Concepts of Human Exposure to Air Pollution, *Environment International*, **7**: 179–196.

Ott, W.R. (1983–1984) Exposure Estimates Based on Computer-Generated Activity Patterns, *Journal of Toxicology: Clinical Toxicology*, **21**: 97–128.

Ott, W. (1985) Total Human Exposure, *Environmental Science and Technology*, **19**(10): 880–886.

Ott, W.R. (1990) Total Human Exposure: Basic Concepts, EPA Field Studies, and Future Research Needs, *Journal of the Air & Waste Management Association*, **40**(7): 966–975.

Ott, W.R. (1995) Human Exposure Assessment: The Birth of a New Science, *Journal of Exposure Analysis and Environmental Epidemiology*, **5**(4): 449–472.

Ott, W.R. and Roberts, J.W. (1998) Everyday Exposure to Toxic Pollutants, *Scientific American*, **278**(2): 86–91.

Ott, W.R., Mage, D.T., and Thomas, J. (1992) Comparison of Microenvironmental CO Concentrations in Two Cities for Human Exposure Modeling, *Journal of Exposure Analysis and Environmental Epidemiology*, **2**(2): 249–267.

Ott, W., Thomas, J., Mage, D., and Wallace, L. (1988) Validation of the Simulation of Human Activity and Pollutant Exposure (SHAPE) Model Using Paired Days from the Denver, CO, Carbon Monoxide Field Study, *Atmospheric Environment*, **22**: 2101–2113.

Ott, W., Wallace, L., Mage, D., Akland, G., Lewis, R., Sauls, H., Rodes, C., Kleffman, D., Kuroda, D., and Morehouse, K. (1986) The Environmental Protection Agency's Research Program on Total Human Exposure, *Environment International*, **12**: 475–494.

Özkaynak, H., Xue, J., Spengler, J., Wallace, L., Pellizzari, I., and Jenkins, P. (1996) Personal Exposure to Particles and Metals: Results from the Particle TEAM Study in Riverside, CA, *Journal of Exposure Analysis and Environmental Epidemiology*, **26**(1): 57–78.

Pellizzari, E.D., Thomas, K.W., Clayton, C.A., Whitmore, R.W., Shores, R.C., Zelon, H.S., and Perritt, R.L. (1993a) *Particle Total Exposure Assessment Methodology (PTEAM)*: Riverside, California Pilot Study, Volume I, NTIS # PB 93-166 957/AS, National Technical Information Service.

Pellizzari, E.D., Thomas, K.W., Clayton, C.A., Whitmore, R.W., Shores, R.C., Zelon, H.S., and Perritt, R.L. (1993b) *Project Summary: Particle Total Exposure Assessment Methodology (PTEAM): Riverside, California Pilot Study, Volume I*, Report No. EPA/600/SR-93/050, U.S. Environmental Protection Agency, Research Triangle Park, NC.

Resnick, R., Halliday, D., and Krane, S. (1992) *Physics*, John Wiley & Sons, New York, NY.

Rodes, C.E., Camens, R.M., and Wiener, R.W. (1991) The Significance and Characteristics of the Personal Activity Cloud on Exposure Assessment Measurements for Indoor Contaminants, *Indoor Air*, **2**: 123–145.

Sexton, K. and Ryan, P.B. (1988) Assessment of Human Exposure to Air Pollution: Methods, Measurements, and Models, in *Air Pollution, the Automobile, and Public Health*, National Academy Press, Washington, DC.

Sexton, K., Selevan, S.G., Wagener, D.K., and Lybarger, J.A. (1992) Estimating Exposures to Environmental Pollutants: Availability and Utility of Existing Databases, *Archives of Environmental Health*, **47**(6): 398–407.

Tardiff, R.G. and Goldstein, B.D. (1991) *Methods for Assessing Exposure of Human and Non-Human Biota*, Scientific Group on Methodologies for the Safety Evaluation of Chemicals (SGOMSEC), John Wiley & Sons, New York, NY.

Thomas, K.W., Pellizzari, E.D., Clayton, C.A., Whitaker, D.A., Shores, R.C., Spengler, J.D., Özkaynak, H., and Wallace, L.A. (1993) Particle Total Exposure Assessment Methodology (PTEAM) Study: Method Performance and Data Quality for Personal, Indoor, and Outdoor Aerosol Monitoring at 178 Homes in Southern California, *Journal of Exposure Analysis and Environmental Epidemiology*, **3**: 203–226.

USEPA (1991) *Air Quality Criteria for Carbon Monoxide*, Report No. EPA/600/8-90/04SF, U.S. Environmental Protection Agency, Office of Research and Development, USEPA, Washington, DC.

USEPA (1992) Guidelines for Exposure Assessment, *Federal Register*, 57(104): 22888–22938.

USEPA (2002) *Framework for Cumulative Risk Assessment*, Report No. EPA/630/P-02/001A, U.S. Environmental Protection Agency, Office of Research and Development, National Center for Environmental Assessment, Washington, DC.

Wallace, L.A. (1987) *The TEAM Study: Summary and Analysis: Volume I*, Report No. EPA/600/6-87/002a, NTIS PB 88-100060, U.S. Environmental Protection Agency, Washington, DC.

Wallace, L.A. (1993) A Decade of Studies of Human Exposure: What Have We Learned? *Risk Analysis*, **13**(2): 135–143.

Wallace, L.A. (1995) Human Exposure to Environmental Pollutants: A Decade of Experience, *Clinical and Experimental Allergy*, **25**: 4–9.

Wallace, L.A., Pellizzari, E.D., Hartwell, T.D., Davis, V., Michael, L.C., and Whitmore, R.W. (1989) The Influence of Personal Activities on Exposure to Volatile Organic Compounds, *Environmental Research*, **50**: 37–55.

Wallace, L.A., Pellizzari, E., Hartwell, T., Sparacino, C., Sheldon, L., and Zelon, H. (1985) Personal Exposures, Indoor-Outdoor Relationships and Breath Levels of Toxic Air Pollutants Measured for 355 Persons in New Jersey, *Atmospheric Environment*, **19**: 1651–1661.

Wallace, L.A., Pellizzari, E.D., Hartwell, T.D., Sparacino, C., Whitmore, R., Sheldon, L., Zelon, H., and Perritt, R. (1987) The TEAM Study: Personal Exposures to Toxic Substances in Air, Drinking Water, and Breath of 400 Residents of New Jersey, North Carolina, and North Dakota, *Environmental Research*, **43**: 290–307.

Wallace, L.A., Pellizzari, E.D., Hartwell, T.D., Whitmore, R., Sparacino, C., and Zelon, H. (1986) Total Exposure Assessment Methodology (TEAM) Study: Personal Exposures, Indoor-Outdoor Relationships, and Breath Levels of Volatile Organic Compounds in New Jersey, *Environment International*, **12**: 369–387.

Wallace, L.A., Pellizzari, E.D., Hartwell, T.D., Whitmore, R., Zelon, H., Perritt, R., and Sheldon, L. (1988a) The California TEAM Study: Breath Concentrations and Personal Exposures to 26 Volatile Compounds in Air and Drinking Water of 188 Residents of Los Angeles, Antioch, and Pittsburgh, CA, *Atmospheric Environment*, **22**(10): 2141–2163.

Wallace, L., Thomas, J., Mage, D., and Ott, W. (1988b) Comparison of Breath CO, CO Exposure, and Coburn Model Predictions in the US EPA Washington-Denver CO Study, *Atmospheric Environment*, **22**(10): 2183–2193.

Weinstein, I.B. (1988) Molecular Cancer Epidemiology: The Use of New Laboratory Methods in Studies on Human Cancer Causation, in *Epidemiology and Health Risk Assessment,* Gordis, L., Ed., Oxford University Press, New York, NY.

Whitmore, R.W., Immerman, F.W., Camann, D.E., Bond, A.E., Lewis, R.G., and Schuam, J.L. (1994) Non-Occupational Exposures to Pesticides for Residents of Two U.S. Cities, *Archives of Environmental Contamination and Toxicology*, **26**: 47–59.

WHO (2004) IPCS Risk Assessment Terminology, Harmonization Project Document No. 1, ISBN 92 4 156267 6, WHO Document Production Services, Geneva, Switzerland.

Zartarian, V.G. (1996) DERM (Dermal Exposure Reduction Model): A Physical-Stochastic Model for Understanding Dermal Exposure to Chemicals, Ph.D. diss., Stanford University, Civil Engineering Department, Environmental Engineering and Science Program, Stanford, CA.

Zartarian, V.G., Özkaynak, H., Burke, J.M., Zufall, M.J., Rigas, M.L., Furtaw Jr., E.J. (2000), A Modeling Framework for Estimating Children's Residential Exposure and Dose to Chlorpyrifos Via Dermal Residue Contact and Non-Dietary Ingestion, *Environmental Health Perspectives*, **108**(6): 505–514.

Zartarian, V.G., Ott, W.R., and Duan, N. (1997) A Quantitative Definition of Exposure and Related Concepts, *Journal of Exposure Analysis and Environmental Epidemiology*, **7**(4): 411–437.

Zartarian, V.G., Bahadori, T., and McKone, T.E. (2004) Feature Article: The Adoption of an Official ISEA Glossary, *Journal of Exposure Analysis and Environmental Epidemiology,* **15**: 1–5.

Zweidinger, R., Erickson, M., Cooper, S., Whitaker, D., Pellizzari, E., and Wallace, L. (1982) *Direct Measurement of Volatile Organic Compounds in Breathing-Zone Air, Drinking Water, Breath, Blood, and Urine*, Report No. NTIS PB-82-186-545, U.S. Environmental Protection Agency, Washington, DC.

3 Probability-Based Sampling for Human Exposure Assessment Studies

Roy Whitmore
RTI International

CONTENTS

3.1 Synopsis 65
3.2 What Is a Sample? 66
3.3 Why Use a Sample? 66
3.4 Does It Matter How the Sample Is Selected? 66
3.5 How Is a Representative Sample Selected? 67
3.6 Example of a Probability-Based Sample 67
3.7 When Should You Use a Probability-Based Sample? 69
3.8 When Is a Probability-Based Sample Not Necessary? 69
3.9 Is Probability-Based Sampling Sufficient to Support Robust Inferences? 70
3.10 Where Do You Begin? 70
3.11 Why Does One Stratify the Sampling Frame? 71
3.12 Are Special Statistical Analysis Techniques Needed? 72
3.13 Additional Examples of Probability-Based Sampling for Exposure Assessment Studies....72
3.14 Conclusion 74
3.15 Questions for Review 74
References 77

3.1 SYNOPSIS

Whenever one studies a sample of people to determine the effects of environmental exposures on them, one usually wants to extrapolate the findings beyond the individuals actually studied and measured. This extrapolation from study subjects to the population represented by them requires application of inferential statistics. A firm foundation for inferential statistics is established by using the scientific method to design and implement the study. The scientific method requires one to explicitly define the population about which one wants to make inferences and to select a sample from that population in such a manner that the probability of being selected into the sample is known for every person selected into the sample. Sampling procedures that result in known probabilities of selection from a specified finite population are referred to as probability-based sampling methods. The purpose of this chapter is to give you a basic understanding of what probability-based sampling methods are, when they should be used, how they are applied, when it may be satisfactory to not use them, and what else is required to support defensible inferences from a sample to the population it represents. For example, it is important that the size of the sample be large enough and that valid measurements be obtained for a high proportion of the people selected into the sample.

3.2 WHAT IS A SAMPLE?

A sample is a subset of the individuals, or units, belonging to a larger group that one would like to characterize or about which one would like to make inferences. The larger group about which one would like to make inferences is usually referred to as the universe, or target population, of interest.

For example, suppose we want to characterize personal exposures to airborne allergens for the residents of Virginia during the upcoming summer. Any subset of the people living in Virginia during that summer would be a sample from the population of interest, but some samples would be better than others. One would need a sample that was representative of various geographic areas of Virginia and representative of all months of the summer. However, *any* subset of the summer residents of Virginia would be a sample from that target population.

3.3 WHY USE A SAMPLE?

A sample is used to learn about the universe from which the sample was selected without observing and measuring *all* units in the universe. However, not all samples provide the same amount of information about the universe from which they were selected.

3.4 DOES IT MATTER HOW THE SAMPLE IS SELECTED?

Probability-based samples are samples for which every unit in the universe has a known, positive probability of being included in the sample. These types of samples ensure representativeness of the universe from which they were selected. Moreover, they allow one to characterize the uncertainty of inferences from the sample. For example, one can calculate the probability that an inference from the sample may be incorrect. If that probability is too high for comfort, one can take a larger sample to reduce the degree of uncertainty.

The inadequacy of sampling methods that might at first seem adequate has been dramatically illustrated in two U.S. presidential election polls: the 1936 and 1948 presidential election polls. Prior to the 1936 election, the *Literary Digest* magazine sent postcards to its subscribers asking whom they would vote for. Based on over 2,000,000 returned postcards, the *Literary Digest* predicted that Alfred E. Landon would defeat Franklin D. Roosevelt by a 57% to 43% margin. However, George Gallop correctly predicted that Roosevelt would win based on a random sample of 300,000 likely voters. The primary lesson learned from this experience was that all members of the population of interest must have a chance of being included in the sample. The readers of the *Literary Digest* were more affluent than the majority of the population, which was still struggling through the Great Depression (1929–1939).

Prior to the 1948 presidential election, Gallop conducted a final poll of likely voters on October 20, about 2 weeks before the November 2 election, and predicted that Thomas Dewey would defeat Harry Truman by a 49.5% to 44.5% margin (with the remainder of the votes going to Henry Wallace and Strom Thurmond). Instead, Truman defeated Dewey by a 49.9% to 45.5% margin. Apparently, voter preferences changed sufficiently during the final 2 weeks of campaigning that the October 20 poll was not valid for making inferences regarding the election outcome. This reinforced the earlier lesson about sampling from the population of interest but with a different twist. As is true in many environmental studies, the outcome of interest was time-dependent. For election polls and for many environmental studies, the population of interest must be carefully defined in terms of both geographic and temporal scope, and the sample selection methods must ensure that the sample is representative in both dimensions because the measurements obtained for study subjects may vary temporally (e.g., by season, by day, or by time of day).

3.5   HOW IS A REPRESENTATIVE SAMPLE SELECTED?

As mentioned above, probability-based sampling is necessary to ensure that a sample is representative of the universe from which it was selected. Probability-based samples use randomization (like flipping a coin or rolling a die) to ensure that a sample is representative of the units on the list from which the sample was selected. The list from which the sample is selected is referred to as the sampling frame. The sampling frame is sometimes a simple list of the units in the population. For example, many European counties maintain population registries that can be used as sampling frames. More often, however, a registry of the population of interest does not exist, and a multistage listing process is used that effectively includes all members of the population of interest. For example, probability-based samples of the residents of the United States are usually selected by first sampling from all counties in the United States, then sampling from all census blocks in the sample counties, and finally sampling from all households in the sample blocks.

The simplest example of a probability-based sample is a simple random sample. A simple random sample is one in which all possible samples of the same size have the same probability of being selected, such as when all units in the sample are determined by independent random draws from the population units. However, simple random samples are rarely used in practice. Instead, the sampling frame is usually partitioned into subsets, called strata, and an independent sample is selected from each subset, or stratum. One reason to partition the population into strata is to guarantee that some of the sample will come from each of the different portions of the population. For example, if we were sampling the summer residents of Virginia, we might want to stratify spatially into four or five geographic regions and temporally by month. That would guarantee that the sample would include persons from all geographic regions of the state and all months of the summer.

Probability-based samples are representative of the units on the sampling frames from which the samples were selected. However, for reasons of practical or cost expediency, the sampling frame from which the sample is selected may not include all units of the population of interest. For example, some probability-based surveys of the U.S. population use probability-based samples of telephone numbers. These samples are only representative of households with telephone service. Moreover, some telephone households may be excluded, such as those on banks of telephone numbers that have only been recently released. The population represented by the sampling frame is often referred to as the survey population, whereas the population about which inferences are desired is the target population. When the survey and target populations are not identical, one needs to carefully consider the potential for bias when making inferences regarding the target population.

Kalton (1983) provides a basic introduction to the concepts of probability-based sampling. Gilbert (1987) and Thompson (1992) provide more in-depth discussions of probability-based sampling methods for environmental studies.

3.6   EXAMPLE OF A PROBABILITY-BASED SAMPLE

In the mid-1990s, the U.S. Environmental Protection Agency (USEPA) conducted a multimedia, multipathway human exposure assessment study, the National Human Exposure Assessment Survey (NHEXAS), in USEPA Region 5[1] as a field test of procedures that could be used in a national study of human exposures to environmental toxicants. A probability-based sampling design was used for the field test because that would be necessary for a defensible national study. The statistical

---

[1] The six states of Minnesota, Wisconsin, Michigan, Illinois, Indiana, and Ohio.

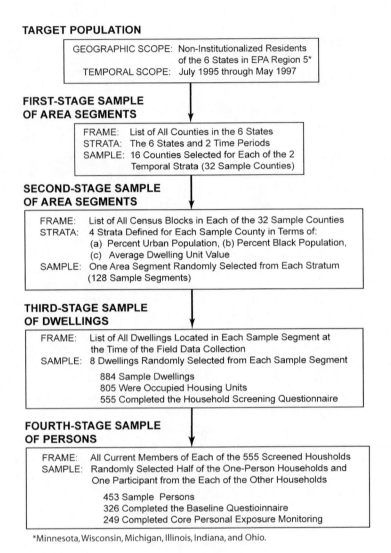

TARGET POPULATION

> GEOGRAPHIC SCOPE: Non-Institutionalized Residents
> of the 6 States in EPA Region 5*
> TEMPORAL SCOPE: July 1995 through May 1997

**FIRST-STAGE SAMPLE
OF AREA SEGMENTS**

> FRAME: List of All Counties in the 6 States
> STRATA: The 6 States and 2 Time Periods
> SAMPLE: 16 Counties Selected for Each of the 2
> Temporal Strata (32 Sample Counties)

**SECOND-STAGE SAMPLE
OF AREA SEGMENTS**

> FRAME: List of All Census Blocks in Each of the 32 Sample Counties
> STRATA: 4 Strata Defined for Each Sample County in Terms of:
> (a) Percent Urban Population, (b) Percent Black Population,
> (c) Average Dwelling Unit Value
> SAMPLE: One Area Segment Randomly Selected from Each Stratum
> (128 Sample Segments)

**THIRD-STAGE SAMPLE
OF DWELLINGS**

> FRAME: List of All Dwellings Located in Each Sample Segment at
> the Time of the Field Data Collection
> SAMPLE: 8 Dwellings Randomly Selected from Each Sample Segment
>
> 884 Sample Dwellings
> 805 Were Occupied Housing Units
> 555 Completed the Household Screening Questionnaire

**FOURTH-STAGE SAMPLE
OF PERSONS**

> FRAME: All Current Members of Each of the 555 Screened Houslholds
> SAMPLE: Randomly Selected Half of the One-Person Households and
> One Participant from the Each of the Other Households
>
> 453 Sample Persons
> 326 Completed the Baseline Questioinnaire
> 249 Completed Core Personal Exposure Monitoring

*Minnesota, Wisconsin, Michigan, Illinois, Indiana, and Ohio.

FIGURE 3.1 Flow chart for the four-stage sampling design for the National Human Exposure Assessment Survey (NHEXAS) field study in USEPA Region 5.

sampling design for the field study is described in detail by Whitmore et al. (1999) and is summarized in Figure 3.1.[2]

The target population for the NHEXAS field test consisted of the noninstitutionalized residents of USEPA Region 5 during the period of field data collection, from July 1995 through May 1997. Members of the sample were selected using a four-stage, probability-based sampling design. The stages of sampling were:

1. First-stage sample of counties
2. Second-stage sample of area segments defined by census blocks within sample counties

---

[2] The study objectives, hypotheses, and exposure study design are described by Pellizzari et al. (1995).

3. Third-stage sample of households within sample segments
4. Fourth-stage sample of persons within sample households

Temporal stratification was implemented for the NHEXAS first-stage sample of counties by selecting two independent samples of 18 counties — one to be used for the first 9 months of data collection and the other to be used for the last 9 months. Four counties were selected into both samples.

For reasons of practical expediency and cost control, the survey population covered by the field test sampling design differed in some known ways from the target population. In particular, the following subpopulations were not included in the field test:

1. People not living in households (e.g., homeless persons and those in prisons, nursing homes, dormitories, etc.)
2. People living on military bases
3. People who were not mentally or physically capable of participating

3.7   WHEN SHOULD YOU USE A PROBABILITY-BASED SAMPLE?

You should use a probability-based sample when you can afford to collect data for more than a minimal number of units in the universe of interest, say more than 20 or 30 persons, and you want to:

1. Make defensible inferences regarding the universe, such as the proportion of people whose personal exposures exceed a specified threshold
2. Quantify the uncertainty of those inferences
3. Do so while making few assumptions about the universe, such as the shape of the distribution of all exposures in the universe

Other types of inference that are less reliant on assumptions when supported by a probability-based sample include the following:

1. Testing hypotheses, such as whether the mean personal exposure of people with low socioeconomic status exceeds that of the rest of the population
2. Estimating relationships between exposures, health effects, and demographic characteristics (e.g., correlations or regression models)

Estimating relationships between variables is often done with data that do not come from a probability-based sample. However, if the relationship between the variables of interest is not the same for all members of the universe, a model based on a probability-based sample protects against getting a biased estimate of the relationships.

3.8   WHEN IS A PROBABILITY-BASED SAMPLE NOT NECESSARY?

When one can only afford to collect data for a small sample, say 20 or fewer persons, or when a small sample is considered sufficient for the purposes of the investigation, a purposively selected sample is usually superior to a probability-based sample. When there will be only a small number of observations, it is important to carefully pick the units to be observed so that they are as representative of the universe of interest as can be achieved with a small sample. One needs to use expert judgment and knowledge of the universe to ensure that the sample units are not unusual (e.g., not mostly large or mostly small units and not just the units most convenient or least expensive to observe and measure).

USEPA (2002) provides some examples where judgmental samples are appropriate. One example is characterization of groundwater contamination beneath a Brownfield site, a site suspected of being contaminated with industrial waste that is being redeveloped. In this case, the high cost of collecting groundwater samples may preclude probability-based sampling.

Another example is determining whether or not the concentration of a contaminant in surface soils exceeds a specified threshold anywhere on a Brownfield site. In this case, samples can be collected from the areas where industrial spills are known to have occurred (e.g., based on visual inspection of the soil). If none of the samples exceeds the threshold, the investigation may be finished. However, if any of the samples exceeds the threshold, then a probability-based sampling design may be necessary to characterize the distribution of soil concentrations of the contaminant throughout the site.

3.9   IS PROBABILITY-BASED SAMPLING SUFFICIENT TO SUPPORT ROBUST INFERENCES?

Unfortunately, selecting a probability-based sample from the population of interest is not sufficient to support defensible inferences regarding the population. You must actually collect data from most, if not all, members of the sample. When the sample units are not people or commercial establishments (e.g., trees, soil, water, or air), data usually can be obtained for most of the sample members. However, when you must get people to agree to let you collect information from them, considerable effort must be devoted to obtaining a good response rate. USEPA (1983) recommends setting the target response rate to be at least 75% for mail and telephone surveys. A reasonably high response rate (at least 50% or better) is necessary to limit the uncertainty due to the possibility that the outcomes being measured may be systematically different for respondents and nonrespondents.

Moreover, not just any data are sufficient. You must use measurement methods that have been carefully developed and tested to ensure that the measurements are accurate and not biased. This principle is equally important for analytical instruments that measure concentrations of contaminants in the environment as well as for questionnaires that attempt to obtain data regarding the attributes and activities of people and businesses.

The size of the sample also is important for limiting uncertainty regarding the inferences made from a probability-based sample. A small probability-based sample could result in inferences with a high level of uncertainty because of the great deal of variability in outcomes that could be obtained with small samples. As the sample size increases, the variability between the results from different samples using the same probability-based sampling design decreases and the precision of the statistical inferences increases. USEPA (2002) provides sample size guidance for simple random sampling designs, which serves as a useful starting point for other sampling designs.

3.10   WHERE DO YOU BEGIN?

The scientific method is used to proceed from research objectives to the study design that best fulfills the requirements of the research. The scientific method for developing a probability-based sampling design is explained as a seven-step process, called the Data Quality Objectives (DQO) process, in USEPA (2000) and in Chapter 2 of Millard and Neerchal (2001). Although these documents describe the process in terms of testing a hypothesis to make a decision, the process is equally applicable to studies for which the objective is to characterize the status and trends of exposures in a population.

The scientific process for developing a probability-based sampling design begins with explicit specification of the goals of the study. The population of interest must be specified, including both the spatial and temporal extent of the population. In addition to overarching study objectives, one must specify specific population parameters to be estimated and, if applicable, specific hypotheses to be tested (e.g., the mean concentration of cotinine in saliva is higher for smokers than for nonsmoking adults). After one has identified the key estimates and hypothesis tests that will drive the inferential needs of the study, one must specify the level of uncertainty that can be tolerated for the estimates or inferences. At this point, it is necessary to consult with a survey statistician to mathematically formulate the precision requirements and determine appropriate classes of statistical sampling designs. Having identified appropriate classes of statistical sampling designs, the cost of data collection must be estimated in terms of cost per sampling unit (e.g., the differential cost for going to one more county, going to one more area in a sample county, or collecting and analyzing data from one more participant in a sample area). A sampling statistician can use the precision requirements, cost estimates, and constraints on study resources (e.g., the research budget) to determine the optimum sampling design (e.g., sampling frames and strata, stages of sampling, and sample sizes) that will either achieve the precision requirements for the least cost or achieve maximum precision within specified cost constraints.

3.11 WHY DOES ONE STRATIFY THE SAMPLING FRAME?

As mentioned earlier, the sampling design for a research study sample is seldom a simple random sample, even if there is a list of the members of the population of interest that can be used as a sampling frame. One usually at least stratifies the sampling frame. A stratified sample is one in which the sampling frame is partitioned into disjoint subsets, called strata, and a separate, statistically independent sample is selected from each stratum.

There are several reasons why sampling frames usually are stratified before sample selection. The reasons include: (a) to improve the representativeness of the sample, (b) to control the precision of estimates, (c) to control costs, and (d) to enable use of different sampling designs for different portions of the population.

The representativeness of a sample can be improved by stratifying the sampling frame because the sampling design then guarantees that a specified number of units will be selected from each stratum. Hence, if there are population subsets (e.g., spatial or temporal domains) that are so important that the sample would be considered to be deficient if it did not include any units from them, they should be sampling strata.

There are two quite different ways that strata often are used to improve the precision of study estimates. If separate estimates are required for various population domains (e.g., defined by age, race, gender, or socioeconomic status), adequate precision for these estimates may require that certain domains receive more that a proportionate sample. In this case, one often defines the domains that need to be oversampled (need more than a proportionate sample) to be sampling strata so that the sample sizes needed for adequate precision can be guaranteed (see, e.g., Ezzati-Rice and Murphy 1995). In a similar manner, if hypotheses will test for differences between two groups (e.g., exposed and unexposed individuals), then the two groups may be defined to be strata so that they receive equal sample sizes, thereby maximizing the likelihood that the hypothesis test will detect any differences between the two groups.

Alternatively, stratification can be used to maximize the precision of estimates of overall population parameters (e.g., overall population means or proportions). In this case, one forms strata that are as homogeneous as possible with respect to the key outcomes of interest in the study. Lacking any firm knowledge of how the outcomes will be distributed across strata, the precision of estimates is then maximized by proportionate allocation of the sample to the strata. However, if one knows ahead of time approximately what the variability of outcome measures will be within strata, precision can be improved for overall population estimates by using higher sampling fractions in the strata with higher variability. Hence, efficient stratification requires knowledge of spatial and temporal variability.

For human exposure assessment surveys, one often is tempted to define strata that contain population members who are expected to experience higher-than-average exposures to the environmental pollutants of interest and oversample these strata. This can produce improved precision for estimates of overall population mean exposures if the stratum sampling rates are proportional to the variability (standard deviations) of the stratum measures of exposure concentrations. However, this practice can lead to loss of precision if the higher exposures are not sufficiently concentrated in the oversampled strata (see, e.g., USEPA 1990a,b; Whitmore et al. 1994; Sexton et al. 2003). High exposure strata can be oversampled with little loss of precision for overall population estimates only if the following two conditions are simultaneously satisfied: (a) the percentage of persons with high exposures is much higher in the oversampled strata and (b) the oversampled strata contain a high proportion (say, 75% or more) of the highly exposed population (see Callahan et al. 1995).

Stratification also can be used to control study costs if the population can be partitioned into strata that have different costs per unit for data collection and analysis. For example, units of the population that are difficult to access and measure could be defined as a separate sampling stratum. In this case, study costs can be reduced by assigning lower sampling fractions to the higher cost strata.

Finally, if different sampling designs are most efficient for different parts of the population, use of different sampling designs is facilitated by stratifying the sample because a statistically independent sample is selected from each stratum. For example, a survey of industrial establishments may require use of two lists that contain quite different types of information for different industrial sectors. Defining each list, or industrial sector, to be a stratum allows one to use entirely different sampling designs for the two strata within the overall probability-based sampling design for the industry.

3.12   ARE SPECIAL STATISTICAL ANALYSIS TECHNIQUES NEEDED?

Use of a probability-based sampling design allows one to compute robust estimates of precision (e.g., standard errors) that usually are larger than one would get from the usual statistical analysis procedures, which typically assume that all observations come from a simple random sample and are independent and identically distributed. The probability-based estimates of standard errors are based on the known sampling design: the stages of sampling, the strata, and the probabilities of selection. Ignoring the sampling design is likely to result in underestimation of the sampling variance and, hence, erroneous claims of statistically significant results.

If the population units do not all have the same probability of selection, the responses from each unit must be weighted inversely to the unit's probability of selection to avoid design bias. For example, if minorities were selected at twice as high a rate as other strata, then each minority person in the sample represents half as many population members as the other sample members. In this case, minority sample members would be assigned a statistical analysis weight that was half as large as that of the other sample members. Ignoring the sampling weights would result in biased estimates.

Several statistical analysis software packages are available to enable proper analysis of data from probability-based samples. An overview of the variance estimation methods is provided by Wölter (1985). A summary of currently available software can be obtained at http://www.hcp.med.harvard.edu/statistics/survey-soft/.

3.13   ADDITIONAL EXAMPLES OF PROBABILITY-BASED SAMPLING
FOR EXPOSURE ASSESSMENT STUDIES

Because of the cost of chemical and physical analysis of environmental samples, most human exposure assessment studies have smaller sample sizes than the NHEXAS and are usually confined to a local geographic area (e.g., a city or county). The carbon monoxide exposure assessment study conducted in Washington, DC, during the winter of 1982–1983 is a good example. The statistical

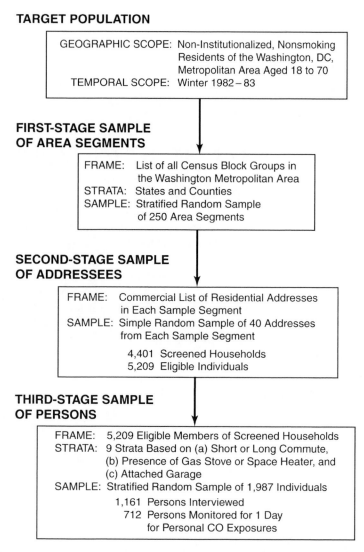

FIGURE 3.2 Flow chart for the three-stage sampling design for the Washington, DC, carbon monoxide personal exposure monitoring field study.

sampling design is described in detail in USEPA (1984) and is summarized in Figure 3.2 and in Akland et al. 1985.

The target population for the CO study consisted of the nonsmoking, noninstitutionalized residents of the Washington, DC metropolitan area during the winter of 1982–1983 who were 18–70 years of age. Members of the sample were selected using a three-stage probability-based sampling design. The stages of sampling were:

1. First-stage sample of area segments defined by census block groups
2. Second-stage sample of residential addresses obtained from a commercial vendor
3. Third-stage sample of persons

The address lists obtained from the vendor did not include some members of the target population. An incomplete sampling frame was used to reduce study costs. In this case, the survey population excluded people with no telephone service and most people whose address was not in the latest published telephone directory.

A bibliography of human exposure assessment studies that have utilized probability-based sampling techniques to make inferences to specified target populations is provided in Table 3.1.

3.14 CONCLUSION

In conclusion, human exposure assessment studies usually are designed to make inferences from a sample of persons to a finite population with specific spatial and temporal bounds. In order to make defensible inferences from the sample to the target population, probability-based sampling methods must be used to select the study subjects. Randomization in the probability-based sampling method protects against biased selection of subjects and facilitates computation of robust, design-based measures of precision. It also makes use of expert knowledge to design an efficient sample. For example, overall sample precision can be maximized by defining sampling strata that have relatively homogeneous values of the outcome measurements.

Use of tested and validated measurement methods ensures that reliable measurements will be obtained from all sample persons. Using survey procedures that result in valid measurements for a high proportion of the sample members protects against nonresponse bias. Finally, having a sufficiently large sample ensures that survey statistics will have adequate precision.

The scientific method, as outlined by the USEPA's Data Quality Objectives process, is recommended for designing and implementing human exposure assessment studies. This process ensures statistically valid inferences with sufficient precision to satisfy the study objectives.

3.15 QUESTIONS FOR REVIEW

1. What is a sampling frame?
2. What is a probability-based sample?
3. How do you determine if probability-based sampling is necessary for a given research study?
4. What is necessary in addition to probability-based sampling to support defensible inferences to the population from which the sample was selected?
5. What are sampling strata?
6. Why would you stratify a sampling frame?
7. What are the basic steps of the Data Quality Objectives process?

TABLE 3.1
Bibliography of Human Exposure Assessment Studies That Used Probability-Based Sampling

Adgate, J.L., Barr, D.B., Clayton, C.A., Eberly, L.E., Freeman, N.C., Lioy, P.J., Needham, L.L., Pellizzari, E.D., Quackenboss, J.J., Roy, A., and Sexton, K. (2001) Measurement of Children's Exposure to Pesticides: Analysis of Urinary Metabolite Levels in a Probability-Based Sample, *Environmental Health Perspectives*, **109**(6): 583–590.

Adgate, J.L., Clayton, C.A., Quackenboss, J.J., Thomas, K.W., Whitmore, R.W., Pellizzari, E.D., Lioy, P.J., Shubat, P., Stroebel, C., Freeman, N.C., and Sexton, K. (2000) Measurement of Multi-pollutant and Multi-pathway Exposures in a Probability-based Sample of Children: Practical Strategies for Effective Field Studies, *Journal of Exposure Analysis and Environmental Epidemiology*, **10**(6), Part 2: 650–661.

Adgate, J.L., Eberly, L.E., Stroebel, C. Pellizzari, E.D., and Sexton, K. (2004) Personal, Indoor, and Outdoor VOC Exposures in a Probability Sample of Children, *Journal of Exposure Analysis and Environmental Epidemiology*, **14** (suppl. 1): S4–S13.

Akland, G.G., Hartwell, T.D., Johnson, T.R., and Whitmore, R.W. (1985) Measuring Human Exposure to Carbon Monoxide in Washington, DC, and Denver, Colorado, During the Winter of 1982–1983. *Environmental Science and Technology*, **19**(10): 911–918.

Becker, K., Schulz, C., Kaus, S., Seiwert, M., and Serfert, B. (2003) German Environmental Survey 1998 (GerES III): Environmental Pollutants in the Urine of the German Population, *International Journal of Hygiene and Environmental Health*, **206**(1): 15–24.

Becker, K., Seiwert, M., Angerer, J., Heger, W., Koch, H.M., Hagorka, R., Rosskamp, E., Schlüter, C., Seifert, B., and Ullrich, D. (2004) DEHP Metabolites in Urine of Children and DEGP in House Dust, *International Journal of Hygiene and Environmental Health*, **207**: 409–417.

Clayton, C.A., Perritt, R.L., Pellizzari, E.D., Thomas, K.W., Whitmore, R.W., Wallace, L.A., Özkaynak, H., and Spengler, J.D. (1993) Particle Total Exposure Assessment Methodology (PTEAM) Study: Distributions of Aerosol and Elemental Concentrations in Personal, Indoor, and Outdoor Air Samples in a Southern California Community, *Journal of Exposure Analysis and Environmental Epidemiology*, **3**(2): 227–250.

Ezzati-Rice, T.M. and Murphy, R.S. (1995) Issues Associated with the Design of a National Probability Sample for Human Exposure Assessment, *Environmental Health Perspectives*, **103** (suppl. 3): 55–59.

Hänninen, O.O., Alm, S., Katsouyanni, K., Künzli, N., Maroni, M., Nieuwinhuijsen, M.J., Saarela, K., Srám, R.J., Zmirou, D, and Jantunen, M. (2004) The EXPOLIS Study: Implications for Exposure Research and Environmental Policy in Europe, *Journal of Exposure Analysis and Environmental Epidemiology*, **14**(6): 440–456.

Hartwell, T.D., Pellizzari, E.D., Perritt, R.L., Whitmore, R.W., and Zelon, H.S. (1987) Comparison of Volatile Organic Levels between Sites and Seasons for the Total Exposure Assessment Methodology (TEAM) Study, *Atmospheric Environment*, **21**(11): 2413–2424.

Hartwell, T.D., Pellizzari, E.D., Perritt, R.L., Whitmore, R.W., Zelon, H.S., Sheldon, L.S., and Sparacino, C.M. (1987) Results from the Total Exposure Assessment Methodology (TEAM) Study in Selected Communities in Northern and Southern California, *Atmospheric Environment*, **21**(9):1995–2004.

Klepeis, N.E., Nelson, W.C., Ott, W.R., Robinson, J.P., Tsang, A.M., Switzer, P., Behar, J.V., Hern, S.C., and Engelmann, W.H. (2001) The National Human Activity Pattern Survey (NHAPS): A Resource for Assessing Exposure to Environmental Pollutants, *Journal of Exposure Analysis and Environmental Epidemiology*, **11**(3): 231–252.

Krause, C., Chutsch, M., Henke, M., Leiske, M., Schulz, C., Schwarz, E., and Seifert, B. (1992) Heavy Metals in the Blood, Urine, and Hair of a Representative Population Sample in the Federal Republic of Germany 1985/86, in: *Proceedings of the Third European Meeting of Environmental Hygiene*, **3**: 159–162.

Olsen, G.W., Logan, P.W., Hansen, K.J., Simpson, C.A., Burris, J.M., Burlew, M.M., Vorarath, P.P., Venkateswarlu, P., Schumpert, J.C., and Mandel, J.H. (2003) An Occupational Exposure Assessment of a Perfluorooctanesulfonyl Fluoride Production Site: Biomonitoring, *Journal of the American Industrial Hygiene Association*, **64**(5): 651–659.

Özkaynak, H., Xue, J., Spengler, J., Wallace, L., Pellizzari, E., Jenkins, P. (1996) Personal Exposure to Airborne Particles and Metals: Results from the Particle TEAM Study in Riverside, California, *Journal of Exposure Analysis and Environmental Epidemiology*, **6**(1): 57–78.

Pang, Y.H., MacIntosh, D.L., Camann, D.E., and Ryan, B. (2002) Analysis of Aggregate Exposure to Chlorpyrifos in the NHEXAS-Maryland Investigation, *Environmental Health Perspectives*, **110**(3): 235–240.

Pellizzari, E.D., Clayton, C.A., Rodes, C.E., Mason, R.E., Piper, L.L., Fort, B., Pfeifer, G., and Lynam, D. (1999) Particulate Matter and Manganese Exposures in Toronto, Canada, *Atmospheric Environment*, **33**(5): 721–734.

TABLE 3.1 (CONTINUED)
Bibliography of Human Exposure Assessment Studies That Used Probability-Based Sampling

Pellizzari, E., Lioy, P., Quackenboss, J., Whitmore, R., Clayton, A., Freeman, N., Waldman, J., Thomas, K., Rodes, C., and Wilcosky, T. (1995) Population-Based Exposure Measurements in USEPA Region 5: A Phase I Field Study in Support of the National Human Exposure Assessment Survey, *Journal of Exposure Analysis and Environmental Epidemiology*, **5**(3): 327–358.

Pellizzari, E.D., Hartwell, T.D., Sparacino, C.M., Sheldon, L.S., and Zelon, H. (1985) Personal Exposures, Indoor-Outdoor Relationships, and Breath Levels of Toxic Air Pollutants Measured for 355 Persons in New Jersey, *Atmospheric Environment*, **19**(10): 1651–1661.

Quackenboss, J.J., Pellizzari, E.D., Shubat, P., Whitmore, R.W., Adgate, J.L., Thomas, K.W., Freeman, N.C.G., Stroebel, C., Lioy, P.J., Clayton, C.A., and Sexton, K. (2000) Design Strategy for Assessing Multi-Pathway Exposure for Children: The Minnesota Children's Pesticide Exposure Study (MNCPES), *Journal of Exposure Analysis and Environmental Epidemiology*, **10**(2): 145–158.

Robertson, G.L., Lebowitz, M.D., O'Rourke, M.K., Gordon, S., and Moschandreas, D. (1999) The National Human Exposure Assessment Survey (NHEXAS) Study in Arizona — Introduction and Preliminary Results, *Journal of Exposure Analysis and Environmental Epidemiology*, **9**(5): 427–434.

Rotko, T., Oglesby, L., Künzli, N., and Jantunen, M. (2000) Population Sampling in European Air Pollution Exposure Study, EXPOLIS: Comparisons Between the Cities and Representativeness of the Samples, *Journal of Exposure Analysis and Environmental Epidemiology*, **10**(4): 355–364.

Schulz, C., Becker, K., Bernigau, W., Hoffmann, K., Krause, C., Nöllke, P., Schwabe, R., and Seiwert, M. (1995) The 1990/92 Environmental Survey in the Old and New States of the Federal Republic of Germany, *Annals of Clinical Laboratory Science*, **25**: 561.

Sexton, K., Adgate, J.L., Church, T.R., Greaves, I.A., Ramachandran, G., Fredrickson, A.L., Geisser, M.S., and Ryan, A.D. (2003) Recruitment, Retention, and Compliance Results from a Probability Study of Children's Environmental Health in Economically Disadvantaged Neighborhoods, *Environmental Health Perspectives*, **111**(5): 731–736.

Thijs, L., Staessen, J., Amery, A., Bruaux, P., Buchet, J.P., Claeys, F., DePlaen, P., DuCoffre, G., Lauwerys, R., Lijnen, P. et al. (1992) Determinants of Serum Zinc in a Random Population Sample of Four Belgian Towns with Different Degrees of Environmental Exposure to Cadmium, *Environmental Health Perspectives*, 251–258.

Trapido, A.S., Mqoqi, N.P., Williams, N.W., Solomon, A., Goode, R.H., Macheke, C.M., Davies, A.J., and Panter, C. (1998) Prevalence of Occupational Lung Disease in a Random Sample of Former Mineworkers, Libode District, Eastern Cape Province, South Africa. *American Journal of Industrial Medicine*, **34**(4): 305–313.

Ullrich, D., Brenske, K.-R., Heinrich, J., Hoffmann, K., Ung, L, and Seifert, B. (1996) Volatile Organic Compounds: Comparison of Personal Exposure and Indoor Air Quality Measurements, in: *Indoor Air '96: Proceedings of the International Conference on Indoor Air Quality and Climate*, 301–306.

Ullrich, D., Gleue, C., Krause, C., Lusansky, C., Nagel, R., Schulz, C., and Seifert, B. (2002) German Environmental Survey of Children and Teenagers 2000 (GerES IV): A Representative Population Study Including Indoor Air Pollutants, in: *Indoor Air '02: Proceedings of the 9th International Conference on Indoor Air Quality and Climate*, **1**: 209–213.

Whitmore, R.W., Byron, M.Z., Clayton, C.A., Thomas, K.W., Zelon, H.S., Pellizzari, E.D., Lioy, P.J., and Quackenboss, J.J. (1999) Sampling Design, Response Rates, and Analysis Weights for the National Human Exposure Assessment Survey (NHEXAS), *Journal of Exposure Analysis and Environmental Toxicology*, **9**(5): 369–380.

Whitmore, R.W., Immerman, F.W., Camann, D.E., Bond, A.E., Lewis, R.G., and Schaum, J.L. (1994) Non-Occupational Exposures to Pesticides for Residents of Two U.S. Cities, *Archives of Environmental Contamination and Toxicology*, **26**: 47–59.

Wilson, N.K., Chuang, J.C., Iachan, R., Lyu, C., Gordon, S.M., Morgan, M.K., Özkaynak, H., and Sheldon, L.S. (2004) Design and Sampling Methodology for a Large Study of Preschool Children's Aggregate Exposures to Persistent Organic Pollutants in Their Everyday Environments, *Journal of Exposure Analysis and Environmental Epidemiology*, **14**(3): 260–274.

REFERENCES

Akland, G.G., Hartwell, T.D., Johnson, T.R., and Whitmore, R.W. (1985) Measuring Human Exposure to Carbon Monoxide in Washington, DC, and Denver, Colorado, during the winter of 1982–1983, *Environmental Science and Technology,* **19**(10): 911–918.

Callahan, M.A., Clickner, R.P., Whitmore, R.W., Kalton, G., and Sexton, K. (1995) Overview of Important Design Issues for a National Human Exposure Assessment Survey, *Journal of Exposure Analysis and Environmental Epidemiology,* **5**(3): 257–282.

Ezzati-Rice, T.M. and Murphy, R.S. (1995) Tissues Associated with the Design of a National Probability Sample for Human Exposure Assessment, *Environmental Health Perspectives,* **103**(suppl. 3): 55–59.

Gilbert, R.O. (1987) *Statistical Methods for Environmental Pollution Monitoring,* Van Nostrand Reinhold, New York, NY.

Kalton, G. (1983) *Introduction to Survey Sampling,* Sage Publications, Beverly Hills, CA.

Millard, S.P. and Neerchal, N.K. (2001) *Environmental Statistics with S-Plus,* CRC Press, Boca Raton, FL.

Pellizzari, E., Lioy, P., Quackenboss, J., Whitmore, R., Clayton, A., Freeman, N., Waldman, J., Thomas, K., Rodes, C., and Wilcosky, T. (1995) Population-Based Exposure Measurements in USEPA Region 5: A Phase I Field Study in Support of the National Human Exposure Assessment Survey, *Journal of Exposure Analysis and Environmental Epidemiology,* **5**(3): 327–358.

Sexton, K., Adgate, J.L., Eberly, L.E., Clayton, C.A., Whitmore, R.W., Pellizzari, E.D., Lioy, P.J., and Quackenboss, J.J. (2003) Predicting Children's Short-Term Exposure to Pesticides: Results of a Questionnaire Screening Approach, *Environmental Health Perspectives,* **111**(1): 123–128.

Thompson, S.K. (1992) *Sampling,* Wiley Interscience, New York, NY.

USEPA (1983) *Survey Management Handbook, Volume I: Guidelines for Planning and Managing a Statistical Survey,* Report No. EPA/230/12-84/002, U.S. Environmental Protection Agency, Office of Policy, Planning, and Evaluation, Washington, DC.

USEPA (1984) *Final Sampling Report for the Study of Personal CO Exposure,* Report No. EPA/600/4-84/034, U.S. Environmental Protection Agency, Environmental Monitoring Systems Laboratory, Research Triangle Park, NC.

USEPA (1990a) *Nonoccupational Pesticide Exposure Study (NOPES): Final Report,* Report No. EPA/600/3-90/003, U.S. Environmental Protection Agency, Office of Research and Development, Washington, DC.

USEPA (1990b) *National Survey of Pesticides in Drinking Water Wells: Phase 1 Report,* Report No. EPA/570/9-90/015, U.S. Environmental Protection Agency, Office of Pesticide Programs, Washington, DC.

USEPA (2000) *Guidance for the Data Quality Objectives Process (EPA QA/G-4),* Report No. EPA/600/R-96/055, U.S. Environmental Protection Agency, Office of Environmental Information, Washington, DC, http://www.epa.gov/quality/qs-docs/g4-final.pdf (accessed on April 25, 2006).

USEPA (2002) *Guidance on Choosing a Sampling Design for Environmental Data Collection (EPA QA/G-5S),* Report No. EPA/240/R-02/005, U.S. Environmental Protection Agency, Office of Environmental Information, Washington, DC, (http://www.epa.gov/quality/qs-docs/g5s-final.pdf (accessed on April 25, 2006).

Whitmore, R.W., Byron, M.Z., Clayton, C.A., Thomas, K.W., Zelon, H.S., Pellizzari, E.D., Lioy, P.J., and Quackenboss, J.J. (1999) Sampling Design, Response Rates, and Analysis Weights for the National Human Exposure Assessment Survey (NHEXAS), *Journal of Exposure Analysis and Environmental Toxicology,* **9**(5): 369–380.

Whitmore, R.W., Immerman, F.W., Camann, D.E., Bond, A.E., Lewis, R.G., and Schaum, J.L. (1994) Non-Occupational Exposures to Pesticides for Residents of Two U.S. Cities, *Archives of Environmental Contamination and Toxicology,* **26**: 47–59.

Wölter, K.M. (1985) *Introduction to Variance Estimation,* Springer-Verlag, New York, NY.

Part II

Inhalation

4 Inhalation Exposure, Uptake, and Dose

Andrea R. Ferro
Clarkson University

Lynn M. Hildemann
Stanford University

CONTENTS

4.1   Synopsis 81
4.2   Introduction 82
4.3   Identifying the Major Pollutants of Concern 82
    4.3.1   Criteria Pollutants 82
    4.3.2   Other Air Pollutants of Concern 83
4.4   Uptake of Pollutants by the Respiratory Tract 83
    4.4.1   Respiratory Tract Regions 84
4.5   Uptake of Gaseous Pollutants 84
    4.5.1   Modeling the Mass Transfer of Gaseous Air Pollutants to the Bloodstream 84
4.6   Uptake of Particulate Pollutants 89
    4.6.1   Size Ranges of Airborne Particles 91
    4.6.2   Health Impacts and Residence Times of Deposited PM 91
4.7   Correlation between Breath and Blood Concentrations and Dose 92
    4.7.1   Example: Uptake of CO by the Bloodstream 92
4.8   Conclusion 96
4.9   Questions for Review 97
References 97

4.1   SYNOPSIS

Uptake of pollutants by the respiratory tract is sequential. Gaseous and particulate pollutants must first sorb or deposit onto the liquid surface of the lungs before they can travel through the liquid layer, come into contact with the lung tissue, and diffuse into the bloodstream. The fate and transport of pollutants that enter the respiratory system are dependent on many factors, including the diffusivity, solubility, reactivity, and size of the pollutants; the airflow characteristics in the respiratory system; and the physiology of the respiratory and vascular systems. Mathematical models can be used to estimate diffusion of pollutants from the respiratory tract to the bloodstream as well as partitioning of pollutants in the body after they have been absorbed.

4.2 INTRODUCTION

Inhalation exposure occurs when a person breathes air that contains pollutants into his or her respiratory system. For some pollutants, such as carbon monoxide (CO) and asbestos, inhalation is the major route of exposure. For other pollutants, such as lead and chloroform, inhalation exposure can contribute to overall exposure in combination with dermal exposure or ingestion. By examining the mechanisms of pollutant uptake by the respiratory system, we can learn how pollutant inhalation exposure contributes to dose and, potentially, health impacts.

As discussed in Chapter 2, the contact boundary for inhalation exposure is selected by the analyst. By selecting the oral/nasal region as the contact boundary, inhalation exposure to airborne pollutants is defined as the concentration in the air at the mouth and nose boundary, with the assumption that this concentration is inhaled by the person. The *intake dose* is the mass of the pollutant that crosses the oral/nasal boundary into the respiratory system. However, not all the intake dose becomes absorbed by the respiratory system; some of the pollutant is exhaled with the breath. The remainder of the pollutant deposits or sorbs (reversibly or irreversibly) to the liquid surface of the respiratory tract. This portion is called the *absorbed dose*. A fraction of the pollutant in the liquid layer can be expelled from the respiratory tract by blowing, sniffling, sneezing, swallowing, coughing, and spitting. Finally, some of the pollutant will transport through this layer and make contact with the tissues in the respiratory tract. This bioavailable portion of the potential dose is called the *effective dose*.

Pollutants follow different transport paths through the liquid layer. Dissolved gaseous pollutants diffuse through the liquid layer to reach the tissue. Soluble chemical species desorb from particle surfaces into the liquid layer and then diffuse through to the tissue. In the pulmonary region, insoluble particles can slowly migrate through the liquid layer and come into contact with the tissue. *Uptake* is the incorporation of the pollutant into the body via chemical and physical mechanisms.

This chapter reviews the major pollutants of concern for inhalation exposure, provides an overview of the respiratory system, and discusses the uptake of gaseous and particulate pollutants in the respiratory system. Several models to estimate the uptake of gaseous and particulate pollutants by the respiratory system are introduced.

4.3 IDENTIFYING THE MAJOR POLLUTANTS OF CONCERN

4.3.1 CRITERIA POLLUTANTS

The 1970 U.S. Clean Air Act Amendments identified six criteria air pollutants for which the U.S. Environmental Protection Agency (USEPA) set National Ambient Air Quality Standards (NAAQS). The NAAQS are periodically updated by the USEPA based on current scientific knowledge about the health effects of the pollutants. The six criteria air pollutants established were particulate matter (PM), nitrogen dioxide (NO_2), ozone (O_3), sulfur dioxide (SO_2), carbon monoxide (CO), and nonmethane hydrocarbons (NMHC). Since 1970, lead was added and NMHC was removed as a criteria air pollutant, and currently the six criteria air pollutants are PM, NO_2, O_3, SO_2, CO, and lead. PM and lead are in particle form, while NO_2, O_3, SO_2, and CO are gases. PM is regulated for two size fractions: PM with an aerodynamic diameter less than 10 μm (PM_{10}) and PM with an aerodynamic diameter less than 2.5 μm ($PM_{2.5}$). PM_{10} is also called *inhalable* PM, $PM_{2.5}$ is called *fine* PM, while the difference between PM_{10} and $PM_{2.5}$ is called *coarse* PM.

Table 4.1 summarizes the target sites and potential pathological effects of inhaled criteria air pollutants. Effects can occur in the respiratory system or elsewhere in the body, as is the case for PM, lead, and CO. Primary NAAQS are also provided in Table 4.1. The reader may obtain the specific instructions for meeting each NAAQS directly from the USEPA (USEPA 2005).

TABLE 4.1
Summary of Uptake and Potential Pathological Effects to Inhaled Airborne Pollutants

Pollutant	Target Site or Organ	Mode of Action and Pathology	Primary Standards	Averaging Times
Particulate Matter (PM_{10})	Respiratory tract	Aggravation of responses to other more toxic pollutants	50 μg m^{-3} (arith. mean) 150 μg m^{-3}	Annual 24-hour
Particulate Matter ($PM_{2.5}$)	Respiratory tract, heart (fine PM)	Possibly pulmonary and systemic inflammation, accelerated hardening of arteries, decreased heart rate variability	15 μg m^{-3} (arith. mean) 65 μg m^{-3}	Annual 24-hour
Lead (Pb)	Kidney, liver, brain, reproductive system	Decreased hemoglobin, anemia, impaired motor and cognitive function, decreased fertility	1.5 μg m^{-3}	Quarterly average
Nitrogen dioxide (NO_2)	Bronchialar and alveolar airways of lung	Irritation, inflammation, fluid accumulation, scarring of tissue, bronchitis	0.053 ppm (100 μg m^{-3}) (arith. mean)	Annual
Ozone (O_3)	Bronchiolar and alveolar distal airways of lung	Irritation, inflammation, difficulty breathing, fibrosis	0.08 ppm 0.12 ppm	8-hour 1-hour
Sulfur dioxide (SO_2)	Bronchial airways	Activation of bronchial receptors, difficulty breathing, bronchitis	0.3 ppm (arith. mean) 0.14 ppm	Annual 24-hour
Carbon monoxide (CO)	Blood and living cells in all organs	Formation of carboxyhemoglobin in red blood cells, limiting circulation of oxygen	9 ppm (10 mg m^{-3}) 35 ppm (40 mg m^{-3})	8-hour 1-hour

(Adapted from Raabe 1999 with NAAQS from USEPA, and additions in PM pathology from Pope et al. 2004 and lead pathology from the U.S. ATSDR 1999.)

4.3.2 OTHER AIR POLLUTANTS OF CONCERN

Because there are currently approximately 60,000 chemicals and 2 million chemical mixtures in commercial use, and more than 100 new chemical compounds synthesized each year (Heinsohn and Kabel 1999), it is infeasible to discuss all potential air pollutants of concern. Additionally, the exposure routes and toxicology of many of these chemicals or potential reaction products of these chemicals are unknown. Instead, this book concentrates on two criteria air pollutants, CO and PM (Chapter 6 and Chapter 8, respectively); two large classes of air pollutants that have received considerable attention, volatile organic compounds and pesticides (Chapter 7 and Chapter 15, respectively); one air pollutant that people have high levels of exposure to, secondhand smoke (Chapter 9); and several classes of persistent pollutants that may have several routes of exposure, such as PCBs, dioxins, furans, and endocrine disruptors (Chapter 16). In this chapter, in addition to considering PM and CO, we will present a model for the respiratory uptake of VOCs into the human body.

4.4 UPTAKE OF POLLUTANTS BY THE RESPIRATORY TRACT

Uptake of pollutants by the respiratory tract is sequential. That is, pollutants move sequentially from the oral/nasal boundary through the respiratory system. After making contact with the surface of the respiratory tract, pollutants move sequentially through the layers of the respiratory tract before reaching the bloodstream or lymphatic system. Each layer the pollutant moves through offers some resistance to the transport of the pollutants. Therefore, a brief discussion of the anatomy and function of the respiratory system components is necessary for understanding pollutant uptake in the respiratory tract.

4.4.1 RESPIRATORY TRACT REGIONS

Figure 4.1 is a schematic of the respiratory tract divided into three major regions. The first region, called the *nasal-pharyngeal* or *extrathoracic* region, includes the nasal passages, the pharynx, and the larynx. This region warms and humidifies the inhaled air and filters out coarse PM in the complex airflow patterns of its passages. The second region, called the *tracheobronchial* or *thoracic* region, includes the trachea, bronchi, and bronchioles. The trachea branches into the left and right main bronchi, which lead to the left and right lung, respectively. The main bronchi then branch into increasingly smaller-in-diameter-and-length bronchi and bronchioles. The tracheo-bronchial region subdivides the airway to reach the third region, called the *pulmonary* or *gas exchange* region, where the gas exchange occurs. The pulmonary region includes partially alveolated respiratory bronchioles, alveolar ducts, and alveoli. The alveoli are tiny sacs, which collectively have an enormous surface area (approximately 100 m^2 during inhalation) to efficiently exchange gases with the bloodstream (Heinsohn and Kabel 1999).

The nasal-pharyngeal and tracheo-bronchial regions are lined with mucus-secreting cells, which form highly viscous *mucus*, and *cilia*, which clear the mucus from the respiratory tract into the digestive tract. The pulmonary region is coated with a liquid film of *surfactant*, which is less viscous than mucus and works to lower the surface tension at the alveolus–air interface. By absorbing and transporting pollutants, these liquid layers play an important role in the uptake of pollutants by the respiratory tract.

4.5 UPTAKE OF GASEOUS POLLUTANTS

Human respiration depends on oxygen (O_2) diffusing from the air in the lungs to the bloodstream; and carbon dioxide (CO_2) diffusing from the bloodstream back out to the lungs.

The mass transfer of O_2 and CO_2 occurs in the pulmonary region of the respiratory tract through several physical layers, and gaseous pollutants follow the same pathway. Figure 4.2 illustrates the sequential diffusion of O_2 and CO_2 in the alveolar region of the lung. To reach the bloodstream from the air in the alveoli, the gases must diffuse through the surfactant layer to reach the lung tissue; through the lung tissue, which is comprised of the alveolar epithelium and the epithelial basement membrane; through the *interstitial space* between the lung and the bloodstream; through the blood vessel wall, which consists of the capillary basement membrane and the capillary endothelium; and into the bloodstream. The interstitial space, which is the space between cells in the body, is filled with *interstitial fluid*.

Figure 4.3 provides a flow diagram for pollutant uptake through the applicable layers for each of the three main regions of the respiratory system. Figure 4.3 shows that the absorption and diffusion of gases is bidirectional. After gases from the inhaled air are absorbed to the mucus or surfactant layer, dissolved gases and volatile species may desorb back out to the air in the lungs. All dissolved species are able to diffuse in either direction; the direction of diffusion is dependent on the concentration gradient between two layers. In the nasal-pharyngeal and tracheobronchial regions, pollutants may or may not reach the bloodstream through the much wider layer of interstitial fluid. Some pollutants, such as insoluble material, may instead travel from the interstitial fluid into the lymphatic system. However, in the pulmonary region, the density of capillaries is extremely high and the thin layer of interstitial fluid between the lung tissue and the capillaries does not present a significant hindrance to the gases reaching the bloodstream.

4.5.1 MODELING THE MASS TRANSFER OF GASEOUS AIR POLLUTANTS TO THE BLOODSTREAM

While the gases are transporting to the bloodstream, they may also chemically react. Figure 4.4 provides a schematic of four-layer diffusion through the bronchial wall for both an inert and a

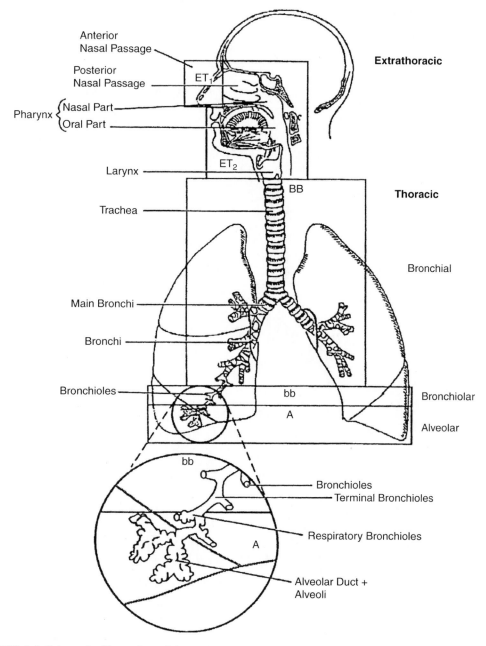

FIGURE 4.1 Schematic illustration of the major regions of the respiratory tract. (Adapted from Figure 1, ICRP 2004.)

reactive gas. Each layer offers some resistance to gaseous transport, decreasing the concentration of species i as it moves from the air ($P_{i,g}$) to the blood ($P_{i,b}$). For a reactive gas, the concentration is further decreased by its reactivity. The mass flow rate of species i through the layers is given by $\dot{m}_i$. The four layers included in the schematic are described as follows:

a. The air layer (g), which represents the difference between the mainstream pollutant concentration inhaled and what is found adjacent to the mucus/surfactant layer

b. the liquid mucus (or surfactant) layer (s), which includes the cilia in the tracheobronchial region
c. the tissue layer (t)
d. the blood layer (b)

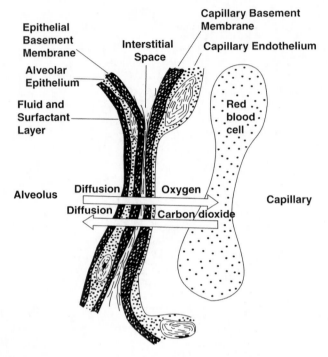

FIGURE 4.2 Gas exchange through the alveolar membrane. (Adapted from Figure 39-9 in Guyton 2000. With permission from Elsevier.)

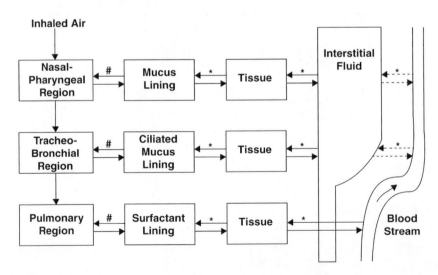

FIGURE 4.3 Sequential pollutant uptake by the respiratory tract. * Denotes a pathway that exists for all dissolved species; # denotes a pathway that exists only for dissolved gases and volatile species.

The mass transfer of gaseous air pollutants from the respiratory tract to the bloodstream can be modeled by representing the individual layers as "resistors" in series. That is, analogous to resistors in an electrical circuit opposing the passage of an electric current, each layer presents resistance to the diffusion of the pollutant through the layer. The gas exchange from the lung to the blood can be described as a 4-resistance mass transfer model for the gas, mucus, tissue, and blood layers (Ultman 1988). By extending the electrical circuit analogy, Ohm's law states that resistances in series are additive. Accordingly, the overall mass transfer coefficient K_m equals the inverse of the total resistance, where the total resistance equals the sum of the resistances in each of the layers:

$$K_m = \frac{1}{\left(\dfrac{RT}{k_m}\right)_g + \left(\dfrac{f}{\alpha_x k_m}\right)_s + \left(\dfrac{f}{\alpha_x k_m}\right)_t + \left(\dfrac{f}{\alpha_x k_m}\right)_b} \qquad (4.1)$$

$$\quad\quad\quad\quad\text{gas}\quad\quad\quad\text{mucus}\quad\quad\quad\text{tissue}\quad\quad\quad\text{blood}$$

where

K_m	= overall mass transfer coefficient [mole m^{-2} Pa^{-1} sec^{-1}]
k_m	= layer-specific mass transfer coefficient, which depends on the pollutant's diffusivity in the layer [m sec^{-1}]
R	= universal gas constant [8.3145 Pa m^3 mole^{-1} K^{-1}]
T	= temperature inside lungs [K; typically 300 K]
f	= fraction of pollutant left unreacted/undissociated in each layer [dimensionless]
α_x	= solubility coefficient, defined as the liquid molar volume divided by the Henry's Law coefficient [mole m^{-3} Pa^{-1}]

$\dfrac{RT}{k_m}$ = resistance in the gas layer [m^2 Pa sec moles^{-1}]

$\dfrac{f}{\alpha_x k_m}$ = resistance in each subsequent layer [m^2 Pa sec moles^{-1}]

The overall mass transfer coefficient can be used to estimate the dose of the pollutant in the blood over time, as given in Equation 4.2. The difference of the partial pressure in the gas layer p_{xg} and the final layer p_{xb} is called the "driving force" of the mass transfer. In other words, as the difference between the partial pressures increases, the mass transfer increases.

$$\frac{Dose}{Time} = K_m S \left(p_{xg} - p_{xb} \right) \qquad (4.2)$$

where

K_m	= overall mass transfer coefficient [mole m^{-2} Pa^{-1} sec^{-1}]
S	= surface area available for transfer [m^2, ~100 m^2 for alveolar area]
p_{xg}	= partial pressure of gaseous pollutant x in gas adjacent to mucus/surfactant layer [Pa]
p_{xb}	= partial pressure of gaseous pollutant x dissolved in final layer (tissue or blood) [Pa]

Higher values for solubility and diffusivity result in a higher overall mass transfer coefficient, due to increases in α_x and k_m. Higher values for reactivity (e.g., for ozone) also result in a higher overall mass transfer coefficient due to a decrease in f; however, the overall mass transfer of the pollutant to the blood is reduced. For example, chemical reaction of the pollutant in the mucus reduces the partial pressure of the pollutant in the mucus and increases the driving force at the initial gas mucus barrier. Subsequent layers have a resulting lower driving force, and the mass flow rate reduces from layer to layer. The difference between the mass flow rate of an inert and a reactive pollutant is illustrated in Figure 4.4.

Table 4.2 (data calculated from Table 6 in Ultman 1988) provides estimates of the resistance for the gas, mucus, and tissue in the terminal bronchioles, deep in the respiratory tract, for SO_2, CO, and O_3. Of the three pollutants, SO_2 is taken up most quickly because its high aqueous solubility results in a very low resistance in the mucus and tissue layers. For SO_2, the resistance of the gaseous layer limits the overall mass transfer. Ozone and CO, on the other hand, have much lower aqueous solubilities than SO_2, resulting in a higher resistance to mass transfer. In addition, the ozone is reactive, which reduces its mass transfer through the tissue to the blood due to reactive losses. Overall, for these three pollutants, CO is taken up most slowly, with the substantial resistance of the tissue layer being most limiting.

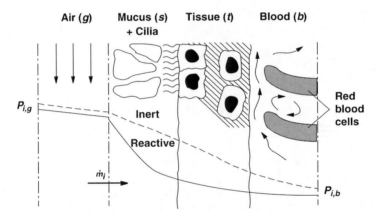

FIGURE 4.4 Schematic diagram of diffusion through the bronchial wall. In the alveoli, the mucus + cilia layer is replaced by a surfactant layer. (From Ultman 1988. With permission.)

TABLE 4.2
Estimates of Resistances (10^{10} m² Pa sec mol^{-1}) in the Terminal Bronchioles

	Gas	Mucus	Tissue	Overall
SO_2	0.05	0.00015	0.0011	0.05
CO	0.071	16.95	120.5	137.0
O_3	0.077	0.91	0.20	1.19

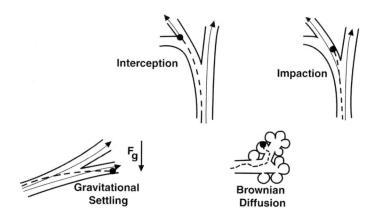

FIGURE 4.5 Illustration of the major physical mechanisms of deposition in the respiratory tract for inhaled airborne particles. (Adapted from Figure 10 in Raabe 1982.)

4.6 UPTAKE OF PARTICULATE POLLUTANTS

Particulate pollutants must first deposit in the lungs before they can travel through the mucus/surfactant and come into contact with the lung tissue. There are four important mechanisms for deposition of inhaled particulate pollutants (particles) in the respiratory tract: interception, impaction, gravitational settling, and Brownian diffusion. These mechanisms are illustrated in Figure 4.5. The relative importance of the mechanisms depends upon the characteristics of the depositing particle. In general, it is assumed that every particle that contacts the mucus or surfactant surface of the respiratory tract remains deposited.

Interception occurs when a particle following an air streamline brushes up against an obstacle and is removed. For particles found in most human environments, interception is not an important deposition mechanism. However, interception can be significant for deposition of particles with elongated shapes (Raabe 1999), such as asbestos and fiberglass particles.

Impaction occurs when a particle with too much inertia to follow a speed or directional change in the air streamlines crosses the streamlines and collides into the surface of the respiratory tract. The respiratory tract has many bends and bifurcations that cause changes in the velocity of the streamlines. Coarse particles, due to their increased inertia, are affected more by inertial impaction than are fine particles. Impaction is the primary removal mechanism for coarse particles in the respiratory tract, most of which occurs in the nasal-pharyngeal and tracheobronchial regions (Raabe 1999).

Gravitational settling occurs when the influence of gravity on a particle causes the particle to cross the air streamlines and settle on the surface of the respiratory tract. The settling velocity, which is a function of the gravity force and the opposing drag force on the particle, is proportional to the square of the aerodynamic diameter of the particle. Gravitational settling is most important where the velocity of the air streamlines has diminished sufficiently to allow the particles to settle, which corresponds approximately with the confluence of the bronchial and pulmonary regions (Raabe 1999).

Deposition due to *Brownian diffusion* occurs when a particle undergoing Brownian motion wanders into the wall of the respiratory tract. Brownian motion is caused by the random collisions of gas molecules against the particle. Therefore, as the particle size decreases, the influence of the gas molecules on the movement of the particle increases. Brownian diffusion is the most important mechanism for deposition of particles smaller than 0.5 μm in diameter in the pulmonary region of the respiratory tract (Raabe 1999).

Figure 4.6 and Figure 4.7 plot the predicted deposition fraction for spherical, insoluble, stable, and unit density (i.e., 1 g cm^{-3}) particles in the nasal/oropharyngeal, tracheobronchial, and

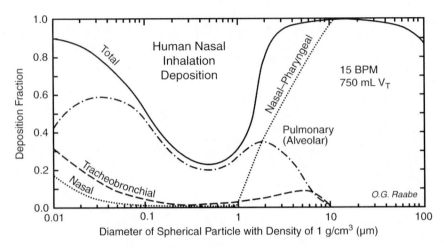

FIGURE 4.6 Inhalation deposition as a function of particle size for nasal breathing. A rate of 15 breaths per minute (BPM) and a tidal volume V_T of 750 mL (at rest) are assumed. (Adapted from Figure 4 in Raabe 1999.)

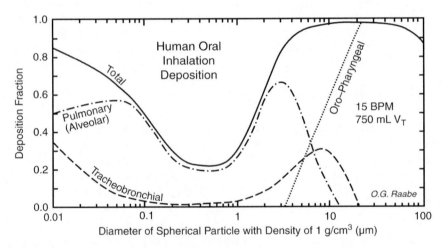

FIGURE 4.7 Inhalation deposition as a function of particle size for oral breathing. A rate of 15 breaths per minute (BPM) and a tidal volume V_T of 750 mL (at rest) are assumed. (Adapted from Figure 5 in Raabe 1999.)

pulmonary regions of the respiration tract. The predictions are for both nasal (Figure 4.6) and oral (Figure 4.7) inhalation and are based on a mathematical model using the respiratory parameters for a typical adult male at rest breathing at a rate of 15 breaths per minute (BPM) and a tidal volume (V_T) of 750 mL, or 16.2 m^3 day^{-1}. This model predicts that for both nose and mouth breathing, larger PM is deposited primarily in the nasal-pharyngeal region due to impaction, and smaller PM is deposited primarily in the pulmonary region due to uptake by Brownian diffusion. A dip in the total deposition fraction is shown for particles between approximately 0.2 and 1 μm. This phenomenon occurs because particles of this size are not affected strongly by the deposition mechanisms. Therefore, most of these particles are exhaled. Mouth breathing decreases impaction and interception in the nasal-pharyngeal region and allows larger PM to penetrate deeper into the lungs. Like the uptake of gaseous pollutants, the uptake of particles in the respiratory system is sequential.

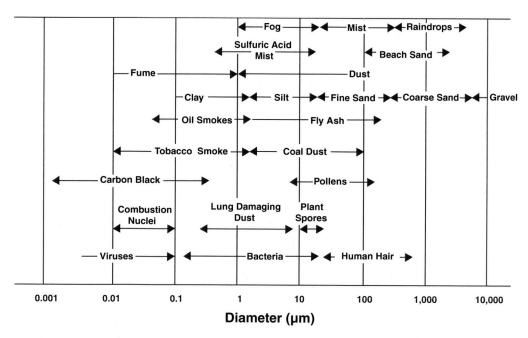

FIGURE 4.8 Approximate size ranges for different types of airborne particles. (Reprinted with permission from the Mine Safety Appliances Company, Pittsburgh, PA.)

4.6.1 SIZE RANGES OF AIRBORNE PARTICLES

Airborne particles range from approximately 1 nm to 10 mm, with different types of particles having different size ranges. Therefore, for each particle type, one can predict where it will tend to deposit in the respiratory tract. As shown in Figure 4.8, coarse particles, which deposit primarily in the nasal-pharyngeal region, tend to be from natural sources, such as windblown dust, sand, or gravel. Fine particles, which deposit primarily in the pulmonary region tend to be from anthropogenic sources, such as combustion of fossil fuel and smoking. Biogenic PM (i.e., bioaerosols) ranges in size from tiny viruses (0.01–0.1 μm) to large pollens (10–100 μm), which deposit primarily in the pulmonary and nasal-pharyngeal regions, respectively.

4.6.2 HEALTH IMPACTS AND RESIDENCE TIMES OF DEPOSITED PM

Many studies have found that ambient particulate matter (PM) concentrations are positively correlated with respiratory and cardiovascular disease (Vedal 1997; Samet et al. 2000; Peters et al. 2001). Particulate matter may be responsible for the mortality of 20,000 persons annually as well as the exacerbation of asthma leading to hundreds of thousands of cases annually. However, scientists still do not understand completely why PM causes health effects.

Many scientists believe that quantifying dose based solely on the mass of PM depositing in the respiratory tract is inadequate. The total number of particles depositing may be important, since each particle depositing in the pulmonary region constitutes a separate insult to the system. The total surface area of deposited particles may be important because the more surface area, the more soluble toxins/carcinogens are potentially available to desorb into the mucus or surfactant. Additionally, chemical characteristics of particles could be important. For example, particle acidity can influence the pH conditions and pH-controlled functions in the lungs. For some particles, shape is a factor for instigating health effects. For example, asbestos is made up of thin, filamentous particles that tend to lodge permanently in the narrow alveoli, causing irritation and sometimes cancer in the lungs. Some scientists argue that the health effects are exacerbated by exposure to the mixture of particles and associated gaseous pollutants.

Health impacts from deposited PM depend on whether the PM is soluble or insoluble and where the PM deposits in the respiratory tract. Soluble PM dissolves into the mucus or surfactant layer and transports through this layer to the tissue. Dissolved species can irritate tissues by changing the pH or other chemical properties. Dissolved species are transported to the rest of the body via the bloodstream or the gastrointestinal (GI) tract. There, the body's natural metabolic processes can activate or deactivate toxins and carcinogens.

Insoluble PM (e.g., lead particles) that deposit in the nasal-pharyngeal region can be cleared by blowing, sniffling, sneezing, swallowing, and spitting. Swallowing can lead to a subsequent potential dose in the GI tract. The main health impacts of insoluble PM in the nasal-pharyngeal region are inflammation leading to nasal congestion. The residence time of insoluble PM in the nasal-pharyngeal region is on the order of an hour.

In the tracheobronchial region, particles are cleared by the action of cilia which move particles in the mucus up out of lungs and into the throat, again leading to a potential subsequent GI tract dose. The main health impact of insoluble PM in the tracheobronchial region is irritation of the bronchi, leading to chest congestion, soreness, and cough. The residence time of insoluble PM in the nasal-pharyngeal region is on the order of a day.

In the pulmonary region, which lacks cilia, particles are engulfed and eventually degraded by special scavenger cells. The main health impact of insoluble PM in the pulmonary region is swelling of alveoli, leading to coughing and shortness of breath. The residence time of insoluble PM in the pulmonary region is on the order of a year (Raabe 1999).

4.7 CORRELATION BETWEEN BREATH AND BLOOD CONCENTRATIONS AND DOSE

4.7.1 EXAMPLE: UPTAKE OF CO BY THE BLOODSTREAM

Due to limited solubility and non-reactivity, high CO levels reach the pulmonary region, but it takes hours for blood CO levels to build up and equilibrate with inhaled CO concentrations. Figure 4.9 plots the percent of blood hemoglobin bound with CO (COHb) as a function of the concentration and duration of inhaled CO, and shows what health effects result from different levels of COHb in the blood (Seinfeld 1986; USEPA 2000). Even at an inhaled concentration of 600 parts per million by volume (ppm), which would eventually induce a coma, no symptoms beyond headache and reduced mental acuity would be felt for the first hour of exposure. Figure 4.9 shows that equilibrium of CO in the blood is approached after approximately 6–8 hours of exposure to a constant CO concentration in the air.

The blood contains a baseline (endogenous) level of COHb due to the generation of small amounts of CO in the body as a metabolic by-product; additional, exogenous COHb is formed due to exposure to inhaled CO. The total amount of COHb present at any time is:

$$[COHb]_{tot} = [COHb]_{bas} + [COHb]_{exo} \tag{4.3}$$

The relationship between the exogenous COHb and the inhaled CO concentration can be modeled as a simple equilibration process as shown in Figure 4.10 (Ott and Mage 1978). The equation representing this approach to equilibrium for $[COHb]_{exo}$ in the blood compartment is shown below as Equation 4.4. The rate of increase of the percent exogenous COHb in the blood compartment is equal to the flow entering the compartment minus the flow leaving the compartment.

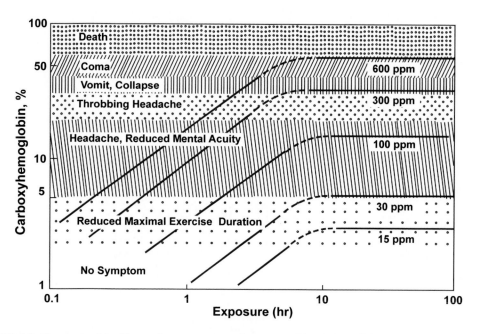

FIGURE 4.9 Human health effects of exposure to carbon monoxide. (Adapted from Figure 2.1 in Seinfeld 1986. With permission from Wiley–Interscience.)

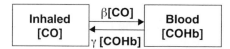

FIGURE 4.10 One-compartment model for relating VOC breath concentration to dose. (Adapted from Wallace et al. 1993. With permission from the Nature Publishing Group and the first author.)

$$\frac{d[COHb]_{exo}}{dt} = \beta[CO] - \gamma[COHb]_{exo} \tag{4.4}$$

where

$[COHb]_{exo}$ = percent hemoglobin that is combined with exogenous CO in the blood [%]
$[CO]$ = concentration of inhaled CO [ppm]
β = rate constant for exogenous CO uptake by the blood [0.06% COHb ppm^{-1} h^{-1}]
γ = rate constant for return of exogenous blood CO to the lungs [0.402 h^{-1}]

The solution to this materials balance is:

$$[COHb]_{tot} = [COHb]_{bas} + [COHb]_{exo,0}\, e^{-\gamma t} + \frac{\beta}{\gamma}[CO]\left(1 - e^{-\gamma t}\right) \tag{4.5}$$

where

$[COHb]_{bas}$ = baseline (endogenous) CO level in blood, constant at 0.5%
$[COHb]_{exo,0}$ = exogenous CO level in blood at time $t = 0$

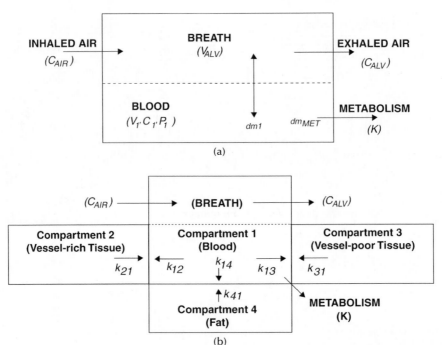

FIGURE 4.11 Two-compartment (a) and four-compartment (b) models for relating VOC breath concentration to dose. (Adapted from Wallace et al. 1993. With permission from the Nature Publishing Group and the first author.)

For uptake of VOCs, Wallace, Pellizzari, and Gordon (1993) described a simple one-compartment model similar to what we used for CO. However, Wallace et al.'s model, described in Figure 4.11a and Equation 4.6, also accounts for bodily metabolism. Because all the uptake of VOCs effectively occurs at the alveoli, the model uses the *alveolar respiration rate*, which is often assumed to be 70% of the measured respiration rate, or about 400 liters hr^{-1}. The remaining inspired airflow fills non-alveolar *dead space*, which is not important to gas exchange.

$$C_{ALV} = \left(C_{ALV}\right)_0 e^{-t/\tau_1} + fC_{AIR}\left(1 - e^{-t/\tau_1}\right) \tag{4.6}$$

where

C_{ALV}	= concentration of the VOC in air of the alveolar region, assumed to be equal to the exhaled air concentration
C_{AIR}	= concentration of the VOC in the air at the mouth and nose boundary, defined earlier as "exposure"
f	= $(\dot{V}_{ALV} + KP_1)/\dot{V}_{ALV}$, the fraction of inhaled contaminant that is exhaled at equilibrium ($f = 1$ for no metabolic degradation; $f < 1$ for some metabolic degradation)
$\dot{V}_{ALV}$	= alveolar respiration rate
K	= first order rate constant for metabolic degradation (value varies with compound)
P_1	= the blood/air partition coefficient defined as C_1/C_{ALV} (value varies with compound)
τ_1	= $P_1 V_1/\dot{V}_{ALV}$, the residence time in the blood compartment
V_1	= volume of the blood
$dm1$	= $\dot{V}_{ALV}(C_{AIR} - C_{ALV})\,dt$, mass of VOC in alveolar air entering or leaving the blood during time dt
dm_{MET}	= $-KC_1\,dt$, mass of VOC lost to metabolism during time dt
C_1	= concentration of the VOC in the blood

TABLE 4.3
Median Calculated Values of a_i in Five Subjects for Four Compartments

Chemical	Blood	Vessel-Rich Tissues	Vessel-Poor Tissues	Fat
Aromatics				
Toluene	0.25	0.15	0.25	0.35
p–Xylene	0.25	0.17	0.21	0.35
o–Xylene	0.29	0.16	0.15	0.36
Ethylbenzene	0.17	0.17	0.22	0.36
Aliphatics				
Hexane	0.35	0.24	0.15	0.20
Decane	0.45	0.25	0.16	0.17
Chlorinated				
1,1,1-Trichloroethane	0.21	0.24	0.19	0.31
Trichloroethylene	0.24	0.19	0.24	0.28
Dichloromethane	0.32	0.25	0.26	0.19

(From Wallace et al. 1997.)

However, VOCs can also partition from the blood into the tissue and fat in other parts of the body, creating reservoirs that will influence the breath concentrations. Therefore, Wallace et al. (1997) developed a 4-compartment model to more accurately represent this (neglecting the initial alveolar burden at t = 0). This model is described in Figure 4.11b and Equation 4.7.

$$C_{ALV} = fC_{AIR}\left(1 - a_1e^{-t/\tau_1} - a_2e^{-t/\tau_2} - a_3e^{-t/\tau_3} - a_4e^{-t/\tau_4}\right) \tag{4.7}$$

where
a_i = fraction of bodily dose in compartment i at equilibrium
τ_i = residence time for pollutant in compartment i

Table 4.3 provides results of the model for nine VOC pollutants. At equilibrium, for most of the VOCs studied, the blood compartment has 20–35% of the VOC dose; the vessel-rich tissue compartment has 15–25%; the vessel-poor compartment has 15–25%; and the fat compartment has 20–35%.

As shown in Table 4.4, the residence times for VOCs differ greatly among the four compartments. The blood compartment has τ_1 = 3–10 minutes (<0.2 h); the vessel-rich tissue compartment has τ_2 = 0.3–2 h; the vessel-poor tissue compartment has τ_3 = 2–5 h; and the fat compartment has τ_4 = 30–90 h. Since the tissue and fat compartments store 60–85% of the VOC dose at equilibrium, and have much longer residence times than the blood compartment, clearance of VOCs from the human body occurs more slowly than what would be predicted by a single-compartment model that focused just on the blood compartment.

As shown by Table 4.5, some VOCs like 1,1,1-trichloroethane appear to be poorly metabolized ($f > 0.9$ corresponds to $K < 4$ h^{-1}), while others, like ethylbenzene, hexane, and decane, are metabolized efficiently ($f < 0.1$ corresponds to $K > 100$ h^{-1}). In general, those VOCs that are most poorly metabolized will be retained in the human body the longest.

TABLE 4.4
Median Residence Times τ_i (hours) in Five Subjects for Four Compartments

Chemical	Blood	Vessel-Rich Tissues	Vessel-Poor Tissues	Fat
Aromatics				
Toluene	0.06	0.64	5.3	84
p–Xylene	0.10	0.59	4.2	51
o–Xylene	0.15	1.9	3.7	64
Ethylbenzene	0.07	0.58	3.3	90
Aliphatics				
Hexane	0.07	0.38	4.7	56
Decane	0.05	0.27	2.0	63
Chlorinated				
1,1,1-Trichloroethane	0.15	0.68	4.8	29
Trichloroethylene	0.09	0.55	4.2	44
Dichloromethane	0.19	0.53	4.7	61

(From Wallace et al. 1997.)

TABLE 4.5
Calculated Values of f for Five Subjects, Mean, and Standard Deviation (SD)

Chemical	Mean	SD
Aromatics		
Toluene	0.162	0.022
p–Xylene	0.080	0.015
o–Xylene	0.063	0.016
Ethylbenzene	0.080	0.008
Aliphatics		
Hexane	0.351	0.076
Decane	0.102	0.019
Chlorinated		
1,1,1-Trichloroethane	0.875	0.072
Trichloroethylene	0.218	0.046
Dichloromethane	0.233	0.031

(Adapted from Wallace et al. 1997.)

4.8 CONCLUSION

In this chapter, we examined the mechanisms of pollutant uptake by the respiratory system, where uptake is the incorporation of the pollutant into the body via chemical and physical mechanisms. Drawing from fate and transport models, chemical equilibrium relationships, and materials balance, we introduced models to predict the pollutant uptake in the major regions of the respiratory tract. We found that for gases, uptake is dependent on solubility, diffusivity, and reactivity; for particles, uptake is primarily dependent on particle size and shape. For our examples, we applied the models to specific pollutants, but we did not study these pollutants in depth. However, the following chapters will further develop human exposure to various classes of pollutants.

4.9 QUESTIONS FOR REVIEW

1. Why does most gas exchange occur in the pulmonary region of the respiratory tract?
2. Explain the sequential uptake of pollutants in the respiratory tract.
3. Derive the relationship between inhaled CO and blood CO levels given in Equation 4.3.
4. Assume that the average concentration of particulate matter (PM) in the air breathed by a person is 25 μg m^{-3}, where 75% of the PM (by mass) is 6 μm particles, 20% is 2 μm particles, and 5% is 0.1 μm particles.
   a. What is the average mass of PM the person would inhale over a 1-day period? Use an inhalation rate of 12 m^3 day^{-1}? [Answer: 300 μg PM day^{-1}.]
   b. If this person is a nose breather, what PM dose would he or she accumulate in each of the 3 regions of the respiratory tract over a 1-day period (in units of mass per day)? Compare this with a mouth breather. [Answer: nose breather: nasal-pharyngeal 201 μg day^{-1}; tracheobronchial 26 μg day^{-1}; pulmonary 51 μg day^{-1}; mouth breather: oro-pharyngeal 79 μg day^{-1}; tracheobronchial 73 μg day^{-1}; pulmonary 120 μg day^{-1}.]
   c. Recall that PM is cleared from different regions of the respiratory tract at different rates. The average amount of PM residing in *each region* of the respiratory tract M can be estimated as $M = F\tau$, where F is the dose found in part (b), and τ is the average time it takes the particles to clear from a given region of the respiratory tract. Compare the total M for the nose breather with the total M for the mouth breather, assuming the particles are insoluble. [Answer: nose breather $M = 19$ g; mouth breather $M = 44$ g.]
5. You are planning to expose volunteers to two VOCs (*p*-xylene and 1,1,1-trichloroethane) for 24 hours in a chamber. Your measurement device has a limit of detection of 1 μg/m^3 for each pollutant. What is the minimum concentration of each pollutant that you must supply to the chamber to assure a final exhaled breath concentration of 1 μg/m^3? Assume equilibrium after 24 hours. Hint: What is f for each pollutant? [Answer: 1/f: 12.5 μg/m^3 for *p*-xylene; 1.1 μg/m^3 for 1,1,1-trichlorethane.]

REFERENCES

Guyton, A.C. (2000) *Textbook of Medical Physiology*, W.B. Saunders, Philadelphia, PA.
Heinsohn, R.J. and Kabel, R.L. (1999) *Sources and Control of Air Pollution*, Prentice-Hall, Upper Saddle River, NJ.
ICRP (1994) International Commission on Radiological Protection Publication 66, *Annals of the ICRP* **24**(1–3): 1–482.
Ott, W.R. and Mage, D.T. (1978) Interpreting Urban Carbon Monoxide Concentrations by Means of a Computerized Blood COHb Model, *Journal of the Air Pollution Control Association*, **28**: 911–916.
Peters, A., Dockery, D.W., Muller, J.E., and Mittleman, M.A. (2001) Increased Particulate Air Pollution and the Triggering of Myocardial Infarction, *Circulation*, **103**: 2810–2815.
Pope, C.A., Burnett, R.T., Thurston, G.D., Thun, M.J., Calle, E.E., Krewski, D., Godleski, J.J. (2004) Cardiovascular Mortality and Long-term Exposure to Particulate Air Pollution — Epidemiological Evidence of General Pathophysiological Pathways of Disease, *Circulation*, **109**: 71–77.
Raabe, O.G. (1982) Deposition and Clearance of Inhaled Aerosols, in *Mechanisms in Respiratory Toxicology, Volume 1*, Witschi, H. and Nettesheim, P., Eds., CRC Press, Boca Raton, FL.

Raabe, O.G. (1999) Respiratory Exposure to Air Pollutants, in *Air Pollutants and the Respiratory Tract*, Swift, D.L. and Foster, W.M., Eds., Marcel Dekker Inc., New York, NY.

Samet, J.M., Dominici, F., Curriero, F.C., Coursac, I., and Zeger, S.L. (2000) Fine Particulate Air Pollution and Mortality in 20 U.S. Cities, 1987–1994. *New England Journal of Medicine*, **343**(24): 1742–1749.

Seinfeld, J.H. (1986) *Atmospheric Chemistry and Physics of Air Pollution*, Wiley Interscience, New York, NY.

Ultman, J.S. (1988) Transport and Uptake of Inhaled Gases, in *Air Pollution, the Automobile, and Public Health*, Watson, A.Y., Bates, R.R., and Kennedy, D., Eds., National Academies Press, Washington, DC.

USEPA (2000) *Air Quality Criteria for Carbon Monoxide*, Report No. EPA 600/P-99/001F, U.S. Environmental Protection Agency, Office of Research and Development, Washington, DC.

USEPA (2005) National Ambient Air Quality Standards, http://www.epa.gov/air/criteria.html (accessed July 12, 2005).

U.S. Department of Health and Human Services, Agency for Toxic Substances and Disease Registry (ATSDR) (1999) *Toxicological Profile for Lead*, CAS#7439-92-1, Atlanta, GA.

Vedal, S. (1997) Ambient Particles and Health: Lines That Divide, *Journal of the Air & Waste Management Association*, **47**: 551–581.

Wallace, L. A., Nelson, W.C., Pellizzari, E.D., and Raymer, J.H. (1997) A Four-Compartment Model Relating Breath Concentrations to Low-Level Chemical Exposures: Application to a Chamber Study of Five Subjects Exposed to Nine VOCs, *Journal of Exposure Analysis and Environmental Epidemiology*, **7**(2): 141–163.

Wallace, L.A., Pellizzari, E.D., and Gordon, S. (1993) A Linear Model Relating Breath Concentrations to Environmental Exposures: Application to a Chamber Study of Four Volunteers Exposed to Volatile Organic Chemicals, *Journal of Exposure Analysis and Environmental Epidemiology*, **3**(1): 75–102.

5 Personal Monitors

Lance A. Wallace
U.S. Environmental Protection Agency (ret.)

CONTENTS

5.1 Synopsis 99
5.2 Introduction 99
5.3 VOC Monitors 100
5.4 Pesticide Monitors 102
5.5 Carbon Monoxide Monitors 103
5.6 Particle Monitors 104
5.7 Newer Monitors for Other Pollutants/Parameters 107
5.8 Conclusion 109
References 110

5.1 SYNOPSIS

This chapter deals with personal air quality monitors, small devices people can carry with them through their daily activities that will sample the air they are breathing. Personal monitors are the "gold standard" for estimating human exposure to air pollutants. In no other way can we measure the changing concentrations as persons go from their houses to their cars to their workplaces over the course of a day. The development of personal monitors for measuring environmental pollutants began in the 1970s and continues today. The chapter provides the seldom-told history of how the U.S. Environmental Protection Agency (USEPA) responded to the National Academy of Sciences recommendation to "foster development" of personal monitors. Early pioneers in developing monitors for volatile organic compounds (VOCs), pesticides, carbon monoxide (CO), and particles are recognized. (Current technologies are also described, such as the new "multipollutant personal monitor" developed at Harvard.) Only brief attention is paid to the results of the large-scale studies using these monitors, because these findings are treated at greater length in accompanying chapters on these pollutants. But the monitors themselves are illustrated profusely.

5.2 INTRODUCTION

Personal air quality monitors have a long history, perhaps beginning with the radiation badge invented early in the twentieth century to protect workers in radiological industries. Until about the late 1960s, personal monitors were used mainly in occupational settings. Development of the monitors was encouraged by legislation such as the Occupational Health and Safety Act of 1970 and the Coal Mine Health and Safety Act of 1969.

Personal monitors for measuring occupational exposure to volatile organic compounds (VOCs) often employed small pumps to pull air across an activated charcoal sorbent. The charcoal was then bathed in a liquid solvent (usually carbon disulfide) and analyzed by gas chromatography with flame ionization detection (GC/FID). Concentration levels were in the parts per million (ppm) range, since the occupational standards for many VOCs were in the neighborhood of 50–100 ppm.

Personal monitors for occupational exposure to particles usually consisted of a cyclone design with a pump and filter. The *cut point* concept associated with particle monitors must be described here. Such monitors work by dividing particles into two groups based on their diameters. The group with smaller diameter is collected on the filter for further analysis. The larger particles are discarded. The diameter separating the two groups is called the cut point. This is the diameter at which exactly half the particles are collected on the filter and half escape. A higher percentage of particles at slightly greater diameters escape collection, and at a high enough diameter none are collected. The cut point in the early occupational monitors was often 3.5 μm to restrict the measured particles to the respirable range. Particle concentrations were in the mg/m^3 range.

When the USEPA was created in 1970, little attention was paid to personal monitors. The Agency concentrated on setting up a nationwide outdoor monitoring system for six "criteria" pollutants, including particles (Total Suspended Particles, or TSP) and several specific gases (CO, ozone, SO$_2$, and NO$_2$). The sixth criteria pollutant was hydrocarbons, or nonmethane hydrocarbons, a non-specific mix of VOCs not including most chlorinated or oxygenated species. The levels of concern for these pollutants were far below the occupational standards and therefore the occupational personal monitors would not be suitable for making environmental measurements. (However, the historical record is not clear on whether USEPA scientists gave much thought to personal monitors at all.)

A few years later, some researchers including M. Granger Morgan at Carnegie Mellon Institute of Technology and Leonard Hamilton at the Brookhaven National Laboratory (BNL) realized that technology was rapidly becoming available that would make personal monitors for environmental concentrations feasible. They sponsored a workshop on personal monitors at BNL that reviewed recent improvements in the field (Morgan and Morris 1976).

At about the same time, the National Academy of Sciences was asked by Congress to review the scientific approach of the USEPA. In 1977, a nine-volume report appeared. Volume IV of this effort was entitled "Environmental Monitoring" (NAS 1977), and one of its major recommendations reads as follows.

> ...the Agency is not currently developing personal air quality monitors...By equipping a controlled sample of people with these portable sampling devices, the concentration of the pollutants to which they are exposed could be measured....There is evidence to challenge a common objection that personal monitors are technologically impracticable (... Morgan and Morris 1976, Wallace 1977). Prototypes have been developed to detect several air pollutants at levels near ambient concentrations (Wallace 1977).

- We recommend that EPA coordinate and support a program to foster the development of small, quiet, accurate, and sensitive personal air pollution monitors for use in conjunction with other methods for measuring human exposure to ambient air quality.

Soon after this recommendation was published, USEPA sponsored a symposium on personal monitors (Mage and Wallace 1979) and set aside a small amount of resources ($250,000) to begin developing personal air quality monitors.

5.3 VOC MONITORS

In the meantime, progress was being made on several other fronts. Small quiet pumps capable of pumping air at 2–4 Lpm were developed. Batteries capable of supplying adequate current to power the pumps for 24 hours or more were developed. However, in the case of the VOCs, the single most important development was the serendipitous discovery of a new sorbent as part of an industrial process in Switzerland. The sorbent was called Tenax™, and had several advantages over activated charcoal. First, it could be thermally desorbed. This had the advantage of not diluting the analytes first with a solvent before reconcentrating them. Second, it gave up its sorbed agents more easily and reproducibly than activated charcoal. Third, it was hydrophobic, which meant that water molecules could not take up all the active sites when monitoring under humid conditions. And fourth, following thermal desorption, it could be reused, an important consideration, since it was quite expensive.

Early exploiters of the new sorbent included Edo Pellizzari of the Research Triangle Institute, and Boguslaw Krotoszynski of IIT Research Institute in Chicago. Pellizzari studied ambient air concentrations of VOCs in petrochemical production areas such as the Louisiana-Texas Gulf of Mexico area (Pellizzari 1977). Krotoszynski studied exhaled breath concentrations of nurses in Chicago (Krotoszynski, Gabriel, and O'Neill 1977; Krotoszynski, Bruneau, and O'Neill 1979). Thus, almost from the first uses of Tenax, the double usage for environmental (air) and body burden (breath) sampling was established.

In 1979, a USEPA initiative committed the Agency to personal monitoring, beginning a program called the Total Exposure Assessment Methodology (TEAM) Studies. The first TEAM study investigated the possibility of measuring personal exposure to four groups of pollutants: VOCs, particles, pesticides, and PAHs. A full-scale 1-year pilot study on nine persons in New Jersey and three in the Research Triangle Park area of North Carolina ensued (Wallace et al. 1982). At the end of this effort, however, only the VOCs were considered feasible targets with the methodology then available. USEPA researchers reviewed progress and research needs in 1981–1982 (Wallace 1981; Wallace and Ott 1982).

The first TEAM study equipped 355 persons in New Jersey with Tenax monitors (Figure 5.1), measuring not only their daytime and overnight exposures to 25 target pollutants, but also their exhaled breath at the end of the 24-hour monitoring period (Wallace et al. 1985). Repeat visits were made in different seasons over a period of 3 years (1981–1983). Smaller comparison studies of 25 persons each were carried out in North Carolina and North Dakota. Another large study took place in Los Angeles, Antioch, and Pittsburg, California in 1984, with 120 persons visited over two seasons (Pellizzari et al. 1987a,b; Wallace 1987; Wallace et al. 1988.) Several new target pollutants were added to bring the total up to 32; the single most prevalent VOC was d-limonene, a natural constituent of citrus fruits and a popular fragrance (lemon scent) in cleaning products.

The TEAM study is now considered a watershed in environmental studies. Not only was it the first large-scale study to focus on personal monitors for environmental measurements, it was also the first to use a probability design. Such a design is the only accepted way to use a small group to extrapolate to a larger population. The 1979–1984 TEAM studies sampled about 500 persons but the total population represented was nearly 500,000 residents of New Jersey and California. Finally, the additional focus on body burden (exhaled breath) measurements, allowed the discovery of the overwhelming importance of smoking in exposing smokers to benzene and other aromatic compounds such as xylenes, ethylbenzene, and styrene (Wallace et al. 1987).

The most surprising finding of the TEAM studies was undoubtedly the importance of indoor sources of VOCs, outweighing the outdoor contribution for essentially all of the compounds studied. In some cases, these indoor sources were easily identified: dry-cleaned clothes as a source of tetrachloroethylene; moth crystals and air fresheners as sources of paradichlorobenzene; volatilization of chloroform from chlorinated water in homes. Persons visiting dry cleaners or gas stations in the past week had significantly higher levels of tetrachloroethylene and benzene in their breath.

The basic findings of the TEAM studies have been replicated repeatedly in studies in Europe. Large-scale studies of 300 homes in the Netherlands (Lebret et al. 1986) and 500 homes in West Germany (Krause et al. 1987) took place shortly after the TEAM studies. Since then, the German national studies have included the entire country and more than 1,000 participants, with VOCs, pesticides, and metals as major targets (Seifert et al. 2000).

Tenax, like all sorbents, collects only a fraction of the VOCs that are present in ambient or indoor air. It cannot measure very volatile organics (VVOCs) because they have a small *breakthrough*

FIGURE 5.1 Tenax cartridge and vest used in the TEAM VOC studies. The vest has a Velcro flap to protect the glass cartridge. The pump is in the pocket of the vest. The flow is 30 cc/min for a nominal 12-hour sample.

volume, the volume of air that can be passed through the monitor before the target chemical begins to emerge from the back end of the sampling cartridge. For example, Tenax cannot measure vinyl chloride and methylene chloride if less volatile organics are also targets, and chloroform is at the edge of measurability. Also, Tenax cannot measure many of the more reactive chemicals such as 1,3-butadiene and those containing oxygen.

To counter these deficiencies, multisorbent sampling cartridges can be used (Hodgson, Girman, and Binenboym 1986). These commonly employ Tenax for the bulk of the target organics, but also include some form of activated carbon to pick up the very volatile and oxygenated chemicals that Tenax cannot collect. Although these were used in the 100-building BASE (Building Assessment Survey and Evaluation) study as fixed indoor monitors, they have not been widely used as personal monitors (USEPA 2005).

5.4 PESTICIDE MONITORS

The next personal monitor to be developed employed polyurethane foam to collect vapor-phase semivolatile organics (SVOCs) including many common pesticides (Lewis and Macleod 1982). A new TEAM study of exposure to pesticides was carried out for 250 residents of Springfield, Massachusetts, and Jacksonville, Florida, (Immerman and Schaum 1990a,b). The same procedures including probability-based selection of participants and use of personal samplers were employed.

Once again the findings were unexpected, with indoor exposures even more dominant than in the case of the VOCs (Lewis et al. 1988). Although five pesticides had been banned as of the time of the study (DDT, aldrin, dieldrin, chlordane, heptachlor) they were all found in quantities such that they presented the highest risk of all the target pesticides. Since DDT had been banned more than a decade before, this was evidence that these pesticides were stable for long periods of time in the environment. For an illustration of the pesticide monitor, see Chapter 15.

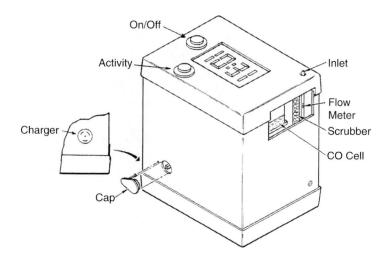

On/Off

Activity

Inlet

Flow Meter

Scrubber

CO Cell

Charger

Cap

FIGURE 5.2 CO monitor/data logger designed by Ott for use in a USEPA study of 1,200 residents of Washington, DC, and Denver, Colorado. The display contains an "activity" button that was pressed by the participant each time he changed microenvironments. This caused the data logger to average the measurements since the last time the activity button was pressed. By reading the time on the display, the participant could write down the location and the time so that each microenvironment was matched to the proper average.

5.5 CARBON MONOXIDE MONITORS

A third personal monitor came on line in the mid-1980s: an electrochemical monitor for carbon monoxide (CO). Electrochemical monitors oxidize CO to CO_2 and collect the free electrons created — the current is proportional to the CO concentration. For optimal use in a field study, it was important to develop a way to monitor not only the concentrations but also the locations where the person was. This problem was solved by attaching a data logger and allowing the participant to punch in code numbers indicating the times he or she entered or left each different microenvironment encountered during the day (Ott et al. 1988) (Figure 5.2). The program then calculated the average concentration associated with each microenvironment. Also, an average was calculated for each hour. At the end of the 24-hour exposure period, a breath sample was collected in a Tedlar® bag and analyzed by a single CO monitor in a central laboratory. This third TEAM study was carried out in Washington, DC, and Denver, Colorado, on more than 1,200 persons.

The study resulted in a rich database with scores of microenvironments ranked according to the CO exposures in each (Akland et al. 1985). CO exposures were shown to depend on car travel, as expected. The breath samples were instrumental in identifying a problem with the CO monitors that had gone undetected; the weakening of the battery toward the end of the 24-hour period resulted in lowering the measured exposures at that time. Thus, the breath samples, together with the model relating breath levels to air levels, were used to adjust the readings of the monitors during the latter part of the monitoring period.

Several electrochemical monitors are available for CO, but one, the Langan monitor, has certain special features of interest. Many of the monitors only supply readouts to the nearest ppm, but CO levels have decreased so much in recent years that monitors capable of reading sub-ppm values are desirable. The Langan has a feature allowing selection of a 10-times scale allowing readings down to 0.1 ppm. Drawbacks include a sensitivity to temperature that must be corrected by a user-specified algorithm.

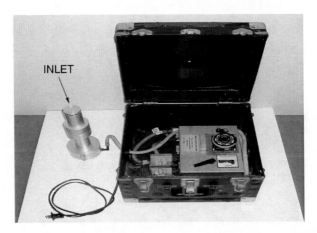

FIGURE 5.3 Harvard impactor (HI), also known as the Black Box. Shown is a PM$_{2.5}$ nozzle. The HI is programmable using the timer to collect data automatically for specified periods up to 2 weeks.

5.6 PARTICLE MONITORS

Meanwhile, the late 1970s and early 1980s saw the development of the Harvard Impactor (HI), and its use in another watershed study, the Harvard 6-City Study. This study, which went on for 20 years, has provided invaluable data on indoor concentrations of particles and associated lung function declines in children and adults. The HI is not a personal monitor, but has had a long useful life as an indoor and outdoor monitor (Figure 5.3). An impactor uses a collector plate to intercept particles too large to follow the sharply curved streamlines within the monitor. The smaller particles deposit on a filter that has been preweighed, and is then weighed again to determine the particle concentration. By contrast, the cyclone monitors widely used in occupational sampling use the walls of the monitor to collect particles that fail to follow helical streamlines within the monitor. An impactor has a sharper *size cut* (the dividing diameter separating the two particle groupings) than a cyclone. The size of particles is very important in determining the depth of penetration into the lung. Therefore the USEPA standards have been set for particles of a specified size. Over the years, the standards have evolved from total suspended particles (TSP) to inhalable particles (particles smaller than 10 μm, or PM$_{10}$) to fine particles (those smaller than 2.5 μm, or PM$_{2.5}$).

In the early days of the 6-City Study, a personal cyclone monitor, with a cut point at 3.5 μm was employed (Dockery and Spengler 1981). Later a new personal impactor was developed in conjunction with Virgil Marple of the University of Minnesota (Marple et al. 1987). The Marple personal exposure monitor (PEM) employs a design that can provide a cut point of any desired magnitude by adjusting the airflow rate and by blocking off a specified number of the 10 holes in the inlet. (The PEM is pictured in Figure 5.4.)

The PEM adjusted to sample PM$_{10}$ was employed in the fourth and last TEAM study, the Particle TEAM (PTEAM) study in Riverside, California (Clayton et al. 1993; Thomas et al. 1993). One hundred seventy-eight Riverside residents wore the monitor for two 12-hour periods (day and night) in the fall of 1990. To ensure comparability between the personal and fixed samples, the PEM was pressed into service as a stationary indoor monitor (SIM) and stationary outdoor monitor (SAM) in each home.

Results of the PTEAM study included discovery of the "personal cloud," a widespread phenomenon in which personal exposures exceed the time-weighted average of indoor and outdoor concentrations (Özkaynak et al. 1996a,b). Another important finding was the importance of cooking, especially frying, in contributing to indoor exposures. Cooking appeared to be second only to smoking as a particle source.

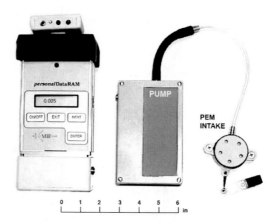

FIGURE 5.4 MIE optical scattering monitor and Marple PEM. The MIE uses a laser at 880 nm wavelength to estimate real-time particle numbers in the range of about 0.1–10 μm. The PEM employs a Teflon filter that can be analyzed for a dozen or so elements using x-ray fluorescence after weighing.

The Ethyl Corporation sponsored a year-long study of personal exposure to manganese (Mn) in Toronto, Canada, and also in Indianapolis, Indiana (Pellizzari et al. 1999, 2001). Mn is produced in auto exhaust when methylcyclopentadienyl manganese tricarbonyl (MMT) is used in gasoline to improve engine performance. MMT has been used in Canada for many years, and is also legal for use in the United States, although little market penetration has occurred due to the opposition of environmental groups and also automobile manufacturers, who fear the "poisoning" of their onboard computer systems by Mn. The Toronto study was made in part to meet USEPA desires for a full-scale study of personal exposure to Mn. For the Toronto and Indianapolis studies, which were carried out by the Research Triangle Institute, a new personal monitor was developed with several interesting features. It had a timer to turn the pump on and off in a 1 minute on, 2 minutes off sequence, allowing measurements to extend over 3 days but requiring battery power to be applied for only 24 hours. There was also a device that kept track of physical activity by measuring electrical capacitance of a transducer kept on the person. This was useful in establishing whether the monitor was actually being worn at the times and places the person reported.

Time series studies in the 1990s established the importance of air pollution in apparently causing morbidity and mortality among sensitive subpopulations (Dockery, Schwartz, and Spengler 1992). However, although much evidence points toward fine particles, a fair amount of evidence suggests other agents, including CO, SO_2, NO_2, ozone, sulfates, nitrates, and coarse particles. Clearly it would be desirable to collect information on personal exposure to as many of these "co-pollutants" or possible confounders as possible. Harvard University School of Public Health accepted this challenge and developed a multipollutant personal monitor (Figure 5.5a and Figure 5.5b). The monitor has a 5.2 Lpm airflow split four ways: 1.8 Lpm each through a PM_{10} and $PM_{2.5}$ nozzle with a Teflon filter; 0.8 Lpm to collect nitrates or sulfates (depending on which is more prevalent in the region of interest) through a denuder onto a sodium carbonate-coated glass fiber filter; and 0.8 Lpm through a $PM_{2.5}$ nozzle onto a quartz filter that can be analyzed for organic carbon and elemental carbon (Demokritou et al. 2001). A clever addition to the monitor's range of pollutants was made by attaching two passive Ogawa badges to the elutriator portion of the monitor, assuring a steady passage of air across the surface of the badges, which can be selected to measure ozone and either SO_2 or NO_2 (again, depending on the pollutant of interest in the region). The monitor was used in a series of panel studies in Atlanta, Boston, and Los Angeles sponsored by the USEPA and also (Los Angeles only) by the California Air Resources Board (Wallace, Williams, and Suggs 2004). It is now commercially available.

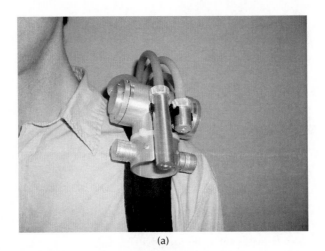

(a)

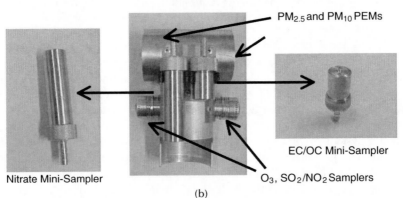

(b)

FIGURE 5.5 Harvard multipollutant monitor, as worn (a) and with parts identified (b). The pump (not shown) has a 5.2 Lpm capacity, with 1.8 Lpm flowing through each of the impactors at the top ($PM_{2.5}$ and PM_{10}). 0.8 Lpm flows through the quartz filter for later analysis of elemental carbon (EC) and organic carbon (OC). The final 0.8 Lpm flows through a sodium carbonate–coated tube to a filter for collection of nitrates or sulfates. Two passive Ogawa badges can be mounted on the vertical elutriator to measure ozone, NO_2, or SO_2.

In the 1990s, personal monitors employing optical scattering to detect and count particles became available. The most widely used of these is the MIE personal DataRAM™, or pDR. This passive monitor uses a laser tuned to 880 nm wavelength. It is most sensitive to particles of about 0.6 μm diameter, but can detect particles as small as about 0.1 μm and as large as 10 μm. Since the MIE is calibrated to a road dust aerosol of specific gravity 2.6, it overestimates the concentration of ambient particles, which typically have a specific gravity of about 1.5–1.7. This means it is an uncertain indicator of the exact mass of the particles of interest; however, the great advantage of having nearly continuous measurements is the ability to determine short-term peaks and identify sources more easily than with a 24-hour integrated sample. The MIE has been used in many studies (Wallace et al. 2003), including the EPA-supported University of Washington study of four Seattle cohorts (Liu et al. 2002, 2003; Allen et al. 2003, 2004) and another EPA-sponsored study in Research Triangle Park, North Carolina (Wallace et al. 2006). In the latter study, participants carried both the gravimetric Marple PEM and the optical-based MIE pDR, allowing a comparison of their respective responses to PM. (Both monitors are pictured in Figure 5.4.) All these studies used the MIE to identify short-term peaks due to indoor sources. When these peaks were then subtracted from the record, the remaining concentrations were assumed to be due to outdoor particles penetrating the home. This separation of indoor particles into those from outdoors and those from indoors is desired by epidemiologists so that they can determine the correlation between

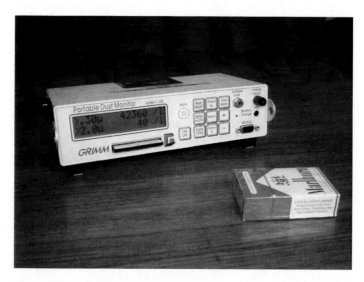

FIGURE 5.6 The Grimm 1.108 laser particle counter provides 1-minute particle counts in 14 size ranges. In this photograph, the display shows a count of 42,360 particles/L in the sizes above 0.3 micrometers and 40 particles/L above 2 micrometers. The instrument stores its readings in a memory card located underneath the display window, and the time series of 1-minute counts for 14 sizes can be downloaded by cable into a personal computer.

exposure to particles of outdoor origin and the measurements of those particles at central sites. Drawbacks of the MIE include a sensitivity to high humidity, which is not important for use as a personal monitor, but requires adjustments when used as an outdoor monitor. Another drawback is a tendency for a zero drift, which is correctable if the drift is positive, but because the manufacturer suppresses negative readings, is not completely correctable if the drift is negative.

Other optical scattering devices are also available, including the Radiance Research instrument and the Grimm monitor (Figure 5.6). The Grimm has the advantage of including a filter along with the optical scattering optics to collect particles for later weighing to compare with the optical scattering estimates. It also measures eight size categories simultaneously.

Although not strictly a personal monitor, the Piezobalance can be mentioned here due to its importance in early studies of environmental tobacco smoke (ETS) (Repace and Lowrey 1980). The Piezobalance employs a piezoelectric crystal electrically driven to oscillate at a certain frequency (Figure 5.7). Particles depositing on the crystal change the frequency. The change is detected by comparing it to another crystal isolated from the airflow. The particles are induced to deposit on the crystal by an electrostatic precipitator. The instrument weighs about 20 pounds, so it is too large to qualify as a personal monitor, but it is portable and quiet and has been employed in many publicly accessible places such as restaurants, bars, and even churches. Because the response of the crystal is not linear beyond a certain threshold, the Piezobalance requires frequent cleaning, up to every half hour in a high-concentration aerosol, a requirement that limits its usefulness.

5.7 NEWER MONITORS FOR OTHER POLLUTANTS/PARAMETERS

In recent years, due to intense interest in particle exposures related to the finding of increased morbidity and mortality, a number of new approaches to particle measurement have been instituted. It remains to be seen which of these will find widespread use in field studies. Following is a brief survey of these newer monitors.

A diffusion charger, which responds to the active surface area rather than the mass or the number of particles, is commercially available (Figure 5.8). (The active surface area is an important

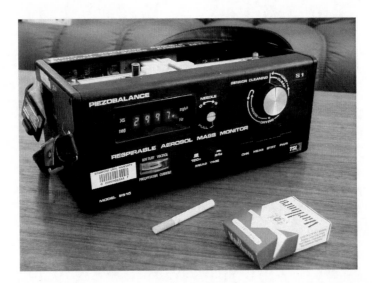

FIGURE 5.7 Kanomax 8510 piezoelectric microbalance (Piezobalance) for measuring particle mass concentration at 1-minute or 2-minute averaging times. In this photograph, the display reads "2997" momentarily to show the user the initial starting frequency of the piezoelectric crystal. An internal size impactor is installed inside the instrument to allow it to measure either $PM_{2.5}$ or $PM_{3.5}$ respirable suspended particulates (RSP) mass concentrations.

feature to measure, since it is surface chemistry that may determine the toxicity of many particles.) The diffusion charger uses a corona discharge to charge all the particles in the sensing volume. A sensitive electrometer measures the resultant current. Since the number of charges attaching to the particle depends on surface area, it gives a nearly direct measure of surface area. Diffusion chargers have been used to measure exposures, in conjunction with health measures, of children with asthma.

Polyaromatic hydrocarbons (PAHs) include a number of carcinogenic agents such as benzo-a-pyrene. Since PAHs have a wide range of volatilities, some exist as gases and others are attached to particle surfaces at normal temperatures. A personal monitor that responds to these *particle-bound* PAHs is the EcoChem PAS CE2000 (Figure 5.9). The PAS 2000 series employs UV emission from a krypton element at an energy of 5.6 eV to ionize PAH residing on particle surfaces. (Since some non-PAH compounds, most importantly black carbon, are also ionizable at this energy, there can be a problem with interferences.) The electrons emitted by the ionized PAHs form a current detectable by the instrument. The analyzer signal is a combined measure of total PAH adsorbed on carbon particles and the underlying black carbon itself. A second major drawback of this instrument is the inability to calibrate it in the field. Thus it must be considered semiquantitative

FIGURE 5.8 Diffusion charging (DC) monitor. The DC monitor provides a measure of the surface area of particles less than 1 micrometer in diameter.

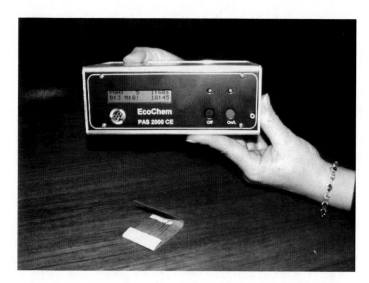

FIGURE 5.9 EcoChem PAS 2000 CE photoionization monitor for measuring the mass concentration of fine particulate polycyclic aromatic hydrocarbons (PPAH). The display shows a current PPAH concentration reading of 5 ng/m^3, the sampling time setting of 60 sec), a code for the battery life remaining (B = 3), a code for the memory used (M = 8), and the clock time of 18:45.

at best. However, it has a near-instantaneous response and is thus highly suited to picking up short-term peaks such as those from combustion in the home, particularly cigarette smoking (Ott et al. 1994). A drawback of later models is a poor reaction to products from cooking, specifically frying using oil or butter. The aerosol from frying produces a depression in the instrument response. The earlier PAS 1000 series, which used a more energetic emission line from mercury vapor (6.7 eV), did not have this problem. Since cooking oil contains PAHs, and the combustion process creates PAHs, it is unfortunate that this problem persists, since it leads to missing an important source of PAHs in the home.

5.8 CONCLUSION

Thanks to several far-sighted individuals on a National Academy of Sciences (NAS) panel, and to an extraordinarily rapid response to their recommendations by the Environmental Protection Agency, the last few decades have seen rapid advances in our knowledge of the actual exposures experienced by several thousand persons representing millions of persons in the population. These advances have included the discovery that many of the most important sources of pollution are small and close to the person (smoking, cooking, using air fresheners or indoor pesticides, wearing or storing dry-cleaned clothes, etc.). Most of these discoveries could not have been made without the incontrovertible evidence supplied by the personal monitors, which provided the "ground truth" to validate (or, more often, invalidate) existing models of pollutant emissions. From being considered "curiosities" at the beginning of their use, the place of personal monitors in studies of human exposure is now secure, so much so that one of the first recommendations of a recent NAS "research portfolio" prepared for the USEPA is to use personal monitors to investigate the relationship between personal exposure to particles and measurements at central monitoring stations. For a fuller discussion of the topics just touched upon in this concluding paragraph, the reader can go to the chapters on volatile organic compounds, pesticides, carbon monoxide, or particles in the remainder of this book.

REFERENCES

Akland, G., Hartwell, T.D., Johnson, T.R., and Whitmore, R. (1985) Measuring Human Exposure to Carbon Monoxide in Washington, DC and Denver, CO during the Winter of 1982–83, *Environmental Science and Technology,* **19**: 911–918.

Allen, R., Larson, T., Sheppard, L., Wallace, L., and Liu, L.-J.S. (2003) Use of Real-Time Light Scattering Data to Estimate the Contribution of Infiltrated and Indoor-Generated Particles to Indoor Air, *Environmental Science and Technology,* **37**: 3484–3492.

Allen, R., Sheppard, L., Liu, L.-J.S., Larson, T., and Wallace, L. (2004) Estimated Hourly Personal Exposures to Ambient and Non-Ambient Particulate Matter Among Sensitive Populations in Seattle, Washington, *Journal of the Air & Waste Management Association,* **54**: 1197–1211.

Clayton, C.A., Perritt, R.L., Pellizzari, E.D., Thomas, K.W., Whitmore, R.W., Özkaynak, H., Spengler, J.D., and Wallace, L.A. (1993) Particle Total Exposure Assessment Methodology (PTEAM) Study: Distributions of Aerosol and Elemental Concentrations in Personal, Indoor, and Outdoor Air Samples in a Southern California Community, *Journal of Exposure Analysis and Environmental Epidemiology,* **3**: 227–250.

Demokritou, P., Kavouras, I., Ferguson, S., and Koutrakis, P. (2001) Development and Laboratory Performance Evaluation of a Personal Multi-Pollutant Sampler for Simultaneous Measurements of Particulate and Gaseous Pollutants, *Aerosol Science and Technology,* **35**: 741–752.

Dockery, D.W. and Spengler, J.D. (1981) Personal Exposure to Respirable Particulates and Sulfates, *Journal of the Air Pollution Control Association,* **31**(2): 153–159.

Dockery, D.W., Schwartz, J., and Spengler, J.D. (1992) Air Pollution and Daily Mortality: Associations With Particulates and Acid Aerosols, *Environment Research,* **59**: 362–373.

Hodgson, A.T., Girman, J.R., and Binenboym, J. (1986) A Multi-Sorbent Sampler for Volatile Organic Compounds in Indoor Air, presented at *79th Annual Meeting of the Air Pollution Control Association, June, Minneapolis, MN*, Pittsburgh, PA, Air Pollution Control Association, Paper No. 86–37.1.

Immerman, F.W. and Schaum, J.L. (1990a) *Nonoccupational Pesticide Exposure Study (NOPES) Project Summary*, Report No. EPA/600/S3-90/003, US Environmental Protection Agency, Research Triangle Park, NC.

Immerman, F.W. and Schaum, J.L. (1990b) *Nonoccupational Pesticide Exposure Study: Final Report*, Report No. EPA/600-3-90-003, U.S. Environmental Protection Agency, Washington, DC.

Krause, C., Mailahn, W., Nagel, R., Schulz, C., Seifert, B., and Ullrich, D. (1987) Occurrence of Volatile Organic Compounds in the Air of 500 Homes in the Federal Republic of Germany, in *Proceedings of the 4th International Conference on Indoor Air Quality and Climate*, Vol. 1, Institute for Soil, Water, and Air Hygiene, Berlin (West), Germany, 102–106.

Krotoszynski, B.K., Bruneau, G.M., and O'Neill, H.J. (1979) Measurement of Chemical Inhalation Exposure in Urban Populations in the Presence of Endogenous Effluents, *Journal of Analytical Toxicology,* **3**: 225–234.

Krotoszynski, B.K., Gabriel, G., and O'Neill, H. (1977) Characterization of Human Expired Air: A Promising Investigation and Diagnostic Technique, *Journal of Chromatographic Science,* **15**: 239–244.

Lebret, E., Van De Weil, H.J., Noij, D., and Boleij, J.S.M. (1986) Volatile Hydrocarbons in Dutch Homes, *Environment International,* **12**(1–4): 323–332.

Lewis, R.G. and Macleod, K.E. (1982) A Portable Sampler for Pesticides and Semivolatile Industrial Organic Chemicals in Air, *Analytical Chemistry,* **54**: 310–315.

Lewis, R.G., Bond, A.E., Johnson, D.E., and Hsu, J.P. (1988) Measurement of Atmospheric Concentrations of Common Household Pesticides: A Pilot Study, *Environmental Monitoring and Assessment,* **10**: 59–73.

Liu L.-J.S., Box, M., Kalman, D., Kaufman, J., Koenig, J., Larson, T., Lumley, T., Sheppard, L., and Wallace, L.A. (2003) Exposure Assessment of Particulate Matter for Susceptible Populations in Seattle, *Environmental Health Perspectives,* **111**: 909–918.

Liu, L.-J.S., Slaughter, C., and Larson, T. (2002) Comparison of Light Scattering Devices and Impactors for Particulate Measurements in Indoor, Outdoor, and Personal Environments, *Environmental Science and Technology,* **36**: 2977–2986.

Mage, D.T. and Wallace, L.A., Eds. (1979) *Proceedings of the National Symposium on Development and Usage of Personal Exposure Monitors in Studies of Human Exposure to Environmental Pollutants,* U.S. Environmental Protection Agency, Washington, DC.

Marple, V.A., Rubow, K.L., Turner, W., and Spengler, J.D. (1987) Low Flow Rate Sharp Cut Impactors for Indoor Air Sampling: Design and Calibration, *Journal of the Air Pollution Control Association,* **37**: 1303–1307.

Morgan, M.G. and Morris, S.C. (1976) *Individual Air Pollution Monitors: An Assessment of National Research Needs. Report of a Workshop On Assessment of Research Needs for Individual Air Pollution Monitors for Ambient Air,* Report No. BNL 50482, Brookhaven National Laboratory, Upton, NY, July 8–10, 1975.

NAS (1977) *Environmental Monitoring: Vol. IV of Analytical Studies for the U.S. Environmental Protection Agency,* National Academy of Sciences, Washington, DC.

Ott, W.R., Rodes, C.E., Drago, R., Williams, C., and Burmann, F.J. (1988) Automated Data-Logging Personal Exposure Monitors for Carbon Monoxide, *Journal of the Air Pollution Control Association,* **36**(8): 883–887.

Ott, W., Wilson, N., Klepeis, N., and Switzer, P. (1994) Real-Time Monitoring of Polycyclic Aromatic Hydrocarbons (PAH) and Respirable Suspended Particles (RSP) from Environmental Tobacco Smoke in a Home, in *Proceedings of the Air and Waste Management Symposium on Measurement of Toxic and Related Air Pollutants,* Research Triangle Park, NC.

Özkaynak, H., Xue, J., Spengler, J.D., Wallace, L.A., Pellizzari, E.D., and Jenkins, P. (1996b) Personal Exposure to Airborne Particles and Metals: Results from the Particle TEAM Study in Riverside, CA, *Journal of Exposure Analysis and Environmental Epidemiology,* **6**: 57–78.

Özkaynak, H., Xue, J., Weker, R., Butler, D., Koutrakis, P., and Spengler, J. (1996a) *The Particle TEAM (PTEAM) Study: Analysis of the Data, Volume III, Final Report,* Contract #68-02-4544, U.S. Environmental Protection Agency, Research Triangle Park, NC.

Pellizzari, E.D. (1977) *The Measurement of Carcinogenic Vapors in Ambient Atmospheres,* Final Report, U.S. Environmental Protection Agency, Research Triangle Park, NC.

Pellizzari, E.D., Clayton, C.A., Rodes, C.E., Mason, R.E., Piper, L.L., Fort, B., Pfeifer, G., and Lynam, D. (1999) Particulate Matter and Manganese Exposures in Toronto, Canada, *Atmospheric Environment,* **31**: 721–734.

Pellizzari, E.D., Clayton, C.A., Rodes, C.E., Mason, R.E., Piper, L.L., Pfeifer, G., and Lynam, D. (2001) Particulate Matter and Manganese Exposures in Indianapolis, Indiana, *Journal of Exposure Analysis & Environmental Epidemiology,* **11**: 423–440.

Pellizzari, E.D., Perritt, R., Hartwell, T.D., Michael, L.C., Sheldon, L.S., Sparacino, C.M., Whitmore, R., Leininger, C., Zelon, H., Handy, R.W., Smith, D., and Wallace, L.A. (1987a) *Total Exposure Assessment Methodology (TEAM) Study: Elizabeth and Bayonne, New Jersey; Devils Lake, North Dakota, and Greensboro, North Carolina: Volume II,* Report No. EPA 600/6-87/002b, NTIS PB 88-10007, U.S. Environmental Protection Agency, Washington, DC.

Pellizzari, E.D., Perritt, R., Hartwell, T.D., Michael, L.C., Whitmore, R., Handy, R.W., Smith, D., Zelon, H., and Wallace, L.A. (1987b) *Total Exposure Assessment Methodology (TEAM) Study: Selected Communities in Northern and Southern California: Vol. III,* Report No. EPA 600/6-87/002c, NTIS PB 88-100086, U.S. Environmental Protection Agency, Washington, DC.

Repace, J.L. and Lowrey, A.H. (1980) Indoor Air Pollution, Tobacco Smoke, and Public Health, *Science,* **208**: 464–472.

Seifert, B., Becker, K., Hoffman, K., Krause, C., and Schulz, C. (2000) The German Environmental Survey 1990/92 (GerESII): A Representative Population Study, *Journal of Exposure Analysis and Environmental Epidemiology,* **10**(2): 115–125.

Thomas, K.W., Pellizzari, E.D., Clayton, C.A., Whitaker, D.A., Shores, R.C., Spengler, J.D., Özkaynak, H., and Wallace, L.A. (1993) Particle Total Exposure Assessment Methodology (PTEAM) 1990 Study: Method Performance and Data Quality for Personal, Indoor, and Outdoor Monitoring, *Journal of Exposure Analysis and Environmental Epidemiology,* **3**: 203–226.

USEPA (2005) Building Assessment, Survey and Evaluation Study (BASE), http://www.epa.gov/iaq/base/ (accessed September 30, 2005).

Wallace, L.A. (1977) *Personal Air Quality Monitors: Supplement III, Vol. IVa of Analytical Studies for the U.S. Environmental Protection Agency,* National Academy of Sciences, Washington, DC.

Wallace, L.A. (1987) *The TEAM Study: Summary and Analysis: Volume I*, Report No. EPA 600/6-87/002a, NTIS PB 88-100060, U.S. Environmental Protection Agency, Washington, DC.

Wallace, L.A. (1981) Recent Progress in Developing and Using Personal Monitors to Measure Human Exposure to Air Pollutants, *Environment International*, **5**: 73-75.

Wallace, L.A. and Ott, W.R. (1982) Personal Monitors: A State-of-the-Art Survey, *Journal of the Air Pollution Control Association*, **32**: 601-610.

Wallace, L.A., Mitchell, H., O'Connor, G.T., Liu, L.-J., Neas, L., Lippmann, M., Kattan, M., Koenig, J., Stout, J.W., Vaughn, B.J., Wallace, D., Walter, M., and Adams, K. (2003) Particle Concentrations in Inner-City Homes of Children with Asthma: The Effect of Smoking, Cooking, and Outdoor Pollution, *Environmental Health Perspectives*, **111**: 1265–1272.

Wallace, L.A., Pellizzari, E., Hartwell, T., Perritt, K., and Ziegenfus, R. (1987) Exposures to Benzene and Other Volatile Organic Compounds from Active and Passive Smoking, *Archives of Environmental Health*, **42**: 272–279.

Wallace, L.A., Pellizzari, E., Hartwell, T., Sparacino, C., Sheldon, L., and Zelon, H. (1985) Personal Exposures, Indoor-Outdoor Relationships and Breath Levels of Toxic Air Pollutants Measured for 355 Persons in New Jersey, *Atmospheric Environment*, **19**: 1651-1661.

Wallace, L.A., Pellizzari, E.D., Hartwell, T.D., Whitmore, R., Perritt, R., and Sheldon, L. (1988) The California TEAM Study: Breath Concentrations and Personal Exposures to 26 Volatile Compounds in Air and Drinking Water of 188 Residents of Los Angeles, Antioch, and Pittsburgh, CA, *Atmospheric Environment*, **22**: 2141–2163.

Wallace, L.A., Williams, R., Rea, A., and Croghan, C. (2006) Continuous Weeklong Measurements of Personal Exposures and Indoor Concentrations of Fine Particles for 37 Health-Impaired North Carolina Residents for Up to Four Seasons, *Atmospheric Environment*, **40**: 399–414.

Wallace, L., Williams, R., and Suggs, J. (2004) *Exposure of High-Risk Subpopulations to Particles: Final Report—APM-21*, Report No. EPA/600/R-03/145, U.S. Environmental Protection Agency, National Exposure Research Laboratory, Research Triangle Park, NC.

Wallace, L.A., Zweidinger, R., Erickson, M., Cooper, S., Whitaker, D., and Pellizzari, E.D. (1982) Monitoring Individual Exposure: Measurement of Volatile Organic Compounds in Breathing-Zone Air, Drinking Water, and Exhaled Breath, *Environment International*, **8**: 269-282.

6 Exposure to Carbon Monoxide

Peter G. Flachsbart
University of Hawai'i at Manoa

CONTENTS

6.1   Synopsis...113
6.2   Introduction...114
6.3   Sources of Carbon Monoxide 115
6.4   Health Effects of Carbon Monoxide...116
6.5   Early Studies of CO Exposure 117
      6.5.1   Surveys of Exposure while Driving in Traffic...117
      6.5.2   Surveys of CO Concentrations on Streets and Sidewalks...........................118
6.6   The Clean Air Act Amendments of 1970 119
6.7   Limitations of Fixed-Site Monitors...120
6.8   Estimates of Nationwide Population Exposure 121
6.9   Estimating Total CO Exposure...122
6.10 Field Surveys of Commercial Microenvironments..123
6.11 Direct and Indirect Approaches to Measure Exposure.....................................125
      6.11.1 Studies Using the Direct Approach...126
      6.11.2 A Study Using the Indirect Approach 129
6.12 Occupational Exposures 130
6.13 Residential Exposures 130
6.14 Recreational Exposures 130
6.15 Population Exposure Models 131
6.16 Activity Patterns 132
6.17 Public Policies Affecting Exposure to Vehicle Emissions 133
      6.17.1 Effects of Motor Vehicle Emission Standards on Unintentional
              Deaths Attributed to Exposure 134
      6.17.2 Effects of Transportation Investments on Commuter Exposure..................134
      6.17.3 Effects of Motor Vehicle Emission Standards on Commuter Exposure 136
6.18 The El Camino Real Commuter Exposure Surveys..136
6.19 International Comparisons of Commuter Exposure 138
6.20 Conclusions...139
6.21 Acknowledgments 140
6.22 Questions for Review 141
References 141

6.1 SYNOPSIS

Incomplete combustion of carbonaceous fuels (i.e., fuels with carbon atoms) can produce significant quantities of carbon monoxide (CO). Exposure to CO occurs during a variety of daily activities such as traveling by motor vehicle in traffic or cooking food over an unvented gas range. Fortunately, reducing CO exposures has been one of the "... greatest success stories in air-pollution control," according to a report published by the National Research Council in 2003. Much of that success is due to the adoption in 1968 of nationwide emission controls on new cars, and to promulgation in 1970 of the National Ambient Air Quality Standards (NAAQS) for CO and several other "criteria" air pollutants. In spite of that success, many people die or suffer the ill effects of high CO exposure every year. In fact, CO is the only regulated air pollutant that appears on death certificates. Accordingly, this chapter first summarizes the principal sources and health effects of CO. It then describes key studies of CO exposure over the last 40 years to show how the goals and methods of these studies have evolved over time. Studies of CO exposure in the 1960s and 1970s essentially pioneered the field of exposure analysis. The earliest studies found that CO concentrations on congested roadways and busy intersections in downtown areas typically exceeded ambient CO levels measured at fixed-site monitors. The U.S. Environmental Protection Agency (USEPA) relies on these monitors to determine compliance with the NAAQS. The chapter reveals typical concentrations of CO that people encounter in their daily lives and identifies factors that affect or contribute to CO exposures as a person performs his or her daily activities. The chapter shows how policies and programs of the Clean Air Act have affected trends in CO exposure over time. The chapter concludes that CO exposure studies are essential for identifying health risks to human populations, for setting and reviewing air quality standards, and for evaluating emission control policies and programs. The chapter recommends that studies of CO exposure are particularly applicable to developing countries that have rapidly growing motor vehicle populations, congested streets and confined spaces in urban areas, and nascent motor vehicle emission control programs.

6.2   INTRODUCTION

The National Research Council (NRC) in Washington, DC recently issued a report titled *Managing Carbon Monoxide Pollution in Meteorological and Topographical Problem Areas*. The report concluded: "CO control has been one of the greatest success stories in air-pollution control. As a result, the focus of United States air quality management has shifted to characterizing and controlling other pollutants, such as tropospheric ozone, fine particulate matter ($PM_{2.5}$), and air toxics." (NRC 2003, p. 149) As evidence for this conclusion, the NRC acknowledged that the number of monitoring stations showing violations of the National Ambient Air Quality Standards (NAAQS) for CO had fallen significantly from the early 1970s when CO monitoring became widespread. Many of the remaining violations occur in areas with meteorological or topographical handicaps. For example, CO violations in Fairbanks, Alaska, have been attributed to stagnant air masses during winter. The atmosphere is more likely to be stable during winter, because there is less solar heating and more frequent ground-level temperature inversions. Other contributing factors are low wind speeds and mountains that hinder dispersion of air pollutants. However, violations are occurring less frequently even in areas with these natural handicaps (NRC 2003).

If most Americans are no longer exposed to unhealthy CO levels, then why study CO exposure? One reason is that CO studies pioneered the field of exposure analysis. The earliest CO exposure studies, which date back to the mid-1960s, focused on tailpipe emissions on urban expressways, because motor vehicles represented the highest percentage of total CO emissions. These studies found that CO exposures on congested roadways and busy intersections typically exceeded ambient levels of CO measured at fixed-site monitors. This problem receded when automakers equipped motor vehicles with catalytic converters to satisfy tailpipe emission standards. Nevertheless, many Americans are still exposed to hazardous and sometimes fatal CO concentrations in their daily activities. Second, CO can be viewed as an indicator of other types of roadway emissions that are relatively stable in the atmosphere. For example, CO concentrations are highly correlated with concentrations of air pollutants such as benzene (a known carcinogen), black carbon, and certain ultra-fine particles (NRC 2003). CO is not an indicator of reactive air pollutants, such as hydrocarbons and nitrogen oxides, which are also emitted by motor vehicles. Third, personal exposure to CO can easily be measured using relatively inexpensive and reliable portable monitors that run on batteries. Certain monitors are capable of very precise and accurate CO measurements, and can store data electronically for later analysis. Finally, CO exposure studies are relevant to developing countries that have seen rapid growth in the use of motor vehicles. Many developing countries mandate less stringent vehicular emission standards than are found in North America, Japan, or Europe. Consequently, these countries have motor vehicle fleets with outdated emission controls.

6.3 SOURCES OF CARBON MONOXIDE

Carbon monoxide is a gaseous by-product that results from incomplete combustion of fuels (e.g., oil, natural gas, coal, kerosene, and wood) and other materials (e.g., tobacco products) that contain carbon atoms. According to the national inventory of air pollutant emissions compiled annually by the U.S. Environmental Protection Agency (USEPA), transportation sources in the United States accounted for nearly 70% of total CO emissions in 2000. Fuel combustion, industrial processes, and miscellaneous sources comprised the remaining 30%. Transportation sources include both on-road motor vehicles (e.g., cars and trucks) and non-road engines and vehicles (e.g., aircraft, boats, locomotives, recreational vehicles, and gasoline-powered lawnmowers). Of particular importance are those sources (such as cars, trucks, and lawnmowers) that release their emissions in close proximity to human receptors (Colvile et al. 2001).

The USEPA's annual emission inventory shows that the relative shares of on-road and non-road sources have shifted during the last two decades. The share of total emissions from on-road vehicles fell from 66.5% in 1980 to 44.3% in 2000, while the share from non-road vehicles increased from 12.3% in 1980 to 25.6% in 2000. This shift can be attributed to tailpipe exhaust emission standards, which have affected all new cars sold in the United States since 1968, and the fact that non-road sources are largely unregulated (USEPA 2003). Although CO emission rates of new cars have fallen over time, due to tighter emission standards imposed by the Clean Air Act (CAA), the number of vehicles and miles driven per vehicle have both been increasing due to population growth and urban sprawl (TRB 1995). Total miles of travel by all types of motor vehicles increased three times faster than population growth in the United States between 1980 and 2000 (Downs 2004).

The USEPA and the State of California have separate models to inventory motor vehicle emissions, because California is allowed to set its own motor vehicle emission standards under the CAA. Both models show that CO emission rates climb substantially when average speeds fall below 15 miles per hour. Since these low speeds often occur during periods of severe traffic congestion, many CO exposure studies focus on commuting activities, particularly during peak periods of travel. Other studies focus on "cold starts" that occur after vehicles have been parked for several hours. Cold starts may elevate CO exposures in homes (Akland et al. 1985) and office buildings (Flachsbart and Ott 1986) with attached garages. After several minutes the engine reaches higher temperatures and CO emissions begin to subside. Higher tailpipe emissions also occur when the driver accelerates the vehicle, runs its air conditioning system, or climbs a hill, because more fuel is needed to achieve extra power. By comparison, diesels emit less CO because excess air is used in the combustion process. CO emissions during conditions of severe fuel enrichment, which are essentially unregulated, can account for 40% of a typical trip's total CO emissions, even though these conditions prevail for only 2% of trip time (Faiz, Weaver, and Walsh 1996). Higher CO emissions also occur if the driver defers vehicle maintenance and repairs, tampers with the catalytic converter, or uses leaded fuel, which renders the converter ineffective (NRC 2000). These actions can spike the exposure of pedestrians, cyclists, and motorists if they are exposed to malfunctioning vehicles on city streets and roads. In 1973, the United States began to phase out lead from gasoline and banned lead additives in commercial gasoline after December 31, 1995. The remaining use of leaded gasoline in U.S. motor vehicles occurs predominantly in rural areas (Walsh 1996). Last but not least, defective exhaust systems can contaminate the passenger compartments of motor vehicles (Amiro 1969) and sustained-use vehicles such as buses, taxicabs, and police cars (Ziskind et al. 1981), and lead to accidental CO poisoning of passengers in the back of pickup trucks (Hampson and Norkool 1992).

6.4 HEALTH EFFECTS OF CARBON MONOXIDE

CO molecules, which have no color, odor, or taste, enter the body through normal breathing (i.e., inhalation exposure). In the lung, the CO molecule passes into the bloodstream through the alveolar and capillary membranes of the lung and blood vessels, respectively. Once in the blood, CO competes with oxygen for attachment to iron sites in red blood cells (hemoglobin). The attraction of hemoglobin (Hb) to CO is about 250 times stronger than it is for oxygen (Burr 2000). The chemical bond between CO and Hb is known as carboxyhemoglobin (COHb). COHb not only reduces the amount of oxygen that can be delivered to organs and tissues, a condition known as hypoxia, it also interferes with the release of oxygen from the blood. This interference occurs because COHb strengthens the bond between hemoglobin and oxygen in the blood. The percentage of COHb in the blood is thus a dosage indicator of CO exposure and a physiological marker that can be linked to various health effects of CO exposure. The percentage of COHb in a person's blood depends not only on the duration of one's exposure to CO concentrations in the air, but also on one's breathing rate, lung capacity, health status, and metabolism. Because of the high affinity of CO and Hb, the elimination of COHb from the body can take between 2 and 6.5 hours depending on the initial level of COHb in the blood (USEPA 2000). Because the elimination of COHb from the body is a slow process, continuous exposure to even low concentrations of CO may increase COHb (Godish 2004).

Everyone on Earth is exposed to background CO concentrations in the ambient air on the order of 120 parts per billion (ppb) by volume in the Northern Hemisphere and about 40 ppb in the Southern Hemisphere. This difference occurs because the Northern Hemisphere is more developed than the Southern Hemisphere and their respective atmospheres are not completely mixed. In addition, metabolism of heme in the blood produces an endogenous level of CO that occurs naturally in the body. As a result, the body of a nonsmoker has a baseline or residual COHb level in the range of 0.3–0.7% and an endogenous breath CO level of 1–2 ppm. This level varies from one person to another due to human variation in basal metabolisms and other metabolic factors (USEPA 2000). Besides exogenous sources of CO, metabolism of many drugs, solvents (e.g., methylene chloride), and other compounds can also elevate COHb levels above baseline levels through endogenous production of CO. If exposure to drugs and solvents continues for several hours, it can prolong cardiovascular stress caused by excess COHb in the blood. The maximum COHb level from endogenous CO production can last up to twice as long as comparable COHb levels caused by exposures to exogenous CO (Wilcosky and Simonsen 1991; ATSDR 1993).

The percentage of COHb in blood can be related to the breath concentration of CO by simultaneously sampling a person's blood for COHb and his or her end-tidal breath for CO concentration. Coburn, Forster, and Kane (1965) developed an equation to predict the percentage of blood COHb in nonsmokers, based on external CO exposure and assumptions about breathing rate, altitude, blood volume, hemoglobin level, lung diffusivity, and endogenous rate of CO production. For example, a nonsmoking adult engaged in light exercise can expect to have COHb levels below 2–3% if exposed to CO levels of less than 25–50 ppm for 1 hour or 4–7% if the same exposure lasted for 8 hours. Since endogenous COHb leads to a breath CO of about 1–2 ppm, a measured breath CO level of 10 ppm corresponds roughly to an exogenous exposure of 9 ppm (under steady-state conditions).

Burr (2000) describes acute, subacute, chronic, and long-term cardiovascular effects of CO exposure in healthy and diseased populations. Severe oxygen deprivation first affects the brain and then the heart. Patients with heart disease, anemia, emphysema or other lung disease are more susceptible to the harmful effects of CO because their bodies are unable to compensate for oxygen deficiencies. Healthy pregnant women, young children, the elderly, and tobacco smokers are more likely to be adversely affected by CO exposure than are other people. COHb levels of 2.4% or higher can induce chest pain in patients with angina, and levels of 2.3–4.3% can affect the performance of people competing in athletic events (USEPA 2000). COHb levels below 5% can result from exposure to high CO concentrations in the ambient air. People working in certain occupations (e.g., chainsaw gas tool operators, firefighters, garage mechanics, forklift operators) can have COHb levels above 5%, which can affect visual perception and learning ability. Baseline COHb concentrations in smokers average around 4% and range from 3–8% for people who smoke one to two packs per day. COHb levels between 5% and 20% can affect vigilance and diminish hand-eye coordination, which can affect a person's ability to drive a vehicle in traffic. Dizziness, fainting and fatigue can occur at COHb levels of 20% (USEPA 2000). Coma, convulsions and death may occur if COHb levels exceed 60% (Burr 2000). A more complete discussion of the health effects of CO appears in reports by Jain (1990), Penney (1996), and Ernst and Zibrak (1998).

6.5 EARLY STUDIES OF CO EXPOSURE

The commercial districts of cities generate large volumes of motor vehicle traffic during business hours. Vehicles often circulate at low speeds with frequent stops and starts at intersections. This traffic pattern can produce relatively high CO emissions particularly during peak travel periods. Tailpipe exhaust gases rise in the atmosphere, because they are warmer and less dense than air. CO spreads through the atmosphere very easily, because it has a lighter molecular weight than air. In open areas, CO concentrations fall rapidly with greater wind speed and distance from sources. Higher CO concentrations may occur in street canyons, however, because tall buildings affect wind patterns. These facts may explain why early studies of exposure focused on activities such as driving in traffic, while other studies measured roadside concentrations attributable to different levels of traffic in urban areas.

6.5.1 SURVEYS OF EXPOSURE WHILE DRIVING IN TRAFFIC

Until the early 1950s, most automotive engineers thought that motor vehicle emissions played a minor role in air pollution. That thinking began to change in November 1950, when Professor Arie J. Haagen-Smit announced results of his laboratory experiments at the California Institute of Technology (Cal Tech). Haagen-Smit's experiments showed how sunlight converted certain gases emitted by motor vehicles and oil refineries, namely oxides of nitrogen and volatile hydrocarbons, into a secondary air pollutant known as ozone (O_3) (Doyle 2000).

Compared to Haagen-Smit's now famous laboratory experiments on ozone formation, his field surveys of CO concentrations while driving in Los Angeles are not as well known. In these surveys, he equipped the passenger cabin of his car with a prototype, continuously recording CO analyzer developed by Dr. P. Hersch. Haagen-Smit placed the instrument next to the dashboard of his car and ran a glass tube from the instrument's CO sensor to the *outside* air through the front window. (The outside measurement can be a good approximation of exposure inside the car, if there is a rapid exchange of air between the passenger cabin and exterior environment.) He made eight 30-mile round-trips, including travel on suburban streets in Pasadena, portions of two interstate freeways, and surface streets near downtown Los Angeles. CO concentrations outside the vehicle averaged 37 ppm and ranged from 23–58 ppm for trips of 40–115 minutes. Average CO concentrations ranged from 38–72 ppm when he drove under 20 miles per hour (mph) in heavy traffic. Ambient CO levels at fixed-site monitors were above 20 ppm during summer and above 30 ppm during the winter season on 50% of days monitored between 1960 and 1964. Thus, Haagen-Smit appears to be the first analyst to observe that CO concentrations on freeways exceeded urban ambient levels, and that these concentrations rose in heavy traffic moving at slow speeds (Haagen-Smit 1966).

Field surveys similar to Haagen-Smit's pioneering effort in Los Angeles were performed shortly thereafter in many U.S. cities. For example, Brice and Roesler (1966) used Mylar™ bags to measure CO, as well as hydrocarbon concentrations, *inside* vehicles moving in traffic in five U.S. cities (Chicago, Cincinnati, Denver, St. Louis, and Washington, DC) between 7 A.M. and 7 P.M. Air samples were also collected at points alongside traffic routes in Chicago, Washington, DC, and Philadelphia. The average CO concentration measured for trips of 20–30 minutes on arterial streets and express-ways ranged from 21 ppm in Cincinnati to 40 ppm in Denver. The average CO levels on high-density traffic routes were 1.3–6.8 times the corresponding CO concentrations measured at fixed monitoring stations. The study concluded that ambient monitoring stations significantly underesti-mated the pollutant exposures of commuters and those working long hours in traffic (e.g., bus drivers, taxicab drivers, policemen, etc.). Besides revealing inadequacies of ambient monitoring, the study provided a significant baseline for comparing the results of later studies of commuter exposure.

At about the same time as the Brice and Roesler study, Lynn et al. (1967) measured commuter CO and hydrocarbon exposures in 14 American cities between April 1966 and June 1967. They used a mobile sampling van and trailer to collect exposures during 30-minute trips. Lynn et al. (1967) attributed variation in the ratio of commuter exposure to ambient concentrations to variation in the location of monitoring stations. After combining and reanalyzing the data for all 14 cities, Ott, Switzer, and Willits (1993a) reported that the average CO concentrations inside test vehicles varied from 28 ppm on routes through city centers to 22 ppm on arterials and 18 ppm on express-ways. The variation in exposure by route could be explained by variation in traffic volume and vehicle speed on each route.

6.5.2 Surveys of CO Concentrations on Streets and Sidewalks

Early studies showed that CO emissions and roadside concentrations can increase dramatically whenever motor vehicles form a queue at street intersections. Therefore, the severity of concentra-tions may partly depend on how much traffic is handled by an intersection and one's distance from it. To test this hypothesis, Ramsey (1966) surveyed 50 intersections over a 6-month period in Dayton, Ohio. Concentrations were 56.1 ± 18.4 ppm (mean ± one standard deviation) for heavy traffic, 31.4 ± 31.5 ppm for moderate traffic, and 15.3 ± 10.2 ppm for light traffic. Ramsey also reported that concentrations were greater at intersections along major arteries somewhat removed from downtown Dayton, and that their mean concentration was 3.4 times the mean concentration of intersections a block away and perpendicular to the axis of the arterial. In a later study, Claggett, Shrock, and Noll (1981) found that CO concentrations at intersections with signals were higher than those measured near freeways that had two to three times greater traffic volumes.

Colucci and Begeman (1969) found that outdoor mean CO concentrations were usually the highest but varied the most (3.5–10 ppm) in commercial areas of Detroit, New York, and Los Angeles. By comparison, outdoor CO levels varied less near freeways (6–8 ppm) and were lowest in residential areas (2.5–5.5 ppm). They also found that outdoor CO concentrations in New York and Los Angeles tended to be higher during summer and autumn when average wind speeds were generally lower. Later studies looked at how CO concentrations varied with distance from sources for a given location. For example, Besner and Atkins (1970) reported that CO concentrations declined with greater distance from an expressway in an open area of Austin, Texas. At 16 feet from the road, CO concentrations ranged from 3.4–6.0 ppm, while at 95 feet concentrations ranged from 2.4 to 3.9 ppm.

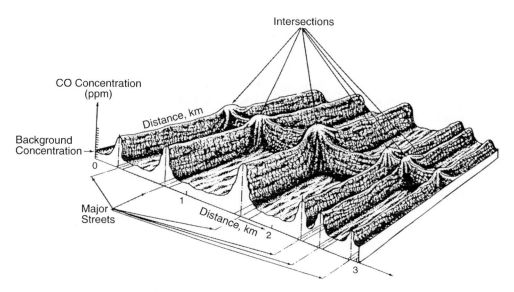

FIGURE 6.1 Model of the spatial variation of CO concentrations at breathing level in an urban area. (From Ott, 1982. With permission from Elsevier.)

These early studies supported the view that CO concentrations at breathing levels were higher in commercial districts of cities, and at intersections and along city streets, but were lower as one moved away from traffic. Figure 6.1 depicts this view of CO concentrations for a portion of a city, based on what was known about the spatial distribution of CO concentrations in the 1970s (Ott 1982). The vertical scale of this figure, which represents CO concentration, would have to be divided by three or four to make the concentrations shown in the figure relevant to the present. The figure also illustrates the superposition principle of CO exposure measurement. This principle holds that the observed CO concentration at a given point in time and space consists of the sum of microenvironmental and background components.

6.6 THE CLEAN AIR ACT AMENDMENTS OF 1970

Most of the early exposure studies were cited in a document titled *Air Quality Criteria for Carbon Monoxide*, published in March 1970 by the National Air Pollution Control Administration (NAPCA) of the U.S. Department of Health, Education, and Welfare (NAPCA 1970). NAPCA, along with several other governmental agencies, became the U.S. Environmental Protection Agency (USEPA) on July 9, 1970, when President Richard Nixon issued an executive order creating the agency. Another significant event of 1970 was congressional approval of amendments to the Clean Air Act (CAA), which required the USEPA to promulgate National Ambient Air Quality Standards (NAAQS) for several air pollutants including CO. Exposure studies frequently refer to the NAAQS for guidance on allowable limits of exposure. The NAAQS include a set of primary standards to protect public health and secondary standards to protect public welfare, such as crop damage from ozone. The NAAQS apply to "criteria" pollutants, because the USEPA must issue air quality criteria for pollutants that may reasonably endanger public health or welfare. Accordingly, the NAAQS set maximum permissible concentrations in ambient air for specified averaging times. The standards include a safety margin to reflect uncertainties in the science of effects of air pollution.

On April 30, 1971, the USEPA promulgated identical primary and secondary NAAQS for CO. In 1985, the USEPA rescinded the secondary standards, because there was no evidence of adverse effects on public welfare due to ambient CO levels. However, the USEPA retained the primary standards, which have remained since 1971 at 9 ppm for an 8-hour average and 35 ppm for a 1-hour average. These standards are designed to keep COHb levels below 2% in the blood of the general public, including probable high-risk groups. These groups include the elderly; pregnant women; fetuses; young infants; and those suffering from anemia or certain other blood, cardiovascular, or respiratory diseases. People at greatest risk from exposures to ambient CO levels are those with coronary artery disease. These people may suffer chest pain during exercise when exposed to COHb levels ≥2.4% (USEPA 2000). Although annual death rates from heart disease have been declining since 1980, heart disease is still the nation's leading cause of death (Arias et al. 2003). Coronary artery disease reduces a person's circulatory capacity, which is particularly critical during exercise when muscles need more oxygen.

In accordance with the CAA, the USEPA must determine whether or not a community complies with the NAAQS based on measurements of ambient air quality made by a nationwide network of fixed-site monitoring stations. This network consists of state and local air monitoring stations (SLAMS), which send data to USEPA's Aerometric Information Retrieval System (now Air Quality System) within 6 months of acquisition (Blumenthal 2005). Several stations within the SLAMS network belong to a network of national air monitoring stations (NAMS) to enable national assessments of air quality. A station is in non-attainment of the NAAQS for CO if it records an ambient concentration that exceeds either the 1-hour or 8-hour standard more than once per year. These stations use the non-dispersive infrared (NDIR) method to measure ambient CO concentrations. Monitoring instruments based on NDIR are large, complex, and expensive, and require an air-conditioned facility for the production of accurate and reliable data. Because NDIR monitors are not portable, they cannot be used to measure CO exposure as a person performs routine daily activities.

The CAA amendments also mandated stringent automobile emission standards to assist in attainment of the NAAQS. When the NAAQS were adopted, highway vehicles accounted for substantial percentages of total national emissions of CO, hydrocarbons, and nitrogen oxides (NO_x). Compared to emissions from new cars sold during the 1970 model year, the CAA amendments required automakers to produce passenger cars that achieved 90% reductions in CO and hydrocarbon emissions by the 1975 model year. By the 1976 model year, manufacturers had to achieve a 90% rollback in NO_x emissions over 1971 levels (Ortolano 1984). Studies persuaded Congress that these emission standards would accomplish ambient air quality goals by 1990 in those areas that had the worst air pollution in the nation (Grad et al. 1975).

Automobile manufacturers viewed the emission standards as "technology forcing," because the technology to achieve them did not exist when the standards were adopted (Ortolano 1984). During the 1970s, the industry's efforts to reduce vehicle emissions were achieved through the use of increasingly elaborate and sophisticated technologies (e.g., the three-way catalytic converter). By the 1981 model year, the CO emission rate of new passenger cars was below the pre-control level (prior to 1968) by 96% (Johnson 1988).

6.7 LIMITATIONS OF FIXED-SITE MONITORS

Several pioneering studies during the 1970s revealed the inability of fixed-site monitors to represent human exposure to CO in certain situations. In one study, Yocom, Clink, and Cote (1971) reported that when make-up air was introduced into an air-conditioned building during morning rush hours (when outdoor CO levels were high), indoor CO concentrations exceeded outdoor levels for the remainder of the day. This finding took on added significance when social scientists reported during the early 1970s that many Americans spent most of their time indoors (Szalai 1972; Chapin 1974). In another study, Wayne Ott collected "walking samples" of CO concentrations on sidewalks along congested streets in downtown San Jose, California, for his doctoral dissertation in civil engineering at Stanford University. He collected samples in large Tedlar™ bags filled by a constant flow pump over a 5-minute period at various times over an 8-hour period. Ott, together with his faculty adviser,

Professor Rolf Eliassen, reported average CO levels ranging from 5.2–14.2 ppm on San Jose's sidewalks. Concurrent CO levels, reported as 1-hour averages at nearby fixed-site monitors, were only 2.4–6.2 ppm (Ott and Eliassen 1973).

A few years later, Cortese and Spengler (1976) did the first survey to determine the CO exposure of "real" people who commuted to and from work. This type of study was made possible by the development of portable electrochemical CO monitors in the early 1970s (USEPA 1991). The research team recruited 66 nonsmoking volunteers who lived in different parts of the metropolitan area of Boston, Massachusetts. The study focused on several travel corridors serving the city's central business district. Each volunteer carried an Ecolyzer monitor attached to a Simpson recorder for 3–5 days between October 1974 and February 1975. The study also estimated COHb levels in the blood based on samples of air in the alveolar sacs of the lung before and after each trip. The study's simultaneous measurement of CO exposure and body burden (% COHb) set a precedent for subsequent studies that involved human participants.

The study reported that the mean of all commuter exposures (11.9 ppm) was about twice the mean concentration measured concurrently at six fixed-site monitors (6 ppm). That was similar to the ratio observed in five cities by Brice and Roesler in the mid-1960s. However, the net mean in-vehicle exposure in Boston was about 42% of the net value reported by Brice and Roesler (1966). Excluding commuters whose cars had "faulty exhaust systems," only 0.5% of 346 sampled CO exposures in the Boston study exceeded the 1-hour CO NAAQS of 35 ppm. Automobile commuters had exposures nearly twice that of transit users, and about 1.6 times that of people who did "split-mode" commuting, which involved both auto and transit. Based on the Boston study, Cortese and Spengler recommended a mobile monitoring program to supplement data from fixed-site monitors.

6.8   ESTIMATES OF NATIONWIDE POPULATION EXPOSURE

The USEPA inherited a fixed-site monitoring program when the agency was established in 1970. Moreover, the Clean Air Act amendments of 1970 did not require measurements of personal exposure to supplement air quality monitoring at fixed sites. There were several proposals to estimate *potential* population exposure to air pollutants during the 1970s (Ott 1982). For example, one estimate simply multiplies the number of days that violations of the NAAQS are observed at county monitoring stations times the county's population. Estimates of exposure using this method are expressed in units of person-days (CEQ 1980).

Knowing that crude estimates of population exposure to CO were potentially inaccurate, the U.S. Public Health Service (PHS) measured the percentage of COHb in the blood of a nationwide sample of 8,405 people between 1976 and 1980. The National Health and Nutrition Examination Survey (NHANES) estimated that 6.4% of those people who never smoked had COHb levels above 2% (Radford and Drizd 1982). This estimate is based on data from a random selection of 3,141 people ranging in age from 12–74 years living in 65 geographic areas of the United States. The estimate was made when ambient CO concentrations were much higher than they are today. As shown by Figure 6.2, the estimated probability distribution of COHb levels appears to be lognormal (Apte 1997). The curve is based on data with a geometric mean (GM) of 0.725% and a geometric standard deviation (GSD) of 2.15%.

The USEPA continues to report the number of Americans who live in areas of the country that are in non-attainment of the NAAQS on an annual basis. The agency's Office of Air Quality Planning and Standards (OAQPS) estimated that 19.130 million Americans residing in 13 counties as of September 2002 (roughly 6.6% of the resident U.S. population) were exposed to ambient CO concentrations that exceeded the 1-hour NAAQS of 35 ppm. Two major metropolitan areas (Los Angeles and Phoenix) accounted for 88.2% of that population at risk (USEPA 2003).

As indicated above, crude estimates of population exposure to CO are made by combining census data on county populations with data on violations of the CO NAAQS recorded by stationary

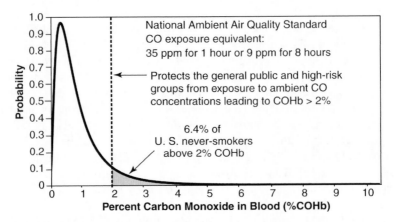

FIGURE 6.2 Estimated probability distribution of carboxyhemoglobin (COHb) levels in blood samples from never-smokers in the United States, 1976–1980.

monitors in each county. Crude estimates of population exposure are based on four assumptions (CEQ 1980):

1. The population does not travel outside the area represented by the fixed-site monitor
2. Air pollutant concentrations measured by fixed-site monitors are representative of the concentrations inhaled by the population throughout the area represented by the monitor
3. The air quality in any one area is only as good as that at the location that had the worst recorded air quality
4. There are no violations in areas of the country (e.g., rural areas) that are not monitored

The early exposure studies (cited previously) challenged the validity of the second assumption regarding the ability of fixed-site monitors to represent the actual CO exposures of people living in cities. Recognizing these studies, the OAQPS developed a risk-analysis framework to support periodic reviews of the NAAQS for CO (Padgett and Richmond 1983; Jordan, Richmond, and McCurdy 1983). This framework gave purpose to subsequent CO exposure studies and stimulated the development of methods and models to estimate total exposure to CO, which is the topic of discussion below.

6.9 ESTIMATING TOTAL CO EXPOSURE

Technical improvements in personal exposure monitors during the 1970s stimulated scholarly interest in how to use and apply them. Fugas (1975) and Duan (1982) advocated that a person's total air pollution exposure could be estimated indirectly based on the following mathematical model:

$$E_i = \sum_{k=1}^{K} c_k(t_{ik}) \tag{6.1}$$

where

E_i = the total integrated exposure of person i over some time period of interest (e.g., 24 hours)

c_k = the air pollutant concentration in microenvironment type k

t_{ik} = the amount of time spent by person i in microenvironment type k

K = the number of microenvironment types encountered by person i over the period of interest

This model states that an individual's total exposure over a given time period can be estimated as the sum of a series of separate exposures resulting from time spent in different types of microenvironments. According to Sexton and Ryan (1988), this model rests on four assumptions: (1) air pollutant concentrations in each microenvironment remain constant for the duration of a person's visit to that microenvironment; (2) the air pollutant concentration and time spent in a microenvironment are independent of each other (i.e., a person does not avoid or leave a microenvironment simply because it is polluted); (3) an adequate number of microenvironments can be identified to characterize a person's total exposure; and (4) the total integrated exposure for a given time period can be related to health effects.

Duan (1982) defined a microenvironment as "a chunk of air space with homogeneous pollutant concentration" (p. 305). Identifying microenvironments is a challenge for analysts who study CO exposure, because CO concentrations not only vary in time and space but can be affected by many factors. For example, CO concentrations in a kitchen will vary depending on whether the kitchen has a gas range, whether it is vented or not, and the duration of its use. In population exposure studies, analysts may ask study participants to keep a diary of time spent in various indoor and outdoor locations and to record certain information about activities (e.g., cooking, smoking behavior) that occur there. Figure 6.3 shows a page from a diary used for this purpose in a study of population exposure in Denver, Colorado. The response categories shown on this page indicate microenvironments relevant to CO exposure.

The concept of integrated air pollutant exposure inspired Ott to develop a computer simulation model to estimate the CO exposure of an urban population. Since Ott's simulation model (described later) needed typical CO concentrations for different microenvironments, he launched two major studies to acquire the data in the late 1970s. These were (1) the field surveys of commercial microenvironments in four California cities (described below), and (2) the El Camino Real commuter exposure surveys on the San Francisco Peninsula (described toward the end of the chapter). At the same time, Ott promoted the emerging science of exposure analysis, which focused on humans as receptors of environmental pollution, to the scientific community (Ott 1985).

6.10 FIELD SURVEYS OF COMMERCIAL MICROENVIRONMENTS

Milton Feldstein, who directed the Bay Area Air Quality Management District in San Francisco during the 1970s, was familiar with Ott's "walking sampling" survey on the streets of downtown San Jose. At the conclusion of that study, Feldstein encouraged Ott to monitor the personal exposures of people, not only as they walked on sidewalks, but also while they shopped in retail stores, ate in restaurants, and did similar activities in the commercial areas of cities. In 1979, Ott proposed this study to Peter Flachsbart at Stanford University's Department of Civil Engineering. The study not only contributed valuable data to Ott's simulation model of population exposure, it also provided insights to the design of two large-scale field surveys of urban population exposure. These larger scale surveys (described later), took place in Denver, Colorado, and Washington, DC.

Knowing the potential for CO exposure along busy streets, Ott and Flachsbart (1982) surveyed locations in the Westwood district of Los Angeles, the financial and Union Square districts of San Francisco, and the central business districts of two suburbs (Palo Alto and Mountain View) of the Bay Area. Using a General Electric electrochemical monitor, they collected 5,000 instantaneous measurements of CO concentrations at 1-minute intervals as they walked the streets of each commercial district. Data collection was cumbersome, because the CO concentration had to be read and recorded from the monitor's digital display, and the time and location of each CO reading

```
┌─────────────────────────────────┬─────────────────────────────────┐
│ TIME FROM MONITOR │1│7│2│3│     │ D.  ONLY IF IN TRANSIT          │
│ A.  ACTIVITY                    │     (1) Start address _____ │
│     driving from office         │         200 Arapahoe St.        │
│     to grocery store            │     (2) End address _____ │
│ _____            │         101 Ellsworth Ave.      │
│                EXAMPLE           │     (3) Mode of travel:         │
│ B.  LOCATION                    │         Walking . . . . . . . 1 │
│     In transit . . . . . . .(1) │         Car . . . . . . . . .(2)│
│     Indoors, residence . . . 2  │         Bus . . . . . . . . . 3 │
│     Indoors, office . . . . 3   │         Truck . . . . . . . . 4 │
│     Indoors, store . . . . 4    │         Train/subway . . . . 5  │
│     Indoors, restaurant . . 5   │         Other . . . . . . . . 6 │
│     Other indoor location . 6   │            Specify: _____  │
│        Specify: _____      │                                 │
│                                 │ E.  ONLY IF INDOORS             │
│     Outdoors, within 10 yards   │     (1) Garage attached to      │
│       of road or street . . 7   │             building?           │
│     Other outdoor location . 8  │         Yes . . . . . . . . . 1 │
│        Specify: _____      │         No . . . . . . . . . 2  │
│                                 │         Uncertain . . . . . . 3 │
│     Uncertain . . . . . . . 9   │     (2) Gas stove in use?       │
│ C.  ADDRESS (if not in transit) │         Yes . . . . . . . . . 1 │
│                                 │         No . . . . . . . . . 2  │
│                                 │         Uncertain . . . . . . 3 │
│                                 │ F.  ALL LOCATIONS               │
│                                 │     Smokers present?            │
│                                 │         Yes . . . . . . . . . 1 │
│                                 │         No . . . . . . . . .(2) │
│                                 │         Uncertain . . . . . . 3 │
└─────────────────────────────────┴─────────────────────────────────┘
```

FIGURE 6.3 Example of a completed page from a diary for an exposure study. (From Johnson 1984.)

had to be recorded manually on a clipboard. During 15 field surveys between November 1979 and July 1980, Ott and Flachsbart visited 588 "commercial settings": 220 indoor settings (e.g., department stores, hotels, office buildings, parking garages, retail stores, restaurants, banks, travel agencies, and theaters); and 368 outdoor settings (e.g., street intersections, sidewalks, arcades, parks, plazas, and parking lots).

The results of their study supported the hypothesis that CO emissions from traffic in commercial areas tend to diffuse into adjacent buildings and stores. Although indoor levels were above zero, they seldom were very high — except for parking garages and certain buildings attached to garages. They found that indoor CO concentrations were relatively stable for nearly 2 hours. This finding was very important, because it suggested that estimates of 1-hour indoor CO exposures could be made based on short visits of only a few minutes to each setting. Thus many settings could be visited during each survey. Excluding 10 parking garages, they found that CO concentrations at 6 of 210 settings (2.9%) equaled or exceeded 9 ppm during brief visits (Ott and Flachsbart 1982; Flachsbart and Ott 1984).

On their first survey of downtown Palo Alto, they observed high CO concentrations (above 9 ppm) on 5 floors of a 15-story office building on University Avenue. They traced the source of the problem to an underground parking garage, where CO levels often exceeded 20 ppm. Based on this observation, they developed a "rapid survey technique" to measure CO concentrations in the building and parking garage during nine visits to the building in 1980. High CO levels accumulated in the garage, because its ventilation system had been shut down periodically to save energy in the wake of high electricity costs triggered by a national oil crisis. A survey of the entire building revealed that CO concentrations diffused into the building from the garage through a stairwell. The door between the garage and stairwell was often kept open. Having identified a potential problem, Flachsbart and Ott told the building manager that the building was "hot." When several tenants threatened legal action, the manager hired an air quality consultant to confirm the problem. In 1984, Flachsbart and Ott resurveyed the building on five different dates. By then, the manager was operating the building's ventilation system on a continuous basis and had installed heavier door closers to make sure the door to the garage would shut automatically after opening. A comparison of average CO concentrations, before and after these actions, showed that levels in the garage had fallen from 40.6 ppm in 1980 to 7.9 ppm in 1984, while typical CO levels in the building fell from 11–12 ppm in 1980 to 1–2 ppm in 1984 (Flachsbart and Ott 1986).

Since the detection of the "hot" building in Palo Alto was accidental, the study could not show how to detect "hot" buildings from among the thousands of office buildings that exist in urban areas. As a result, the prevalence of "hot" buildings has never been determined on an urban, state, or national scale. Such a survey would probably require an amendment of the Clean Air Act to enable routine monitoring of *indoor* air quality in the United States. Until then, the Palo Alto study is noteworthy, because it illustrates how portable monitors can be used to survey a high-rise office building quickly once elevated CO concentrations are detected, to determine the source of high concentrations, and to evaluate solutions to reduce those concentrations.

6.11 DIRECT AND INDIRECT APPROACHES TO MEASURE EXPOSURE

The advent of microelectronics during the 1970s enabled the initial development of reliable, compact personal exposure monitors. This development stimulated new thinking about how to measure exposure in the field. Most exposure analysts recognize direct and indirect methods for measuring an individual's exposure to an air pollutant (Sexton and Ryan 1988; Mage 1991).

In the direct approach, personal monitors are distributed ideally to a representative, probability sample of a human population. Population exposure parameters cannot be estimated from a "convenience sample," such as the Boston study of volunteer commuters (described earlier), because such a sample may not represent the population from which it was drawn. Personal monitors must be calibrated with gases of known concentration, before and after their distribution to the public, to assure that the CO measurements are accurate. As one can imagine, distributing calibrated monitors to a large sample of participants on a daily basis creates logistical problems. Participants must be provided with instructions on how to record their exposures as they perform their daily activities during the next day or two. Subjects may be asked to use a diary (Figure 6.3) to record pertinent information about these activities. This information (e.g., presence of a wood-burning fireplace or people smoking) may shed light on circumstances that affect high exposures at certain times of the day. Many exposure analysts believe that the direct approach provides the most accurate estimate of population exposure, because it surveys the actual exposures of people doing their daily activities. However, the direct approach is expensive, because it requires substantial amounts of time and labor to gather data from large samples of urban populations.

The intrusiveness and expense of the direct approach motivated some researchers to favor the indirect approach of exposure measurement. In that approach, trained technicians use calibrated portable monitors to measure pollutant concentrations in specific microenvironments. That information is then combined in mathematical models with separate estimates of time spent in those microenvironments from activity surveys of a population. This approach can lead to errors in estimates of population exposure, because measurements of CO concentrations for different microenvironments and time spent in them are made separately.

Besides choosing between the direct and indirect approaches, the exposure analyst must consider several types of personal monitors. Small, passive CO monitors are usually placed near a person's oral/nasal cavity close to where inhalation exposure actually occurs. Larger monitors with air pumps usually come with a shoulder strap so they can be carried. In commuter studies, monitors are set or mounted somewhere inside the vehicle. Using data from portable monitors, one can plot CO concentrations as a function of time and location for a particular activity, such as commuting. From this information, one can determine the average CO concentration to which a person has been exposed for a given time period.

6.11.1 Studies Using the Direct Approach

In the early 1980s, the USEPA funded a pilot study and two large-scale surveys of urban population exposure based on the direct approach. In the pilot study, Ziskind, Fite, and Mage (1982) asked nine residents of Los Angeles to record their exposures and corresponding activities manually using diaries. Because this process was cumbersome and potentially affected the amount of time doing the activity, the USEPA funded the development in the early 1980s of automated data-logging personal exposure monitors (PEMs). These instruments measured and stored CO concentrations, as well as the time spent doing activities associated with those concentrations (Ott et al. 1986). Figure 6.4 shows an example of a CO monitor carried by a woman shopping in a grocery store.

The Los Angeles pilot study provided useful insights on how to design field surveys of population exposure using the direct approach. For example, the study showed that there was greater variation in CO exposure from person to person than from day to day for any one person. This suggested that subsequent studies should try to sample more people rather than to sample the same person over many days. Akland et al. (1985) described how the USEPA used this insight in designing two large-scale field surveys of population exposure that occurred in the fall of 1982 and winter of 1983. Study participants consisted of 454 residents of Denver, Colorado, and 712 residents of Washington, DC. In each city, the target population consisted of non-institutionalized, nonsmoking residents ages 18–70 who lived in the metropolitan area. Akland et al. (1985) estimated the size of the target population to be 500,000 people in Denver and about 1.2 million people in Washington. Study participants carried a CO PEM and diary for 48 hours in Denver and for 24 hours in Washington.

The goal of these studies was to estimate the percentage of the urban population that was exposed to CO concentrations in excess of the NAAQS for CO. The studies showed that over 10% of the daily maximum 8-hour personal exposures in Denver exceeded the NAAQS of 9 ppm, and about 4% did so in Washington. By comparison, the CO concentrations at fixed-site monitors exceeded the 8-hour NAAQS for CO (9 ppm) only 3% of the time in Denver, and never exceeded the 9 ppm standard in Washington, during the survey period (Akland et al. 1985). These results raised further doubts as to the ability of fixed-site monitors to represent the total CO exposure of urban populations. The studies also showed that the end-expired breath CO levels (after correcting the measured breath concentrations for the influence of room air CO concentrations) were in excess of 10 ppm, which was roughly equivalent to about 2% COHb, in about 12.5% of the Denver participants and about 10% of the Washington participants.

The two studies also looked at factors that contributed to higher levels of exposure. Of 10 different microenvironments, both studies found that parking garages had the highest average CO

FIGURE 6.4 Lady carrying a personal exposure monitor while shopping in a grocery store. (Courtesy of U.S. Environmental Protection Agency.)

concentrations, but the shortest duration of exposure. The Denver study found that the average PEM indoor mean exposure (unadjusted for cofactors) was increased 2.59 ppm (134%) by gas stove operation, 1.59 ppm (84%) by tobacco use from smokers other than study participants, and 0.41 ppm (22%) by attached garages. Table 6.1 shows CO concentrations for selected microenvironments of the Denver study. The table indicates that higher CO exposures occurred for travel by motor vehicle (motorcycle, bus, car, and truck) than for pedestrian or bicycle modes of travel. It also shows that high indoor concentrations (above the 8-hour NAAQS of 9 ppm) occurred in public garages, service stations or vehicle repair facilities. In the Washington study, those who commuted 6 hours or more per week had higher average CO exposures than those who commuted fewer hours per week. Also, certain occupations increased one's CO exposure in the Washington study. People who drove trucks, buses or taxis, as well as people who worked as automobile mechanics, garage workers, and policemen had a mean CO exposure (22.1 ppm) that was three times greater than those who did not work with gasoline-powered automobiles (6.3 ppm). Of the two cities, Denver had higher average CO concentrations than did Washington for all microenvironments. This difference was later attributed to Denver's higher altitude and colder winter climate (Ott, Mage, and Thomas 1992).

Since the Denver–Washington surveys of 1982–1983, there have been only a few small-scale studies of CO exposure using the direct approach. Recognizing the higher exposures of certain types of commuters, Shikiya et al. (1989) measured in-vehicle concentrations of CO and several air toxics (e.g., benzene) in the Los Angeles metropolitan area. The researchers selected a random sample of 140 nonsmokers who commuted from home to work in privately owned vehicles during peak commuting hours during the summer of 1987 and winter of 1988. Trips averaged 33 minutes and driving patterns and ventilation conditions were not controlled. Based on 192 samples, in-vehicle CO concentrations averaged 6.5 ppm in summer and 10.1 ppm in winter.

TABLE 6.1
CO Concentrations of Selected Microenvironments in Denver,
CO, 1982–1983 (Listed in Descending Order of Mean CO
Concentration)

Microenvironment	n	Mean[a] (ppm)	Standard Error (ppm)
In-Transit			
Motorcycle	22	9.79	1.74
Bus	76	8.52	0.81
Car	3,632	8.10	0.16
Truck	405	7.03	0.49
Walking	619	3.88	0.27
Bicycling	9	1.34	1.20
Outdoor			
Public garages	29	8.20	0.99
Residential garages or carports	22	7.53	1.90
Service stations or vehicle repair facilities	12	3.68	1.10
Parking lots	61	3.45	0.54
Other locations	126	3.17	0.49
School grounds	16	1.99	0.85
Residential grounds	74	1.36	0.26
Sports arenas, amphitheaters	29	0.97	0.52
Parks, golf courses	21	0.69	0.24
Indoor			
Public garages	116	13.46	1.68
Service stations or vehicle repair facilities	125	9.17	0.83
Other locations	427	7.40	0.87
Other repair shops	55	5.64	1.03
Shopping malls	58	4.90	0.85
Residential garages	66	4.35	0.87
Restaurants	524	3.71	0.19
Offices	2,287	3.59	0.002
Auditoriums, sports arenas, concert halls	100	3.37	0.48
Stores	734	3.23	0.21
Healthcare facilities	351	2.22	0.23
Other public buildings	115	2.15	0.30
Manufacturing facilities	42	2.04	0.39
Homes	21,543	2.04	0.02
Schools	426	1.64	0.13
Churches	179	1.56	0.25

[a] An observation was recorded whenever a person changed a microenvironment, and on every clock hour; thus each observation had an averaging time of 60 min or less.

Source: Johnson (1984) as reported in U.S. Environmental Protection Agency (1991).

The studies by Shikiya et al. (1989) and Cortese and Spengler (1976) involved real commuters. A comparison of these studies, which were performed 13 years apart, shows the effect of stricter CO regulations. To factor out the effect of ambient CO concentrations, an analyst compares their net mean in-vehicle CO concentrations, which is the estimated mean in-vehicle CO concentration during commuting minus the mean ambient CO concentration, as recorded concurrently at an appropriate fixed-site monitor. The net value in Boston (7.4 ppm) in 1974–1975 was 51% higher than the net value in Los Angeles (4.9 ppm) in 1987–1988, which are the respective years in which these studies were performed. The difference in net values can be attributed to the prevalence of emission controls on motor vehicles, which differed for the two studies. By the 1980 model year, half of all passenger cars in use in the United States had catalytic and other types of emission controls (MVMA 1990). Hence, most cars in the Boston study probably lacked these emission controls while most cars in the Los Angeles study probably had them.

6.11.2 A Study Using the Indirect Approach

In the early 1980s, a significant CO exposure study occurred in Honolulu, which is located on the Island of Oahu in Hawai'i. Honolulu has no major smokestack industries except for a few refineries located in an industrial area on the southwest corner of the island. Generally, prevailing "trade winds" from the northeast blow most air pollutants in Honolulu out to sea. Occasionally, southerly winds trap air pollutants against the Ko'olau Mountains and create a stagnant air mass. The city is thus an ideal location for doing microenvironmental exposure studies, because ambient CO levels are generally low and satisfy the NAAQS. Several Honolulu "walking surveys" revealed high CO concentrations on the street level of the Ala Moana Shopping Center, less than a mile west of Waikiki Beach (Flachsbart and Brown 1985, 1986).

Several factors contributed to what appeared to be a significant CO exposure problem at the Ala Moana Center. First, it had 155 business outlets that attracted 40 million people including many tourists each year. Second, it had an attached structure with 7,800 parking spaces on several decks. CO emission rates were high, because the posted speed limit for driveways in the structure was 15 mph for the safety of pedestrians. Third, one deck of the parking structure functioned as a lid on the exhaust emissions of cars at the street level of the structure. Fourth, many of the 94 outlets at street level kept their doors open during business hours to attract customers. This allowed CO concentrations from the parking area and internal driveways to diffuse into many retail outlets.

Flachsbart and Brown (1989) devised an indirect method to estimate the CO exposures of employees working in retail stores at the center's street (semi-enclosed) level. Using a portable monitor, they measured CO concentrations and counted employees who worked at 25 retail outlets. Data collection occurred during three periods of the day (10 A.M.–12 noon, 2–4 P.M., and 6–8 P.M.) at 5-day intervals between early November 1981 and late March 1982. Based on 30 days of sampling, they estimated that between 24.5% and 36.1% of employee exposures could have exceeded the 8-hour NAAQS of 9 ppm, and between 2.2% and 2.4% of employee exposures could have exceeded the 1-hour NAAQS of 35 ppm. By comparison, the vast majority (88.5%) of the 8-hour CO averages at the nearest fixed-site monitor, located 3 miles east of the shopping center, were less than or equal to 1 ppm during the study period. The survey on December 21, 1981, showed that CO concentrations in 10 of 25 stores (40%) exceeded the 1-hour NAAQS of 35 ppm. The average CO levels during visits to these 10 stores on that date ranged from a low of 36.3 ppm to a high of 86.7 ppm (Flachsbart and Brown 1985, 1989). CO concentrations on that scale are sufficient to trigger actions by public health officials if they have a mandate to act.

The Ala Moana study showed the potential of portable monitors to reveal CO exposure problems in a specific population, much like the study of the 15-story office building in downtown Palo Alto by Flachsbart and Ott (1986). Both studies also revealed a gap in existing environmental laws and regulations, which do not protect the public from high CO exposures on private property. The Honolulu study also revealed a gap in occupational standards. The Pollution Investigation and Enforcement branch of the Hawai'i State Department of Health (DOH) acknowledged that the Ala Moana study revealed a potential CO exposure problem for retail workers. However, the DOH branch of Occupational Safety and Health found no technical violation of occupational standards for CO, because the occupational CO standards were 200 ppm for 1 hour and 50 ppm for 8 hours at that time. Unlike the NAAQS, which are designed to protect the most sensitive class of the general population, occupational standards for CO assume a healthy young male employee working in an industrial setting. However, it is not clear that these standards fully protect everyone who works at the Ala Moana Shopping Center, because many of its employees are women and older adults.

6.12   OCCUPATIONAL EXPOSURES

The direct studies of urban populations in Denver and Washington, and the indirect study of shopping center employees in Honolulu, both found higher exposures among occupations that involved use of or proximity to gasoline-powered motor vehicles. To put these studies in perspective, a national survey found that 3.5 million workers in the private sector potentially are exposed to CO from motor exhaust, a figure greater than that for any other physical or chemical agent (Pedersen and Sieber 1988). In 1992, there were 900 work-related CO poisonings resulting in death or illness in private industry as reported by the U.S. Bureau of Labor Statistics in a publication by the National Institute for Occupational Safety and Health (NIOSH 1996). Three risk factors affect industrial occupational exposure: (1) the work environment is located in a densely populated area that has high background (i.e., ambient CO concentrations); (2) the work environment produces CO as a product or by-product of an industrial process, or the work environment tends to accumulate CO concentrations that may result in occupational exposures; and (3) the work environment involves exposure to methylene chloride, which is metabolized to CO in the body. Proximity to fuel combustion of all types elevates CO exposure for certain occupations: airport employees; auto mechanics; small gasoline-powered tool operators (e.g., users of chainsaws); parking garage or gasoline station attendants; policemen; taxi, bus, and truck drivers; tollbooth and roadside workers; warehouse workers and forklift operators (USEPA 1991, 2000).

6.13   RESIDENTIAL EXPOSURES

Residential exposures are an important component of total daily exposure to air pollution, because people spend a substantial portion of their day at home often in close proximity to significant sources of CO emissions. Wilson, Colome, and Tian (1993a,b) and Colome, Wilson, and Tian (1994) reported CO exposures for a random sample of California homes with gas appliances. Of surveyed homes, 13 of 286 homes (4.5%) had indoor CO concentrations above the NAAQS of 9 ppm for 8 hours, and 8 of 282 homes (2.8%) had outdoor CO concentrations above the standard. They found that a small percentage of California homes would still have indoor CO problems even if outdoor CO levels complied with federal ambient standards. They traced higher net indoor CO levels (indoor minus outdoor CO concentrations) to space heating with gas ranges and gas-fired wall furnaces, use of gas ranges with continuous gas pilot lights, small home volumes, and cigarette smoke. Other factors contributing to high CO concentrations included malfunctioning gas furnaces; automobile exhausts leaking into homes from attached garages and carports; improper use of gas appliances (e.g., gas fireplaces); and improper installation of gas appliances (e.g., forced-air unit ducts).

6.14   RECREATIONAL EXPOSURES

Potentially dangerous CO exposures also occur in both outdoor and indoor recreational settings. The CO exposure of cycling as a travel mode has been studied and compared to the exposure of motorists in two European countries. In England, Bevan et al. (1991) reported that the mean CO exposure of cyclists in Southhampton was 10.5 ppm, based on 16 runs over two 6-mile routes that took an average of 35 minutes to complete. Note that the CO exposures of European cyclists may not be comparable to cyclists' exposures in the United States, because installation of catalytic converters on new cars in Europe occurred in 1988, about 13 years after their introduction in the United States (Faiz, Weaver and Walsh 1996). In the Netherlands, Van Wijnen et al. (1995) compared exposures of volunteers serving as both car drivers and cyclists on several routes in Amsterdam during winter and spring. For a given route, the mean personal 1-hour CO concentrations were always higher for car drivers than for cyclists regardless of when sampling occurred during the year. However, the analysts reported that a volunteer inhaled 2.3 times more air per minute on average as a cyclist than as a car driver. When adjusted for variation in breathing rate, the range in median 1-hour averaged uptakes of CO of cyclists (2.4–3.2 mg) approached that of car drivers (2.7–3.4 mg).

Significant quantities of CO can accumulate over short periods of time in poorly ventilated indoor arenas used for sporting events. The CO is emitted by several sources including ice resurfacing machines and ice-edgers during skating events; gas-powered radiant heaters used to heat viewing stands; and motor vehicles at motocross, monster-truck, and tractor-pull competitions. Some of these sporting events and competitions involve large numbers of spectators and motor vehicles with no emission controls. Several U.S. studies are briefly discussed here to show the scale of the problem.

Lee et al. (1994) reported that CO concentrations measured inside six enclosed rinks in the Boston area during a 2-hour hockey game ranged from 4–117 ppm, whereas outdoor levels were about 2–3 ppm. The alveolar CO concentrations of hockey players increased by an average of 0.53 ppm per 1 ppm of CO exposure over 2 hours. Fifteen years earlier, Spengler, Stone, and Lilley (1978) found CO levels ranging from 23 to 100 ppm in eight enclosed rinks in the Boston area, which suggested that CO levels in ice arenas had not changed very much.

Boudreau et al. (1994) reported CO levels for three indoor sporting events (i.e., monster-truck competitions, tractor pulls) in Cincinnati, Ohio. The CO measurements were taken before and during each event at different elevations in the public seating area of each arena with most readings obtained at the midpoint elevation where most people were seated. Average CO concentrations over 1–2 hours ranged from 13 to 23 ppm before the event and from 79 to 140 ppm during the event. Measured CO concentrations were lower at higher seating levels. The ventilation system was operated maximally, and ground-level entrances were completely open.

6.15 POPULATION EXPOSURE MODELS

Under the Clean Air Act, the USEPA has a statutory requirement to perform a periodic review of the criteria that support the NAAQS. Each review must be based on the latest information published in scientific, peer-reviewed literature. Following the 1991 review, the USEPA's Office of Air Quality Planning and Standards (OAQPS) estimated that fewer than 0.1% of the nonsmoking population with cardiovascular diseases would experience a COHb level of $\geq 2.1\%$ when exposed to CO levels at current ambient standards in the absence of indoor sources (USEPA 1992). That estimate influenced Carol Browner, who was the USEPA administrator under President Clinton, to retain the existing NAAQS for CO in 1994. The next review of the criteria in 2000 supported the conclusions of the review in 1991. The review in 2000 also found that "there is not a good estimate of CO exposure distribution for the current population" (USEPA 2000, p. 7-10).

The OAQPS estimate of population exposure in 1992 was derived from an exposure model. Such models are important because it is impossible to measure the exposure of every person in a population on a real-time basis. Models of human exposure are empirically derived mathematical relationships, theoretical algorithms, or hybrids of these two. To support policy decisions related to the setting of ambient and emission standards, the USEPA supported development of four general population exposure models: (1) the Simulation of Human Activity and Pollutant Exposure (SHAPE) model, (2) the NAAQS Exposure Model (NEM), (3) the probabilistic NEM for CO (pNEM/CO), and (4) the Air Pollutants Exposure Model (APEX). These models rest on Duan's (1982) theory for estimating total human exposure to air pollution as previously discussed.

The SHAPE model used a stochastic approach to simulate the exposure of an individual over a 24-hour period (Ott 1983–1984). The model replicates a person's daily activity pattern by sampling from probability distributions representing the chance of entry, time of entry, and time spent in 22 different microenvironments. Transition probabilities determine a person's movement from one microenvironment to another. The model assumes that microenvironmental concentrations reflect the contribution of an ambient concentration and a component representing CO sources within each microenvironment. Because SHAPE relies on field surveys of representative populations, the data requirements of the model are fairly extensive. The SHAPE model can estimate the frequency distribution of maximum standardized exposures to CO for an urban population and the cumulative frequency distribution of maximum exposures for both 1-hour and 8-hour periods, thereby allowing estimates of the proportion of the population that is exposed to CO concentrations above the NAAQS. An evaluation of SHAPE by Ott et al. (1988), using survey data from the aforementioned Denver population study, showed that the observed and predicted arithmetic means of the 1-hour and 8-hour maximum average CO exposures were in close agreement; however, SHAPE over-predicted low-level exposures and under-predicted high-level exposures.

Unlike SHAPE, which uses diary data from a probability sample of a population, NEM aggregates people into cohorts. The NEM has evolved over time from deterministic to probabilistic versions. As described elsewhere (Johnson and Paul 1983; Paul and Johnson 1985), the deterministic version of NEM simulates movements of selected groups (cohorts) of an urban population through a set of exposure districts or neighborhoods and through different microenvironments. Cohorts are identified by district of residence and, if applicable, district of employment, as well as by age-occupation group and activity pattern subgroup. The NEM uses empirical adjustment factors for indoor and in-transit microenvironments, and accumulates exposure over 1 year. Although the deterministic NEM was able to estimate central tendencies in total exposure accurately, it did less well estimating the associated uncertainty caused by variation in time spent in various microenvironments (Sexton and Ryan 1988) or variation in microenvironmental concentrations (Akland et al. 1985). Paul, Johnson, and McCurdy (1988) discussed improvements in the deterministic version of NEM.

Subsequently, the USEPA developed the probabilistic NEM for CO (pNEM/CO). The model enables the USEPA to evaluate alternative ambient standards for CO by establishing distributions of personal exposures to CO when the alternative CO standard is met. The USEPA evaluated the predictions of pNEM/CO against observed CO exposure data for subjects of the Denver study. That evaluation concluded that there was relatively close agreement between simulated and observed exposures for CO concentrations near the average exposure, within the range of 6–13 ppm for the 1-hour standard and within 5.5–7 ppm for the 8-hour standard. However, the model over-predicted lower exposures and under-predicted higher exposures for both standards (Law et al. 1997).

More recently, the USEPA developed the Air Pollutants Exposure Model (APEX). Like its predecessors, APEX includes a dosimetry module to estimate the percentage of COHb in the blood of an individual person for a calendar year. Richmond et al. (2003) describe an application of APEX to estimate adult exposure to CO in the Los Angeles, Orange, and San Bernardino counties of California. Fixed-site monitors in the region had recorded ambient CO concentrations in excess of the 1-hour standard of 35 ppm in 2002 (USEPA 2003). Model inputs included ambient air quality data for the metropolitan region in 1997, data on two major indoor sources of CO (i.e., passive smoking and gas stoves), demographic data from the 2000 census, and diary data from the USEPA's Consolidated Human Activity Database (CHAD). The ability of APEX to make accurate estimates of population exposure to CO must still be determined.

6.16 ACTIVITY PATTERNS

Population exposure models require extensive data on human activity patterns. To supply that data, the USEPA authorized and supported the National Human Activity Pattern Survey (NHAPS). This survey collected 24-hour diary data on human activities and their locations from a sample of 9,386 U.S. residents between October 1992 and September 1994 (Klepeis, Tsang, and Behar 1996; Klepeis et al. 2001). To enable projections to the larger U.S. population, the sample was weighted by the 1990 U.S. census data to account for disproportionate sampling of certain population groups defined by age and gender. Results were analyzed across a dozen subgroups defined by various character-istics of respondents: gender, age, race, education, employment, census region, day of week, season, asthma, angina, and bronchitis/emphysema. The weighted results showed that, on average, 86.9% of a person's day was spent indoors (68.7% at residential locations), 7.2% of the day was spent in or near vehicles, and 5.9% of the day was spent in outdoor locations.

The study also reported unweighted descriptive statistics and percentiles for both the full population and various subpopulations (i.e., people who actually did certain activities or who spent time in certain microenvironments) (Tsang and Klepeis 1996). The findings on activities that contribute to elevated CO exposures are of particular relevance. Of all respondents, 38.3% reported having a gas range or oven at home, and 23.7% said that the range or oven had a burning pilot light. When asked about motor vehicle use, 10% of 6,560 people (7.0% of the total sample) spent more than 175 minutes per day inside a car, and 10% of 1,172 people (1.2% of the total sample) spent more than 180 minutes inside a truck or van. Of those who were inside a car and knew they had angina (n = 154 respondents), 10% of them spent more than 162 minutes per day inside a car. The survey also asked about sources of household pollutants. Of 4,723 respondents to this question, 10.5% were exposed to solvents, 10.4% to open flames, and 8.4% to "gasoline-diesel"-powered equipment; 6.3% of these respondents were in a garage or indoor parking lot; and 5.7% reported that someone smoked cigarettes at home. Only 1.8% of 4,663 respondents reported having a kerosene space heater at home.

Workers have adjusted their commuting behavior during the past 25 years in response to growing traffic congestion and social trends. The decennial census collected travel time data for the first time in 1980. Census reports showed that the nation's average commuting time of 21.7 minutes in 1980 increased only 40 seconds to 22.4 minutes in 1990 (Pisarski 1992), but then jumped to 25.5 minutes in 2000 (U.S. Census Bureau 2003). Although the number of workers who commuted 45 minutes or more increased from 10.9 million in 1980 to 13.9 million in 1990, the mean travel time of this commuter cohort actually decreased slightly from 59.6 minutes in 1980 to 58.5 minutes in 1990. One reason for this is that more people were taking their morning commute from home to work during the "shoulder hours" from 6–7 A.M. or from 8–9 A.M. than during the "peak hour" from 7–8 A.M. In 1990, the "shoulder hours" accounted for about 37% of worker trip starts, whereas the "peak hour" accounted for only 32% of trip starts (Pisarski 1992). Flachsbart (1999c) showed that a commuter's travel time and CO exposure for a trip from home to work was related to trip departure time. In his study of a coastal artery in Honolulu, travel during off-peak hours reduced both travel time and CO exposure compared to travel during peak hours.

Increasingly, more people are working, shopping, and entertaining themselves at home to avoid traffic. This trend is possible due to the advent of the "information superhighway" (i.e., broadband, two-way communications facilitated by personal computers and the Internet). The percentage of people working at home increased from 2.3% in 1980 (Pisarski 1992) to 3.3% in 2000 (U.S. Census Bureau 2003). The overall impact of this trend on commuter CO exposures has not been studied.

6.17 PUBLIC POLICIES AFFECTING EXPOSURE TO VEHICLE EMISSIONS

The Clean Air Act (CAA) amendments of 1970, 1977, and 1990 articulated three approaches for regulating motor vehicle emissions. The foremost approach requires companies that sell cars and trucks for the U.S. market to produce vehicles that emit fewer air pollutants, and oil companies to sell fuels that achieve the same goal. The second approach relies on state governments to implement programs that maintain the effectiveness of motor vehicle emission control systems and reduce the use of motor vehicles. The third approach requires that new investments in transportation systems support achievement of the NAAQS (Howitt and Altshuler 1999). This section discusses the effect of the Clean Air Act on CO exposure.

6.17.1 EFFECTS OF MOTOR VEHICLE EMISSION STANDARDS ON UNINTENTIONAL DEATHS ATTRIBUTED TO EXPOSURE

Cobb and Etzel (1991) reported that the unintentional, CO-related, annual death rate per 100,000 people in the United States declined from 0.67 in 1979 to 0.39 in 1988, based on death certificate reports compiled by the National Center for Health Statistics. Motor vehicle exhaust gas accounted for 6,552 deaths, or 56.7% of the total 11,547 unintentional, CO-related deaths that occurred during the 10-year period. The highest death rates occurred among males, blacks, the elderly, and residents of northern states. Monthly variation in death rates indicated a seasonal pattern, with January fatalities routinely about 2–5 times higher than in July. The study speculated that declining death rates could be attributed to industry compliance with motor vehicle CO emission standards of the CAA amendments. Table 6.2 shows data to support the study's conclusion. The investigators argued that tighter CO emission standards enabled cars to emit exhaust into an enclosed space for a longer period of time before CO concentrations could accumulate to toxic levels. The findings of the study by Cobb and Etzel (1991) are noteworthy even though mortality is not a health effect used by USEPA to set the NAAQS for CO.

6.17.2 EFFECTS OF TRANSPORTATION INVESTMENTS ON COMMUTER EXPOSURE

Since the mid-1960s, major construction projects intended to expand highway capacities have been opposed in some metropolitan areas in the United States. Opponents claimed that these projects promoted urban sprawl and induced motor vehicle travel that raised regional air pollutant emissions. To address these concerns, the 1990 CAA amendments stipulated that transportation actions (plans, programs, and projects) cannot create new NAAQS violations, increase the frequency or severity of existing NAAQS violations, nor delay attainment of the NAAQS (U.S. Code 1990). Pursuant thereto, the USEPA promulgated its Transportation Conformity Rule (TCR). Complementary provisions of the 1991 Intermodal Surface Transportation Efficiency Act (ISTEA) offered financial incentives to improve air quality under the Congestion Management and Air Quality (CMAQ) improvement program (Ortolano 1997). Under CMAQ, metropolitan planning organizations (MPOs) were offered federal funds to improve air quality by implementing TCMs. Examples of TCMs include programs to promote car- and van-pooling, flexible working schedules, special lanes for high occupancy vehicles (HOVs), and parking restrictions. USEPA's Office of Mobile Sources (OMS) did an extensive study of how TCMs have changed travel activity, including the number of trips, vehicle miles of travel, vehicle speed, travel time, and the extent to which commuters have shifted travel from peak to off-peak periods. Using an emission factors model (i.e., MOBILE5), OMS estimated how much TCMs would change the average speeds and CO emissions of motor vehicles (USEPA 1994).

By comparison, the direct effect of TCMs on commuter exposure to CO has received only limited study. Flachsbart (1989) hypothesized that priority (with-flow and contra-flow) lanes could be effective in reducing exposure to CO, because these lanes enable commuters to travel at higher speeds than commuters in unrestricted lanes. In theory, priority lanes provide a speed advantage during the line-haul phase of one's trip to compensate for the extra time needed to collect passengers at the origin of the trip. This speed advantage could translate to lower CO exposure for people using priority lanes. To test this hypothesis, Flachsbart recruited several volunteers who used the Kalaniana'ole Highway, a coastal artery in Honolulu. All volunteers commuted from their homes in East Honolulu to the Manoa campus of the University of Hawai'i during the 1981–1982 academic

TABLE 6.2
Motor Vehicle CO Emission Standards, In-Vehicle CO Exposures, and Unintentional CO-Related Annual Death Rates

Year	New Passenger Car CO Emission Standard Federal (g/mile)	California (g/mile)	Net Mean In-Vehicle CO Concentration (ppm)	CO Exposure Study Location	United States CO-Related Annual Death Rates per 10^5 People
Pre-control	84.0	84.0			
~1965	84.0	84.0	12	Los Angeles, CA	
1966	84.0	51.0	17.5	5 U.S. cities	
1968	51.0	51.0			
1970	34.0	34.0			
1972	28.0	34.0			
1973	28.0	34.0	11.5	Los Angeles, CA	
1974	28.0	34.0			
1974–75	15.0	9.0	7.4	Boston, MA	
1975–77	15.0	9.0			
1978	15.0	9.0	10.3	Washington, DC	
1979	15.0	9.0	9.7	Los Angeles, CA	0.67
1980–81	7.0	9.0	8.3	Santa Clara Co., CA	0.55
1981	3.4	7.0	5.2	Denver, CO	0.58
1981	3.4	7.0	4.3	Los Angeles, CA	0.58
1981	3.4	7.0	2.9	Phoenix, AZ	0.58
1981	3.4	7.0	2.9	Stamford, CT	0.58
1981–82	3.4	7.0	9.5	Honolulu, HI	
1982	3.4	7.0			0.56
1982–83	3.4	7.0	1.4	Denver, CO	
1982–83	3.4	7.0	1.8	Washington, DC	
1983	3.4	7.0	9.4	Washington, DC	0.53
1984	3.4	7.0			0.49
1985	3.4	7.0			0.49
1986	3.4	7.0			0.44
1987	3.4	7.0			0.39
1987–88	3.4	7.0	4.9	Los Angeles, CA	
1988	3.4	7.0	8.4	Raleigh, NC	0.39
1989–90	3.4	7.0			
1991–92	3.4	7.0	~3.6	Santa Clara Co., CA	
1992	3.4	7.0	<3.0	New Jersey suburbs of New York City	

Source: From Flachsbart, 1999. With permission from Elsevier.

year. During weekdays a crew from the state Department of Transportation coned off priority lanes on a 4-mile segment of the artery for the exclusive use of express buses, carpools, and vanpools between 6 and 8 A.M. Volunteers using these modes of travel entered designated lanes within 15 minutes of each other to facilitate valid comparisons of their travel times and CO exposures.

The results of this study show the potential of express lanes to reduce commuter exposure to CO. Compared with average CO concentrations inside passengers cars on the highway's unrestricted lanes, average CO concentrations were 61% less for passengers of express buses (n = 28 trips), about 28% less for those using non-bus, high-occupancy vehicles (HOVs) (n = 20 trips), and about 18% less for people in carpools (n = 52 trips). One explanation for these results is that faster vehicles in priority lanes created more air turbulence, which dispersed air pollutants surrounding the vehicle. Of course, users of priority lanes also reduced their total trip exposure to CO, because these lanes were very effective in reducing total travel time. The differences in CO exposure between lanes existed even though the priority lanes were frequently downwind of large numbers of slower vehicles emitting high levels of CO in congested lanes. However, differences in CO exposure by travel lane also could have been caused by differences in vehicle type and ventilation, both of which could not be controlled by study design.

6.17.3 Effects of Motor Vehicle Emission Standards on Commuter Exposure

The percentage of in-use passenger cars in the United States with catalytic and other types of emission controls increased from 50.0% in 1980 to 90.3% by 1989 (MVMA 1990). To study the effect of this trend on the CO exposure of commuters, Flachsbart (1995) reviewed 16 in-vehicle studies performed in various cities in the United States between the mid-1960s and early 1990s. In each study, trips lasted for an hour or less. Table 6.2 summarizes the results of these studies, and Figure 6.5 shows evidence of downward trends in both the mean in-vehicle CO concentrations (top line) and the concurrently measured ambient CO concentrations (bottom line) of these studies. These lines do not imply that CO concentrations can be inferred from points on the lines themselves, or that relationships exist between results for different cities. If one assumes that the results of this "meta-analysis" (i.e., study of studies) are representative of typical CO exposures for commuters in large U.S. cities, then CO exposures fell approximately 90% for commuters over a 27-year period.

In comparing the methodology of the 16 studies, Flachsbart (1995) found that typical in-vehicle CO exposures varied by study approach (direct vs. indirect) and even by researcher. For the more common type of study (i.e., the indirect approach), typical exposures varied by city, season of the year, roadway type and location, travel mode, and the ventilation settings of the test vehicle. The two studies of an arterial highway (El Camino Real) in Northern California by Ott, Switzer, and Willits (1993b) were noteworthy exceptions to this observation, because they adhered to a standardized protocol to facilitate comparisons in exposure over one decade. (Figure 6.5 identifies these two studies as #8 in 1980, and #15 in 1991.) Flachsbart (1995) recommended that future studies use standard methods similar to those used by Ott, Switzer, and Willits (1993b) to "improve their potential to assess the effectiveness of public policies in curtailing automotive emissions over the long term" (Flachsbart 1995, p. 493). Based on that recommendation, Ott resurveyed the study site on El Camino Real in 2001–2002, and recruited Flachsbart to analyze and compare the data from all three studies. The next section discusses the results of that comparison.

6.18 THE EL CAMINO REAL COMMUTER EXPOSURE SURVEYS

The three studies of El Camino Real, which together span the period between 1980 and 2002, show compelling evidence of the effectiveness of California's motor vehicle emission control program. This program includes a set of tailpipe exhaust emission standards, which have affected all new cars sold in the state since 1966, and an inspection and maintenance (I/M) program (known as Smog Check), which was implemented in 1984 to regulate emissions of in-use vehicles. The El Camino Real study differs from many other exposure studies, because it adopted a longitudinal design to observe the cumulative effect of progressively tighter emission standards and the I/M program over two decades. In this study, field surveys of the highway occurred during 15-month periods in 1980–1981, 1991–1992, and 2001–2002. During each period, Wayne Ott measured his personal CO exposure inside an automobile during round-trips on a 5.9-mile segment of the highway

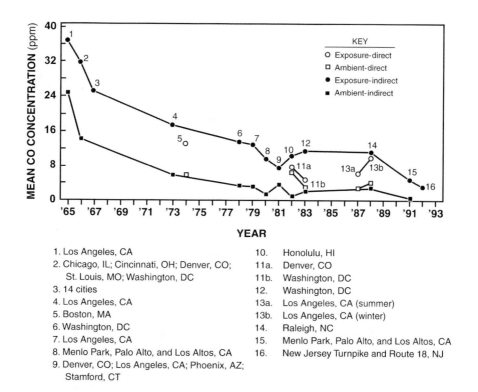

FIGURE 6.5 Trends in ambient CO concentrations and in-vehicle CO exposures in the United States, 1965–1992. (From Flachsbart, 1995. With permission from Macmillan Publ.)

using a standardized protocol to control for confounding variables. Fifty trips from each period — for a total of 150 trips — were matched by date, day of the week, and starting time to facilitate comparisons over two decades (Flachsbart, Ott, and Switzer 2003).

The net CO concentration of each trip was obtained by subtracting the background CO level from the average CO concentration for the entire trip. The mean net CO concentration (1.7 ppm) for 2001–2002 was 35.4% of the corresponding value (4.8 ppm) for 1991–1992 and 17.5% of the value (9.7 ppm) for 1980–1981. Figure 6.6 shows that the median (1.5 ppm) of the observed CO concentrations for the 50 trips in 2001–2002 was only slightly below the predicted range in median values (1.55–1.78 ppm) for 2002–2003. This prediction was based on the roadway emission estimate of the STREET model, as reported by Yu, Hildemann, and Ott (1996).

Flachsbart, Ott, and Switzer (2003) attributed the results of their study to the adoption of progressively tighter tailpipe CO emission standards on new motor vehicles sold in the state since 1966, and to the gradual replacement of older cars with newer models. California's cold-temperature CO standard implemented in 1996 appeared to reduce high CO exposures that were observed during the late fall and winter of 1980–1981. The lack of sharp peaks in CO measurements taken during trips in 2001–2002 indicated fewer "high-emitting" vehicles on the highway relative to comparable measurements taken during previous surveys. This result was attributed to both the I/M program and tougher "durability standards" on emission controls that were phased in on new cars sold in 1993 and 1994.

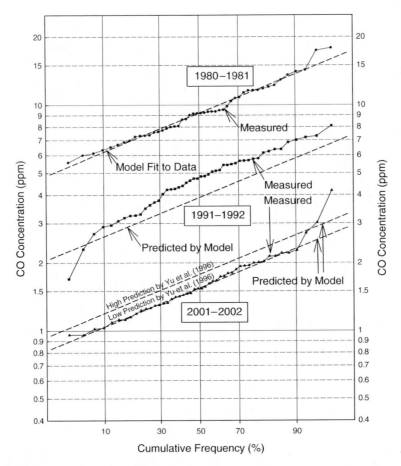

FIGURE 6.6 Logarithmic probability plots of net in-vehicle CO concentrations for three time periods. (From Flachsbart, Ott, and Switzer 2003.)

6.19 INTERNATIONAL COMPARISONS OF COMMUTER EXPOSURE

By comparing studies in different countries, Flachsbart (1999a,b) suggested that commuter exposure to motor vehicle emissions may be an indicator of a country's level of economic development. Similar studies of CO exposure for three modes of travel (i.e., automobile, diesel bus, and rail transit) were performed in Washington, DC (Flachsbart et al. 1987) and in Mexico City (Fernandez-Bremauntz and Ashmore, 1995a,b) about 8 years apart. Each study used the indirect approach to measure commuter CO exposure during peak periods of travel and gathered concurrent ambient CO levels from fixed-site monitors. Table 6.3 reports the net mean CO concentrations (i.e., ambient CO levels subtracted from in-vehicle exposure) by travel mode for each study. Notice that net mean CO concentrations varied significantly, not only by travel mode but also by country. Variation by travel mode could have occurred due to variation in the traffic volume of sampled routes, to variation in the height of the portable monitor above the roadway, and to variation in the vehicle ventilation system of each mode.

The variation in exposure by country can partly be explained by comparing the history of automotive emission standards in each country. The United States initiated nationwide emission standards on new passenger cars in 1968, adopted progressively tighter controls throughout the 1970s (see Table 6.2), and required catalytic converters on all new cars beginning in the 1975

TABLE 6.3
Typical Net Mean CO Concentrations by Travel Mode in Washington, DC, and Mexico City[a,b]

Travel Mode	Washington, DC, U.S. (1983)		Mexico City, Mexico (1991)	
	Net Mean CO Concentrations (ppm)	Averaging Times (minutes)	Net Mean CO Concentrations (ppm)	Averaging Times (minutes)
Automobile	7–12	34–69	37–47	35–63
Diesel bus	2–6	82–115	14–27	40–99
Rail transit	0–3	27–48	9–13	39–59

[a] "Typical" means do not include outlier values that can be attributed to unusual circumstances.

[b] Net mean CO concentration = mean in-vehicle CO concentration – mean ambient CO concentration.

Source: From Flachsbart, 1999. With permission from Elsevier.

model year (Johnson 1988). Mexico adopted a tailpipe CO emission standard of 47.0 grams/mile for the 1975 model year, when the U.S. standard was 15.0 grams/mile, and reached parity with the 1981 U.S. standard of 3.4 grams/mile by the 1993 model year (Faiz, Weaver, and Walsh 1996). The Washington exposure study occurred in early 1983, about 2 years after the 3.4 grams/mile standard took effect in the United States, and the Mexico City exposure study occurred in 1991, 2 years before the same standard took effect in Mexico. Hence, the vehicle fleet in Washington, DC, at the time of its exposure study in 1983 was under tighter emission standards than the fleet in Mexico City at the time of its exposure study in 1991.

6.20 CONCLUSIONS

In accordance with the Clean Air Act, the USEPA must determine whether or not a community complies with National Ambient Air Quality Standards (NAAQS). This determination is based on measurements of ambient air quality made by a nationwide network of fixed-site monitoring stations. Using personal exposure monitors, early studies revealed that measurements at fixed-site monitors were not sufficient to estimate the total CO exposure of urban populations in U.S. cities. In response to these findings, the USEPA's Office of Air Quality Planning and Standards (OAQPS) developed a risk-analysis framework to enable more formal treatment of the uncertainties associated with scientific estimates of population exposure. This framework benefited from progress made in the science and technology of personal exposure monitoring devices and in the development of new methods for measuring exposure as people performed their daily activities.

Over the years, the USEPA also supported the development of several large-scale, population exposure models (e.g., SHAPE, NEM, pNEM/CO, and APEX) at considerable expense to assist the agency in its periodic review and evaluation of the NAAQS for CO. Each review required an accurate estimate and assessment of population exposure to CO. To provide data for these models, the USEPA supported several direct and indirect studies of population exposure to CO at considerable expense. These studies included the landmark Denver–Washington population exposure studies of 1982–1983. Compared to crude methods, these models improved the accuracy of the

USEPA's estimates of population exposure. Even so, the NAAQS for CO have not changed since their promulgation in April 1971. Future adjustments to the NAAQS for CO may depend on more refined estimates of exposure in sensitive populations.

It is not likely that we shall see another large-scale survey in the United States of urban population exposure to CO. The surveys of population exposure in Denver and Washington were undertaken when ambient CO levels were higher than they are today and the ability of fixed-site CO monitors to represent population exposure was in question. Noting the steady decline in ambient CO concentrations in many cities nationwide, the National Research Council declared in 2003 that the control of CO has been a success and that air quality management in the United States has shifted to other air pollutants such as tropospheric ozone and fine particulate matter. In fact, CO exposure has not been the sole focus of most commuter exposure studies that have been performed in the United States since the mid-1980s. The trend has been to measure exposure to CO, together with exposures to air toxics (e.g., benzene) and particles. Given these trends, the purpose and methods of CO exposure studies will likely change in the future. At best, we may see a few indirect studies of exposure in the United States that build on existing knowledge of the sources of CO emissions and human activity patterns, and take advantage of the portability of personal monitors. These studies could discover new microenvironments with elevated CO concentrations or revisit known microenvironments (e.g., buildings attached to parking garages, indoor sport arenas) to establish trends in exposure.

Over the years there has been a series of unrelated commuter exposure studies in the United States undertaken by many researchers using different methodologies. Viewed collectively, these studies suggest that substantial reductions in commuter CO exposure can be attributed to the Clean Air Act amendments of 1970, 1977, and 1990. The strongest evidence of this trend comes from a long-term study of commuter exposure on an arterial highway (El Camino Real) in northern California. The study quantified a significant reduction in commuter CO exposure over two decades, which could be attributed to California's motor vehicle emission control program. This implies that all past measurements of commuter CO exposure (i.e., those made prior to 2000) may no longer be indicative of current CO exposures. It also implies that large-scale models (e.g., the new APEX model) of urban population exposure are trying to provide estimates of a moving target.

Studies of exposure are particularly applicable to mega-cities in developing countries that have rapidly growing motor vehicle populations, congested streets and confined spaces in urban areas, and nascent motor vehicle emission control programs. Studies of population exposure would enable these countries to determine what fraction of total exposure to CO and other mobile source pollutants can be attributed to commuting activities. Microenvironmental studies of commuter exposure would enable them to evaluate the effectiveness of emission control programs and transportation measures designed to reduce exposure to mobile source emissions. These countries should first establish baseline information on commuter exposure, as the United States did in the mid-1960s before implementing tailpipe emission standards for new passenger cars in 1968. Periodically, exposure should then be remeasured using a standard data collection protocol to assess progress and facilitate comparisons over time. Ideally, the measurement protocol should take advantage of modern personal monitors, which are capable of measuring CO and other vehicle-related air pollutants continuously over time in the field, storing these measurements automatically using portable data loggers, and transferring data directly to computer files for analysis. Moreover, the protocol should represent typical travel patterns, recognize that exposure can vary over time and space, and adhere to good principles of research design.

6.21 ACKNOWLEDGMENTS

Several people assisted the author in preparing this chapter. Lance Wallace and Doug Eisinger provided thoughtful advice and comments on a previous draft. Wayne Ott and Michael Apte rendered invaluable assistance in preparing figures. The author expresses his gratitude to each of them but assumes responsibility for any factual errors that may still remain.

Portions of text of this chapter are reprinted, with permission, from: *Chemosphere—Global Change Science,* Vol. 1, M.A.K. Khalil, J.P. Pinto, and M.J. Shearer, Eds., P.G. Flachsbart, "Human Exposure to Carbon Monoxide from Mobile Sources," pages 301–329, copyright (1999b), published by Elsevier; *Urban Traffic Pollution*, D. Schwela and O. Zali, Eds., P.G. Flachsbart, "Chapter 4: Exposure to Exhaust and Evaporative Emissions from Motor Vehicles," pages 89–132, copyright (1999a), published by the World Health Organization; and *Journal of Exposure Analysis and Environmental Epidemiology,* **5**(4), Peter G. Flachsbart, "Long-Term Trends in United States Highway Emissions, Ambient Concentrations, and In-Vehicle Exposure to Carbon Monoxide in Traffic," pages 473–495, copyright (1995), published by Macmillan Publishers Ltd.

6.22 QUESTIONS FOR REVIEW

1. What are the principal sources of CO emissions? Under what conditions or circumstances do these motor vehicles emit higher amounts of CO?
2. What constitutes exposure to CO for human beings? What groups of people are at greater risk from exposure to ambient CO concentrations in urban areas?
3. What are the NAAQS for CO? How does the USEPA determine whether a community is in compliance with the NAAQS? What carboxyhemoglobin (COHb) level in the blood of the general population are the NAAQS for CO designed to prevent?
4. What assumptions are necessary to estimate population exposure based on fixed-site monitors? Explain how CO exposure studies have challenged the validity of some of these assumptions.
5. If ambient CO concentrations for a community comply with the NAAQS, give several reasons to study CO exposure in that community.
6. How do direct studies differ from indirect studies of CO exposure? Give examples of microenvironmental studies. What are the pros and cons of each type of study?
7. What factors contribute to elevated CO exposures as a person performs his or her daily activities? What types of microenvironments, travel modes, and daily activities increase CO exposure?
8. Describe gaps in the ability of the Clean Air Act as amended to protect public health based on your knowledge of CO exposure studies.
9. What is net CO exposure? Why is it necessary to compute net exposures in making comparisons of exposure over time and space?
10. Why study trends in CO exposure over time? What can a trend study of CO exposure reveal about motor vehicle emission control programs?
11. Explain why commuter CO exposure varies among developed and developing countries. Assess the potential for a commuter CO exposure study in a mega-city of a developing country.

REFERENCES

Akland, G.G., Hartwell, T.D., Johnson, T.R., and Whitmore, R.W. (1985) Monitoring Human Exposure to Carbon Monoxide in Washington, D.C., and Denver, Colorado, during the Winter of 1982–1983, *Environmental Science and Technology,* **19**(10): 911–918.

Amiro, A. (1969) CO Presents Public Health Problems, *Journal of Environmental Health,* **32**: 83–88.

Apte, M.G. (1997) A Population-Based Exposure Assessment Methodology for Carbon Monoxide: Development of a Carbon Monoxide Passive Sampler and Occupational Dosimeter, Ph.D. diss., Department of Environmental Health Sciences, University of California, Berkeley, CA.

Arias, E., Anderson, R.N., Kung, H.-C., Murphy, S.L., and Kochanek, K.D. (2003) Deaths: Final Data for 2001, *National Vital Statistics Reports*, Center for Health Statistics, Hyattsville, MD, **52**(3).

ATSDR (1993) Methylene Chloride Toxicity, Agency for Toxic Substances and Disease Registry, *American Family Physician*, **47**(5): 1159–1166.

Besner, D. and Atkins, P. (1970) The Dispersion of Lead and Carbon Monoxide from a Heavily Travelled Expressway, presented at the 63rd Annual Meeting of the Air Pollution Control Association, St. Louis, MO.

Bevan, M.A.J., Proctor, C.J., Baker-Rogers, J., and Warren, N.D. (1991) Exposure to Carbon Monoxide, Respirable Suspended Particulates, and Volatile Organic Compounds while Commuting by Bicycle, *Environmental Science & Technology*, **25**(4): 788–791.

Blumenthal, D. (2005) The Use of Real-Time Air Quality Data in Daily Forecasting and Decision-Making, *EM: The Magazine for Environmental Managers,* September: 18.

Boudreau, D.R., Spadafora, M.P., Wolf, L.R., and Siegel, E. (1994) Carbon Monoxide Levels during Indoor Sporting Events — Cincinnati, 1992–1993, [reprint] from *MMWR* 43(2): 21–23, *JAMA, Journal of the American Medical Association*, **271**(6): 419.

Brice, R.M. and Roesler, J.F. (1966) The Exposure to Carbon Monoxide of Occupants of Vehicles Moving in Heavy Traffic, *Journal of the Air Pollution Control Association,* **16**(11): 597–600.

Burr, M.L. (2000) Combustion Products, in *Indoor Air Quality Handbook*, Spengler, J.D., Samet, J.M., and McCarthy, J.F., Eds., McGraw-Hill, New York, NY, 29.3–29.25.

CEQ (1980) *Environmental Quality — 1980: The Eleventh Annual Report of the Council on Environmental Quality*, Council on Environmental Quality, U.S. Government Printing Office, Washington, DC.

Chapin, F.S. Jr. (1974) *Human Activity Patterns in the City*, Wiley-Interscience, New York, NY.

Claggett, M., Shrock, J., and Noll, K. (1981) Carbon Monoxide near an Urban Intersection, *Atmospheric Environment*, **15**(9): 1633–1642.

Cobb, N. and Etzel, R.A. (1991) Unintentional Carbon Monoxide-Related Deaths in the United States, 1979 through 1988, *JAMA: Journal of the American Medical Association*, **266**(5): 659–663.

Coburn, R.F., Forster, R.E., and Kane, P.B. (1965) Considerations of the Physiological Variables That Determine the Blood Carboxyhemoglobin Concentration in Man, *Journal of Clinical Investigation,* **44**: 1899–1910.

Colome, S.D., Wilson, A.L., and Tian, Y. (1994) *California Residential Indoor Air Quality Study, Volume 2: Carbon Monoxide and Air Exchange Rate, an Univariate and Multivariate Analysis*, GRI-93/0224.3, Gas Research Institute, Chicago, IL.

Colucci, J. and Begeman, C. (1969) Carbon Monoxide in Detroit, New York, and Los Angeles Air, *Environmental Science and Technology,* **3**(1): 41–47.

Colvile, R.N., Hutchinson, E.J., Mindell, J.S., and Warren, R.F. (2001) The Transport Sector as a Source of Air Pollution, *Atmospheric Environment,* **35**(9): 1537–1565.

Cortese, A.D. and Spengler, J.D. (1976) Ability of Fixed Monitoring Stations to Represent Personal Carbon Monoxide Exposure, *Journal of the Air Pollution Control Association,* **26**(12): 1144–1150.

Downs, A. (2004) *Still Stuck in Traffic: Coping with Peak-Hour Traffic Congestion*, Brookings Institution Press, Washington, DC.

Doyle, J. (2000) *Taken for a Ride: Detroit's Big Three and the Politics of Pollution*, Four Walls Eight Windows, New York, NY.

Duan, N. (1982) Models for Human Exposure to Air Pollution, *Environment International*, **8**: 305–309.

Ernst, A. and Zibrak, J.D. (1998) Carbon Monoxide Poisoning, *New England Journal of Medicine*, **339**(22): 1603–1608.

Faiz, A., Weaver, C.S., and Walsh, M.P. (1996) *Air Pollution from Motor Vehicles: Standards and Technologies for Controlling Emissions*, The World Bank, Washington, DC.

Fernandez-Bremauntz, A.A. and Ashmore, M.R. (1995a) Exposure of Commuters to Carbon Monoxide in Mexico City, I. Measurement of In-Vehicle Concentrations, *Atmospheric Environment*, **29**(4): 525–532.

Fernandez-Bremauntz, A.A. and Ashmore, M.R. (1995b) Exposure of Commuters to Carbon Monoxide in Mexico City, II. Comparison of In-Vehicle and Fixed-Site Concentrations, *Journal of Exposure Analysis and Environmental Epidemiology*, **5**(4): 497–510.

Flachsbart, P.G. (1989) Effectiveness of Priority Lanes in Reducing Travel Time and Carbon Monoxide Exposure, *Institute of Transportation Engineers Journal*, **59**(1): 41–45.

Flachsbart, P.G. (1995) Long-Term Trends in United States Highway Emissions, Ambient Concentrations, and In-Vehicle Exposure to Carbon Monoxide in Traffic, *Journal of Exposure Analysis and Environmental Epidemiology*, **5**(4): 473–495.

Flachsbart, P.G. (1999a) Exposure to Exhaust and Evaporative Emissions from Motor Vehicles, in *Urban Traffic Pollution*, Schwela, D. and Zali, O., Eds., World Health Organization, Geneva, Switzerland, 89–132.

Flachsbart, P.G. (1999b) Human Exposure to Carbon Monoxide from Mobile Sources, *Chemosphere: Global Change Science*, **1**: 301–329.

Flachsbart, P.G. (1999c) Models of Exposure to Carbon Monoxide inside a Vehicle on a Honolulu Highway, *Journal of Exposure Analysis and Environmental Epidemiology*, **9**(3): 245–260.

Flachsbart, P.G. and Brown, D.E. (1985) *Surveys of Personal Exposure to Vehicle Exhaust in Honolulu Microenvironments*, Department of Urban and Regional Planning, University of Hawai'i at Manoa, Honolulu, HI.

Flachsbart, P.G. and Brown, D.E. (1986) A Seasonal Study of Personal Exposure to CO in Indoor and Outdoor Microenvironments of Honolulu, presented at the 79th Annual Meeting of the Air Pollution Control Association, Minneapolis, MN.

Flachsbart, P.G. and Brown, D.E. (1989) Employee Exposure to Motor Vehicle Exhaust at a Honolulu Shopping Center, *The Journal of Architectural and Planning Research*, **6**(1): 19–33.

Flachsbart, P.G. and Ott, W.R. (1984) *Field Surveys of Carbon Monoxide in Commercial Settings Using Personal Exposure Monitors,* Report No. EPA 600/4-84-019, Office of Monitoring Systems and Quality Assurance, U.S. Environmental Protection Agency, Washington, DC.

Flachsbart, P.G. and Ott, W.R. (1986) A Rapid Method for Surveying CO Concentrations in High-Rise Buildings, *Environment International,* **12**: 255–264.

Flachsbart, P.G., Mack, G.A, Howes, J.E., and Rodes, C.E. (1987) Carbon Monoxide Exposures of Washington Commuters, *Journal of the Air Pollution Control Association*, **37**(2): 135–142.

Flachsbart, P., Ott, W., and Switzer, P. (2003) Long-Term Trends in Exposure to Carbon Monoxide on a California Arterial Highway, presented at the 96th Annual Meeting and Exhibition of the Air and Waste Management Association, San Diego, CA.

Fugas, M. (1975) Assessment of Total Exposure to an Air Pollutant, presented at the International Symposium on Environmental Sensing and Assessment, Las Vegas, NV.

Godish, T. (2004) *Air Quality*, 4th ed., CRC Press, Boca Raton, FL.

Grad, F.P., Rosenthal, A.J., Ruckett, L.R., Fay, J.A., Heywood, J., Kain, J.F., Ingram, G.K., Harrison, D. Jr., and Tietenberg, T. (1975) *The Automobile and the Regulation of Its Impact on the Environment*, University of Oklahoma Press, Norman, OK.

Haagen-Smit, A.J. (1966) Carbon Monoxide Levels in City Driving, *Archives of Environmental Health*, **12**(5): 548–551.

Hampson, N.B. and Norkool, D.M. (1992) Carbon Monoxide Poisoning in Children Riding in the Back of Pickup Trucks, *JAMA: Journal of the American Medical Association,* **267**: 538–540.

Howitt, A.M. and Altshuler, A. (1999) The Politics of Controlling Auto Air Pollution, in *Essays in Transportation Economics and Policy: A Handbook in Honor of John R. Meyer*, Gomez-Ibanez, J., Tye, W.B., and Winston, C., Eds., Brookings Institution Press, Washington, DC, 223–255.

Jain, K.K. (1990) *Carbon Monoxide Poisoning*, Warren H. Green, Inc., St. Louis, MO.

Johnson, J.H. (1988) Automotive Emissions, in *Air Pollution, the Automobile, and Public Health*, Watson, A.Y., Bates, R.R., and Kennedy, D., Eds., National Academy Press, Washington, DC., 39–75.

Johnson, T. (1984) *A Study of Personal Exposure to Carbon Monoxide in Denver, Colorado*, Report No. EPA-600/4/84-014, U.S. Environmental Protection Agency, Environmental Monitoring Systems Laboratory, Research Triangle Park, NC.

Johnson, T. and Paul, R.A. (1983) *The NAAQS Exposure Model (NEM) Applied to Carbon Monoxide,* Report No. EPA 450/5-83-003, Office of Air Quality Planning and Standards, U.S. Environmental Protection Agency, Research Triangle Park, NC.

Jordan, B.C., Richmond, H.M., and McCurdy, T. (1983) The Use of Scientific Information in Setting Ambient Air Standards, *Environmental Health Perspectives*, **52**: 233–240.

Klepeis, N.E., Tsang, A.M., and Behar, J.V. (1996) *Analysis of the National Human Activity Pattern Survey (NHAPS) Respondents from a Standpoint of Exposure Assessment,* Report No. EPA 600/R-96/074, Office of Research and Development, U.S. Environmental Protection Agency, Washington, DC.

Klepeis, N.E., Nelson, W.C., Ott, W.R., Robinson, J.P., Tsang, A.M., Switzer, P., Behar, J.V., Hern, S.C., and Engelmann, W.H. (2001) The National Human Activity Pattern Survey (NHAPS): A Resource for Assessing Exposure to Environmental Pollutants, *Journal of Exposure Analysis and Environmental Epidemiology,* **11**(3): 231–252.

Law, P.L, Lioy, P.J., Zelenka, M.P., Huber, A.H., and McCurdy, T.R. (1997) Evaluation of a Probabilistic Exposure Model Applied to Carbon Monoxide (pNEM/CO) using Denver Personal Exposure Monitoring Data, *Journal of the Air and Waste Management Association,* **47**(4): 491–500.

Lee, K., Yanagisawa, Y., Spengler, J.D., and Nakai, S. (1994) Carbon Monoxide and Nitrogen Dioxide Exposures in Indoor Ice Skating Rinks, *Journal of Sport Science,* **12**: 279–283.

Lynn, D.A., Ott, W.R., Tabor, E.C., and Smith, R. (1967) Present and Future Commuter Exposures to Carbon Monoxide, presented at the 60th Annual Meeting of the Air Pollution Control Association, Cleveland, OH.

Mage, D. (1991) A Comparison of the Direct and Indirect Methods of Human Exposure, in *New Horizons in Biological Dosimetry: Proceedings of the International Symposium on Trends in Biological Dosimetry,* Gledhill, B.L. and Mauro, F., Eds., Wiley-Liss, New York, NY, 443–454.

MVMA (1990) *MVMA Motor Vehicle Facts & Figures '90,* Motor Vehicle Manufacturers Association of the United States, Detroit, MI.

NAPCA (1970) *Air Quality Criteria for Carbon Monoxide,* National Air Pollution Control Administration, Public Health Service, Environmental Service, Report No. AP-62, U.S. Department of Health, Education, and Welfare, Washington, DC.

NIOSH (1996) *Preventing Carbon Monoxide Poisoning from Small Gasoline-Powered Engines and Tools [NIOSH Alert],* Report No. 96-118, Public Health Service, National Institute for Occupational Safety and Health, U.S. Department of Health and Human Services, Cincinnati, OH.

NRC (2000) *Modeling Mobile-Source Emissions,* National Research Council, National Academies Press, Washington, DC.

NRC (2003) *Managing Carbon Monoxide Pollution in Meteorological and Topographical Problem Areas,* National Research Council, National Academies Press, Washington, DC.

Ortolano, L. (1984) *Environmental Planning and Decision Making,* John Wiley & Sons, New York, NY.

Ortolano, L. (1997) *Environmental Regulation and Impact Assessment,* John Wiley & Sons, New York, NY.

Ott, W.R. (1982) Concepts of Human Exposure to Air Pollution, *Environment International,* **7**: 179–196.

Ott, W.R. (1983–1984) Exposure Estimates Based on Computer Generated Activity Patterns, *Journal of Toxicology — Clinical Toxicology,* **21**(1&2): 97–128.

Ott, W.R. (1985) Total Human Exposure: An Emerging Science Focuses on Humans as Receptors of Environmental Pollution, *Environmental Science and Technology,* **19**(10): 880–886.

Ott, W.R. and Eliassen, R. (1973) A Survey Technique for Determining the Representativeness of Urban Air Monitoring Stations with Respect to Carbon Monoxide, *Journal of the Air Pollution Control Association,* **23**(8): 685–690.

Ott, W.R. and Flachsbart, P.G. (1982) Measurement of Carbon Monoxide Concentrations in Indoor and Outdoor Locations Using Personal Exposure Monitors, *Environment International,* **8**: 295–304.

Ott, W.R., Mage, D.T., and Thomas, J. (1992) Comparison of Microenvironmental CO Concentrations in Two Cities for Human Exposure Modeling, *Journal of Exposure Analysis and Environmental Epidemiology,* **2**(2): 249–267.

Ott, W.R., Rodes, C.E., Drago, R.J., Williams, C., and Burmann, F.J. (1986) Automated Data-Logging Personal Exposure Monitors for Carbon Monoxide, *Journal of the Air Pollution Control Association,* **36**(8): 883–887.

Ott, W.R., Switzer, P., and Willits, N. (1993a) *Carbon Monoxide Exposures inside an Automobile Traveling on an Urban Arterial Highway: Final Report on the 1980-81 Field Study,* SIMS Technical Report No. 150, Department of Statistics, Stanford University, Stanford, CA.

Ott, W.R., Switzer, P., and Willits, N. (1993b) Trends of In-Vehicle CO Exposures on a California Arterial Highway over One Decade, presented at the 86th Annual Meeting and Exhibition of the Air and Waste Management Association, Denver, CO.

Ott, W.R., Thomas, J., Mage, D.T., and Wallace, L. (1988) Validation of the Simulation of Human Activity and Pollutant Exposure (SHAPE) Model Using Paired Days from the Denver, CO, Carbon Monoxide Field Study, *Atmospheric Environment,* **22**(10): 2101–2113.

Padgett, J. and Richmond, H. (1983) The Process of Establishing and Revising National Ambient Air Quality Standards, *Journal of the Air Pollution Control Association*, **33**(1): 13–16.

Paul, R.A. and Johnson, T. (1985) *The NAAQS Exposure Model (NEM) Applied to Carbon Monoxide: Addendum,* Report No. EPA 450/5-85-004, Office of Air Quality Planning and Standards, U.S. Environmental Protection Agency, Research Triangle Park, NC.

Paul, R.A., Johnson, T., and McCurdy, T. (1988) Advancements in Estimating Urban Population Exposure, presented at the 81st Annual Meeting of the Air Pollution Control Association, Dallas, TX.

Pedersen, D.H. and Sieber, W.K. (1988) *National Occupational Exposure Survey. Volume III: Analysis of Management Interview Responses,* Publication No. 89-103, National Institute of Occupational Safety and Health, U.S. Department of Health and Human Services, Cincinnati, OH.

Penney, D.G., Ed. (1996) *Carbon Monoxide*, CRC Press, Boca Raton, FL.

Pisarski, A.E. (1992) *New Perspectives in Commuting Based on Early Data from the 1990 Decennial Census and the 1990 Nationwide Personal Transportation Study,* Report No. FHWA-PL/92-026, Federal Highway Administration, U.S. Department of Transportation, Washington, DC.

Radford, E.P. and Drizd, T.A. (1982) *Blood Carbon Monoxide Levels in Persons 3–74 Years of Age: United States, 1976-80,* PHS 82-1250, Public Health Service, National Center for Health Statistics, U.S. Department of Health and Human Services, Hyattsville, MD.

Ramsey, J. (1966) Concentrations of Carbon Monoxide at Traffic Intersections in Dayton, Ohio, *Archives of Environmental Health,* **13**(1): 44–46.

Richmond, H., Langstaff, J., Johnson, T., Capel, J., Rosenbaum, A., and McCurdy, T. (2003) EPA's Air Pollutants Exposure Model (APEX): Estimation of Carbon Monoxide Population Exposure in Los Angeles — A Case Study, presented at the 13th Annual Conference of the International Society of Exposure Analysis, Stresa on Lago Maggiore, Italy.

Sexton, K. and Ryan, P.B. (1988) Assessment of Human Exposure to Air Pollution: Methods, Measurements, and Models, in *Air Pollution, the Automobile, and Public Health*, Watson, A.Y., Bates, R.R., and Kennedy, D., Eds., National Academy Press, Washington, DC, 207–238.

Shikiya, D.C., Liu, C.S., Kahn, M.I., Juarros, J., and Barcikowski, W. (1989) *In-Vehicle Air Toxics Characterization Study in the South Coast Air Basin,* Office of Planning and Rules, South Coast Air Quality Management District, El Monte, CA.

Spengler, J.D., Stone, K.R., and Lilley, F.W. (1978) High Carbon Monoxide Levels Measured in Enclosed Skating Rinks, *Journal of the Air Pollution Control Association*, **28**(8): 776–779.

Szalai, A., Ed. (1972) *The Use of Time: Daily Activities of Urban and Suburban Populations in Twelve Countries*, Mouton and Co., The Hague, The Netherlands.

TRB (1995) *Expanding Metropolitan Highways: Implications for Air Quality and Energy Use*, Transportation Research Board, National Academy Press, Washington, DC.

Tsang, A.M. and Klepeis, N.E. (1996) *Descriptive Statistics Tables from a Detailed Analysis of the National Human Activity Pattern Survey (NHAPS) Data,* Report No. EPA 600/R-96/148, Office of Research and Development, U.S. Environmental Protection Agency, Washington, DC.

U.S. Census Bureau (2003) *Statistical Abstract of the United States: 2003*, U.S. Department of Commerce, Washington, DC.

U.S. Code (1990) Clean Air Act, as amended by PL 101-549, November 15, 1990. U. S. C. §§42 7401-7671q.

USEPA (1991) *Air Quality Criteria for Carbon Monoxide*, Report No. 600/8-90/045F, Office of Health and Environmental Assessment, Environmental Criteria and Assessment Office, U.S. Environmental Protection Agency, Research Triangle Park, NC.

USEPA (1992) *Review of the National Ambient Air Quality Standards for Carbon Monoxide: 1992 Reassessment of Scientific and Technical Information*, Report No. EPA 452/R-92-004, Office of Air Quality Planning and Standards, U.S. Environmental Protection Agency, Research Triangle Park, NC.

USEPA (1994) *Methodologies for Estimating Emission and Travel Activity Effects of TCMs*, Report No. EPA 420-R-94-002, Office of Mobile Sources, U.S. Environmental Protection Agency, Washington, DC.

USEPA (2000) *Air Quality Criteria for Carbon Monoxide,* Report No. EPA 600/P-99/001F, Office of Research and Development, National Center for Environmental Assessment, U.S. Environmental Protection Agency, Washington, D.C, http://www.epa.gov/NCEA/pdfs/coaqcd.pdf (accessed March 22, 2004).

USEPA (2003) *National Air Quality and Emissions Trends Report, 2003 Special Studies Edition*, Report No. EPA 454/R-03-005, Office of Air Quality Planning and Standards, U.S. Environmental Protection Agency, Research Triangle Park, NC, http://www.epa.gov/oar/aqtrnd03 (accessed May 24, 2004).

Van Wijnen, J.H., Verhoeff, A.P., Jans, H.W.A., and Van Bruggen, M. (1995) The Exposure of Cyclists, Car Drivers and Pedestrians to Traffic-Related Air Pollutants, *International Archives of Occupational and Environmental Health*, **67**: 187–193.

Walsh, M. (1996) EPA Bans Leaded Gasoline, *Car Lines*, **96**(2): 17–18.

Wilcosky, T.C. and Simonsen, N.R. (1991) Solvent Exposure and Cardiovascular Disease, *American Journal of Industrial Medicine*, **19**: 569–586.

Wilson, A.L., Colome, S.D., and Tian, Y. (1993a) *California Residential Indoor Air Quality Study, Volume 1: Methodology and Descriptive Statistics,* GRI-93/0224.1, Gas Research Institute, Chicago, IL.

Wilson, A.L., Colome, S.D., and Tian, Y. (1993b) *California Residential Indoor Air Quality Study, Volume 1: Methodology and Descriptive Statistics, Appendices,* GRI-93/0224.2, Gas Research Institute, Chicago, IL.

Yocom, J.E., Clink, W.L., and Cote, W.A. (1971) Indoor/Outdoor Air Quality Relationships, *Journal of the Air Pollution Control Association*, **21**(5): 251–259.

Yu, L., Hildemann, L., and Ott, W. (1996) A Mathematical Model for Predicting Trends in Carbon Monoxide Emissions and Exposures on Urban Arterial Highways, *Journal of the Air and Waste Management Association*, **46**(5): 430–440.

Ziskind, R.A., Fite, K., and Mage, D.T. (1982) Pilot Field Study: Carbon Monoxide Exposure Monitoring in the General Population, *Environment International*, **8**: 283–293.

Ziskind, R.A., Rogozen, M., Carlin, T., and Drago, R. (1981) Carbon Monoxide Intrusion into Sustained-Use Vehicles, *Environment International*, **5**: 109–123.

7 Exposure to Volatile Organic Compounds

Lance A. Wallace
U.S. Environmental Protection Agency (ret.)

Sydney M. Gordon
Battelle Memorial Institute

CONTENTS

7.1   Synopsis 147
7.2   Introduction 148
7.3   Human Exposure 150
    7.3.1   Air 150
        7.3.1.1   Benzene 154
        7.3.1.2   *para*-Dichlorobenzene 155
        7.3.1.3   Tetrachloroethylene 156
        7.3.1.4   Carbon Tetrachloride 157
        7.3.1.5   Formaldehyde 157
        7.3.1.6   1,3-Butadiene 157
    7.3.2   Drinking Water 158
7.4   Discussion 158
7.5   Conclusion 159
7.6   Appendix — Measurement Methods 160
    7.6.1   Air 160
        7.6.1.1   Air Sampling 160
        7.6.1.2   Analysis 163
        7.6.1.3   Real-Time (Simultaneous Sampling and Analysis) Techniques 163
    7.6.2   Body Fluids 165
        7.6.2.1   Breath 165
        7.6.2.2   Blood 166
        7.6.2.3   Urine 167
7.7   Questions for Review 167
References 168

7.1 SYNOPSIS

Volatile organic compounds (VOCs) surround us at all hours of the day. Each breath we take contains some hundreds of these compounds, many of which our bodies must metabolize or excrete to remain healthy. Both outdoor and indoor sources contribute to our exposure, but for a large percentage of these compounds, it is a small indoor or personal source right under our nose that is the largest contributor. These facts were first demonstrated on a large national scale by the Total Exposure Assessment Methodology (TEAM) studies of the early 1980s, and this historical effort is described below. The TEAM studies benefited from the availability of new sorbents, stronger batteries, and miniaturized pumps to allow small personal monitors to measure 12-hour exposures to a target list of some 32 VOCs. Because of the extraordinarily small concentrations of most VOCs, methods of sampling and analysis have had to meet extraordinary demands for sensitivity and stability. Progress in developing such methods is continuing today, and this chapter provides a very thorough coverage of these methods.

VOCs generally "prefer" to be in the air, but can also be found in drinking water and in our bodies (breath, blood, urine, and fat cells, particularly breast milk). A large number of studies of VOC concentrations outdoors, indoors, and in these biological media are discussed. Major sources of VOCs (automobile exhaust, secondhand smoke, building materials, consumer products, chlorinated water, air fresheners) are identified and several individual VOCs with the greatest cancer risk are characterized in full: benzene, chloroform, *para*-dichlorobenzene, formaldehyde, tetrachloroethylene, and carbon tetrachloride. Less frightening but possibly causing far more economic harm is the apparent involvement of VOCs in reducing worker productivity through eye, nose, and throat irritation, headaches, and other apparently "minor" but debilitating symptoms. Sometimes these symptoms escalate into full-blown syndromes such as sick building syndrome (SBS) or multiple chemical sensitivity (MCS), although the involvement of VOCs has not been proven due to the lack of adequate double-blind studies.

7.2 INTRODUCTION

Volatile organic compounds (VOCs) comprise some thousands of chemicals, many of which are in wide use as paints, adhesives, solvents, fragrances, and other ingredients in processes and consumer products (Table 7.1). VOCs have great economic importance. Many chemicals with the highest annual production figures are VOCs. They also sometimes occur as *unwanted* ingredients or impurities — for example, benzene in gasoline, or formaldehyde in pressed wood products.

As their name implies, VOCs are so volatile that under normal conditions they are found overwhelmingly in the gaseous state. They may occur in liquids, often volatilizing from those liquids when given the chance (for example, chloroform from treated water, methyl tert-butyl [MTBE] from groundwater). They may also be in the form of solids (e.g., naphthalene and *para*-dichlorobenzene, used as mothballs and bathroom deodorants) that *sublime* (go from solid to gas without an intervening liquid stage) at room temperature.

Human exposure to most VOCs is mainly through inhalation; a small number of VOCs are in drinking water as contaminants. Some VOCs may travel in groundwater or through soil from hazardous waste sites, landfills, or gasoline spills to inhabited areas.

Two main health effects are of interest: cancer and acute irritative effects (eye, nose, throat, and skin irritation, headaches, difficulty concentrating, etc.). The latter may have a greater economic effect than the former because of reduced productivity of workers (Fisk and Rosenfeld 1997).

Some VOCs are considered to be human carcinogens (benzene, vinyl chloride, formaldehyde). Others are known animal carcinogens and may be human carcinogens (methylene chloride, trichloroethylene, tetrachloroethylene, chloroform, *p*-dichlorobenzene, 1,3-butadiene). Others are mutagens (α–pinene) or weak animal carcinogens (limonene).

Many common VOCs have well-documented health effects, often neurobehavioral, at high (occupational) concentrations. A recent study of benzene exposures in the shoemaking industry in China showed clear reductions in white blood cells even for the lowest exposures (about 0.57 ppm), well under the U.S. occupational standard of 1 ppm (Lan, Zhang, and Li 2004). Acute effects at lower environmental concentrations are often difficult to observe under controlled conditions, although Mølhave and coworkers (Mølhave and Møller 1979; Mølhave 1982, 1986; Mølhave, Bach, and Pedersen 1984, 1986) were able to observe some subjective effects such as reported headache

TABLE 7.1
Common Volatile Organic Chemicals and Their Sources

Chemicals	Major Sources of Exposure
Acetone	Cosmetics
Alcohols (ethanol, isopropanol)	Spirits, cleansers
Aromatic hydrocarbons (toluene, xylenes, ethylbenzene, trimethylbenzenes)	Paints, adhesives, gasoline, combustion sources
Aliphatic hydrocarbons (octane, decane, undecane)	Paints, adhesives, gasoline, combustion sources
Benzene	Smoking, auto exhaust, passive smoking, driving, refueling automobiles, parking garages
Butylated hydroxytoluene (BHT)	Urethane-based carpet cushions
Carbon tetrachloride	Fungicides, global background
Chloroform	Showering, washing clothes, dishes
p-Dichlorobenzene	Room deodorizers, moth cakes
Ethylene glycol, Texanol	Paints
Formaldehyde	Pressed wood products
Furfural	Cork parquet flooring
Methylene chloride	Paint stripping, solvent use
Methyl-tert-butyl ether (MTBE)	Gasoline, groundwater contaminant
Phenol	Vinyl flooring, cork parquet flooring
Styrene	Smoking
Terpenes (limonene, α-pinene)	Scented deodorizers, polishes, cigarettes, food, beverages, fabrics, fabric softeners
Tetrachloroethylene	Wearing/storing dry-cleaned clothes
Tetrahydrofuran	Sealer for vinyl flooring
1,1,1-Trichloroethane	Aerosol sprays, solvents, many consumer products
Trichloroethylene	Cosmetics, electronic parts, correction fluid

using a mixture of 22 common VOCs at a total concentration of 5 mg/m³, which is high but is sometimes encountered in new or renovated buildings. The U.S. Environmental Protection Agency (USEPA) later confirmed these findings (Otto et al. 1990). Irritation from VOCs is thought to be mediated through the trigeminal nerve, or "common chemical sense" (Bryant and Silver 2000; Cometto-Muñiz 2001).

Despite the difficulty of observing effects under controlled conditions, a very common worldwide phenomenon is reported increases in symptoms of large numbers of workers following occupation of a new or renovated building (Berglund, Berglund, and Lindvall 1984; Sundell et al. 1990; Preller et al. 1990). This phenomenon has come to be known as sick building syndrome (SBS) and is characterized by multiple symptoms: eye irritation, stuffy nose, sore throat, headaches, skin rashes, and difficulty concentrating (Mølhave 1987; Ten Brinke et al. 1998). Since such new or renovated buildings almost always have very high levels of VOCs for a period of 6 months or more after completion, SBS has been thought to be a possible effect of VOC exposure.

A similar, more serious, syndrome, multiple chemical sensitivity (MCS), has also been suggested to be a result of VOC or pesticide exposure, either chronic or following a single massive dose. A comprehensive review of MCS is found in Ashford and Miller (1997). The International Programme on Chemical Safety (IPCS) has recommended double-blind controlled chamber studies to determine if symptoms can be reproducibly created by exposures to VOCs (IPCS 1996). Although several such studies have been carried out, results are mixed (Fiedler and Kipen 2001). A recent study (Joffres, Sampalli, and Fox 2005) determined that sensitive patients take longer to adapt to the conditions prior to the testing, which may account for some of the negative findings.

A less serious but possibly very costly result of VOC exposure may be reduced productivity resulting from minor ailments such as headache and eye irritation. The total annual cost of poor indoor air quality has been estimated to be in the neighborhood of 100 billion dollars (Fisk and Rosenfeld 1997).

This chapter concentrates on exposure in air, with a brief discussion of exposure in drinking water. VOCs can also be absorbed through the skin, and this is the subject of a separate chapter on dermal exposure from baths and showers. Because of the crucial importance of measurement methods in detecting low-level VOCs, there is an Appendix on measurement methods in air, water, and biological media (breath, blood, urine). Since new methods are often developed and described as part of studies of human exposure, such exposure studies are also described and referenced in this Appendix when appropriate.

7.3 HUMAN EXPOSURE

7.3.1 AIR

Between 1979 and 1987, the USEPA carried out the TEAM studies to measure personal exposures of the general public to VOCs in several geographic areas in the United States (Pellizzari et al. 1987a,b; Wallace 1987). About 20 target VOCS were included in the studies, which involved about 750 persons, representing 750,000 residents of the areas. Each participant carried a personal air quality monitor containing 1.5 g Tenax. A small battery-powered pump pulled about 20 L of air across the sorbent over a 12-hour period. Two consecutive 12-hour personal air samples were collected for each person. Concurrent outdoor air samples were also collected in the participants' backyards. In the studies of 1987, fixed indoor air samplers were also installed in the living rooms of the homes.

The initial TEAM pilot study (Wallace et al. 1982) in Beaumont, TX, and Chapel Hill, NC, indicated that personal exposures to about a dozen VOCs exceeded outdoor air levels, even though Beaumont, TX, has major oil producing, refining, and storage facilities. These findings were supported by a second pilot study in Bayonne–Elizabeth, NJ (another major chemical manufacturing and petroleum refining area) and Research Triangle Park, NC (Wallace et al. 1984a). A succeeding major study of 350 persons in Bayonne–Elizabeth (Wallace et al. 1984b) and an additional 50 persons in a nonindustrial city and a rural area (Wallace et al. 1987a) reinforced these findings. A second major study in Los Angeles, CA, and in Antioch–Pittsburg, CA (Wallace et al. 1988) with a follow-up study in Los Angeles in 1987 (Wallace et al. 1991a) added a number of VOCs to the list of target chemicals with similar results.

Major findings of these TEAM studies included the following:

- Personal exposures exceeded median outdoor air concentrations by factors of 2 to 5 for nearly all prevalent VOCs. The difference was even larger (factors of 10 or 20) when the maximum values were compared, despite the fact that most of the outdoor samples were collected in areas with heavy industry (New Jersey) or heavy traffic (Los Angeles).
- Major sources are consumer products (bathroom deodorizers, moth repellents); personal activities (smoking, driving); and building materials (paints and adhesives). In the United States, one chemical (carbon tetrachloride) has been banned from consumer products and exposure is thus limited to the global background of about 0.7 $\mu g/m^3$.
- Traditional sources (automobiles, industry, petrochemical plants) contributed only 20–25% of total exposure to most of the target VOCs (Wallace 1987). No difference in exposure was noted for persons living close to chemical manufacturing plants or petroleum refineries.

A more recent study of personal exposure to VOCs was carried out on 450 persons in six cities in Europe as part of the EXPOLIS study (Saarela et al. 2003, Edwards et al. 2005). In every city, indoor home VOC concentrations were greater than outdoor levels, with ratios generally ranging from 1 to 3. Personal exposures were also greater than indoor air levels for some compounds, particularly aromatics and alkanes, leading the authors to posit exposures in traffic as likely contributors. The most common VOCs included toluene and xylenes among the aromatics and limonene and α-pinene among the terpenes, similar to most other studies.

Son, Breysse, and Yang (2003) used passive badges to measure personal, indoor, and outdoor concentrations of 10 target VOCs for 30 persons in Seoul, South Korea, and 30 in a smaller city of Asan. Average indoor, outdoor, and personal exposures to benzene in Asan were 20–23 $\mu g/m^3$, and 40–43 $\mu g/m^3$ in Seoul. These are several times higher than in the United States. Benzene levels were increased in homes with smoking and homes that used mosquito coils (incense).

A "new" VOC of considerable interest and concern has arisen as a result of attempts to reduce the carbon monoxide emitted from incomplete combustion in automobiles. To improve combustion, oxidizers are required to be added to gasoline in some areas of the United States. One of the most popular of these is methyl-*tert*-butyl ether (MTBE), added to gasoline in amounts as high as 17%. MTBE appears to have some serious toxic effects, and complaints have been received from residents of some (but not all) of the areas where it has been added to gasoline. Additionally, enough time has passed for it to have become one of the most common contaminants of groundwater. Several studies have documented the human exposure resulting from refueling autos (Lindstrom and Pleil 1996, Lioy et al. 1994). Based on these findings, the USEPA recently banned MTBE as a gasoline additive.

Three large studies of VOCs, involving 300–800 homes, were carried out in the 1980s in the Netherlands (Lebret et al. 1986), West Germany (Krause et al. 1987) and the United States (Wallace 1987). Observed concentrations were remarkably similar for most chemicals, indicating similar sources in these countries. One exception is chloroform, present at typical levels of 1–4 $\mu g/m^3$ in the United States but not found in European homes. This is to be expected, since the likely source is volatilization from chlorinated water (Wallace et al. 1982; Andelman,1985a,b); the two European countries do not chlorinate their water.

Major findings of these indoor air studies include the following:

- Indoor levels in homes and older buildings (>1 year) are typically several times higher than outdoor levels. Sources include dry-cleaned clothes, cosmetics, air fresheners, and cleaning materials.
- New buildings (<1 month) have levels of some VOCs (aliphatics and aromatics) 100 times higher than outdoor, falling to 10 times outdoor about 2–3 months later. Major sources include paints and adhesives.
- About half of 750 homes in the United States had total VOC levels (obtained by integrating the total ion current response curve of the gas chromatography/mass spectrometry profile) greater than 1 mg/m^3, compared to only 10% of outdoor samples (Wallace, Pellizzari, and Wendel 1991b).
- More than 500 different VOCs were identified in four buildings in Washington, DC, and Research Triangle Park, NC (Sheldon et al. 1988a).

One study (Wallace et al. 1989) involved seven volunteers undertaking about 25 activities suspected of causing increased VOC exposures; a number of these activities (using bathroom deodorizers, washing dishes, cleaning an auto carburetor) resulted in 10–1,000-fold increases in 8-hour exposures to specific VOCs.

One study of 12 California office buildings (Daisey et al. 1994) found some chemicals to be emitted primarily by indoor sources (cleaning solvents, building materials, bioeffluents) and others to be likely intrusions from outdoor air (e.g., motor vehicle emissions).

Early studies of organics indoors were carried out in the 1970s in the Scandinavian countries (Johansson 1978; Mølhave and Møller 1979; Berglund, Johanssen and Lindvall 1982a,b). Mølhave (1982) showed that many common building materials used in Scandinavian buildings emitted organic gases. Early U.S. measurements were made in 9 Love Canal residences (Pellizzari, Erickson, and Zweidinger 1979); 34 Chicago homes (Jarke and Gordon 1981); and in several buildings (Hollowell and Miksch, 1981; Miksch, Hollowell and Schmidt 1982).

Hundreds of VOCs have been identified in environmental tobacco smoke (Bi et al. 2005; Daisey, Mahanama, and Hodgson 1998; Higgins, Griest, and Olerich 1983; Higgins 1987; Guerin, Higgins, and Jenkins 1987; Jermini, Weber, and Grandjean 1976; Löfroth et al., 1989; Hodgson et al. 1996; Gundel, Hansen, and Apte 1997; Singer, Hodgson, and Nazaroff 2003), which contaminates about 60% of all U.S. homes and workplaces (Repace and Lowrey 1980, 1985; Miller, Branoff, and Nazaroff 1998). Among these are several human carcinogens, including benzene and 1,3-butadiene. Environmental tobacco smoke (ETS) and other indoor combustion sources such as kerosene heaters (Traynor et al. 1990) and woodstoves (Highsmith, Zweidinger, and Merrill 1988) may emit both volatile and semivolatile organic compounds.

Later studies also investigated building materials (Sheldon et al. 1988a,b) but added cleaning materials and activities such as scrubbing with chlorine bleach or spraying insecticides (Wallace et al. 1987c) and using adhesives (Girman et al. 1986) or paint removers (Girman, Hodgson, and Wind 1987). Knöppel and Schauenburg (1987) studied VOC emissions from 10 household products (waxes, polishes, detergents); 19 alkanes, alkenes, alcohols, esters, and terpenes were among the chemicals emitted at the highest rates from the 10 products. All of these studies employed either headspace analysis or chambers to measure emission rates.

Other studies estimated emission rates from measurements in homes or buildings. For example, Wallace (1987) estimated emissions from a number of personal activities (visiting dry cleaners, pumping gas) by regressing measurements of exposure or breath levels against the specified activities. Girman and Hodgson (1987) extended their chamber studies of paint removers to a residence, finding similar (high ppm) concentrations of methylene chloride in this more realistic situation.

The U.S. National Aeronautics and Space Agency (NASA) has measured organic emissions from about 5,000 materials used in space missions (Nuchia 1986). Perhaps 3,000 of these materials are in use in general commerce (Özkaynak et al. 1987). The chemicals emitted from the largest number of materials included toluene (1,896 materials), methyl ethyl ketone (1,261), and xylenes (1,111).

A 41-day chamber study (Berglund, Johansson, and Lindvall 1987) of aged building materials taken from a "sick" preschool indicated that the materials had absorbed about 30 VOCs, which they re-emitted to the chamber during the first 30 days of the study. Only 13 of the VOCs originally present in the first days of the study continued to be emitted in the final days, indicating that these 13 were the only true components of the materials. This finding has significant implications for remediating "sick buildings." Even if the source material is identified and removed, weeks may be needed before re-emission of organics from sinks in the building stops.

Emission rates of most chemicals in most materials are greatest when the materials are new. For "wet" materials such as paints and adhesives, most of the total volatile mass may be emitted in the first few hours or days following application (Tichenor and Mason 1987; Tichenor et al. 1990). USEPA studies of new buildings indicated that 8 of 32 target chemicals measured within days after completion of the building were elevated 100-fold compared to outdoor levels: xylenes, ethylbenzene, ethyltoluene, trimethylbenzenes, decane, and undecane (Sheldon et al. 1988b). The half-lives of these chemicals varied from 2–6 weeks; presumably some other nontarget chemicals, such as toluene, would have shown similar behavior. The main sources were likely to be paints and adhesives. Thus new buildings would be expected to require about 6 months to a year to decline to the VOC levels of older buildings.

For dry building materials, such as carpets and pressed wood products, emissions are likely to continue at low levels for longer periods. Formaldehyde from pressed wood products may be slowly emitted with a half-life of several years (Breysse, 1984). According to several recent studies, 4-phenylcyclohexene (4-PC), a reaction product occurring in the styrene-butadiene backing of carpets, is the main VOC emitted from carpets after the first few days. 4-PC is likely to be largely responsible for the new carpet odor (Hodgson 1999).

A recent study of VOC emissions from building materials included three common materials: paint, carpets, and vinyl flooring (Hodgson 1999). For the paints, the dominant VOCs were a solvent component (ethylene glycol or propylene glycol) and Texanol, a coalescent aid. The carpets emitted lower levels of VOCs, but all emitted 4-PC. Two types of carpet cushions were tested. The urethane-based cushions all emitted butylated hydroxytoluene (BHT) and a mixture of unsaturated hydrocarbons, whereas the synthetic fiber cushions emitted alkanes primarily. The vinyl flooring emitted n-tridecane and phenol. The associated seam sealer emitted tetrahydrofuran and cyclohexanone, and the adhesive emitted toluene.

This study also tested the effectiveness of several procedures that have been suggested for reducing exposures. Increased ventilation on the days following application of the paints decreased concentrations on those days, but succeeding days saw a rise to higher levels. Similarly, heating immediately following application reduced concentrations only temporarily, with no indication that a permanent decrease had been achieved. Airing out the carpet assembly materials reduced the total exposure to some VOCs, but not 4-PC or BHT in any significant amount. Long-term emissions of Texanol from the paints were considerably reduced, but emissions of BHT from carpets tended to increase over time. BHT and TXIB emissions from vinyl flooring tended to remain constant over the 12 weeks of the study. The authors concluded that most of these procedures had limited value, and that selection of low-emitting materials showed the most promise for reducing exposures.

A major category of human exposure to toxic and carcinogenic VOCs is room air fresheners and bathroom deodorants. Since the function of these products is to maintain an elevated indoor air concentration in the home or the office over periods of weeks (years with regular replacement), extended exposures to the associated VOCs are often the highest likely to be encountered by most (nonsmoking) persons. The main VOCs used in these products are para-dichlorobenzene (widely used in public restrooms), limonene, and α-pinene. The first is carcinogenic to two species (NTP 1986), the second to one (NTP 1988), and the third is mutagenic. Limonene (lemon scent) and α-pinene (pine scent) are also used in many cleaning and polishing products, which would cause short-term peak exposures during use, but which might not provide as much total exposure as the air freshener. Recently it has been found that these terpenes can react with ozone to form large numbers of ultrafine particles and also hydroperoxides (Fan et al. 2005; Weschler and Shields 1999).

Awareness is growing that most exposure comes from these small nearby sources. In California, Proposition 65 focuses on consumer products, requiring makers to list carcinogenic ingredients. The USEPA carried out a "shelf survey" of solvents containing just six VOCs, finding some thousands of consumer products containing the target chemicals (USEPA 1987a; Sack et al. 1992). Environmental tobacco smoke (ETS) was declared a known human carcinogen by the USEPA in 1991; smoking has been banned from many public places and many private workplaces during the last few years.

However, an unintended result of increased consumer awareness of VOC emissions from building materials may be the replacement of some volatile and odorous chemicals with less volatile but longer-lasting chemicals of unknown toxicological properties. For example, a study of 51 renovated homes in Germany with complaints included a number in which the complaints had only begun 2 years after renovation (Reitzig et al. 1998). Upon investigation, a number of "new" VOCs were found, including longifolene, phenoxyethanol, and butyldiglycolacetate. These may represent a class of less traditional compounds that have been added to building materials to replace the "bad actors" identified by toxicological and carcinogenic studies; however, these compounds may themselves have toxic properties that will emerge following new studies. They have the property that, instead of being emitted in large quantities shortly following application of the surface coating, they are emitted in smaller quantities at first but tend to keep a steady emission rate for much longer periods of time.

Outdoor air levels of many of the most common VOCs, even in heavily industrialized areas or areas with high densities of vehicles, are usually considerably lower than indoor levels (Wallace 1987). This fact has not been fully recognized or incorporated into regulations. For example, in the United States, the 1990 reauthorization of the Clean Air Act continues to deal only with outdoor air ("air external to buildings"), while adding 189 toxic chemicals to the list of those to be regulated. Many of these chemicals, which include common solvents and household pesticides, have been shown to be far more prevalent and at higher concentrations in homes than outdoors. A partial exception to this general rule was noted in a Harvard University study of the heavily industrialized Kanawha Valley in West Virginia, where outdoor levels of chloroform were quite high at times (Cohen et al. 1989, 1990).

For many hydrocarbons, the major source of outdoor air levels is gasoline vapor or auto exhaust (Sigsby, Tejada, and Ray 1987; Zweidinger et al. 1988). For example, about 85% of outdoor air benzene in the United States is from mobile sources and only about 15% from stationary combustion sources. A recent study of VOCs in 43 Chinese cities indicated that in 10 cities, traffic appeared to be the major source, while in 25 other cities, coal combustion for heating businesses and homes appeared to be major (Barletta et al. 2005). Exposure to certain aromatics (benzene, toluene, xylenes, ethylbenzene) while inside the automobile can exceed ambient levels by a factor of six or so (SCAQMD 1989). Tollbooth workers are in the midst of traffic and experience high exposures, although the booth provides considerable protection if they do not have to lean out to accept the tolls (Sapkota, Williams, and Buckley 2005).

Although risk assessment remains at a primitive level, several estimates of carcinogenic risk to VOCs have basically agreed on the VOCs found to have the highest risk: benzene, chloroform, and *para*-dichlorobenzene (Tancrede et al. 1987; McCann et al. 1987; Wallace 1991a). Recent studies have added formaldehyde as a possible major carcinogen (Hauptmann et al. 2004; IARC, 2005). Carbon tetrachloride and tetrachloroethylene are estimated to lie about an order of magnitude below the other three, and recently 1,3-butadiene has been added to this list at about this position. We consider these in turn.

7.3.1.1   Benzene

The major source of benzene exposure for the 50 million smokers remaining in the United States is their smoking — calculations indicate that close to 90% of their total exposure to benzene is through mainstream cigarette smoke (Higgins, Griest, and Olerich 1983; Wallace et al. 1987b; Wallace 1990). Also, smokers are exposed to about 6–10 times more benzene daily than nonsmokers (Wallace et al. 1987b). Even for nonsmokers, those exposed to smokers in the home get about 10% of their exposure from secondhand smoke (SHS) (Krause et al. 1987; Wallace 1990).

A second major source of exposure is from automobile exhaust. However, in the past decade, the amount of benzene in gasoline has dropped from 5 to 1%, and outdoor concentrations have fallen accordingly, from about 6 μg/m³ to about 2–3 μg/m³ in many cities (Wilson, Colome, and Tian 1993; Wallace 1996). 24-hour average benzene levels have been measured every 12th day at about 20 sites throughout the state of California since 1986. Statewide average annual values fluctuated between 5 and 7 μg/m³ until 1993 and 1994, when they dropped to about 4 μg/m³. The decline continued over the next decade, reaching levels of about 2 μg/m³. The decline may be due to one or more of several factors: (1) the 50% reduction in hydrocarbon emissions mandated for new cars; (2) the Stage II vapor recovery controls recently in effect; (3) a reduction in benzene content in gasoline down to the 1% mandated in the 1990 Clean Air Act Amendments. Benzene from automobiles accounts for about 40% of the average nonsmoker's exposure.

Although industry, oil refineries, and chemical plants are feared by some as major sources of exposure to benzene, in fact only about 6% of the nonsmoker's exposure is due to these stationary sources, well below even the exposure from SHS. Even in Valdez, Alaska, where the offshore tankers facility was feared to cause increased benzene exposure, it was found to account for only about 11% of the total nonsmoking exposure (Goldstein et al. 1992).

Two notable studies focusing on benzene exposure in the home include a nationwide Canadian study (Fellin and Otson 1993) measuring 24-hour indoor air concentrations of benzene in 754 randomly selected homes and a study of 173 homes in Avon, U.K. (Brown and Crump 1996). The largest study of in-vehicle benzene exposure was a 200-trip study (SCAQMD 1989) of Los Angeles commuters, which found an average benzene exposure of 13 ppb (40 µg/m³) for commuters during rush hour, on the order of 5 times the concentration measured at a fixed outdoor site. More recently, a study (Weisel et al. 1992) of benzene levels in two passenger vehicles during typical commutes in the New Jersey-New York area resulted in measured exposures of 9–12 µg/m³ in suburban and turnpike conditions, and 26 µg/m³ in the Lincoln Tunnel. The authors stated that the concentrations during the commutes to New York City were about 10 times the ambient background concentration measured the same day in suburban New Jersey.

Gammage, White, and Gupta (1984) and McClenny et al. (1986) reported finding gasoline vapor in homes with attached garages. This could arise from evaporative emissions following parking, or from storage of gasoline in the garage. A study of four homes with attached garages (Thomas et al. 1993) showed that 3 of the homes received extensive emissions from gasoline vapors or exhaust in the garage. A larger more recent study of 15 homes with attached garages (Batterman et al. 2006) confirmed these findings, determining that much of the exposure was due to paints and other solvents stored in the garage. These studies suggest that the attached garage may be one of the most important sources of VOC exposures, due to the large number of homes with attached garages, the widespread use of them for storage of gasoline, solvents, and paints, and (often) the lack of good insulation for the door from the garage to the house.

From the above considerations, we can construct a nationwide benzene exposure budget apportioning the observed benzene exposures to the most important sources. The results indicate that smoking accounts for roughly half of the exposure, with the remaining half split fairly evenly between personal activities (driving, visiting gas stations, parking "hot" cars in attached garages) (~30%) and the traditional outdoor sources (~20%) (Figure 7.1, bottom pie chart).

On the other hand, emissions present a very different picture. The traditional sources — motor vehicles and industry — account for 99.9% of the total emissions, compared to 0.1% from cigarettes (Figure 7.1, top pie chart).

These findings have important implications for our regulatory and control strategies. For example, if emissions from all stationary sources were reduced by a Draconian 50%, the total reduction in population exposure would be an unnoticeable 2% (50% × 15% × 20%). The same total effect (although affecting different people) could be achieved by reducing the average benzene content of cigarettes by 4% (from 57 to 55 µg/m³). The idea of trading in exposure rather than emissions is described in Roumasset and Smith (1990), based on earlier work by Smith (1988a,b).

7.3.1.2 *para*-Dichlorobenzene

Results from six TEAM study cities showed that *para*-dichlorobenzene p-DCB was almost exclusively an indoor air pollutant, outweighing outdoor air by more than 20 to 1 (Wallace 1987). Assuming that one-third of homes contain p-DCB, we may calculate that users of these products are increasing their exposures by factors of roughly 60 compared to nonusers.

This chemical has two major uses: to mask odors and to kill moths. Both uses require that the chemical maintain a high concentration in the home for periods of months or even years. A large number of American homes may contain high levels of p-DCB. Many schools, offices, hotels and other places with public restrooms also use *p*-DCB to mask odors.

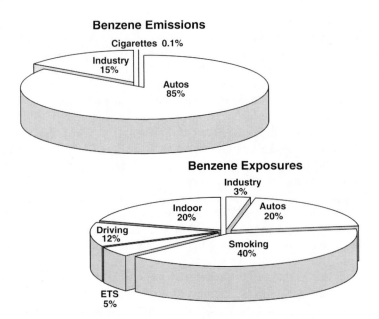

FIGURE 7.1 (Top) Emissions of benzene are dominated by auto exhaust. Emissions from cigarettes are negligible by comparison. (Bottom) Exposures to benzene are dominated by active smoking. Even passive smoking (ETS) accounts for more exposure to benzene than all of American industry.

About 12 million pounds annually are used to kill moths. An estimated 25% of American households contain mothballs, moth crystals, or moth cakes formed from nearly pure p-DCB. About 70% of TEAM study homes in Baltimore and Los Angeles reported using air fresheners or bathroom deodorants. *para*-Dichlorobenzene accounts for a fraction of the air freshener market (perhaps 10%). Assuming 25% of homes have p-DCB moth repellents and an additional 7% have p-DCB air fresheners, we may calculate that about a third of the 85 million homes in the United States contain p-DCB.

In 1986, following a 2-year test of male and female rats and mice, the National Toxicology Program announced that p-DCB caused several different types of malignant tumors in both sexes of the mice and in male rats (NTP 1986). Traditionally, when a chemical causes cancer in two different species of mammals, it is considered a probable human carcinogen. In this case, because the tumors occurred in the male rat kidney and the mouse liver, both of which have been questioned for their relevance to human cancer, p-DCB has been provisionally classified as a possible human carcinogen.

7.3.1.3 Tetrachloroethylene

Early TEAM studies showed that tetrachloroethylene levels were higher among employed people, suggesting that exposure to one's own or to coworkers' dry-cleaned clothes could be important. A later TEAM study (Pellizzari et al. 1984b; Thomas, Pellizzari, and Cooper 1991) indicated that tetrachloroethylene levels in homes increase by factors of 100-fold (to levels exceeding 100 $\mu g/m^3$) following the introduction of dry-cleaned clothes into the home. (The study also indicated that indoor air levels decrease when the clothes are removed from the home and increase when they are put back, thus supporting the notion that "airing out" the clothes on a balcony or patio before introducing them into the home can be effective in reducing exposure.) The same study showed that wearing the clothes also increased personal exposure. Finally, a small but noticeable source of exposure occurs during the few minutes the clothes are being picked up at the dry cleaning shop; earlier TEAM studies (Pellizzari et al. 1984b) indicated that levels in dry cleaning shops varied between 10,000 and 20,000 $\mu g/m^3$. Thus a 5-minute exposure would provide as much tetrachloroethylene as 5 days of normal exposure. The "exposure budget" for tetrachloroethylene is shown in Figure 7.2, bottom pie chart.

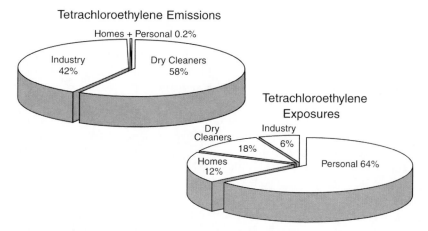

FIGURE 7.2 (Top) Tetrachloroethylene emissions are split mainly among dry cleaning shops and other industry uses, particularly as a solvent (degreaser). (Bottom) Tetrachloroethylene exposures are dominated by wearing or storing dry-cleaned clothes. Only a fraction of exposure is due to regulated emissions from dry cleaning shops or industry.

More recently, a series of studies of indoor air and body burden of persons living in the same building with a dry cleaner have documented large and chronic exposures (Schreiber 1992, 1993; Schreiber et al. 1993, 2002). Because tetrachloroethylene is highly lipophilic and long-lived in the human body, it tends to concentrate preferentially in breast milk. Schreiber documented high concentrations in breast milk of persons living in apartments above dry cleaning shops in New York City.

The dry cleaning shop is considered the major source of outdoor tetrachloroethylene (Figure 7.2. top pie chart). However, these emissions account for no more than 20% of total exposure. Thus, reducing emissions from dry cleaning shops by 50% would result in a barely noticeable 10% reduction in exposure. The same reduction might be achievable if people hung their dry cleaning outside for an 8-hour period before taking it into the house.

7.3.1.4 Carbon Tetrachloride

This is interesting as the exception that proves the rule. It is the only compound of those discussed for which human exposure is entirely driven by outdoor concentrations. This is due to its banning from all consumer products by the Consumer Product Safety Commission. The global background is about 0.7 μg/m^3.

7.3.1.5 Formaldehyde

This highly reactive compound is found at high levels in homes (particularly mobile homes) with a high percentage of pressed wood or particleboard products. It is also produced by certain automobile fuels (e.g., methanol), but the indoor concentrations far outweigh outdoor concentrations as a source of exposure.

7.3.1.6 1,3-Butadiene

Recent toxicological studies show that this compound is an unusually broad animal carcinogen, causing a number of different types of cancer. A major source is cigarette smoke, with about as much 1,3-butadiene produced in sidestream smoke as benzene (400–500 μg) (Gordon et al. 2002). Another source of more recent concern is exhaust from automobiles using alternative fuels.

7.3.2 DRINKING WATER

In areas that chlorinate their drinking water, chlorination by-products such as chloroform and other trihalomethanes (THMs) contaminate the finished water (Rook 1974, 1976, 1977; Bellar and Lichtenberg 1974; IARC 1991). The discovery of chloroform in the blood of New Orleans residents (Dowty et al. 1975) led to the Safe Drinking Water Act (SDWA) of 1974, which set a limit of 100 μg/L for total THMs in finished water supplies.

A series of nationwide surveys of THM levels in water supplies have been carried out in the United States since the passage of the SDWA. The National Organics Reconnaissance Survey of 80 treatment plants (Symons et al. 1975) indicated that about 20% exceeded the THM standard of 100 μg/L. A more recent survey of 727 utilities (McGuire and Meadow 1988) representing more than half of consumers indicated that only about 3.6% of water supplies surveyed continued to exceed the standard.

In succeeding years, it has become more evident that THM exposure occurs in ways other than drinking chlorinated water. Since treated water is normally used for all other household purposes, volatilization from showers, baths, and washing clothes and dishes is an important route of exposure. Studies of experimental laboratory showers by Andelman (1985a,b, 1990; Andelman, Meyers, and Wilder 1986; Andelman, Wilder, and Meyers 1987) resulted in the estimate that chloroform exposure from inhaling the volatilized chloroform from a typical 8-minute shower could range from 0.1–6 times the exposure from drinking water from the same supply, depending on the amount of water ingested per day. Other studies of full-scale showers corroborate these conclusions, with estimates of exposure equivalent to ingestion of 1.3–3.7 L/day of tap water at the same concentrations as the water in the shower (Hodgson et al. 1988; Jo, Weisel, and Lioy 1990a,b; McKone 1987; McKone and Knezovich 1991; Wilkes et al. 1990). Inhalation of airborne chloroform during the rest of the day was also found to be comparable to ingestion, based on measurements of 800 persons and homes in the TEAM study. Weisel and Chen (1993) found that water kept in hot water heaters overnight increased the chloroform levels by 50%. Wallace (1997) summarized the scientific literature on human exposure to chloroform through all routes (inhalation, ingestion, dermal absorption). Corsi and Howard (1998) studied volatilization of five compounds from baths, showers, washing machines, and dishwashers.

Shapiro et al. (2004) used historical chromatograms from 413 wells to estimate the increased contamination of groundwater by VOCs such as tetrachloroethylene. Pre-1940 water had 6.7% of samples with perchloroethylene (PCE) exceeding equilibrium with atmospheric concentrations compared to 68% of samples of post-2000 recharged water. Lindstrom et al. (1994) investigated a case of benzene contamination through groundwater.

7.4 DISCUSSION

For each of the chemicals discussed above, the "traditional" sources of emissions (mobile sources, industry) have accounted for only 2–20% of total human exposure. This same conclusion has been documented for a number of other volatile organic chemicals: styrene, xylenes, ethylbenzene, trimethylbenzenes, chloroform, trichloroethylene, α-pinene, limonene, decane, undecane, etc. (Wallace 1987; Sheldon et al. 1988a,b). For most of these chemicals, the major sources of exposure have been identified (personal activities, consumer products, building materials), but cannot be regulated under existing environmental authorities.

This situation has led to a peculiar split in the perception of risk. The public perceives indoor air pollution as considerably less risky than, say, hazardous waste sites (Smith 1988a,b), whereas experts at the USEPA put indoor air and consumer product exposure at the top of the list of health risks, with hazardous waste sites near the bottom (USEPA 1987b, 1989). Nonetheless, the amount of resources devoted to these two problems reflect the public perception, not that of the experts.

How can this situation be rectified? A continuing process of consumer information and media attention may ultimately result in greater public awareness of the problem. Some steps to reduce exposure can be taken by the public without waiting for cumbersome government attempts at regulation. Other actions, such as setting up consensus guidelines, can be taken by professional organizations — e.g., ASHRAE (ventilation requirements), ASTM (standardized testing for organic emissions from building materials). Information on the economic impacts of indoor air pollution may ultimately convince employers to improve their employees' working conditions. Market forces may also play a role — manufacturers may find substitute chemicals or processes leaving less residue in their products, if the public demands it.

We have seen that in every case the major sources of exposure to our four example chemicals have been small but nearby: cigarettes (benzene); air fresheners (p-dichlorobenzene); dry-cleaned clothes (tetrachloroethylene); and the shower (chloroform). These sources are different from those that have usually been implicated. While these findings indicate that we have been (and still are) pursuing the "wrong" (i.e., less important) sources, they also point the way to a more efficient control of exposure to these carcinogens. For example, the exposure of children (and that of fetuses) to benzene from their mothers' smoking may be reduced by warning women of this risk. Advising consumers of air freshener ingredients and their possible human carcinogenicity could reduce exposures, particularly to p-dichlorobenzene. "Airing out" dry-cleaned clothes and ventilating bathrooms while showering could reduce exposure to tetrachloroethylene and chloroform, respectively, at very little cost.

However, these consumer-oriented recommendations can be effective only if they can be efficiently communicated. At present, few of these findings appear to be general knowledge. Organizations such as the American Lung Association have been in the forefront in trying to communicate this information. The USEPA has completed a booklet on these and other sources of indoor air pollution aimed at the consumer (USEPA 1995). It would also be helpful if medical organizations could join in the communication of this information to doctors and through them to their patients. A recent decision has been made by the California Air Resources Board to ban the use of p-dichlorobenzene in air fresheners as of January 2006 (CARB 2005).

7.5 CONCLUSION

Probably the one central finding of all of these studies has been the following: The major sources of exposure to all chemical groups studied have been small and close to the person — usually inside his or her home.

This finding is so much at odds with the conventional wisdom — that the major sources are industry, autos, urban areas, incinerators, landfills, and hazardous waste sites — that it seems safe to say that most decision makers have not yet grasped its import. For example, if these studies are correct, it makes little sense to spend millions of dollars a year monitoring the outdoor air for VOCs, since so little of our exposure is provided by that route. Similarly, the present allocation of research for outdoor air vs. indoor air (about $50 million to $2 million) appears out of balance. Moreover, the main control options — national air quality standards or emission controls — obviously would need to be replaced by more innovative approaches. One such approach — exposure trading, in which the "currency" is units of human exposure rather than the more common units of source emissions — has been outlined (Roumasset and Smith 1990).

Long-term health effects may include cancer. However, with the exception of benzene, a human carcinogen, most other prevalent VOCs are not known to cause human cancer, and risk estimates are highly uncertain. However, the possible acute effects of VOCs include reduced productivity, which alone has been estimated to cost the nation on the order of scores of billions of dollars annually. If SBS or MCS ultimately prove to be caused or exacerbated by VOC exposure, even higher costs would be attributable to indoor VOC sources.

7.6 APPENDIX — MEASUREMENT METHODS

The measurement of human exposure to VOCs, whether in air or body fluids (e.g., exhaled breath, blood, or urine), requires the use of suitable samplers, sorbents, and analytical techniques. A valuable and comprehensive review of sampling methods for VOCs is provided by Lewis and Gordon (1996).

7.6.1 AIR

7.6.1.1 Air Sampling

VOCs in air are sampled either by batch techniques, (i.e., by concentration or capture) (Rafson 1998) or by continuous, real-time sampling and detection (Lewis and Gordon 1996). Sorbents are used to collect VOCs by the concentration technique; whole-air sample capture is achieved using a suitable container. The sampling mode can be active (i.e., using a pump or vacuum-assisted critical orifice) or passive (i.e., by thermal diffusion to the sorbent), and the sample can be taken over an extended period (time-integrated sampling), or it may be continuous, sequential, or instantaneous (grab sampling) (Otson and Fellin 1992).

Solid sorbents and evacuated containers are the two most widely used procedures for collecting low levels of VOCs from air. For indoor or personal exposure monitoring, actively pumped sorbent cartridges or passive badges are preferred for collecting compounds of interest. Evacuated containers are usually limited to fixed-site sampling, and samples are generally collected by allowing the pre-evacuated container to fill to near-atmospheric pressure by using a suitable flow controller to maintain constant air flow.

7.6.1.1.1 Concentration (Sorbent) Sampling

Solid sorbents used to sample VOCs fall into three major categories: macroreticular and porous resins (Tenax®, Porapak®, Chromosorb®, XAD™), inorganic sorbents (silica gel, alumina, Florisil®, molecular sieves), and activated charcoal and carbon-based sorbents such as graphitized carbon blacks (Carbotrap®, Carbopack®), and carbon molecular sieves (Carbosieve S-III®, Carboxens®, Spherocarb®) (Lewis and Gordon 1996).

Activated Charcoal
Activated charcoal, a porous nonpolar sorbent that irreversibly absorbs low molecular weight VOCs, was first developed for use in occupational sampling. Because it irreversibly sorbs these compounds, a solvent (generally carbon disulfide) is used to recover the *analytes* (target chemicals). Solvent extraction suffers from the disadvantage that the collected compounds are diluted in a relatively large volume of liquid. As a result, when detection limits in the low- or sub-part-per-billion (ppb) range are required, as in environmental (non-occupational) applications, solvent desorption is unsuitable. In contrast, with thermal desorption, the sorbent is simply heated in an inert gas stream and the entire sample is transferred to the analytical system, thus enhancing the sensitivity of the analysis.

Although activated charcoal also collects water from the air, it has nevertheless found increasing acceptance as a sorbent for passive sampling. A number of studies have reported the use of activated charcoal badges (3M OVM-3500 badges) for personal monitoring by diffusive air sampling (Chung et al. 1999a,b). Notable amongst these are a large Canadian indoor air survey of 757 randomly selected homes in which samples were collected passively and analyzed for average 24-hour concentrations of 26 VOCs (Fellin and Otson 1994); a German environmental survey in which personal exposures of 113 participants to 74 VOCs were measured following seven days of passive badge sampling (Hoffmann et al. 2000); the National Human Exposure Assessment Survey (NHEXAS) in Arizona in which passive samplers were used to collect fixed indoor and outdoor air samples at 170 homes and analyzed for three primary VOCs (Gordon et al. 1999); and a recent study in which OVM-3500 badges were used to determine 2-day average concentrations of 15

VOCs that were measured concurrently in indoor, outdoor, and personal air samples taken in the breathing zone of 71 participants in three urban areas over three seasons (Sexton et al. 2004).

Advantages of these passive badges over actively pumped samplers for monitoring personal exposures to hazardous air pollutants include their ease of use, small size, low weight, and reduced burden on participants. However, careful studies of the performance of these badges in sampling indoor, outdoor, and personal air for specific VOCs found that they underestimated the delivered concentration compared to active samplers for most VOCs targeted (Chung et al. 1999a,b; Gordon et al. 1999). Nonetheless, the negative bias was usually less than 25%, suggesting that the passive badges can be effectively used over typical non-occupational concentration ranges and environmental conditions (Chung et al. 1999b).

7.6.1.1.2 Tenax

Tenax, a porous polymeric resin, has several advantages over activated charcoal for the collection of VOCs in air. It has a low affinity for water, low background contamination when carefully cleaned, and high thermal stability (up to 250°C), allowing thermal desorption and combined gas chromatography/mass spectrometry (GC/MS) for VOC recovery and analysis (Gordon 1988; Rothweiler, Wäger, and Schlatter 1991; Jurvelin et al. 2001). Although expensive, Tenax can be cleaned and reused many times. Significant drawbacks are its inability to quantitatively retain very volatile organic compounds (e.g., vinyl chloride and methylene chloride) and artifact formation of several compounds (e.g., benzaldehyde, acetophenone, and phenol) (Coutant, Lewis, and Mulik 1986).

7.6.1.1.3 Multisorbent Systems

Samplers containing two or more sorbents have been developed in an effort to overcome the limited compound sampling range achievable with Tenax, and obtain more reliable recoveries (Heavner, Ogden, and Nelson 1992; Hodgson, Girman, and Binenboym, 1986). Highly volatile compounds tend to break through Tenax, resulting in erroneous concentration estimates. Multisorbent systems that use two or more sorbents in tandem extend the collection of VOCs to a broader range of volatilities and chemical types. For example, in a recent study of spacecraft air, a sorbent combination was used to trap and desorb moderately volatile compounds (on Tenax-TA) and very volatile compounds (on Carbotrap, Carboxen 569, or Carbosieve S-III) (Matney et al. 2000). The Tenax-TA/Carboxen 569 combination gave the best overall recoveries for the 10-component gas mixture tested, which included the highly volatile compounds methanol and Freon 12.

Multisorbent cartridges have also been evaluated for the adsorption and thermal desorption-GC/MS analysis of a wide range of VOCs in air (Pankow et al. 1998). The study targeted 87 analytes, including halogenated alkanes and alkenes, ethers, alcohols, nitriles, esters, ketones, aromatics, a disulfide, and a furan, with volatilities ranging from that of Freon 12 to that of 1,2,3-trichlorobenzene. The eight most volatile compounds were characterized using a Carbotrap B/Carboxen 1000 combination, while the remaining compounds were sampled with a cartridge containing relatively more Carbotrap B and relatively less Carboxen 1000. No breakthrough was evident using these combinations, and excellent recoveries were obtained. Multisorbent cartridges were also used in the USEPA's 100 office building assessment survey and evaluation (BASE) study of VOCs in indoor and outdoor air (Girman et al. 1999). At each building, samples were taken at three indoor locations and one outdoor location, and the samples were analyzed for 57 VOCs by GC/MS.

7.6.1.1.4 Solid-Phase Microextraction (SPME)

SPME, developed by Pawliszyn in 1989, is a simple, efficient, and solvent-free sample concentration technique that has been widely used for sampling gases, liquids, or solids (Zhang, Yang, and Pawliszyn 1994; Vas and Vékey 2004). The SPME technique makes use of a fused silica fiber that is coated with a thin polymer film and is contained in a specially designed syringe, whose needle protects the fiber when it is inserted into a septum. To collect a sample, the SPME fiber is lowered into the gas atmosphere by depressing the syringe plunger. Compounds partition into the polymeric coating of the fiber until equilibrium is established (usually in a matter of minutes).

After withdrawing the plunger and retracting the fiber into the needle, the needle is introduced into the injector of a GC, where the compounds are thermally desorbed and analyzed. The method has the advantage that it does not require special equipment, unlike some methods for gaseous sample analysis, and each fiber can be used repeatedly. The SPME technique with GC or GC/MS has been used to analyze air samples for VOCs (Vas and Vékey 2004; Jia, Koziel and Pawliszyn 2000). A recent application has been the use of SPME coupled with GC/MS for the sensitive detection of microbial VOCs produced by indoor mold (Wady et al. 2003).

7.6.1.1.5 Capture (Whole-Air) Sampling

In the capture technique, samples are collected as they exist in the atmosphere being sampled (Rafson 1998). For this purpose, gas tight syringes, air (Tedlar®) bags, or metal canisters followed by direct injection or cryogenic trapping for GC/MS analysis are generally used.

Sampling with evacuated electropolished stainless steel canisters for later laboratory analysis has become one of the most widely used sampling approaches for trace-level measurements of VOCs in air (Lewis and Gordon 1996; McClenny et al. 1991). The extensive field use of canisters has confirmed that samples captured in this way produce data of acceptable quality, as measured by precision and accuracy (both within ± 25%) for most target VOCs at low ppb levels (Evans et al. 1992).

Canister samples are collected by allowing the air to enter an evacuated container either directly without any flow restriction (grab sampling) or through the use of a suitable critical orifice or a pump to fill the canister to a pressure of a few atmospheres (time-integrated sampling) (McClenny et al. 1991). For analysis, an aliquot (ca. 50–200 mL) of air is withdrawn from the canister and cryofocused into a high-resolution GC column attached to a mass spectrometer, flame ionization detector, or other suitable detector (Lewis and Gordon 1996; Winberry, Murphy, and Riggin 1990). Detection limits are generally less than 1 $\mu g/m^3$ for most VOCs of interest. When equipped with a flow controller and used in the passive sampling mode, canister sampling can be easily adjusted to provide a 24-hour sample. Using a timer and solenoid-controlled valving system, sampling can be extended up to 1 week (Lewis and Gordon 1996; McClenny et al. 1991).

Compared with sorbent sampling, a major advantage of whole-air sampling with canisters is that a large sample can be collected (canisters typically range in size from ~1–6 liters), which is then available for multiple analyses, thus reducing analytical errors. Canisters also have relatively good storage stability, especially for nonpolar VOCs, greatly reduce problems due to contamination and artifact formation, and have no breakthrough limitations (Lewis and Gordon 1996).

Major disadvantages associated with the use of canisters include their cost and bulkiness, the co-collection of water with air samples, and the potential reaction or adsorption of more polar compounds on surfaces (Lewis and Gordon 1996). Much of the early canister sampling work was done using SUMMA-polished stainless steel containers; careful studies have shown that many nonpolar VOCs are stable for at least 7 days, some for as many as 30 days, in humidified SUMMA-passivated canisters (Lewis and Gordon 1996; McClenny et al. 1991). To maintain sample integrity and minimize analyte losses due to surface reactions or adsorption, new surface deactivation technologies have been developed as alternatives to the SUMMA polishing technique (Shelow et al. 1994). The Silcosteel and Sulfinert processes, which bond micron-thick layers of silica directly onto inner stainless steel canister surfaces, have been used to investigate the storage stability of selected volatile sulfur compounds (Sulyok et al. 2001) and to analyze for the presence of these compounds in exhaled breath at low ppb levels without loss of the target analytes (Ochiai et al. 2001).

7.6.1.1.6 Comparison of Sampling Methods

No single method of sampling VOCs in the atmosphere or indoors has become a standard or reference method. In the United States, the two preferred methods are Tenax and evacuated canisters. These two methods were compared under controlled conditions in an unoccupied house (Spicer et al. 1986). Ten chemicals were injected at nominal levels of about 3, 9, and 27 $\mu g/m^3$. The results showed that the two methods were in excellent agreement, with precisions of better than 10% for all chemicals at all spiked levels.

In Europe, the two most common methods are Tenax and activated charcoal. One study employing both methods side by side (Skov et al. 1990) found consistently higher levels of total VOC on the charcoal sorbent. The difference may be due to very volatile organics such as pentane and isopentane, which are collected by charcoal but which break through Tenax readily.

The sorbent methods lend themselves to personal monitoring — a small battery-powered pump is worn for an 8- or 12-hour period to provide a time-integrated sample. Until recently, the whole-air methods employed bags or canisters that were too bulky or heavy to be used as personal monitors; however, a small 1-L canister sampler has been modified for use as a personal sampler (Lindstrom and Pleil 1996; Rossner et al. 2004).

7.6.1.2 Analysis

Samples are usually analyzed by first separating the components by GC. Three common detection methods are flame ionization (FID), electron capture (ECD), and mass spectrometry (MS). Only GC/MS can unambiguously identify many chemicals in a complex mixture. By breaking chemicals into fragments and then uniquely identifying those fragments, MS often provides a means of even distinguishing among coeluting chemicals (i.e., chemicals that emerge from the GC column at the same time), something that is not feasible with either GC/FID or GC/ECD. However, because chemicals are identified in mass spectrometry by comparing their mass spectral patterns to existing libraries, which, although large, are incomplete, even GC/MS identifications are often tentative or incorrect.

The most widely used analytical methods for measuring trace levels of VOCs in air are those that have been established by the USEPA (Winberry et al. 1993). The USEPA TO (Toxic Organic) Compendium Methods for sampling and analyzing ambient VOCs are summarized in Table 7.2 (Winberry, Murphy, and Riggin 1990).

7.6.1.3 Real-Time (Simultaneous Sampling and Analysis) Techniques

A major disadvantage of the integrated air sampling methods described above is that each sample collected must be returned to the laboratory for later analysis before any information about the composition of the sample is available. Thus a field study requires a large number of canisters or sorbent cartridges, creating logistical difficulties. Each sample is analyzed separately, at considerable cost per analysis. These factors have combined to create severe restrictions on the number of samples investigators have been able to collect in field and chamber studies.

7.6.1.3.1 MS-Based Methods

Several MS-based methods have been developed recently that allow direct real-time analysis of VOCs in air (Lewis and Gordon 1996). Here, a key requirement is the replacement of the GC in conventional GC/MS systems with a mass spectrometer, which serves as both the separation device and detector. When operated in the tandem mass spectrometric mode, i.e., MS/MS, the masses are selected in the first stage, then are dissociated, and the fragments so formed are identified and quantified in the second stage. To achieve mass separation and identification in this case requires a soft ionization technique to resolve the components of a mixture. Suitable ionization techniques include atmospheric pressure chemical ionization (APCI) (Carroll et al. 1974), atmospheric sampling glow discharge ionization (ASGDI) (Goeringer et al. 1994; Brinkman et al. 1998), selected ion flow tube (SIFT) (Smith and Španěl 1996; Smith et al. 1998), and proton transfer reaction (PTR) (Hansel et al. 1998; Lindinger, Hansel, and Jordan 1998). All of these methods, which yield protonated molecular ions by proton transfer reactions largely from H_3O^+, are rapid and effective to perform because the sample of air containing the chemicals of interest is introduced directly

TABLE 7.2
USEPA Compendium Methods for VOCs in Air

Compendium Method No.	Sampling & Analysis Procedure	Applicability
TO-1	Tenax sorption with thermal desorption-GC/MS analysis	Nonpolar VOCs with boiling points in 80–200°C range
TO-2	Carbon molecular sieve adsorption with GC/MS analysis	Very volatile nonpolar organics with boiling points in −15 to 120°C range
TO-3	Cryogenic trapping with GC/FID or GC/ECD analysis	Nonpolar VOCs with boiling points in −10 to 200°C range
TO-5	Dinitrophenylhydrazine (DNPH) liquid impinger sampling with high performance liquid chromatography (HPLC)/UV analysis	Aldehydes and ketones
TO-6	Aniline/toluene midget liquid impinger sampling with HPLC analysis	Phosgene
TO-7	Porous polymer (Thermosorb/N) adsorption with GC analysis	N-Nitrosodimethylamine
TO-8	Sodium hydroxide midget liquid impinger sampling with HPLC analysis	Phenol and methyl phenols (cresols)
TO-11A	Adsorbent cartridge sampling with HPLC analysis	Formaldehyde, other aldehydes, and ketones
TO-12	Cryogenic preconcentration and flame ionization detection (FID)	Nonmethane organic compounds
TO-14A	Evacuated, passivated canister sampling with GC/MS analysis	Nonpolar VOCs
TO-15	Evacuated, passivated canister sampling with GC/MS analysis	Polar/nonpolar VOCs
TO-16	Real-time monitoring by Fourier transform infrared spectroscopy (FTIR)	VOCs
TO-17	Real-time or sorbent sampling with MS or standard GC detectors	VOCs

into the instrument. When combined with tandem mass spectrometry, the high sensitivity and selectivity of MS/MS generally allow the air sample to be introduced without prior treatment or concentration (Busch, Glish, and McLuckey 1988).

APCI-MS/MS, which may be regarded as the precursor to the ASGDI ion trap mass spectrometer (ASGDI-ITMS), SIFT-MS, and PTR-MS methods, has been used for the detection of a wide variety of chemicals at trace levels in real time. Examples include the determination of amines in indoor air from steam humidification (Edgerton, Kenny, and Joseph 1989), and the continuous measurement of dimethyl sulfide in air at low parts-per-trillion (ppt) levels (Kelly and Kenny 1991).

Several years of EPA-sponsored development resulted in a system consisting of a compact (2 cubic foot), lightweight (<50 lb) enhanced ion trap mass spectrometer (ITMS) equipped with a sensitive atmospheric sampling glow discharge ionization source (Goeringer et al. 1994; Brinkman et al. 1998). In this ASGDI-ITMS system, trace-level chemicals in the air enter the ionization source, where they are immediately ionized and, through the application of special filtered noise fields, compounds of interest are isolated according to mass. This technique, which also provides a means for performing multistage MS (i.e., MS/MS), allows continuous real-time measurements with time resolutions as short as 6 seconds and detection limits in the low ppb range (Goeringer et al. 1994; Brinkman et al. 1998; Gordon et al. 1998; Gordon et al. 2002).

SIFT-MS and PTR-MS are two similar chemical ionization methods that use a quadrupole mass spectrometer to monitor protonated VOCs in air at trace levels. Both techniques have been extensively used and discussed in recent review papers (Smith et al. 1998; Lindinger et al. 1998; Hewitt, Hayward, and Tani 2003). Since its introduction, the sensitivity of the PTR-MS system has been improved by almost two orders of magnitude, so that now online monitoring of VOCs at concentrations of a few ppt is possible (Hansel et al. 1998).

The mass spectra produced by SIFT-MS and PTR-MS are generally quite simple because of the limited fragmentation caused by chemical ionization. However, some VOCs still show significant fragmentation despite the use of this soft ionization technique. To resolve this problem, Prazeller et al. (2003) have recently coupled the PTR drift tube to an ITMS. The advantage of using an ITMS is that it can operate in the MS/MS mode and possibly distinguish between fragments as well as between isomers and other isobars.

7.6.1.3.2   Olfactory Analysis

Because of the complexity of most indoor and outdoor air samples, the cost and inaccuracy of measurement methods, and the almost complete lack of knowledge of the relationship of measured VOC levels within complex mixtures to resulting health effects, an alternative to chemical measurement methods has arisen in recent years — use of a trained panel to judge the possible health or comfort effects of an air sample directly. This method, pioneered by Dravnieks (Dravnieks and Prokop 1975; Dravnieks and Jarke 1980) at Illinois Institute of Technology in Chicago, and extended to broad use in Europe by Ole Fänger of Denmark (Fänger 1987) employs panels of 6–10 persons who have been previously trained by sampling known mixtures of odorous compounds. When exposed to a test atmosphere, the judges provide an instantaneous estimate of its pollution potential, measured in units called decipols. One decipol is equivalent to the amount of pollution (body odor) produced by one person in a room ventilated at one air change per hour. The method is capable of predicting how persons will react to air of a given quality, and also of estimating the relative contribution of various sources (e.g., ventilation system, office machines, employees) to indoor air quality.

7.6.2   Body Fluids

For the biological monitoring of VOCs, the approaches that have proved to be the most useful include analysis of exhaled breath, blood, and urine (Heinrich-Ramm et al. 2000; Draper 2001). Favorable Henry's Law constants make headspace (static or dynamic) and SPME useful techniques for the analysis of unchanged VOCs in blood and urine. Techniques similar to those described above are also used to sample exhaled breath. Coupled with appropriate pharmacokinetic models including knowledge of how the chemical partitions between blood, air, and fat, the measurement of a single body fluid, such as breath or blood, can give an indication of the total body burden (Wallace 2001).

7.6.2.1   Breath

Breath analysis is relatively simple since the matrix is less complex than that of blood or urine. Time-integrated breath sampling methods (Wallace et al. 1996) include collection in Tedlar® bags (Wallace et al. 1987a), concentration through sorbent cartridges (Wallace et al. 1987b; Gordon et al. 1988; Pellizzari, Wallace, and Gordon 1992) or on SPME fibers (Grote and Pawliszyn 1997; Prado, Marin, and Periago, 2003) and capture in electropolished canisters (Thomas et al. 1991; Pleil and Lindstrom 1995a,b). Detection limits using samples collected in sorbent cartridges or evacuated canisters are usually well below 1 $\mu g/m^3$.

Early efforts to characterize the distribution and initial decay of a chemical from the body through breath measurements were hampered by the relatively short half-lives associated with the initial decay and the long collection and cycle times that were achievable with the sorbent-based sampling methods (Wallace 2001). As a result, a new canister-based method was developed with significantly shorter collection (i.e., ~1 minute) and turnaround (i.e., ~5 minutes) times (Raymer et al. 1990; Thomas, Pellizzari, and Cooper 1991). In addition, the method is able to sample about 98% alveolar air with no restrictions or training needed for the person supplying the exhaled air. This procedure was used in several subsequent field studies and chamber studies on a large number of nonpolar VOCs (Raymer et al. 1991, 1992).

Recently, a single-breath canister method was developed (Pleil and Lindstrom 1995a,b, 1997). The subject breathes directly into a 1-L evacuated canister through a straw-like attachment; after wasting the first (dead space) portion of his or her breath, the subject opens the valve on the canister to allow collection of the second (alveolar) portion. The cycle time is approximately 1 minute. This method allows for immediate collection following exposure, thus defining the maximum breath (and therefore, blood) concentration attained during the exposure period. However, like all discrete breath sampling methods, this one has the drawback of having a high cost per sample, limiting the number of samples that can be collected in any one field study.

Adaptation of the APCI-MS/MS and ASGDI/ITMS real-time instruments, originally developed for monitoring VOCs in air, to the needs of breath analysis, was undertaken in large part to overcome the time-resolution limitations and relatively high costs associated with the discrete sampling methods (Benoit et al. 1985; Gordon, Kenny, and Kelly 1992; Gordon 1998). In one of the earliest exposure-related real-time studies, the APCI-MS/MS instrument was used to measure the elimination of 1,1,1-trichloroethane from the breath of an occupationally exposed machine-shop worker, in which samples were taken continuously at 5.5-second intervals over a 20-minute period shortly after exposure ceased (Gordon, Kenny, and Kelly 1992). The data from this experiment were evaluated in terms of a two-compartment model, and biological residence times were derived.

This study, as well as more recent work, makes use of a specially designed breath inlet system that allows continuous real-time analysis when interfaced with any of these mass spectrometers (Gordon, Kenny, and Kelly 1992; Gordon 1998; Gordon et al. 2002). The undiluted breath sample is vacuum-extracted at a constant rate from the breath inlet by the vacuum in the interface and flows into the mass spectrometer without any attention from the subject. The volume of the breath inlet is roughly one-fifth the mean value of the adult tidal volume, so that each breath exhalation effectively displaces the previous breath sample and a steady gas flow is maintained into the analyzer. This ensures that unit resolution is achieved between individual breath exhalations, while at the same time producing a constant and undiluted sample for analysis. The ASGDI/ITMS system with the breath interface was used recently to determine, by real-time breath analysis, the effect of bathwater temperature on dermal exposure to chloroform (Gordon et al. 1998), and the suitability of some volatile smoke carcinogens as breath biomarkers for active and passive smoking (Gordon et al. 2002).

Both the SIFT-MS and PTR-MS methods have also been used extensively for breath analysis. Applications have included measurement of exposure to solvents and other chemicals (Smith et al. 1998; Lindinger et al. 1998; Wilson et al. 2002), quantification of recent smoking behavior and exposure to environmental tobacco smoke (Jordan et al. 1995; Senthilmohan et al. 2001; Lirk et al. 2004), measurement of VOC distribution coefficients in human blood and urine (Wilson et al. 2003), and measurement of breath VOCs for possible applications in metabolic disorders and cancer screening (Rieder et al. 2001).

7.6.2.2 Blood

Measurements of toxic VOCs in blood at occupational levels have been made for many years (Brugnone et al. 1989). Measurements at environmental levels — typically in the ng/L range — are more recent (Ashley et al. 1992, 1996). In the analytical method developed at the Centers for Disease Control and Prevention (CDC) to monitor blood VOC levels in the general population (Ashley et al. 1992), the VOCs are extracted from a 10-mL blood sample by a purge and trap method, then are separated by capillary GC and measured by high-resolution (magnetic sector) mass spectrometry. Although extremely selective and sensitive (low ppt levels) for measuring 32

VOCs in blood, this method requires considerable operator interaction, which limits sample through-put. Pleil, Fisher, and Lindstrom (1998) compared their own blood method to results from their alveolar breath method on subjects exposed to trichloroethylene.

An alternative method that has gained increasing attention is headspace SPME (Schimming et al. 1999, Cardinali et al. 2000). In this approach, a SPME fiber is inserted into the headspace above the blood sample and the VOCs bind to the phase on the outside of the fiber shaft. This fiber is then inserted into the GC inlet, the VOCs are thermally desorbed, and analysis is accomplished using a benchtop quadrupole MS. This simple, rapid, sensitive (~50 ppt), and easily automated method is a viable alternative to the high resolution MS method described above, and makes blood VOC analysis more readily available and affordable to both researchers and analytical laboratories (Cardinali et al. 2000).

7.6.2.3 Urine

Because of their relatively high water solubility, short-chain alcohols and ketones (e.g., acetone, 2-butanone, methanol, etc.) are easily excreted unchanged in the urine and are measurable in urine (Heinrich-Ramm et al. 2000).

Compared to VOCs in blood, which are generally rapidly eliminated (Wallace, Pellizzari, and Gordon 1993a; Wallace et al. 1997), half-lives are significantly longer for VOC excretion in urine. Furthermore, the literature on the measurement of VOCs in urine for biomonitoring purposes is limited largely because of (a) the relatively small fraction of lipophilic VOCs that are excreted in the urine; (b) the increased analytical challenge associated with the detection of these low levels; (c) the high probability of VOC loss during the pre-analysis phase; and (d) the more complex sample handling requirements (Heinrich-Ramm et al. 2000; Ikeda 1999). Analytical methods that have been developed and used successfully include a thermal desorption-GC method for nonpolar VOCs (benzene, toluene, ethylbenzene, and xylenes) (Hung, Lee, and Chen 1998), headspace-SPME/GC-MS for a variety of VOCs (Guidotti et al. 2001; de Martinis and Martin 2002; Prado, Garrido, and Periago 2004; Krämer et al. 2004), and a headspace/SIFT-MS measurement of VOC distribution coefficients in urine (Wilson et al. 2003).

7.7 QUESTIONS FOR REVIEW

1. Which three VOCs present the highest carcinogenic risk? [Answer: Benzene, chloroform, *para*-dichlorobenzene. Name the single major source of each. [Answer: Smoking, water chlorination, moth cakes/air fresheners.]
2. True or false — residence near chemical plants, oil refineries, and hazardous waste sites significantly increases human exposure to VOCs. [Answer: False.] Name the analysis that led to this conclusion. [Answer: The TEAM study comparing 150 persons living near such sites in New Jersey to 150 living farther away.]
3. Which carcinogenic VOC is an exception to the rule of thumb that indoor sources outweigh outdoor sources? [Answer: Carbon tetrachloride.] Why are there no indoor sources any more? [The Consumer Product Safety Commission banned it from consumer products.] Does this answer provide an indication of possible effective regulatory actions on *para*-dichlorobenzene? [Answer: It could be banned from sale as an air freshener, or at least labeled for consumer protection.]
4. What type of collection device is best suited to the collection of personal air samples? [Answer: Passive sampling badge.]
5. What are two major reasons why sorbent tubes or canisters are not the preferred systems for the collection of air samples in which the concentrations of the contaminants change rapidly? [Answer: Collection with sorbent tubes or canisters takes time, during which

rapid changes in concentration may be missed; collection and subsequent analysis of multiple samples becomes very expensive.]

6. Pick the VOC you found most interesting in the chapter. Why is it interesting? What could regulatory agencies do to reduce exposure to this VOC? What could you do?

REFERENCES

Andelman, J.B. (1985a) Human Exposures to Volatile Halogenated Organic Chemicals in Indoor and Outdoor Air, *Environmental Health Perspectives*, **62**: 313–318.

Andelman, J.B. (1985b) Inhalation Exposure in the Home to Volatile Organic Contaminants of Drinking Water, *The Science of the Total Environment*, **47**: 443–460.

Andelman, J.B., Meyers, S.M., and Wilder, L.C. (1986) Volatilization of Organic Chemicals from Indoor Uses of Water, in *Chemicals in the Environment*, Lester, J.N., Perry, R., and Sterritt, R.M., Eds., Selper Ltd., London, U.K., 323–330.

Andelman, J.B., Wilder, L.C., and Meyers, S.M. (1987) Indoor Air Pollution from Volatile Chemicals in Water, in *Proceedings of the 4th International Conference on Indoor Air Quality and Climate*, Volume 1, Institute for Soil, Water and Air Hygiene, Berlin (West), 37–42.

Andelman, J.B. (1990) Total Exposure to Volatile Organic Compounds in Potable Water, in *Significance and Treatment of Volatile Organic Compounds in Water Supplies*, Ram, N.M., Christman, R.F., and Cantor, K.P., Eds., Lewis Publishers, Chelsea, MI.

Ashford, N. and Miller, C. (1998) *Chemical Exposures: Low Levels and High Stakes*, 2nd ed., Van Nostrand Reinhold, New York, NY.

Ashley, D.L., Bonin, M.A., Cardinali, F.L., McCraw, J.M., Holler, J.L., Needham, L.L., and Patterson, D.G. (1992) Determining Volatile Organic Compounds in Human Blood from a Large Sample Population by Using Purge and Trap Gas Chromatography-Mass Spectrometry, *Analytical Chemistry*, **64**: 1021–1029.

Ashley, D.L., Bonin, M.A., Cardinali, F.L., McCraw, J.M., and Wooten, J.V. (1996) Measurement of Volatile Organic Compounds in Human Blood, *Environmental Health Perspectives*, **104**(suppl. 5): 871–877.

Barletta, B., Meinardi, S., Rowland, F.S., Chan, C.Y., Wang, X., Zou, S., Chan, L.Y., and Blake, D.R. (2005) Volatile Organic Compounds in 43 Chinese Cities, *Atmospheric Environment*, **39**: 5979–5990.

Batterman, S., Hatzivasilis, G., and Jia, C. (2006) Concentrations and Emissions of Gasoline and Other Vapors from Residential Vehicle Garages, *Atmospheric Environment*, **40**: 399–414.

Bellar, T.A. and Lichtenberg, J.J. (1974) Determining Volatile Organics at Microgram-per-Liter Levels by Gas Chromatography, *Journal of the American Water Works Association*, **66**: 739–744.

Benoit, F.M., Davidson, W.R., Lovett, A.M., Nacson, S., and Ngo, A. (1985) Breath Analysis by API/MS — Human Exposure to Volatile Organic Solvents, *International Archives of Occupational and Environmental Health*, **55**: 113–120.

Berglund, B., Berglund, U., and Lindvall, T. (1984) Characterization of Indoor Air Quality and "Sick Buildings," *ASHRAE Transactions*, **90** (Part 1): 1045–1055.

Berglund, B., Johansson, I., and Lindvall, T. (1982a) A Longitudinal Study of Air Contaminants in a Newly Built Preschool, *Environment International*, **8**: 111–115.

Berglund, B., Johansson, I., and Lindvall, T. (1982b) The Influence of Ventilation on Indoor/Outdoor Air Contaminants in an Office Building, *Environment International*, **8**: 395–399.

Berglund, B., Johansson, I., and Lindvall, T. (1987) Volatile Organic Compounds from Building Materials in a Simulated Chamber Study, in *Proceedings of the 4th International Conference on Indoor Air Quality and Climate*, Volume 1, Institute for Soil, Water and Air Hygiene, Berlin (West), 16–21.

Bi, X., Sheng, G., Feng, Y., Fu, J., and Xie, J. (2005) Gas and Particulate-phase Specific Tracer and Toxic Organic Compounds in Environmental Tobacco Smoke, *Chemosphere*, **61**(10): 1512–1522.

Breysse, P. A. (1984) Formaldehyde Levels and Accompanying Symptoms Associated with Individuals Residing in Over 1000 Conventional and Mobile Homes in the State of Washington, in *Indoor Air: Sensory and Hyperreactivity Reactions to Sick Buildings*, Volume 3, Berglund, B., Lindvall, T., and Sundell, J., Eds., Swedish Council for Building Research, Stockholm, Sweden, 403–408.

Brinkman, M.C., Callahan, P.J., Delitala, C., Kenny, D.V., Loggia, M.C., Porcelli, M., and Gordon, S.M. (1998) Direct Analysis of Trace Organic Compounds in Air by Ion Trap Mass Spectrometry, in *Air Pollution VI*, Brebboa, C.A., Ratto, C.F., and Power, H., Eds., WIT Press/Computational Mechanics Publications, Southampton, U.K., 99–108.

Brown, V.M. and Crump, D.R. (1996) Volatile Organic Compounds, in *Indoor Air Quality in Homes: Part I. the Building Research Establishment Indoor Environment Study*, Berry, R.W., Brown, V.M., Coward, S.K.D., Crump, D.R., Gavin, M., Grimes, C.P., Higham, D.F., Hull, A.V., Hunter, C.A., Jeffery, I.G., Lea, R.G., Llewellyn, J.W., and Raw, G.J., Eds., Construction Research Communications, London, England.

Brugnone, F., Perbellini, L., Faccini, G.B., and Pasini, F. (1989) Benzene in the Breath and Blood of Normal People and Occupationally Exposed Workers, *American Journal of Industrial Medicine,* 16: 385–399.

Bryant, B. and Silver, W.L. (2000) Chemesthesis: The Common Chemical Sense, in *The Neurobiology of Taste and Smell,* 2nd ed., Finger, T.E., Silver, W.L., and Restrepo, D., Eds., Wiley-Liss, New York, NY, 73–100.

Busch, K.L., Glish, G.L., and McLuckey, S.A. (1988) *Mass Spectrometry/Mass Spectrometry: Techniques and Applications of Tandem Mass Spectrometry,* VCH Publishers, Inc., New York, NY, 333.

CARB (2005) Rulemaking document 2004-06-24, Air Resources Board, State of California, www.arb.ca.gov/regact/conprod/execsum.pdf (accessed September 30, 2005).

Cardinali, F.L., Ashley, D.L., Wooten, J.V., McCraw, J.M., and Lemire, S.W. (2000) The Use of Solid-Phase Microextraction in Conjunction with a Benchtop Quadrupole Mass Spectrometer for the Analysis of Volatile Organic Compounds in Human Blood at the Low Parts-Per-Trillion Level, *Journal of Chromatographic Science,* 38: 49–54.

Carroll, D.I., Dzidic, I., Stillwell, R.N., Horning, M.G., and Horning, E.C. (1974) Subpicogram Detection System for Gas Phase Analysis Based upon Atmospheric Pressure Ionization (API) Mass Spectrometry, *Analytical Chemistry,* 46: 706–710.

Chung, C-W., Morandi, M.T., Stock, T.H., and Afshar, M. (1999a) Evaluation of a Passive Sampler for Volatile Organic Compounds at ppb Concentrations, Varying Temperatures, and Humidities with 24-h Exposures: Part 1. Description and Characterization of Exposure Chamber System, *Environmental Science and Technology,* 33: 3661–3665.

Chung, C-W., Morandi, M.T., Stock, T.H., and Afshar, M. (1999b) Evaluation of a Passive Sampler for Volatile Organic Compounds at ppb Concentrations, Varying Temperatures, and Humidities with 24-h Exposures: Part 2. Sampler Performance, *Environmental Science and Technology,* 33: 3666–3671.

Cohen, M.A., Ryan, P.B., Yanagisawa, Y., and Hammond, S.K. (1990) Validation of a Passive Sampler for Indoor and Outdoor Concentrations of Volatile Organic Compounds, *Journal of the Air & Waste Management Association,* 40: 993–997.

Cohen, M.A., Ryan, P.B., Yanagisawa, Y., Spengler, J.D., Özkaynak, H., and Epstein, P.S. (1989) Indoor/Outdoor Measurements of Volatile Organic Compounds in the Kanawha Valley of West Virginia, *Journal of the Air Pollution Control Association,* 39: 1086–1093.

Cometto-Muñiz, J.E. (2001) Physicochemical Basis for Odor and Irritation Potency of VOCs, *Indoor Air Quality Handbook*, Spengler, J.D., Samet, J., and McCarthy, J.F., Eds., McGraw-Hill, New York, NY, 20.1–20.21.

Corsi, R. and Howard, C. (1998) *Volatilization Rates from Water to Indoor Air, Phase II*, EPA Final Report for Grant No. CR 824228-01, U.S. Environmental Protection Agency, Washington, DC.

Coutant, R.W., Lewis, R.G., and Mulik, J.D. (1986) Modification and Evaluation of a Thermally Desorbable Passive Sampler for Volatile Organic Compounds in Air, *Analytical Chemistry,* 58: 445–448.

Daisey, J.M., Hodgson, A.T., Fisk, W. J., Mendell, M.J., and Ten Brinke, J. (1994) Volatile Organic Compounds in 12 California Office Buildings: Classes, Concentrations, and Sources, *Atmospheric Environment,* 28(22): 3557–3562.

Daisey, J.M., Mahanama, K.R.R., and Hodgson, A.T. (1998) Toxic Volatile Organic Compounds in Simulated Environmental Tobacco Smoke: Emissions Factors for Exposure Assessment, *Journal of Exposure Analysis and Environmental Epidemiology,* 8(3): 313-334.

De Martinis, B.S. and Martin, C.C. (2002) Automated Headspace Solid-Phase Microextraction and Capillary Gas Chromatography Analysis of Ethanol in Postmortem Specimens, *Forensic Science International,* 128: 115–119.

Dowty, B., Carlisle, D., Laseter, J.L., and Storer, J. (1975) Halogenated Hydrocarbons in New Orleans Drinking Water and Blood Plasma, *Science,* **187**: 75–77.

Draper, W.M. (2001) Biological Monitoring: Exquisite Research Probes, Risk Assessment, and Routine Exposure Measurement, *Analytical Chemistry,* **73**: 2745–2760.

Dravnieks, A. and Jarke, F. (1980) Odor Threshold Measurement by Dynamic Olfactometry: Significant Operational Variables, *Journal of the Air Pollution Control Association,* **30**: 1284–1289.

Dravnieks, A. and Prokop, W.H. (1975) Source Emission Odor Measurement by a Dynamic Forced-Choice Triangle Olfactometer, *Journal of the Air Pollution Control Association,* **25**: 28–35.

Edgerton, S.A., Kenny, D.V., and Joseph, D.W. (1989) Determination of Amines in Indoor Air from Steam Humidification, *Environmental Science and Technology,* **23**: 484–488.

Edwards, R.D., Schweizer, C., Jantunen, M., Lai, H.K., Bayer-Oglesby, L., Katsouyanni, K., Nieuwenhuijsen, M., Saarela, K., Sram, R., and Kunzli, N. (2005) Personal Exposures to VOC in the Upper End of the Distribution — Relationships to Indoor, Outdoor, and Workplace Concentrations, *Atmospheric Environment,* **39**: 2299–2307.

Evans, G.F., Lumpkin, T.A., Smith, D.L., and Somerville, M.C. (1992) Measurements of VOCs from the TAMS Network, *Journal of the Air & Waste Management Association,* **42**: 1319–1323.

Fan, Z., Weschler, C.J., Han, I.-K., and Zhang, J. (2005) Co-Formation of Hydroperoxides and Ultra-Fine Particles during the Reactions of Ozone with a Complex VOC Mixture under Simulated Indoor Conditions, *Atmospheric Environment,* **39**: 5171–5182.

Fänger, O. (1987) A Solution to the Sick-Building Mystery, in *Indoor Air '87, Proceedings of the 4th International Conference on Indoor Air,* Seifert, B., Ed., Institute for Water, Soil and Air Hygiene, Berlin, **4**: 49–57.

Fellin, P. and Otson, R. (1993) Seasonal Trends of Volatile Organic Compounds (VOCs) in Canadian Homes, in *Indoor Air '93: Proceedings of the 6th International Conference on Indoor Air Quality and Climate,* Jaakola, J.J.K., Ilmarinen, R., and Seppänen, O., Eds., Helsinki University of Technology, Espoo, Finland, **1**: 339–343.

Fellin, P. and Otson, R. (1994) Assessment of the Influence of Climatic Factors on Concentration Levels of Volatile Organic Compounds (VOCs) in Canadian Homes, *Atmospheric Environment,* **28**: 3581–3586.

Fiedler, N. and Kipen, H.M. (2001) Controlled Exposures to Volatile Organic Compounds in Sensitive Groups, *Annals of the New York Academy of Sciences,* **923**: 24–37.

Fisk, W.J. and Rosenfeld, A.H. (1997) Estimates of Improved Productivity and Health from Better Indoor Environments, *Indoor Air,* **7**: 158–172.

Gammage, R.B., White, D.A., and Gupta, K.C. (1984) Residential Measurements of High Volatility Organics and Their Sources, in *Indoor Air, Chemical Characterization and Personal Exposure,* Swedish Council for Building Research, Stockholm, Sweden, **4**: 157–162.

Girman, J.R., Hadwen, G.E., Burton, L.E., Womble, S.E., and McCarthy, J.F. (1999) Individual Volatile Organic Compound Prevalence and Concentrations in 56 Buildings of the Building Assessment Survey and Evaluation (BASE) Study, in *Proceedings of Indoor Air 1999, the 8th International Conference on Indoor Air Quality and Climate,* Edinburgh, Scotland, August 8-13, 1999, **2**: 460–465.

Girman, J.R., Hodgson, A.T., Newton, A.W., and Winkes, A.W. (1986) Volatile Organic Emissions from Adhesives with Indoor Applications, *Environment International,* **12**: 317–321.

Girman, J. R., Hodgson, A.T., and Wind, M.L. (1987) Considerations in evaluating emissions from consumer products, *Atmospheric Environment,* **21**: 315–320.

Girman, J.R. and Hodgson, A.T. (1987) *Exposure to Methylene Chloride from Controlled Use of a Paint Remover in a Residence,* Report No. LBL 23078, paper presented at 80th Annual Meeting of the Air Pollution Control Association in New York, NY, June 21–26, 1987, Lawrence Berkeley Laboratory, Berkeley, CA.

Goeringer, D.E., Asano, K.G., McLuckey, S.A., Hoekman, D., and Stiller, S.W. (1994) Filtered Noise Field Signals for Mass-Selective Accumulation of Externally Formed Ions in a Quadrupole Ion Trap, *Analytical Chemistry,* **66**: 313–318.

Goldstein, B.D., Tardiff, R.G., Baker, S.R., Hoffnagle, G.F., Murray, D.R., Catizone, P.A., Kester, R.A., and Caniparoli, D.G. (1992) *Valdez Air Health Study,* Alyeska Pipeline Service Co., Anchorage, AK.

Gordon, S., Wallace, L., Pellizzari, E., and O'Neill, H. (1988) Breath Measurements in a Clean Air Chamber to Determine Washout Times for Volatile Organic Compounds at Normal Environmental Concentrations, *Atmospheric Environment,* **22**: 2165–2170.

Gordon, S.M. (1988) Sampling of Volatile Organic Compounds in Ambient Air, in *Advances in Air Sampling*, American Conference of Governmental Industrial Hygienists, Lewis Publishers, Chelsea, MI, 133–142.

Gordon, S.M. (1998) *Development and Application of a Compact Mobile Ion Trap (Tandem) Mass Spectrometer System for Real-Time Measurement of Volatile Organics in Air and Breath*, Cooperative Agreement CR 822062, Final Report, U.S. Environmental Protection Agency, Washington, DC.

Gordon, S.M., Callahan, P.J., Nishioka, M.G., Brinkman, M.C., O'Rourke, M.K., Lebowitz, M.D., and Moschandreas, D.J. (1999) Residential Environmental Measurements in the National Human Exposure Assessment Survey (NHEXAS) Pilot Study in Arizona: Preliminary Results for Pesticides and VOCs, *Journal of Exposure Analysis and Environmental Epidemiology, 9*: 456–470.

Gordon, S.M., Kenny, D.V., and Kelly, T.J. (1992) Continuous Real-Time Breath Analysis for the Measurement of Half-Lives of Expired Volatile Organic Compounds, *Journal of Exposure Analysis and Environmental Epidemiology, (Suppl 1)*: 41–54.

Gordon, S.M., Wallace, L.A., Brinkman, M.C., Callahan, P.J., and Kenny, D.V. (2002) Volatile Organic Compounds as Breath Biomarkers for Active and Passive Smoking, *Environmental Health Perspectives, 110*: 689–698.

Gordon, S.M., Wallace, L.A., Callahan, P.J., Kenny, D.V., and Brinkman, M.C. (1998) Effect of Water Temperature on Dermal Exposure to Chloroform, *Environmental Health Perspectives, 106*(6): 337–345.

Grote, C. and Pawliszyn, J. (1997) Solid-Phase Microextraction for the Analysis of Human Breath, *Analytical Chemistry, 69*: 587–596.

Guerin, M.R., Higgins, C.E., and Jenkins, R.A. (1987) Measuring Environmental Emissions from Tobacco Combustion: Sidestream Cigarette Smoke Literature Review, *Atmospheric Environment, 21*: 291–297.

Guidotti, M., Onorati, B., Lucareli, E., Blasi, G., and Ravaioli, G. (2001) Determination of Chlorinated Solvents in Exhaled Air, Urine, and Blood of Subjects Exposed in the Workplace Using SPME and GC-MS, *American Clinical Laboratory, 20*: 23–26.

Gundel, L.A., Hansen, A.D.A., and Apte, M.G. (1997) Real-time Measurement of Environmental Tobacco Smoke by Ultraviolet Absorption, presented at Annual Meeting of the American Association for Aerosol Research, Denver, CO, Oct. 13–17.

Hansel, A., Jordan, A., Warneke, C., Holzinger, R., and Lindinger, W. (1998) Improved Detection Limit of the Proton-Transfer Reaction Mass Spectrometer: On-Line Monitoring of Volatile Organic Compounds at Mixing Ratios of a Few pptv, *Rapid Communications in Mass Spectrometry, 12*: 871–875.

Hauptmann, M., Lubin, J.H., Stewart, P.A., Hayes, R.B., and Blair, A. (2004) Mortality from Solid Cancers among Workers in Formaldehyde Industries, *American Journal of Epidemiology, 159*: 1117–1130.

Heavner, D.L., Ogden, M.W., and Nelson, P.R. (1992) Multisorbent Thermal Desorption/Gas Chromatography/Mass Selective Detection Method for the Determination of Target Volatile Organic Compounds in Indoor Air, *Environmental Science and Technology, 26*: 1737–1746.

Heinrich-Ramm, R., Jakubowski, M., Heinzow, B., Molin Christensen, J., Olsen, E., and Hertel, O. (2000) Biological Monitoring for Exposure to Volatile Organic Compounds (VOCs), *Pure & Applied Chemistry, 72*: 385–436.

Hewitt, C.N., Hayward, S., and Tani, A. (2003) The Application of Proton Transfer Reaction-Mass Spectrometry (PTR-MS) to the Monitoring and Analysis of Volatile Organic Compounds in the Atmosphere, *Journal of Environmental Monitoring, 5*: 1–7.

Higgins, C.E. (1987) Organic Vapor Phase Composition of Sidestream and Environmental Tobacco Smoke from Cigarettes, in *Proceedings of the 1987 EPA/APCA Symposium on Measurement of Toxic and Related Air Pollutants*, 140–151.

Higgins, C., Griest, W.H., and Olerich, G. (1983) Applications of Tenax Trapping to Cigarette Smoking, *Journal of the Association of Official Analytical Chemists, 66*: 1074–1083.

Highsmith, V.R., Zweidinger, R.B., and Merrill, R.G. (1988) Characterization of Indoor and Outdoor Air Associated with Residences Using Woodstoves: A Pilot Study, *Environment International, 14*: 213–219.

Hodgson, A.T. (1999) *Common Indoor Sources of Volatile Organic Compounds: Emission Rates and Techniques for Reducing Consumer Exposures*, Report No. 95-302, California Air Resources Board, Sacramento, CA.

Hodgson, A.T., Daisey, J.M., Mahanama, K.R.R., Ten Brinke, J., and Alevantis, L.E. (1996) Use of Volatile Tracers to Determine the Contribution of Environmental Tobacco Smoke to Concentrations of Volatile Organic Compounds in Smoking Environments, *Environment International, 22*(3): 295–307.

Hodgson, A.T., Girman, J.R., and Binenboym, J. (1986) *A Multi-Sorbent Sampler for Volatile Organic Compounds in Indoor Air*, Report LBL-21378, Lawrence Berkeley Lab, Berkeley, CA.

Hodgson, A.T. et al. (1988) Evaluation of Soil–Gas Transport of Organic Chemicals into Residential Buildings: Final Report. Lawrence Berkeley Lab Report LBL-25465, Berkeley, CA.

Hoffmann, K., Krause, C., Seifert, B., and Ullrich, D. (2000) The German Environmental Survey 1990/92 (GerES II): Sources of Personal Exposure to Volatile Organic Compounds, *Journal of Exposure Analysis and Environmental Epidemiology, 10*: 115–125.

Hollowell, C.D. and Miksch, R.R. (1981) Sources and Concentrations of Organic Compounds in Indoor Environments, *Bulletin of the New York Academy of Medicine, 57*: 962–977.

Hung, I.F., Lee, S.A., and Chen, R.K. (1998) Simultaneous Determination of Benzene, Toluene, Ethylbenzene, and Xylenes in Urine by Thermal Desorption-Gas Chromatography, *Journal of Chromatography B, 706*: 352–357.

IARC (1991) *Monographs on the Evaluation of Carcinogenic Risks to Humans, Vol. 52, Chlorinated Drinking-Water, Chlorination By-Products; Some Other Halogenated Compounds; Cobalt and Cobalt Compounds*, International Agency for Research on Cancer, Lyon, France.

IARC (2005) *Monographs on the Evaluation of Carcinogenic Risks to Humans, Vol. 88, Formaldehyde, 2-Butoxyethanol, and 1-Tert-Butoxy-2-Propanol*, International Agency for Research on Cancer, Lyon, France.

Ikeda, M. (1999) Solvents in Urine as Exposure Markers, *Toxicology Letters, 5*: 99–106.

IPCS (1996) Conclusions and Recommendations of a Workshop on Multiple Chemical Sensitivities (MCS), *Regulatory Toxicology and Pharmacology, 24*: S1888–S189.

Jarke, F.H., and Gordon, S.M. (1981) Recent Investigations of Volatile Organics in Indoor Air at Sub-ppb Levels. Paper No. 81-57.2, presented at the 74th Annual Meeting of the Air Pollution Control Association, Pittsburgh, PA.

Jermini, C., Weber, A., and Grandjean, E. (1976) Quantitative Determination of Various Gas-Phase Components of the Sidestream Smoke of Cigarettes in Room Air, *International Archives of Occupational and Environmental Health, 36*: 169–181.

Jia, M.Y., Koziel, J., and Pawliszyn, J. (2000) Fast Field Sampling/Sample Preparation and Quantification of VOCs in Indoor Air by SPME and Portable GC, *Field Analytical Chemistry Technology, 4*: 73–84.

Jo, W.K, Weisel, C.P., and Lioy, P.J. (1990a) Routes of Chloroform Exposure and Body Burden from Showering with Contaminated Tap Water, *Risk Analysis, 10*: 575–580.

Jo, W.K., Weisel, C.P. and Lioy, P.J. (1990b) Chloroform Exposure and the Health Risk Associated with Multiple Uses of Chlorinated Tap Water, *Risk Analysis, 10*: 581–585.

Joffres, M.R., Sampalli, T., and Fox, R. (2005) Physiologic and Symptomatic Responses to Low-Level Substances in Individuals with and without Chemical Sensitivities: A Randomized Controlled Blinded Pilot Booth Study, *Environmental Health Perspectives, 113*(9): 1178–1183.

Johansson, I. (1978) Determination of Organic Compounds in Indoor Air with Potential Reference to Air Quality, *Atmospheric Environment, 12*: 1371–1377.

Jordan, A., Hansel, A., Holzinger, R., and Lindinger, W. (1995) Acetonitrile and Benzene in the Breath of Smokers and Non-Smokers Investigated by Proton Transfer Reaction Mass Spectrometry (PTR-MS), *International Journal of Mass Spectrometry and Ion Processes, 148*: L1–L3.

Jurvelin, J., Edwards, R., Saarela, K., Laine-Ylijoki, J., De Bortoli, M., Oglesby, L., Schläpfer, K., Georgoulis, L., Tischerova, E., Hänninen, O., and Jantunen, M. (2001) Evaluation of VOC Measurements in the EXPOLIS Study, *Journal of Environmental Monitoring, 3*: 159–165.

Kelly, T.J. and Kenny, D.V. (1991) Continuous Determination of Dimethylsulfide at Part-per-Trillion Concentrations in Air by Atmospheric Pressure Chemical Ionization Mass Spectrometry, *Atmospheric Environment, 25A*: 2155–2160.

Knöppel, H. and Schauenburg, H. (1987) Screening of Household Products for the Emission of Volatile Organic Compounds, *Environment International, 15*: 443–447.

Krämer Alkalde, T., de Carmo Ruaro Peralba, M., Alcaraz Zini, C., and Bastos Caramão, E. (2004) Quantitative Analysis of Benzene, Toluene, and Xylenes in Urine by Means of Headspace Solid-Phase Microextraction, *Journal of Chromatography A, 1027*: 37–40.

Krause, C., Mailahn, W., Nagel, R., Schulz, C., Seifert, B., and Ullrich, D. (1987) Occurrence of Volatile Organic Compounds in the Air of 500 Homes in the Federal Republic of Germany, in *Proceedings of the 4th International Conference on Indoor Air Quality and Climate*, Institute for Soil, Water, and Air Hygiene, Berlin (West), Germany, **1**: 102–106.

Lan, Q., Zhang, L., and Li, G. (2004) Hematotoxicity in Workers Exposed to Low Levels of Benzene, *Science,* **306**: 1774–1776.

Lebret, E., Van de Weil, H.J., Noij, D., and Boleij, J.S.M. (1986) Volatile Hydrocarbons in Dutch Homes, *Environment International,* **12**: 323–332.

Lewis, R.G. and Gordon, S.M. (1996) Sampling for Organic Chemicals in Air, in *Principles of Environmental Sampling*, 2nd ed., Keith, L.H., Ed., American Chemical Society, Washington DC, 401–470.

Lindinger, W., Hansel, A., and Jordan, A. (1998) On-Line Monitoring of Volatile Organic Compounds at pptv Levels by Means of Proton-Transfer-Reaction Mass-Spectrometry (PTR-MS): Medical Applications, Food Control and Environmental Research, *International Journal of Mass Spectrometry and Ion Processes,* **173**: 191–241.

Lindstrom, A.B., Highsmith, V.R., Buckley, T.J., Pate, W.J., and Michael L. (1994) Gasoline-Contaminated Ground Water As a Source of Residential Benzene Exposure: A Case Study, *Journal of Exposure Analysis and Environmental Epidemiology,* **4**(2): 183–196.

Lindstrom, A.B. and Pleil, J.D. (1996) Alveolar Breath Sampling and Analysis to Assess Exposures to Methyl Tertiary Butyl Ether (MTBE) during Motor Vehicle Refueling, *Journal of the Air & Waste Management Association,* **46**: 676–682.

Lioy, P.J., Weisel, C.P., Jo, W., Pellizzari, E., and Raymer, J.H. (1994) Microenvironmental and Personal Measurements of Methyl-Tertiary Butyl Ether Associated with Automobile Use Activities, *Journal of Exposure Analysis and Environmental Epidemiology,* **4**(4): 427–441.

Lirk, P., Bodrogi, F., Deibl, M., Kahler, C.M., Colvin, J., Moser, B., Pinggera, G., Raifer, H., Rieder, J., and Schobersberger, W. (2004) Quantification of Recent Smoking Behaviour Using Proton Transfer Reaction-Mass Spectrometry (PTR-MS), *Wien Klinische Wochenschrift,* **116**: 21–25.

Löfroth, G., Burton, B., Forehand, L., Hammond, S.K., Seila, R., Zweidinger, R., and Lewtas, J. (1989) Characterization of Environmental Tobacco Smoke, *Environmental Science and Technology,* **23**: 610–614.

Matney, M.L., Beck, S.W., Limero, T.F., and James, J.T. (2000) Multisorbent Tubes for Collecting Volatile Organic Compounds in Spacecraft Air, *American Industrial Hygiene Association Journal,* **61**: 69–75.

McCann, J., Horn, L., Girman, J., and Nero, A.V. (1987) *Potential Risks from Exposure to Organic Carcinogens in Indoor Air,* Report No. LBL-22473, Lawrence Berkeley Lab, Berkeley, CA.

McClenny, W.A., Lumpkin, T.A., Pleil, J.D., Oliver, K.D., Bubacz, D.K., Faircloth, J.W., and Daniels, W.H. (1986) Canister-Based VOC Samplers, in *Proceedings of the EPA/APCA Symposium on Measurement of Toxic Air Pollutants*, Air Pollution Control Association, Pittsburgh, PA.

McClenny, W.A., Pleil, J.D., Evans, G.F., Oliver, K.D., Holdren, M.W., and Winberry, W.T. (1991) Canister-Based Method for Monitoring Toxic VOCs in Ambient Air, *Journal of the Air & Waste Management Association,* **41**: 1308–1318.

McGuire, M.J. and Meadow, R.G. (1988) AWWARF Trihalomethane Survey, *Journal of the American Water Works Association,* **80**: 61.

McKone, T.E. (1987) Human Exposure to Volatile Organic Compounds in Household Tap Water: The Indoor Inhalation Pathway, *Environmental Science and Technology,* **21**: 1194–1201.

McKone, T.E. and Knezovich, J. (1991) The Transfer of Trichloroethylene (TCE) from a Shower to Indoor Air: Experimental Measurements and Their Implications, *Journal of the Air & Waste Management Association,* **40**: 282–286.

Miksch, R.R., Hollowell, C.D, and Schmidt, H.E. (1982) Trace Organic Chemical Contaminants in Office Spaces, *Environment International,* **8**: 129–137.

Miller, S.L., Branoff, S., and Nazaroff, W.W. (1998) Exposure to Toxic Air Contaminants in Environmental Tobacco Smoke: An Assessment for California Based on Personal Monitoring Data, *Journal of Exposure Analysis and Environmental Epidemiology,* **8**(3): 287–312.

Mølhave, L. (1982) Indoor Air Pollution Due to Organic Gases and Vapours of Solvents in Building Materials, *Environment International,* **8**(1–6): 117–127.

Mølhave, L. (1986) Indoor Air Quality in Relation to Sensory Irritation Due to Volatile Organic Compounds, *ASHRAE Transactions,* **92**: 2954.

Mølhave, L. (1987) The Sick Buildings — A Sub-Population among the Problem Buildings? in *Indoor Air '87: Proceedings of the 4th International Conference on Indoor Air Quality and Climate, August 17–21, 1987*, Institute For Water, Soil, and Air Hygiene, West Berlin, 469–474.

Mølhave, L. and Møller, J. (1979) The Atmospheric Environment in Modern Danish Dwellings: Measurements in 39 Flats, in *Indoor Climate*, Fänger, P.O. and Valbjørn, O., Eds., Danish Building Research Institute (SBI), Hørsholm, Denmark, 171–186.

Mølhave, L., Bach, B., and Pedersen, O.F. (1984) Human Reactions during Controlled Exposures to Low Concentrations of Organic Gases and Vapours Known as Normal Indoor Air Pollutants, in *Indoor Air: Sensory and Hyperreactivity Reactions to Sick Buildings*, Berglund, B., Lindvall, T., and Sundell, J., Eds., Swedish Council for Building Research, Stockholm, Sweden, **3**: 431–436.

Mølhave, L., Bach, B., and Pedersen, O.F. (1986) Human Reactions to Low Concentrations of Volatile Organic Compounds, *Environment International*, **12**(1-4): 167–175.

NTP (1986) *Technical Report on the Toxicity and Carcinogenesis of 1,4-Dichlorobenzene (CAS #106-46-7) in F344/N Rats and B6C3F1 Mice (Gavage Study)*, NTP Technical Report #319, Board Draft.

NTP (1988) *Technical Report on the Toxicity and Carcinogenesis of D-Limonene (CAS #5989-27-5) in F344/N Rats and B6C3F1 Mice (Gavage Study)*, NTP Technical Report #347, NIH Publication #88-2802, Bethesda, MD.

Nuchia, E. (1986) *MDAC — Houston Materials Testing Database Users' Guide*, McDonnell Douglas Corp., Software Technology Development Laboratory, Houston, TX.

Ochiai, N., Takino, M., Daishima, S., and Cardin, D.B. (2001) Analysis of Volatile Sulphur Compounds in Breath by Gas Chromatography-Mass Spectrometry Using a Three-Stage Cryogenic Trapping Preconcentration System, *Journal of Chromatography B*, **762**: 67–75.

Otto, D., Mølhave, L., Rose, G., Hudnell, H.K., and House, D. (1990) Neurobehavioral and Sensory Irritation Effects of Controlled Exposure to a Complex Mixture of Volatile Organic Compounds, *Neurotoxicology and Teratology*, **12**: 649–652.

Otson, R. and Fellin, P. (1992) Volatile Organics in the Indoor Environment: Sources and Occurrence, in *Gaseous Pollutants: Characterization and Cycling*, Nriagu, J.O., Ed., Wiley & Sons, New York, NY, 335–421.

Özkaynak, H., Ryan, P.B., Wallace, L.A., Nelson, W.C., and Behar, J.V. (1987) Sources and Emission Rates of Organic Chemical Vapors in Homes and Buildings, in *Proceedings of the 4th International Conference on Indoor Air Quality and Climate*, Institute for Soil, Water, and Air Hygiene, Berlin (West), Germany, **1**: 3–7.

Pankow, J.F., Luo, W., Isabelle, L.M., Bender, D.A., and Baker, R.J. (1998) Determination of a Wide Range of Volatile Organic Compounds in Ambient Air Using Multisorbent Adsorption/Thermal Desorption and Gas Chromatography/Mass Spectrometry, *Analytical Chemistry*, **70**: 5213–5221.

Pellizzari, E.D., Erickson, M.D., and Zweidinger, R. (1979) *Formulation of a Preliminary Assessment of Halogenated Organic Compounds in Man and Environmental Media*, U.S. Environmental Protection Agency, Washington, DC.

Pellizzari, E.D., Perritt, K., Hartwell, T.D., Michael, L.C., Whitmore, R., Handy, R.W., Smith, D., and Zelon, H. (1987a) *Total Exposure Assessment Methodology (TEAM) Study: Elizabeth and Bayonne, New Jersey, Devils Lake, North Dakota, and Greensboro, North Carolina, Vol. II*, U.S. Environmental Protection Agency, Washington, DC.

Pellizzari, E.D., Perritt, K., Hartwell, T.D., Michael, L.C., Whitmore, R., Handy, R.W., Smith, D., and Zelon, H. (1987b) *Total Exposure Assessment Methodology (TEAM) Study: Selected Communities in Northern and Southern California, Vol. III*, U.S. Environmental Protection Agency, Washington, DC.

Pellizzari, E.D., Sheldon, L.S., Perritt, K., Hartwell, T.D., Michael, L.C., Whitmore, R., Handy, R.W., Smith, D., and Zelon, H. (1984b) *Total Exposure and Assessment Methodology (TEAM): Dry Cleaners Study*, EPA Contract No. 68-02-3626, U.S. Environmental Protection Agency, Office of Research and Development, Washington, DC.

Pellizzari, E.D., Wallace, L.A., and Gordon, S.M. (1992) Elimination Kinetics of Volatile Organics in Humans Using Breath Measurements, *Journal of Exposure Analysis and Environmental Epidemiology*, **2**(3): 341–356.

Pleil, J.D. and Lindstrom, A.B. (1995a) Collection of a Single Alveolar Exhaled Breath for Volatile Organic Compounds Analysis, *American Journal of Industrial Medicine*, **27**: 109–121.

Pleil, J.D. and Lindstrom, A.B. (1995b) Measurement of Volatile Organic Compounds in Exhaled Breath as Collected in Evacuated Electropolished Canisters, *Journal of Chromatography B*, **665**: 271–279.

Pleil, J.D. and Lindstrom, A.B. (1997) Exhaled Human Breath Measurement Method for Assessing Exposure to Halogenated Volatile Organic Compounds, *Clinical Chemistry*, **43**: 723–730.

Pleil, J.D., Fisher, J.W., and Lindstrom, A.B. (1998) Comparison of Human Blood and Breath Levels of Trichlorethylene from Controlled Inhalation Exposure, *Environmental Health Perspectives*, **106**(9): 573–580.

Prado, C., Garrido, J., and Periago, J.F. (2004) Urinary Benzene Determination by SPME/GC-MS: A Study of Variables by Fractional Factorial Design and Response Surface Methodology, *Journal of Chromatography B*, **804**: 255–261.

Prado, C., Marin, P., and Periago, J.F. (2003) Application of Solid-Phase Microextraction and Gas Chromatography-Mass Spectrometry to the Determination of Volatile Organic Compounds in End-Exhaled Breath Samples, *Journal of Chromatography A*, **1011**: 125–134.

Prazeller, P., Palmer, P.T., Boscaini, E., Jobson, T., and Alexander, M. (2003) Proton Transfer Reaction Ion Trap Mass Spectrometer, *Rapid Communications in Mass Spectrometry*, **17**: 1593–1599.

Preller, L., Zweers, T., Brunekreef, B., and Boleij, J. (1990) Sick Leave Due to Work-Related Health Complaints among Office Workers in the Netherlands, in *Indoor Air '90: Proceedings of the 5th International Conference on Indoor Air Quality and Climate, Toronto, Canada, 29 July–3 August 1990*, Walkinshaw, D., Ed., Canada Mortgage and Housing Association, Ottawa, Canada, **1**: 227–230.

Rafson, H.J., Ed. (1998) *Odor and VOC Control Handbook*, McGraw-Hill, NY.

Raymer, J.H., Thomas, K.W., Cooper, S.D., and Pellizzari, E.D. (1990) A Device for Sampling Human Alveolar Breath for the Measurement of Expired Volatile Organic Compounds, *Journal of Analytical Toxicology*, **14**: 337–344.

Raymer, J.H., Pellizzari, E.D., Thomas, K.W., and Cooper, S.D. (1991) Elimination of Volatile Organic Compounds in Breath after Exposure to Occupational and Environmental Microenvironments, *Journal of Exposure Analysis and Environmental Epidemiology*, **1**: 439–451.

Raymer, J.H., Pellizzari, E.D., Thomas, K.W., Kizakevich, P., and Cooper, S.D. (1992) Kinetics of Low-Level Volatile Organic Compounds in Breath, II: Relationship between Concentrations Measured in Exposure Air and Alveolar Air, *Journal of Exposure Analysis and Environmental Epidemiology*, **2**(supp. 2): 67–83.

Reitzig, M., Mohr, S., Heinzrow, B., and Knöppel, H. (1998) VOC Emissions after Building Renovations: Traditional and Less Common Indoor Air Contaminants, Potential Sources, and Reported Health Complaints, *Indoor Air*, **8**(1): 91–102.

Repace, J.L. and Lowrey, A.H. (1980) Indoor Air Pollution, Tobacco Smoke, and Public Health, *Science*, **208**: 464–472.

Repace, J.L. and Lowrey, A.H. (1985) A Quantitative Estimate of Non-Smokers' Lung Cancer Risk from Passive Smoking, *Environment International*, **11**: 3–22.

Rieder, J., Lirk, P., Ebenbichler, C., Gruber, G., Prazeller, P., Lindinger, W., and Amann, A. (2001) Analysis of Volatile Organic Compounds: Possible Applications in Metabolic Disorders and Cancer Screening, *Wien Klinische Wochenschrift*, **113**: 181–185.

Rook, J.J. (1974) Formation of Haloforms during Chlorination of Natural Waters, *Journal of Water Treatment and Examination*, **23**: 234.

Rook, J.J. (1976) Haloforms in Drinking Water, *Journal of the American Water Works Association*, **68**: 168–172.

Rook, J.J. (1977) Chlorination Reactions of Fulvic Acids in Natural Waters, *Environmental Science and Technology*, **11**: 478–482.

Rossner, A., Warner, S.D., Vyskocil, A., Tardif, R., and Farant, J.P. (2004) Performance of Small Evacuated Canisters Equipped with a Novel Flow Controller for the Collection of Personal Air Samples, *Journal of Occupational and Environmental Hygiene*, **1**: 173–181.

Rothweiler, H., Wäger, P.A., and Schlatter, C. (1991) Comparison of Tenax TA and Carbotrap for Sampling and Analysis of Volatile Organic Compounds in Air, *Atmospheric Environment*, **25B**: 231–235.

Roumasset, J.A. and Smith, K.R. (1990) Exposure Trading: An Approach to More Efficient Air Pollution Control, *Journal of Environmental Economics and Management*, **18**: 276–291.

Saarela, K., Tirkkonen, T., Laine-Ylijoki, J., Jurvelin, J., Nieuwenhuijsen, M.J., and Jantunen, M. (2003) Exposure of Population and Microenvironmental Distributions of Volatile Organic Compound Concentrations in the EXPOLIS Study, *Atmospheric Environment*, **37**: 5563–5575.

Sack, T.M., Steele, D.H., Hammerstrom, K., and Remmers, J. (1992) A Survey of Household Products for Volatile Organic Compounds, *Atmospheric Environment*, **26A**(6): 1063–1070.

Sapkota, A., Williams, D., and Buckley, T.J. (2005) Tollbooth Workers and Mobile Source-Related Hazardous Air Pollutants: How Protective Is the Indoor Environment?, *Environmental Science and Technology*, **39**: 2936–2943.

SCAQMD (1989) *In-Vehicle Characterization Study in the South Coast Air Basin*, South Coast Air Quality Management District, Los Angeles, CA.

Schimming, E., Levsen, K., Köhme, C., and Schürmann, W. (1999) Biomonitoring of Benzene and Toluene in Human Blood by Headspace-Solid-Phase Microextraction, *Fresenius Zeitschrift Analytische Chemie*, **363**: S.88–91.

Schreiber, J. (1992) An Exposure and Risk Assessment Regarding the Presence of Tetrachloroethylene in Human Breast Milk, *Journal of Exposure Analysis and Environmental Epidemiology*, **2**(supp. 2): 15–26.

Schreiber, J. (1993) An Assessment of Tetrachloroethylene in Breast Milk, *Risk Analysis*, 13(5): 515–524.

Schreiber, J., House, S., Prohonic, E., Smead, G., Hudson, C., Styk, M., and Lauber, J. (1993) An Investigation of Indoor Air Contamination in Residences above Dry Cleaners, *Risk Analysis*, **13**(3): 335–344.

Schreiber, J.S., Hudnell, H.K., Geller, A.M., House, D.E., Aldous, K.M., Force, M.S., Langguth, K., Prohonic, E.J., and Parker, J.C. (2002) Apartment Residents' and Day Care Workers' Exposures to Tetrachloroethylene and Deficits in Visual Contrast Sensitivity, *Environmental Health Perspectives*, **110**(7): 655–664.

Senthilmohan, S.T., McEwan, M.J., Wilson, P.F., Milligan, D.B., and Freeman, C.G. (2001) Real Time Analysis of Breath Volatiles Using SIFT-MS in Cigarette Smoking, *Redox Reports*, **6**: 185–187.

Sexton, K., Adgate, J.L., Ramachandran, G., Pratt, G.C., Mongin, S.J., Stock, T.H., and Morandi, M.T. (2004) Comparison of Personal, Indoor, and Outdoor Exposures to Hazardous Air Pollutants in Three Urban Communities, *Environmental Science and Technology*, **38**: 423–430.

Shapiro, S.D., Busenberg, E., Focazio, M.J., and Plummer, L.N. (2004) Historical Trends in Occurrence and Atmospheric Inputs of Halogenated Volatile Organic Compounds in Untreated Ground Water Used as a Source of Drinking Water, *The Science of the Total Environment*, **321**: 201–217.

Sheldon, L.S., Handy, R.W., Hartwell, T.D., Whitmore, R.W., Zelon, H.S., and Pellizzari, E.D. (1988a) *Indoor Air Quality in Public Buildings, Vol. I*, Report No. EPA 600/6-88/009a, U.S. Environmental Protection Agency, Research Triangle Park, NC.

Sheldon, L.S., Eaton, C., Hartwell, T.D., Zelon, H.S., and Pellizzari, E.D. (1988b) *Indoor Air Quality in Public Buildings, Vol. II,* Report No. EPA 600/6-88/009b, U.S. Environmental Protection Agency, Research Triangle Park, NC.

Shelow, D., Silvis, P., Schuyler, A., Stauffer, J., Pleil, J.D., and Holdren, M.W. (1994) Deactivating SUMMA® Canisters for Collection and Analysis of Polar Volatile Organic Compounds in Air, presented at the 1994 EPA/A&WMA Symposium on Measurement of Toxic and Related Air Pollutants, Durham, NC.

Sigsby, J.E., Tejada, S., and Ray, W. (1987) Volatile Organic Compound Emissions from 46 In-Use Passenger Cars, *Environmental Science and Technology*, **21**: 466–475.

Singer, B.C., Hodgson, A.T., and Nazaroff, W.W. (2003) Gas-Phase Organics in Environmental Tobacco Smoke: 2. Exposure-Relevant Emission Factors and Indirect Exposures from Habitual Smoking, *Atmospheric Environment*, **37**: 5551–5561.

Skov, P., Valbjorn, O., Pedersen, B.V., and the Danish Indoor Climate Study Group (1990) Influence of Indoor Climate on the Sick Building Syndrome in an Office Environment, *Scandinavian Journal of Work and Environmental Health*, **16**: 363–371.

Smith, D. and Španěl, P. (1996) The Novel Selected-Ion Flow Tube Approach to Trace Gas Analysis of Air and Breath, *Rapid Communications in Mass Spectrometry*, **10**: 1183–1198.

Smith, D., Španěl, P., Thomson, J.M., Rajan, B., Cocker, J., and Rolfe, P. (1998) The Selected Ion Flow Tube Method for Workplace Analysis of Trace Gases in Air and Breath: Its Scope, Validation and Applications, *Applied Occupational and Environmental Hygiene*, **13**: 817–823.

Smith, K.R. (1988a) Air Pollution: Assessing Total Exposure in the United States, *Environment*, 30(8): 10–38.

Smith, K.R. (1988b) Exposure Trading, *Environment*, October 1988.

Son, B., Breysse, P., and Yang, W. (2003) Volatile Organic Compounds Concentrations in Residential Indoor and Outdoor and Its Personal Exposure in Korea, *Environment International*, **29**: 79–85.

Spicer, C.W., Holdren, M.W., Slivon, L.E., Coutant, R.W., Graves, M.E., Shadwick, D.S., McClenny, W.A., Mulik, J.D., and Fitz-Simons, T.R. (1986) Intercomparison of Sampling Techniques for Toxic Organic Compounds in Indoor Air, in *Proceedings of the 1986 EPA/APCA Symposium on the Measurement of Toxic Air Pollutants*, Hochheiser, S. and Jayanti, R.K.M., Eds., Air Pollution Control Association, Pittsburgh, PA, 45–60.

Sulyok, M., Haberhauer-Troyer, C., Rosenberg, E., and Grasserbauer, M. (2001) Investigation of the Storage Stability of Selected Volatile Sulfur Compounds in Different Sampling Containers, *Journal of Chromatography A*, **917**: 367–374.

Sundell, J., Lonnberg G., Wall, S., Stenberg, B., and Zingmark, P.A. (1990) The Office Illness Project in Northern Sweden. Part III: A Case-Referent Study of Sick Building Syndrome (SBS) in Relation to Building Characteristics and Ventilation, in *Indoor Air '90: Proceedings of the 5th International Conference on Indoor Air Quality and Climate,* Vol. 4, Walkinshaw, D., Ed., Canada Mortgage and Housing Association, Ottawa, Canada, 633–638.

Symons, J.M., Bellar, T.A., Carsell, J.K., Demarco, J., Kropp, K.L., Robeck, G.G., Seeger, D.R., Slocum, C.J., Smith, B.L., and Stevens, A.A. (1975) National Organics Reconnaissance Survey for Halogenated Organics, *Journal of the American Water Works Association*, **67**: 708–729.

Tancrede, M., Wilson, R., Zeise, L., and Crouch, E.A. (1987) The Carcinogenic Risk of Some Organic Vapors Indoors: A Theoretical Survey, *Atmospheric Environment,* **21**: 2187–2205.

Ten Brinke, J., Selvin, S., Hodgson, A.T., Fisk, W.J., Mendell, M.J., Koshland, C.P., and Daisey, J.M. (1998) Development of New Volatile Organic Compound (VOC) Exposure Metrics and Their Relationship to Sick Building Syndrome Symptoms, *Indoor Air,* **8**(3): 140–152.

Thomas, K.W., Pellizzari, E.D., and Cooper, S.D. (1991) A Canister-Based Method for Collection and GC/MS Analysis of Volatile Organic Compounds in Human Breath, *Journal of Analytical Toxicology*, **15**: 54-59.

Thomas, K.W., Pellizzari, E.D., Clayton, C.A., Perritt, R.L., Dietz, R.N., Goodrich, R.W., Nelson, W.C., and Wallace, L.A. (1993) Temporal Variability of Benzene Exposure for Residents in Several New Jersey Homes with Attached Garages or Tobacco Smoke, *Journal of Exposure Analysis and Environmental Epidemiology,* **3**: 49–73.

Thomas, K.W., Pellizzari, E.D., Perritt, R.L., and Nelson, W.C. (1991) Effect of Dry-Cleaned Clothes on Tetrachloroethylene Levels in Indoor Air, Personal Air, and Breath for Residents of Several New Jersey Homes, *Journal of Exposure Analysis and Environmental Epidemiology,* **1**: 475–490.

Tichenor, B.A. and Mason, M.A. (1987) Organic Emissions from Consumer Products and Building Materials to the Indoor Environment, *Journal of the Air Pollution Control Association,* **38**: 264–268.

Tichenor, B.A., Sparks, L.E., White, J.B., and Jackson, M.D. (1990) Evaluating Sources of Indoor Air Pollution, *Journal of the Air & Waste Management Association,* **41**: 487–492.

Traynor, G.W., Apte, M.G., Sokol, H. A., Chuang, J.C., Tucker, W.G., and Mumford, J.L. (1990) Selected Organic Pollutant Emissions from Unvented Kerosene Space Heaters, *Environmental Science & Technology,* **24**: 1265–1270.

USEPA (1987a) *Household Solvent Products: A Shelf Survey with Laboratory Analysis*, Report No. EPA-OTS 560/5-87-006, U.S. Environmental Protection Agency, Washington, DC.

USEPA (1987b) *Unfinished Business: A Comparative Assessment of Environmental Problems, Vol. I, Overview,* NTIS # PB-88-127048, U.S. Environmental Protection Agency, Washington, DC.

USEPA (1989) *Comparing Risks and Setting Environmental Priorities: Overview of Three Regional Projects,* U.S. Environmental Protection Agency, Washington, DC.

USEPA (1995) *The Inside Story: A Consumer's Guide to Indoor Air Pollution*, U.S. Environmental Protection Agency, Washington, DC.

Vas, G. and Vékey, K. (2004) Solid-Phase Microextraction: A Powerful Sample Preparation Tool Prior to Mass Spectrometric Analysis, *Journal of Mass Spectrometry*, **39**: 233–254.

Wady, L., Bunte, A., Pehrson, C., and Larsson, L. (2003) Use of Gas Chromatography-Mass Spectrometry/Solid Phase Microextraction for the Identification of MVOCs from Moldy Building Materials, *Journal of Microbiological Methods*, **52**: 325–332.

Wallace, L.A. (1987) *The TEAM Study: Summary and Analysis: Volume I*, Report No. EPA 600/6-87/002a, NTIS PB 88-100060, U.S. Environmental Protection Agency, Washington, DC.

Wallace, L.A. (1990) Major Sources of Exposure to Benzene and Other Volatile Organic Compounds, *Risk Analysis*, **10**: 59–64.

Wallace, L.A. (1991) Comparison of Risks from Outdoor and Indoor Exposure to Toxic Chemicals, *Environmental Health Perspectives*, **95**: 7–13.

Wallace, L.A. (1996) Environmental Exposure to Benzene: An Update, *Environmental Health Perspectives*, **104**(Supplement 6): 1129–136.

Wallace, L.A. (1997) Human Exposure and Body Burden for Chloroform and Other Trihalomethanes, *Critical Reviews in Environmental Science and Technology*, **27**: 113–194.

Wallace, L.A. (2001) Human Exposure to Volatile Organic Pollutants: Implications for Indoor Air Studies. *Annual Review of Energy and the Environment*, **26**: 269–301.

Wallace, L.A., Buckley, T., Pellizzari E.D., and Gordon, S. (1996) Breath Measurements as VOC Biomarkers: EPA's Experience in Field and Chamber Studies, *Environmental Health Perspectives*, **104** (suppl) 5, 861–869.

Wallace, L.A., Nelson, W.C., Pellizzari, E.D. and Raymer, J.H. (1997) A Four Compartment Model Relating Breath Concentrations to Low-Level Chemical Exposures: Application to a Chamber Study of Five Subjects Exposed to Nine VOCs, *Journal of Exposure Analysis and Environmental Epidemiology*, **7**(2): 141–163.

Wallace, L.A., Nelson, W.C., Ziegenfus, R., Pellizzari, E. (1991a) The Los Angeles TEAM Study: Personal Exposures, Indoor-Outdoor Air Concentrations, and Breath Concentrations of 25 Volatile Organic Compounds, *Journal of Exposure Analysis and Environmental Epidemiology*, **1**(2): 37–72.

Wallace, L.A., Pellizzari, E.D., and Gordon, S. (1993a) A Linear Model Relating Breath Concentrations to Environmental Exposures: Application to a Chamber Study of Four Volunteers Exposed to Volatile Organic Chemicals, *Journal of Exposure Analysis and Environmental Epidemiology*, **3**(1): 75–102.

Wallace, L.A., Pellizzari, E.D., Hartwell, T.D., Davis, V., Michael, L.C., and Whitmore, R.W. (1989) The Influence of Personal Activities on Exposure to Volatile Organic Compounds, *Environmental Research*, **50**: 37–55.

Wallace, L.A., Pellizzari, E.D., Hartwell, T., Perritt, K., and Ziegenfus, R. (1987b) Exposures to Benzene and Other Volatile Organic Compounds from Active and Passive Smoking, *Archives of Environmental Health*, **42**: 272–279.

Wallace, L.A., Pellizzari, E.D., Hartwell, T., Rosenzweig, R., Erickson, M., Sparacino, C., and Zelon, H. (1984a) Personal Exposure to Volatile Organic Compounds: I. Direct Measurement in Breathing-Zone Air, Drinking Water, Food, and Exhaled Breath, *Environmental Research*, **35**: 293–319.

Wallace, L.A., Pellizzari, E.D, Hartwell, T., Sparacino, C., Sheldon, L., and Zelon, H. (1984b) Personal Exposures, Indoor-Outdoor Relationships and Breath Levels of Toxic Air Pollutants Measured for 355 Persons in New Jersey, *Atmospheric Environment*, **19**: 1651–1661.

Wallace, L.A., Pellizzari, E.D., Hartwell, T.D., Sparacino, C., Whitmore, R., Sheldon, L., Zelon, H., and Perritt, R. (1987a) The TEAM Study: Personal Exposures to Toxic Substances in Air, Drinking Water, and Breath of 400 Residents of New Jersey, North Carolina, and North Dakota, *Environmental Research*, **43**: 290–307.

Wallace, L.A., Pellizzari, E.D., Hartwell, T.D., Whitmore, R., Perritt, R. and Sheldon, L. (1988) The California TEAM Study: Breath Concentrations and Personal Exposures to 26 Volatile Compounds in Air and Drinking Water of 188 Residents of Los Angeles, Antioch, and Pittsburgh, CA, *Atmospheric Environment*, **22**: 2141–2163.

Wallace, L.A., Pellizzari, E., Leaderer, B., Hartwell, T., Perritt, R., Zelon, H., and Sheldon, L. (1987c) Emissions of Volatile Organic Compounds from Building Materials and Consumer Products, *Atmospheric Environment*, **21**: 385–393.

Wallace, L.A., Pellizzari, E. and Wendel, C. (1991b) Total Volatile Organic Concentrations in 2700 Personal, Indoor, and Outdoor Air Samples Collected in the US EPA TEAM Studies, *Indoor Air*, **4**: 465–477.

Wallace, L. A., Zweidinger, R., Erickson, M., Cooper, S., Whitaker, D., and Pellizzari, E.D. (1982) Monitoring Individual Exposure: Measurement of Volatile Organic Compounds in Breathing-Zone Air, Drinking Water, and Exhaled Breath, *Environment International*, **8**: 269-282.

Weisel, C.P. and Chen, W.J. (1993) Exposure to Chlorination By-Products from Hot Water Uses, *Risk Analysis* **14**: 101–106.

Weisel, C.P., Lawryk, N.J., and Lioy, P.J. (1992) Exposure to Emissions from Gasoline within Automobile Cabins, *Journal of Exposure Analysis and Environmental Epidemiology*, **2**(1): 79–96.

Weschler, C.J. and Shields, H.C. (1999) Indoor Ozone/Terpene Reactions as a Source of Indoor Particles, *Atmospheric Environment*, **33**: 2301–2312.

Wilkes, C.R., Small, M.J., Andelman, J.B., Giardino, N.J., and Marshall, J. (1990) Air Quality Model for Volatile Constituents from Indoor Uses of Water, in *Indoor Air '90: Proceedings of the 5th International Conference on Indoor Air Quality and Climate, Toronto, Canada,* Walkinshaw, D.S., Ed., Canada Mortgage and Housing Association, Ottawa, Canada, **2**: 783–788.

Wilson, A.L., Colome, S.D., and Tian, Y. (1993) *California Residential Indoor Air Quality Study, Volume I: Methodology and Descriptive Statistics,* Integrated Environmental Services, Irvine, CA.

Wilson, P.F., Freeman, C.G., McEwan, M.J., Allardyce, R.A., and Shaw, G.M. (2003) SIFT-MS Measurement of VOC Distribution Coefficients in Human Blood Constituents and Urine, *Applied Occupational and Environmental Hygiene,* **18**: 759–763.

Wilson, P.F., Freeman, C.G., McEwan, M.J., Milligan, D.B., Allardyce, R.A., and Shaw, G.M. (2002) In Situ Analysis of Solvents on Breath and Blood: A Selected Ion Flow Tube Mass Spectrometric Study, *Rapid Communications in Mass Spectrometry,* **16**: 427–432.

Winberry, W.T., Jr., Forchard, L., Murphy, N.T., Ceroli, A., Phinnay, B., and Evans, A. (1993) *Methods for Determination of Indoor Air Pollutants: EPA Methods,* Noyes Data Corporation, Park Ridge, NJ.

Winberry, W.T., Jr., Murphy, N.T., and Riggin, R.M. (1990) *Methods for Determination of Toxic Organic Compounds in Air: EPA Methods,* Noyes Data Corporation, Park Ridge, NJ, http://www.epa.gov/ttn/amtic/airtox.html (accessed on September 27, 2005).

Zhang, Z., Yang, M., and Pawliszyn, J. (1994) Solid Phase Microextraction: a Solvent-Free Alternative in Sample Preparation, *Analytical Chemistry,* **66**: 844A–845A.

Zweidinger, R.B., Sigsby, J.E., Tejada, S.B., Stump, F.D., Dropkins, D.L., and Ray, W.D. (1988) Detailed Hydrocarbon and Aldehyde Mobile Source Emissions from Roadway Studies, *Environmental Science and Technology,* **22**: 956–962.

8 Exposure to Particles

Lance A. Wallace
U.S. Environmental Protection Agency (ret.)

Kirk R. Smith
University of California

CONTENTS

8.1 Synopsis 181
8.2 Introduction 182
8.3 Particle Characteristics 183
8.4 Particle Exposures 184
8.5 Studies of the Exposure of High-Risk Subpopulations in the Developed World 186
8.6 Exposure to Particles of Outdoor Origin 187
        8.6.1 Separating Indoor from Outdoor Particles in the Measurements 187
        8.6.2 The Infiltration Factor 188
        8.6.3 The Ambient Exposure Factor 191
8.7 Total Particle Exposure 192
8.8 Reduction of Exposure 192
8.9 Summary 194
8.10 Questions for Review 194
8.11 Answers to Questions for Review 195
References 195

8.1 SYNOPSIS

The World Health Organization (WHO) estimates that about 800,000 premature deaths occur each year due to urban outdoor particle air pollution around the world. A further 1.6 million premature deaths are attributed to indoor pollution from solid fuel use for cooking and heating in developing countries. Studies in many cities have shown that more deaths occur on days of high air pollution, and the responsible pollutant is most often identified as particles. Fine particles are created by combustion processes, and can evade the body's defenses and penetrate the lungs and bloodstream. The precise mechanisms by which particles cause morbidity and mortality are uncertain, and the critical toxic components of the particles are also unclear. Exposure studies are helping to reduce this uncertainty by determining the relationship between personal exposure and the concentrations measured at central site monitors, particularly for those people at highest risk. Exposure studies have also identified the most important sources of particles: smoking, heating, cooking, and traffic. This chapter reports on the basic studies that have provided most of our knowledge about the sources of indoor particles and the extent of human exposure. It provides the basic mathematical relationship between outdoor and indoor concentrations of particles, and presents recent data showing the range of "protectiveness" of homes with respect to outdoor pollution. "Tight" homes that reduce the penetration of outdoor particles unfortunately increase the concentrations of indoor-generated particles, so other means of reducing exposure to particles are considered. It is shown that certain high-performance air cleaners can cut particle levels in homes by about a factor of two. This would be a useful protective step to take for persons at risk, such as those with chronic respiratory or cardiovascular problems.

In the developing world, the particle emissions from solid fuels used for cooking and heating, the extended cooking period with women and children in close proximity to the fire, the poor ventilation in many homes, and the lack of stoves with vented chimneys are a combination that produces high indoor exposures. Some countries are taking steps to reduce the toll of sickness and death by providing cleaner fuels, improved stoves, chimneys, and information to their citizens.

8.2 INTRODUCTION

Invisible particles in the air we breathe have recently been identified as probably the greatest cause of premature death and ill health of all environmental pollutants. Studies in hundreds of cities around the world show higher mortality and morbidity occurring on days when outdoor particle concentrations are slightly higher than normal. Estimates of annual particle-related deaths in the United States are in the neighborhood of 20,000, which would make airborne outdoor particles the greatest of all environmental causes of death in the United States (not counting behaviors such as smoking and diet). Morbidity, including loss of productivity and increased childhood asthma, is estimated to cost the nation on the order of $100 billion yearly (Fisk and Rosenfeld 1997).

In the developing world, there is also a great toll from urban ambient particles, some 400,000 premature deaths per year in India and China alone, for example (Cohen et al. 2004; International Scientific Oversight Committee, 2004). An even greater toll is caused by poorly vented indoor cooking and heating with household solid fuels such as wood, crop residues, and coal. Recent estimates put the toll at about 1.6 million premature deaths, albeit with an uncertainty of about 50% (Smith, Mehta, and Maeusezahl-Feuz 2004). Although there are many pollutants in the smoke generated by solid fuels, as in other combustion situations, the best single indicator of health risk is thought to be small particles. Taken together, the impact of indoor and outdoor particle exposures is thought to be nearly 2.4 million premature deaths, greater than that of any other environmental pollutant globally. Occupational particle exposures add another 240,000. For comparison, the next largest type of environmental pollution, due to poor water and sanitation, is thought to be responsible for about 1.7 million premature deaths annually (WHO 2002).

The principal source of these findings for the effect of outdoor particles on short-term risk has been epidemiological studies, both cohort studies comparing effects across different cities (e.g., Dockery et al. 1993) and time-series studies of daily deaths in cities as a function of air pollution (Dockery, Schwartz, and Spengler 1992; Schwartz, Dockery, and Neas 1996). Such time-series studies use each person as his own control and are thus not plagued by most confounders (such as smoking) encountered in more traditional cross-sectional studies. However, both types of studies depend on the assumption that changes in personal exposure to outdoor-generated particles are highly correlated with changes in concentrations measured at central sites. For exposure scientists, this assumption can be tested with properly designed studies, and indeed such studies have recently been carried out and will be examined toward the end of this chapter.

Although outdoor particle levels have greatly decreased in much of the developed world in the last 100 years, our present understanding is that typical current particle concentrations are not protective against these health effects, and that relatively common fluctuations in particle concentrations in even the cleanest areas are associated with measurable increases in mortality in many cities. For example, although typical annual average levels in U.S. and European cities of PM_{10} (particles less than 10 micrometers [μm] in aerodynamic diameter) are 10–40 μg/m^3, daily changes of 10 μg/m^3 are associated with an approximate increase of 0.5% in mortality in 20 major U.S. cities (Samet et al. 2000).

Although the relatively few indoor concentration and exposure studies done in developing country households provide evidence that particle exposures in these settings are often great, they are not yet of sufficient number, quality, and representativeness to be used for health risk estimates. Indeed, this is to a large extent true of indoor exposure information across the world, whether in developed or developing countries, because of the difficulty, intrusiveness, and expense of doing such measurements.

In the households of the developing world, therefore, health impact information comes from epidemiological studies using less precise indicators of exposure, such as whether or not solid fuels are used or whether a child has been carried on the mother's back during cooking. Although such crude indicators undoubtedly lead to some exposure misclassification, the results have been remarkably consistent in studies in different parts of the world for a wide range of health effects (Smith 1987, 1988b, 1993, 1996, 2000; Chen et al. 1990; Smith et al. 1993, 2000; Bruce et al. 2000). The most well-established risks are for acute lower respiratory infections in children and chronic obstructive lung disease and lung cancer in adult women. Multiple studies have also shown effects, however, for tuberculosis, adverse pregnancy outcomes, cataracts, and other cancers. Impacts on heart disease are also suspected, as they have been demonstrated for outdoor particle pollution (Smith, Mehta, and Maeusezahl-Feuz 2004).

8.3 PARTICLE CHARACTERISTICS

Airborne particles are produced by several mechanisms, the most important of which are combustion and mechanical grinding or resuspension. Combustion processes, whether burning natural gas, oil, coal, or wood, can produce particles of all sizes, ranging from invisible particles scarcely larger than a molecule to visible soot and ash, although most mass is usually concentrated in sizes between about 0.1 and 1 μm. Grinding processes or windblown dust generally result in larger particles, which settle out more quickly than the small combustion particles.

The behavior of particles is determined largely by their size. The smallest particles are called *nanoparticles*, and are less than 50 nanometers (nm), or 0.05 micrometers (μm) in diameter. (Medium-size molecules may be a few nanometers in diameter.) These are created in a few milliseconds from the gases produced by incomplete combustion. They may consist of many different hydrocarbons, including acetylene and benzene. Once the first benzene rings are formed in the flame others can be attached to create polycyclic aromatic hydrocarbons (PAHs), some of which cause cancer in rodents, and probably humans as well. Gas stoves create particles with peak numbers at about 10 nm in diameter, whereas electric toasters create particles peaking at about 30 nm (Wallace 2000b; Wallace and Howard-Reed, 2002). These nanoparticles quickly coagulate into *ultrafine* particles (less than 100 nm in diameter), and these may further coagulate into *accumulation mode* particles between 0.1 and 1 μm in diameter.

The U.S. Environmental Protection Agency (USEPA) defined *fine* particles as all those less than 2.5 μm ($PM_{2.5}$), and *inhalable* particles as all those less than 10 μm (PM_{10}). The range between 2.5 and 10 μm became known as *coarse* particles.

Particle size governs how fast particles will fall to the ground (outdoors) or deposit on walls, ceilings, or floors (indoors). The smallest (ultrafine) particles will quickly deposit on nearby surfaces or on larger particles because of rapid incessant jittering from being struck by air molecules (*Brownian motion*). The larger coarse particles will also settle out quickly, due to their large mass (a 10-μm particle is 1 million times heavier than the largest ultrafine [0.1 μm] particle). But the accumulation mode (0.1–1 μm) will stay airborne for long periods of time.

Particle size also determines how deeply they will penetrate into the lungs. The ultrafine particles can penetrate all levels of the lungs and *brachioles* (small branchings of the lungs) down to the *alveolar sacs* (where oxygen is exchanged with the blood), whereas coarse particles may be filtered out by the nasal passages.

Composition is also important. Fine particles in the eastern part of the United States consist largely of sulfates (typically 40–50% of the total mass), most produced by the great coal-burning electric power plants of the Midwest. Another large portion consists of carbon particles, either *elemental carbon* (or soot, about 5–10% of the mass) or *organic* carbon (typically 20–40% of the mass). Windblown soil, or *crustal* material, makes up the remainder of the $PM_{2.5}$ particles in the east, and perhaps half of the total PM_{10}. In the western part of the country, sulfates are not so prevalent, but nitrates from agricultural feedlots and mobile sources often form a major portion of the particles found outdoors. Woodsmoke and other biomass smokes also contribute significantly to outdoor as well as indoor particle exposures in many parts of the world (Naeher et al. 2005). Although woodsmoke particles are different in composition from particles due to combustion of fossil fuels, there is no definitive evidence indicating a difference in effect (WHO 2006).

Despite considerable knowledge about how particles are produced, what they are made of, and which particles deposit where in the lungs, we still do not know what produces the observed health effects. Among the many hypothesized causal agents are the ultrafines (which are more toxic than larger particles), metals (which undergo chemical reactions), fine particles (which deposit deep in the lungs), coarse particles (which include mold spores and other biological aerosols), and combustion particles (which include highly reactive oxygenated species). Some studies can be found that give support to each of these hypotheses, and none (so far) have been found that rule out any of these hypotheses. In general, however, control methods focused on PM_{10} and $PM_{2.5}$ with no further distinctions have greatly reduced health risks.

8.4   PARTICLE EXPOSURES

The first and still the most productive of all studies of exposure to particles was the Harvard 6-City Study (Dockery and Spengler 1981) begun in the 1970s and continuing to the present day. Thousands of persons have had their homes measured for particle levels and their lung functions tested. Early findings showed a decrease in lung function associated with particle exposure. The decrease was faster for persons at higher exposures, such as those living in homes with smokers. Homes with smokers averaged 48 $\mu g/m^3$ $PM_{2.5}$ compared to 17 $\mu g/m^3$ in homes without smokers (Neas et al. 1994). This increase of about 30 $\mu g/m^3$ produced by smoking has been confirmed by many U.S. studies (e.g., Clayton et al. 1993; Miller and Nazaroff 2001). Models of nonsmokers' exposure to environmental tobacco smoke have also been developed (e.g., Klepeis et al. 2003).

Following the time series studies associating *daily* mortality with air pollution, the 6-City Study was revisited to determine whether mortality increased *across cities* with higher outdoor pollution (Dockery et al. 1993). Indeed this was found, adding evidence of chronic health effects from long-term particle exposure to the acute effects identified by the time series studies. A more recent revisit of the 6-City Study used tracer metals to apportion particle mixtures to several sources, concluding that combustion-related particles were associated with increased mortality but that crustal (soil-related) particles were not (Laden et al. 2000).

A second important study aimed at determining the impact of indoor combustion sources on particle levels was sponsored by the Department of Energy in two New York State counties. Four indoor combustion sources were considered: smoking, gas stoves, wood stoves or fireplaces, and kerosene heaters. Of the four, only smoking was found to increase concentrations significantly (Koutrakis, Briggs, and Leaderer 1992).

A third large-scale study was the particle total exposure assessment methodology (PTEAM) study in Riverside, CA, in 1989–1990 (Clayton et al. 1993; Özkaynak et al. 1996). PTEAM was the first particle study with a probability-selection design. (Probability-weighted selection is the only accepted way to extrapolate from the monitored group to a larger population. See Chapter 3 by Whitmore in this book.) One hundred seventy-eight persons represented the entire population of Riverside (130,000). At this time, the USEPA had revised its National Ambient Air Quality

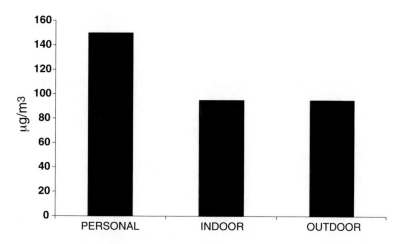

FIGURE 8.1 Observed mean personal exposures, indoor and outdoor concentrations of PM_{10} during the 12-hour daytime period in the PTEAM Study. The personal cloud may be defined as the difference between the personal exposures and indoor concentrations.

Standard (NAAQS) to focus on PM_{10}. Therefore the PTEAM study included personal monitoring for PM_{10}, although indoor and outdoor monitoring included both PM_{10} and $PM_{2.5}$.

One major finding of the PTEAM study was the discovery that personal PM_{10} exposures were nearly always greater than either indoor or outdoor concentrations (Figure 8.1). This was interpreted as being due to personal activities, and could be visualized as a cloud of particles surrounding the person: the "personal cloud" effect, or sometimes more graphically referred to as the "Pigpen effect" after the beloved character in the Peanuts comic strip perpetually surrounded by a cloud of dust.

The second major finding of the PTEAM study was that cooking appeared to be the second most important single indoor source of particles. While smoking appeared to add about 30 $\mu g/m^3$ to indoor concentrations, cooking seemed to add 10–15 $\mu g/m^3$ to the PM_{10} total.

Concurrently in the developing world, interest began to focus on cooking and heating fuels. The concept of the "household energy (or fuel) ladder" (Office of Technology Assessment 1990; Smith et al. 1994) has been used to frame the fashion in which societies progress from solid biofuels such as wood and cow dung to liquid fuels such as kerosene and to liquefied petroleum gas to natural gas for household cooking and heating. Although individual households often use more than one fuel, it is clear that each step up the ladder in population represents energy efficiency increases and pollution decreases. Besides the fuel itself, however, several other characteristics have led to greatly increased indoor air pollution, experienced mostly by women and children. One of these is the traditional method of cooking, which often involves long periods of cooking, burning, and being near the fire. Another is the lack of ventilation in many homes, as a means of conserving heat or simply because of poor construction due to poverty. And a third is the lack of chimneys to vent the combustion particles. Finally, infants are often close to their mothers, perhaps on her back in a sling, as she carries out the cooking task. Thus this form of particle pollution is gender- and age-specific, singling out women and children as its major target population, although also exposing men and older children.

Early studies also showed that simple improved solid-fuel stoves with chimneys often did not lower exposures significantly over time because of maintenance and construction problems and because the smoke was not eliminated but merely removed from the room into the immediate environment of the house (Reid et al. 1986; Ramakrishna et al. 1989). Further work documented this "neighborhood pollution," which often surrounds communities of solid fuel–using households (Smith et al. 1994). This is at a scale intermediate between indoor and ambient, complicating exposure assessment because the pollution is poorly indicated by measurements at these other scales.

Providing evidence of the serious effects of these exposures on these populations, meta-analyses of published epidemiological studies examining the impact of using solid fuels in simple stoves without good ventilation show that women cooking for many years bear about a 220% excess of chronic obstructive pulmonary disease (relative risk = 3.2; confidence interval: 2.3–4.8) and young children experiencing such conditions have a 130% increase in lower respiratory infections (relative risk = 2.3; confidence interval: 1.9–2.7), which are the chief cause of childhood mortality in the world (Smith, Mehta, and Maeusezohl-Feuz 2004).

8.5 STUDIES OF THE EXPOSURE OF HIGH-RISK SUBPOPULATIONS IN THE DEVELOPED WORLD

The studies described above were mostly concerned with healthy persons, although the 6-City Study did include many children with asthma. However, the mortality studies suggested that the persons most sensitive to particle exposure were those with respiratory or cardiovascular problems. Therefore it became important to study such high-risk subpopulations, since their activities and resulting exposures might be quite different than those of healthy persons. One hypothesis was that they would be less active, stay at home more, perform fewer particle-generating activities in their home, and thus be more likely to be affected by changes in outdoor air pollution than healthy persons.

Previously, due to study designs involving short-term monitoring of a large number of people, the relation between exposure and outdoor concentrations had been compared on a *cross-sectional* basis; that is, single-day exposures of *different people* were compared to outdoor concentrations. These cross-sectional correlations were typically very low. For example, the Pearson correlation r between personal PM_{10} exposure and outdoor concentrations in the PTEAM study was only 0.4, indicating that only about 16% of the variation in measured exposures was due to outdoor concentrations (Özkaynak et al. 1996). However, the correlation more relevant to the time-series studies was that between *each individual's* exposure and the outdoor concentration *over a number of days*: the *longitudinal* correlation.

Therefore in the mid-1990s a series of studies were launched of these sick, usually elderly, persons. Early studies were sponsored by the Electric Power Research Institute and the American Petroleum Institute, in Nashville and Boston (Rojas-Bracho, Suh, and Koutraki, 2000). Two even earlier studies with a longitudinal component were the PTEAM pilot study, which followed 18 people over 5–7 days (Spengler et al. 1989) and a study in Phillipsburg, NJ, of 14 persons monitored for 14 days (Lioy et al. 1990). All four of these studies found that only about half the monitored subjects had moderate longitudinal correlations (coefficients >0.5) of personal exposure with outdoor concentrations (Wallace 2000a).

Additional studies were carried out in Europe, with similar findings (Janssen et al. 1997, 1998, 1999, 2000). More studies in the United States and Canada followed, continuing to find much the same results (Abt et al. 2000; Ebelt et al. 2000; Keeler et al. 2002; Long et al. 2001; Wallace et al. 2003; Williams et al. 2003).

In 1996, as part of an evaluation of USEPA's research on particle air pollution, the National Academy of Sciences recommended that further attempts be made to relate personal particle exposure to outdoor concentrations for high-risk subpopulations (NRC-NAS 1998). By this time, the USEPA had once again changed the focus of its particle regulations, selecting fine particles ($PM_{2.5}$) as a new target, while retaining the older regulation on PM_{10}. The Academy recommended that the USEPA carry out research to answer the following question:

What are the quantitative relationships between concentrations of particulate-matter and gaseous co-pollutants measured at stationary outdoor air-monitoring sites, and the contributions of these concentrations to actual personal exposures, especially for potentially susceptible subpopulations and individuals? (p. 46)

Congress responded by appropriating funds for use in mounting such studies, and the USEPA sponsored four new studies on personal exposure to particles, focusing mainly on $PM_{2.5}$ (USEPA 2002, 2004). The studies were carried out during 1998–2003 by the Harvard University School of Public Health, the University of Washington, the New York University School of Medicine, and the National Exposure Research Laboratory of USEPA. All studies used the same set of questionnaires and the same core measurements (personal exposure, indoor and outdoor measurements of particles and associated gases) carried out over a number of days in different seasons. Five cities were included: Atlanta, Boston, Los Angeles, New York, and Seattle. Subjects included adults with respiratory or cardiovascular disease and children with asthma.

A major goal of the studies was to relate personal $PM_{2.5}$ exposure to measurements made outdoors, and the studies were remarkably consistent in finding that 40–50% of respondents had moderate longitudinal correlations of 0.5 or better over periods of 10 to 14 days. However, even for these moderate correlations, less than half of them were significantly different from zero (USEPA 2004).

8.6 EXPOSURE TO PARTICLES OF OUTDOOR ORIGIN

These correlations between exposure to particles from *all sources* (both indoor and outdoor) and concurrent outdoor concentrations are *not* the correlations of interest to regulators interested in tracing the influence of outdoor particles. To them, the relevant correlation is that between exposure to particles *of outdoor origin* and outdoor concentrations. To see this, assume that both indoor and outdoor-generated particles can cause mortality. Then the observed daily mortality will be due partly to both sources. However, the *variations* in that daily mortality, which are correlated with *variations* in outdoor particle concentrations, must be due to the outdoor-generated particles only. (This conclusion depends on *no* correlation of indoor-generated particles with outdoor concentrations. But this seems reasonable, since it is unlikely, for example, that people smoke more cigarettes or cook more often on days with higher pollution.) From this point of view, the possible toxicity of indoor particles is *irrelevant* to the question of the relation of personal exposure to outdoor-generated particles.

On the other hand, to those concerned with *health*, as opposed to those concerned with *outdoor regulation*, the health effects of indoor-generated particles are also of interest. For these persons, it matters if indoor particles are toxic, because their risk affects the types of controls that might be needed to reduce exposure. For example, certain actions to reduce exposure to outdoor particles, such as tightening homes and closing windows, lead directly to *higher* concentrations of indoor-generated particles, whereas other actions (using exhaust fans and high-quality air filters) lead to lower concentrations of *both* indoor-generated and outdoor-generated particles.

Consequently, the USEPA-sponsored studies included measurements that allowed for a separation of the indoor-generated particles from outdoor particles (USEPA 2004).

8.6.1 Separating Indoor from Outdoor Particles in the Measurements

Imagine that the indoor space is a single well-ventilated volume, with outdoor particles entering uniformly from all points of the building envelope. Then the indoor concentration will change from moment to moment according to (1) the penetration of outdoor particles at a certain concentration; (2) the creation of particles from indoor sources; and (3) the loss of particles either through depositing on walls and floors or leaving the house through cracks or open windows. These considerations lead to the following differential equation:

$$\frac{dC_{in}}{dt} = Pa\,C_{out} - (a + k)C_{in} + \frac{S}{V} \tag{8.1}$$

where

C_{in} = indoor number or mass concentration (cm^{-3} or µg/m^3)
C_{out} = outdoor number or mass concentration
P = penetration coefficient across building envelope
A = air exchange rate (h^{-1})
V = volume of building (cm^3 or m^3)
S = indoor-generated source strength (h^{-1} or µg h^{-1})
k = total decay rate of particles (h^{-1})

This *mass balance equation* is considered to be applicable to all particle sizes, with all terms except air exchange rate and building volume considered to be functions of particle size. The further assumption is made that the entire house is a single well-mixed zone, with instantaneous mixing of particles throughout the house, and that the measured air exchange rate in one room applies to the entire house. Finally, we assume that the averaging time over which the equation is to be evaluated is sufficiently long that transient terms due to short-term changes in the outdoor concentration are negligible compared to the long-term average concentrations. Under these assumptions, the solution to the mass balance equation is

$$C_{in} = \frac{Pa}{a+k} C_{out} + \frac{S}{V(a+k)} \tag{8.2}$$

where the various terms are averages over the integration period.

A more complete derivation including physical processes such as coagulation and filtration may be found in Nazaroff and Cass (1989). See also Chapter 18 in this book on mass balance modeling of indoor sources to predict indoor air quality.

8.6.2 THE INFILTRATION FACTOR

The coefficient of the outdoor concentration is sometimes called the infiltration factor:

$$F_{inf} = \frac{Pa}{a+k} \tag{8.3}$$

The infiltration factor determines the indoor–outdoor relationship, and to a large degree, the personal–outdoor relationship. This is expected to vary by household and residence characteristics. For example, a tightly built house may have a lower penetration coefficient than a drafty house. A house with a large surface/volume ratio area (e.g., many carpets, rugs, or fibrous wall hangings) may have higher deposition rates (Lai and Nazaroff 2000). Use of fans or filters may also increase particle deposition rates (Thatcher and Layton 1995; Howard-Reed, Wallace, and Emmerich 2003; Wallace, Emmerich, and Howard-Reed, 2004a). Open windows will increase F_{inf} by increasing the air exchange rate (Howard-Reed, Wallace, and Ott 2002; Wallace, Howard-Reed, and Emmerich 2002) and possibly by redirecting infiltrating particles through the open window ($P = 1$) rather than through the rest of the building envelope ($P < 1$) (Liu and Nazaroff 2001). Use of air conditioning has been shown to lower the infiltration factor, either because of reducing air exchange rates by shutting windows or increasing deposition rates by recirculation of indoor air through ductwork (Howard-Reed, Wallace, and Emmerich 2003; Sarnat, Koutrakis, and Suh 2000; Lai Burne, and Goddard 1999; Thornburg et al. 2004). For example, a recent study known as RIOPA (Residential Indoor, Outdoor, and Personal Air) found that the infiltration factor was much lower in Houston,

TX, than in Los Angeles, CA, or Elizabeth, NJ, most likely due to the extensive use of air conditioning in the Texas homes (Meng et al. 2005).

The infiltration factor is the crucial term that must be calculated for each house to distinguish the measured indoor concentration into the portions due to indoor-generated and outdoor-generated particles. How can this term be calculated? Recall from Equation 8.2 that we (generally) have no information on the value of the indoor-generated particle source term S. However, if we can measure something that has no indoor sources, we can determine the infiltration factor directly from a measurement of the indoor and outdoor concentrations. A number of studies indicate that there are few or no indoor sources of sulfur (Sarnat et al. 2002; Wallace and Williams 2005). If so, and if sulfur particles (which often make up a substantial fraction of $PM_{2.5}$) behave like $PM_{2.5}$, then the $PM_{2.5}$ infiltration factor is simply equal to the ratio of the indoor and outdoor sulfur (S) concentrations:

$$F_{inf} = \frac{S_{in}}{S_{out}} \tag{8.4}$$

In the USEPA study of 36 homes in North Carolina, measurements included x-ray fluorescence analyses of the particle filters to measure sulfur. The $PM_{2.5}$ infiltration factor calculated in this way ranged from 0.26–0.89 (Figure 8.2). The median value was 0.59, and the interquartile range was 0.48–0.69 (Wallace and Williams 2005). These results provide one of the first indications of the range of the infiltration factor across a set of homes. (Previous studies had estimated the *average* $PM_{2.5}$ or PM_{10} infiltration factor for the homes monitored, but were unable to estimate the factor for individual homes.)

In the University of Washington studies, the special measurements included both the sulfur measurements in a subset of filters and use of a continuous optical scattering device both indoors and outdoors to provide a way of detecting short-term indoor peaks. By subtracting those peaks from the record, the remaining indoor concentrations were assumed to be due to outdoor penetration

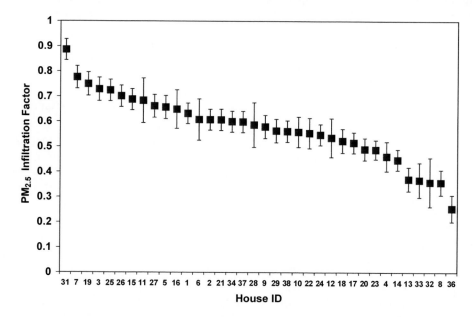

FIGURE 8.2 Indoor/outdoor sulfur ratios (infiltration factors) by home averaged across all seasons in the North Carolina study. Error bars are 95% confidence intervals.

of particles. This method resulted in an estimate of the $PM_{2.5}$ infiltration factor for 44 homes from 0.2–1.0 (Allen et al. 2003, 2004), with a mean value of 0.66. The investigators noted a seasonal difference in F_{inf}, with a mean (SD) value of 0.79 (0.18) in the non-heating season and 0.53 (0.16) in the heating season.

Although attempts have been made to estimate the infiltration factor using regression of indoor $PM_{2.5}$ on outdoor concentrations (USEPA 2004) it appears that this approach cannot provide reliable results. The reason may be the (likely) large variation of the indoor-generated particle source strengths from day to day, contrary to the assumption of a single value applicable to all days.

What about some of the other parameters affecting partitioning of particles into indoor and outdoor sources? The three parameters forming the infiltration factor are P, a, and k. Only one of these, the air exchange a, can be measured in most cases. We rely on theory and experiment (in chambers, experimental rooms, or instrumented homes) to provide values for the penetration coefficient P and the deposition rate k. Theory suggests that the penetration coefficient P is very close to 1 for the accumulation-mode particles making up the bulk of the $PM_{2.5}$ fraction (Liu and Nazaroff 2001). P begins to fall off for particles larger than 1 μm, and also falls off for those smaller than 0.1 μm. Typical values of the penetration coefficient averaged over all particle sizes up to 2.5 μm seem to range from about 0.7 to about 0.9, although some studies have found values indistinguishable from unity (Özkaynak et al. 1996; Thatcher and Layton 1995).

The deposition rate k depends on several factors, including the characteristics of the room surfaces and volume (k will be larger for fully furnished rooms with carpets or other irregular surfaces [Thatcher et al. 2002]); the airflow (k will be larger for higher air movement and turbulence); use of a central furnace or air conditioning fan (k will be larger with the fan on because the particles are forced through the ductwork where they can deposit more readily [Howard-Reed, Wallace, and Ott 2002]); and the use of a filter (many typical membrane filters hardly affect k, but HEPA filters and electrostatic precipitators can greatly increase deposition of particles [Howard-Reed, Wallace, and Emmerich 2003; Wallace, Emmerich, and Howard-Reed 2004a]).

A recent study directly measured the distribution of deposition rates across all particle sizes from ultrafines to fine particles; the increased rate when a central fan is operating; and the much larger rate when an efficient air filter is mounted in a central air system (Figure 8.3) (Wallace, Emmerich, and Howard-Reed 2004a). The study has a very important message for persons at risk from high concentrations of particles — the use of the best filters (such as duct-mounted electrostatic precipitators) can reduce exposures by 45–65%. Moreover, this reduction works for all types of particles, both outdoor and indoor generated, unlike simply tightening the house, which reduces exposure to outdoor particles but *increases* exposure to indoor particles.

Although P and k are often unknown or extremely difficult to calculate in field studies, it appears that the "lumped" determining factor F_{inf} for identifying particles of outdoor origin is relatively insensitive to the variation of P and k and therefore we can simply use our estimated values of F_{inf} without worrying about the values of P and k (Ott, Wallace, and Mage 2000).

Finally, the value of the indoor-generated source strength S is known for only a few sources and from only a few studies. The PTEAM study gave an estimate for S of 20 mg/cigarette and 6 mg/hour of cooking. Wallace, Emmerich, and Howard-Reed (2004b) estimated a value for S for cooking at 10^{14} particles (mostly ultrafine) per hour. Apart from these estimates, we have little knowledge of the magnitude and variation of S across different indoor sources. For example, the PTEAM study found about 25% of the indoor particle concentrations were due to unknown sources. One guess for a major indoor source for PM_{10} is resuspension from clothes or from walking, sitting, etc. in homes, but no study has provided an estimate of the source strength associated with resuspension. Even vented combustion devices such as gas clothes dryers have been implicated as possible sources of ultrafine particles (Wallace 2005).

Recently, some success in determining sources of both indoor-generated and outdoor-generated particles has been found in studies employing factor analysis, particularly positive matrix factorization (PMF) (Hopke et al. 2003). PMF usually depends on identification of 12 to 20 elements

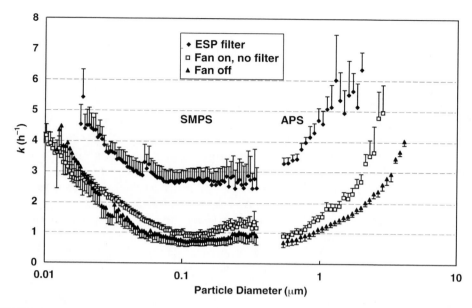

FIGURE 8.3 Deposition rates of particles under natural conditions (no fan); using a fan in a central heating or cooling system; and using an electrostatic precipitator (ESP) mounted in the return air duct. The fan alone increases the deposition rate of the particles by a moderate amount, but the addition of the filter forces the particles out of the air at much higher rates. Using this graph, it was determined that the ESP could reduce particle levels in the entire house by 45–65%. Persons at risk from exposure to particles could thus reduce their exposure while in their homes by more than a factor of two. Error bars are 1 standard deviation (shown only above or below the point estimate to reduce confusion). The SMPS is a scanning mobility particle sizer capable of measuring ultrafine particles; the APS is an aerodynamic particle sizer able to measure larger particles up to 2.5 μm in diameter.

on a filter sample, which are then analyzed statistically to detect correlations. The correlations can be further studied to extract profiles fitting certain sources. For example, a factor containing mostly Na and Cl might be associated with sea salt, whereas one containing vanadium might be associated with home heating oil. In one example of this approach, Yakovleva, Hopke, and Wallace (1999) were able to estimate that 20% of the indoor concentrations and 30% of the personal cloud observed in the PTEAM study were due to resuspended soil.

8.6.3 THE AMBIENT EXPOSURE FACTOR

With the calculation of the infiltration factor for individual homes, we have almost reached our goal of splitting personal exposure into its indoor and outdoor sources. However, the infiltration factor tells us the fraction of the outdoor particles penetrating indoors, but not the fraction that contributes to personal exposure as a person moves through his or her daily life — at home, at work, outdoors, in a vehicle. Since some of these *microenvironments* are difficult to monitor (at the office, inside one's car), it is more difficult to determine the *ambient exposure factor* F_{exp} (that portion of the outdoor concentration contributing to personal exposure) than the infiltration factor. However, something that helps us here is to remember that the average person in the United States spends 89% of his or her time indoors, and about 67% indoors at home. Therefore the infiltration factor determines a large fraction of the ambient exposure factor. It may be that the infiltration factor is almost sufficient to estimate the ambient exposure factor. In fact, the sulfur determinations in the North Carolina study provide a direct measure of the ambient exposure factor:

$$F_{\exp} = \frac{S_{pers}}{S_{out}} \qquad\qquad (8.5)$$

where
  S_{pers} = personal exposure to sulfur
  S_{out} = outdoor sulfur concentration

The results showed that the ambient exposure factor was slightly smaller than the infiltration factor for nearly all participants, and could be estimated very well from a knowledge of the infiltration factor alone, with R^2 values exceeding 90% (Wallace and Williams 2005). Theoretically, we expect the ambient exposure factor to be slightly higher than the infiltration factor for most persons, since for the 11% of the time they spend outdoors or in their vehicles they are exposed to the full outdoor particle concentrations rather than a fraction on the order of 0.66. However, the finding of a *smaller* ambient exposure factor is supported by the PTEAM finding that persons commuting and spending time in an office had lower concentrations than persons staying at home. The reason may be that air in office buildings, department stores, etc. is conditioned and recycled through filters, thus lowering one's exposure to particles of outdoor origin.

8.7   TOTAL PARTICLE EXPOSURE

With our ability to calculate the infiltration and ambient exposure factors, and with a knowledge of source strengths or at least of typical indoor concentrations, and finally some idea of the magnitude of the personal cloud, we can make an estimate of personal exposures to particles from all sources for certain common cases. For the 30% or so of homes in the United States that have a smoker, indoor concentrations will be about 30 $\mu g/m^3$ above normal background levels. Since about half of the counties in the United States are expected to exceed the outdoor $PM_{2.5}$ standard of 15 $\mu g/m^3$, that level will be close to an average outdoor concentration. Multiplying that by about 0.6, the average infiltration factor found in several studies, we find that the background level of $PM_{2.5}$ due to outdoor particles in many homes may be close to 9–10 $\mu g/m^3$. Cooking might add 5 $\mu g/m^3$. Resuspension seems to be important for PM_{10}, probably less so for $PM_{2.5}$. Thus a smoker's home might typically have an indoor $PM_{2.5}$ concentration of 40–45 $\mu g/m^3$ while a nonsmoker's home might have a concentration closer to 15 $\mu g/m^3$. For personal exposure, one must add the personal cloud, which has a value on the order of 2–4 $\mu g/m^3$ for $PM_{2.5}$ (Williams et al. 2003; Liu et al. 2003). (For PM_{10}, it is much higher at about 35 $\mu g/m^3$ [Özkaynak et al. 1996; Pellizzari et al. 1999]). A pictorial representation of these different contributors to personal exposure, based on the PTEAM results, is provided as Figure 8.4.

8.8   REDUCTION OF EXPOSURE

Since most particle exposure occurs in the home, reducing in-home concentrations is the most efficient way to reduce total exposure. In the developing world, particularly in rural areas, the most important source of in-home exposure is cooking in unventilated homes. A number of programs have been initiated by national governments and nongovernmental organizations (NGOs) to introduce improved stoves, with mixed success to date. The most successful program by far, Chinese National Improved Stove Program (NISP) was able to facilitate the introduction of more than 180 million improved biomass stoves from the early 1980s to the mid-1990s. Like all such programs around the world during this period, the focus was on fuel savings, but the NISP-approved stoves were required also to have working chimneys (Sinton et al. 2004). Although studies have shown that particle levels decreased as a result, they still do not reach the levels specified in the Chinese standards for indoor air quality (IAQ) in households (150 $\mu g/m^3$ PM_{10} for 24 hours). In addition,

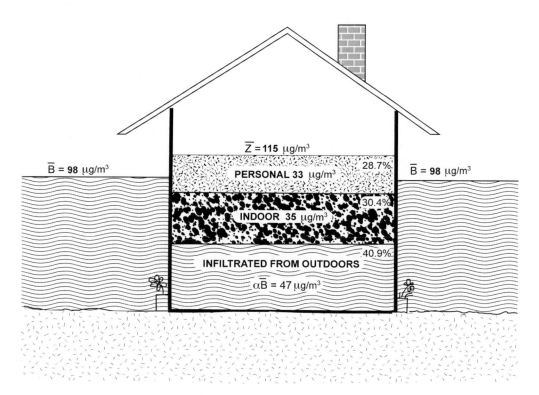

HOUSE OF PARTICLES

PTEAM PM$_{10}$ PERSONAL EXPOSURES

$\overline{Z}$ = 115 μg/m³

$\overline{B}$ = 98 μg/m³   PERSONAL 33 μg/m³   28.7% $\overline{B}$ = 98 μg/m³

30.4%

INDOOR 35 μg/m³

40.9%

INFILTRATED FROM OUTDOORS

$\alpha\overline{B}$ = 47 μg/m³

FIGURE 8.4 Assuming an infiltration factor α for PM$_{10}$ of about 0.48, the basic contributors to personal exposure in the PTEAM study were as shown. Homes with and without smokers were combined to provide an overall average. Homes with smokers would have a higher indoor contribution and homes without smokers a lower one. (Based on analyses of PTEAM data in Ott, Wallace, and Mage 2000.)

relatively little attention has been given to coal stoves, which are becoming more common in rural areas as the economy expands. In addition, millions of people are exposed to high levels of F, As, and other toxic contaminants as well as combustion products from household use of "poisonous coals."

In housing typical of developed countries, households can be tightened to reduce outdoor particle penetration — however, this has the effect of increasing indoor-generated particle concentrations. Source removal (e.g., banning smoking indoors) is highly effective for certain sources. Since the greatest source of particles in nonsmoking homes is often the stove, installing and using an effective range hood is important. (However, hoods are often inefficient, ducts get blocked, and in general, constant vigilance is required.) A portable or duct-mounted air filter reduces both indoor-generated and outdoor-infiltrating particles. Unfortunately, however, there is great opportunity to buy worthless products. Many "ion generators" or "ozone cleaners" not only do not reduce particle concentrations in any significant amount, but also may increase the level of harmful ozone in the home. The most effective type of air cleaner is the duct-mounted electrostatic precipitator, which can reduce concentrations by 45–65% (Riley et al. 2002; Wallace, Emmerich, and Howard-Reed 2004a). Although these electrostatic precipitators (ESPs) also create some ozone, since the ozone is released in the duct it is likely to react immediately with the duct walls and not escape into the home. Portable air cleaners with HEPA filters can also be effective in cleaning single rooms, provided they have sufficiently powerful fans (300 cfm is a reasonable lower threshold airflow rate).

8.9 SUMMARY

In the last two decades, combustion-related airborne particles have been recognized as the single greatest environmental contributor to sickness and death. Considerable strides have been made in understanding the most important sources (smoking, vehicle emissions, and, in developing countries, cooking or heating with biofuels) and effective countermeasures (exhaust fans, electrostatic precipitators, and, in developing countries, better stoves and chimneys). However, a decade of intense research on the toxicology of particles has so far failed to identify the causal mechanism.

Exposure studies have been central in identifying sources (e.g., the Harvard 6-City Study quantifying the effect of smoking, the PTEAM study quantifying the effect of cooking and the extent of the PM_{10} personal cloud, and the studies in the developing world identifying the cooking or heating behaviors and the effect of home construction in producing the health effects so cruelly focused on women and children). Exposure studies have also answered the question posed by the National Academy of Science regarding the relation of personal exposure to particles of ambient origin with central-site measurements: the relationship is strong (significant) for less than half the population. As the determinants of exposure are becoming better understood, as well as personal exposure being recognized as a better indicator of health effect than area concentrations, there are now also discussions for using exposure more directly in regulation (Smith 1996; Krzyzanowski 2004). Finally, exposure studies have shown that indoor-generated particles produce approximately equal exposures as outdoor-generated particles (for nonsmokers). This puts a new emphasis on determining toxicity of indoor-generated particles, with possible important repercussions on future control strategies.

8.10 QUESTIONS FOR REVIEW

1. Describe one main finding from each of the three large-scale studies of particle exposure described in the chapter:
   a. The Harvard 6-City Study
   b. The New York State 2-County Study
   c. The Particle TEAM Study in Riverside
2. Why are longitudinal correlations of personal exposure with outdoor concentrations more relevant to the time-series epidemiological studies than cross-sectional correlations?
3. Describe the two main terms on the right-hand side of Equation 8.2. What is the effect on each term of
   a. Increasing the air exchange rate a?
   b. Increasing the deposition rate k?
   c. Based on your answers to a and b above, which would be better to do to decrease indoor particle concentrations?
4. What is the approximate average size of the infiltration factor F_{inf}? The ambient exposure factor F_{exp}? Why does theory predict a higher value for F_{exp} than for F_{inf}? Why do exposure studies show a *lower* value for F_{exp} than for F_{inf}?
5. The outdoor National Ambient Air Quality Standard for $PM_{2.5}$ is 15 $\mu g/m^3$. Use what you know about the infiltration factor to estimate the average person's exposure to particles of outdoor origin when the standard is just achieved. Compare this level to the exposure expected (a) from having a smoker in the house; (b) from cooking.
6. Agencies such as the Department of Energy (DOE) and USEPA recommend that people build energy-efficient houses and close their windows on days of high outdoor pollution. What is wrong with this advice?
7. Compare the annual mortality due to particle air pollution in the developed world vs. that in the developing world.

8.11 ANSWERS TO QUESTIONS FOR REVIEW

1. a. The importance of smoking in elevating indoor particle concentrations.
   b. The relative unimportance of other combustion sources, such as fireplaces.
   c. The importance of cooking; the importance of personal activities; and the "personal cloud" in contributing to PM_{10} exposure.

2. The time-series epidemiology studies use each person as his or her own control, thus eliminating the confounding aspects of smoking, spatial variation, etc. These confounders are also eliminated in the longitudinal correlations (calculated for each individual) but remain in the cross-sectional correlations (calculated for all individuals at once at a point in time).

3. The first term represents the infiltration of outdoor air. The second term represents the contribution of indoor sources.
   a. An increase in the contribution of outdoor air; a decrease in the contribution of the indoor source.
   b. A decrease in both terms.
   c. It is *generally* better to increase k (by using, for example, an effective air cleaner) than to increase a (by, for example, opening windows), since the latter allows more outdoor air particles to enter. On the other hand, for situations in which the indoor source dominates the particle concentration, as, for example, when food is burned or cigarette smoking occurs, it is clear that effecting a large and immediate increase in a by opening many windows is preferable to the slower approach of using the air cleaner.

4. About 0.5–0.7 for $PM_{2.5}$. About the same. Because persons spend a fraction of their time (about 13%) either outside or in vehicles, which normally have higher particle concentrations than indoor levels. [Since exposure studies rarely measure concentrations in schools, office buildings or other places with filtered recirculated air, and these locations often have lower concentrations than residences, extrapolating the home indoor particle concentration to the concentrations in these locations results in overestimating the true F_{exp}.]

5. Multiplying by, say, 0.6, results in an estimated indoor concentration of 9 µg/m³. Since F_{exp} is similar to F_{inf}, but adding the personal cloud (about 2 µg/m³ for $PM_{2.5}$) will give an exposure on the order of 10 µg/m³. (a) Add 30 µg/m³ from smoking to the 10 from outdoor air to get about 40 µg/m³; (b) Add 10–15 µg/m³ from cooking to the 10 from outdoor air to get about 20–25 µg/m³.

6. Both the energy-efficient homes and closing windows reduces the air exchange a, which reduces infiltration of outdoor air particles but increases the contribution of indoor sources.

7. The opening sentence of this chapter mentions about 800,000 deaths due to urban air pollution and 1.6 million deaths due to use of solid fuel for heating and cooking, mostly in the developing world. Since many cities associated with the first type of pollution are in the developing world, the ratio of deaths in the developing world to deaths in the developed world is greater than 2:1.

REFERENCES

Abt, E., Suh, H.H., Catalano, P.J., and Koutrakis, P. (2000) Relative Contribution of Outdoor and Indoor Particle Sources to Indoor Concentrations, *Environmental Science and Technology*, **34**: 3579–3587.

Allen, R., Larson, T., Sheppard, L., Wallace, L., and Liu, L.-J.S. (2003) Use of Real-Time Light Scattering Data to Estimate the Contribution of Infiltrated and Indoor-Generated Particles to Indoor Air, *Environmental Science and Technology*, **37**: 3484–3492.

Allen, R., Sheppard, L., Liu, L.-J.S., Larson, T., and Wallace, L. (2004) Estimated Hourly Personal Exposures to Ambient and Non-Ambient Particulate Matter among Sensitive Populations in Seattle, Washington, *Journal of the Air and Waste Management Association*, **54**(9): 1197–1211.

Bruce, N., Perez-Padilla, R. and Albalak, R. (2000) Indoor Air Pollution in Developing Countries: A Major Environmental and Public Health Challenge, *WHO Bulletin*, **78**(9): 1078–1092.

Chen, B.H., Hong, C.J., Pandey, M.R., and Smith, K.R. (1990) Indoor Air Pollution and its Effects in Developing Countries, *WHO Statistical Quarterly*, **43**(3): 127–138.

Clayton, C.A., Perritt, R.L., Pellizzari, E.D., Thomas, K.W., Whitmore, R.W., Wallace, L.A., Özkaynak, H., and Spengler, J.D. (1993) Particle Total Exposure Assessment Methodology (PTEAM) Study: Distributions of Aerosol and Elemental Concentrations in Personal, Indoor, and Outdoor Air Samples in a Southern California Community, *Journal of Exposure Analysis and Environmental Epidemiology, 3*: 227–250.

Cohen, A.J., Anderson, H.R., Ostro, B., Pandey, K.D., Krzyzanowski, M., Kuenzli, N., Gutschmidt, K., Pope, C.A., Romieu, I., Samet, J.M., and Smith, K.R. (2004) Mortality Impacts of Urban Air Pollution, in *Comparative Quantification of Health Risks: Global and Regional Burden of Disease Due to Selected Major Risk Factors*, Ezzati, M., Rodgers A.D., Lopez, A.D., and Murray, C.J.L., Eds., World Health Organization, Geneva, Switzerland 1353–1433.

Dockery, D.W. and Spengler, J.D. (1981) Personal Exposure to Respirable Particulates and Sulfates, *Journal of the Air Pollution Control Association, 31*(2): 153–159.

Dockery, D.W., Pope, C.A., III, Xu, X., Spengler, J.D., Ware, J.H., Fay, M.E., Ferris, B.G., Jr., and Speizer, F.E. (1993) An Association Between Air Pollution and Mortality in Six U.S. Cities, *New England Journal of Medicine*, **329**: 1753–1759.

Dockery, D.W., Schwartz, J., and Spengler, J.D. (1992) Air Pollution and Daily Mortality: Associations with Particulates and Acid Aerosols, *Environment Research*, **59**: 362–373.

Ebelt, S.T., Petkau, A.J., Vedal, S., Fisher, T.V., and Brauer, M. (2000) Exposure of Chronic Obstructive Pulmonary Disease Patients to Particulate Matter: Relationship between Personal and Ambient Air Concentrations, *Journal of the Air and Waste Management Association*, **50**: 1081–1094.

Fisk, W.J. and Rosenfeld, A.H. (1997) Estimates of Improved Productivity and Health from Better Indoor Environments, *Indoor Air*, 7: 158–172.

Hopke, P.K., Ramadan, Z., Paatero, P., Norris, G.A., Landis, M.L., Williams, R.W., and Lewis, C.W. (2003) Receptor Modeling of Ambient and Personal Exposure Samples: 1998 Baltimore Particulate Matter Epidemiology-Exposure Study, *Atmospheric Environment*, **37**(23): 3289–3302.

Howard-Reed, C., Wallace, L.A., and Emmerich, S.J. (2003) Effect of Ventilation Systems and Air Filters on Decay Rates of Particles Produced by Indoor Sources in an Occupied Townhouse, *Atmospheric Environment*, **37**(38): 5295–5306.

Howard-Reed, C.H., Wallace, L.A., and Ott, W.R. (2002) The Effect of Opening Windows on Air Change Rates in Two Homes, *Journal of the Air and Waste Management Association*, **52**: 147–159.

International Scientific Oversight Committee (2004) *Health Effects of Outdoor Air Pollution in Developing Countries of Asia: A Literature Review*, Special Report 15, Health Effects Institute, Boston, MA. http://www.healtheffects.org/Pubs/SpecialReport15.pdf (accessed October 5, 2005).

Janssen, N.A.H., De Hartog, J.J., Hoek, G., Brunekreef, B., Lanki, T., Timonen, K.L., Pekkanen, J. (2000) Personal Exposure to Fine Particulate Matter in Elderly Subjects: Relation between Personal, Indoor, and Outdoor Concentrations, *Journal of the Air and Waste Management Association*, **50**: 1133–1143.

Janssen, N.A.H., Hoek, G., Brunekreef, B., Harssema, H., Mensink, I., and Zuldhof, A. (1998) Personal Sampling in Adults: Relation among Personal, Indoor, and Outdoor Air Concentrations, *American Journal of Epidemiology*, **147**: 537–547.

Janssen, N.A.H., Hoek, G., Harssema, H., and Brunekreef, B. (1997) Childhood Exposure to PM_{10}: Relation between Personal, Classroom, and Outdoor Concentrations, *Occupational and Environmental Medicine*, **54**: 888–894.

Janssen, N.A.H., Hoek, G., Harssema, H., and Brunekreef, B. (1999) Personal Exposure to Fine Particles in Children Correlates Closely with Ambient Fine Particles, *Archives of Environmental Health*, **54**: 95–101.

Keeler, G.J., Dvonch, J.T., Yip, F.Y., Parker, E.A., Israel, B.A., Marsik, F.J., Morishita, M., Barres, J.A., Robins, T.G., Brakefield-Caldwell, W., and Sam, M. (2002) Assessment of Personal and Community-Level Exposures to Particulate Matter among Children with Asthma in Detroit, Michigan, as Part of Community Action Against Asthma (CAAA), *Environmental Health Perspectives,* **110**(supp. 2): 173–181.

Klepeis, N.E., Apte, M.G., Gundel, L.A., Sextro, R.G., and Nazaroff, W.W. (2003) Determining Size-Specific Emissions Factors for Environmental Tobacco Smoke Particles, *Aerosol Science and Technology,* **37**: 780–790.

Koutrakis, P., Briggs, S.L.K., and Leaderer, B.P. (1992) Source Apportionment of Indoor Aerosols in Suffolk and Onondaga Counties, New York, *Environmental Science and Technology,* **26**: 521–527.

Krzyzanowski, M. (Ed.) (2004) Role of Human Exposure Assessment in Air Quality Management WHO-JRC-ECA Workshop, Bonn, October 2002, EUR 21052 EN, European Communities, Brussels, http://www.euro.who.int/document/e79501.

Laden, F., Neas, L.M., Dockery, D.W., and Schwartz, J. (2000) Association of Fine Particulate Matter from Different Sources with Daily Mortality in Six U.S. Cities, *Environmental Health Perspectives,* **108**(10): 941–947.

Lai, A.C.K., Burne, M.A., and Goddard, A.J.H. (1999) Measured Deposition of Aerosol Particles on a Two-Dimensional Ribbed Surface in a Turbulent Duct Flow, *Journal of Aerosol Science,* **30**(9): 1201–1214.

Lai, A.C.K. and Nazaroff, W.W. (2000) Modeling Indoor Particle Deposition from Turbulent Flow onto Smooth Surfaces, *Journal of Aerosol Science,* **31**: 463–476.

Lioy, P.J., Waldman, J.M., Buckley, T., Butler, J. and Pietarinen, C. (1990) The Personal, Indoor and Outdoor Concentrations of PM-10 Measured in an Industrial Community during the Winter, *Atmospheric Environment,* **24B**: 57–66.

Liu, D.L. and Nazaroff, W.W. (2001) Modeling Pollutant Penetration across Building Envelopes, *Atmospheric Environment,* **35**: 4451–4462.

Liu, L.-J.S., Box, M., Kalman, D., Kaufman, J., Koenig, J., Larson, T., Sheppard, L., Slaughter, C., Lewtas, J., and Wallace, L.A. (2003) Exposure Assessment of Particulate Matter for Susceptible Populations in Seattle, WA, *Environmental Health Perspectives,* **111**(7): 909–918.

Long, C.M., Suh, H.H., Catalano, P.J. and Koutrakis, P. (2001) Using Time- and Size-Resolved Particulate Data to Quantify Indoor Penetration and Deposition Behavior, *Environmental Science and Technology,* **35**(10): 2089–2099.

Meng, Q.Y., Turpin, B.J., Korn, L., Weisel, C.P., Morandi, M., Colome, S., Zhang, J., Stock, T., Spektor, D., Winer, A., et al. (2005) Influence of Ambient (Outdoor) Sources on Residential Indoor and Personal PM$_{2.5}$ Concentrations: Analyses of RIOPA Data, *Journal of Exposure Analysis and Environmental Epidemiology,* **15**: 17–28.

Miller, S.L. and Nazaroff, W.W. (2001) Environmental Tobacco Smoke Particles in Multizone Indoor Environments, *Atmospheric Environment,* **35**: 2053–2067.

Naeher, L.P., Smith, K.R., Brauer, M., Chowdhury, Z., Simpson, C., Koenig, J.Q., Lipsett, M., and Zelikoff, J.T. (2005) Critical Review of the Health Effects of Woodsmoke, Air Health Division, Health Canada, Ottawa.

National Research Council-National Academy of Science (NRC–NAS) (1998) *Research Priorities for Airborne Particulate Matter I: Immediate Priorities and a Long-Range Research Portfolio,* National Academies Press, Washington, DC.

Nazaroff, W.W. and Cass, G.R. (1989) Mathematical Modeling of Indoor Aerosol Dynamics, *Environmental Science and Technology,* **23**: 157–166.

Neas, L.M, Dockery, D.W., Ware, J.H., Spengler, J.D., Ferris Jr., B.G., and Speizer, F.E. (1994) Concentration of Indoor Particulate Matter as a Determinant of Respiratory Health in Children, *American Journal of Epidemiology,* **139**: 1088–1099.

Office of Technology Assessment (1990) Biomass Energy Technologies for Developing Countries, U.S. Congress, Washington, DC.

Ott, W., Wallace, L., and Mage, D. (2000) Predicting Particulate (PM$_{10}$) Personal Exposure Distributions Using a Random Component Superposition Statistical Model, *Journal of the Air and Waste Management Association,* **50**: 1390–1406.

Özkaynak, H., Xue, J., Spengler, J.D., Wallace, L.A., Pellizzari, E.D., and Jenkins, P. (1996) Personal Exposure to Airborne Particles and Metals: Results from the Particle TEAM Study in Riverside, CA, *Journal of Exposure Analysis and Environmental Epidemiology*, **6**: 57–78.

Pellizzari, E.D., Clayton, C.A., Rodes, C.E., Mason, R.E., Piper, L.L., Fort, B., Pfeifer, G., and Lynam, D. (1999) Particulate Matter and Manganese Exposures in Toronto, Canada, *Atmospheric Environment*, **31**: 721–734.

Ramakrishna, J., Durgaprasad, M.B., and Smith, K.R. (1989) Cooking in India: The Impact of Improved Stoves on Indoor Air Quality, *Environment International*, **15**(1–6): 341–352.

Reid, H.F., Smith, K.R., and Sherchand, B. (1986) Indoor Smoke Exposure from Traditional and Improved Cookstoves: Comparisons among Rural Nepali Women, *Mountain Research and Development*, **6**(4): 293–304.

Riley, W.J., McKone, T.E., Lai, A.C., and Nazaroff, W.W. (2002) Indoor Particulate Matter of Outdoor Origin: Importance of Size-Dependent Removal Mechanisms, *Environmental Science and Technology*, **36**(2): 200–207.

Rojas-Bracho, L., Suh, H.H., and Koutrakis, P. (2000) Relationships among Personal, Indoor, and Outdoor Fine and Coarse Particle Concentrations for Individuals with COPD, *Journal of Exposure Analysis and Environmental Epidemiology*, **10**: 294–306.

Samet, J.M., Dominici, F., Curriero, F.C., Coursac, I., and Zeger, S.L. (2000) Fine Particulate Air Pollution and Mortality in 20 U.S. Cities, 1987–1994. *New England Journal of Medicine*, **343**(24): 1742–1749.

Sarnat, J.A., Koutrakis, P. and Suh, H.H. (2000) Assessing the Relationship between Personal Particulate and Gaseous Exposures of Senior Citizens Living in Baltimore, MD, *Journal of the Air and Waste Management Association*, **50**(7): 1184–1198.

Sarnat, J.A., Long, C.M., Koutrakis, P., Coull, B.A., Schwartz, J., and Suh, H.H. (2002) Using Sulfur as a Tracer of Outdoor Fine Particulate Matter, *Environmental Science and Technology*, **36**: 5305–5314.

Schwartz, J., Dockery, D.W., and Neas, L.M. (1996) Is Daily Mortality Associated Specifically with Fine Particles?, *Journal of the Air and Waste Management Association*, **46**: 927–939.

Sinton, J.E., Smith, K.R., Peabody, J.W., Liu, Y., Zhang, X., Edwards, R., and Gan, Q. (2004) An Assessment of Programs to Promote Improved Household Stoves in China, *Energy for Sustainable Development*, **8**(3): 33–52.

Smith, K.R. (1987) *Biofuels, Air Pollution and Health*, Plenum Press, New York.

Smith, K.R. (1993) Fuel Combustion, Air Pollution, and Health in Developing Countries, *Annual Review of Energy and Environment*, **18**: 529–566.

Smith, K.R. (1996) *The Potential of Human Exposure Assessment for Air Pollution Regulation*, Report No. WHO/EHG/95.09, Office of Global and Integrated Environmental Health, World Health Organization, Geneva, Switzerland.

Smith, K.R. (2000) National Burden of Disease in India from Indoor Air Pollution, *Proceedings of the National Academy of Sciences*, **97**(24): 13286–13293.

Smith, K.R., Aggarwal, A.L., and Dave, R.M. (1983) Air Pollution and Rural Biomass Fuels in Developing Countries: A Pilot Village Study in India and Implications for Research and Policy, Atmospheric Environment, **17**(11): 2343–2362.

Smith, K.R., Apte, M.G., Ma, Y., Wathana, W., and Kulkarni, A. (1994) Air Pollution and the Energy Ladder in Asian Cities, *Energy, the International Journal*, **19**(5): 587–600.

Smith, K.R., Ghuhua, G., Kun, H., and Daxiong, Q. (1993) One Hundred Million Improved Cookstoves in China: How Was It Done? *World Development*, **21**(6): 941–961.

Smith, K.R., Mehta, S., and Maeusezahl-Feuz, M. (2004) Indoor Smoke from Household Solid I. Fuels, in *Comparative Quantification of Health Risks: Global and Regional Burden of Disease Due to Selected Major Risk Factors*, Volume 2, Ezzati, M., Rodgers, A.D., Lopez, A.D., and Murray, C.J.L., Eds., World Health Organization, Geneva, 1435–1493.

Spengler, J.D., Özkaynak, H., Ludwig, J., Allen, G., Pellizzari, E.D., and Wiener, R. (1989) Personal Exposures to Particulate Matter: Instruments and Methodologies of PTEAM, in *Proceedings of EPA/AWMA Symposium on Measurement of Toxic and Related Air Pollutants*, Air and Waste Management Association, VIP-13, Pittsburgh, PA, 449–463.

Thatcher, T. and Layton, D. (1995) Deposition, Resuspension, and Penetration of Particles within a Residence, *Atmospheric Environment*, **29**(13): 1487–1497.

Thatcher, T.L., Lai, A.C.K., Moreno-Jackson, R., Sextro, R.G., and Nazaroff, W.W. (2002) Effects of Room Furnishings and Air Speed on Particle Deposition Rates Indoors, *Atmospheric Environment,* **36**: 1811–1819.

Thornburg, J.W., Rodes, C.E., Lawless, P.A., Stevens, C.D., and Williams, R.W. (2004) A Pilot Study of the Influence of Residential HAC Duty Cycle on Indoor Air Quality, *Atmospheric Environment,* **38**(11): 1567–1577.

USEPA (2002) *Preliminary Particulate Matter Mass Concentrations Associated with Longitudinal Panel Studies: Assessing Human Exposures of High-Risk Subpopulations to Particulate Matter,* Report No. EPA/600/R-01/086, U.S. Environmental Protection Agency, National Exposure Research Laboratory, Office of Research and Development, Washington, DC.

USEPA (2004) *Exposure of High-Risk Subpopulations to Particles, Final Report (APM-21),* Report No. EPA/600/R-03/145, U.S. Environmental Protection Agency, National Exposure Research Laboratory, Office of Research and Development, Washington, DC.

Wallace, L.A. (2000a) Correlations of Personal Exposure to Particles with Outdoor Air Measurements: A Review of Recent Studies, *Aerosol Science and Technology,* **32**: 15–25.

Wallace, L.A. (2000b) Real-time Monitoring of Particles, PAH, and CO in an Occupied Townhouse, *Applied Occupational and Environmental Hygiene,* **15**(1): 39–47.

Wallace, L.A. (2005) Ultrafine Particles from a Vented Gas Clothes Dryer, *Atmospheric Environment,* **39**: 5777–5786.

Wallace, L.A. and Howard-Reed, C.H (2002) Continuous Monitoring of Ultrafine, Fine, and Coarse Particles in a Residence for 18 Months in 1999-2000, *Journal of the Air and Waste Management Association,* **52**(7): 828–844.

Wallace, L.A. and Williams, R. (2005) Use of Personal-Indoor-Outdoor Sulfur Concentrations to Estimate the Infiltration Factor, Outdoor Exposure Factor, Penetration Coefficient, and Deposition Rate for Individual Homes, *Environmental Science and Technology,* **39**: 1707–1714.

Wallace, L.A., Emmerich, S.J., and Howard-Reed, C. (2004a) Effect of Central Fans and In-Duct Filters on Deposition Rates of Ultrafine and Fine Particles in an Occupied Townhouse, *Atmospheric Environment,* **38**(4): 405–413.

Wallace L.A., Emmerich, S.J., and Howard-Reed, C. (2004b) Source Strengths of Ultrafine and Fine Particles Due to Cooking with a Gas Stove, *Environmental Science and Technology,* **38**(8): 2304–2311.

Wallace, L.A., Howard-Reed, C.H., and Emmerich, S.J. (2002) Continuous Measurements of Air Change Rates in an Occupied House for One Year: The Effect of Temperature, Wind, Fans, and Windows, *Journal of Exposure Analysis and Environmental Epidemiology,* **12**: 296–306.

Wallace, L.A., Mitchell, H., O'Connor, G.T., Liu, L.-J., Neas, L., Lippmann, M., Kattan, M., Koenig, J., Stout, J.W., Vaughn, B.J., Wallace, D., Walter, M., and Adams, K. (2003) Particle Concentrations in Inner-City Homes of Children with Asthma: the Effect of Smoking, Cooking, and Outdoor Pollution, *Environmental Health Perspectives,* **111**: 1265–1272.

WHO (2002) *World Health Report: Comparing Risks,* World Health Organization, Geneva, Switzerland.

WHO (2006) *Global Air Quality Guidelines: Updates for Particles SO$_2$, NO$_2$, and Ozone,* World Health Organization, Geneva, Switzerland.

Williams, R.W., Suggs, J., Rea, A., Leovic, K., Vette, A., Sheldon, L., Rodes, C., and Thornburg, J. (2003) The Research Triangle Park Particulate Matter Panel Study: PM Mass Concentration Relationships, *Atmospheric Environment,* **37**(38): 5349–5363.

Yakovleva, E., Hopke, P., and Wallace, L. (1999) Positive Matrix Factorization in Determining Sources of Particles Measured in EPA's Particle TEAM Study, *Environmental Science and Technology,* **33**: 3645–3652.

9 Exposure to Secondhand Smoke

James L. Repace
Tufts University and Repace Associates, Inc.

CONTENTS

9.1 Synopsis...201
9.2 Introduction..202
9.3 Pollutants from SHS...203
9.4 Smoking Prevalence and Trends 209
9.5 Workplace Exposure...210
9.6 Determining Exposure..210
    9.6.1 Biomarkers for SHS: Body Fluid Cotinine 210
    9.6.2 Misclassification Problems...210
9.7 Secondhand Smoke as Microenvironmental Air Pollution....................................212
    9.7.1 SHS Concentrations..212
    9.7.2 Smoking Rates and Emissions 213
    9.7.3 A Person's Daily Exposure to Particulate Air Pollution 214
9.8 Atmospheric Tracers for SHS: Nicotine and RSP...215
9.9 Time-Averaged Models for SHS Concentrations 217
    9.9.1 The Active Smoking Count...218
    9.9.2 The Habitual Smoker Model (Equation 9.2 with Defaults)..........................220
9.10 Time-Varying SHS Concentrations...221
9.11 Applications — SHS in Naturally and Mechanically Ventilated Buildings 223
    9.11.1 RSP from SHS in Homes...223
    9.11.2 Predicting RSP from SHS in Mechanically Ventilated Buildings 223
    9.11.3 SHS in the Hospitality Industry..223
    9.11.4 The Delaware Air Quality Survey...224
9.12 Applications — RSP and CO from SHS in a Vehicle 225
9.13 Applications — Dosimetry: Translation of Exposure into Dose via
    Cotinine Analysis 226
9.14 Future Issues...228
9.15 Acknowledgments 229
9.16 Questions for Review 229
References 231

9.1 SYNOPSIS

Secondhand smoke (SHS) has been estimated to cause as much as 2.7% of all deaths in the United States annually. Its adverse health effects have been estimated to cost more than $25 billion annually in California alone. SHS is a source of at least 172 toxic substances in indoor air. A major pollutant emitted by SHS is respirable suspended particles (RSP). RSP concentrations in indoor microenvironments where smoking occurs greatly exceed the levels encountered in smoke-free environments, outdoors, and in vehicles on busy highways. SHS appears to contribute the overwhelming majority of carcinogenic particle-bound polycyclic aromatic hydrocarbons in the air of most buildings where smoking occurs. SHS-carbon monoxide levels measured in pubs are in the range that produces acute cardiovascular effects. This chapter discusses the toxic constituents of SHS, and the prevalence of nonsmokers' exposure, as well as factors determining exposure and dose. The microenvironments of greatest importance are those where the population spends the most time: at home and in the workplace. In-vehicle exposure is also of concern due to the high concentrations observed. Use of a personal exposure monitor to estimate relative contributions of smoking, cooking, and diesel exhaust to a person's RSP exposure is illustrated. How such personal exposures combine into a population distribution is illuminated, and the major U.S. field study of SHS dose in the population is deconstructed. Field studies and controlled measurements of SHS concentrations in homes and workplaces are reviewed. SHS emission and removal rates are discussed in the context of the time-averaged mass-balance model for estimating concentrations in naturally and mechanically ventilated buildings, and examples for homes and bars are given. SHS-RSP can be related to SHS-nicotine, which when inhaled is metabolized into the SHS biomarker, cotinine. National surveys have shown that nearly half of nonsmokers who report no SHS exposure have detectable levels of cotinine in their body fluids. Despite growing trends toward indoor public and workplace smoking bans, SHS exposure continues for half of all children at home in the United States, and for most bar and casino workers.

9.2 INTRODUCTION

Cigarette smoking proliferated after World War I, but not until the 1964 U.S. Surgeon General's Report on Smoking (Surgeon General 1964) was the public made aware that smoking could kill. At that time there was little understanding that indoor air pollution from the tobacco smoke exhaled by smokers and emitted from the burning ends of cigarettes, pipes, and cigars — i.e., secondhand smoke (SHS), also known as environmental tobacco smoke (ETS) (Figure 9.1) — was for nonsmokers actually the most significant source of human air pollution exposure. Moreover, as smoking was then wrongly but widely regarded as a "personal choice," public health advice was largely limited to smoking cessation. Regulation of tobacco products in the United States became difficult as the tobacco industry quietly got Congress to exempt its products from every conceivable federal law under which their emissions could possibly be regulated (Repace 1981). Yet, exposure to tobacco smoke at the levels encountered in smoking damages nearly every organ in the human body and caused an estimated 440,000 excess deaths per year in 2004 (Surgeon General 2004). Thus, tobacco smoke is a seriously toxic substance, for which no safe level of exposure has been identified.

Research into indoor air pollution from SHS proliferated in the 1970s, resulting in a cascade of lengthy, authoritative, peer-reviewed reports by national and international environmental, occupational, and public health authorities. The Surgeon General (1986), the National Academy of Sciences (National Research Council 1986), the International Agency for Research on Cancer (IARC 1987, 2004), the National Institute for Occupational Safety and Health (NIOSH 1991), the U.S. Environmental Protection Agency (USEPA 1992), the Occupational Safety & Health Administration (OSHA 1994), the National Cancer Institute (NCI 1993, 1998, 1999), the California EPA (CalEPA 1997, 2005), and the National Toxicology Program (NTP 2000) variously concluded that nonsmokers' exposure to SHS causes fatal heart disease; lung, breast, and nasal sinus cancer; asthma induction and aggravation; middle ear infection; sudden infant death syndrome; and respiratory impairment; as well as irritation of the mucous membranes of the eyes, nose, and throat. SHS is now widely accepted as the third leading preventable health hazard after active smoking and alcohol (Surgeon General 2004); nevertheless it continues to be a widespread indoor pollutant in many homes, workplaces, and public access buildings in the United States and abroad.

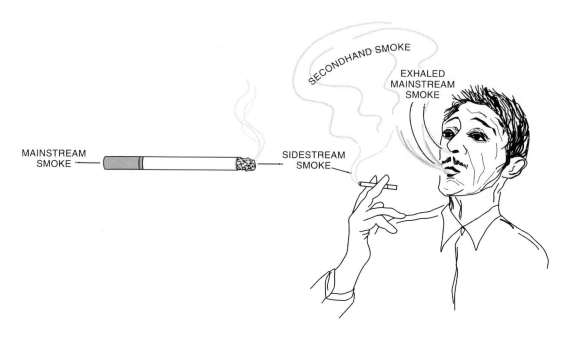

FIGURE 9.1 Mainstream smoke is inhaled by the smoker during puffing, and to a small extent diffuses through the cigarette paper. Between puffs, the smoker emits exhaled mainstream smoke, and the burning end of the cigarette emits sidestream smoke. The combination of sidestream (~90%) and exhaled mainstream smoke (~10%) is called secondhand smoke. An older term for secondhand smoke is environmental tobacco smoke.

9.3 POLLUTANTS FROM SHS

The tobacco smoke aerosol is a mixture of more than 4,000 chemical by-products of tobacco combustion, 500 of which are in the gas phase (Hoffmann and Hoffmann 1998). Of these SHS by-products, 172 are known toxic substances, many of which are regulated — except in the non-industrial indoor air environment, where most exposure takes place. SHS includes 3 criteria air pollutants and 33 hazardous air pollutants regulated under the Clean Air Act, 47 pollutants that are classified as hazardous wastes whose disposal in solid or liquid form is regulated by the Resource Conservation and Recovery Act, 67 known human or animal carcinogens, and 3 industrial chemicals regulated under the Occupational Health and Safety Act (Table 9.1). Nevertheless, although widely regarded as a major nuisance and worse by nonsmokers due to its irritating properties and health hazards, until the 1990s there were few successful attempts to regulate SHS exposure.

In the mid-1990s the U.S. Centers for Disease Control discovered that most nonsmokers of all ages had tobacco combustion products in their blood (Pirkle et al. 1996). By the late 1990s SHS had been linked to a wide variety of fatal and nonfatal diseases (Table 9.2), with credible estimates of the toll from passive smoking reaching as high as 60,000 U.S. nonsmoker deaths per year from all known or suspected causes (Wells 1999; CalEPA 1997). By comparison, for the period 1990–1994, the estimated total annual average death toll from active smoking was 431,000 persons, or 19.5% of all U.S. deaths (Centers for Disease Control 1997) and 8.8% of all the world's deaths annually (Brandt and Richmond 2004). Thus, in the United States, SHS pollution may be responsible for as many as 2.7% of all U.S. deaths annually.

TABLE 9.1
172 Toxic Substances in Tobacco Smoke, Including 33 Hazardous Air Pollutants (HAPs), 47 Chemicals Restricted as Hazardous Waste (HW), 67 Known Human or Animal Carcinogens, and 3 EPA Criteria Pollutants (CP)

Compound(s)[a]
Known Human Carcinogen (H)
Probable Human Carcinogen (HP)
Animal Carcinogen (A)
	Toxic = T, from references 1, 2, 3, 4, 5, 6, or 9;
	hazardous wastes = HW, reference 7;
	hazardous air pollutants = HAP reference 8;
	OSHA = Regulated Workplace Carcinogen,
	from reference 10

1. 1,1-Dimethylhydrazine *HP*	T[4,9] HW, HAP
2. 1,3 Butadiene	T[1,9] HAP
3. 1-Methylindole	T[5]
4. 2,6-Dimethylaniline *A*	T[9]
5. 2-Naphthylamine *A, H*	T[4,9] HW, OSHA
6. 2-Nitropropane *A*	T[4,9] HW
7. 2-Toluidine *A*	T[4,9] HW, HAP
8. 3-Vinylpyridine	T[4]
9. 4-(Methylnitrosamino)-1-(3-pyridil)-1-butanone (NNK) *A*	T[4,9]
10. 4,4-Dichlorostilbene	T[5]
11. 4-Aminobiphenyl *A, H*	T[4,9] HW, HAP, OSHA
12. 5-Methylchrysene *A*	T[4,9]
13. 7H-Dibenzo(c,g)carbazole *A*	T[4] HW
14. 9-Methylcarbazole	T[5]
15. AaC* *A*	T[9]
16. Acetaldehyde *A*	T[4,9] HW, HAP
17. Acetamide *A*	T[9]
18. Acetone	T[4]
19. Acetonitrile	T[1]
20. Acrolein	T[4] HW, HAP
21. Acrylonitrile	T[4,9] HW, HAP
22. Acrylymide *A*	T[9]
23. Alkylcatechols	T[5]
24. Ammonia	T[1]
25. Anabasine	T[3]
26. Aniline	T[1] HAP
27. Anthracenes (5)	T[2]
28. Antimony	T[2,5] HAP
29. Arsenic *H*	T[4,9] HW, HAP
30. Aza-arenes *A*	T[9]
31. Benz(a)anthracene *A*	T[4,9] HW
32. Benzene *A,H*	T[4,9] HW, HAP
33. Benzo(a)pyrene *A,H*	T[4,9] HW
34. Benzo(b)fluoranthene *A*	T[4,9] HW
35. Benzo(b)furan *A*	T[9]
36. Benzo(j)fluoranthene *A*	T[4,9]
37. Benzo(k)fluoranthene *A, HP*	T[4,9] HW
38. Benzofurans (4) *A*	T[2]
39. Beryllium *H*	T[9]
40. Butyrolactone	T[6]
41. Cadmium *H*	T[4,9] HW, HAP
42. Caffeic acid *A*	T[9]
43. Carbon monoxide	T[4]
44. Carbonyl sulfide	T[4]

TABLE 9.1 (CONTINUED)
172 Toxic Substances in Tobacco Smoke, Including 33 Hazardous Air Pollutants (HAPs), 47 Chemicals Restricted as Hazardous Waste (HW), 67 Known Human or Animal Carcinogens, and 3 EPA Criteria Pollutants (CP)

Compound(s)[a]
Known Human Carcinogen (H)
Probable Human Carcinogen (HP)
Animal Carcinogen (A)
	Toxic = T, from references 1, 2, 3, 4, 5, 6, or 9;
	hazardous wastes = HW, reference 7;
	hazardous air pollutants = HAP, reference 8;
	OSHA = Regulated Workplace Carcinogen,
	from reference 10

45. Catechol	T[4,9] HAP
46. Chromium VI *H*	T[4,9] HW, HAP
47. Chrysene	T[4] HW
48. Cobalt	T[9]
49. Cresols (all 3 isomers)	T[5] HW, HAP
50. Crotonaldehyde	T[4] HW
51. Cyanogen	T[2] HW, HAP
52. DDD	T[5,2] HW
53. DDE *A*	T[9]
54. DDT *A*	T[5,2,9] HW
55. Dibenz(a,h)acridine *A*	T[4,9] HW
56. Dibenz(a,j)acridine *A*	T[4,9] HW
57. Dibenz(a,h)anthracene *A*	T[4,9] HW
58. Dibenzo(a,e)pyrene *A*	T[9]
59. Dibenzo(a,i)pyrene *A*	T[4]
60. Dibenzo(a,l)pyrene *A*	T[4,9] HW
61. Dibenzo(c,g)carbazole *A*	T[9]
62. Dimethylamine	T[2,6]
63. Di(2-ethylhexyl)phthalate *A*	T[9]
64. Endosulfan	T[5] HW
65. Endrin	T[5,2] HW
66. Ethylbenzene	T[1,12]
67. Ethyl Carbamate *A*	T[4,9] HAP
68. Ethylene Oxide *A,H*	T[9]
69. Fluoranthenes (5)	T[2]
70. Fluorenes (7)	T[2] HAP
71. Formaldehyde	T[1,9] HW
72. Furan	T[2,9]
73. Glu-P-1* *A*	T[9]
74. Glu-P-2* *A*	T[9]
75. Hydrazine	T[4] HW, HAP
76. Hydrogen cyanide	T[4] HW
77. Hydrogen sulfide	T[1]
78. Hydroquinone	T[5,2] HAP
79. Indeno(1,2,3-c,d)pyrene *A*	T[4,9]
80. Indole	T[2]
81. IQ* *A, HP*	T[9]
82. Isoprene	T[2,9]
83. Lead [210] *A, H*	T[5,9] HW, HAP
84. Limonene	T[2]
85. Maleic hydrazide	T[5] HW
86. Manganese	T[5,2] HAP
87. Mercury	T[5,2] HW, HAP
88. Methanol	T[1] HAP

TABLE 9.1 (CONTINUED)
172 Toxic Substances in Tobacco Smoke, Including 33 Hazardous Air Pollutants (HAPs), 47 Chemicals Restricted as Hazardous Waste (HW), 67 Known Human or Animal Carcinogens, and 3 EPA Criteria Pollutants (CP)

Compound(s)[a] Known Human Carcinogen (H) Probable Human Carcinogen (HP) Animal Carcinogen (A)	Toxic = T, from references 1, 2, 3, 4, 5, 6, or 9; hazardous wastes = HW, reference 7; hazardous air pollutants = HAP, reference 8; OSHA = Regulated Workplace Carcinogen, from reference 10
89. Methyl formate	T[1]
90. Methyl chloride	T[3]
91. Methylamine	T[1]
92. Methyleugenol	T[9]
93. Naphthalene	T[1] HAP
94. Nickel A,H	T[4,9] HAP
95. Nicotine	T[4] HW
96. Nitric oxide	T[4]
97. Nitrobenzene A, HP	T[9]
98. Nitrogen dioxide	T[4]
99. Nitromethane A	T[9]
100. N-Nitrosodi-n-propylamine A	T[9]
101. N'-Nitrosoanabasine A	T[4]
102. N-Nitrosodiethanolamine A	T[4,9] HW
103. N-Nitrosodiethylamine A	T[4,9] HW
104. N-Nitrosodimethylamine A	T[4,9] HAP, HW, OSHA
105. N-Nitrosodi-n-butylamine A	T[9]
106. N-Nitrosoethylmethylamine A	T[4,9]
107. N-Nitrosomorpholine A	T[4] HAP
108. N'-Nitrosonornicotine A	T[4,9] HW
109. N-Nitrosopiperidine A	T[9]
110. N-nitrosopyrrolidine A	T[4,9]
111. NAT* A	T[9]
112. NNN* A	T[4]
113. Nornicotine	T[3]
114. Octane	T[1,12]
115. o-Toluidine	T[4] HW
116. Palmitic acid	T[2]
117. Parathion	T[5] HW
118. Phenol	T[2] HW, HAP
119. Phenols (volatile)	T[4] HW
120. PhIP A	T[9]
121. Picolines (3)	T[3]
122. Polonium[210] A, H	T[4] HAP
123. Propionic acid	T[1]
124. Propylene oxide A	T[9]
125. Pyrenes (6)	T[2]
126. Pyridine	T[1] HW
127. Quinolines (7) A	T[2,9]
128. Resorcinol	T[5] HW
129. Styrene	T[1,9,12] HAP
130. Toluene	T[1] HAP
131. Trp-P-1* A	T[9]
132. Trp-P-2* A	T[9]

TABLE 9.1 (CONTINUED)
172 Toxic Substances in Tobacco Smoke, Including 33 Hazardous Air Pollutants (HAPs), 47 Chemicals Restricted as Hazardous Waste (HW), 67 Known Human or Animal Carcinogens, and 3 EPA Criteria Pollutants (CP)

Compound(s)[a] Known Human Carcinogen (*H*) Probable Human Carcinogen (*HP*) Animal Carcinogen (*A*)	Toxic = T, from references 1, 2, 3, 4, 5, 6, or 9; hazardous wastes = HW, reference 7; hazardous air pollutants = HAP, reference 8; OSHA = Regulated Workplace Carcinogen, from reference 10
133. Urethane	T[5,2]
134. Vinyl chloride *H*	T[4,9] HW, HAP
135. Xylenes (3)	T[1,12]
136. PM$_{2.5}$	CP
137. Nitrogen dioxide	CP
138. Carbon monoxide	CP

[a] From Tables 5, 6, 7, 8, or 9 in Reference 4 or 5.

* Abbreviations: (all PAHs): AaC, 2-amino-9H-pyrido[2,3-b]indole; IQ, 2-amino-3-methylimidazo[4,5,b]quinoline; Glu-P-1, 2-amino-6-methyl[1,2,-a:3′,2″-d]imidazole; Glu-P-2, 2-aminodipyrido[1,2-a:3′2″-d]imidazole; Phlp, 2-amino-1-methyl-6-phenylimidazo[4,5-b]pyridine; Trp-P-1, 3-amino-1,4,-dimethyl-5Hpyrio[4,3-b]indole; Trp-2, 3 amino-1-methyl-5H-pyrido[4,3,-b]indole. NAT, *N′*-nitrosoanatabine; NNN, *N′*-nitrosonornicotine.

REFERENCE SOURCES FOR TABLE 9.1

1. *NIOSH Pocket Guide to Chemical Hazards* (1994) U.S. Department of Health & Human Services, Centers for Disease Control & Prevention, June 1994.
2. Sax, N.I. (1984) *Dangerous Properties of Industrial Materials*, 6th ed., Van Nostrand Reinhold, New York.
3. *The Merck Index — An Encyclopedia of Chemicals, Drugs, and Biologicals,* 11th ed. (1989) Budavari, S., O'Neill, M.J., Smith, A., and Heckelman, P.E., Eds., Merck & Co., Rahway, NJ.
4. *Reducing the Health Consequences of Smoking, 25 Years of Progress. A Report of the Surgeon General, 1989* (1989) USDHHS, Rockville, MD.
5. *Smoking and Health, A Report of the Surgeon General, 1979.* USDHEW, Washington, DC.
6. Wynder, E. & Hoffmann, D. (1967) *Tobacco and Tobacco Smoke*, Academic Press, New York.
7. Appendix VIII, Part 261, 40 CFR Part 268 Subpart A Sec. 268.2(b).
8. Section 112 Hazardous Air Pollutants (2005) US EPA Unified Air Toxics, www.epa.gov/ttn/uatw/188polls. html.
9. Hoffmann, D. and Hoffmann, I. 1998. Chemistry and Toxicology, in *Smoking and Tobacco Control Monograph 9*. National Institutes of Health, National Cancer Institute, Bethesda, MD.
10. Thirteen OSHA-Regulated Carcinogens. Appendix B in *NIOSH Pocket Guide to Chemical Hazards* (1994) U.S. Department of Health & Human Services, Centers for Disease Control & Prevention, June 1994.
11. *IARC Monographs on the Evaluation of Carcinogenic Risks to Humans Volume 83 Tobacco Smoke and Involuntary Smoking* (2004) Table 1.14 Carcinogens in Cigarette Smoke. World Health Organization International Agency for Research on Cancer, Lyon, France.
12. Wallace, L.A., Pellizzari, E., Hartwell, T., Perritt, K., and Ziegenfus, R. (1987) Exposures to Benzene and Other Volatile Organic Compounds from Active and Passive Smoking, *Archives of Environmental Health* 42: 272–279.

Note: The substances in Table 9.1 are all listed as constituents of tobacco smoke. Although few of them have actually been reported as measured in secondhand smoke, all of them have been measured in mainstream, and to a lesser extent, sidestream, smoke. Secondhand smoke consists of fresh and aged exhaled mainstream and sidestream smoke, and mainstream smoke is formed in the same burning cone as sidestream. Generally, sidestream and secondhand smoke contain greater total quantities of given chemicals (e.g., more NO$_2$ and more NNK), and are more toxic than mainstream smoke, which is formed at a higher temperature, and is also filtered by the tobacco rod and the cigarette filter.

TABLE 9.2
Health Effects Associated with Exposure to Secondhand Smoke

I. Effects Causally Associated with SHS Exposure

Developmental Effects

Fetal growth: low birth weight and decrease in birth weight
Sudden Infant Death Syndrome (SIDS)

Respiratory Effects

Acute lower respiratory tract infections in children
 (e.g., bronchitis and pneumonia)
Asthma induction and exacerbation in children and adults
Chronic respiratory symptoms in children
Eye and nasal irritation in adults
Middle ear infections in children

Carcinogenic Effects

Lung cancer
Nasal sinus cancer
Breast cancer

Cardiovascular Effects

Heart disease mortality
Acute and chronic coronary heart disease morbidity
Altered vascular properties

II. Effects with Suggestive Evidence of a Causal Association with ETS Exposure

Reproductive and Developmental Effects

Spontaneous abortion, intrauterine growth retardation
Adverse impact on cognition and behavior
Allergic sensitization
Decreased pulmonary function growth
Adverse effects on fertility or fecundability
Menstrual cycle disorders

Cardiovascular and Hematological Effects

Elevated stroke risk in adults

Respiratory Effects

Exacerbation of cystic fibrosis
Chronic respiratory symptoms in adults

Carcinogenic Effects

Cervical cancer
Brain cancer and lymphomas in children
Nasopharyngeal cancer
All cancers — adult and child

Source: Data from CalEPA (1997); CalEPA (2005).

9.4 SMOKING PREVALENCE AND TRENDS

Smoking prevalence varies by historical time period, by age group, and by group characteristics. During the period prior to 1990, two-thirds of U.S. children grew up in households with one or more smokers, and it was nearly impossible to find a job in a workplace where smoking was prohibited (Repace 1985). U.S. smoking trends show that cigarette smoking prevalence peaked in 1965 at 42.4%, with a notable race and gender disparity, and has steadily declined since (Figure 9.2). In 2001, 46.2 million adults (22.8%) in the U.S. were current cigarette smokers — 25.2% of men and 20.7% of women. Current U.S. smoking prevalence varies markedly by state, from a low of 13% in Utah to a high of 31% in Kentucky, and has become increasingly concentrated in lower income and less educated segments of the populace. Smoking prevalence also varies by age group; in the past decade, prevalence rates have been declining in all age groups except among persons 18 to 24, among whom prevalence has increased, from 23% in 1991 to 27% in 2000. Smoking is also higher among veterans than in the general population, with the prevalence of smoking among Vietnam War veterans at a very high 47%. Smoking prevalence also ranges from 50% to over 80% among persons with psychiatric or substance-abuse disorders, with one study estimating that such persons may account for 44% of all cigarettes smoked in the U.S. (Schroeder 2004); very high smoking prevalence is also common among the prison population. The declining prevalence of smoking and the proliferation of smoke-free workplace laws (in six states at this writing) has produced a reduction in population exposure to SHS, as evidenced by a 70% decline in blood (serum) cotinine, a biomarker for SHS exposure, from a median of 0.20 ng/ml in 1988–1991 to 0.059 ng/ml in 2003 (Pirkle et al. 1996; Centers for Disease Control 2003).

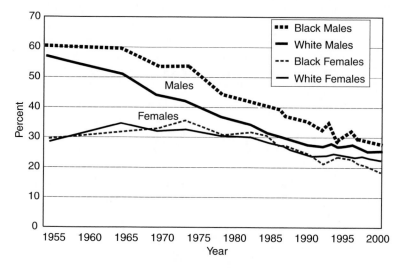

FIGURE 9.2 Trends in the prevalence of cigarette smoking among adults by race and gender in the U.S. 1955–2001. (From Shopland D., personal communication, 2003; National Cancer Institute. [1955 Data are based on the Census Bureau's Current Population Survey, data from 1965–2000 are based on estimates from the National Health Interview Survey, 1965–2001.])

9.5 WORKPLACE EXPOSURE

Workplace smoking policies are important determinants of exposure. While over three-quarters of white-collar workers are covered by smoke-free policies, including 90% of teachers, just 43% of the U.S.'s 6.6 million food preparation and service occupations workers benefit from this level of protection, and of these, only 13% of bartenders can avoid breathing smoke on the job (Shopland et al. 2004). Furthermore, under intense lobbying by tobacco-allied interests, 20 state legislatures have passed intentionally weak laws preempting local efforts to enact strict clean indoor air laws, which particularly limit smoking restrictions in the hospitality industry (Shopland et al. 2004, Schroeder 2004). Only six U.S. states ban smoking in all workplaces; of the remaining 44, a patchwork of local laws provides a mixture of bans, ventilation, air cleaning, designated smoking areas, and no restrictions. Moreover, even in California, one of the states with the most complete workplace restrictions, as of 2000, half of smokers continued to smoke inside the home.

9.6 DETERMINING EXPOSURE

9.6.1 BIOMARKERS FOR SHS: BODY FLUID COTININE

Microenvironmental air pollution contributes to total exposure as follows. The total inhaled SHS exposure of a person over a lifetime is the time-weighted sum over all of the microenvironments visited during each day of life, taken over all days lived, of the product of three factors: the concentrations of pollutants a person encounters in each microenvironment; the person's respiration rate during the exposure; and the duration of that person's exposure in each microenvironment. All individuals in the population have their own exposure profiles (Repace, Ott, and Wallace 1980). For secondhand smoke, an inhaled exposure of the atmospheric SHS marker, nicotine, would be absorbed through the lung and converted by the liver into a dose of its metabolites, of which cotinine is one, by the pharmacokinetics of the individual (Repace and Lowrey 1993; Repace et al. 1998; Benowitz 1999). What does this dose distribution look like? Figure 9.3 shows the distribution of serum cotinine, which reflects the total body burden of SHS, for a representative sample of all non-tobacco-using individuals in the U.S. population from 1989–1991. Figure 9.3 then reflects the U.S. population's cross-sectional exposure profiles, incorporating the concentrations of SHS nicotine to which each member of the population was exposed, duration of exposure, and the respiration rates of those nonsmokers during exposure, as modulated by individuals' metabolisms which combine to transform inhaled SHS exposure into actual dose of cotinine (Repace and Lowrey 1993; Repace et al. 1998; Benowitz 1999). In 1990, an estimated 25.5% of the adult population were current cigarette smokers (MMWR 1992), and workplace smoking restrictions were just beginning to be enforced in offices, but not in restaurants or bars.

9.6.2 MISCLASSIFICATION PROBLEMS

How much exposure to SHS do people receive? An obvious, but not necessarily accurate, way to find out is to ask a nonsmoker "Does your spouse smoke?" Assessing the level of exposure by questionnaire is highly subjective, since it depends upon an individual's sensitivity to SHS. The health risks of SHS have often been assessed in the epidemiological literature by comparing disease incidence in nonsmokers who live with smokers and who therefore are presumed to be SHS exposed, to that of nonsmokers who live with nonsmokers, and who are therefore presumed to be unexposed to SHS. There has been scant appreciation of how poor an exposure assessment such self-reports often yield. The problem of people not comprehending their true smoke exposure is illustrated by the Third National Health and Nutrition Examination Survey (NHANES III), the first nationwide survey in which both questionnaires and measurements of serum cotinine were combined. A major finding of the 1989–1991 NHANES III survey was that exposure of the U.S. nonsmoking population to SHS was pandemic, with 87.3% manifesting detectable levels of cotinine in the blood, despite

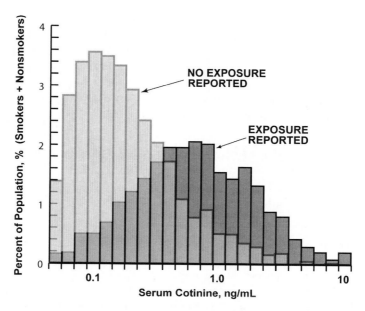

FIGURE 9.3 Distribution of non-tobacco users' blood (serum) cotinine ($n = 6,056$), reflecting SHS dose in a national probability sample of the U.S. population during 1989–1991 adapted from Pirkle et al. (1996). 88% of this nonsmoking population sample has detectable cotinine levels, showing widespread public exposure, but only 40% actually report SHS exposure on the NHANES III questionnaire. The intersection (medium gray) of the distribution of those reporting SHS exposure at work or at home (dark gray) with that of those reporting no exposure in those locations (light gray) indicates that many persons who report having no SHS exposure actually have exposures greater than those who do report having SHS exposure. (Note that the values on the horizontal axis represent the antilogs of the cotinine distribution.)

only 40% actually reporting SHS exposure. The geometric mean of those reporting exposure was much higher than those who did not (0.925 ng/ml vs. 0.132 ng/ml) (Pirkle et al. 1996). However, the frequency distribution of cotinine for the group reporting "no exposure" (histogram on the left side of Figure 9.3) overlapped with the frequency distribution for the group reporting exposure to SHS (histogram on right side of Figure 9.3). In this overlap zone, some nonsmokers who reported "no exposure to SHS" had higher cotinine levels than some nonsmokers who reported "exposure to SHS." Persons who reported no exposure but had high cotinine levels apparently were exposed without knowing it.

In a health effects study, an epidemiologist with only the questionnaire results available might erroneously assume that the persons who said they were not exposed had zero exposure. The overlap between the two groups, based on measured cotinine in their blood, reveals a serious statistical problem called "misclassification." For many people in this population, the questionnaire alone gives erroneous information about who was exposed and who was not. Because differences in observable health effects may be small, misclassification reduces the apparent magnitude of the health effect induced by SHS, as well as its statistical significance. On the other hand, misclassification can also work to increase apparent risk, as when smokers do not respond accurately about their smoking status; tobacco industry consultants (e.g., Lee and Forey 1996) have argued that this explains the observed risk elevation in epidemiological studies of passive smoking; however, the USEPA (1992) and others (Wells et al. 1998) persuasively rejected these arguments, concluding that this effect is a small fraction of the observed risk. This problem could be addressed (e.g., Repace, Hughes, and Benowitz in press) by directly measuring each person's cotinine or by asking respondents to wear a personal monitor to directly measure their exposure (see Chapter 6). Unfortunately, such measurements are rarely made in epidemiological studies.

Pirkle et al. (1996) reported that for children aged 2 months to 11 years, 43% lived in a home with at least one smoker. Both home and workplace environments contributed significantly to SHS exposure. Since the survey was based on a large national statistical sample with measured blood levels (n = 10,642 persons), the authors' conclusions are significant: "The high proportion of the population with detectable serum cotinine levels indicates widespread exposure to environmental tobacco smoke in the U.S. population." Since that time, U.S. nonsmokers' exposures have declined with the proliferation of smoking restrictions (Centers for Disease Control 2003).

9.7 SECONDHAND SMOKE AS MICROENVIRONMENTAL AIR POLLUTION

9.7.1 SHS CONCENTRATIONS

High U.S. smoking prevalence, peaking at 57% of men and 34% of women in 1955, coupled with scant restrictions on indoor smoking until the late 1980s, led to widespread nonsmokers' exposures during this era (Schroeder 2004). In the late 1970s, two-thirds of homes, most offices, most factories and other workplaces, as well as trains, buses, ships, aircraft, theaters, college classrooms, doctors' offices, hospitals, waiting rooms, restaurants, bars, public buildings, stores, supermarkets, and small businesses, as well as social events, including parties, weddings, and dances, were smoke filled (Repace and Lowrey 1980, 1982; Repace 1985). A 1978 investigation of the range and nature of the nonsmoking public's exposure to SHS concluded that nonsmokers were exposed to significant air pollution burdens from smoking (Repace and Lowrey 1980), for two basic reasons. First, people on average spent 90% of their time indoors (Repace and Lowrey 1980), and, second, respirable suspended particulates (RSP) in microenvironments where smoking occurred greatly exceeded the concentrations of RSP encountered in smoke-free environments. Figure 9.4 shows ~20-minute-average Piezobalance measurements of RSP levels in 17 smoking (alphabetically labeled data) and 13 nonsmoking microenvironments (unlabeled) as a function of the density of burning cigarettes (D_s). The measurements shown in Figure 9.4 are of $PM_{3.5}$, also called respirable suspended particles (RSPs), an older generic term widely used in much of the SHS measurement literature, which includes the closely related fine particles (FP), or $PM_{2.5}$, a regulated (outdoor) air pollutant since 1997. Both fresh and aged SHS consist of FP. Both RSP and FP consist of fine solid or liquid particles that remain airborne for extended periods, are able to penetrate deep into the lung when breathed, and have slow pulmonary clearance times. (At the time of the study in Figure 9.4, the only regulated outdoor particulate air pollutant was total suspended particulate, TSP, which includes the nonrespirable particulate fraction between PM_{10} and PM_{50}. TSP was superseded by PM_{10}, which remains regulated, since it is inhaled by mouth breathers, but $PM_{2.5}$ appears to be the more serious health threat.) Indoor microenvironments, such as in homes, offices, restaurants, bars, theaters, lodge halls, bowling alleys, bingo games, and other commonly frequented venues, including emergency rooms at hospitals, all measured in Figure 9.4, were found polluted with RSP to a much greater extent than indoor nonsmoking environments such as homes, churches, and libraries or outdoors on city streets, or even in vehicles on busy commuter highways. The essential findings of this field study were that the nonsmoking microenvironment RSP levels ranged from about 20–60 $\mu g/m^3$, while the smoking microenvironment RSP levels ranged from about 90–700 $\mu g/m^3$, and that the RSP concentration differences between the smoking and nonsmoking microenvironments were due to tobacco smoke. As shown by the linear regression equation in Figure 9.4 (r^2 = 0.50), 50% of the variance in RSP levels is explained by the smoker density. Such levels of indoor air pollution, although ubiquitous in that era, were unpleasant for many nonsmokers; to place these measurements into a nonsmoker's perspective, Junker et al. (2001) reported an odor acceptability threshold of 1 $\mu g/m^3$ for RSP from SHS, and an irritation threshold of 4.4 $\mu g/m^3$. Viewed from the perspective of early 21st century air quality health advisories for fine-particle pollution ($PM_{2.5}$), if the $PM_{3.5}$ air quality levels of RSP in these typical 20th century smoking microenvironments had

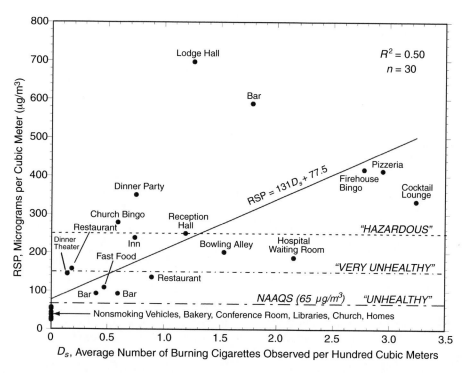

FIGURE 9.4 Respirable particle mass concentration (RSP) vs. smoker density for 17 smoking microenvironments and 13 nonsmoking microenvironments in the Washington, DC metro area in the Spring of 1978 (data adapted from Repace and Lowrey 1980; 1982). The locations with active smoking included restaurants, bars, nightclubs, bingo games, a bowling alley, and a hospital waiting room. The smoke-free locations ($D_s = 0$) included 5 homes, 2 libraries, a church, a bagel bakery, and a conference room in an office building. Most visits lasted for about 20 minutes. Environmental Protection Agency health-based short-term air quality advisories for $PM_{2.5}$ in the outdoor air are shown for comparison as "Unhealthy," "Very Unhealthy," and "Hazardous." (Data from Repace and Lowrey 1980, 1982.)

been interpreted using the current outdoor air criteria, they would have been judged variously as *Unhealthy (Code Red)*, *Very Unhealthy,* or *Hazardous*, as shown by the horizontal dashed lines in Figure 9.4.

9.7.2 Smoking Rates and Emissions

The concentrations of SHS to which the population are exposed will depend upon the smoking prevalence and the number of cigarettes, pipes, and cigars smoked by each smoker per unit time in each of the microenvironments that the nonsmoking population frequents. Because the vast majority of smokers smoke cigarettes, the treatment given here will focus on cigarettes. To illuminate the smoking rates that contributed to the nonsmokers' exposures shown in Figure 9.4 and doses shown in Figure 9.3, consider the following. In 1994, among U.S. adults aged ≥18 years, there were 48.0 million adult current smokers, of whom 25.3 million were men, and 22.7 million were women. These persons smoked 485 billion cigarettes, and thus smoked 10,104 cigarettes per smoker per year, or ~28 cigarettes per smoker per day (MMWR 1996; USDA 1995; Maxwell 1995). If one assumes a 16-hour smoking day, this is equivalent to about 1.8 cigarettes per smoker-hour.

RSP is the major pollutant emitted by burning tobacco products, and cigars emit three times as much RSP as cigarettes (Repace, Ott, and Klepeis 1990; Klepeis, Ott, and Repace 1999). Figure 9.5 shows a histogram of the SHS RSP emissions of the top 50 brands of cigarettes (Nelson 1994; Martin et al. 1997). These data are useful for showing the range in RSP emissions of cigarettes,

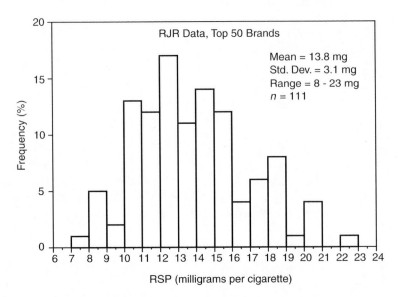

FIGURE 9.5 Histogram of SHS RSP emission factors for the top 50 brands of cigarettes, representing 65.3% of the U.S. cigarette market (plus a University of Kentucky research cigarette K1R4F), when two of each brand were smoked for 11 minutes each by human smokers in 1994. Data were digitized from a plot presented by RJ Reynolds Tobacco Company in testimony before the Occupational Safety and Health Administration (OSHA) hearing on OSHA's proposed Indoor Air Rule in Washington, DC in January 1995 (Nelson 1994). Other details from this study were published by Martin et al. (1997), which reported an overall sales-weighted RSP emission average of 13.67 mg/cigarette (Standard Error, 0.4106 mg/cigarette), $n = 100$.

explaining one source of the variation in RSP levels in field studies of the impact of smoking on indoor air. Note, however, that the actual variation in emissions may be smaller because only a few prominent brands dominate sales. In 2003, the top five cigarette brands alone accounted for two thirds of total U.S. cigarette consumption. Marlboro is the most popular brand, with sales greater than the five leading competitors combined. The market share for Marlboro was 37.5%, followed by Newport (8.0%), Doral (6.3%), Camel (6.1%), Winston (4.6%), and Basic (4.4%) (Maxwell 1995). The range in RSP emissions of cigarettes is determined primarily by the differences in mass when smoked identically.

9.7.3 A Person's Daily Exposure to Particulate Air Pollution

Epidemiological studies (studies of the causation of disease) have linked increases in daily average outdoor inhalable (PM_{10}) exposures to increased morbidity and mortality, effects that are stronger for finer particles ($PM_{2.5}$) (Holgate et al. 1999). There is recent evidence that even shorter-term $PM_{2.5}$ exposures can have cardiopulmonary health effects (Pope and Dockery 1999). For this reason, particulate air pollution is regulated under the Clean Air Act, and $PM_{2.5}$ is subject to an annual average health-based National Ambient Air Quality Standard (NAAQS) of 15 $\mu g/m^3$ in the outdoor air, and a 24-hour NAAQS of 65 $\mu g/m^3$. RSP from SHS is at least as toxic as outdoor RSP, and likely far more toxic, as illustrated by Table 9.1.

If a person carries a personal air pollution monitor for 24 hours as he or she visits various microenvironments, a picture emerges of that person's total daily exposure. Figure 9.6 (Repace, Ott, and Wallace 1980) shows one of the first experiments designed to investigate this issue:

> On October 16, 1979, I awoke to a new day in my home, commuted 25 miles from my residence in suburban Maryland to the Environmental Protection Agency (EPA) headquarters in Washington, DC, where I had a private smoke-free office. During my waking hours, I carried along a portable, battery-powered TSI Model 3500 Piezobalance (see Chapter 6) to measure RSP (each datum represents at least

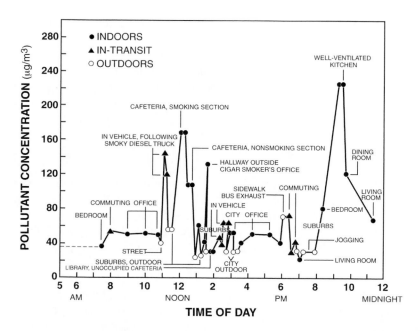

FIGURE 9.6 A nonsmoker's 24-hour exposure to RSP on October 16, 1979, illustrating a person's air pollution exposure in the metropolitan Washington, DC area, on one day of his life. A series of these daily patterns comprise an individual's lifetime exposure.

ten 2-minute $PM_{3.5}$ averages). I next traveled by car to the Goddard Space Flight Center in suburban Maryland, partly stuck in traffic behind a smoky diesel truck. I dined in the Goddard cafeteria, followed by a short tour of various buildings. The RSP level in the smoking section of the cafeteria was 55% greater than in the nonsmoking section, and 8.5 times greater than levels outdoors. On my return trip to Washington, RSP levels were much lower in the absence of diesel exhaust. Another period of a few hours in my smoke-free office was followed by a second encounter with diesel bus exhaust as I walked along the city sidewalk. My commute back to my suburban residence was followed by jogging outdoors, then dinner. Despite a powerful kitchen ceiling exhaust fan, roasting chicken in the electric oven caused high particulate levels in the kitchen, which penetrated into the dining and living rooms.

The data in Figure 9.6 show that I spent 84% of my time indoors, 9% in transit, and 7% outdoors. My total 24-hour integrated exposure for the day was 1428 microgram-hours per cubic meter (μg-h/m^3), equivalent to a 24-hour average exposure concentration of 59.5 μg/m^3. Contributions to my total RSP exposure break down into 82% from indoor microenvironments, 10% from in-transit microenvironments, and 8% from outdoors. (Note that even though high levels of RSP due to diesel exhaust were encountered in transit, the contribution to total exposure was still little more than that expected from the fraction of time spent traveling.) On this day, indoor RSP levels are generally higher than outdoors; diesel exhaust produces higher RSP levels than gasoline-powered vehicles; both cooking and smoking produced the highest RSP levels of the day. More recent field studies of personal exposure to particulate matter, such as the USEPA particle total exposure assessment methodology (PTEAM) studies (Özkaynak et al. 1996) have confirmed that smoking, cooking, and other indoor activities are major sources of human exposure to particulate matter.

9.8 ATMOSPHERIC TRACERS FOR SHS: NICOTINE AND RSP

What is the best atmospheric marker for SHS, a mixture of over 4,000 chemicals? Nicotine, a semi-volatile organic compound unique to tobacco smoke, is the most commonly used atmospheric tracer for SHS. Hammond (1999) has reviewed a number of field measurements of nicotine concentrations in homes, offices, restaurants, etc. These "opportunistic" (that is, not necessarily representative) samples using both personal or area monitors, with relatively long averaging times from an 8-hour work shift to a day or a week or more, show that workplaces with smoke-free policies generally have measured nicotine levels below 1 $\mu g/m^3$. On the other hand, mean nicotine concentrations in workplaces that allow smoking generally range from 2–6 $\mu g/m^3$ in offices, 3–8 $\mu g/m^3$ in restaurants, and from 1–6 $\mu g/m^3$ in blue-collar workplaces, compared to 1–3 $\mu g/m^3$ in the homes of smokers (Hammond 1999). However, workplace concentrations have been highly variable, with some studies reporting concentrations more than 10–100 times the average home levels. For example, bars and discos have had measured median nicotine concentrations from SHS ranging from 19 $\mu g/m^3$ to 122 $\mu g/m^3$ (Nebot et al. 2005).

Field-deployable exposure monitors for nicotine use either active pump-and-filter methods, with minimum averaging times of about 15 minutes, or passive techniques with minimum averaging times of about a day. Both methods require chemical analysis. Repace et al. (1998) developed a model for the prediction of time-averaged nicotine concentrations based upon smoker densities and design ventilation rates, making it possible to generalize measured nicotine data if smoker densities are also measured.

Although RSP is not unique to SHS, measurement of RSP concentrations from SHS is important because the fine-particulate fraction of RSP, $PM_{2.5}$, is a regulated outdoor air pollutant with health-based national air quality standards, and daily air quality advisories are widely reported in the media in qualitative terms, using an Air Quality Index (AQI). Available real-time RSP monitors have averaging times of minutes or seconds, and do not require laboratory analysis, (see Chapter 6). Therefore, these monitors are useful for real-time studies of the growth and decay of individual cigarette smoke concentrations, as well as for short-term field surveys of SHS, as will be described later in this chapter. Cigarettes (and pipes, and cigars as well) have major emission components in the $PM_{2.5}$ size and range (Klepeis et al. 2003; National Research Council 1986). A number of studies have examined the SHS-RSP emissions of tobacco products. Martin et al. (1997) reported that the average gravimetric SHS-RSP yield was 13.7 mg/cigarette (standard error [SE] 0.4 mg/cigarette, $n = 100$) for the 50 top-selling U.S. cigarette brand styles in 1991. Martin et al. (1997) measured "RSP" both using a Piezobalance and a 1-μm pore gravimetric filter, but did not otherwise characterize the particle size distribution. The yield estimated by Piezobalance data was slightly lower, at 11.55 mg/cigarette (SE 0.36 mg/cigarette, $n = 100$). The SHS was generated by human smokers in a test chamber, who smoked each cigarette for 11 minutes. This value is in good agreement with the data presented in Figure 9.5, and with the ~14 mg/cigarette value ($PM_{2.5}$) obtained by regression analysis from the USEPA's Riverside PTEAM study (Wallace 1996). Dr. W.R. Ott (personal communication 2003) observed smoking times for 33 people in a Las Vegas casino reporting a mean of 9.25 minutes (SD = 2.3 min). Assuming a smoking time of 9.25 minutes and a cigarette emission factor of 1.43 mg/min, as reported by Klepeis, Ott, and Switzer (1996) for smoking lounges in San Francisco and San Jose, CA, airports, this yields an estimated RSP ($PM_{3.5}$) yield of 13.2 mg/cigarette. Repace and Lowrey (1980) observed a smoking rate of 9.8 minutes per cigarette in seven smokers, suggesting little change in cigarette smoking duration over a period of 2 decades. Industry studies report a range of about 8 to 23 mg/cigarette for RSP emissions from cigarettes (Figure 9.5), so that controlled experiments performed with individual cigarettes may differ. It is reasonable to assume a default emission rate of approximately14 mg/cigarette and a smoking duration of ~10 min/cigarette.

For SHS nicotine, Martin et al. (1997) reported that the average measured SHS nicotine yield for the 50 top-selling U.S. cigarette brand styles in 1991 was 1,585 $\mu g/cigarette$ (standard error, SE 42.21 $\mu g/cigarette$, $n = 100$), yielding an RSP-to-nicotine emission ratio of 8.6 for SHS from these cigarettes. Daisey (1999) in an excellent discussion of atmospheric tracers for SHS, observes that nicotine can be used to estimate RSP exposures provided that smoking occurs regularly in the microenvironment, that the system is in a steady state, and that the sampling time is longer than the characteristic times for removal processes. Under these conditions, the ratio of RSP to nicotine from SHS is approximately 10:1 (Nagda et al. 1989, Leaderer and Hammond 1991, Daisey 1999). This 10:1 ratio permits body fluid cotinine (e.g., Figure 9.3) which is derived from atmospheric nicotine, to be related to that portion of the RSP air pollution exposures of the population that are due to SHS using pharmacokinetic models (Repace and Lowrey 1993; Repace, Hughes, and Benowitz in press; Repace, Al-Delaimy, and Bernert 2006).

9.9 TIME-AVERAGED MODELS FOR SHS CONCENTRATIONS

Models for SHS concentrations are important for the prediction of human exposures in indoor air quality, epidemiological, or forensic investigations, to generalize field measurements (Repace 1987), to evaluate putative SHS control measures such as ventilation or air cleaning, as well as to debunk SHS junk science (Repace 2004a; Ott 1999; Repace and Lowrey 1995). As described in Chapter 18, mathematical models have been developed for predicting indoor air concentrations for a variety of sources, including cigarettes. These models are derived from the mass-conservation law of physics. The only other required assumption is that SHS concentrations be reasonably spatially uniform at any instant of time, as when air motion and convection cause the smoke in a room to mix rapidly with the air. The spatial variation of the concentration in a room may not be uniform while a point source is emitting, because concentrations usually are higher very close to the source (McBride et al. 1999), but concentrations in a room, or even a home, often become spatially homogeneous soon after the source stops emitting (Klepeis, Nelson, Ott et al. 2003; see Chapter 18 of this book).

Consider a particle emitted from a point source in a well-mixed room with emission rate $g(t)$ and air exchange rate a with pollutant-free outdoor air and a deposition rate k on the room surfaces (Ott, Langan, and Switzer 1992). The exchange with outdoor air is due to natural or mechanical ventilation and is measured as the number of room air changes per unit time. The deposition rate is also measured in number of (equivalent) air changes per unit time. The total air exchange rate is then $\phi = a + k$. If the initial concentration is $x(t) = 0$ in the room at time $t = 0$, then the following expression based on the mass balance model (from Chapter 5) gives the relationship between the mean concentration over averaging time $T = (t_f - t_0)$, (where t_f is the end of the averaging period) the mean emission rate over time T, and the instantaneous concentration $x(T)$, in typical units:

$$\overline{x(T)} = \frac{\overline{g(T)}}{\phi v} - \frac{\tau}{T} x(T) \tag{9.1}$$

where

$\overline{x(T)}$	= average indoor concentration over time T ($\mu g/m^3$)
$\overline{g(T)}$	= average source emission rate over time T ($\mu g/hr$)
$x(t)$	= instantaneous concentration at time t ($\mu g/m^3$)
v	= volume of the well-mixed room (m^3)
ϕ	= total loss rate of particles due to both air exchange and deposition (h^{-1})
τ	= mean residence time $1/\phi$ (h)

As discussed in Chapter 5, the last term on the right-hand side of Equation 9.1 includes the ratio of the residence time to the averaging time τ/T, which generally becomes small as the averaging time becomes large relative to the residence time.

Example 1. A typical home with the exterior doors and windows closed might have an air exchange rate of $a = 0.5$ h^{-1}, and a deposition rate $k = 0.3$ h^{-1}, which corresponds to a mean residence time of $\tau = 1/\phi = 1/(0.8$ $h^{-1}) = 1.25$ hours. If the averaging period were $T = 10$ hours, then the ratio would be $\tau/T = 1.25/10 = 0.125$. Thus, in many practical situations, the rightmost term in Equation 9.1 is negligible, and setting this term equal to zero gives the following approximate relationship for the average concentration as a function of the average source emission rate:

$$\overline{x(T)} \cong \frac{\overline{g(T)}}{\phi v} \qquad (9.2)$$

Example 2. Consider a single cigarette smoked for 10 minutes in a room beginning at time t = 0, with a constant RSP emission rate of 1.4 mg/min. Assume an air exchange rate of a = 0.5 h^{-1}, a deposition rate of k = 0.39 h^{-1} (Özkaynak et al. 1996) so that ϕ = 0.89 h^{-1}, and a room volume of v = 43 m^3 (the size of a small bedroom). To calculate the mean concentration in the room over an 8-hour period beginning at t = 0, we first calculate the mean emission rate over the 8-hour period, $\overline{g(8)}$ = (10 min) (1.4 mg/min)/(8 hr) = 1.75 mg/hr. Substituting this average emission rate into the numerator of Equation 9.2 and using ϕ = $a + k$ gives:

$$\overline{x(8)} \cong \frac{1.75 \text{ mg/hr}}{(0.5 \text{ hr}^{-1} + 0.39 \text{ hr}^{-1})(43 \text{ m}^3)} = 45.7 \times 10^{-3} \text{ mg/m}^3 = 45.7 \text{ μg/m}^3 \qquad (9.3)$$

9.9.1 THE ACTIVE SMOKING COUNT

Some investigators have formally named the number of cigarettes smoked over a specific time period as the "Active Smoking Count (ASC)" (Ott, Switzer, and Robinson 1996). One can consider either the instantaneous ASC $n(t)$ or the expected value, or average, ASC of $\overline{n}_{cig}$. Using this notation, the average emission rate will be $\overline{g(T)} = \overline{n}_{cig} \overline{g}_{cig} /T$, where $\overline{g}_{cig}$ is the average emission rate per cigarette. How is the ASC measured? As an example of the timed observation of natural smoking activity patterns of 8 smokers, consider Harry's Hofbrau, a sports tavern in Redwood City, California, sampled in 1995 (Figure 9.7; W.R. Ott personal communication 2005). Averaged on a minute-by-minute basis, the mean number of cigarettes (ASC) being smoked at any one time in Figure 9.7 is $\overline{n}_{cig}$ = 2.213 active smokers. In the typical field study where the investigator is recording room sizes, counting cigarettes and people in a large crowd, measuring concentrations, and trying to look inconspicuous, it is difficult to time individual smokers. Accordingly, Repace and Lowrey (1980) developed an approximate method for estimation of the ASC: they assumed the estimated 1978 U.S. national average smoking rate of 2 cigarettes per hour, found empirically that it took about 10 minutes to smoke a cigarette, and concluded that a typical smoker might be expected to spend one third of the hour actively smoking. Based on these assumptions, Repace and Lowrey (1980) demonstrated that an investigator could arrive at an empirical estimate of $\overline{n}_{cig}$ inside a bar or tavern with smokers by counting the number of actively burning cigarettes approximately every 10 minutes while walking around the location, and use this value to estimate the number of smokers present. For example, in the real-life tavern smoking pattern of Figure 9.7, if the number of burning cigarettes is counted every 10 minutes beginning at 8:40 P.M., the result is 3, 4, 2, 2, 2, and 2, for an average of 2.5 cigarettes (close to the more accurate minute-by-minute calculation of 2.2 above); if this is multiplied by 3, it yields an estimated 7.5 smokers present. If 6 of the 8 smokers were counted present for the full observation period; and the remaining 2 (who were not) are counted as being present for a half-cycle of smoking, then an average of 7 smokers would be estimated present over the full period. Thus, in crowded field studies, such as in stand-up bars where the ASC cannot be measured with a fine time resolution, the empirical method of Repace and Lowrey (1980) can be used to provide an estimate of the ASC, as well as an estimate of the total number of smokers present. A better, but often infeasible, alternative would be to collect and count the number of cigarette butts smoked during the averaging time T, and estimate the ASC by assuming an average cigarette burning time, e.g., 10 minutes. Ott, Switzer, and Robinson (1996) found in 52 visits to

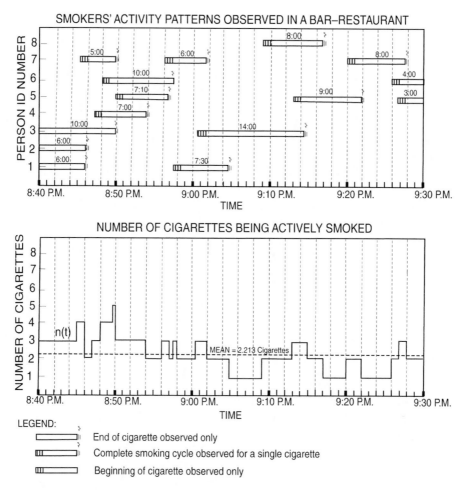

FIGURE 9.7 The smoking time-activity pattern for 8 smokers recorded using a stopwatch and a diary in Harry's Hofbrau in Redwood City, CA, on December 21, 1996 (top). (From W.R. Ott, personal communication, 2005.) The upper figure shows the number of minutes each smoker smoked and the time at which the cigarette was smoked. Smoker #1 departed after smoking two cigarettes, and smoker #8 arrived about 9 P.M. The lower part of the graph shows the actual physical observations in surveys of smoking: $n(t)$, the instantaneous number of cigarettes being smoked. The mean number of cigarettes being smoked over the 50-minute observation period is $n_{ave} = 2.213$ cigarettes.

the Oasis Bar in Redwood City, CA, the measured average RSP concentration from SHS had a correlation coefficient $R^2 = 0.61$ with the active smoking count, similar to the value of $R^2 = 0.50$ derived from the Washington, DC area studies of Repace and Lowrey (1980, 1982) shown in Figure 9.4, and the value of $R^2 = 0.54$ derived by Repace (2004a) for 8 establishments in metropolitan Wilmington, DE.

The ASC concept can also be used to derive a simple model for estimating RSP concentrations based on the number of smokers using certain default assumptions. In many applications, regulators, risk assessors, and investigators making field measurements under real-world uncontrolled conditions are in need of simple models for the prediction of SHS concentrations or the assessment of ventilation rates from measurements of concentrations. The Habitual Smoker Model, based upon Equation 9.2, was developed for these purposes.

9.9.2 THE HABITUAL SMOKER MODEL (EQUATION 9.2 WITH DEFAULTS)

In order to be able to relate observations of burning cigarettes, room sizes, and the design ventilation rates of the American Society of Heating, Refrigerating, and Ventilating Engineers (ASHRAE 1973 to 1989 et seq.) to measured concentrations of RSP from SHS, Repace and Lowrey (1980, 1982, 1985) introduced the Habitual Smoker Model (HSM). The HSM is simply derived from Equation 9.2 as follows. Assume an RSP emission rate of 1.4 mg/min-cigarette, a cigarette smoking time of 10 min/cigarette, and an average smoking rate of two cigarettes per smoker per hour. Then, over a 1-hour averaging period, if an average of $\bar{n} = 1$ burning cigarette is observed, an average of 6 cigarettes per hour will have been smoked (1 cig/ten minutes)(6 10-minute periods/hour), and Equation 9.2 is written as: $\bar{x}(T) = \bar{g}(T)/\phi v = \bar{x}(1\,h) = \bar{g}(1\,h)/\phi v = [(6\,cig/h)\,(1.4\,mg/min\text{-}cig)\,(10\,min/cig)]/[\phi v] = (84\,mg/cig\text{-}h/\phi v)$ in units of milligrams per cubic meter per cigarette. If the RSP mass emission is expressed in units of micrograms and the space volume in units of hundred cubic meters, then the average number of burning cigarettes per 1-hour averaging time yields a concentration of $\bar{x}(1\,h) = [\bar{n}\bar{g}(1\,h)/\phi(100v)](1000\,\mu g/mg) = 840(\bar{n}/v')/\phi$, in units of micrograms per cubic meters, where v' is defined as the space volume expressed in practical units of hundred cubic meters.

If we define ($\bar{n}/v') = D_s$ as the active smoker density (in units of burning cigarettes per hundred cubic meters) and $\phi = a + k$ as before, then

$$\bar{x} = 840\,D_s/\phi\ [\mu g/m^3] \tag{9.4}$$

If we further assume that *all* smokers have identical smoking patterns (2 cigarettes per hour, defined as "habitual smokers") then the number of habitual smokers N_{hs} is three times the ASC, or expressed in terms of smoker densities, the habitual smoker density $(N_{hs}/v') = D_{hs} = 3D_s$. With this substitution, $\bar{x} = 840(D_{hs}/3\phi)$, or

$$\bar{x} = 280\,D_{hs}/\phi\ [\mu g/m^3] \tag{9.5}$$

where D_{hs} is in units of habitual smokers per 100 m^3, and ϕ is in units of inverse hours (i.e., in air changes per hour from ventilation and its equivalent for deposition). For modeling purposes for existing smoking-permitted commercial premises such as restaurants, bars, and casinos, design ventilation rates q (usually expressed in units of cubic feet per minute per person [P] or liters per second per person) are recommended by ASHRAE Standard 62, Ventilation for Acceptable Indoor Air Quality (ASHRAE pre-2004 versions), where q is related to a by the equation $q = av/P$, in consistent units. While q and a are readily measurable, deposition rates k are not as well understood, and may vary depending upon room characteristics such as surface area, surface type, ventilation type, and perhaps airflow rates in the room. I have long used Equation 9.4 and Equation 9.5 with the default assumption $\phi = 1.3a$ (Repace and Lowrey 1980, 1982; Repace 2004a). In summary, the Habitual Smoker Model has been used when it has not been possible to make measurements of a, k, and $\bar{g}$, as for example, in risk assessment or in forensic applications where the SHS exposure takes place in the past or future and default assumptions are required. Using this assumption, the chamber data of Leaderer, Cain, Isseroff, and Berglund (1984) were predicted reasonably accurately by Repace and Lowrey (1982). Equation 9.2 should always be used with exact parameter values whenever available.

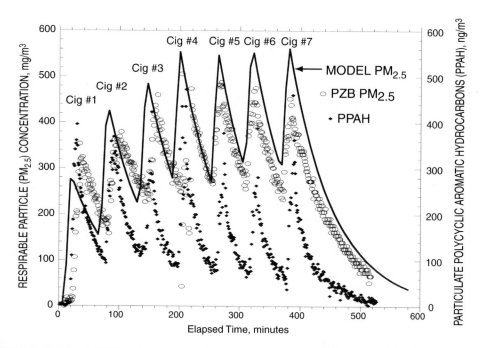

FIGURE 9.8 $PM_{2.5}$ and PPAH sidestream smoke concentrations for a series of seven Marlboro 100 medium cigarettes smoldered for 10 minutes each at the approximate rate of 1 per hour. The experiment was conducted in a closed 41 m^3 bedroom with total decay rates $\phi_{PM2.5}$ = 0.8 h^{-1}, and ϕ_{PPAH} = 1.45 h^{-1}, in a single-family home in Redwood City, CA. These measurements were made using a Model 3511 Piezobalance for RSP and an EcoChem PAS2000CE for PPAH. The model (Ott, Langan, and Switzer 1992; see also Chapter 18) assumes a 14 mg/cigarette RSP emission strength and uses the measured decay rate for RSP.

9.10 TIME-VARYING SHS CONCENTRATIONS

The equations given above are time-averaged equations that average over a typically fluctuating concentration of SHS during the smoking period, as distinct from time-dependent models (Ott, Klepeis, and Switzer 2003). This fluctuation is illustrated by the following controlled experiment. Figure 9.8 shows the concentration as a function of time (i.e., the *time series*) for RSP and particle-bound polycyclic aromatic hydrocarbons (PPAH) concentrations for a series of 7 Marlboro cigarettes smoldered (i.e., producing sidestream smoke only) at the approximate rate of 1 cigarette per hour, in a 41 m^3 bedroom of a detached home in Redwood City, CA, with the windows and bedroom door closed. These measurements were made using the Piezobalance for RSP and the EcoChem PPAH monitor described in Chapter 6 of this book. This experiment illustrates how PPAH and RSP concentrations rise and fall as cigarettes burn and are put out; it also illustrates the faster rate at which PPAH are removed from the room air relative to RSP from SHS, shows how this faster rate results in the PPAH oscillations stabilizing about the mean more quickly than RSP does, and the PPAH decaying back to background more rapidly after the last cigarette is smoked. The removal rate for each class of SHS compound may be calculated from the slope of the final decay curve when it is plotted on a semi-logarithmic graph.

What are PPAH? PPAH are part of a broader class of gas- and particulate-PAH, a group of more than 100 different molecules formed during the incomplete combustion of organic material such as tobacco, fossil fuels, or wood, and are usually found in complex mixtures such as soot or tar (see also Chapter 14). PPAH are generally formed in the gas phase with subsequent transition to the particulate form. PAH have at least two benzene rings sharing a common border; 2- and 3-ring PAH generally exist in the vapor phase in the atmosphere, whereas 5- and 6-ring PAH are

TABLE 9.3
Carcinogenic PPAH, IARC Status, Amount in Cigarette Smoke[a]

Particulate Phase PAH (PPAH) with Four or More Rings[b]	IARC Carcinogen in Lab Animals (A) Humans (H)	Amount Measured in Mainstream Smoke (MS) (ng/cig)[c]	Amount Measured in Sidestream Smoke (SS) or SHS (ng/cig)[c]	Reference
Benz(a)anthracene	Sufficient (A)	20–70	412	Hoffmann and Hoffmann (1998); Gundel et al. (1995); IARC (2004)
Benzo(b)fluoranthene	Sufficient (A)	4–22	132	Hoffmann and Hoffmann (1998); Gundel et al. (1995); IARC (2004)
Benzo(j)fluoranthene	Sufficient (A)	6–21	32	Hoffmann and Hoffmann (1998), Gundel et al. (1995); IARC (2004)
Benzo(k)fluoranthene	Sufficient (A)	6–12		Hoffmann and Hoffmann (1998), IARC Monographs (2004)
Benzo(a)pyrene	Sufficient (A,H)	20–40 8.5–11.6	74	Hoffmann and Hoffmann (1998) Gundel et al. (1995); IARC (2004)
Dibenzo(a,e)pyrene	Sufficient	present		IARC (2004)
Dibenzo(a,i)pyrene	Sufficient (A)	1.7–3.2		Hoffmann and Hoffmann (1998)
Dibenz(a,h)anthracene	Sufficient (A)	4		Hoffmann and Hoffmann (1998)
Indeno(1,2,3-cd)pyrene	Sufficient (A)	4–20	51	Hoffmann and Hoffmann (1998), Hecht (2003)
5-methylchrysene	Sufficient (A)	ND–0.6		Hoffmann and Hoffmann (1998)
All PPAH in SS machine-smoked 1R4F Univ. of KY research cigarette	–	–	1,067	Gundel et al. (1995)
All PPAH in SHS plus exhaled MS human-smoked Camel, Merit, Winston, Benson & Hedges cigarettes	–	–	13,500	Rogge, Hildemann, Mazurek et al. (1994)
All PPAH SHS plus exhaled MS human-smoked Marlboro Lite 100s (0.7 g smoked)	–	–	13,260	Repace (abstract, 2004)

[a] Measured by EcoChem PAS 2000CE monitor.
[b] Wynder and Hoffmann (1967).
[c] Nanograms per cigarette; blank cells indicate no data available; IARC = International Agency for Research on Cancer.

predominantly found in the particle phase, and 4-ring compounds exist in both phases. Some PAH are potent locally acting carcinogens in laboratory animals, inducing cancers of the upper respiratory tract when inhaled. Excess cancers in workers are caused by PAH from coke-oven emissions in iron and steel foundries, and PAH are regulated as Hazardous Air Pollutants under Section 112 of the U.S. Clean Air Act. PPAH are particle-bound PAH, which generally consist of 4 or more ring PAH compounds. Table 9.3 shows 10 4+-ring PPAH that have been quantified in MS, SS, and in SHS from human smokers and that are known animal or human carcinogens. Several of the individual PPAH compounds listed in Table 9.3 have been measured in indoor atmospheres at levels ranging from 0.3–2 ng/m^3 (IARC 2004; Hecht 2003; EcoChem 2005).

9.11 APPLICATIONS — SHS IN NATURALLY AND MECHANICALLY VENTILATED BUILDINGS

Time-activity patterns show that the population spends about 88% of its time at home and at work. What are typical SHS concentrations in homes and workplaces? Can time-averaged models help us to understand those levels?

9.11.1 RSP FROM SHS IN HOMES

In the early 1980s, the Harvard 6-City Study (Dockery and Spengler 1981), collected RSP in 55 homes in six U.S. cities; they found that in the average home, the annual mean RSP increased overall by 0.88 µg/m^3 per cigarette, and in fully air-conditioned homes (which presumably have lower air exchange rates), by 2.11 µg/m^3 per cigarette. Leaderer et al. (1990) and Leaderer and Hammond (1991), in the New York State Energy Research and Development Authority study in the winter of 1986, measured RSP and nicotine in 96 New York State homes with detectable nicotine concentrations, whose mean air exchange rates a for all homes averaged 0.54 h^{-1}, and whose mean volumes v averaged 353 m^3. A mean daily cigarette usage of 14.2 cigarettes was reported per home, and 24-hour, 7-day averages for SHS-RSP and SHS-nicotine were 29 µg/m^3 (SD = 25.9 µg/m^3) and 2.2 µg/m^3 (SD = 2.43 µg/m^3), respectively. Wallace (1996) observed that in comparison of the three large-scale RSP in-home studies (Harvard 6-City, New York State, and Riverside PTEAM) that the estimates of smoking contributions to indoor fine particle concentrations in homes with smokers was 25–47 µg/m^3. Wilson et al. (1996) has collected a large database of air exchange rates in California homes, which show median air exchange rates in the range between 0.5 h^{-1} and 1 h^{-1}, depending upon climate and season. Models incorporating measured values for home volume, air exchange rate, and cigarette source strength predict measured data reasonably well (Repace, Ott, Klepeis, and Wallace 2000).

9.11.2 PREDICTING RSP FROM SHS IN MECHANICALLY VENTILATED BUILDINGS

For mechanically ventilated premises, since 1973, ASHRAE Standard 62, Ventilation for Acceptable Indoor Air Quality, has prescribed ventilation rates based primarily upon controlling carbon dioxide levels from human metabolism. ASHRAE Standard 62-1999 specified that maintaining a ventilation rate of 7.5 L/s per person (15 cfm/P) will yield a steady-state CO_2 concentration of 700 parts per million (ppm) above the outdoor background level. These "design rates" are then based on human occupancy, with default values for occupants per unit occupiable floor area given based on the maximum expected occupancy. Per-occupant design mechanical ventilation rates for commercial buildings based on default occupancy and building type have been prescribed by ASHRAE since 1973. This approach makes it possible to estimate air exchange rates for offices, restaurants, bars, etc. based on the design ventilation rates and default occupancy in the various versions of the standard issued in 1973, 1975, 1981, 1989, 1999, and 2000–2004 when these standards were annually updated. If the smoking prevalence for the historical period obtained from the Centers for Disease Control is assumed, applying the same using a default ceiling height, the smoker density can similarly be estimated. These parameters can then be input into the HSM model to estimate the RSP, PPAH, nicotine, or carbon monoxide concentrations from SHS, or into the active smoking model (ASM) model to estimate particle loss rates ϕ based on measurements of SHS concentration, ceiling height, and smoker density. (Estimating air exchange rates a to compare with the ASHRAE standard requires a further estimate of the deposition rate k.)

9.11.3 SHS IN THE HOSPITALITY INDUSTRY

Now we turn to a topic that has generated more heat than light: control of SHS in the hospitality industry. While occupational and public health authorities have recommended that exposure to SHS be eliminated, bans on smoking in the hospitality industry have met with determined resistance, particularly from owners of bars and casinos, who have aggressively promoted ventilation alternatives to smoking bans. To appreciate the level of control required, one can ask, What level of SHS should be expected in a bar? The design ventilation standard, ASHRAE Standard 62-2001 recommends a ventilation rate of 15 L/second per occupant at a design maximum occupancy of 100 persons per thousand square feet of occupiable floor space, unchanged from ASHRAE Standard 62-1989, but 40% lower than the 25 L/second–occupant of ASHRAE Standard 62-1981. [Note that ASHRAE Standard 62-2004 only recommends ventilation rates for nonsmoking premises]. By contrast, U.S. smoking prevalence was 33% in 1981; by 2001, the median adult current smoking prevalence was 23.4% (range: 13.3–30.9%) for the 50 U.S. states and DC (MMWR 2003), a decline of 29%. As Equation 9.2 and Equation 9.4 show, unless the cigarette RSP emission rate has changed dramatically, or bars are making widespread use of unusually effective ventilation or filtration, the SHS concentration in the hospitality industry might be about the same in 2002 as it was 20 years earlier. The following study investigates this issue.

9.11.4 THE DELAWARE AIR QUALITY SURVEY

This study investigated air pollution in the hospitality industry before and after a smoking ban (Repace 2004a). This real-time study of indoor, outdoor, and in-transit air quality measured RSP and PPAH in the Wilmington, DE metropolitan area in 2002–2003. The Wilmington Study illustrates the collection and analysis of field data measured to assess the impact of secondhand smoke on human exposures in the hospitality industry, as well as the use of the mass-balance model to analyze and generalize such data (Repace 2004a). Using concealed real-time data-logging monitoring equipment (the MIE 1200AN nephelometer and the EcoChem PAS 2000CE photoelectric aerosol sampler), coupled with a time-activity pattern diary, I assessed air quality outdoors, in transit, and in eight hospitality venues (a casino, six bars, and a pool hall) on two Friday evenings: first on November 15, 2002, under conditions of unrestricted smoking (Figure 9.9, top) and then on January 24, 2003, 2 months after a statewide smoke-free workplace law (Figure 9.9, bottom). The mean indoor concentration of RSP dropped from 230 µg/m³ to 24.5 µg/m³ and indoor PPAH fell from 134 ng/m³ to 4.6 ng/m³. After subtracting the measured outdoor levels, the reduction due to the smoking ban was 92% for RSP and 85% for PPAH. This cross-sectional study yielded similar results to that observed in 52 visits to a tavern in the longitudinal study of Ott, Switzer, and Robinson (1996), who found a 90% reduction in measured RSP after a smoking ban. The correlation of RSP and PPAH, adjusted by subtracting outdoor levels, during smoking in this study was $R^2 = 0.55$ (Figure 9.10).

How do the measured RSP concentrations averaged over the six bars compare with the value predicted for a bar using the defaults in the HSM model? Assuming a default ceiling height of 10 ft, a bar at maximum occupancy would have 100 persons per 10,000 ft³ (283 m³) of space volume. In 2002, the State of Delaware had a smoking prevalence of 23% (MMWR 2003). Thus, for a bar with a Delaware average smoking prevalence, at maximum occupancy, there would be an expected 23 smokers per 283 m³, for an expected habitual smoker density $D_{hs} = 8$ habitual smokers per hundred cubic meters. The design air exchange rate is calculated from the ASHRAE Standard as: $a = (100 \text{ occupants})(15 \text{ L/s-occ})(3600 \text{ s/h})(1\text{m}^3/1000\text{L})/283 \text{ m}^3 = 19 \text{ h}^{-1}$. Because ventilation rates are not regulated, operational rates may be much lower than the design values. Applying the HSM yields: SHS $= 280 D_{hs}/(a + k) = (280)(8)/(19 + k)$. Assuming k is small compared to a, we estimate an SHS concentration of 118 µg/m³ in a bar ventilated according to ASHRAE design criteria. Assuming a non-SHS RSP background of 10 µg/m³, (about the PM$_{2.5}$ average for all U.S. counties outdoors in 2002) the predicted total RSP concentration in an ASHRAE Standard–ventilated bar at maximum occupancy and Delaware average smoking prevalence would be about 128 µg/m³. For the six bars measured in the Delaware Study, the measured mean concentrations averaged 109 µg/m³ (SD = 83 µg/m³) indoors before the smoking ban and 11 µg/m³ (SD = 9 µg/m³) post-ban.

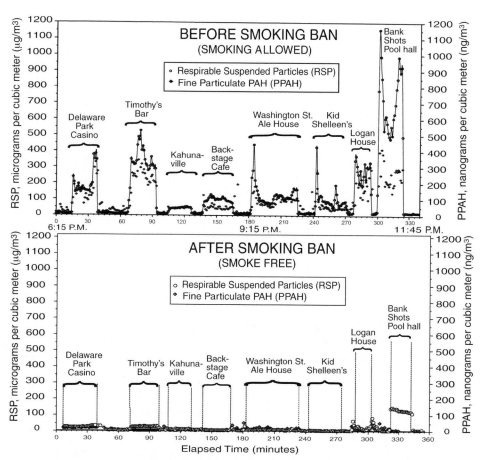

FIGURE 9.9 A real-time study of RSP and PPAH air pollution in a casino, six bars, and a pool hall before and after a smoking ban. An MIE pDR 1200 AN active-mode nephelometer for $PM_{3.5}$ and the EcoChem PAS 2000CE monitor for PPAH, assessed air quality in the 8 hospitality venues indoors, outdoors and in-transit. Top: Before the ban, on November 15, 2002, under conditions of unrestricted smoking. Bottom: This study was repeated on January 24, 2003, 2 months after a statewide smoke-free workplace law. (From Repace 2004a. With permission).

Using my rule-of-thumb default $\phi = 1.3a$ yields SHS = 98 $\mu g/m^3$, which when added to the 11 $\mu g/m^3$ background, yields 109 $\mu g/m^3$.

9.12 APPLICATIONS — RSP AND CO FROM SHS IN A VEHICLE

Ott, Langan, and Switzer (1992) measured RSP from SHS in a 1986 Mazda 626 4-door sedan at 20 miles per hour with the windows closed and in which 3 Marlboro filter cigarettes had been smoked by a smoker at a rate of ~3 cigarettes in 40 minutes, or 4.5 cigarettes per hour. The car's air exchange rate was measured at $a = 7.27$ h^{-1}, and its volume was $v = 3.7$ m^3. RSP deposition rates were not measured. These investigators found amazingly high peak RSP levels of 3000 $\mu g/m^3$ and valleys of about 1000 $\mu g/m^3$ between cigarettes. Ott, Langan, and Switzer (1992) also reported peak carbon dioxide (CO) levels at 12 ppm with valleys between cigarettes at 6 ppm compared to a measured roadway CO background of 1.5 ppm. Also applying the sequential cigarette exposure model (SCEM) to this experiment, the authors calculated an emission rate of 88 mg CO/cigarette. To place the measured concentration in perspective, the CO level in this car is higher than that observed by Otsuka et al. (2001), who showed that exposure to secondhand smoke carbon monoxide (SHS-CO) at levels of 6.02 ppm for 30 minutes induces acute endothelial dysfunction of the

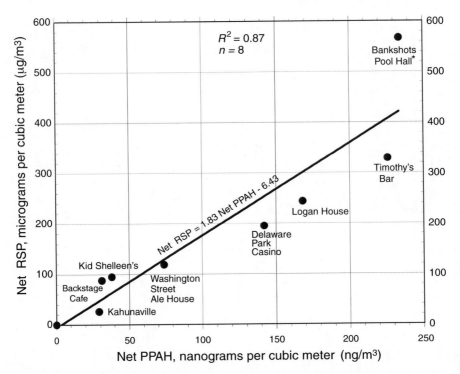

FIGURE 9.10 Net RSP and PPAH (indoor–outdoor levels) for eight Wilmington, DE locations (Figure 9.9, top) are correlated during smoking. Net PPAH is about 0.05% of net RSP. *(Data calculated from pre-ban RSP and post-ban indoor RSP [Figure 9.9, bottom] to account for an apparent non-SHS indoor source.) (From Repace 2004a, with permission.)

coronary circulation in nonsmokers. Thus, these observed SHS-CO levels are physiologically significant. Mulcahy and Repace (2002) reported 6.36 ppm median SHS-CO levels in 14 Galway, Ireland pubs. Recently, the Centers for Disease Control cautioned persons at risk of cardiovascular disease to avoid exposure to SHS (Pechacek and Babb 2004).

What do the peaks and valleys generated by CO emissions of multiple smoldered cigarettes look like? Figure 9.11 illustrates the CO emissions measured in the same Silicon Valley experiment of Figure 9.8. Using the SCEM model of Ott, Langan, and Switzer (1992), these data fit a SHS CO emission rate of 6.5 mg/min, or 65 mg CO/cigarette.

9.13 APPLICATIONS — DOSIMETRY: TRANSLATION OF EXPOSURE INTO DOSE VIA COTININE ANALYSIS

The concentration of one major SHS atmospheric marker can be estimated from the concentration of another (e.g., nicotine, RSP, and CO), and therefore are all relatable to the major body-fluid biomarker, cotinine, because it is derived from nicotine (Repace and Lowrey 1993; Repace et al. 1998; Repace, Al-Delaimy, and Bernert 2006). This allows a greater understanding of how the dose distribution in Figure 9.3 comes about. The reader is referred to the excellent article by Benowitz (1999) for a detailed discussion of nicotine and cotinine pharmacokinetics. Because cotinine has a 19-hour half-life, Figure 9.3 essentially represents a daily dose "snapshot" of the cross-sectional SHS exposure for the U.S. population for the time period 1988–1991, and illustrates a two-order of magnitude spread in SHS exposure among the population. Can the models and data given in this chapter shed any light on this distribution? As part of a risk assessment of passive smoking–induced lung cancer, Repace and Lowrey (1985) modeled the average SHS-RSP exposure of

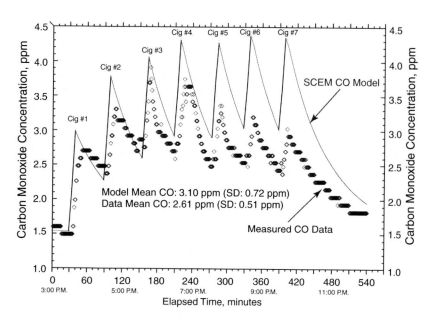

FIGURE 9.11 The simultaneous CO emissions for the seven smoldered cigarettes experiment shown in Figure 9.8. Using the SCEM model of Ott, Langan, and Switzer (1992), these data fit a SHS CO emission rate of 6.5 mg/min, or 65 mg CO/cigarette (Data from Ott and Repace 2003). The room volume is 41 m^3 and the air exchange rate is 0.82 h^{-1}.

a typical person in the U.S. population using the HSM, incorporating time-activity pattern studies, exposure probabilities for home and workplace exposure derived from surveys, coupled with respiration rates for home and work activities, using the default $\phi = 1.3a$. Using these methods, we assumed that the typical person was exposed only at home or at work, which covered 88% of a person's time. Based on the survey-grounded assumption that 86% of the U.S. population was exposed to SHS either at work or at home, or in both microenvironments, Repace and Lowrey (1985, 1993) estimated that the average American circa 1985 inhaled 1430 micrograms of SHS-RSP daily. Repace and Lowrey (1993), assuming a 10:1 ratio for RSP to nicotine in SHS, translated the modeled SHS-RSP inhaled exposure for this hypothetical person into an estimated daily nicotine exposure of 143 micrograms daily, and by developing a pharmacokinetic model translating inhaled nicotine into its biomarker, cotinine, estimated that the typical U.S. adult nonsmoker was exposed to 1 nanogram of cotinine per milliliter (ng/ml) for the most-exposed persons, 10 ng/ml. When updated with an improved 78% nicotine-to-cotinine conversion efficiency (Benowitz 1999; Repace et al. 1998), these values are reduced by a factor of 0.86, to 0.86 ng/ml and 8.6 ng/ml, respectively.

The NHANES III national cotinine dosimetry survey shown in Figure 9.3 later found an 88% SHS exposure probability. When compared to the NHANES III "reported home or work ETS exposure" distribution of Figure 9.3, the cotinine doses estimated by Repace and Lowrey (1993) yield estimates strikingly consistent with both the median and upper extreme values observed for those exposed at work or at home for the U.S. population in 1989–1991 (Pirkle et al. 1996; Repace, Al-Delaimy, and Bernert 2006).

What exposure scenarios might lead some unfortunate nonsmokers to be at the upper end of the SHS dose distribution of Figure 9.3 (dark gray)? A serum cotinine value of 8 ng/ml corresponds to saliva cotinine value of about 9.3 ng/ml (Repace et al. 1998). In the early 1990s, London pub workers had measured *median* salivary cotinine levels of 8 ng/ml (Jarvis, Foulds, and Feyerabend 1992). By 2000 a sample ($n = 44$) of London bar-worker salivary cotinines manifested mean values of 6.16 ng/ml, seven times higher than the mean values for a sample of all English households of 0.86 ng/ml (Jarvis 2001). This suggests that bar staff may populate the upper dose levels of the cotinine distribution. What daily average SHS air pollution concentration does a serum cotinine value at the upper extreme represent? The Repace et al. (1998) model equates a saliva cotinine concentration of 9.36 ng/ml to an 8-hour time-weighted average (TWA) SHS nicotine exposure concentration of 109 $\mu g/m^3$, which in turn equates to an estimated 8-hour TWA SHS-RSP concentration of 1090 $\mu g/m^3$ (Repace and Lowrey 1993). Levels this high are consistent with the most extreme SHS values encountered in workplace exposures in the hospitality industry shown in Figure 9.4 and Figure 9.9.

Application of such models permits forensic risk assessment. For example, Repace (2004b) analyzed a cotinine dosimetry study of flight attendants conducted by the National Cancer Institute (Mattson et al. 1989) on several Air Canada flights in 1989, and estimated that the median serum cotinine dose of typical flight attendants in aircraft cabins, at 2.88 ng/ml, was about 6-fold that of the average U.S. worker and about 14-fold that of the average person at that time. As there are in excess of 3,000 flight attendant SHS personal-injury lawsuits currently in litigation, such estimates are indispensable for assisting the court in placing such exposures in perspective. If a flight attendant develops a disease associated with SHS, exposure or dose models permit incomparably better evaluation of risk probabilities than is possible simply by saying Ms. X worked as a flight attendant and was exposed to SHS in aircraft cabins for 20 hours per week. Application of such models also permits improvements in exposure assessment in passive smoking epidemiology, which has been based almost wholly on spousal smoking status, which ignores concentration, duration, and respiration rates, and relates only tangentially to activity patterns, and thus is a weak surrogate for exposure (Repace, Al-Delaimy, and Bernert 2006).

9.14 FUTURE ISSUES

Secondhand smoke pollution will continue to be a major public health problem for many people throughout much of the world for many decades. To place this into perspective, in 1986, 50 million U.S. cigarette smokers smoked 584 billion cigarettes, burning an estimated 424,330 metric tons of tobacco in indoor microenvironments, with another estimated 29,700 metric tons burned by pipe and cigar smokers (Repace and Lowrey 1990). Fourteen years later, in 2000, the number of cigarette smokers had declined only by 7% to 46.5 million. Table 9.4 gives the number of U.S. smokers, cigarettes consumed by them, and estimated smoking rates for various years from 1965–2000. The number of smokers has been slowly declining since its peak in 1983. Both the number of cigarettes consumed and the cigarette smoking rate peaked in 1979 and have also slowly declined since. The decline in average cigarette smoking rate may actually reflect fewer opportunities to smoke because of the increased prevalence of workplace and home smoking bans rather than a decline in personal smoking rate when an individual is at liberty to smoke freely; in turn, the reduced smoking prevalence may facilitate the growth of home and workplace smoking bans. However 18 state legislatures currently preclude total workplace smoking bans (American Nonsmokers Rights 2005), and this decline may occur largely outside those states.

Half of all U.S. children still grow up in homes with smokers. In 2003, 94% of locations measured in seven countries in Latin America were observed to have detectable levels of SHS nicotine (Navas-Acien et al. 2004). The California Environmental Protection Agency (CalEPA 2005) estimates that indoor air pollution from SHS in California — even after banning all workplace smoking — still causes more than $25 billion in associated healthcare costs. More research is needed in several areas. Smoking in multifamily dwellings continues to be a significant problem due to infiltration of tobacco smoke from one unit to another; little research has been done on the concentrations and mitigation, but this is a major complaint expressed to local health departments and has been the subject of litigation.

Outgassing of deposited SHS tars is another major concern of nonsmokers moving into homes or apartments formerly occupied by heavy smokers, as is re-emission from surfaces in elementary school multipurpose rooms time-shared with bingo games. Johansson, Olander, and Johansson

TABLE 9.4
Estimated Number of U.S. Smokers, Cigarettes Smoked and Smoking Rates, 1965–2000, for 8 Available Years

Year	1965	1970	1974	1979	1983	1988	1995	2000
Millions of U.S. smokers[a]	50.1	48.1	48.9	51.1	53.5	49.4	47.0	46.5
Billions of cigarettes consumed[b]	521.1	534.2	594.5	621.8	603.6	560.7	482.2	430
Cigarettes smoked per smoker per day[c]	28.5	30.4	33.3	33.3	30.9	31.0	28.1	25.3

[a] CDC TIPS: Number (in millions) of adults 18 years and older who are current, former, or never smokers, overall and by sex, race, Hispanic origin, age, and education, National Health Interview Surveys, selected years — United States, 1965–2000.

[b] Federal Trade Commission sales by manufacturers to wholesalers and retailers within the U.S. and Armed forces personnel stationed outside the U.S. National Center for Chronic Disease Prevention and Health Promotion, Centers for Disease Control and Prevention.

[c] Leap-year adjusted.

(1993) reported that the surfaces in a room where cigarette smoking occurs become secondary particulate-phase pollution sources. Singer et al. (2004) found that lower volatility gas-phase SHS compounds (nicotine, 3-ethenylpyridine, phenols, cresols, naphthalene, and methylnapthalenes) outgassed from surfaces where they had sorbed, suggesting widespread contamination of buildings and air-handling systems with deposited SHS. Such outgassing contributes to indirect exposure to SHS contaminants even when smokers are not present or not smoking in microenvironments with daily smoking, particularly where infants are concerned. Further research is needed into the range of deposition and re-emission rates, and persistence of SHS tars on room surfaces. Smoking in hotels exposes guests in lobbies and maids cleaning smokers' rooms to unknown levels of outgassed SHS, which as Table 9.1 shows, contains a wide variety of regulated carcinogens. Smoking in outdoor cafes also continues to expose waitstaff to SHS; preliminary research into this problem shows that outdoor levels of SHS-RSP and SHS-PPAH in the proximity of smokers may be as high as indoor microenvironments (Klepeis, Ott, and Switzer 2004; Repace 2004 abstract). SHS in prisons and psychiatric institutions has been little studied. Additional studies of cotinine dose in specific worker groups, such as bar and casino workers, and its relationship to SHS exposure and ventilation and air cleaning would be useful to evaluate claims of efficacy for these putative SHS control measures relative to smoke-free workplaces. Finally, there will be an expected 1 billion tobacco deaths worldwide in the 21st century if current global smoking patterns continue — a tenfold increase of the 20th century toll of 100 million (CRC 2002; WHO 2005). Based on the ratio of estimated U.S. passive smoking to active smoking deaths (Surgeon General 2004; CalEPA 2005), one out of eight of those deaths will be nonsmokers exposed to SHS.

9.15 ACKNOWLEDGMENTS

This work is supported by the Flight Attendant Medical Research Institute's William Cahan Distinguished Professor Award. The author is grateful to W.R. Ott and L.A. Wallace for helpful discussions and suggestions.

9.16 QUESTIONS FOR REVIEW

1. A bitter epidemiological dispute developed when Enstrom and Kabat (2003), funded by the tobacco industry, published a re-analysis and 40-year follow-up of the American Cancer Society's (ACS) Cancer Prevention Study I cohort that concluded that "the results

do not support a causal relation between environmental tobacco smoke (ETS) and tobacco-related mortality." Thun (2003) of ACS responded that Enstrom and Kabat's (2003) analysis was "misleading science" and was fundamentally flawed in major part because of exposure misclassification of CPS I subjects. Enstrom and Kabat replied that in this cohort, in 1959, "the majority of female never smokers married to never smokers were not exposed to ETS." The U.S. adult smoking prevalence in 1965 was 42.4%, 37% in 1974, and by 1990 had declined to 25.5%. If Figure 9.3 represents the exposure distribution of the nonsmoking U.S. population in 1990, who is more likely to be correct? Why?

2. Air quality control authorities recommend that citizens remain indoors during outdoor air pollution alerts. Is this likely to be good advice for nonsmokers who live with smokers?

3. It is clear that smoke-free workplace laws can reduce indoor air pollution in casinos, bars, and restaurants virtually to outdoor levels. Opponents of such laws argue that ventilation can produce acceptable indoor air quality. In Figure 9.9, the SHS-RSP level in Timothy's Bar averages 337 $\mu g/m^3$ above the 10 $\mu g/m^3$ outdoor background concentration at an active smoker density of 1.44 burning cigarettes per 100 m^3. Holding the smoker density and outdoor background concentration constant, how many air changes per hour would it take to reduce the total indoor RSP to 15 $\mu g/m^3$, the level of the Annual National Ambient Air Quality Standard which defines clean air for RSP? Is this value possible to attain at the 18 h^{-1} air exchange rate recommended by the ASHRAE Standard 62-2001? The bar's existing particle removal rate was calculated using the ASM model to be $\phi = 0.87$ h^{-1}. The actual indoor RSP level in the bar after the ban was 24 $\mu g/m^3$. How much ventilation would it take to attain the NAAQS in this case? [Answer: In the case of a 10 $\mu g/m^3$ background particle concentration, one would need to attain a level of $(15-10) = 5$ $\mu g/m^3$ of SHS-RSP, which results in the equation $(337-10)(0.87/\phi) = 5$ $\mu g/m^3$, and solving for ϕ gives a particle decay rate of $\phi = a + k = 75.2$ h^{-1}. Assuming the particle deposition rate k is small compared with the air change rate a, the ventilation rate would need to be about 4 times the level recommended by ASHRAE. In the case of a 24 $\mu g/m^3$ background level, attainment of the NAAQS would be impossible.]

4. One of the most common complaints for local health departments is infiltration of SHS from a neighboring apartment. If a nonsmoker resides in a mid-level apartment in a three or more story building, how many apartments might surround the nonsmoker? What is the probability of having at least one smoker in a neighboring apartment if the smoking prevalence in the building is 25%? About one half to two thirds of the air in a multifamily dwelling appears to infiltrate from neighboring apartments in older buildings. What methods can you think of to minimize or eliminate this infiltration? [Answer 1: The nonsmoker's apartment will typically be surrounded by four neighboring apartments. If two people reside in each neighboring apartment, of the eight neighbors, chances are that at least two smoke. Therefore, one to two neighboring apartments will have a smoker. Answer 2: To minimize infiltration, plugging cracks behind electrical outlets and around plumbing pipes, plus positively pressurizing the nonsmoker's apartment or depressurizing the wall between apartments are possible engineering controls. Smoking and nonsmoking buildings may be possible future policies.]

5. Numerous flight attendants have filed lawsuits against the tobacco industry and against individual airlines over their exposure to SHS. An important question arising in such litigation is what are SHS exposures like on a typical aircraft? Assume that a Boeing 747 flying from the U.S. to Europe has a cabin volume $v = 790$ m^3 and an air exchange rate $a = 14.7$ h^{-1}. Assume a complement of 288 passengers, with an average smoking prevalence of 13.7%, and a smoking rate of 1.5 cigarettes per hour. Using the mass-balance model, estimate the well-mixed SHS-RSP level. Compare your result with the average of four measurements taken on eight international flights sampling four locations

on the aircraft: $(133 + 36 + 21 + 11)/4 = 50$ µg/m³, by Nagda et al. 1989). Assume that flight attendants moving about the aircraft will approximate this average. [Answer: Using the HSM of Equation 9.3, the habitual smoker density is $D_{hs} = 100(\{1.5$ cigarettes/h$\}/\{2$ cigarettes/h$\})(0.137$ smoking prevalence)(288 passengers) / (790 m³) = 3.74 HS/100 m³. Assume k small compared to a. Thus, Equation 9.3 becomes: ETS-RSP = 280(3.74/14.7) = 71 µg/m³. With the default of $\phi = 1.3a$, ETS-RSP = 55 µg/m³.]

6. Table 9.1 shows a list of 172 toxic compounds in SHS for which SHS atmospheric markers are surrogates. Fowles and Dybing (2004) used toxicological methods, which in principle provide a plausible and objective framework for estimating cancer risk, applied to a list of toxic substances in tobacco smoke such as those in Table 9.1. They found that this method underestimates observed cancer risk in smokers by a factor of 5. Can you think of a reason why? [Answer: This may be due to synergy between chemicals. Other reasons include omission of some chemicals; unknown toxicity of many chemicals; inadequacy of risk assessment methods; difficulty in extrapolating from animals to humans and from low dose to high dose, etc.]

7. Average house volumes in the U.S. were about 350 m³ in the late 1990s. For an air exchange rate of 0.6 h⁻¹, a $PM_{2.5}$ deposition rate of 0.4 h⁻¹, and 1 smoker spending 8 waking hours in the home smoking 2 cigarettes per hour with a $PM_{2.5}$ emission rate of 14 mg/cigarette, calculate the 24-hour average $PM_{2.5}$ concentration. How does your answer compare with the observed range in three large studies of 25 to 45 µg/m³? [Answer: 26.7 µg/m³.]

REFERENCES

American Nonsmokers Rights Foundation (2005) States and Municipalities with 100% Smokefree Laws in Workplaces, Restaurants, or Bars Currently in Effect as of October 4, 2005, American Nonsmokers Rights Foundation, Berkeley, CA, http://www.no-smoke.org/goingsmokefree.php?dp=d13%7Cp140.

ASHRAE (1989) *Ventilation for Acceptable Indoor Air Quality, ASHRAE Standard 62-1989,* American Society of Heating, Refrigerating, and Air Conditioning Engineers, Atlanta, GA. Accessed March 10, 2005.

ASHRAE (2004) *Ventilation for Acceptable Indoor Air Quality, ASHRAE Standard 62-2004,* American Society of Heating, Refrigerating, and Air Conditioning Engineers, Atlanta, GA (accessed June 8, 2005).

Benowitz, N.L. (1999) Biomarkers of Environmental Tobacco Smoke Exposure, *Environmental Health Perspectives,* **107**(2): 349–355.

Brandt, A.M. and Richmond, J.B. (2004, January 15) Tobacco Pandemic, *Washington Post,* p. A21.

CalEPA (1997) Health Effects of Exposure to Environmental Tobacco Smoke, Final Report, Office of Environmental Health Hazard Assessment, California Environmental Protection Agency.

CalEPA (2005) Proposed Identification of Environmental Tobacco Smoke as a Toxic Air Contaminant, as Approved by the Scientific Review Panel on June 24, 2005, California Air Resources Board, California Environmental Protection Agency, Sacramento, CA.

Centers for Disease Control (CDC) (1997) Smoking-Attributable Mortality and Years of Potential Life Lost, *Morbidity and Mortality Weekly Report,* **46**(20): 449.

Centers for Disease Control (CDC) (2003, January) *Second National Report on Human Exposure to Environmental Chemicals,* Department of Health and Human Services, Centers for Disease Control and Prevention, National Center for Environmental Health, Division of Laboratory Sciences, NCEH Pub. No. 02-0716, Atlanta, GA.

CRC Cancer Stats (2002) U.K. Smoking and Cancer September 2002, Cancer Research U.K.

Daisey, J.M. (1999) Tracers for Assessing Exposure to Environmental Tobacco Smoke: What Are They Tracing? *Environmental Health Perspectives,* **107**(2): 319–327.

Dockery, D.W. and Spengler, J.D. (1981) Indoor-Outdoor Relationships of Respirable Sulfates and Particles, *Atmospheric Environment,* **15**: 335–343.

EcoChem (2005) *EcoChem Analytics Users' Guide,* Edition 1.2, http://www.ecochem.biz (accessed March 10, 2005).

Enstrom, J.E. and Kabat, G.C (2003) Letter, *British Medical Journal, 3*: 588–589.

Fowles, J. and Dybing, E. (2004) Application of Toxicological Risk Assessment Principles to the Chemical Constituents of Tobacco Smoke, *Tobacco Control, 12*: 424–430.

Gundel, L.A., Mahanama, K.R.R., and Daisey, J.M. (1995) Semivolatile and Particulate Aromatic Hydrocarbons in Environmental Tobacco Smoke: Cleanup, Speciation, and Emission Factors, *Environmental Science and Technology, 29*: 1607–1614.

Hammond, S.K. (1999) Exposure of U.S. Workers to Environmental Tobacco Smoke, *Environmental Health Perspectives, 107*(2): 329–340.

Hecht, S.S. (2003) Carcinogen Derived Biomarkers: Applications in Studies of Human Exposure to Secondhand Tobacco Smoke, *Tobacco Control, 13*(suppl. I): i48–i56.

Hoffmann, D. and Hoffmann, I. (1998) Chemistry and Toxicology, in *Cigars: Health Effects and Trends, Smoking and Tobacco Control Monograph 9,* National Institutes of Health, National Cancer Institute, Bethesda, MD.

Holgate, S.T., Samet, J.M., Koren, H.S., and Maynard, R.L. (1999) *Air Pollution and Health,* Academic Press, London.

IARC (1987) *Environmental Carcinogens — Selected Methods of Analysis — Volume 9, Passive Smoking,* IARC Scientific Publications No. 81, O'Neill, I.K., Brunnemann, K.D., Dodet, B., and Hoffmann, D., Eds., International Agency for Research on Cancer, World Health Organization, United Nations Environment Programme, Lyon, France.

IARC (2004) IARC Monographs on the Evaluation of Carcinogenic Risks to Humans, *Tobacco Smoke and Involuntary Smoking, Volume 83,* International Agency for Research on Cancer, Lyon, France.

Jarvis, M. J. (2001) *Quantitative Survey of Exposure to Other People's Smoke in London Bar Staff,* Department of Epidemiology and Public Health, University College, London.

Jarvis, M.J., Foulds, J., and Feyerabend, C. (1992) Exposure to Passive Smoking among Barstaff, *British Journal of Addiction, 87*: 111–113.

Johansson, J., Olander, L., and Johansson, R. (1993) Long-Term Test of the Effect of Room Air Cleaners on Tobacco Smoke, in *Proceedings of the 6th International Conference on Indoor Air Quality and Climate, Vol. 6, Thermal Environment, Building Technology, Cleaning,* Helsinki, Finland, July 4–8.

Junker, M.H., Danuser, B., Monn, C., and Koller, T. (2001) Acute Sensory Response of Nonsmokers at Very Low Environmental Tobacco Smoke Concentrations in Controlled Laboratory Settings, *Environmental Health Perspectives, 109*: 1045–1052.

Klepeis, N.E. (1999) Validity of the Uniform Mixing Assumption: Determining Human Exposure to Environmental Tobacco Smoke, *Environmental Health Perspectives, 107*(2): 357–363.

Klepeis, N.E., Apte, M.G., Gundel, L.A., Sextro, R.G., and Nazaroff, W.W. (2003) Determining Size-Specific Emission Factors for Environmental Tobacco Smoke Particles, *Aerosol Science and Technology, 37*: 780–790.

Klepeis, N.E., Ott, W.R., and Repace, J.L. (1999) The Effect of Cigar Smoking on Indoor Levels of Carbon Monoxide and Particles, *Journal of Exposure Analysis and Environmental Epidemiology, 9*: 1–14.

Klepeis, N.E., Ott, W.R., and Switzer, P. (1996) A Multiple-Smoker Model for Predicting Indoor Air Quality in Public Lounges, *Environmental Science and Technology, 30*: 2813–2820.

Klepeis, N.E., Ott, W.R., and Switzer, P. (2004) Real-Time Monitoring of Outdoor Environmental Tobacco Smoke Concentrations — A Pilot Study, University of California, San Francisco, Contract No. 3317SC, for the Tobacco Control Section, Department of Health Services, State of California.

Leaderer, B.P., Cain, W.S., Isseroff, R., and Berglund, L.G. (1984) Ventilation Requirements in Buildings — II. Particulate Matter and Carbon Monoxide from Cigarette Smoking, *Atmospheric Environment, 18*(1): 99–106.

Leaderer, B.P. and Hammond, S.K. (1991) An Evaluation of Vapor-Phase Nicotine and Respirable Suspended Particle Mass as Markers for Environmental Tobacco Smoke, *Environmental Science & Technology, 25*: 770–777.

Leaderer, B., Koutrakis, P., Briggs, S., and Rizzuto, J. (1990) Impact of Indoor Sources on Residential Aerosol Concentrations, in *Proceedings of the 5th International Conference on Indoor Air Quality & Climate,* Toronto, Canada, July 29–August 3.

Lee, P.N. and Forey, B.A. (1996) Misclassification of Smoking Habits as a Source of Bias in the Study of Environmental Tobacco Smoke and Lung Cancer, *Statistics in Medicine, 15*: 581–605.

Martin, P., Heavner, D.L., Nelson, P.R., Maiolo, K.C., Risner, C.H., Simmons, P.S., Morgan, W.T., and Ogden, M.W. (1997) Environmental Tobacco Smoke (ETS): A Market Cigarette Study, *Environment International,* **23**: 75–90.

Mattson, M.E., Boyd, G., Byar, D., Brown, C., Callahan, J.F., Corle, D., Cullen, J.W., Greenblatt, J., Haley, N.J., and Hammond, K. (1989) Passive Smoking on Commercial Air Flights, *Journal of the American Medical Association,* **261**: 867–872.

Maxwell, J.C. (1995) The Maxwell Report: Sales Estimates for the Cigarette Industry, Richmond, VA, John C. Maxwell, Jr., February.

McBride, S.J., Ferro, A.R., Ott, W.R., Switzer, P., and Hildemann, L.M. (1999) Investigations of the Proximity Effect for Pollutants in the Indoor Environment, *Journal of Exposure Analysis & Environmental Epidemiology,* **9**: 602–621.

MMWR (1992) Cigarette Smoking among Adults — United States, 1990 *Morbidity & Mortality Weekly Report,* **41**(20): 354–355, 361–362.

MMWR (1996) Cigarette Smoking among Adults — United States, 1994, *Morbidity & Mortality Weekly Report,* **45**(27): 588–590.

MMWR (2003) State-Specific Prevalence of Current Cigarette Smoking among Adults —United States, 2003, *Morbidity & Mortality Weekly Report,* **53**(44): 1035–1037.

Mulcahy, M. and Repace, J.L. (2002) Passive Smoking Exposure and Risk for Irish Bar Staff, in *Proceedings the 9th International Conference on Indoor Air Quality and Climate,* Monterey, CA.

Nagda, N.L., Fortmann, R.C., Koontz, M.D., Baker, S.R., and Ginevan, M.E. (1989) Airliner Cabin Environment: Contaminant Measurements, Health Risks and Mitigation Options, Report No. DOT P-15-89-5, U.S. Department of Transportation, Washington, DC.

National Cancer Institute (1993) *Respiratory Health Effects of Passive Smoking: Lung Cancer and Other Disorders,* The Report of the U.S. Environmental Protection Agency, National Cancer Institute Smoking and Tobacco Control Monograph 4, NIH Publication No. 93-3605, National Institutes of Health, Bethesda, MD.

National Cancer Institute (1998) *Cigars — Health Effects and Trends,* Smoking and Tobacco Control Monograph 9, National Institutes of Health, National Cancer Institute, Bethesda, MD.

National Cancer Institute (1999) *Health Effects of Exposure to Environmental Tobacco Smoke,* The Report of the California Environmental Protection Agency, National Cancer Institute Smoking and Tobacco Control Monograph 10, NIH Publication No. 99-4645, National Institutes of Health, Bethesda, MD.

National Institute for Occupational Safety and Health (1991) *Environmental Tobacco Smoke in the Workplace, Lung Cancer and Other Health Effects,* Current Intelligence Bulletin No. 54, U.S. Department of Health and Human Services, National Institute for Occupational Safety and Health, Cincinnati, OH.

National Research Council (1986) *Environmental Tobacco Smoke — Measuring Exposures and Assessing Health Effects,* National Academy Press, Washington, DC.

National Toxicology Program (2000) *9th Report on Carcinogens,* U.S. Department of Health and Human Services, National Institute of Environmental Health Sciences, Research Triangle Park, NC.

Navas-Acien, A., Peruga, A., Breysse, P., Zavaleta, A., Blanco-Marquizo, A., Pitarque, R., Acuña, M., Jiménez-Reyes, K., Colombo, V.L., Gamarra, G., Stillman, F.A., and Samet, J. (2004) Secondhand Tobacco Smoke in Public Places in Latin America, 2002–2003, *Journal of the American Medical Association,* **291**(22): 2741–2745.

Nebot, M., López, M.J., Gorini, G., Neuberger, M., Axelsson, S.M., Pilali, M., Fonseca, C., Abdennbi, K., Hackshaw, A., Moshammer, H., Laurent, A.M., Salles, J., Georgouli, M., Fondelli, M.C., Serrahima, E., Centrich, F., and Hammond, S.K. (2005) Environmental Tobacco Smoke Exposure in Public Places of European Cities, *Tobacco Control,* **14**: 60–63.

Nelson, P. (1994) Testimony of R.J. Reynolds Tobacco Company, OSHA Docket No. H-122, Comment 8-266, Indoor Air Quality, Proposed Rule, U.S. Occupational Safety & Health Administration, Washington, DC.

OSHA (1994) U.S. Department of Labor, Occupational Safety and Health Administration, 29 CFR Parts 1910, 1915, 1926, and 1928, *Indoor Air Quality, Proposed Rule,* Federal Register 59, No. 65, Tuesday, April 5, 1994, 15968–16039.

Otsuka, R., Watanabe, H., Hirata, K., Tokai, K., Yoshiyama, M., Takeuchi, K., and Yoshikawa, J. (2001) Acute Effects of Passive Smoking on the Coronary Circulation in Healthy Young Adults, *Journal of the American Medical Association,* **286**: 436–441.

Ott, W.R. (1999) Mathematical Models for Predicting Air Quality from Smoking Activity, *Environmental Health Perspectives,* **107**(2): 375–381.

Ott, W.R. and Repace, J.L. (2003) *Modeling and Measuring Indoor Air Pollution from Multiple Cigarettes Smoked in Residential Settings,* Poster at the International Society for Exposure Analysis, Stresa, Italy, September 21–25.

Ott, W.R., Klepeis, N.E., and Switzer P. (2003) Analytical Solutions to Compartmental Air Quality Models with Application to Environmental Tobacco Smoke Concentrations Measured in a House, *Journal of the Air & Waste Management Association,* **53**: 918–936.

Ott, W.R., Langan, L., and Switzer, P. (1992) A Time Series Model for Cigarette Smoking Activity Patterns: Model Validation for Carbon Monoxide and Respirable Particles in a Chamber and an Automobile, *Journal of Exposure Analysis and Environmental Epidemiology,* **2**(2): 175–200.

Ott, W., Switzer, P., and Robinson, J. (1996) Particle Concentrations Inside a Tavern Before and After Prohibition of Smoking: Evaluating the Performance of an Indoor Air Quality Model, *Journal of the Air & Waste Management Association,* **46**: 1120–1134.

Özkaynak, H., Xue, J., Spengler, J., Wallace, L.A., Pellizzari, E.D., and Jenkins, P. (1996) Personal Exposure to Airborne Particles and Metals: Results from the Particle TEAM Study in Riverside California, *Journal of Exposure Analysis and Environmental Epidemiology,* **6**: 57–78.

Pechacek, T.F. and Babb, S. (2004) Commentary: How Acute and Reversible Are the Cardiovascular Risks of Secondhand Smoke? *British Medical Journal,* **328**: 980–983.

Pirkle, J.L., Flegal, K.M., Bernert, J.T., Brody, D.J., Etzel, R.A., and Maurer, K.R. (1996) Exposure of the Population to Environmental Tobacco Smoke: The Third National Health and Nutrition Examination Survey, 1988 to 1991, *Journal of the American Medical Association,* **275**: 1233–1240.

Pope, C.A. and Dockery, D.W. (1999) Epidemiology of Particle Effects, in *Air Pollution and Health,* Holgate, S.T., Samet, J.M., Koren, H.S., and Maynard, R.L, Eds., Academic Press, London.

Repace, J.L. (1981) The Problem of Passive Smoking, *Bulletin of the New York Academy of Medicine,* **57**: 936–946.

Repace, J.L. (1985) Risks of Passive Smoking, in *To Breathe Freely, Risk, Consent, and Air, Maryland Studies in Public Philosophy,* Gibson, M., Ed., Center for Philosophy and Public Policy, University of Maryland, Rowman and Allenheld, Totowa, N.J.

Repace, J.L. (1987) Indoor Concentrations of Environmental Tobacco Smoke: Field Surveys, in *Environmental Carcinogens — Selected Methods of Analysis — Volume 9, Passive Smoking,* IARC Scientific Publications, No. 81, O'Neill, I.K., Brunnemann, K.D., Dodet, B., and Hoffmann, D., Eds., International Agency for Research on Cancer, World Health Organization, United Nations Environment Programme, Lyon, France.

Repace, J.L. (2002) Letter, Effects of Passive Smoking on Coronary Circulation, *Journal of the American Medical Association,* **287**(3).

Repace, J.L. (2004a) Respirable Particles and Carcinogens in the Air of Delaware Hospitality Venues before and after a Smoking Ban, *Journal of Occupational and Environmental Medicine,* **46**: 887–905.

Repace, J.L. (2004b) Flying the Smoky Skies: Secondhand Smoke Exposure of Flight Attendants, *Tobacco Control,* **13**(suppl. I): i8–i19.

Repace, J.L. (2004) Indoor/Outdoor PAH Carcinogen Pollution on a Cruise Ship in the Presence and Absence of Tobacco Smoking, presented at the 14th Annual Conference of the International Society of Exposure Analysis, Philadelphia, PA, October 17–21.

Repace, J.L. and Lowrey, A.H. (1980) Indoor Air Pollution, Tobacco Smoke, and Public Health, *Science,* **208**: 464–474.

Repace, J.L. and Lowrey, A.H. (1982) Tobacco Smoke, Ventilation, and Indoor Air Quality, *ASHRAE Transactions,* **88**(I): 895-914.

Repace, J.L. and Lowrey, A.H. (1985) A Quantitative Estimate of Nonsmokers' Lung Cancer Risk from Passive Smoking, *Environment International,* **11**: 3–22.

Repace, J.L. and Lowrey, A.H. (1990) Risk Assessment Methodologies in Passive Smoking-Induced Lung Cancer, *Risk Analysis,* **10**: 27–37.

Repace, J.L. and Lowrey, A.H. (1993) An Enforceable Indoor Air Quality Standard for Environmental Tobacco Smoke in the Workplace, *Risk Analysis,* **13**: 463–475.

Repace, J.L. and Lowrey, A.H. (1995) A Rebuttal to Tobacco Industry Criticism of an Enforceable Indoor Air Quality Standard for Environmental Tobacco Smoke, *Risk Analysis,* **15**: 7–13.

Repace, J.L., Al-Delaimy, W.K., and Bernert, J.T. (2006) Correlating Atmospheric and Biological Markers in Studies of Secondhand Tobacco Smoke Exposure and Dose in Children and Adults, *Journal of Occupational and Environmental Medicine,* **48**: 181–194.

Repace, J., Hughes, E., and Benowitz, N. (in press) Exposure to Secondhand Smoke Air Pollution Assessed from Bar Patrons' Urine Cotinine, *Nicotine and Tobacco Research.*

Repace, J.L, Jinot, J., Bayard, S., Emmons, K., and Hammond, S.K. (1998) Air Nicotine and Saliva Cotinine as Indicators of Passive Smoking Exposure and Risk, *Risk Analysis,* **18**: 71–83.

Repace, J.L., Ott, W.R., and Klepeis, N.E. (1998) Indoor Air Pollution from Cigar Smoke, in *Cigars — Health Effects and Trends: Smoking and Tobacco Control Monograph 9,* National Institutes of Health, National Cancer Institute, Bethesda, MD.

Repace, J.L., Ott, W.R., Klepeis, N.E., and Wallace, L.A. (2000) Predicting Environmental Tobacco Smoke Concentrations in California Homes, Paper 5E-04p, Session: Environmental Tobacco Smoke: Determining Concentrations & Assessing Exposures, in *10th Annual Conference of the International Society of Exposure Analysis,* Monterey, CA, October 24–27.

Repace, J.L., Ott, W.R., and Wallace, L.A. (1980) Total Human Exposure to Air Pollution, Paper 80-61.6, presented at the 73rd Annual Meeting of the Air Pollution Control Association, Montreal, Quebec, June 22–27.

Rogge, W.F., Hildemann, L.F., Mazurek, M.A., and Cass, G.R. (1994) Sources of Fine Organic Aerosol, 6. Cigarette Smoke in the Urban Atmosphere, *Environmental Science & Technology,* **26**: 1375–1388.

Schroeder, S.A. (2004) Tobacco Control in the Wake of the 1998 Master Settlement Agreement, *New England Journal of Medicine,* **350**(3): 293–301.

Shopland, D.R., Anderson, C.M., Burns, D.M., and Gerlach, K.K. (2004) Disparities in Smoke-Free Workplace Policies among Food Service Workers, *Journal of Occupational and Environmental Medicine,* **46**: 347–356.

Singer, B.C., Revzan, K.L., Toshifumi, H., Hodgson, A.T., and Brown, N.J. (2004) Sorption of Organic Gases in a Furnished Room, *Atmospheric Environment,* **38**: 2483–2494.

Surgeon General (1964) *Smoking and Health: Report of the Advisory Committee to the Surgeon General of the Public Health Service,* U.S. Department of Health and Education, and Welfare, Public Health Service, Washington, DC.

Surgeon General (1986) *The Health Consequences of Involuntary Smoking, A Report of the Surgeon General,* U.S. Department of Health and Human Services, Public Health Service, Washington, DC.

Surgeon General (2004) *The Health Consequences of Smoking: A Report of the Surgeon General,* U.S. Department of Health and Human Services, Public Health Service, Washington, DC.

Thun, M.J. (2003) Letters, *British Medical Journal,* **3**: 588–589.

USDA (1995) *Domestic Cigarette Sales, 1994,* U.S. Department of Agriculture, December 1995.

USEPA (1992) *Health Effects of Passive Smoking: Assessment of Lung Cancer in Adults, and Respiratory Disorders in Children,* Report No. EPA/600/6-90/006F, U.S. Environmental Protection Agency, Office of Research and Development, Office of Health and Environmental Assessment, Washington, D.C.

Wallace, L.A. (1996) Indoor Particles: A Review, *Journal of the Air & Waste Management Association,* **46**: 98–126.

Wallace, L.A., Pellizzari, E., Hartwell, T., Perritt, K., and Ziegenfus, R. (1987) Exposures to Benzene and Other Volatile Organic Compounds from Active and Passive Smoking, *Archives of Environmental Health,* **42**: 272–279.

Wells, A.J. (1999) Total Mortality from Passive Smoking, *Environment International,* **25**: 515–519.

Wells, A.J., English, P.B., Posner, S.F., Wagenknecht, L.E., and Perez-Stable, E.J. (1998) Misclassification Rates for Current Smokers Misclassified as Smokers, *American Journal of Public Health,* **88**: 1503–1509.

WHO (2005) Press Release, Global Cancer Rates Could Increase by 50% to 15 Million by 2020, http://www.who.int/mediacentre/news/releases/2003/pr27/en/ (accessed March 10, 2005).

Wilson, A.L., Colome, S.D., Tian, Y., and Becker, E.W. (1996) California Residential Air Exchange Rates and Residence Volumes, *Journal of Exposure Analysis and Environmental Epidemiology,* **6**: 311–326.

10 Intake Fraction

Julian D. Marshall
University of British Columbia

William W. Nazaroff
University of California

CONTENTS

10.1 Synopsis 237
10.2 Introduction 238
10.3 Background 238
10.4 What Is Intake Fraction?...239
10.5 Typical Intake Fraction Values 240
10.6 Estimating Intake Fraction Values Using a One-Compartment Model 241
10.7 The Use of Intake Fraction in Prioritizing Emission Reduction Efforts.........244
     10.7.1 Particulate Matter from On-Road Sources.......................................245
     10.7.2 Self-Pollution...246
10.8 Using Intake Fraction When Considering Environmental Justice Concerns.........246
10.9 Acknowledgments...247
10.10 Questions for Review..247
References 249

10.1 SYNOPSIS

Intake fraction, a metric that summarizes the emission-to-inhalation relationship, facilitates comparisons among sources in terms of their exposure potential. For a given emission source and pollutant, intake fraction is the cumulative mass inhaled by the exposed population divided by the cumulative emissions. One way to estimate the environmental health impact of a pollution source or source class is as the product of three terms: emission rate (mass per time), intake fraction (mass inhaled per mass emitted), and toxicity (health impact per mass inhaled). In the ideal situation, one would know all three terms for all major emission sources. However, important insight can be gained even without complete information. For example, if two sources are identical except that the intake fraction is twice as high for source A as for source B, then the health benefit per mass emission reduction is expected to be twice as large for A as for B.

Intake fraction is a metric not a method. Values of the intake fraction can be determined from models or from measurements. Typical values for the intake fraction are as low as 0.1 per million for releases to outdoor air in remote rural areas, roughly 10 per million for releases near ground level in urban areas, and roughly 5,000 per million for indoor releases in occupied buildings. Thus, releases to indoor air have roughly 500 times as great an intake fraction as for outdoor releases to urban air. In other words, a gram released indoors while painting your living room is, from the standpoint of population exposure, roughly equivalent to half a kilogram released into the urban atmosphere from the paint plant.

This chapter illustrates the use of a simple model, the one-compartment box model, to estimate intake fraction values and compare values among types of sources. Examples are included of how one might compare intake fraction values for two sources and then use this information to prioritize emission reductions. For example, since intake fraction values are expected to be higher in urban areas than in rural areas, all else being equal, the health benefits attributable to an emission reduction are expected to be greater if that emission reduction occurs in an urban area than if it occurs in a rural area. As another example, because people are, on average, closer to on-road emissions than to other ambient sources, emission reductions targeted at on-road sources will have a greater health impact per mass emission reduction than reductions targeted at other ambient sources. As a third example, "self-pollution" of school buses, whereby a small fraction of emissions migrate inside the same vehicle that generated the pollution, has the potential to greatly increase intake fractions. Mitigating self-pollution represents "low-hanging fruit" in terms of exposure control: the mass of pollution inhaled could be significantly reduced even without reducing emissions.

Intake fraction can be useful in a variety of situations where a summary of the emission-to-inhalation pathway is needed. It can be used in cost-effectiveness analyses to compare emission reduction options in terms of the cost per gram inhaled rather than the cost per gram emitted. Because it can be disaggregated according to who inhales the pollution, intake fraction may also be useful when considering environmental justice concerns related to the distribution of exposure concentrations among the population.

10.2 INTRODUCTION

The effectiveness of air pollution control measures may be evaluated in terms of changes in emissions rates, using measures such as tons per year. Indirectly, the effects of such reductions may be observed through changes in ambient air concentrations as measured at ambient monitoring stations. It might be assumed that decreases in ambient air concentrations cause commensurate decreases in human exposure. However, this is not necessarily the case, because personal exposures can vary substantially from what ambient air monitors indicate. For example, measured ambient benzene concentrations decreased in the area of Los Angeles, California, from 1989 to 1997 by a factor of four, from 4 to 1 ppb. However, exposure concentrations were calculated to have decreased by only a factor of three, from 6 to 2 ppb (Fruin et al. 2001). In this case, only about half of the exposure reductions occurred because of reductions in ambient air concentrations. Other contributions came from reduced exposure to environmental tobacco smoke and from decreased benzene concentrations in cars and garages, improvements that would not be detected at ambient monitoring stations.

This chapter presents ideas about how to prioritize emission reductions based on their effectiveness in reducing exposures. These include considering the location of the emissions source, the surrounding population densities, and the factors affecting dilution of the emissions.

10.3 BACKGROUND

There are many sources of environmental pollution. Important environmental health goals include identification of sources, estimation of personal and population exposures, and effective prioritization of emission reductions. Scientific and engineering analyses are needed to determine which sources to mitigate and by how much. Because the most important reason for regulating air pollution is to reduce its adverse effects on public health, environmental health impact is a logical basis for prioritizing emission reductions. While this chapter focuses on pollutant emissions to air, similar approaches are applicable to groundwater, surface water, and soil pollution.

One way to estimate the environmental health impact of an air pollution source or source class is as the product of three terms: emission rate (mass per time), intake fraction (mass inhaled per

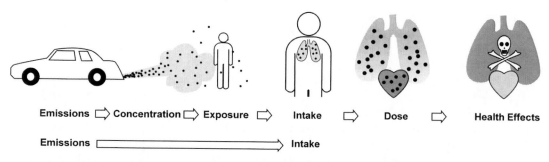

Emissions $\Rightarrow$ Concentration $\Rightarrow$ Exposure $\Rightarrow$ Intake $\Rightarrow$ Dose $\Rightarrow$ Health Effects

Emissions $\longrightarrow$ Intake

FIGURE 10.1 The air-pollution emission-to-effects paradigm. (After Smith 1993.)

mass emitted), and toxicity (health impact per mass inhaled). In the ideal situation, one would know all three terms for all significant emission sources. However, as we describe below, one can gain important insight even without complete information. This chapter focuses on the second term in this relationship (intake fraction).

10.4   WHAT IS INTAKE FRACTION?

Intake fraction summarizes in a compact and transparent form the relationship between emissions and inhalation of these emissions. Intake fraction is useful in connecting emissions to effects because mass inhaled is a much better indicator of potential adverse health impacts than either mass emitted or airborne concentration.

The emission-to-effects relationship involves a series of causally related steps. As illustrated in Figure 10.1, emissions are transported and transformed to generate pollutant concentrations that generally vary in space and time. Human encounters with concentrations constitute exposures, and inhalation of pollutants results in intake. Pollutant transfer into the body of an exposed individual leads to doses to physiological targets, such as organs, which in turn can elevate the risk of adverse health effects. Intake fraction quantitatively summarizes an important portion of this chain of events by describing the emission-to-intake relationship as a single number.

Intake fraction should be understood to be a metric not a method. Like emissions and concentrations, intake fraction can be determined through several different methods. Investigations that generate intake fraction results can range from simple to complex and can rely on modeling or on experimental measurement.

Intake fraction for a primary pollutant is the total mass inhaled from an emission source divided by the total mass emitted from that source. The emission source evaluated in the denominator can be a single emitter, such as an industrial stack or a cigarette, or a broad source class, such as motor vehicles or household cleaning products. When considering an entire population, the value of the numerator would be the cumulative mass inhaled by all exposed individuals. When considering a subpopulation or an individual, the value in the numerator would be the mass inhaled by that subpopulation or individual. Mass inhaled can be determined as the average intake rate multiplied by exposure duration.

$$\text{intake fraction} = \frac{\text{mass inhaled}}{\text{mass emitted}}$$

This expression can be evaluated in terms of cumulative intake per unit emissions for a release episode. In this case, both numerator and denominator would have units of mass. Or, for processes that continuously emit pollutants, the intake fraction can be evaluated as the ratio of the time-averaged inhalation rate to the time-averaged emission rate. In this case, both numerator and denominator would have units of mass per time. Intake rate (mass per time) can be evaluated as exposure concentration (mass per volume) times breathing rate (volume per time). In all cases, the intake fraction is a dimensionless ratio that reflects the fraction of pollution released to the environment that is taken in by an exposed population.

Intake fraction depends on many parameters that influence the emission-to-intake relationship, such as whether the emission occurs indoors or outdoors. Therefore, intake fraction can vary with location and over time. For example, if two outdoor emission sources emit the same mass of pollution, but one source is in a densely populated urban area while the other is in a rural area, the first source will have a higher associated intake fraction because there are more people in the vicinity of the emissions. For a given indoor or outdoor emission source, the intake fraction varies with the pollutant's removal rate from the environment of concern. For ground-level outdoor emissions, intake fraction is smaller during periods of rapid mixing and dispersion than during stagnant air conditions. For indoor emissions into a building of a given size, intake fraction is smaller for a high air-exchange rate (a "leaky" building) than for a low air-exchange rate (a "tight" building).

One important attribute of intake fraction is that it can be applied to groups of pollutants, rather than only to specific species. For example, if two pollutants are emitted from the same source and have the same fate and transport characteristics, then their intake fraction values will be the same, even if their chemical composition and mass emission rates differ. Consider emissions from passenger vehicles in a specific urban environment. To the extent that $PM_{2.5}$ from gasoline-powered motor vehicles behaves like a conserved (nonreacting) pollutant, then its intake fraction would be similar to the intake fraction of other conserved pollutants from motor vehicles, such as carbon monoxide. Similarly, the indoor fate and transport of acrylonitrile and 1,3-butadiene, two toxic chemicals found in environmental tobacco smoke (ETS), are similar, indicating that their intake fractions associated with ETS would be similar. This characteristic offers the promise of efficiency in the use of intake fraction. One can envision that as more studies of intake fraction are completed, a compendium of intake fraction results could be compiled that would provide useful guidance about expected values for sources not yet assessed.

10.5 TYPICAL INTAKE FRACTION VALUES

Intake fraction values vary over several orders of magnitude. Three important factors affecting intake fraction are the size of the exposed **population**, the **proximity** between the emission source and the exposed population, and the **persistence** of the pollutant in the air parcel. These three factors are informally known as "the three P's." Typical intake fraction values for some release categories are presented in Figure 10.2.

For outdoor releases in rural or urban areas, intake fraction values are typically in the range 0.1–100 per million. An intake fraction of 1 per million means that for every kilogram emitted, 1 mg is collectively inhaled. This intake fraction value also means that to reduce inhalation intake by 1 mg would require reducing emissions by 1 kg. Intake fraction values are much higher for indoor releases than for outdoor releases because dilution and dispersion rates are lower indoors than outdoors. A typical intake fraction difference between indoor and outdoor releases can be as large as three orders of magnitude. This result leads to the "rule of 1,000," which states that from an inhalation intake standpoint, 1 gram of emissions to indoor environments is roughly equivalent to 1000 grams of emissions to outdoor environments.

The term "intake fraction" was first introduced in the literature in 2002 (Bennett et al. 2002a). However, there is a much longer history of the idea of quantitatively relating pollutant emissions to inhalation intake, as reviewed by Bennett et al. (2002a) and by Evans et al. (2002). Smith and colleagues have explored the policy implications of intake fraction and related concepts (e.g., Smith and Edgerton 1989; Smith 1995, 2002). Although not yet large, the literature on intake fraction is diverse, addressing primary and secondary pollutants, inhalation and other intake pathways, and varied sources such as motor vehicles, power plants, and dry cleaners (see Table 10.1).

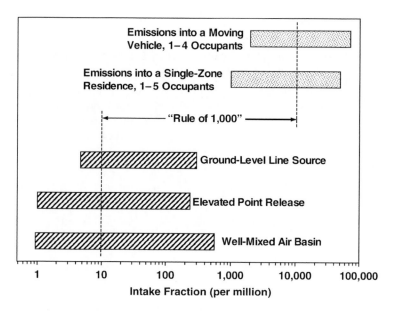

FIGURE 10.2 Typical intake-fraction values for nonreactive air pollutants emitted from different source classes. (Data from Lai, Thatcher, and Nazaroff 2000.) The upper two bars represent indoor emissions and the lower three bars correspond to outdoor emissions. Within each category, considerable variability is possible, depending largely on population exposed, their proximity, and the persistence of the pollutants. The difference in central tendency between the bars is ~3 orders of magnitude, as illustrated by the arrow labeled "rule of 1,000."

Broadly, there are two approaches for quantifying the emission-to-intake relationship: models and measurements. During the past several decades, much work in air-quality engineering has developed and used these approaches to understand emission-to-airborne concentration relationships. The methods developed and the results obtained can also be used to inform the emission-to-intake relationship. Models range from simple, one-compartment representations of a household or an urban area, to complex, three-dimensional urban airshed models. Measurement methods include experiments involving the deliberate release of a tracer gas as well as utilization of "tracers-of-opportunity" (i.e., chemical compounds that act as a "fingerprint" for an emission source).

The following factors have been found to significantly influence intake:

- Whether a release occurs within a confined space (indoors) or into an open environment
- Population density and size of the exposed population in the vicinity of the release
- Meteorological conditions controlling air dispersion, such as wind speed and mixing height, or analogously for indoor releases, the ventilation rate of a building
- Pollutant persistence, depending on the rate of mechanisms such as deposition
- Dominant exposure pathway (inhalation, dermal absorption, ingestion) and emission media (air, soil, surface water, groundwater)
- Transformations such as bioaccumulation (for ingestion) and secondary formation (for air pollutants)

10.6 ESTIMATING INTAKE FRACTION VALUES USING A ONE-COMPARTMENT MODEL

In this section we demonstrate how a one-compartment box model can be used to estimate inhalation intake fraction. The one-compartment model is straightforward, produces reasonable quantitative intake fraction estimates, and provides insight about the dependence of intake fraction on key

TABLE 10.1
Recently Published Studies Reporting Evaluations of Intake Fractions

Sources	Pollutants	Media	Pathways	Methods	References
Dry cleaners	Perchloroethylene	Air	Inhalation	Dispersion model, box model	Evans, Thompson, and Hattis (2000)
General	Organic pollutants	Air, water, soil, food	Inhalation and ingestion	Multimedia model	Bennett et al. (2002b)
General	Benzene, carbon tetrachloride, benzo[a]pyrene, and dioxin	Air, water, soil, food	Inhalation and ingestion	Multimedia model	MacLeod et al. (2004)
General	PCDDs/DFs and PCBs	Air, water, soil, food	Inhalation and ingestion	Multimedia model, data analysis	Hirai et al. (2004)
General	Semivolatile organics	Air, water, soil, food	Ingestion	Multimedia model	Lobscheid, Maddalena, and McKone (2004)
General air emissions	Primary air pollutants	Air	Inhalation	Box models, dispersion models	Lai, Thatcher, and Nazaroff (2000)
General air emissions	PCDDs/DFs	Air, water, soil, food	Ingestion	Multimedia model, data analysis	Margni et al. (2004)
Motor vehicles	Primary pollutants (CO, benzene)	Air	Inhalation	Data analysis of tracers of opportunity	Marshall et al. (2003)
Motor vehicles	Primary pollutants	Air	Inhalation	Models	Marshall, Teoh, and Nazaroff (2005)
Motor vehicles: school buses	Primary pollutants	Air	Inhalation	Tracer gas	Marshall and Behrentz (2005)
Power plants, motor vehicles	Primary and secondary PM	Air	Inhalation	Dispersion model	Levy, Wolff, and Evans (2002)
Power plants, motor vehicles	Primary pollutants	Air	Inhalation	Dispersion model	Nigge (2001)
Power plants	SO_2, sulfate	Air	Inhalation	Dispersion model	Hao et al. (2003)
Power plants	Primary and secondary PM	Air	Inhalation	Dispersion model	Levy et al. (2003)
Power plants	Primary and secondary PM	Air	Inhalation	Dispersion model	Li and Hao (2003)
Power plants	SO_2, primary and secondary PM	Air	Inhalation	Dispersion model	Zhou et al. (2003)
Power plants, distributed electricity generation	Primary pollutants ($PM_{2.5}$, formaldehyde)	Air	Inhalation	Dispersion model	Heath, Hoats, and Nazaroff (2003)

Note: PCDDs = polychlorinated dibenzo-dioxins; DFs = dibenzo-furans; PCBs = polychlorinated biphenyls.

parameters. We begin by applying this model to the indoor release of a nonreactive pollutant in a household, and then apply it to similar conditions in an urban area.

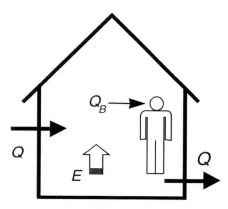

FIGURE 10.3 Schematic for estimating intake fraction for indoor release of a nonreactive pollutant. The indoor environment is treated as well mixed, meaning that the pollutant concentration is assumed to be uniform throughout the indoor air.

For the case of a residence, as illustrated in Figure 10.3, the parameters needed to determine intake fraction are the ventilation rate of the building (Q, units: m^3 h^{-1}), the number of individuals exposed (P, units: person), and the average volumetric breathing rate (Q_B, units: m^3 h^{-1} person^{-1}). The ventilation rate, Q, can be represented as the product of the house volume (V, units: m^3) and the air exchange rate (X, units: h^{-1}). If the pollutant release rate (E, units: μg h^{-1}) and the ventilation rate are constant, and the pollutant is nonreactive and well mixed throughout the indoor air, then the steady-state concentration of the contaminant attributable to the indoor release is simply

$$C = \frac{E}{Q} . \tag{10.1}$$

With P individuals in the space, each breathing at an average volumetric rate Q_B, the cumulative mass inhalation rate of pollutants owing to the indoor source, I, is

$$I = CPQ_B = \frac{EPQ_B}{Q} . \tag{10.2}$$

The intake fraction is the intake rate, I, normalized by the emission rate, E:

$$iF = \frac{PQ_B}{Q} = \frac{PQ_B}{VX} . \tag{10.3}$$

In this model, a pollutant emitted inside can have one of two fates: either it is inhaled, or it exits via air exchange. Intake fraction is the probability that an emitted pollutant is inhaled rather than removed by ventilation. Although Equation 10.3 was derived for steady-state conditions with constant emissions and ventilation, the same relationship holds for the intake fraction of certain episodic releases (Nazaroff accepted).

To generate quantitative estimates of intake fraction, we next consider typical values for each of the terms in Equation 10.3. Estimates in the literature of population-average breathing rate, Q_B, vary. Commonly used values for daily-average conditions (units: m^3 person^{-1} d^{-1}) are 12 (Layton 1993; USEPA 1997), 15 (Marty et al. 2002), and 17 (OEHHA 1996). Air-exchange rates vary among houses and over time. In tropical climates, homes can be constructed with large designed openings for natural ventilation. On a windy day, the air exchange rate in such a building may be of the order of 10 h^{-1}. In contrast, a modern building in a cold climate may be well sealed, having an air exchange rate as low as of the order of 0.3 h^{-1}. The median value for single family homes in the United States is 0.5 h^{-1} (Murray and Burmaster 1995). Of course, the number of occupants in a residence can vary widely. In the United States, there are an average of three persons living in each occupied household (U.S. Census 2004). The median household size is 160 m^2 (U.S. Census 2004); assuming a ceiling height of 2.4 m, this corresponds to a volume of approximately 400 m^3. Building volume and indoor population can be significantly larger for nonresidential buildings, such as shopping centers, restaurants, and offices, than for residences. However, in temperate climates the ventilation rate per occupant (XV/P) is of consistent magnitude across major building classes. Consequently, intake fraction values are expected to be consistent in magnitude across different building types.

Using the values $Q_B = 12$ m^3 person^{-1} d^{-1}, $P = 3$, $X = 12$ d^{-1} ($= 0.5$ h^{-1}), and $V = 400$ m^3, the intake fraction for releases into a residence is estimated to be 0.75%, or 7,500 per million. That is, in this case people would inhale 7.5 mg per gram of pollutant emitted into an indoor environment. This estimate assumed constant occupancy of the indoor space. Taking into account that people only spend about two thirds of their time indoors in their own residences (Klepeis et al. 2001), a better estimate for the central tendency of intake fraction in U.S. residences might be 0.5% or 5,000 per million. This correction would only be expected to apply for certain emission sources that emit continuously, such as furnishings that emit volatile organic compounds. Many important sources, such as cooking, cleaning, and cigarette smoke, only occur when people are present. For these sources, the time-activity correction of two thirds may not apply.

As a comparison, the box model can also be applied to an urban area. The air inflow rate, Q, can be estimated as the product of the wind speed (u, units: m s^{-1}), the height of the atmospheric mixing layer (H, units: m), and the width of the urban area (W, units: m). For the United States, a typical value for the term uH is 42 million m^2 d^{-1}, and the population-weighted median value for the term P/W in U.S. metropolitan areas is 43 people m^{-1} (Marshall, Teoh, and Nazaroff 2005). The ratio of these two groups of terms indicates that a typical value of outdoor "ventilation" of urban areas per person in the United States is $Q/P \sim 10^6$ m^3 person^{-1} d^{-1}. Combining this result in Equation (10.3) with $Q_B = 12$ m^3 person^{-1} d^{-1} yields typical intake fraction for outdoor urban releases of 12 per million, which is roughly 400 times smaller than the value of 5,000 per million estimated for indoor releases. We can see from this comparison that the "rule of 1,000" is only an approximation, providing a magnitude estimate of the difference in intake fractions between indoor and outdoor releases. In large urban areas, the difference between indoor and outdoor releases can be less than 1,000. In small urban areas or in rural areas, the difference can be more than 1,000. In this particular comparison, the factor of 400 reflects the difference in per-capita ventilation rates that is available to cause dilution of pollutant emissions, comparing the amount of ventilation that is provided to buildings with the amount that nature provides through winds that ventilate urban air basins.

10.7 THE USE OF INTAKE FRACTION IN PRIORITIZING EMISSION REDUCTION EFFORTS

There are many situations in which one can use intake fraction to help prioritize among air pollution control options, even without full information about emissions and toxicity. Each of the next four paragraphs presents a situation wherein different pieces of information are available. In each case, we assume that there are two emission sources, and that the question at hand is how to prioritize between these two sources as the target of emissions reduction policies. We further assume that the pollutant of concern exhibits a linear, no-threshold dose-response relationship. This assumption, which is commonly applied in public health protection for carcinogens and certain other pollutants, means that the adverse health impact associated with a source scales in direct proportion to the intake, regardless of how that intake is distributed within the population. Not all important air pollution problems can be treated in this way. The intake fraction can be useful even when this treatment is not possible, but a more sophisticated assessment is required.

1. When all three terms — emissions, intake fraction, and toxicity — are known, one can estimate the overall health impact from the two sources. The source with the higher health impact would be identified as a higher priority for control. If information about the costs of control technologies is also known, then one could prioritize emission reductions based on a cost-effectiveness analysis. In this case, one might seek to maximize the reduction in adverse health effects per unit cost (Smith and Edgerton 1989, Smith 1995).

2. When only emissions and intake fractions are known, one could prioritize emission source reductions based on total emissions, but using the intake fraction values as multipliers. For example, considering a specific pollutant, if the intake fraction is two times as great for emission source A as for emission source B, an emission reduction of 1 kg from A could be given the same policy "priority" as an emission reduction of 2 kg from B. Assuming pollutants have the same toxicity regardless of the emission source (which is usually the case), then comparisons between sources can be made assuming that inhalation intake is a suitable proxy for adverse effects.

3. When only intake fractions and control costs are known, one can carry out certain cost-effectiveness analyses. For example, if control costs per kg *emitted* are the same for emission sources A and B, but the intake fraction is larger for A, then the control cost per kg *inhaled* is less for A than for B.

4. Finally, when only intake fractions are known, one can compare sources that are similar. For example, comparing natural-gas power plants in different locations, one could prioritize for control the emissions location with the higher intake fraction. This intake fraction difference could be attributable to a variety of factors, including the size of the exposed population, proximity between the population and emissions, and meteorology.

In prioritizing among the thousands of emission sources, it may be useful to group sources into broad classes, such as indoor vs. outdoor, stationary vs. mobile, urban vs. rural, or — for motor vehicles — diesel-powered vs. gasoline-powered. The adverse impact for each source class could be estimated as proportional to the product of the total mass emissions times an emission-weighted estimate of intake fraction. Information on toxicity could be incorporated into this approach independently, to the extent that it is available.

Below are two examples of how intake fraction might be used to prioritize emission source reductions: on-road vs. off-road sources and emissions sources with self-pollution.

10.7.1 PARTICULATE MATTER FROM ON-ROAD SOURCES

People in urban areas typically spend some time in or near vehicles each day. Particulate matter (PM) concentrations are several times higher, on average, in vehicles than in buildings for three reasons: (1) vehicles are an important source of PM emissions, (2) the in-vehicle environment is closer than the indoor environment to vehicle emissions, and (3) buildings offer more protection than vehicles against outdoor PM. These near-source (in-vehicle) exposures increase the intake fraction associated with on-road emissions as compared with off-road emissions.

For example, focusing on urban diesel PM emissions, one can use published data to estimate the typical intake fraction differences between on-road sources and other sources. Recent measurements for diesel vehicles in California suggest a factor of 4–14 difference between in-vehicle and nearby ambient concentrations (Fruin, Winer, and Rodes 2004). The result is that the ~6% of time (80 minutes per day) spent in vehicles (Klepeis et al. 2001) contributes ~25–54% of total exposure to diesel PM, rather than 6%. If it is assumed that 25% of California's diesel PM emissions are from on-road sources (CARB 2000), then on-road sources contribute ~39–63% (rather than 25%) to total diesel PM exposure. On average, on-road sources are estimated to contribute between 1.9 and 5.1 times more diesel PM inhalation per unit emissions than other sources. Thus, from an exposure standpoint, on-road diesel particle emissions should be given a "weighting" of ~2 to 5 relative to other diesel sources. The width of this range reflects uncertainty in diesel PM concentrations in vehicles and in the ambient environment.

10.7.2 SELF-POLLUTION

Combustion sources often possess an exhaust system to deliver emissions to ambient air. The exhaust manifold of a car conducts effluents from the engine to the tailpipe; wood stoves emit their effluents through a chimney that runs from the fireplace to the rooftop. Generally, exhaust systems work well but not perfectly, and a small fraction of emissions can enter the indoor or in-vehicle environment. Such leaks lead to a condition known as "self-pollution."

Intake fractions for pollutant releases into moving vehicles are comparable to those for residences (Lai, Thatcher, and Nazaroff 2000). Because of the "rule of 1,000," even a small amount of self-pollution — on the order of 0.1% of emissions or higher — can significantly increase the intake fraction associated with sources like motor vehicle exhaust and residential wood combustion.

School buses are a good example of the high impacts possible from self-pollution. Tracer-gas experiments indicate that the self-pollution intake fraction is significant for school buses in California, to a degree that depends on the age of the bus. The self-pollution intake fraction reported by Marshall and Behrentz (2005) was 70 per million for the oldest bus investigated (model year: 1975), and averaged 20 per million for the remaining five buses (model years between 1985 and 2002). These values are larger than typical outdoor intake fractions (e.g., a typical value of 12 per million was derived above). On the basis of this work it is estimated that in a typical U.S. urban area, the cumulative mass of a school bus's emissions inhaled by the roughly 40 students on that bus is larger than the cumulative mass of pollution from that school bus inhaled by all of the other exposed individuals. Addressing bus self-pollution could markedly reduce the inhalation intake of diesel PM, even before bus emission reductions are achieved. More broadly, controlling self-pollution wherever it occurs offers the potential to be "low-hanging fruit" for effectively achieving exposure reduction.

10.8 USING INTAKE FRACTION WHEN CONSIDERING ENVIRONMENTAL JUSTICE CONCERNS

Understanding and addressing distributional issues related to air pollution exposure is an important aspect of air quality management. Air pollution control policies need not only reduce the total health impact of emissions, but also ensure that the distribution of burden among the population is not unfair or unjust. Throughout most of this chapter, intake fraction has been based on the total population intake. However, intake fraction can be estimated for population subgroups, even down to the level of individuals. So, for example, if the population is divided into a set of groups, the total population intake fraction can be considered as the sum of the partial intake fractions associated with each group. One indicator of environmental justice would be the degree to which partial intake fractions per capita are consistent among different demographic indicators. Heath, Hoats, and Nazaroff (2003) considered how intake fraction for an electricity-generation station depends on the station's location. They presented, for specific locations, the percentage of the downwind population that is white vs. non-white, and the percentage of total intake that occurs in the white and non-white populations. In three of the five case studies, non-white populations were significantly more exposed than white populations. For example, for a hypothetical small-scale electricity generator located in downtown Los Angeles, 32% of the exposed (i.e., downwind) population is non-white, but this population would receive 69% of the total intake (Heath, Hoats, and Nazaroff 2003).

There are several attributes that can be used to divide the population into groups that could be relevant in considering air pollution exposure aspects of environmental justice, including ethnicity, gender, neighborhood, income, age, and health status. As information emerges about different degrees of susceptibility of demographic subgroups to air pollution exposure, intake fraction analyses could also be conducted to highlight the levels of exposure that these subgroups encounter in relation to the other parts of the population. By doing so, additional control efforts could be targeted at protecting those who are most vulnerable to air pollution.

10.9 ACKNOWLEDGMENTS

The Research Division of the California Air Resources Board provided partial funding for the work described in this chapter. The contents are the opinions of the authors and do not represent the views of the Air Resources Board. Dr. Scott Fruin provided helpful review comments and suggestions on a draft of this work. Figure 10.1 was generated by The Linus Group.

10.10 QUESTIONS FOR REVIEW

1. What is meant by the "three P's" and the "rule of 1,000"? [Answer: The three P's are *population, proximity,* and *persistence*. These are three variables that have a large influence on intake fraction. The *rule of 1,000* says that typical intake fraction values are roughly 1,000 times larger for indoor releases than for outdoor releases. This means that indoor releases are roughly 1,000 times more efficient at delivering their dose to people than are outdoor releases.]

2. The health risk attributable to an emission source can be estimated as the product of three parameters. One of these three parameters is intake fraction. What are the other two parameters, and what are the units on each of the three terms? [Answer: The three terms are emissions (units are mass, or mass per time); intake fraction (mass inhaled per mass emitted, effectively dimensionless); and toxicity (health impact per mass inhaled).]

3. The outdoor intake fraction value in the text (12 per million) is based on the value P/W = 43 people m^{-1}, where P is the population and W is the width of the metropolitan area. Using equations given in this chapter, perform a similar calculation: estimate intake fraction values in the following table using the given values for P and W. We have used the square root of the land area as an estimate of W.

Area	Population (thousands)	Land Area (km²)	P/W (people m⁻¹)	Intake Fraction (mg inhaled per kg emitted)
Los Angeles, CA	12,400	5,800	163	
Sacramento, CA	1,510	990	48	
Rural Nevada	170	280,000	0.3	
Area where you live				

How well does the one-compartment intake fraction estimate for Los Angeles compare with the published estimate of 48 per million (Marshall et al. 2003)? [Answer: The results are presented in the table below. Note the wide variation between sparsely populated rural Nevada and the populous Los Angeles area. The estimate of 58 per million for Los Angeles is ~20% larger than the published value of 48 per million. Given the considerable simplification inherent in the one-compartment model, this level of agreement is considered good.]

Area	Population (thousands)	Land Area (km²)	P/W (people m⁻¹)	Intake Fraction (mg inhaled per kg emitted)
Los Angeles, CA	12,400	5,800	163	58
Sacramento, CA	1,510	990	48	17
Rural Nevada	170	280,000	0.3	0.1
Area where you live				

4. The one-compartment model presented in this chapter can be modified to account for pollutant removal mechanisms other than advection (or ventilation). For example, pollutant removal via deposition can be characterized in terms of a deposition velocity (v_d, units: m s⁻¹) onto a surface of area A (units: m²). The one-compartment intake fraction equation, accounting for deposition, is

$$\text{iF} = \frac{PQ_B}{Q + Av_d}$$

Consider an urban area for which the following values apply: P = 1 million people; Q_B = 15 m³ person⁻¹ d⁻¹; $Q = 10^{12}$ m³ d⁻¹ (i.e., 1 trillion m³ d⁻¹); A = 700 million m². Assume deposition velocities (units: cm s⁻¹) for fine particulate matter (PM) and for coarse PM are 0.03 and 3, respectively. What are the intake fractions for fine PM and coarse PM, and how do these values compare to the intake fraction for a non-depositing conserved pollutant? (*Hint:* A non-depositing pollutant has a deposition velocity of zero.) [Answer: Intake fractions values (units: per million) for fine PM and coarse PM are 14.7 and 5.3, respectively. These values are 2% and 64% less, respectively, than the intake fraction for a non-depositing pollutant (15.0 per million).]

5. A specific household contains two combustion sources: an incense stick and a fireplace. Assume that in 1 hour, 3 grams of incense (about 2 incense sticks) are burned and 3 kg of wood are burned. Assume that 0.3% of the incense mass is emitted as fine particles and that 0.03% of the wood mass is emitted as fine particles. Finally, assume that all incense emissions are to the indoor space, while only 0.5% of the wood emissions enter the indoor space as self-pollution. (The remaining 99.5% of the wood emissions exit via the chimney.) What are the total mass emission rates of fine particles *to the indoor space* for the two sources, in units of mg per hour of use? Using an intake fraction of 1%, what is the total mass of incense particle emissions and of wood particle emissions inhaled, in units of mg per hour of use? What steps could be taken to reduce mass inhalation values? [Answer: The indoor mass emission rates (units: mg per hour of use) are 9 and 4.5, respectively, for the incense emissions and for the wood emissions. The mass inhalation rates (units: mg per hour of use) are 0.09 and 0.045, respectively, for the incense emissions and for the wood emissions. To reduce the intake rate, one can reduce the mass emission rate or the intake fraction. Steps that would reduce the mass emission rate include using low-smoke incense; altering the wood combustion conditions to reduce particle formation rates; and, reducing the mass of fuel combusted. Steps to reduce the intake fraction include separating the emissions and the people (e.g., go in the other room and shut the door); increasing the ventilation rate (e.g., crack open a

window); and, reducing the self-pollution rate (e.g., fully enclose the wood fire to ensure that all of the smoke exits via the chimney rather than into the room).]

6. The "rule of 1,000" indicates that indoor sources are roughly 1,000 times more potent than outdoor sources at delivering inhalation intake to people. Similarly, calculations in the section "Particulate Matter from On-Road Sources" indicate that on-road PM emissions are approximately two to five times more effective than off-road PM emissions at delivering inhalation intake to people. In other words, the intake fraction for PM is two to five times larger for on-road emissions than for off-road emissions. A similar calculation indicates that the intake fraction for nonreactive gas molecules, such as benzene and carbon monoxide, is ~25% higher for on-road emissions than for off-road emissions. Derive this ~25% difference, using the following assumptions:

- 66% of emissions are on-road, and the remainder are off-road
- People spend 80 minutes per day in a vehicle
- Concentrations are 4 times higher in-vehicle than not-in-vehicle
- Buildings do not offer protection against outdoor concentrations (i.e., the indoor concentration attributable outdoor pollution is equal to the outdoor concentration)
- Note that because we are taking the ratio of two intake fractions, the ~25% value does not depend on the ambient concentration, the volume of air breathed per day per person, or the total emission rate.

[Answer: Let C = the ambient concentration, Q_B = the volume of air breathed per day, and E = the total emission rate. Note that 80 minutes per day is $(80/1440) = 5.6\%$ of the day.

Mass inhaled per day = Mass inhaled during driving + Mass inhaled not during driving
= $Q_B(5.6\%)(4C) + Q_B(94.4\%)(C) = (1.17)Q_BC$
Mass inhaled per day attributable to off-road sources = $0.34Q_BC$
Mass inhaled per day attributable to on-road sources = $1.17Q_BC - 0.34Q_BC = 0.83\ Q_BC$
Intake fraction for on-road sources = intake/emissions = $0.83Q_BC/0.66E$
Intake fraction for off-road sources = intake/emissions = $0.34\ Q_BC/0.34E = Q_BC/E$
Ratio of these intake fraction values = $(0.83Q_BC/0.66E)/(Q_BC/E) = 0.83/0.66 = 1.26$

Thus, on-road emissions are ~25% more effective than off-road emissions at delivering inhalation intake.]

REFERENCES

Bennett, D.H., McKone, T.E., Evans, J.S., Nazaroff, W.W., Margni, M.D., Jolliet, O., and Smith, K.R. (2002a) Defining Intake Fraction, *Environmental Science & Technology,* **36**: 206A–211A.

Bennett, D.H., Margni, M.D., McKone, T.E., and Jolliet, O. (2002b) Intake Fraction for Multimedia Pollutants: A Tool for Life Cycle Analysis and Comparative Risk Assessment, *Risk Analysis,* **22**: 905–918.

CARB (2000) *Risk Reduction Plan to Reduce Particulate Matter Emissions from Diesel-Fueled Engines and Vehicles,* Planning and Technical Support Division, Sacramento, California Air Resources Board, Sacramento, CA, http://www.arb.ca.gov/diesel/documents/rrpapp.htm (accessed April 15, 2006).

Evans, J.S., Thompson, K.M., and Hattis, D. (2000) Exposure Efficiency: Concept and Application to Perchloroethylene Exposure from Dry Cleaners, *Journal of the Air & Waste Management Association,* **50**: 1700–1703.

Evans, J.S., Wolff, S.K., Phonboon, K., Levy, J.I., and Smith, K.R. (2002) Exposure Efficiency: An Idea Whose Time Has Come? *Chemosphere,* **49**: 1075–1091.

Fruin, S.A., St. Denis, M.J., Winer, A.M., Colome, S.D., and Lurmann, F.W. (2001) Reductions in Human Benzene Exposure in the California South Coast Air Basin, *Atmospheric Environment,* **35**: 1069–1077.

Fruin, S.A., Winer, A.M., and Rodes, C.E. (2004) Black Carbon Concentrations in California Vehicles and Estimation of In-Vehicle Diesel Exhaust Particulate Matter Exposures, *Atmospheric Environment,* **38**: 4123–4133.

Hao, J.M., Li, J., Ye, X.M., and Zhu, T.L. (2003) Estimating Health Damage Cost from Secondary Sulfate Particles — A Case Study of Hunan Province, China, *Journal of Environmental Sciences,* **15**: 611–617.

Heath, G.A., Hoats, A.S., and Nazaroff, W.W. (2003) *Air Pollution Exposure Associated with Distributed Electricity Generation,* CARB Contract No. 01-341, Department of Civil and Environmental Engineering, University of California, Berkeley, CA, http://ftp.arb.ca.gov/research/apr/abstracts/01–341.htm (accessed April 15, 2006).

Hirai, Y., Sakai, S., Watanabe, N., and Takatsuki, H. (2004) Congener-Specific Intake Fractions for PCDDs/DFs and Co-PCBs: Modeling and Validation, *Chemosphere,* **54**: 1383–1400.

Klepeis, N., Nelson, W., Ott, W., Robinson, J., Tsang, A., Switzer, P., Behar, J., Hern, S., and Engelmann, W. (2001) The National Human Activity Pattern Survey (NHAPS): A Resource for Assessing Exposure to Environmental Pollutants, *Journal of Exposure Analysis and Environmental Epidemiology,* **11**: 231–252.

Lai, A.C.K., Thatcher, T.L., and Nazaroff, W.W. (2000) Inhalation Transfer Factors for Air Pollution Health Risk Assessment, *Journal of the Air & Waste Management Association,* **50**: 1688–1699.

Layton, D.W. (1993) Metabolically Consistent Breathing Rates for Use in Dose Assessments, *Health Physics,* **64**: 23–36.

Levy, J.I., Wolff, S.K., and Evans, J.S. (2002) A Regression-Based Approach for Estimating Primary and Secondary Particulate Matter Intake Fractions, *Risk Analysis,* **22**: 895–904.

Levy, J.I., Wilson, A.M., Evans, J.S., and Spengler, J.D. (2003) Estimation of Primary and Secondary Particulate Matter Intake Fractions for Power Plants in Georgia, *Environmental Science & Technology,* **37**: 5528–5536.

Li, J. and Hao, J.M. (2003) Application of Intake Fraction to Population Exposure Estimates in Hunan Province of China, *Journal of Environmental Science and Health Part A — Toxic/Hazardous Substances & Environmental Engineering,* **38**: 1041–1054.

Lobscheid, A.B., Maddalena, R.L., and McKone, T.E. (2004) Contribution of Locally Grown Foods in Cumulative Exposure Assessments, *Journal of Exposure Analysis and Environmental Epidemiology,* **14**: 60–73.

MacLeod, M., Bennett, D.H., Perem, M., Maddalena, R.L., McKone, T.E., and Mackay, D. (2004) Dependence of Intake Fraction on Release Location in a Multimedia Framework: A Case Study of Four Contaminants in North America, *Journal of Industrial Ecology,* **8**: 89–102.

Margni, M., Pennington, D.W., Amman, C., and Jolliet, O. (2004) Evaluating Multimedia/Multipathway Model Intake Fraction Estimates Using POP Emission and Monitoring Data, *Environmental Pollution,* **128**: 263–277.

Marshall, J.D. and Behrentz, E. (2005) Vehicle Self-Pollution Intake Fraction: Children's Exposure to School Bus Emissions, *Environmental Science & Technology,* **39**: 2559–2563.

Marshall, J.D., Riley, W.J., McKone, T.E., and Nazaroff, W.W. (2003) Intake Fraction of Primary Pollutants: Motor Vehicle Emissions in the South Coast Air Basin, *Atmospheric Environment,* **37**: 3455–3468.

Marshall, J.D., Teoh, S.K., and Nazaroff, W.W. (2005) Intake Fraction of Nonreactive Vehicle Emissions in US Urban Areas, *Atmospheric Environment,* **39**: 1363–1371.

Marty, M.A., Blaisdell, R.J., Broadwin, R., Hill, M., Shimer, D., and Jenkins, M. (2002) Distribution of Daily Breathing Rates for Use in California's Air Toxics Hot Spots Program Risk Assessments, *Human and Ecological Risk Assessment,* **8**: 1723–1737.

Murray, D.M. and Burmaster, D.E. (1995) Residential Air Exchange Rates in the United States: Empirical and Estimated Parametric Distributions by Season and Climatic Region, *Risk Analysis,* **15**: 459–465.

Nazaroff, W.W. Inhalation Intake Fraction of Pollutants from Episodic Indoor Emissions, in *Building and Environment* (accepted).

Nigge, K.M. (2001) Generic Spatial Classes for Human Health Impacts, Part I: Methodology. *International Journal of Life Cycle Assessment,* **6**: 257–264.

OEHHA (1996) Technical Support Document for Exposure Assessment and Stochastic Analysis, California Office of Environmental Health Hazards Assessment, http://www.oehha.org/airhot_spots/final-Stoc.html (accessed April 15, 2006).

Smith, K.R. (1993) Fuel Combustion, Air Pollution Exposure, and Health: The Situation in Developing Countries, *Annual Review of Energy and the Environment,* **18**: 529–566.

Smith, K.R. (1995) *The Potential of Human Exposure Assessment for Air Pollution Regulation*, Report No. WHO/EHG/95.9, Collaborating Centre for Studies of Environmental Risk and Development, World Health Organization, Geneva.

Smith, K.R. (2002) Place Makes the Poison: Wesolowski Award Lecture (1999) *Journal of Exposure Analysis and Environmental Epidemiology,* **12**: 167–171.

Smith, K.R. and Edgerton, S.A. (1989) Exposure-Based Air Pollution Control Strategies: Proposed Three-Way Integration, in *Proceedings of the EPA/A&WMA Specialty Conference on Total Exposure Assessment Methodology,* Las Vegas, NV, 630–641.

U.S. Census (2004) American Housing Survey for the United States: 2003, Report No. H150/03, Washington DC, http://www.census.gov/prod/2004pubs/H150-03.pdf (accessed April 15, 2006).

USEPA (1997) *Exposure Factors Handbook*, Report No. EPA-600-P-95-002Fa, Office of Research and Development, National Center for Environmental Assessment, U.S. Environmental Protection Agency, Washington, DC, http://www.epa.gov/ncea/pdfs/efh/front2.pdf (accessed April 15, 2006).

Zhou, Y., Levy, J.I., Hammitt, J.K., and Evans, J.S. (2003) Estimating Population Exposure to Power Plant Emissions Using CALPUFF: A Case Study in Beijing, China, *Atmospheric Environment,* **37**: 815–826.

Part III

Dermal Exposure

11 Dermal Exposure, Uptake, and Dose

Alesia C. Ferguson
University of Arkansas for Medical Sciences

Robert A. Canales
Harvard University

James O. Leckie
Stanford University

CONTENTS

11.1   Synopsis 256
11.2   Introduction 256
11.3   Importance of Dermal Exposure and Dose..............................256
11.4   Defining Dermal Exposure and Dose 257
11.5   The Human Skin...259
     11.5.1   General Skin Structure..259
     11.5.2   General Skin Function 260
     11.5.3   The Function and Structure of the Stratum Corneum (SC)............260
     11.5.4   Shedding and Hydration in the Stratum Corneum....................261
11.6   Factors Affecting Dermal Dose...262
11.7   Mechanisms and Pathways for Dermal Exposure 263
11.8   Direct Methods for Measuring Dermal Exposure 265
     11.8.1   Surrogate Skin Techniques...265
     11.8.2   Removal Techniques 266
     11.8.3   Fluorescent Tracer Techniques...................................266
     11.8.4   Surface Sampling Techniques.....................................267
11.9   Dermal Exposure Examples 269
11.10   Direct Techniques for Measuring Absorption...........................270
     11.10.1 *In Vitro* Methods...271
     11.10.2 *In Vivo* Methods 272
11.11   Highlighted Dermal Dose Examples...273
11.12   Conclusion 274
11.13   Questions for Review 275
Glossary of Terms 276
References 278

11.1 SYNOPSIS

The dermal route of exposure and dose to toxic agents has gained recognition over the last decade as being important for certain population groups and certain classes of chemicals. This chapter introduces the field of dermal exposure and dose, and all its related components, such as the human skin, the factors affecting percutaneous absorption, the mechanisms of exposure, techniques used in the field for directly measuring dermal exposure, and the *in vitro* and *in vivo* techniques used for measuring percutaneous absorption. Some studies that demonstrate the amounts of chemical agents that contact the skin surface and types of chemicals that are absorbed through the skin are also presented in this chapter. Lastly, this chapter discusses the relative benefits of the direct and indirect techniques for determining dermal exposure and dose. Some effort has been made to describe the chemical and physical structure of the human skin. This type of knowledge can potentially provide insights for predicting the complex interactions between the skin and a toxic agent. Ultimately we wish to accurately predict the amount of a toxic agent that contacts the skin surface, and the mass that is able to cross the skin barrier into the dermal vasculature (i.e., blood vessels of the skin). By understanding the dermal route and developing reliable tools to quantify exposure and dose, the relative importance of the dermal route in a total health risk assessment can be established. Words that may be unfamiliar to the reader, especially biological terms related to the human skin, are in italics where first encountered, and found in the glossary at the end of the chapter. Happy reading!!

11.2 INTRODUCTION

Human exposure and intake dose to environmental toxins has long been recognized as occurring via the inhalation and ingestion routes. More recently, the dermal route of exposure and dose has gained recognition as being an important route to study for certain classes of chemicals (e.g., metals, polychlorinated biphenyls [PCB], polycyclic aromatic hydrocarbons [PAH], and pesticides) and for certain high-risk, susceptible groups of individuals (e.g., children and pesticide handlers). However, much remains to be understood about the scenarios under which chemicals contact and adhere to the skin surface, the types of chemicals that are able to penetrate a mostly impermeable barrier (i.e., the skin), and how the process of dermal penetration occurs. In general, we are interested in dermal exposure and dose for the following reasons: (1) the local effect of chemicals (e.g., corticosteroids) applied to the skin for dermatology; (2) the transport of chemicals through the skin for systemic effects (e.g., nicotine patches); (3) the surface effects of chemicals (e.g., sunscreens); (4) the effect of chemicals applied to target deeper tissues (e.g., nonsteroidal anti-inflammatory drugs (NSAIDs) for muscle inflammation); and (5) the unwanted exposure to and *absorption* of harsh agents (e.g., environmental exposures to industrial solvents, pesticides, allergens) (Roberts and Walters 1998). In most of these cases, we want to enhance the absorption rate of chemicals by altering the barrier function of the skin via chemical enhancement, *iontophoresis*, or *phonophoresis,* for example (Rosado and Rodriques 2003). However, in the case of unwanted environmental exposures, we want to understand the extent of chemical loading on the skin and the absorption process so we can consequently develop methods of reducing that exposure and dose.

11.3 IMPORTANCE OF DERMAL EXPOSURE AND DOSE

The five components of the complete human health risk model have been introduced in Chapter 1. As mentioned, the two components that have not been developed extensively are total human exposure analysis and dosage estimation (Zartarian and Leckie 1998). Risk estimation can therefore be improved with more robustness in *aggregate* or multiroute exposure and dose estimations. The Food Quality Protection Act (FQPA) of 1996 requires the U.S. Environmental Protection Agency (USEPA) to quantify aggregate exposure for chemical health risks — in particular, the health risk posed by pesticides (U.S. Congress 1996). This importance for aggregate exposure assessment has motivated researchers to explore new and refined methodologies for measuring and modeling dermal exposure and dose (Fenske 2000), especially given that the dermal route has been understudied relative to the ingestion and inhalation routes. For the dermal route, quantifying exposure and dose has been challenging because loading onto, removal from, and uptake into the skin varies in space and time (Zartarian et al. 2000). Special interest has also been given to population groups that might receive enhanced exposure through the dermal route because of their unique activity patterns, environmental conditions, or increased biological susceptibility. These groups may include children, pregnant women and their fetuses, pesticide applicators, industrial workers, hairdressers, metal workers, furniture workers, and food handlers.

A common exposure scenario where the dermal route is of special interest is the exposure of children to pesticides in and around the residential environment, including schools and daycare environments (Schmidt 1999; Wilson, Chuang, and Lyu 2001; Wilson et al. 2003, 2004). Up to 90% of households use products containing pesticides (see Chapter 15) providing ample opportunity for exposure through regular use, as well as through misuse and accidents. Homeowners used an estimated 74 million lbs of conventional pesticides in 1995; at the time, 939 million lbs were used for agriculture (Aspelin 1997). Home and garden consumption of pesticides was even higher in 1997 at 137 million lbs (62 million kg), while 946 million lbs (429 million kg) was consumed for agriculture (see Chapter 15). Pesticide use poses a variety of health issues for the general public. Illnesses like brain cancer, childhood leukemia, immune system disorders, and learning disabilities have been linked to long-term exposure to pesticides (Sinclair 1995). Short-term acute health issues include skin rashes, headaches, dizziness, and even death. In 1993, for example, there were 140,000 acute pesticide exposures reported nationwide, and 93% of those exposures took place in the home. Children under age 6 accounted for over half of all reported exposures (Grossman 1995). When it comes to exposure to pesticides through the dermal exposure route, young children are possibly at greater risk when compared to adults because of their unique activity patterns (e.g., crawling on floors, carpets, and in sandboxes in and around the home where pesticides may be present), their larger surface area to body weight ratio, and their developing organs. Children in agricultural communities are a special susceptible group with possible increased exposure to multiple pesticides due to their proximity to agricultural fields and contaminants tracked from farms to the home by their parents (Simcox et al. 1995; Fenske et al. 2000). There is also concern for children's health due to their possibly increased dermal exposure to heavy metals, such as lead, found in house dust (Roels et al. 1980), and arsenic found in a chrominated copper arsenate (CCR) material used to coat decks and play structures (Hemond and Solo-Gabriele 2004). A number of initiatives stemming from the 1996 FQPA and the National Research Council's report *Pesticides in the Diets of Infants and Children* (NRC 1993) have been implemented to protect children, including but not limited to: (1) the Federal Executive Order of 21st April 1997, "Protection of Children from Environmental Risks and Safety Risks"; (2) the creation of the Centers for Children's Environmental Heath and Disease Prevention Research established by the USEPA, Centers for Disease Control and Prevention (CDC), and the National Institute of Environmental Health Sciences (NIEHS) (O'Fallon, Collman, and Dearry 2000); (3) the development of the Children's Health Act; (4) the Strategy for Research on Environmental Risks to Children; (5) the creation of the *Child-Specific Exposure Factors Handbook*; (6) and the Guidance for Assessing Cancer Susceptibility from Early-Life Exposure (Williams, Holicky, and Paustenbach 2003).

11.4 DEFINING DERMAL EXPOSURE AND DOSE

Figure 11.1 illustrates the relationship between dermal exposure and dermal dose and the contact boundary, the skin. Dermal exposure and dose are intimately related, with both processes occurring simultaneously. Dermal exposure occurs at the surface and dermal dose through the skin. Dermal exposure occurs when the human skin (i.e., exposure boundary) contacts a chemical (e.g., pesticide),

Relationship between Dermal Exposure and Dermal Dose

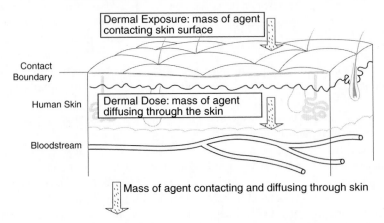

FIGURE 11.1 Relationship between dermal exposure and dermal dose. For dermal exposure, a mass of agent comes into contact with the surface of the skin. If conditions allow, the agent begins to diffuse through the skin barrier towards the bloodstream for dermal dose. Any mass of agent that enters the skin is called potential dose, while any mass that enters the bloodstream is called actual dose. Once in the bloodstream, the agent is distributed throughout the rest of the body.

physical (e.g., building), or biological (e.g., bacteria) agent present in an environmental carrier medium, such as air, liquid, or soil (Zartarian 1996). Chemical exposure is specifically defined as the contact of an exposure boundary (i.e., skin, mouth, nasal passage) of a human target with a pollutant concentration (Duan, Dobbs, and Ott 1990; Zartarian, Ott, and Duan 1997; see Chapter 2). An environmental carrier medium is often called the *vehicle* of transport, and a chemical agent is always associated with one or more vehicles (e.g., water, air, soil, chemical formulation).

A dermal exposure analysis attempts to determine how much of an agent comes into contact with the skin via a carrier medium and how much of the agent remains on the skin surface. The magnitude of an individual's dermal exposure is dictated by: (1) the duration and frequency of contact with surfaces and objects in the environment (i.e. personal activity patterns), (2) the concentrations of chemicals on the surfaces and objects contacted, and (3) the transfer rates of chemicals from surfaces and objects to the skin during the contact events.

A dermal dose analysis attempts to determine the mass of a chemical agent that has penetrated the target via the exposure/contact boundary, the skin (see Chapter 2). *Percutaneous* absorption is another commonly used term for dermal dose, especially in the medical field. Technically, when the pollutant mass of the agent enters the skin, but has not yet entered the bloodstream beneath the epidermis, it is called the potential dose; it becomes an actual dose (or absorbed dose) after entering the bloodstream. There is reason to believe that most of the chemical that enters or is retained in the skin will eventually be systemically absorbed with time, unless there is permanent binding or metabolism of the chemical in the upper layers of the skin. From a dermal exposure estimate of the mass of an agent on the skin surface, calculations can be made of the mass of the agent that enters the skin over time. The dermal intake dose estimate is equal to or less than exposure mass at the skin surface, and is affected by skin, chemical, and environmental factors. The delivered portion of the absorbed dose (i.e., dose to a target tissue or organ as determined by the processes of distribution, metabolism, and elimination) can ultimately result in a health effect.

11.5 THE HUMAN SKIN

11.5.1 General Skin Structure

The dermal exposure boundary, the skin, is a complex, multilayer, multipathway, biological membrane that requires special consideration in order to understand the process of dermal absorption. Being the largest organ of the body, the skin surface area of a male adult is close to 20,000 cm^2, while the mass of the skin accounts for over 10% of our total body mass at around 7 kg for a 65-kg adult (Roberts and Walters 1998). With a thickness between 0.5 to over 4 mm, depending on the area of the body, the skin consists of an outer epidermis, an inner dermis layer, and an underlying subcutaneous layer (Figure 11.2). The epidermis is a stratified, *squamous, keratinized*, thin, *avascular* layer that consists of four to five cell layers (*stratum germinativum, stratum spinosum, stratum granulosum, stratum lucidum, stratum corneum*), depending on the body site. The last four layers of the epidermis are commonly lumped together and called the viable epidermis (VE), while the outermost layer, the *stratum corneum* (SC) or *horny layer*, is considered separately as the most impermeable barrier to absorption of chemicals due to its dried and hardened structure. The dermis, on the other hand, consists mainly of dense irregular connective tissue surrounded by *collagen*, elastic, and *reticular fibers* embedded in an *amorphous* ground substance. The dermis consists of two layers, the *papillary layer* and the *reticular layer* (Figure 11.2). In the papillary layer, the *collagen* and *elastin* fibers are folded in ridges or papillae, which extend into the epidermis (most noticeable in the palms of hands and soles of the feet; Figure 11.2). These undulating papillae increase the surface contact between the epidermis and dermis, facilitating the diffusion of nutrients, growth factors, and xenobiotics for the avascular epidermis.

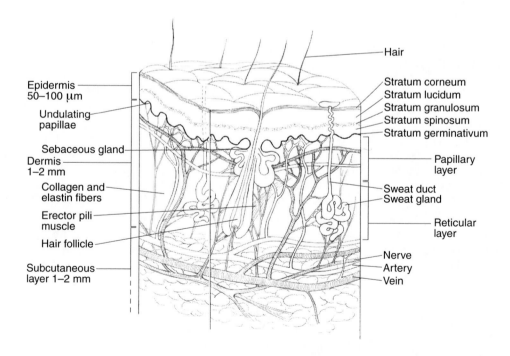

FIGURE 11.2 Complex structure of the skin. The human skin is the largest organ of the body, exhibiting a complex heterogeneous structure. The three main layers of the human skin are the epidermis, dermis, and subcutaneous layers. These layers possess varied structural and physiochemical features that allow the skin to metabolize compounds, detect pain and touch, regulate temperature, protect itself from ultraviolet radiation, give mechanical support, and regulate sweat and sebum secretion.

The subcutaneous layer contains *adipose* tissue, nerve fibers, and blood vessels supplying the skin *capillaries* and *lymphatics*. There is an increase in thickness of the layers of the skin from the surface of the skin toward the bloodstream. Additionally, the skin's thickness varies throughout the body as a function of body site, age, and sex and is an important factor when considering dermal absorption (Whitton and Everall 1973). Appendages on the skin surface are no more than 0.1% of total body surface area, and include sweat ducts (400 glands per cm²), hair follicles (300–500 per cm²), and nail (Roberts and Walters 1998), through which chemicals can be absorbed. *Sebaceous* glands are found along the shaft of hair follicles, and there are one to two glands per hair follicle, which vary in size from 200 to 2,000 μm in diameter.

11.5.2 General Skin Function

Acting like a protective waterproof layer, the skin has many functions, including regulation of body temperature, metabolism, drug biotransformation, and use of sensory nerve endings to detect changes in environmental conditions (Monteiro-Reviere 1996; Chuong et al. 2002). The epidermis allows the exchange of warmth, air, and fluids, is resistant to damage, provides mechanical support, prevents bacterial invasion and evaporation, and contains melanin for skin color and ultraviolet protection. The dermis gives the skin its strength and elastic properties. Sweat and sebaceous glands that originate in the dermis also provide the acid mantle (pH of 5.5), which is a natural film of sebum and sweat that protects the skin from outside attack and may have some effect on the adherence of chemicals to the surface of the skin. The dermis also contains countless tiny blood vessels. These blood vessels: (1) feed the outermost epidermal layer above the dermis; (2) absorb substances applied to skin; and (3) contain important enzymes to break down or inactivate toxic substances (e.g., esterase enzymes). Feeding, excretion, heat exchange, metabolism, and insulation are found in the subcutaneous layer. The subcutaneous layer also anchors the dermis to the underlying muscle or bone.

11.5.3 The Function and Structure of the Stratum Corneum (SC)

Because the SC has unique structural characteristics and is believed to be the main barrier to absorption (USEPA 1992), additional discussion of this layer is warranted. Originally considered a disorganized, nonfunctional layer before the 1950s, the SC is now recognized as being a meta-bolically active, compartmentalized layer (Kligman 1964). In general, the functions of the SC are to retain body fluids, to prevent the disruption of living cells by water or harsh environmental chemicals, and to protect tissues from fatal drying and osmotic damage from bathing. While performing these functions, the SC must still remain thin enough to be flexible and plastic (Kligman 1964). The SC provides a stable environment; it contains insoluble cell membranes, matrix-embedded fibers, specialized desmosome junctions between cells, and intercellular cement. The desmosome junctions between cells are keratin, *cytoskeletal* structures that attach cells creating a tissue very resistant to shearing forces (Roberts and Walters 1998). The SC's importance in the prevention of chemical absorption can be illustrated by considering the oral mucosa, which lacks a SC. The oral mucosa is highly penetrable; drugs are often administered by this route for quick absorption (Kligman 1964).

The physical appearance of the SC has often been called a brick and mortar structure, with the brick being the protein *corneocytes*, and the mortar being the lipid intercellular regions between the corneocytes (Figure 11.3). The SC cells are non-nucleated, fused, flattened, squamous cells filled with keratin fibers. The protein corneocyte cells comprise 99% of the SC and are stacked almost vertically in 15 to 25 layers, making the SC layer 10 to 20 μm thick (Roberts and Walters 1998). The SC cells are thicker in areas of the body that are subjected to frequent direct interaction with the physical environment (e.g., the palms of hands and soles of feet). A typical SC corneocyte cell is 0.8–1.0 μm thick and 25–45 μm in diameter (30 times as wide as thick; Mershon 1975).

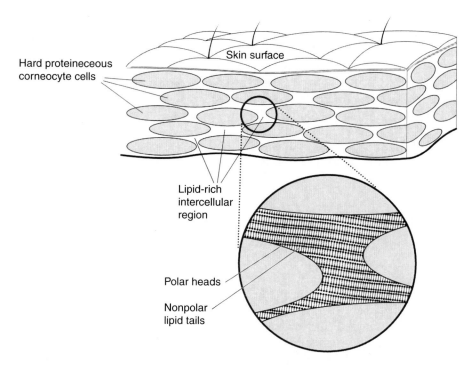

FIGURE 11.3 Brick and mortar structure of the SC and bilayer structure of the intercellular lipids. The stratum corneum is the topmost layer of the viable epidermis and is comprised mostly of protein-rich corneocyte cells that are surrounded by a lipid-rich intercellular region. The corneocytes tend to be stacked vertically, and slowly make their way to the surface of the skin in a shedding process called desquamation. The SC content is 70% protein, 15% water, and another 15% is lipids and other materials.

Each SC cell is made of about 70% insoluble bundled keratins and about 20% lipid encased in a cell envelope (Roberts and Walters 1998). Macromolecular protein fibers dominate in the envelope as well as in the contents (Kligman 1964). The lipid corneocyte envelope resists the passage of water and other small polar molecules in and out of the corneocyte, while playing a role in SC cohesion (Moghimi, Barry, and Williams 1999)

The intercellular region is about 150 Å wide in most tissues but may reach 400 Å between keratinized cells (Mershon 1975), and is many times greater than the internal or external surface of the epidermis. The intercellular lipids are arranged in *lamellar* sheets, which are paired bilayers formed from fused lamellar granule disks (magnified in Figure 11.3). In essence, the polar regions of lipids are attracted to each other and dissolved in an aqueous layer while the hydrocarbon, lipid, non-polar regions mirror each other on the other side of the bilayer (Moghimi, Barry, and Williams 1999). No single component provides the barrier properties in the SC intercellular domain, but experiments establish that the order and physical state of the lamellar structure as described above is essential. Apart from the continuous lipid bilayers of the SC intercellular domain that provide a *diffusional* barrier, a geometrical barrier also exists due to the tortuous pathway around corneocyte cells that chemicals are believed to mostly travel to get across the SC layer.

11.5.4 Shedding and Hydration in the Stratum Corneum

The entire epidermis is said to possess a cohesion gradient; as the cells move upward to the surface of the skin they acquire a tough envelope, and the cytoplasm in the cells becomes progressively packed with more and more consolidated fibrous keratin. The maximum cohesion gradient is reached in the lower SC; the cohesion gradient is reversed in the upper SC, and cell units are able to be destroyed as forces holding them together become weak as *desquamation* takes place (Kligman 1964). Corneocyte cells take 1–6 weeks to reach the surface, resulting in complete renewal of the SC in 15 days. Adults may shed 3.5 g/m²-day from the palm and about 0.2 g/m²-day from the upper arm (Kligman 1964). Chemicals may be lost from the skin surface in the process of shedding.

The density of the SC is about 1.4 g/cm³ in the dry state; hydration is only about 15–20%, compared with the rest of body at 70% hydration. Water may be lost by sweating or by diffusion, where this loss is dependent on conditions of airflow, temperature, humidity, surface characteristics (e.g., *occlusion* and immersion) and thickness of the SC. The average adult has a total transepidermal water loss of 85–170 mL/day under normal conditions. Due to sweating, water loss can increase to 300–500 mL/day (Idson 1978). The general belief is that increased hydration causes an increase in skin absorption.

11.6 FACTORS AFFECTING DERMAL DOSE

Once a chemical agent contacts the skin surface, there is potential for the chemical agent to be absorbed through the skin. Although the SC is said to be the rate-limiting barrier to absorption of chemicals applied to the skin (Monteiro-Riviere 1996; Wertz 1996) permeation can occur, mainly through the intercellular lipid domain of the SC (Malkinson 1964), and mostly for nonpolar compounds. In addition, hair follicles and sebaceous and sweat glands provide possible channels through the skin, especially for ions and large polar molecules; some chemicals therefore bypass the rate-limiting SC barrier (Moghimi, Barry, and Williams 1999). Polar compounds (e.g., organic ions) are also hypothesized to travel by special polar pathways such as keratinized protein cell remnants and polar head regions of the lipid domain (Sznitowska and Berner 1995).

In general, three major variables may account for the chemical rate and amount of penetration through the skin: (1) the concentration of chemical applied, (2) the partition coefficient of the chemical between SC and vehicle, and (3) the diffusivity of the chemical within the vehicle and within the skin. These three variables are in turn controlled by other chemical factors, skin factors, and environmental factors (Figure 11.4). The chemical-related factors affecting absorption through the skin include a chemical's lipid and water solubility, molecular size, volatility, and chemical configuration (Malkinson 1964). The skin-related factors include the physiology of the skin (e.g., metabolism, natural psychological changes in blood flow), the anatomy of the skin (e.g., variation in number of cell layers and thickness of layers due to anatomy location or aging skin), and the condition of the skin (e.g., disease, damage, physical injury, skin exposure) (Jackson 1990). The mechanical properties of the skin (e.g., elasticity) are also important in the skin's protective function (Marks 1983). Some environmental factors affecting absorption include exposure conditions (e.g., fluctuation in chemical mass loadings, occlusion and residence time on skin surface), temperature, humidity, and vehicle properties. Occlusion, humidity, perspiration, and high external temperatures can lead to an increase in chemical absorption by causing an increase in skin hydration (Chang and Riviere 1991; Jewell et al. 2000; Poet et al. 2000; Zhai and Maibach 2001; Pendlington et al. 2001). Increased perspiration, heart rate, body temperature and circulation, on occasion caused by vigorous exercise, can also increase dermal absorption (Williams, Aston, and Krieger 2004).

Reactive vehicles added to pesticide formulations such as *organic, aprotic* solvents (e.g., dimethyl sulfoxide, DMSO), and surfactants (e.g., sodium dodecyl sulfate) can affect solute *permeability* by damaging the skin lipids or increasing the chemical agent's ability to spread over the skin (Scheuplein and Blank 1971; Schaefer, Zesch, and Stuttgen 1982; Menzel 1995). Studies reflect that the permeability coefficient of a chemical for absorption through the skin correlates with its percentage saturation in the vehicle, its partitioning from the vehicle to the epidermis, and its likely new diffusion rate along the epidermal pathway caused by vehicle changes to the epidermis (Hilton et al. 1994; Roy, Manoukian, and Combs 1995; Sartorelli et al. 1997; Selim et al. 1999; Jepson and McDougal 1999; Riviere et al. 2001; Rosado et al. 2003).

Three Main Factors Affecting Percutaneous Absorption

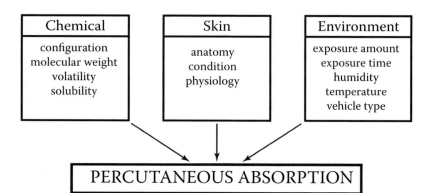

FIGURE 11.4 Compound, skin, and environmental factors, in addition to the complex reactions between all three, affect the percutaneous absorption of the chemical. Insufficient data exist today to clarify the exact relationship between each factor and percutaneous absorption. In general, relatively small lipophilic, nonvolatile chemicals are able to penetrate the skin more readily, absorption is increased through skin that is thin or damaged, and increased temperature and humidity and harsh vehicles increase the rate of absorption.

Significant skin dynamics affecting the absorption of chemicals include skin surface properties and metabolism. Skin surface properties may affect the adherence of chemical agents and vehicles and consequently the mass transfer or diffusion of the agent through the skin. For example, moisture on the skin may increase the adherence of a *soil particle* that has a pesticide concentration. Skin surface properties may vary from person to person (e.g., by age, sex, or race), from one anatomical site to the next (e.g., torso, palm of hand, scrotum) or because of differences in environmental conditions. A few researchers in the cosmetic industry have attempted to define various surface properties of the skin such as roughness, scaliness, skin surface pH, and hydration (Moghimi, Barry, and Williams 1999; Randeau et al. 2001; Eberlein-Konig, Spiegel, and Przybilla 1996; Eberlein-Konig et al. 2000; and Manuskiatti, Schwindt, and Maibach 1998).

Metabolism is the process where chemicals are converted to other chemicals by the reaction and interaction with enzymes. Viable skin contains most of the metabolizing enzymes; xenobiotics-metabolizing enzymes, for example, are thought to be mostly in the *basal* layer of the epidermis (Jewell et al. 2000). *Acetylation*, hydrolysis, alcohol oxidation, and reduction are some of the metabolic processes that have been demonstrated in the skin (Bronaugh et al. 1999). Bashir and Maibach (1999) claim that metabolic enzymes in the skin primarily act on *lipophilic* chemicals and convert them to *hydrophilic* chemicals that are less active and can be excreted via the kidneys. In analyzing the absorption rate and reservoir tendencies of chemicals, consideration must be given to the metabolic rates and metabolic by-products of the chemicals in the skin.

11.7 MECHANISMS AND PATHWAYS FOR DERMAL EXPOSURE

Exposure is a complex phenomenon involving a number of potential chemical agents (e.g., chlorpyrifos, lead) and the vehicles or medium (e.g., water, air, soil, xylene), in which they are immersed,

Factors to Consider in the Transfer of Agent to the Human Skin

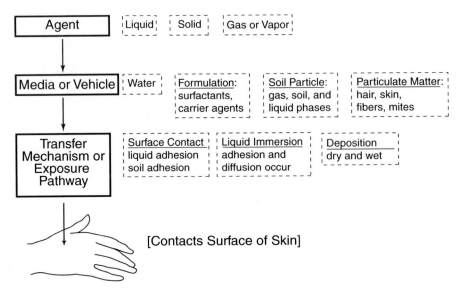

FIGURE 11.5 Conceptual model of exposure pathways. Liquid, solid, or gaseous chemical agents reside in our environment in a number of possible vehicles or media such as water, air, soil or a chemical formulation. There are a number of exposure pathways or transfer mechanisms by which agent and vehicle contact the surface of skin, and these are facilitated by our individual activity patterns. When we contact surfaces, for example, the agent and vehicle adhere to the skin surface.

that contact the skin surface through one or more exposure pathways. An exposure pathway is the means by which a source of agent contacts a human target. Each chemical agent may be present in any one or more of the following forms: pure liquid, pure solid, pure gas, dissolved or suspended in a liquid vehicle, or combined in a solid vehicle (Figure 11.5). To simplify the study of exposure, one chemical agent at a time in potentially different vehicle phases is typically studied.

According to Fenske (1993a), there are typically three exposure pathways: immersion, deposition, or surface contact of the human skin (Figure 11.5). Immersion is possible in residential scenarios with contaminated water (i.e., swimming or taking a bath) or in chemicals volatilizing from contaminated water (e.g., chloroform in hot showers). The extent of the exposure is a function of the chemical concentration, exposed skin area, and duration of contact.

Deposition results when an agent is transported to the skin as a vapor or aerosol. Large aerosols may be intentionally created, for instance, during airless spray painting or pesticide application (Brouwer et al. 2000). Aerosols may also occur incidentally, for example, from emissions of a nearby source or from dislodged pesticide residues from foliage. The extent of exposure due to deposition is a function of the skin loading rate (or deposition velocity) and the exposed skin area (Fenske 1993a).

The surface contact pathway, also referred to as the indirect contact pathway, occurs when skin touches a contaminated surface (including skin and clothing), resulting in the transfer (i.e., adhesion) of chemical residue from the surface to the skin. Exposure from this pathway is often estimated as the product of a skin loading rate and the exposed skin area. The skin loading rate is in turn due to the concentration of the chemical on the surface, contact frequency or duration, and residue transferability. Additional factors, such as contact pressure and motion, a chemical's affinity for the surface of the skin, regional differences in skin composition and conditions, work practices, and hygienic behavior, also affect the skin loading rate (Fenske 1993a). Surface contact can represent the primary exposure pathway in several occupational settings, including agricultural work (e.g., cleaning equipment or picking crops). Children are also more likely to be exposed by the surface contact pathway due to their play activities on previously sprayed surfaces (Zartarian 1996). Although not typically considered "pathways," removal mechanisms are methods by which mass can be removed from the skin surface. These removal mechanisms may include dislodgment, washing or decontamination, surface contact, vaporization, absorption through the skin, and mouthing.

11.8 DIRECT METHODS FOR MEASURING DERMAL EXPOSURE

Direct measurements of dermal exposure (to mostly non- and semi-volatile contaminants) are typically considered as skin exposure sampling techniques or personal sampling techniques (Cohen Hubal et al. 2000). Exposure sampling techniques primarily fall into three categories: (1) use of surrogate skin, (2) chemical removal, and (3) fluorescent tracers (McArthur 1992). Surface sampling is frequently considered another dermal exposure sampling technique and its applicability is discussed below. Some of these methods have been tailored to assess dermal exposure to volatile chemicals, such as the use of charcoal cloth surrogate skins (Cohen and Popendorf 1989). It is important for sampling to be representative by considering the varying factors in designing a sampling strategy and capturing the variability of dermal exposure. This implies, on a detailed level, that all skin surfaces with the potential for dermal exposure should be identified and measured (Fenske 1993a). In some cases the important portions of the body may not be obvious, and there may be a need to review the potential fate and transport of the contaminant. On a larger scale, sampling should be representative of the population of interest (e.g., a children's study should give consideration to a child's microenvironments and play and eating behaviors), and feasible for the study group (e.g., intentional exposure to toxic chemicals may not be practical for children). For all skin exposure sampling techniques not only is consistency in the application method important for interpreting the dermal exposure measurements, but the analytical procedures for extraction and analysis of the contaminant mass are also crucial for data quality.

11.8.1 SURROGATE SKIN TECHNIQUES

Surrogate skin techniques involve placing a collection medium against the skin prior to an exposure and analyzing the collection material for its chemical content after the exposure. Ultimately the realism of this method depends on the ability of the surrogate skin to mimic the adherence of contaminants to actual skin. Both overestimates and underestimates of dermal exposure have been reported (McArthur 1992; Whitmore et al. 1994). Two approaches have been used: *patch samplers* covering a small surface area of skin, and *garment samplers* covering larger regions. Patch samplers can be considered spot or grab samples of the mass of contaminant on a body region (Fenske 1993a), and typically consist of several layers of surgical gauze and a cellulose paper backing (McArthur 1992). In typical studies, between 3 and 8% of the body surface is represented by patches (depending on the number of patches). Exposure is then calculated by extrapolating the mass of contaminant on the patch to the surface area of the anatomical region (Soutar et al. 2000). Patch sampling can serve as a simple and cost-effective method for a preliminary investigation of dermal exposure if the limitations of spatial variability in mass loading are taken into account (Fenske 1993a).

Use of garment samplers or dosimeters (e.g., cotton, nylon, and leather gloves; T-shirts; plastic boots; and entire disposable overalls) may help to overcome some of the limitations of patch sampling by collecting chemical mass over entire anatomical regions (McArthur 1992). Absorbent, cotton gloves, for instance, are used to estimate hand exposure during contact with equipment and in harvesting or spraying of crops (Whitmore et al. 1994). Whole-body samplers have been used to assess occupational and residential pesticide exposure (Krieger and Dinoff 1996; Krieger et al.

2000), worker exposure to antifungal products during wood treatment (Fenske 1993a; Ross et al. 1990), exposure during spraying for coating boats, mixing and diluting in laboratories, and exposure during wiping of biocides in hospitals (Hughson and Aitken 2004). However, garment samplers may be cumbersome to use and few standard methods exist as to the absorption properties of the media.

11.8.2 REMOVAL TECHNIQUES

Washing, rinsing, or wiping can remove chemicals deposited on skin after an exposure event. During bag rinsing, one hand is immersed in a solvent or water-surfactant solution contained in a plastic bag and the hand is shaken vigorously for a fixed time or a fixed number of shakes. Hand washes are similar in manner, except the subject is asked to wash his or her hands in a routine fashion or a set procedure (Brouwer, Foeniger, and Van Hemmen 2000). The remaining liquid after the rinse or wash is collected and analyzed for chemical mass. While the wash or rinse procedure could be standardized to ensure operator independence, a standard approach has not been widely adopted, and few sampling efficiency tests have been conducted. Factors that may play a role in standardization of such procedures include the elapsed time between contamination and start of sampling, skin loading, wash time, and type of solvent or solution (Brouwer et al. 2000). Hand rinses have predominantly been conducted in the study of children's exposure to residential contaminants (Lioy et al. 2000; Pellizzari et al. 2003; Shalat et al. 2003; Freeman et al. 2004).

Skin wiping, which could theoretically be applied to a larger skin area than just the hands, is defined formally as "the removal of contaminant mass from the skin by providing manually an external force to a medium that equals or exceeds the force of adhesion over a defined surface area" (Brouwer, Foeniger, and Van Hemmen 2000). Materials that have been used in skin wiping procedures include cotton fabric wipes, cotton balls, commercial wetting sponges, and cotton surgical pads (Bradman et al. 1997; Lewis et al. 2001; Fenske et al. 2002b; Wilson et al. 2003, 2004). Wetting agents include alcohols, surfactants, and water. Unlike hand washing or rinsing methods, which could be standardized, skin wiping procedures are inherently operator-dependent and include several components of variability (Fenske 1993a). Factors of concern include the number of contacts or passes over the wiped area, applied pressure, efficiency in wiping hard-to-reach areas (e.g., around fingers, mass that has already penetrated the skin), and the surface area of the media utilized (McArthur 1992; Fenske et al. 1998; Brouwer, Foeniger, and Van Hemmen 2000).

11.8.3 FLUORESCENT TRACER TECHNIQUES

Fluorescent tracers provide a means of directly and non-invasively assessing dermal exposure by quantifying deposition of fluorescent materials on the skin. Although some substances have a natural fluorescence, a suitable tracer is typically added to the source to provide the fluorescent property. Video images of the skin prior to exposure are captured, and a standard curve relating the intensity of fluorescence to a quantifiable chemical concentration is developed (Fenske 1993a,b). Post-exposure (i.e., after the exposure event to the source of interest) the skin is then held under a long-wave ultraviolet light, and images capture the exposed body parts. An indication of relative exposure can be obtained by comparing pre- and post-exposure images. A quantitative estimate of exposure can then be obtained by comparing post-exposure images to the developed standard curves.

The fluorescent tracer technique has primarily been applied to assess mixers' and applicators' dermal exposure to pesticides (Fenske et al. 2002a; Fenske 1990). Methods have also been developed to simulate children's exposure to pesticides in the residential setting (Fenske 1993a) and to investigate exposure to nonvolatile chemicals during spray painting (Brouwer et al. 2000) and to pharmaceuticals in a hospital (Kromhout et al. 2000). As illustrated in these last two examples, qualitative use of tracers can also serve to identify transfer processes of agents, discover dermal exposure patterns, recognize variability of exposure among worker groups, evaluate worker practices and hygiene, and help educate and train workers. This technique has limitations, however. Use of a tracer, for instance, requires the introduction of a foreign substance into the chemical of interest that could be incompatible with the agent. Also, the tracer may not perfectly mimic the properties (e.g., volatilization, transfer, binding) of the agent (Fenske 1993a). The tracer methodology is also limited by the number of persons for whom measurements can be made at any one time, which often leads to a small sample size and restricts estimates of variability in dermal exposure. Finally, quantitative assessments using fluorescent tracers could be relatively costly (Cherrie et al. 2000).

11.8.4 SURFACE SAMPLING TECHNIQUES

A fourth method for assessing dermal exposure is to sample a contaminated surface with a material that can be analyzed to measure residues that are available for uptake by human skin during a contact event. Since the mass on the skin is not analyzed, this technique is an indirect or potential estimate of dermal exposure. Often the goal is to quantify what are called removal, transferable, or dislodgeable residues. Thus the interest is not in estimating the mass of contaminant on the surface but those residues that are likely to be transferred to human skin.

Instruments and techniques vary for determining dislodgeable residues. The simplest is wipe sampling, consisting of surgical gauze pads similar to those that may be used as surrogate skin for direct exposure assessment, moistened with water or a solvent solution (McArthur 1992; Fenske et al. 1990, 2002b; Lu and Fenske 1998; Lewis et al. 2001). Specific devices for sampling via wipes, such as the Lioy-Wainman-Weisel (LWW) wipe sampler (Lioy et al. 2000) have also been developed. Lioy et al. (2000) have also used a surface press sampler called the Edwards and Lioy (EL) sampler method. The Dow drag sled is another device that is used by dragging a weighted block through a fixed surface area (Camann et al. 1994). A removable piece of denim cloth is on the underside of the block and collects the contaminant mass, while a weight is placed on the top of the block to simulate pressure exerted upon contact (Figure 11.6). Another device is the California cloth roller, which calls for a weighted foam covered roller to be rolled over a sheet of cotton/polyester cloth, which is placed over the sampling area (Camann et al. 1994; Krieger et al. 2000;

FIGURE 11.6 Dow sled (Dow Chemical Co., MI). The Dow drag sled is a surface sampling technique used to determine mass of dislodgeable chemical residue. The weight is used to mimic the pressure dynamics during contact, while the cloth collects the residue. 1. Weight (3.6 kg); 2. Sled (7.6 cm square × 2–4 cm high, typically made of wood cover with aluminum foil); 3. Collection medium (undyed denim cloth, 8 cm × 10 cm); 4. Drag line (e.g., fishing line); 5. Metal starting platform for sled; 6. Guide ruler. (Courtesy of Dr. Robert Lewis, USEPA.)

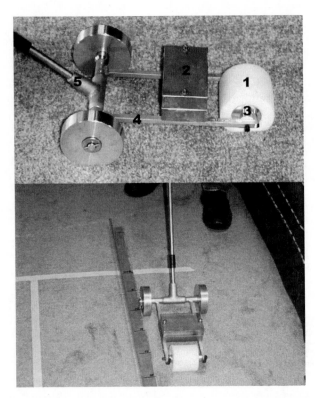

FIGURE 11.7 PUF roller (California Department of Pesticide Regulation, Sacramento, CA). The PUF roller is a surface sampling technique used to determine mass of dislodgeable chemical residue. The foam is rolled over a sampling area to collect the residue. The weight attempts to mimic the dynamics of human skin during contact with floor surfaces. 1. Collection medium sampling material; 2. Weights; 3. Roller (PVC pipe, 13 cm dia. × 63 cm long); 4. Support bar; 5. Handles. (Courtesy of Dr. Robert Lewis, USEPA.)

Williams, Aston, and Krieger 2004). The most widely used methodology, however, utilizes the polyurethane foam (PUF) roller (Figure 11.7). This instrument consists of a weighted aluminum or stainless steel roller with a polyurethane foam ring that is rolled over a specified area. The sampling material (e.g., denim, cotton/polyester cloth, PUF) is removed from each instrument and its contents are analyzed in the laboratory (Camann et al. 1994; Whitmore et al. 1994; Lewis et al. 2001).

In addition to the surface sampling techniques mentioned above, other surface sampling techniques have also been developed and implemented, more specifically for removing dust and soil particles (i.e., vacuum-dislodgeable residues) from carpets and floors (Roberts et al. 1991; Whitmore et al. 1994; Simcox et al. 1995; Bradman et al. 1997; Mukerjee et al. 1997; Lu and Fenske 1998; Lewis et al. 2001; Fenske et al. 2002b). By removing particles from deep within carpets and surfaces, these techniques (e.g., high volume surface sampler: HVS3) better represent maximum, potential chemical exposure.

One obvious issue when using surface sampling methods is selecting the appropriate technique. For the case of estimating the mass on a surface, the best technique could be determined by conducting a series of experiments in which a known mass is deposited on surfaces, and each instrument is evaluated for its efficiency in quantifying the mass. In this approach, an efficiency of 100% is the goal. However, for determining dermal exposure, we are more interested in simulating the transfer of contaminants to human skin. Therefore an estimate of the transfer efficiency of contaminants to skin is necessary. The best technique, then, involves the instrument and method-ology whose efficiency is consistently closest to that of skin.

11.9 DERMAL EXPOSURE EXAMPLES

A number of studies have been conducted in the past to determine normalized measures of chemical transferability (e.g., residue transfer coefficient and soil adherence) under various exposure scenar-ios. These studies have used one or more of the sampling techniques (i.e., removal techniques, surrogate skin techniques, or surface sampling techniques) previously mentioned. The final mea-surements can be used to make some direct assessment of the dermal exposure at that point in time due to a specific contact (e.g., a hand press tells the amount of residue or soil that might adhere to the human skin following contact with a surface), or over a specific measurement period (e.g. a dermal patch tells us the amount of residue that adheres to the skin for the time period the patch is worn). In order for the measurement of soil adherence to be considered a direct exposure measurement, the mass of chemical in the soil must first be known or determined. Often it is assumed that the skin surface is exposed to the entire chemical mass in the soil adhered to the skin. Many times these residue transfers and soil adherence measurements are used in models (i.e., an indirect exposure analysis) to assess long-term dermal exposure, given information on a person's long-term contact activity and the environmental chemical concentrations to which they are exposed.

Determining residue transfer coefficients (i.e., fractional mass transferred from a surface) requires knowledge of the original mass of residue on a surface and the mass of residue transferred to the collection device. Lack of reproducibility and consistency of techniques (e.g., hand press vs. drag sled) and exposure scenarios makes comparison difficult across studies, even for one chemical. For example, for chlorpyrifos, residue transfer coefficients have varied from 0.03 to 35% across past studies (Roberts and Camann 1989; Hsu et al. 1990; Ross et al. 1990, 1991; Williams, Aston, and Krieger 2004).

Soil adherence to the skin is determined by factors such as our activity (e.g., farming, soccer), skin properties (e.g., skin moisture, amount of hair follicles), and the properties of the soil (i.e., moisture content of the soil, particle size) (USEPA 1997). Subsequently penetration into the skin is dependent on the mass distribution of soil on the skin surface, properties of the soil, properties of the soil-bound chemical agent, properties of the skin, the residence time of the soil on skin, mass of chemical in the *soil matrix*, and environmental conditions. Soil to skin adherence has been estimated from staged (e.g., orchestrated hand presses) and unstaged (e.g., removing adhering mass from hands after normal activities) experiments. Again comparison across studies is difficult due to the variability in experimental techniques and exposure scenarios (e.g., various soil types, activities and locations, various sampling methodologies). Results of a few soil adherence studies have found that soil will adhere to the skin across various body parts at a rate of 0.01 to as much as 10 mg/cm^2 skin (Lepow et al. 1975; Que Hee et al. 1985; Driver, Konz, and Whitmyre 1989; Kissel, Richter, and Fenske 1996; Kissel et al. 1998; Holmes et al. 1999).

For a closer look at how direct exposure measurement techniques are applied, two field studies are presented below. In the Minnesota Children's Pesticide Exposure Study (MNCPES) of 102 children (aged 3–13), house dust levels of selected insecticides and a herbicide were measured using surface wipe (LWW: Lioy Wainman Weisel) and press samplers (EL: Edwards Lioy) and compared to hand rinses and urine metabolites (Lioy et al. 2000). The hand rinses were collected by placing a child's hand in a plastic bag containing 150 mL of 2-propanol and agitating the child's hand in the solution to remove any chemical residue. For chlorpyrifos, the only chemical detected in all samples, the median level measured across the study group was 0.07 ng/cm^2 for the EL surface press sampler, 0.07 ng/cm^2 for the EL carpet press sampler, 0.06 ng/cm^2 and 0.43 ng/cm^2 for the LWW surface wipe samplers (two rooms, average based on number of sample was 0.34 ng/cm^2), 0.42 ng/cm^2 for the hand rinse, and 0.03 ng/hand and 6.9 $\mu g/g$ for the urine metabolites. As authors Lioy et al. expected, the values from hand rinses were greater than those from the EL press sampler; the hand rinses represent more than one contact, whereas the EL sampler contacted the surface of the carpet or floor once. The EL sampler was consistent across the floor and carpet surfaces, giving similar measurements. Although the LWW wipe sampler was designed to consistently collect the entire residue on a surface, results logically showed that the location of sampling was important in determining the concentration level. This is attributable to the varied deposition rate of chemical residue after application varied across rooms. In this study, the chlorpyrifos levels measured in hand rinse and dust samples did not correlate with the urine metabolite measurements. In general, the contact dynamics for the EL, LWW, hand rinses and even interpretation from the metabolite collection were too variable for any meaningful comparison. This study highlights the difficulty of drawing conclusions using the varied direct dermal exposure measurement techniques.

In a study to evaluate the efficiency of chemical protective clothing for farmworkers in the citrus groves of Central Florida to the organophosphorus insecticide Ethion, Fenske et al. (2002b) utilized the fluorescent tracer and the patch sampling techniques. Both techniques were used to determine the loading on the skin surface following Ethion penetration into the clothes. Two traditional woven garments (i.e., cotton work shirt/workpants and cotton/polyester overalls) were compared with two nonwoven polypropylene and polyester/wood pulp fabrics. Eight replicate exposures of each garment were conducted. For the tracer technique, a premeasured bag of the fluorescent tracer was mixed with the Ethion formulation. Workers used this mixture in the field under their normal application process and pre- and post-video images of the workers' hands, neck, forearms, upper arms, upper torsos, and lower torsos were compared. For the patch technique, four 103.2 cm^2 square alpha-cellulose patches were pinned to the protective clothing of each worker on the inside and outside of both thighs. After the spray events, these patches were removed and sent for analysis.

The fluorescent tracer video-imaging method proved useful in detecting exposures through garment openings (e.g., neck and sleeves) and, therefore, in comparing the four garment types. However, this technique was not accurate in determining concentration levels. The patch technique, on the other hand, was far more sensitive in determining low levels of tracer on the skin. For the patch technique, the percent penetration was calculated by comparing the inner patch sample value to the outer patch sample value. The mean penetration ranged from 4.7–7.2% and did not differ significantly among the four garment types. The fluorescent tracer video-imaging technique revealed that penetration variability within each garment group was high. This was due to the varying amount of pesticide and tracer ultimately reaching the garment (i.e., individual technique, spray time, and deposition rates varied for each field application). The tracer measurements were then normalized based on the head exposure of the workers who had no protective gear. The normalized tracer measurement showed that the forearms had the highest exposures for all garments and that the woven garments had less exposure than the nonwoven garments. These differences were due not only to penetration but also to deposition through the sleeves.

11.10 DIRECT TECHNIQUES FOR MEASURING ABSORPTION

Various experimental techniques have been developed in the last century for the assessment of percutaneous absorption of xenobiotics, and can be classified into *in vitro* and *in vivo* methods. *In vitro* test methods use animal organs to assess chemical uptake, whereas *in vivo*, clinical methods use live animals (even humans) to assess chemical uptake. Various species have been used to study chemical diffusion through human skin. Guinea pigs, pigs, and monkeys appear to exhibit similar skin permeability characteristics to those of humans. Data from *in vitro* studies are often used to predict *in vivo* conditions for a number of reasons. *In vitro* methods, for example, offer measurement data under conditions where the chemicals are too toxic to test ethically on live animals, especially humans (Bronaugh 1996, 2000). Also, measurement data on chemical absorption rates through the skin can be deduced more directly from *in vitro* studies using just the skin. In contrast, for *in vivo* studies, absorption rates through the skin are deduced from excretion rates in urine/feces or expired

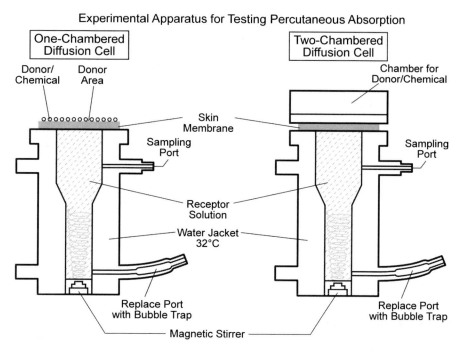

FIGURE 11.8 One- and two-chamber diffusion cells. Two types of experimental chambers are typically used to measure the diffusion rate of chemicals through skin. In the one-chambered diffusion cell, a finite amount of chemical is placed on the donor area, left open to the environment, and allowed to diffuse through the membrane into the receptor solution below. In the two-chambered diffusion cell, the above chamber contains an infinite dose of the penetrating chemical to maintain constant concentration. The chemical is then allowed to diffuse into the receptor chamber below. (Courtesy of Bayard Colyear, Stanford medical illustrator. [Based on drawings originally created by Alesia Ferguson.])

air, and as a result complete assessment requires sufficient understanding of uptake, distribution, excretion of the chemical agent in the body, and time scales for all processes. Nevertheless, the interpretation of *in vitro* measurement data for understanding dermal absorption for a particular human exposure scenario relies on the use of appropriate methodologies (e.g., experimental apparatus, skin thickness and type, finite dose vs. infinite dose of chemical, and accounting for biochemical changes such as desquamation; Franz 1978).

11.10.1 *In Vitro* Methods

In vitro experiments are often carried out in diffusion cells (e.g., Franz cells; Franz 1975) that consist of two or more chambers in which the skin is always positioned above a receptor chamber. Two main types of diffusion cell designs are used to assess percutaneous absorption. These are the one-chambered and two-chambered basic cell types (Figure 11.8). In the one-chambered cell, the skin membrane is open to the environment, and relatively small amounts of a penetrating substance are applied to the skin-air side, not unlike what would be expected in most everyday exposure scenarios (e.g., application of a cream with finite dose of an active ingredient). On the other side is a chamber containing receptor fluid into which the penetrating substance diffuses. This receptor fluid is continually stirred, and samples are extracted through a side arm for the determination of absorption rates. In the two-chambered cell type, the skin membrane is exposed on both sides to two chambers of solutions (Figure 11.8). The above chamber contains a solution of the penetrating chemical in infinite dose to maintain constant concentration while the chamber below contains the receptor fluid for collecting the penetrating chemical. Both chambers can be stirred continually.

This condition is conducive for studying steady-state diffusion through cell membranes. Note that larger volumes of solution containing infinite dose can also be used in the one chamber cell to determine steady-state absorption kinetics (Bronaugh 1996). Occasionally in the two-chambered design, the chambers are laid out horizontally, instead of vertically as in Figure 11.8 (Tojo, Chiang, and Chien 1987).

For very volatile chemicals, consideration is given to their evaporation rate from *in vitro* diffusion chambers. Some chambers have short walls that may prevent air currents that increase evaporation, and some chambers have been built to collect and account for the evaporating material (Diembeck et al. 1999). Temperature should be kept at *in vivo* conditions (30–32°C). This can be achieved by using water jackets or temperature-controlled incubators. The amount of chemical in the receptor fluid should be less than 10% of the chemical's saturation point to prevent any chance of back penetration into the skin (Diembeck et al. 1999). Additionally, consideration of the properties of the diffusing chemical and, in particular, the chemical's hydrophilic and lipophilic properties, should help dictate the appropriate receptor fluid to be used in the *in vitro* study. The type of membrane used in an absorption study should adequately mimic the properties of human skin to be a useful assessment of *in vivo* absorption. When an actual skin membrane is used in the diffusion cell, it can be either whole, split, or epidermal sheets and will vary with the intended use of the study. For risk assessment purposes, what diffuses through and what is retained in the skin at the end of the exposure period should all be considered as being systemically absorbed. Given sufficient time, the mass retained in the skin will eventually make its way to the bloodstream (Bronaugh and Hood 1999; Bronaugh et al. 1999). Sometimes, however, the chemical is preferentially soluble in skin and may remain there for permanent binding or metabolism (Yourick et al. 2004). It is also possible that some of the chemical can resurface, be wiped off, or vaporize.

11.10.2 *IN VIVO* METHODS

Skin viability might be an important concern in percutaneous absorption, so *in vivo* methods are sometimes used. The chemical and physical properties of the vehicle, chemical agent, and the SC will reveal the initial absorption in the skin. However, the vitality of the complete skin structure will determine the metabolism, distribution, and excretion of the chemical in the layers of the skin, in the systemic circulation, and in the body (Wester and Maibach 1999). Numerous small animals are used for *in vivo* studies (especially where the chemical of interest is too toxic for humans). During *in vivo* studies there are a variety of ways, some more direct and precise than others, for obtaining an estimation of the amount absorbed through the skin. After application of a radioactivity-labeled chemical under study, the radioactivity of a chemical in the blood or excreta can be used as an estimation of the amount of chemical absorbed. Estimation by these methods, however, does not account for metabolism within the skin. Surface recovery and surface disappearance can also be used to estimate the amount of chemical absorbed after topical application (Bronaugh et al. 1999). Surface recovery is based on the assumption that what is not recovered topically is absorbed, whereas surface disappearance requires appropriate instrumentation to analyze the mass of the chemical that has moved into the skin from the surface. A technique for surface disappearance, successive tape stripping to remove layers of skin, determines the amount of chemical in the SC after a short exposure period and consequently an extrapolation is made between SC reservoir and percutaneous absorption (Rougier 1999; Bronaugh et al. 1999). Other less reliable means of measuring percutaneous absorption involve the use of a biological/pharmacological response in the body to the presence of a chemical. Newer techniques for measuring the absorption of chemicals into layers of the skin and sometimes into the bloodstream, with application for *in vivo* and *in vitro* analysis, include the isolated perfused porcine skin flap (IPPSF) (Chang, Dauterman, and Riviere 1994), microdialysis (McDonald and Lunte 2003) and various spectroscopy and fluorescence microscopy techniques (Yu et al. 2003; Notinger and Imhof 2004).

11.11 HIGHLIGHTED DERMAL DOSE EXAMPLES

There are a number of experimental studies that illustrate what classes of chemicals are able to penetrate the skin and to what extent (e.g., Wester et al. 1983, 1993; Iwata, Moriya, and Kobayashi 1987; Bucks, Wester, and Mobayen 1989; Barber et al. 1992; Moody et al. 1992; Wester and Maibach 1992; Ademola et al. 1993; Chang and Riviere 1991; Chang, Dauterman, and Riviere 1994; Wester 1995; Baynes, Halling, and Riviere 1997; Benech-Kieffer et al. 2000; Zhao, Singh, and Singh 2001; Turner, Saeed, and Kelsey 2004). Chemicals that have been previously studied include metals, pesticides, steroids, alcohols, and some other organics that are used or encountered systematically in the immediate environment (e.g., cosmetics, jet fuel, vinyl chloride). Studies show that because the SC consists of 75–80% moderately lipophilic materials, the chemicals that effectively permeate the skin are those compounds that are also moderately lipophilic but also possess hydrophilic regions (Rozman and Klaassen 2001).

In one particular study experimenters progressively increased the hydrophilicity of progesterone (a lipophilic steroid) by incorporating one or more hydroxy groups at different positions in the skeletal structure (Tojo, Chiang, and Chien 1987). Then they measured the steady-state permeation of these different chemicals across intact skin (SC + VE) and stripped skin (VE only) of the hairless mouse in order to evaluate the diffusivity and solubility of the chemical in the two layers of the skin. Table 11.1 shows the results from this study. Progesterone, a lipophilic chemical, diffuses across the intact skin more rapidly than hydrocortisone, a hydrophilic chemical; permeation rate of progesterone was 2.37 µg/cm²-h compared to 0.15 µg/cm²-h for hydrocortisone. On the other hand, once the SC is removed, the chemical diffuses across the hydrophilic VE much more slowly than hydrocortisone; permeation rate of progesterone was 3.62 µg/cm²-h compared to 7.26 µg/cm²-h for hydrocortisone. Therefore, according to this study, although the SC presents the primary resistance to diffusion, lipophilic chemicals are able to penetrate this barrier more easily. In addition, the solubility of the penetrating chemical in the membrane is the most important parameter for controlling the permeation rate; permeability steadily increases with chemical solubility in the skin. The position of the added OH groups in the skeletal structure also plays a role in the permeation rate of the chemical.

In another experimental study, the absorption of four pesticides with different properties was evaluated in the IPPSF (Chang, Dauterman, and Riviere 1994). The IPPSF is an anatomically intact

TABLE 11.1
The Estimated Absorption Profile for Progesterone Derivative

Drug Name	Permeation Rate (µg/cm²h) Intact	Stripped	Lag Time (h) Intact	Stripped	Solubility in PEG Solution (µg/mL)	Diffusivity cm²/s SC ($\times 10^{11}$)	VE ($\times 10^8$)	Solubility (cm²/s) SC	VE
1. Progesterone	2.37	3.62	5.49	1.55	208	3.73	4.09	47.70	0.87
2. Prog. + OH	4.73	12.10	3.90	1.20	1380	6.62	5.28	32.40	2.26
3. Prog. + OH	0.57	5.13	6.64	0.96	653	1.42	6.60	12.50	0.79
4. Prog. + OH	0.29	1.92	4.55	1.03	121	3.42	6.15	2.82	0.32
5. Prog. + 2OH	0.97	8.27	5.40	1.45	1840	3.84	4.37	7.97	1.94
6. Prog. + 2OH	0.31	2.44	4.16	0.99	457	4.03	6.40	2.46	0.39
7. Prog. + 3OH, Hydrocortisone	0.15	7.26	5.50	1.44	2590	3.89	4.40	1.09	1.69

Prog.: Progesterone, PEG: polyethylene glycol

(From Tojo, Chiang, and Chien 1987. With permission.)

TABLE 11.2
The Absorption Dynamics for Four Pesticides in the IPPSF

Compounds	Carbaryl	Lindane	Malathion	Parathion
Molecular Weight (g/mol)	201.23	290.85	330.4	291.27
Vapor Pressure (mPa)	0.181	5.6	0.45	0.76
Water Solubility (ppm @200^{0}C)	100	6.6–11	145	24
Henry's Law (Pa-m3/mol @250^{0}C)	0.00028	0.183	0.0014	0.024
Log Octanol–Water Partition Coefficient (log K_{ow})	2.29–3.46	3.6–3.72	2.7	3.83
Percent Dose 8 Hours after Application				
Vascular	0.856	0.481	0.233	2.96
Skin Transitional	0.688	0.513	0.965	5.803
Skin Reservoir	10.84	4.257	34.194	57.066
Surface	37.836	14.451	19.195	0.235
Cumulative Effluent	8.378	2.189	1.197	2.392
Lost from Surface	41.404	78.109	44.245	31.546
Sum of All Compartments	50.219	19.702	54.427	66.062

(From Chang, Dauterman, and Riviere 1994. With permission.)

viable skin preparation that maintains microcirculation and metabolism in the skin. A four-compartment linear *pharmacokinetic* model was used to stimulate the absorption dynamics, where the estimated parameters of the model were based on the cumulative flux of radioactivity appearing in the venous perfusate for each chemical. Table 11.2 shows the results for the simulated IPPSF compartment amounts for the four pesticides after 8 hours, along with some pertinent chemical and physical properties of the chemicals. Carbaryl had the highest absorption into the bloodstream with 8.4% in the cumulative effluent. This high absorption is likely due to its smaller molecular weight and reasonably high octanol-water partition coefficient. The total amount absorbed is also relatively high for carbaryl at over 50%. Lindane, on the other hand, had the lowest total amount absorbed at only 20% because most of the chemical evaporates from the surface due to its high vapor pressure and consequent tendency to be airborne. Malathion and parathion had the highest total absorbed of 54% and 66%, respectively. However, most of their mass is retained in the skin reservoir. These larger chemicals diffuse more slowly through the skin toward the bloodstream.

11.12 CONCLUSION

Dermal exposure and dose depend on contaminants in the environment, human behavior, and the dynamics at the skin contact boundary. Since the concentration of agent, the distribution in the environment, and human activity patterns are highly variable, it is likely that within-person and between-person dermal exposures will vary widely (Fenske 1993a). Skin condition and the dynamics between skin, chemical agent, and vehicle can also cause some variation in the mass of chemical agent absorbed into the body. Therefore, it is important to understand the many factors that affect dermal exposure and dose and develop appropriate tools and protocols for their measurement or estimation.

One of the focal points of this chapter was in describing the direct ways to measure dermal exposure (i.e., sampling techniques and strategies) and dermal dose (i.e., *in vitro* and *in vivo* techniques), with some specific study examples given. However, indirect methods for determining personal exposure and dose also play a valuable role, especially in cases where the direct approach may be infeasible. For example, it is difficult to get chemical concentrations from mobile sources, children may not tolerate personal monitors or patches, and monitoring a statistically representative group may be expensive (Zartarian, Ott, and Duan 1997; Ott 1990; Cohen Hubal et al. 2000). With the indirect approach for dermal exposure, activities or time allocations are collected separately from environmental concentrations. The two, along with exposure factors, are then reconstructed in a model estimating exposure (Ott 1990; Cohen Hubal et al. 2000). (Some discussion of aggregate models can be found in Chapter 20.) For dermal dose, mass loadings on the skin surface are combined with skin, chemical, and environmental factors in a diffusion model to estimate dose. These modeling approaches are sometimes desirable to overcome the difficulties of the direct approach, to permit simulations and predictions, and to serve as a tool for understanding how explanatory variables (e.g., activities, concentrations, transfer factors, chemical and skin properties) affect exposure and dose (Ott 1990; Ritter 1998). Biological monitoring (measuring metabolites of contaminants in blood and urine) has been used as a direct means of measuring the mass of a contaminant cumulatively absorbed through the inhalation, ingestion, and dermal routes. Through interpretation with other data (e.g., knowledge of contaminant properties, chemical distribution and elimination processes in the body, supporting direct dermal exposure measurements), biological monitoring has also been used to indirectly measure dermal exposure and dermal dose (McArthur 1992; Williams, Aston, and Krieger 2004). Ultimately, with any approach for obtaining dermal exposure and dose estimates, we come closer to understanding the relative importance of the dermal route for certain classes of chemicals (e.g., pesticides, volatile organics) and certain populations (e.g., children, farmworkers). Developing strategies to reduce risk and the health problems associated with toxic chemical agents should be the goal of any exposure and dose estimate.

11.13 QUESTIONS FOR REVIEW

1. Under what exposure scenario might the dermal route of exposure be more important relative to the ingestion and inhalation pathway?
2. What types of human activities affect the magnitude of dermal exposure and dose? Are these activities different than the activities that would affect the magnitude of inhalation and ingestion exposure?
3. What layer of the skin is considered the main barrier to absorption and why? What is the viable epidermis and how is it differentiated from the epidermis?
4. Explain the difference between absorbed dose and potential dose. For risk assessment, what type of dose is most practical to consider and why?
5. Name some of the differences in structure and function between the epidermis and dermis layers of the human skin. How far into the skin does a chemical diffuse before it is considered absorbed?
6. Name the pathways or routes that a chemical can enter the skin and diffuse toward the bloodstream. Which pathway is most significant for dermal absorption and why?
7. What are some of the most important factors affecting dermal absorption?
8. What are the pathways for dermal exposure? What pathway is most significant for children and why?
9. How representative are surface sampling techniques for use as measurements of dermal exposure?
10. What types of chemicals are likely to have a relatively large skin absorption and why? Does the vehicle (i.e., environmental media) that the chemical is housed in play a role in the chemical's absorption rate?
11. When are indirect exposure techniques valuable as measurements for dermal exposure and dose?

GLOSSARY OF TERMS

Absorption The movement and uptake of substances (liquids and solutes) into cells or across tissues such as skin, intestine and kidney tubules, by way of diffusion or osmosis.

Acetylation The metabolic process of introducing the acetyl radical (CH_3CO) into a chemical.

Adipose Fat or tissue containing fat cells.

Amorphous Any non-crystalline solid in which the atoms and molecules are not organized in a definite lattice pattern; such solids include glass, plastic, and gel.

Avascular Not associated with or supplied by blood vessels.

Basal Cell Population The innermost cells of the deeper epidermis of the skin.

Capillaries The smallest vessels that contain oxygenated blood. The capillaries, allowing red blood cells to travel in single file, are responsible for delivering oxygen to the tissues on a cellular level.

Collagen The principal protein of the skin, tendons, cartilage, bone, and connective tissue.

Corneocytes Otherwise known as cornified cells, hardened cells of the SC possessing a large fraction of keratin.

Cytoskeletal A three-dimensional network of microtubules and filaments that provides internal support for the cells, anchors internal cell structures, and functions in cell movement and division.

Desmosome A type of junction that attaches one cell to its neighbor.

Desquamated A natural process in the skin where the hardened, dry cells of the upper stratum corneum peel off in the form of scales; shedding of the stratum corneum cells.

Diffusion The spontaneous movement of molecules or other particles in solution, owing to their random thermal motion, to reach a uniform concentration throughout the solvent; a process requiring no addition of energy to the system.

Elastin Fibers Fibers made of a glycoprotein, called elastin, which allows them to recover size and shape after deformation, they are randomly coiled and cross-linked to form elastic fibers that are found in connective tissue.

Fluorescence Microscopy Light microscopy in which the specimen is irradiated at wavelengths that excite fluorochromes, which allows measurement of chemical presence.

Garment Samplers A surrogate skin technique to cover large areas of the body and collect chemical residue during contact events for the assessment of dermal exposure.

Horny Layer Another name given to the hard, dry upper layer of the epidermis, the stratum corneum.

Hydrophilic A material that readily absorbs moisture is hygroscopic and has strongly polar groups that readily interact with water.

In Vitro Outside the living body and in an artificial environment.

In Vivo In the living body of a plant or animal.

Iontophoresis The introduction of an ionized substance (such as a drug) through intact skin by the application of a direct electric current.

Keratin A hard structure produced by the skin (found in hair, nails, and skin).

Keratin Fibers Matured keratin: a hard protein structure that is also the primary constituent of hair and nails.

Lag Time A general term for the delay between an action and a desired effect; for absorption, the time it takes to achieve steady-state equilibrium in the concentration of the chemical (i.e., establish a concentration gradient).

Lamellar Any thin, platelike structure.

Lipophilic Having an affinity for fat, pertaining to or characterized by fat attractiveness.

Lymphatic System System composed of small thin channels similar to blood vessels. They do not carry blood but collect and carry tissue fluid from the body to ultimately drain back into the bloodstream.

Mass Loading The mass of soil per area deposited on the skin surface (grams soil/area skin) or mass of residue per area deposited on the skin surface (grams chemical /area skin).

Microdyalysis Involves the insertion of probes within the lower skin layers, which are perfused with a phosphate buffer solution and sampled at intervals, to measure the chemical concentration that has resulted from diffusion of a chemical into the skin.

Occlusion Covering of an area to prevent evaporation of water or exposure to other environmental conditions.

Organic Pertaining to carbon-based compounds produced by living plants, animals, or by synthetic processes.

Papillary Layer The first layer of the dermis that contains nerve fibers, blood vessels, and loose connective tissue with fine collagen and elastin fibers folded in ridges or papillae, which extend into the epidermis.

Patch Samplers A surrogate skin technique used to collect chemical residues over small areas of the skin for dermal exposure.

Percutaneous Administered, removed, or absorbed by way of the skin, as an injection, needle biopsy, or transdermal drug.

Permeability Capable of being permeated or passed through; yielding passage; passable; penetrable; used especially for substances that allow the passage of fluids; for example, wood is permeable to oil; glass is permeable to light.

Pharmacokinetic The action of drugs or toxic agents in the body over a period of time, including the kinetic processes of absorption, distribution, localization in tissues, biotransformation and excretion, and their appropriate rate constants.

Phonophoresis Process in which an ultrasound machine is used to push medicine deep into joints or muscles by producing heat that swells the blood vessels and increases blood flow and transport of chemicals.

Porcine Of or derived from swine.

Reticular Layer Second layer of the dermis that has no defined boundary and contains denser connective tissue and many thick collagen fibers.

Sebaceous Glands Glands in the skin that secrete sebum (an oily substance) at the surface of the skin.

Soil Matrix An enclosure within which soil originates or develops; the array or arrangement of soil particles.

Soil Particle Bonded soil matter, soil matter having finite mass and internal structure but negligible dimensions.

Spectroscopy Involves decomposing a substance and analyzing the masses of electrically charged particles using an optical instrument.

Squamous Scaly or platelike.

Stratified When a body is arranged in different parts or in separate layers or groups.

Stratum Corneum The first layer of the epidermis, considered the most impenetrable barrier to absorption, consisting mostly of hard corneocyte cells.

Stratum Germinativum The final bottom layer of the epidermis, consists of cuboidal shaped cells that possess large nuclei and distinct cell content (e.g., ribosomes for keratin production).

Stratum Granulosum A layer of granular cells lying immediately above the stratum germinativum in most parts of the epidermis.

Stratum Lucidum A thin, somewhat translucent layer of cells lying above the stratum granulosum and under the stratum corneum especially in thickened parts of the epidermis.

Stratum Spinosum The layers of prickle cells over the layer of the stratum germinativum capable of undergoing mitosis — also called *prickle cell layer.*

Surface Roughness A measure of how rugged (i.e., the heights of hills and valleys) the surface of the skin is.

Vehicle The chemical(s) or medium that the agent is present in (e.g., liquid formulation, soil matrix, air).

REFERENCES

Ademola, J.I., Sedik, L.E., Wester, R.C., and Maibach, H.I. (1993) *In Vitro* Percutaneous Absorption and Metabolism in Man of 2-Chloro-4-Ethylamino-6-Isopropylamine-s-Triazine (Atrazine), *Archives of Toxicology*, **67**: 85–91.

Aspelin, A.L. (1997) *Pesticide Industry Sales and Usage: 1994 and 1995 Market Estimates*, Office of Pesticide Programs, U.S. Environmental Protection Agency, Washington, DC.

Barber, E.D., Teetsel, N.M., Kolberg, K.F., and Guest, D. (1992) A Comparative Study of the Rates of *in Vitro* Percutaneous Absorption of Eight Chemicals Using Rat and Human Skin, *Fundamental and Applied Toxicology*, **19**: 493–497.

Bashir, S.J. and Maibach, H.I. (1999) Cutaneous Metabolism of Xenobiotics, in *Percutaneous Absorption: Drugs, Cosmetics, Mechanism, Methodologies*, Bronaugh, R.L. and Maibach, H.I., Eds., Marcel Dekker, Inc., New York, NY, 65–80.

Baynes, R.E., Halling, K.B., and Riviere, J.E. (1997) The Influence of Diethyl-m-toluamide (DEET) on the Percutaneous Absorption of Permethrin and Carbaryl, *Toxicology and Applied Pharmacology*, **144**: 332–339.

Benech-Kieffer, F., Wegrich, P., Schwarzenbach, R., Klecak, G., Weber, T., Leclaire, J., and Schaefer, H. (2000) Percutaneous Absorption of Sunscreens In Vitro: Interspecies Comparison, Skin Models and Reproducibility Aspects, *Skin Pharmacology and Applied Skin Physiology*, **13**: 324–335.

Bradman, M.A., Harnly, M.E., Draper, W., Seidel, S., Teran, S., Wakeham, D., and Neutra, R. (1997) Pesticide Exposures to Children from California's Central Valley: Results of Pilot Study, *Journal of Exposure Analysis and Environmental Epidemiology*, **7**(2): 217–234.

Bronaugh, R.L. (1996) Methods for *in-Vitro* Percutaneous Absorption, in *Dermatotoxicology*, Marzulli, W.F. and Maibach, H., Eds., Taylor and Francis, Washington, DC, 317–324.

Bronaugh, R.L. (2000) *In-Vitro* Percutaneous Models, *Annals of the New York Academy of Sciences*, **919**: 188–191.

Bronaugh, R.L. and Hood, H.L. (1999) Will Cutaneous Levels of Absorbed Material Be Systemically Absorbed?, in *Percutaneous Absorption: Drugs, Cosmetics, Mechanism, Methodologies*, Bronaugh, R.L. and Maibach, H.I., Eds., Marcel Dekker, Inc., New York, NY, 235–239.

Bronaugh, R.L., Kraeling, M.E.K., Yourick, J.J., and Hood, H.L. (1999) Cutaneous Metabolism during *In-Vitro* Percutaneous Absorption, in *Percutaneous Absorption: Drugs, Cosmetics, Mechanism, Methodologies*, Bronaugh, R.L. and Maibach, H.I., Eds., Marcel Dekker, Inc., New York, NY, 57–63.

Brouwer, D.H., Foeniger, M.F., and Van Hemmen, J. (2000) Hand Wash and Manual Skin Wipes, *Annals of Occupational Hygiene*, **44**(7): 501–510.

Brouwer, D.H., Lansink, C.M., Cherrie, J.W., and Van Hemmen, J.J. (2000) Assessment of Dermal Exposure during Airless Spray Painting Using a Quantitative Visualization Technique, *Annals of Occupational Hygiene*, **44**(7): 543–549.

Bucks, D.A.W., Wester, R.C., and Mobayen, M.M. (1989) *In Vitro* Percutaneous Absorption and Stratum Corneum Binding of Alachlor: Effect of Formulation Dilution with Water, *Toxicology and Applied Pharmacology*, **100**: 417–423.

Camann, D.E., Harding, H.J., Geno, P.W., and Lewis, R.G. (1994) Relationship among Drag Sled, PUF Roller, and Hand Press Transfer of Pesticide Residues from Floors, in *Measurement of Toxic and Related Air Pollutants, Proceedings of the EPA A&WMA International Symposium*, Air & Waste Management Assoc., Pittsburgh, PA, Durham, NC, 832–835.

Chang, S.K. and Riviere, J.E. (1991) Percutaneous Absorption of Parathion In-Vitro in Porcine Skin: Effects of Dose Temperature, Humidity, and Perfusate Composition on Absorptive Flux, *Fundamental and Applied Toxicology*, **17**: 494–504.

Chang, S.W.P., Dauterman W.C., and Riviere, J.E. (1994) Percutaneous Absorption, Dermatopharmacokinetics and Related Bio-Transformation Studies of Carbaryl, Lindane, Malathion, and Parathion in Isolated Perfused Porcine Skin, *Toxicology*, **91**: 269–280.

Cherrie, J.W., Brouwer, D.H., Roff, M., Vermeulen, R., and Kromhout, H. (2000) Use of Qualitative and Quantitative Fluorescence Techniques to Assess Dermal Exposure, *Annals of Occupational Hygiene*, **44**(7): 519–522.

Chuong, C.M., Nickoloff, B.J., Elias, P.M., Goldsmith, L.A., Macher, E., Maderson, P.A., Sundberg, J.P., Tagami, H., Plonka, P.M., Thestrup-Pedersen, K., Bernard, B.A., Schroder, J.M., Dotto, P., Chang, C.H., Williams, M.L., Feingold, K.R., King, L.E., Kligman, A.M., Rees, J.L., and Christophers, E. (2002) What Is the True Function of Skin? *Experimental Dermatology*, **11**(2): 159–187.

Cohen, B.M. and Popendorf, W. (1989) A Method for Monitoring Dermal Exposure to Volatile Chemicals, *American Industrial Hygiene Association*, **50**(4): 216–223.

Cohen Hubal, E.A., Sheldon, L.S., Burke, J.M., McCurdy, T.R., Berry, M.R., Rigas, M.L., Zartarian, V.G., and Freeman, N.C.G. (2000) Children's Exposure Assessment: A Review of Factors Influencing Children's Exposure, and the Data Available to Characterize and Assess That Exposure, *Environmental Health Perspectives*, **108**: 475–502.

Diembeck, W., Beck, H., Benech-Kieffer, F., Courtellemont, P., Dupois, J., Lovell, W., Paye, M., Spengler, J., and Steiling, W. (1999) Test Guidelines for In-Vitro Assessment of Dermal Absorption and Percutaneous Penetration of Cosmetic Ingredients, *Food and Chemical Toxicology*, **37**: 191–205.

Driver, J.H., Konz, J.J., and Whitmyre, G.K. (1989) Soil Adherence to Human Skin, *Bulletin of Environmental Contamination and Toxicology*, **43**: 814–820.

Duan, N., Dobbs, A., and Ott, W. (1990) Comprehensive Definitions of Exposure and Dose to Environmental Pollution, Report No. 159, Stanford Department of Applied Earth Sciences, Department of Statistics, Stanford University, Stanford, CA, 1–32.

Eberlein-Konig, B., Schafer, T., Huss-Marp, J., Darsow, U., Mohrenschlager, M., Herbert, O., Abeck., D., Kramer, U., Behrendt, H., and Ring, J. (2000) Skin Surface PH, Stratum Corneum Hydration, Trans-Epidermal Water Loss and Skin Roughness Related to Atopic Eczema and Skin Dryness in a Population of Primary School Children, *Dermato-Venereologica*, **80**: 188–191.

Eberlein-Konig, B., Spiegel, A., and Przybilla, B. (1996) Change of Skin Roughness due to Lowering Air Humidity in a Climate Chamber, *Acta Dermato-Venereologica*, **76**: 447–449.

Fenske, R.A. (1990) Non-Uniform Dermal Deposition Patterns during Occupational Exposure to Pesticides, *Archives of Environment, Contaminant and Toxicology*, 19: 322–337.

Fenske, R.A. (1993a) Dermal Exposure Assessment Techniques, *Annals of Occupational Hygiene*, **37**(6): 687–706.

Fenske, R.A. (1993b) *Fluorescent Tracer Evaluation of Protective Clothing Performance*, Report No. EPA/600/R-93/143, Risk Reduction Engineering Laboratory, U.S. Environmental Protection Agency, Office of Research and Development, Cincinnati, OH.

Fenske, R.A. (2000) Dermal Exposure: A Decade of Real Progress, *Annals of Occupational Hygiene*, **44**(7): 489–491.

Fenske, R.A., Birnbaum, S.G., Methner, M.M., and Nigg, H.N. (2002a) Fluorescent Tracer Evaluation of Chemical Protective Clothing during Pesticide Applications in Central Florida Citrus Groves, *Journal of Agricultural Safety and Health*, **8**(3): 319–331.

Fenske, R.A., Black, K.G., Elkner, K.P., Lee, C., Methner, M.M., and Soto, R. (1990) Potential Exposure and Health Risks of Infants Following Indoor Residential Pesticide Application, *American Journal of Public Health*, **80**(6): 689–693.

Fenske, R.A., Lu, C., Barr, D., and Needham, L. (2002b) Children's Exposure to Chlorpyrifos and Parathion in Agricultural Community in Central Washington State, *Environmental Health Perspectives*, **110**(5): 549–553.

Fenske, R.A., Lu, C., Simcox, N.J., Loewenherz, C., Touchstone, J., Moate, T.F., Allen, E.H., and Kissel, J.C. (2000) Strategies for Assessing Children's Organophosphorus Pesticide Exposures in Agricultural Communities, *Journal of Exposure Analysis and Environmental Epidemiology*, **10**: 662–671.

Fenske, R.A., Schulter, C., Lu, C., and Allen, E.H. (1998) Incomplete Removal of the Pesticide Captan from Skin by Standard Handwash Exposure Assessment Procedures, *Bulletin of Environmental Contaminant Toxicology*, **61**: 194–201.

Franz, T.J. (1975) Percutaneous Absorption, On the Relevance of *In-Vitro* Data, *The Journal of Investigative Dermatology*, **64**: 190–195.

Franz, T.J. (1978) The Finite Dose Technique as a Valid *In-Vitro* Model for the Study of Percutaneous Absorption in Man, *Current Problems in Dermatology*, **7**: 58–68.

Freeman, N.C.G., Hore, P., Black, K., Jimenez, M., Sheldon, L., Tulve, N., and Lioy, P.J. (2004) Contributions of Children's Activities to Pesticide Hand Loadings Following Residential Pesticide Applications, *Journal of Exposure Analysis and Environmental Epidemiology*, March Online Publication: 1–8.

Grossman, J. (1995) Dangers of Household Pesticides, *Environmental Health Perspectives*, **103**(6): 550–554.

Hemond, H.F. and Solo-Gabriele, H.M. (2004), Children's Exposure to Arsenic from CCA-Treated Wooden Decks and Playground Structures, *Risk Analysis*, **24**(1): 51–64.

Hilton, J., Wollen, B.H., Scott, R.C., Auton, T.R., Trebilock, KL., and Wilks, M.F. (1994) Vehicle Effects on *In-Vitro* Percutaneous Adsorption through Rat and Human Skin, *Pharmaceutical Research*, **11**(10): 1396–1400.

Holmes, K.K., Shirai, J.H., Richter, K.Y., and Kissel, J. (1999) Field Measurement of Dermal Soil Loadings in Occupational and Recreational Activities, *Environmental Research*, Section A, **80**: 148–157.

Hsu, J.P., Camann, D.E., Schattenberg, H., III, Wheeler, B., Villalobos, K., Kyle., M., Quarderer, S., and Lewis, R.G. (1990) New Dermal Exposure Sampling Technique, *US EPA Air & Waste Management Association International Symposium on Measurement of Toxic and Related Air Pollutants*, Raleigh, NC, April 30–May 4, published in AWMA Publication VIP-17, Pittsburgh, PA, 489–497.

Hughson, G.W. and Aitken, R.J. (2004), Determination of Dermal Exposures during Mixing, Spraying and Wiping Activities, *Annals of Occupational Hygiene*, **48** (3): 245–255.

Idson, B. (1978) Hydration and Percutaneous Absorption, *Current Problems in Dermatology*, **7**: 132–141.

Iwata, Y., Moriya, Y., and Kobayashi, T. (1987) Percutaneous Absorption of Aliphatic Compounds, *Cosmetics and Toiletries*, **100**: 53–68.

Jackson, E.M. (1990) Toxicological Aspects of Percutaneous Absorption, *Cosmetics and Toiletries*, **105**: 135–147.

Jepson, G.W. and McDougal, J.N. (1999) Predicting Vehicle Effects on the Dermal Absorption of Halogenated Methanes Using Physiologically Based Modeling, *Toxicological Sciences*, **48**: 180–188.

Jewell, C., Heylings, J., Clowes, H.M., and Williams, F.M. (2000) Percutaneous Absorption and Metabolism of Dinitrochlorobenzene In-Vitro, *Archives of Toxicology*, **74**: 356–365.

Kissel, J.C., Richter, K.Y., and Fenske, R.A. (1996) Field Measurement of Dermal Soil Loading Attributable to Various Activities: Implications or Exposure Assessment, *Risk Analysis*, **16**(1): 115–125.

Kissel, J.C., Shirai, J.H., Richter, K.Y., and Fenske, R.A. (1998) Empirical Investigation of Hand-to-Mouth Transfer of Soil, *Bulletin of Environmental Contamination and Toxicology*, **60**: 379–386.

Kligman, A.M. (1964) The Biology of the Stratum Corneum, in *The Epidermis,* Montagna, W. and Lobitz, W.C., Eds., Academic Press, New York, NY, 387–452.

Krieger, R.I. and Dinoff, T.M. (1996) Human Disodium Octaborate Tetrahydrate Exposure Following Carpet Flea Treatments Is Not Associated with Significant Dermal Absorption, *Journal of Exposure and Environmental Epidemiology*, **6**(3): 279–288.

Krieger, R.I., Bernard, C.E., Dinoff, T.M., Fell, L., Osimitz, T.G., Ross, J.H., and Thongsinthusak, T. (2000) Biomonitoring and Whole Body Cotton Dosimetry to Estimate Potential Human Dermal Exposure to Semi-Volatile Chemicals, *Journal of Exposure Analysis and Environmental Epidemiology*, **10**(1): 50–57.

Kromhout, H., Hoek, F., Uitterhoeve, R., Huijbers, R., Oermars, R.F., Anzion, R., and Vermeulen, R. (2000) Postulating a Dermal Pathway for Exposure to Anti-Neoplastic Drugs among Hospital Workers, Applying a Conceptual Model to the Results of Three Workplace Surveys, *Annals of Occupational Hygiene*, **44**(7): 551–560.

Lepow, M.L., Bruckman, L., Gillete, M., Markowitz, S., Robino, R., and Kapish, J. (1975) Investigations into Sources of Lead in the Environment of Urban Children, *Environmental Research*, **10**: 415–426.

Lewis, R.G., Fortune, C.R., Blanchard, F.T., and Camann, D.E. (2001) Movement and Deposition of Two Organophosphorus Pesticides within a Residence after Interior and Exterior Applications, *Journal of the Air and Waste Management Association*, **5**: 339–351.

Lioy, P.J., Edwards, R.D., Freeman, N., Gurunathan, S., Pellizzari, E., Adgate, J.L., Quakenboss, J., and Sexton, K. (2000) House Dust Levels of Selected Insecticides and a Herbicide Measured by the EL and LWW Samplers and Comparisons to Hand Rinses and Urine Metabolites, *Journal of Exposure Analysis and Environmental Epidemiology*, **10**: 327–340.

Lu, C. and Fenske, R.A. (1998) Air and Surface Chlorpyrifos Residues Following Residential Broadcast and Aerosol Pesticide Applications, *Environmental Science and Technology*, **32**(10): 1386–1390.

Malkinson, F.D. (1964) Permeability of the Stratum Corneum, in *Epidermis,* Montagna, W. and Lowitz, W.C., Eds., Academic Press, New York, NY, 435–452.

Manuskiatti, W., Schwindt, D.A., and Maibach, H.I. (1998) Influence of Age, Anatomic Site and Race on Skin Roughness and Scaliness, *Dermatology,* **196**: 401–407.

Marks, R. (1983) Mechanical Properties of the Skin, in *Biochemistry and Physiology of the Skin,* Goldsmith, L.A., Ed., Oxford Press, U.K. 1237–1254.

McArthur, B. (1992) Dermal Measurement and Wipe Sampling Methods: A Review, *Applied Occupational and Environmental Hygiene,* **7**(9): 599–605.

McDonald, S. and Lunte, C. (2003) Determination of the Dermal Penetration of Esteron Components Using Microdialysis Sampling, *Pharmaceutical Research,* **20**(11): 1827–1834.

Menzel, E. (1995) Assessment of Delipidization as an Enhancing Factor in Percutaneous Penetration, *Exogenous Dermatology,* **22**: 189–194.

Mershon, M.M. (1975) Barrier Surfaces of the Skin, in *Applied Chemistry at Protein Interfaces: A Symposium at the 166th Meeting of the American Chemical Society,* Chicago, IL, Baier, R.E., Ed., ACS Advanced Chemical Series, 41–73.

Moghimi, H.R., Barry, B.W., and Williams, A.C. (1999) Stratum Corneum and Barrier Performance: A Model Lamellar Structural Approach, in *Percutaneous Absorption: Drugs, Cosmetics, Mechanism, Methodologies,* Bronaugh, R.L. and Maibach, H.I., Eds., Marcel Dekker, Inc., New York, NY, 515–553.

Monteiro-Riviere, N. (1996) Anatomical Factors Affecting Barrier Function, in *Dermatotoxicology,* Marzulli, W.F. and Maibach, H., Eds., Taylor and Francis, Washington, DC, 3–17.

Moody, R.P., Wester, R.C., Melendres, J.L., and Maibach, H.I. (1992) Dermal Absorption of the Phenoxy Herbicide 2,4-D Dimethylamine in Humans: Effect of DEET and Anatomic Site, *Journal of Toxicology and Environmental Health,* **36**: 241–250.

Mukerjee, S., Ellenson, W.D., Lewis, R.G., Stevens, R.K., Somerille, M.C., Shadwick, D.S., and Willis, R.D. (1997) An Environmental Scoping Study in the Lower Rio Grande Valley of Texas-III. Residential Microenvironmental Monitoring for Air, House Dust, and Soil, *Environment International,* **23**(5): 657–673.

Notingher, I. and Imhof, R.E. (2004) Mid-Infrared In-Vivo Depth-Profiling of Topical Chemicals on Skin, *Skin Research and Technology,* **10**: 113–121.

NRC (1993) *Pesticides in the Diets of Infants and Children,* National Research Council, National Academy Press, Washington, DC.

O'Fallon, L.R., Collman, G.W., and Dearry, A. (2000) The National Institute of Environmental Health Sciences' Research Program on Children's Health, *Journal of Exposure Analysis and Environmental Epidemiology,* **10**: 630–637.

Ott, W.R. (1990) Total Human Exposure: Basic Concepts, EPA Field Studies, and Future Research Needs, *Journal of Air and Waste Management Association,* **40**(7): 966–973.

Pellizzari, E.D., Smith, D.J., Clayton, A., Quackenboss, J.J. (2003) Assessment of Data Quality for the NHEXAS-Part II: Minnesota Children's Pesticide Exposure Study (MNCPES), *Journal of Exposure Analysis and Environmental Epidemiology,* **13**: 465–479.

Pendlington, R.U., Whittle, E., Robinson, J.A., and Howes, D. (2001) Fate of Ethanol Topically Applied to Skin, *Food and Chemical Toxicology,* **39**: 169–174.

Poet, T.S., Corley, R.A., Thrall, K.D., Edwards, J.A., Tanoojo, H., Weitz, K.K., Hui, X., Maibach, H.I., and Wester, R.C. (2000) Assessment of the Percutaneous Absorption of Trichloroethylene in Rats and Humans Using MS/MS Real-Time Breath Analysis and Physiologically Based Pharmacokinetic Modeling, *Toxicological Sciences,* **56**: 61–72.

Que Hee, S.S., Peace, B., Clack, C.S., Boyle, J.R., Bornschein, R.L., and Hammond, P.B. (1985) Evolution of Efficient Methods to Sample Lead Sources, Such as House Dust, in Homes of Children, *Environmental Research,* **38**: 77–95.

Randeau, M., Kurdian, C., Sirvent, A., and Girard, F. (2001) Characterization of Skin Surface: Comparison of SIA(r) and PRIMOS Methods, *Cosmetics and Toiletries Manufacture Worldwide,* Basel, http://www.ctmw.com/art02.htm (accessed March 2002).

Ritter, S.K. (1998) Modeling Dermal Exposure, *Chemical and Engineering News,* **76**(41): 37–39.

Riviere, J.E., Qiao, G., Baynes, R.E., Brooks, J.D., and Mumtaz, M. (2001) Mixture Component Effects on the *In Vitro* Dermal Absorption of Pentachlorophenol, *Archives of Toxicology,* **75**: 329–334.

Roberts, J.W. and Camann, D.E. (1989) Pilot Study of a Cotton Glove Press Test for Assessing Exposure to Pesticides in House Dust, *Bulletin of Environmental Contamination and Toxicology*, **43**: 717–724.

Roberts, J.W., Budd, W.T., Roby, M.G., Bond, A.E., Lewis, R.G., Wiener, R.W., and Camann, D. E. (1991) Development and Field Testing of a High Volume Sampler for Pesticides and Toxics in Dust, *Journal of Exposure and Environmental Epidemiology*, **1**: 143–155.

Roberts, M. and Walters, K. A. (1998) The Relationship between Structure and Barrier Function of Skin, in *Dermal Absorption and Toxicity Assessment*, Roberts, M. and Walters, K.A., Eds., Marcel Dekker, New York, NY, **91**: 1–42.

Roels, H.A., Buchet, J., Lauwerys, R.R., Bruaux, P., Claeys-Thoreau, F., Lafontaine, A., and Verduyn, G. (1980) Exposure to Lead by the Oral and the Pulmonary Routes of Children Living in the Vicinity of a Primary Lead Smelter, *Environmental Research*, **22**: 81–94.

Rosado, C. and Rodriques, L.M. (2003) Solvent Effects in Permeation Assessed *in Vivo* by Skin Surface Biopsy, *BMC Dermatology*, **3**(5). http://www.biomedcentral.com/1471–5945/3/5.

Rosado, C., Cross, S.E., Pugh, W.J., Roberts, M.S., and Hadgraft, J. (2003) Effect of Vehicle Pretreatment on the Flux, Retention, and Diffusion of Topically Applied Penetrants *in Vitro*, *Pharmaceutical Research*, **20**(9): 1502–1507.

Ross, J., Fong, H.R., Thongsinthusak, T., Margetich, S., and Krieger, R. (1991) Measuring Potential Dermal Transfer of Surface Pesticide Residue Generated from Indoor Fogger Use: Using the CDFA Roller Method Interim Report II, *Chemosphere*, **22**(9-10): 975–984.

Ross, J., Thongsinthusak, T., Fong, H.R., Margetich, S., and Krieger, R. (1990) Measuring Potential Dermal Transfer of Surface Pesticide Residue Generated from Indoor Fogger Use: An Interim Report, *Chemosphere*, **20**(3/4), 349–360.

Rougier, A. (1999) In-Vivo Percutaneous Absorption: A Key Role for Stratum Corneum/Vehicle Partitioning, in *Percutaneous Absorption: Drugs, Cosmetics, Mechanisms, Methodology*, Bronaugh, R.L. and Maibach, H.I., Eds., Marcel Dekker, New York, NY, 193–211.

Roy, S.D., Manoukian, E., and Combs, D. (1995). Absorption of Transdermally Delivered Ketorlac Acid in Humans, *Journal of Pharmaceutical Sciences*, **84**(1): 49–52.

Rozman, K.K. and Klaassen, C.D. (2001) Absorption, Distribution, and Excretion of Toxicants, in *Casarett & Doull's Toxicology: The Basic Science of Poisons*, Klaassen, C.D., Ed., McGraw-Hill, New York, NY, 107–132.

Sartorelli, P.A., Bussani, R., Novelli, M.T., Orsi, D., and Sciarra, G. (1997) In-Vitro Percutaneous Penetration of Methyl-Parathion from a Commercial Formulation through the Human Skin, *Occupational and Environmental Medicine*, **54**: 524–525.

Schaefer, H., Zesch, A., and Stuttgen, G. (1982) *Skin Permeability*, Springer-Verlag, London, U.K.

Scheuplein, R.J. and Blank, I.H. (1971) Permeability of the Skin, *Physiological Reviews*, **51**(4): 702–747.

Schmidt, C.R. (1999) A Closer Look at Chemical Exposures in Children, *Environmental Science and Technology*, **33**(3): 72A–75A.

Selim, S., Preiss, F.J., Gabriel, K.L., Jonkman, J.H.G., and Osiimitz, T.G. (1999) Absorption and Mass Balance of Piperonyl Butoxide Following an 8-hr Dermal Exposure in Human Volunteers, *Toxicology Letters*, **107**: 207–217.

Shalat, S.L., Donnelly, K.C., Freeman, N.C.G., Calvin, J.A., Ramesh, S., Jimenez, M., Black, C.C., Needham, L.L., Barr, D.B., and Ramirez, J. (2003) Non-Dietary Ingestion of Pesticides by Children in an Agricultural Community on the U.S./Mexico Border: Preliminary Results, *Journal of Exposure Analysis and Environmental Epidemiology*, **13**: 42–50.

Simcox, N.J., Fenske, R.A., Wolz, S.A., Lee, I., and Kalman, D.A. (1995) Pesticides in Household Dust and Soil: Exposure Pathways for Children of Agricultural Families, *Environmental Health Perspectives*, **103**(12): 1126–1133.

Sinclair, W. (1995) Pesticide Health Effects Studies, in *Allergy, Asthma, Immunology*. http://www.chem-tox.com/pesticides.

Soutar, A., Semple, S., Aitken, R.J., and Robertson, A. (2000) Use of Patches and Whole Body Sampling for the Assessment of Dermal Exposure, *Annals of Occupational Hygiene*, **44**(7): 511–518.

Sznitowska, M. and Berner, B. (1995) Polar Pathway for Percutaneous Absorption, *Exogenous Dermatology*, **22**: 164–170.

Tojo, K., Chiang, C.C., and Chien, Y.W. (1987) Drug Permeation across the Skin: Effect of Penetrant Hydrophilicity, *Journal of Pharmaceutical Sciences*, **76**(2): 123–126.

Turner, P., Saeed, B., and Kelsey, M.C. (2004) Dermal Absorption of Isopropyl Alcohol from a Commercial Hand Rub: Implications for its Use in Hand Decontamination, *Journal of Hospital Infection*, **56**: 287–290.

USEPA (1992) *Dermal Exposure Assessment: Principles and Applications*, Interim Report, EPA/600/8-91/011B, Exposure Assessment Group, U.S. Environmental Protection Agency, Washington, DC.

USEPA (1997) *Exposure Factors Handbook, Volume 1: General Factors*, Report No. EPA/600/P-95/002Fa, U.S. Environmental Protection Agency, Washington, DC.

United States Congress (1996) Food Quality Protection Act of 1996: Report Together with Additional View (to accompany H.R. 1627) (Including Cost Estimate of the Congressional Budget Office), Rept. 104th Congress, 2d Session, House of Representatives; 104–669, Washington, DC.

Wertz, P.W. (1996) The Nature of the Epidermal Barrier: Biochemical Aspects, *Advanced Drug Delivery Review*, **18**(3): 283–294.

Wester, R.C. (1995) Twenty Absorbing Years, *Pharmacology*, **22**: 112–123.

Wester, R.C., and Maibach, H.I. (1992) In Vitro Percutaneous Absorption of Cadmium from Water and Soil into Human Skin, *Fundamental and Applied Toxicology*, **19**: 1–5.

Wester, R.C., and Maibach, H.I. (1999) In-Vivo Methods for Percutaneous Absorption Measurements, *Percutaneous Absorption: Drugs-Cosmetics-Mechanisms-Methodology*, Bronaugh, R.L. and Maibach, H.I., Eds., Marcel Dekker, New York, NY, 215–227.

Wester, R.C., Maibach, H.I., Bucks, D.A.W., and Guy, R.H. (1983) Malathion Percutaneous Absorption after Repeated Administration to Man, *Toxicology and Applied Pharmacology*, **68**: 116–119.

Wester, R.C., Maibach, H.I., Sedik, L., Melendres, J., and Wade, M. (1993) In-Vivo and In-Vitro Percutaneous Absorption and Skin Decontamination of Arsenic from Water and Soil, *Fundamental and Applied Toxicology*, **20**: 336–340.

Whitmore, R.W., Immerman, F.W., Camann, D.E., Bond, A.E., Lewis, R.G., and Schaum, J.L. (1994) Non-Occupational Exposures to Pesticides for Residents of Two U.S. Cities, *Archives of Environmental Contamination and Toxicology*, **26**: 47–59.

Whitton, J.T. and Everall, J.D. (1973) The Thickness of the Epidermis, *British Journal of Dermatology*, **89**: 467–476.

Williams, P.R.D., Holicky, K.C., and Paustenbach, D.J. (2003) Current Methods for Evaluating Children's Exposures for Use in Health Risk Assessment, *Journal of Children's Health*, **1**(1): 41–98.

Williams, R.L., Aston, L.S., and Krieger, R.I. (2004) Perspiration Increased Human Pesticide Absorption Following Surface Contact during an Indoor Scripted Activity Program, *Journal of Exposure Analysis and Environmental Epidemiology*, **14**(2): 129–136.

Wilson, N.K., Chuang, J.C., Iachan, R., Lyu, C., Gordon, S.M., Morgan, M.K., Özkaynak, H., and Sheldon, L. S. (2004) Design and Sampling Methodology for a Large Scale Study of Preschool Children's Aggregate Exposures to Persistent Organic Pollutants in Their Everyday Environments, *Journal of Exposure Analysis and Environmental Epidemiology*, **14**: 260–274.

Wilson, N.K., Chuang, J.C., and Lyu, C. (2001) Levels of Persistent Organic Pollutants in Several Child Day Care Centers, *Journal of Exposure Analysis and Environmental Epidemiology*, **11**: 449–458.

Wilson, N.K., Chuang, J.C., Lyu, C., Menton, R., and Morgan, M.K. (2003) Aggregate Exposures of Nine Preschool Children to Persistent Organic Pollutants at Day Care and at Home, *Journal of Exposure Analysis and Environmental Epidemiology*, **13**(3): 187–202.

Yourick, J.J., Koenig, M.L., Yourick, D.L., and Bronaugh, R.L. (2004) Fate of Chemicals in Skin after Dermal Application: Does the *in Vitro* Skin Reservoir Affect the Estimate of Systemic Absorption? *Toxicology and Applied Pharmacology*, **195**(3): 309–320.

Yu, B., Kin, K.H., So, P.T.C., Blankschtein, D., and Langer, R. (2003) Evaluation of Fluorescent Probe Surface Intensities as an Indicator of Transdermal Permeant Distributions Using Wide-Area Two Photon Fluorescence Microscopy, *Journal of Pharmaceutical Sciences*, **2**(12): 2354–2365.

Zartarian, V.G. (1996) A Physical-Stochastic Model for Understanding Dermal Exposure to Chemicals, Ph.D. diss., Department of Civil Engineering, Stanford University, Stanford, CA.

Zartarian, V.G and Leckie, J.O. (1998) Dermal Exposure: The Missing Link, *Environmental Science and Technology*, **3**(3): 134A–137A.

Zartarian, V.G., Ott, W.R., and Duan, N. (1997) A Quantitative Definition of Exposure and Related Concepts, *Journal of Exposure Analysis and Environmental Epidemiology*, **7**(4): 411–437.

Zartarian, V.G., Özkaynak, H., Burke, J.M., Zufall, M.J., Rigas, M.L., and Furtaw, E.J. (2000) A Modeling Framework for Estimating Children's Residential Exposure and Dose to Chlorpyrifos via Dermal Residue Contact and Non-Dietary Ingestion, *Environmental Health Perspectives*, **108**(6): 505–514.

Zhai, H. and Maibach, H.I. (2001) Effects of Skin Occlusion on Percutaneous Absorption: An Overview, *Skin Pharmacology and Applied Skin Physiology*, **14**: 1–10.

Zhao, K., Singh, S., and Singh, J. (2001) Effect of Methane on the *in Vitro* Percutaneous Absorption of Tamoxifen and Skin Reversibility, *International Journal of Pharmaceutics*, **219**: 177–181.

12 Dermal Exposure to VOCs while Bathing, Showering, or Swimming

Lance A. Wallace
U.S. Environmental Protection Agency (ret.)

Sydney M. Gordon
Battelle Memorial Institute

CONTENTS

12.1 Synopsis..285
12.2 Introduction...286
12.3 Parameters Affecting Dermal Absorption 287
12.4 Indirect Measures of Dermal Absorption 292
    12.4.1 Showers and Baths 292
    12.4.2 Swimmers 292
12.5 Direct Measures of Dermal Absorption Using Continuous Breath Measurements...........292
12.6 Conclusion 295
12.7 Questions for Review 296
References 296

12.1 SYNOPSIS

We often say to measure exposure, measure the concentration in a medium next to the appropriate body surface. For inhalation and ingestion, this is relatively straightforward. But dermal absorption involves the skin, the body's largest and most extended organ. How is one to determine the appropriate body surface in many situations? For example, in showering some portions of the body are receiving occasional "hits" from small droplets of water, other portions of the body may be constantly exposed as the water rivulets cascade downward, still other parts (e.g., the hair and skull) may be completely dry and unexposed. This chapter deals with how various exposure assessors have dealt with the problem of measuring dermal absorption. Although some efforts have been ingenious, we will see that the problem remains unsolved.

We begin with the history of how first indoor air and then the shower or bath were identified as providing major sources of exposure, both inhalation and dermal, to volatile organic compounds (VOCs), but most particularly chloroform. After a brief detour to investigate how chloroform gets into our water, we review several models of exposure through the skin through contact with contaminated water. The central barrier to exposure is the stratum corneum (SC), which can be viewed as providing a resistance to molecular diffusion through the skin. Different chemicals have different diffusion rates based on their molecular characteristics. A crucial parameter is the lag time

τ_{SC}, the time (roughly speaking) that it takes from the beginning of exposure to the point at which the chemical is pouring out of the inner surface of the SC at its maximum rate. This is important because if the lag time is large, then exposure during a 10-minute shower will not be important, but if it is small then dermal exposure may be significant. In the 1990s, two models were extant predicting very different values (29 minutes and 12 minutes) for the lag time. We shall see that when the crucial experiment was done, both models were shown to be wrong.

We then continue the history of studies of showers and baths, and follow their extension into studies of exposure of swimmers. We conclude with a relatively recent advance in measuring exhaled breath continuously as a means of determining the lag time directly. The experimental lag time turned out to be smaller (6–9 minutes) than either model had predicted, thus indicating that dermal exposure was more important than previously thought. The same study determined that the temperature of bath water was extraordinarily important in determining exposure, with very hot bath water providing 30 times the chloroform uptake as merely warm bath water. One hypothesis to explain this increase is that blood flow is greatly increased to the extremities under hot conditions. This hypothesis turns out to be useful in explaining the different findings of the early swimming pool studies, which estimated widely different dermal uptakes due to different levels of activity in the swimmers studied.

12.2 INTRODUCTION

In 1985, the U.S. Environmental Protection Agency's (USEPA) total exposure assessment methodology (TEAM) study reported chloroform levels in homes that were four to five times higher than those outside (Pellizzari et al. 1987a,b; Wallace et al. 1985). Mean personal exposures to airborne chloroform for the 800 participants in the TEAM Study generally ranged from 1 to 4 $\mu g/m^3$. Simultaneous outdoor air measurements were much lower, generally contributing less than 15% of the indoor air to total airborne exposure. This was the first time that such measurements had been made, and it excited intense interest as to the source of the chloroform. The TEAM Study analysts guessed that the cause was the use of treated water in the home. If about 1,000 L/day of water gave up half its chloroform, it would be sufficient to reach the levels that were measured. In fact, the average use of water by each person in a household is approximately 500 L/day, and later studies found that indeed water gives up roughly half of its chloroform to the air, so for a two- or three-person household, the conditions agree well with observations.

The hypothesis was tested in a small study in which apartment residents were asked to use dishwashers, clothes washing machines, and take showers and baths while several 8-hour integrated measurements were taken of the air in the apartment (Wallace et al. 1989). The results confirmed that water use had the capability of raising chloroform concentrations in a small apartment as high as 40 $\mu g/m^3$ for short (8-hour) periods, compared to typical outdoor levels of about 1 $\mu g/m^3$.

These findings led to many further studies on exposure while showering, bathing, or swimming. Andelman (1985a,b) verified that inhalation exposure during showers might be comparable to ingestion of 1 to 6 liters of drinking water a day. McKone (1987) developed a three-compartment model (shower, rest of bathroom, rest of house) for indoor air concentrations of chloroform due to showering and checked his model against the TEAM Study results. However, all of these early studies ignored the possible contribution of dermal absorption to total body burden of chloroform.

The first studies to take up this problem were those by Jo, Weisel, and Lioy (1990a,b), Little (1992), McKone (1992), and Weisel, Jo, and Lioy (1992). Their experimental and theoretical studies indicated that the dose of chloroform absorbed through the skin during a shower might be comparable to the dose inhaled. Later, studies of indoor swimming pools (Lévesque et al. 1995; Lindstrom, Pleil, and Berkoff 1997) indicated that both inhalation and dermal absorption can provide substantial amounts of chloroform.

In most of these studies, the focus was on chloroform, as the most prevalent volatile organic compound (VOC) in treated water. Therefore we first review how it is that chloroform got into our water in the first place. For a full review of human exposure to chloroform through all routes, see Wallace (1997).

Chloroform and other trihalomethanes (THMs) are created when water is chlorinated. In the chlorination process, chlorine is added as a disinfectant to raw water at treatment plants. Hypochlorous acid is formed and reacts with organic precursors (e.g., humic or fulvic acids), forming chloroform (Rook 1976, 1977; Schnoor et al. 1979; Amy, Chadik, and Chowdhury 1987).

Chloroform is prevalent in tap water throughout much of the country. About half of the U.S. population uses chlorinated surface water, and another 25% consumes chlorinated groundwater (Jolley 1983). As the water ages in the distribution system, the reaction process creating chloroform continues and the levels at the tap are thus higher than the levels leaving the treatment plant. Water temperature also affects the reaction rate, and one study (Weisel and Chen 1994) showed that water stored overnight in the hot water heater in homes will increase its chloroform content by a substantial amount.

Chloroform concentrations in ground and surface water have been measured in a series of surveys (Symons et al. 1975; Brass et al. 1977, 1981; Westrick, Mello, and Thomas 1983; Krasner et al. 1989). In the largest of these surveys (McGuire and Meadow 1988), median values of total THMs from 727 treatment plants ranged from a low of 30 µg/L in the winter quarter to 44 µg/L in the summer. Chloroform accounted for the largest fraction (about 40% of the overall median level of 39 µg/L). Note that levels at the tap will be higher because of aging in the distribution system.

Chloroform and other THMs cause cancer in rats and mice (NCI 1976; NTP, 1985, 1987, 1989a,b). This fact, and the discovery of chloroform in the blood of residents of New Orleans (later reported in Dowty et al. 1975), led to the Safe Drinking Water Act (SDWA) of 1974, which set a maximum allowed level of 100 µg/L for the total THM content of drinking water (USEPA 1979, 1984). Later, a number of epidemiological studies suggested that chlorinated water causes bladder cancer, and possibly rectal cancer, in humans (Cantor, Hoover, and Hartge 1987; Morris et al. 1992; Vena et al. 1993). Both Cantor and Vena found a dose-response relationship: persons with higher fluid intake had progressively higher risks of developing bladder cancer. It is not known whether the main cancer-causing agent resides in the volatile (e.g., the THMs) or nonvolatile (e.g., MX, a chlorinated hydrofuranone) component of the tap water. However, Vena et al. (1993) found a stronger relationship with total fluid intake than with unheated tap water intake, suggesting the nonvolatile component as the one with greater carcinogenic activity.

With this brief introduction, we next turn to a theoretical examination of the major factors affecting dermal absorption.

12.3 PARAMETERS AFFECTING DERMAL ABSORPTION

Dermal absorption of several VOCs has been measured in humans by immersion of hand or thumb in the undiluted liquid (Stewart and Dodd 1964; Hake and Stewart 1977), or immersion of the arm in the solvent vapor (Corley, Markham, and Banks 1997; Giardino et al. 1999). Studies in absorption of high concentration solvents by guinea pigs and hairless mice have also been carried out.

One general approach is to consider the permeability K_p of a compound through the SC:

$$K_p = K_{sw} D/L \tag{12.1}$$

where K_{sw} is the skin-water partition coefficient, D is the diffusion coefficient of the chemical, and L is the thickness of the SC. Sometimes K_{sw} is replaced by the octanol-water partition coefficient K_{ow}.

Flynn (1990) published a set of permeability coefficients for 97 compounds, which have been used in many subsequent studies to develop models of dermal absorption. Patel, Ten Berg, and Cronin (2002) extended the set of chemicals to 143. Fitzpatrick, Corish, and Hayes (2004) have published the most recent review of models based on this equation. Many of these models estimate the permeability coefficient for a given compound from the following relation developed by Potts and Guy (1992) from the Flynn (1990) database:

$$\log K_p = 0.71 \log K_{ow} - 0.006 \text{ MW} - 6.3 \qquad (12.2)$$

where MW is the molecular weight of the chemical.

The equation is an example of a quantitative structure-activity relationship (QSAR) in which some physicochemical attributes of a compound are used to predict other, usually more complex, parameters. Although Patel et al. (2002) developed a more complex equation involving two additional parameters based on his extended dataset, Fitzpatrick et al. (2004) were able to obtain nearly as good a fit using the same two parameters as Potts and Guy:

$$\log K_p = 0.781 \log K_{ow} - 0.0115 \text{ MW} - 2.19 \qquad (12.3)$$

Scheuplein and Blank (1971) provide a comprehensive review of skin permeability. They find that the SC provides the bulk of the resistance to skin penetration by low-molecular-weight compounds. They define a skin permeability coefficient as the ratio of the flux $[ML^{-2}T^{-1}]$ of chemical across the SC to the concentration $[ML^{-3}]$ in the medium — thus the permeability coefficient has the dimensions $[LT^{-1}]$ and is often reported as cm/h. Assuming a constant concentration in a medium touching the skin, they define a "lag time" during which the concentration in blood reaches steady-state equilibrium. (Also during this time, the SC "fills up" with the chemical of interest, acting as a source for a time after contact is broken off.) If the lag time is longer than the time in contact with the medium, it will not be possible to determine a true permeability coefficient, but only an effective permeability coefficient for the time in question.

Bogen, Colston, and Machicao (1992) studied absorption of dilute concentrations of chloroform, trichloroethylene, and tetrachloroethylene in hairless guinea pigs. Six experiments on chloroform at concentrations of 19 to 52 ppb provided a mean permeability constant of 0.13 (SD = 17%) mL water cleared of chloroform per square cm exposed per hour: 0.13 mL/cm²-h. (A 70-kg person has a body area of about 18,000 cm²; about 80% of that area may be considered submerged during showers or baths.) The question arises whether the permeability constants measured in the animal studies could be applied to humans. From the data of Jo, Weisel, and Lioy (1990a), and assuming an alveolar ventilation rate for a 70-kg adult of 378 L/h, with 80% of the total skin surface area of 18,000 cm² immersed during the shower, the authors calculated a human permeability constant for chloroform of 0.16 mL/cm²-h, close to the value of 0.13 mL/cm²-h calculated for the guinea pigs. This leads to an estimate that a 20-minute bath would result in dermal absorption of chloroform equivalent to ingesting 0.76 L (0.6 L if the guinea pig value is used) of the same tap water. Both the Bogen et al. (1992) and Jo et al. (1990a) results indicate that dermal absorption during showers may be a significant contributor to total chloroform exposure, on the order of ingesting 0.3 to 0.4 liters of tap water for a 10-minute shower.

A more detailed kinetic model of dermal absorption has been presented by Brown and Hattis (1989) and Shatkin and Brown (1991). The model includes two skin compartments (SC and viable epidermis), with a large number of physiological parameters describing diffusion, blood flow, partition coefficients, fat content, and elimination rate constants. Shatkin and Brown applied their model to the Jo et al. (1990a) conditions (10-minute shower at 24.5 µg/L chloroform) and arrived at a prediction of 0.003 mg chloroform absorbed, about a factor of 5 below the value calculated by Jo et al. (1990b). Other influential models of skin absorption have been published by Cleek and Bunge (1993) and Wester and Maibach (1989).

McKone and Howd (1992) and the USEPA (1992) published models of dermal exposure relying on correlations of measured skin permeability with measured octanol-water partition coefficients and with molecular weight to predict skin permeability for a number of chemicals. The theories underlying the two models differ somewhat, as do the estimates of the parameters. McKone and Howd find about three times higher values of skin permeability than the USEPA model.

McKone (1992) combined a shower exposure model with a physiologically based pharmaco-kinetics (PBPK) model to account for both dermal and inhalation exposures. He was unable to justify a significant effect from the second compartment (viable epidermis) included in the Shatkin and Brown model, and therefore included only the SC in his model. McKone adjusted three parameters of the model (the skin permeability coefficient, the thickness of the hydrated portion of the SC, and the skin-water partition coefficient) by fitting to the Jo et al. (1990a) data. The values determined for these coefficients were 0.06 cm/h, 0.0025 cm, and 10, respectively. By comparison, the USEPA (1992) model values are 0.0089 cm/h, 0.0010 cm, and 24. Thus the McKone model estimates a considerably higher skin permeability coefficient than does the USEPA (1992) model. McKone's value for the true skin permeability coefficient leads to an estimate for the effective skin permeability coefficient of 0.2 cm/h, agreeing with the Chinery and Gleason (1993) best value of 0.2 cm/h and comparable with Bogen's estimate of 0.13–0.16 cm/h. McKone finds that the USEPA model values underestimate the Jo et al. (1990a) data by about 50%. McKone also calculates a lag time for the SC to achieve steady state as predicted by the USEPA (1992) model (29 minutes) and McKone's model (12 minutes). McKone calculates the cumulative metabolized dose from inhalation and from dermal absorption resulting from a normal 10-minute shower to be approximately 1/3 (for each pathway) of the dose resulting from ingestion of 1L of the tap water. In absolute terms, the amount of chloroform metabolized in μg per μg/L in tap water is 0.17 as a result of dermal uptake and 0.24 as a result of inhalation.

The concept of the lag time is illustrated in Figure 12.1, which illustrates the flux through the SC for a person exposed to a constant level of chloroform in water. The chloroform flux through

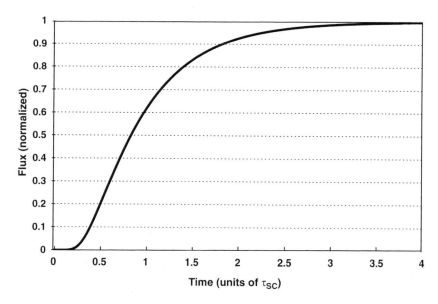

FIGURE 12.1 Exact infinite-series solution for the flux across the inner surface of the SC ($x = L$) assuming a constant concentration at the outer surface ($x = 0$) and a concentration of zero at the inner surface ($x = L$). In units of the time constant τ_{SC}, the flux can be said to reach its equilibrium value at about $\tau_{SC} = 3$. The flux reaches a value of 0.635 of its final equilibrium value when $\tau_{SC} = 1$.

the inner surface of the SC as a function of time from exposure will show no change for a certain period while it diffuses through the SC, and will then start to follow an S-shaped curve until it approaches an asymptote at equilibrium (Figure 12.1). This portion of the curve was drawn according to a solution of the diffusion equation for the concentration $C(x,t)$ in the SC as a function of distance from the outer surface $(0 < x < L)$:

$$\frac{\partial C(x,t)}{\partial t} = -D \frac{\partial^2 C(x,t)}{\partial x^2} \tag{12.4}$$

where D is the diffusion coefficient of the chemical agent of interest.

The solution pictured in Figure 12.1 for the concentration at the inner surface $(x = L)$ of the SC as a function of time beginning with the first exposure to a constant concentration is based on the solution developed by Carslaw and Jaeger (1959) from the theory of heat conduction in solids. Here the SC is visualized as an infinite flat plate of thickness L, where the two sides are held at constant (different) temperatures. (In the present case, the equivalent of the temperatures are the chloroform concentration in the bathwater and the chloroform concentration at the inner surface of the SC. The latter concentration is considered to be zero since the blood in the capillaries is constantly removing the chloroform.) The time constant τ_{SC} is defined as $L^2/6D$, where D is the diffusion coefficient. This turns out to be the time it takes to achieve a concentration approximately (but not exactly) equal to $(1 - 1/e)$ of the asymptotic level. (The exact value is 0.635.) At about 3 time constants, close to 99% of the equilibrium value has been reached (exact value: 98.6%). Thus the "lag time" described above is equal to $3\tau_{SC}$.

The solution for the concentration $C(x,t)$ in the SC as a function of distance from the outer surface $(0 < x < L)$ and time consists of a simple linear function of x plus an infinite series of products of sine waves and exponential functions of time. After sufficient time to reach equilibrium, the infinite series goes to zero and all that is left is the function of x:

$$C(x,t) = P_{sw} \, C_w \, (1 - x/L) \tag{12.5}$$

where P_{sw} is the skin-water partition coefficient and C_w is the concentration in the water.

This function is a simple straight line sloping downward from the concentration in the water (modified by the partition coefficient) at the skin-water interface $(x = 0)$ to a value of zero at the other (inner) surface of the SC $(x = L)$.

Figure 12.1 provided the picture of the buildup of the flux crossing the inner surface of the SC from the moment of first exposure; Figure 12.2 shows the decay of the flux crossing that same inner surface from the moment exposure ceases.

Although this is an elegant solution for the problem of the concentration in the SC and the flux across its inner surface, we cannot measure either of these quantities directly. Therefore the lower curve shown in Figure 12.3 presents what we can measure: the breath concentration as a function of time for persons exposed to a constant concentration in water. This breath concentration curve is linked to the flux curve shown above it. The time until the flux has achieved 0.635 of its final concentration is shown as τ_{SC} in the upper curve. The time to the first appearance of chloroform in the breath is shown as T_0 in the lower curve. The breath concentration begins to increase and after a time T_1 attains a maximum rate of increase which is maintained for a period of time. The time to first attain this maximum rate of increase in the breath corresponds to the time taken for the flux to reach its asymptote. The relationship between these various times can be stated as $T_0 + T_1 = 3\tau_{SC}$.

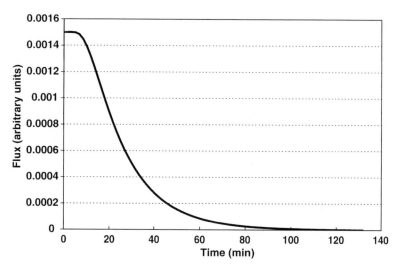

FIGURE 12.2 Exact infinite-series solution for the decay of the flux across the inner surface of the SC when exposure ceases. The time scale on the x-axis depends on the choices made for the width L of the SC and the value D of the diffusion coefficient.

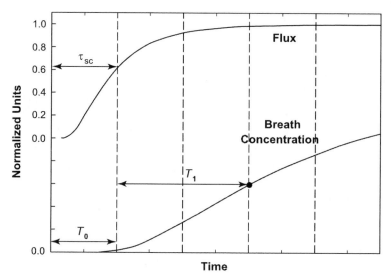

FIGURE 12.3 Relation between flux through SC and breath concentration. τ_{SC} is the time constant for diffusion through the SC (time to reach a value of $L^2/6D = 0.635$ of the equilibrium flux, where L is the thickness of the SC and D is the diffusion coefficient of the compound of interest). T_0 is the time to first appearance of the compound in the breath (about 2–3 minutes for chloroform). T_1 is the time from the first appearance to the maximum rate of increase in the breath concentration (about 4 to 6 minutes for chloroform). The maximum rate of increase is achieved at the time the flux reaches a maximum (i.e., the "lag time" = $3\,\tau_{SC}$). Thus $T_0 + T_1 = 3\,\tau_{SC}$. (From Gordon, Wallace, Callahan et al. 1998. With permission.)

12.4 INDIRECT MEASURES OF DERMAL ABSORPTION

12.4.1 SHOWERS AND BATHS

Jo, Weisel, and Lioy (1990a,b) measured exhaled breath of volunteers following showers with and without wet suits to estimate the fraction of total dose absorbed through the skin. Six individuals took 13 showers with and without wearing rubber raincoats and boots. During the normal shower, they were exposed both by inhalation and by skin absorption, but the other condition allowed only inhalation exposure. Breath samples were taken both before and after each shower. Breath levels following both types of showers increased linearly with the tap water concentration, but the breath levels were roughly double following the normal shower (Jo, Weisel, and Lioy 1990a). The authors interpreted this as evidence that skin absorption during showers (0.22 µg/kg/day for water with a chloroform level of 24.5 µg/L) would be roughly equivalent to the inhalation exposure (0.24 µg/kg/day). For a shower providing 40 µg chloroform through inhalation, another 40 µg would then be absorbed through the skin.

12.4.2 SWIMMERS

A route of exposure that will be unimportant for most persons, but possibly very important for persons swimming for long periods of time in indoor pools, has been a subject of interest. The results from the studies by Aggazzotti et al. (1987, 1990, 1993), Weisel and Shepard (1994), Jo (1994), Lahl et al. (1981), Lévesque et al. (1995), and Lindstrom, Pleil, and Berkoff (1997) suggest that at least 25% and perhaps up to 75% of the total exposure to chloroform from swimming in indoor pools is due to dermal absorption. Weisel and Shepard (1994) calculate that an adult swimming 1 hour/day, 3 days/week in a pool where the chloroform air concentration is 100 µg/m^3 would receive a weekly chloroform dose of 210 µg from inhalation. This estimate may be low, since they assume a respiration rate of only 1 m^3/h (although a swimmer almost certainly is breathing far more than that), with an absorption factor of 70%. They assume an equivalent dermal dose, resulting in a weekly dose due to swimming of 420 µg. The same individual ingesting 2 L/day of water at a chloroform concentration of 25 µg/L would have a weekly dose of 350 µg. They also calculate a weekly dose of 180 µg from a daily shower. Thus the dose from swimming 3 hours per week at an indoor pool is comparable to that from ingestion of drinking water and from inhalation and dermal absorption from showers.

12.5 DIRECT MEASURES OF DERMAL ABSORPTION USING CONTINUOUS BREATH MEASUREMENTS

A more direct method of measuring dermal absorption in humans was developed at Battelle under the sponsorship of the USEPA. The method involves a device for analyzing breath VOCs continuously, rather than collecting discrete samples for later laboratory analysis (Gordon, Kenny, and Kelly 1992; Gordon et al. 1996). The limiting factor in the latter process is cost; each sample costs about $300 to analyze and therefore only a few samples are typically taken during exposures that may last several hours and decay periods that may last many more hours. The typical number of samples taken may be on the order of a dozen. With the continuous breath analyzer, the entire period can be monitored and the equivalent of hundreds of measurements can be made.

The current version consists of a direct sampling system (glow discharge ionization source) and a compact quadrupole ion trap mass spectrometer that uses filtered noise fields for enhanced specificity and sensitivity, and is capable of operation in the full tandem (MS/MS) mass spectrometric mode. For breath analysis, a specially designed breath inlet system is attached to the glow discharge source and provides a constant source of exhaled air for the mass spectrometer. This direct breath sampling/mass spectrometric approach offers a powerful means of extracting VOCs directly from the breath matrix, and eliminates the pre-concentration step that normally precedes exhaled air analysis by conventional gas chromatography/mass spectrometry (GC/MS).

The first full-scale study using this device focused on dermal absorption of chloroform while bathing (Gordon et al. 1998). Ten subjects bathed in chlorinated water while breathing pure air through a face mask (Figure 12.4). Their breath was measured semi-continuously (12-second intervals) for the 30 minutes in the bathtub and an additional 30 minutes out of the tub while still wearing the face mask to prevent inhalation exposure from the elevated air chloroform concentrations following the bath (Figure 12.5). Seven of the subjects bathed at three water temperatures, nominally 30, 35, and 40°C.

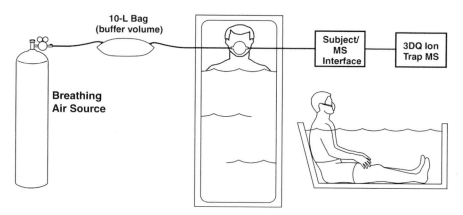

FIGURE 12.4 Experimental setup for sampling exhaled breath samples from a subject immersed in water. The hydrotherapy tub is shown in plan and side views. Pure air from the cylinder passes through a 10-L buffer volume to the face mask and exhaled breath is channeled through the mass spectrometer. (From Gordon, Wallace, Callahan et al. 1998. With permission.)

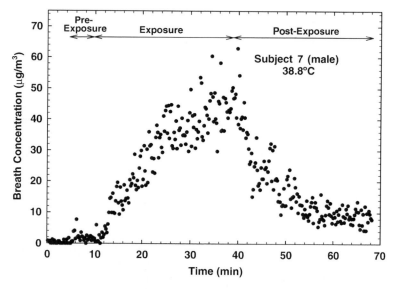

FIGURE 12.5 Breath concentration of chloroform due to dermal absorption only while exposed in a bathtub to a water concentration of 91 μg/L. Exposure began at $T = 10$ minutes and the initial increase in the breath occurred within 2–3 minutes after that. About 90% of the body surface area was under water. (From Gordon, Wallace, Callahan et al. 1998. With permission.)

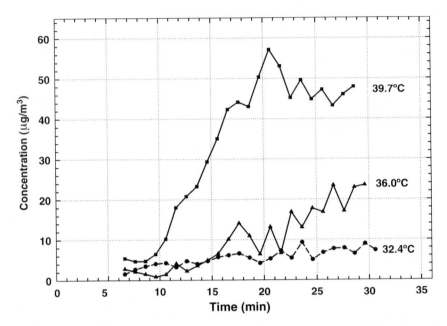

FIGURE 12.6 Effect of water temperature on exhaled breath concentrations of chloroform plotted as a function of time for one subject during dermal-only exposure while bathing. The total dermal absorption at the highest temperature was 30 times that at the lowest temperature. (From Gordon, Wallace, Callahan et al. 1998. With permission.)

The most dramatic result of this study was the powerful influence of bathwater temperature on chloroform uptake — the average uptake at the highest temperature was 30 times that at the lowest temperature (Figure 12.6). One possible explanation for this effect is the heat-dissipating mechanisms of the body (Gordon et al. 1998). At low bath or shower water temperatures, the capillaries closest to the skin surface experience greatly lessened blood flow. This effectively forces the chloroform to diffuse across a greater distance to reach the blood. Also, the lower temperature of the SC may cause a low speed of diffusion, a process strongly dependent on temperature (Glasstone et al. 1941). At high water temperatures, increased blood flow to the skin results in a shorter effective diffusion length and an increased speed of diffusion. If blood flow changes to the skin are responsible for these large changes in dose, we should expect to find temperature-related blood flow changes of at least this order of magnitude. In fact, measurements of blood flow in the extremities have shown even larger factors of 80–600 for blood flow through the fingers on changing from a cold to a hot environment (Mountcastle 1980). A complete PBPK model was prepared to account for the temperature effect by use of increased blood flow (Corley 1998; Corley, Gordon, and Wallace 2000).

This finding of a strong temperature/blood flow effect helps to explain certain results obtained in previous studies. For example, Jo, Weisel, and Lioy (1990a,b) found that for six subjects taking 13 showers with and without wet suits, the amount of chloroform exhaled during normal showers was about twice the amount exhaled during the showers with wet suits (inhalation only), suggesting that the dermal route contributed about 50% of the total dose. However, a study of swimmers by Lévesque et al. (1995) found that only about 25% of the exposure was due to dermal absorption. Finally, in a study of swimmers undergoing rigorous training, Lindstrom, Pleil, and Berkoff (1997) concluded that the dermal pathway was responsible for about 75% of the total exposure to chloroform. These different findings all suggest that the temperature/blood flow relation is paramount in determining dermal absorption. In the first case, all showers were at high temperatures, leading to a large dermal effect. In the Lévesque et al. (1995) swimming pool study, water temperatures were lower, leading to reduced dermal absorption. (Since all the swimmers in this study were exercising for part of the time, a study of persons not exercising much in swimming pools would be expected to produce even less dermal absorption.) Although the water temperature in the Lindstrom, Pleil, and Berkoff (1997) experiment was also lower than that of a normal shower, the rigorous aerobic training resulted in greatly increased blood flow and therefore in increased dermal absorption.

The second important finding from Gordon et al. (1998) was the first measured estimate of the "lag time," or time to reach equilibrium flux. Recall that the USEPA 1992 model estimated 29 minutes for this quantity and the McKone (1993) model estimated 12 minutes. Gordon et al. (1998) estimated this value at about 6–9 minutes. This increases the importance of dermal absorption as a route of exposure for short exposures such as occur in the bath or shower.

Recently it has been discovered that methyl-*tert*-butyl ether (MTBE) is the single most common VOC contaminant in the nation's groundwater (Squillace et al. 1996). MTBE was required by the USEPA to be added to gasoline in order to reduce VOC emissions and the formation of ozone. However, with the finding that it is contaminating groundwater, the USEPA reversed itself and banned MTBE from use. In the meantime, the groundwater will remain contaminated for some decades. Since about half the population is served by groundwater supplies, this could conceivably be a source of concern. Therefore the USEPA's Region 9 sponsored a study of exposure to MTBE while bathing or showering. The study, recently completed and not yet available in peer-reviewed form, was carried out using the same experimental setup as the Gordon et al (1998) paper describes. The final USEPA reports (Gordon 2005a,b) show that MTBE dermal-only exposures while bathing were far lower, and showed little temperature effect, compared to chloroform. However, inhalation exposures, as tested in chamber studies, were expected to be significant while bathing. A recent chamber study involving placing one arm in water highly contaminated with MTBE also developed estimates of the rate of dermal absorption (Prah et al. 2004).

12.6 CONCLUSION

Considerable progress has been made on several fronts in understanding dermal absorption from exposures to water in baths, showers, and swimming pools. Early field studies, such as TEAM, identified chlorinated water as a source of chloroform in indoor air and isolated showers and baths as a major contributor. Quite quickly after that, a series of experimental and theoretical studies attempted to estimate the dermal absorption coefficient associated with exposure in showers and baths. Other studies extended the effort to swimming pools. Certain puzzling results were evident in the mid-1990s, as the dermal absorption contribution appeared to vary widely from study to study.

On the modeling front, an application of a QSAR technique to a set of measured physicochemical characteristics of nearly 100 compounds provided reasonable estimates of the magnitudes of the permeability coefficients for many compounds with known octanol-water partition coefficients. A recent (Fitzpatrick et al. 2004) study extended these results to nearly 150 compounds and developed refined (but rather similar) estimates for the fundamental coefficients in the model.

Near the end of the 1990s, new measurement methods, such as the continuous breath measurement device, were developed to directly determine the time it takes for certain VOCs such as chloroform and MTBE to reach the bloodstream of human subjects. These chamber studies have led to the discovery that temperature and blood flow are crucial parameters that were missing from earlier PBPK models and now have been added to more recent models. They were also conclusive in explaining the ill-understood results of the earlier field studies of swimmers.

Thus there has been a fruitful back and forth relationship between field studies, modeling studies, and chamber studies, with each type sometimes taking the lead in discoveries that then helped the other.

12.7 QUESTIONS FOR REVIEW

1. Graph Equation 12.5 using arbitrary units, for x ranging from 0 to L. Label one side of the graph (which?) as outside the body and the other as inside the body. Use the graph to help you visualize the answers to the next two questions.

2. After you step out of your tub following a long, satisfying bath, the chloroform contained in the stratum corneum (SC) will diffuse in two directions — into and out of the body. Use simple reasoning based on the gradient of chloroform concentration in the SC (use your graph of Equation 12.5) to calculate roughly how much goes in each direction. [Answer: Since it was a "long" bath, the solution to the diffusion equation is just the steady-state linear solution: Equation 12.5. Assume all the chloroform in the outer half of the SC diffuses outward and all in the inner half diffuses inward. Then integration of the concentration (i.e., calculating the area under the curve for the two halves) provides the answer: only 2/7 of the total amount in the SC reaches your bloodstream.]

3. The answer to Question 1 is a slight underestimate of the amount that diffuses inward. Why? [Answer: Not every molecule in each half of the SC will migrate to the nearest boundary. Some in the outer half will reach the bloodstream; some in the inner half will reach the skin surface. Since there are more molecules just to the left (outside) of the center line than there are to the right (inside), more molecules in the outer half of the SC will reach the bloodstream than those molecules in the inner half that reach the surface of the skin.]

4. Three studies of swimmers and bathers concluded that dermal absorption of chloroform was 25%, 50%, or 75% of the total dermal + inhalation exposure. Two of the studies were of swimmers and one was of persons taking a shower. We know now that hot water will increase the dermal absorption of chloroform, probably by increasing blood flow to the skin. True or false: the shower study must have been the one with the 75% estimate. [False. Not only heat but also exertion will increase blood flow, and one of the two studies of swimmers was of Olympic-level swimmers in training situations.]

5. Looking at Equation 12.2, describe qualitatively what happens to skin permeability as larger molecules are considered. [Answer: Permeability goes down, because the diffusion of larger molecules is slower.]

6. If a risk-averse person asked you how to reduce dermal absorption of chloroform in the shower, what could you tell him or her? [Answer: Uptake could be reduced by a factor of 30 by lowering the temperature of the water from 40°C to 30°C. Chloroform concentrations in hot water would be reduced somewhat (but not by a factor of 30!) if the temperature of the hot water heater was lowered.]

REFERENCES

Aggazzotti, G., Fantuzzi, G., Righi, E., Tartoni, P., Cassinadri, T., and Predieri, G. (1993) Chloroform in Alveolar Air of Individuals Attending Indoor Swimming Pools, *Archives of Environmental Health,* **48**: 250-254.

Aggazzotti, G., Fantuzzi, G., Tartoni, P.L., and Predieri, G. (1990) Plasma Chloroform Concentrations in Swimmers Using Indoor Swimming Pools, *Archives of Environmental Health,* **45**: 175–179.

Aggazzotti, G., Predieri, G., Fantuzzi, G., and Benedetti, A. (1987) A Headspace Gas Chromatographic Analysis for Determining Low Levels of Chloroform in Human Plasma, *Journal of Chromatography,* **416**: 125–130.

Amy, G.L., Chadik, P.A., and Chowdhury, Z.K. (1987) Developing Models for Predicting Trihalomethane Formation Potential and Kinetics, *Journal of the Air & Waste Management Association,* **79**: 89.

Andelman, J.B. (1985a) Human Exposures to Volatile Halogenated Organic Chemicals in Indoor and Outdoor Air, *Environmental Health Perspectives,* **62**: 313–318.

Andelman, J.B. (1985b) Inhalation Exposure in the Home to Volatile Organic Contaminants of Drinking Water, *The Science of the Total Environment,* **47**: 443–460.

Bogen, K.T., Colston, B.W., and Machicao, L.K. (1992) Dermal Absorption of Dilute Aqueous Chloroform, Trichloroethylene, and Tetrachloroethylene in Hairless Guinea Pigs, *Fundamental and Applied Toxicology,* **18**: 30–39.

Brass, A.J., Feige, M.A., Halloran, T., Mello, J.W., Munch, D., and Thomas, R.F. (1977) The National Organic Monitoring Survey: Samplings and Analyses for Purgeable Organic Compounds, in *Drinking Water Quality Enhancement through Source Protection*, Pojasek, R.B., Ed., Ann Arbor Science, Ann Arbor, MI, 393–416.

Brass, H.J., Weisner, M.J., and Kingsley, B.A. (1981) *Community Water Supply Survey: Sampling and Analysis for Purgeable Organics and Total Organic Carbon, Draft*, American Water Works Association Annual Meeting, Water Quality Division.

Brown, H.S. and Hattis, D. (1989) The Role of Skin Absorption as a Route of Exposure to Volatile Organic Compounds in Household Tap Water: A Simulated Kinetic Approach, *Journal of the American College of Toxicology,* **8**: 839–851.

Cantor, K.P., Hoover, R., and Hartge, P. (1987) Bladder Cancer, Drinking Water Sources, and Tap Water Consumption: A Case-Control Study, *Journal of the National Cancer Institute,* **79**: 1269–1279.

Carslaw, H.S. and Jaeger, J.C. (1959) *Conduction of Heat in Solids*, Clarendon Press, Oxford, U.K.

Chinery, R.L. and Gleason, K.A. (1993) A Compartmental Model for the Prediction of Breath Concentration and Absorbed Dose of Chloroform after Exposure While Showering, *Risk Analysis,* **13**: 51–62.

Cleek, R.L. and Bunge, A. (1993) A New Method for Estimating Dermal Absorption from Chemical Exposure, 1. General Approach, *Pharmacological Research,* **10**(4): 497–506.

Corley, R. (1998) *Modeling Human Dermal Absorption of Chloroform Following Bathwater Exposures*, Final Report, EPA Contract # 68-D4-0023, U.S. Environmental Protection Agency, National Exposure Research Laboratory, Research Triangle Park, NC.

Corley, R.A., Gordon, S.M., and Wallace, L.A. (2000) Physiologically Based Pharmacokinetic Modeling of the Temperature-Dependent Dermal Absorption of Chloroform by Humans Following Bath Water Exposures, *Toxicological Sciences,* **53**: 13–23.

Corley, R.A., Markham, D.A., and Banks, C. (1997) Physiologically Based Pharmacokinetics and the Dermal Absorption of 2-Butoxyethanol Vapor by Humans, *Fundamental and Applied Toxicology,* **39**: 120–130.

Dowty, B., Carlisle, D., Laseter, J.L., and Storer, J. (1975) Halogenated Hydrocarbons in New Orleans Drinking Water and Blood Plasma, *Science,* **187**: 75–77.

Fitzpatrick, D., Corish, J., and Hayes, B. (2004) Modeling Skin Permeability in Risk Assessment — The Future, *Chemosphere,* **55**: 1309–1314.

Flynn, G.L. (1990) Physicochemical Determinants of Skin Absorption, in *Principles of Route-to-Route Extrapolation for Risk Assessment*, Gerrity, T.R. and Henry, C.J., Eds., Elsevier, New York, NY.

Giardino, N.J., Gordon, S.M., Brinkman, M.C., Callahan, P.J., and Kenny, D.V. (1999) Real-Time Breath Analysis of Vapor Phase Uptake of 1,1,1-Trichloroethane through the Forearm: Implications for Daily Absorbed Dose of Volatile Organic Compounds at Work, *Applied Occupational and Environmental Hygiene,* **14**: 719–727.

Glasstone, S., Laidler, K.J., and Eyring, H. (1941) Viscosity and Diffusion, in *The Theory of Rate Processes: The Kinetics of Chemical Reactions, Viscosity, Diffusion and Electrochemical Phenomena*, McGraw-Hill, New York, NY, 477–511.

Gordon, S. (2005a) *Inhalation Exposure to Methyl Tert-Butyl Ether (MTBE) Using Continuous Breath Analysis*, Final Report, EPA Contract # 68-D-99-011, Report No. EPA/600/R-05/095. U.S. Environmental Protection Agency, National Exposure Research Laboratory, Research Triangle Park, NC.

Gordon, S. (2005b) *Human Exposure to Methyl Tert-Butyl Ether (MTBE) While Bathing with Contaminated Water*, Final Report, EPA Contract # 68-D-99-011, Report No. EPA/600/R-05/094. U.S. Environmental Protection Agency, National Exposure Research Laboratory, Research Triangle Park, NC.

Gordon, S.M., Kenny, D.V., and Kelly, T.J. (1992) Continuous Real Time Breath Analysis for the Measurement of Half Lives of Expired Volatile Organic Compounds, *Journal of Exposure Analysis and Environmental Epidemiology,* Supp. 1: 41–54.

Gordon, S.M., Callahan, P.J., Kenny, D.V., and Pleil, J.D. (1996) Direct Sampling and Analysis of Volatile Organic Compounds in Air by Membrane Introduction and Glow Discharge Ion Trap Mass Spectrometry with Filtered Noise Fields, *Rapid Communications in Mass Spectrometry,* **10**: 1038–1046.

Gordon, S.M., Wallace, L.A., Callahan, P.J., Kenny, D.V., and Brinkman, M.C. (1998) Effect of Water Temperature on Dermal Exposure to Chloroform, *Environmental Health Perspectives,* **106**: 337–345.

Hake, C.L. and Stewart, R.D. (1977) Human Exposure to Tetrachloroethylene: Inhalation and Skin Contact, *Environmental Health Perspectives,* **21**: 231–238.

Jo, W.K. (1994) Chloroform in the Water and Air of Korean Indoor Swimming Pools Using Both Sodium Hypochlorite and Ozone for Water Disinfection, *Journal of Exposure Analysis and Environmental Epidemiology,* **4**: 491–502.

Jo, W.K., Weisel, C.P., and Lioy, P.J. (1990a) Routes of Chloroform Exposure and Body Burden from Showering with Contaminated Tap Water, *Risk Analysis,* **10**: 575–580.

Jo, W.K., Weisel, C.P., and Lioy, P.J. (1990b) Chloroform Exposure and the Health Risk Associated with Multiple Uses of Chlorinated Tap Water, *Risk Analysis,* **10**: 581–585.

Jolley, R.L., Ed. (1983) *Water Chlorination: Environmental Impact and Health Effects,* 4th ed., Ann Arbor Scientific Publications, Ann Arbor, MI.

Krasner, S.W., Mcguire, M.J., Jacangelo, J.G., Patania, N.L., Reagan, K.M., and Aieta, E.M. (1989) The Occurrence of Disinfection By-Products in U.S. Drinking Water, *Journal of the American Water Works Association,* **81**(8): 41–52.

Lahl, U., Batjer, K., Van Duszelin, J., Gabel, B., and Thiemann, W. (1981) Distribution and Balance of Volatile Halogenated Hydrocarbons in the Water and Air of Covered Swimming Pools Using Chlorine for Water Disinfection, *Water Research,* **15**: 803–814.

Lévesque, B., Ayotte, P., Leblanc, A., Dewailly, É., Prud'Homme, D., Lavoie, R., Sylvain, A., and Levallois, P. (1995) Evaluation of Dermal and Respiratory Chloroform Exposures in Humans, *Environmental Health Perspectives,* **102**: 1082–1087.

Lindstrom, A.B., Pleil, J.D., and Berkoff, D.C. (1997) Alveolar Breath Sampling and Analysis to Assess Trihalomethane Exposures during Competitive Swimming Training, *Environmental Health Perspectives,* **105**: 636–642.

Little, J.C. (1992) Applying the Two-Resistance Theory to Contaminant Volatilization in Showers, *Environmental Science and Technology,* **26**(7): 1341-1349.

Mcguire, M.J. and Meadow, R.G. (1988) AWWARF Trihalomethane Survey, *Journal of the American Water Works Association,* **80**: 61.

McKone, T.E. (1987) Human Exposure to Volatile Organic Compounds in Household Tap Water: The Indoor Inhalation Pathway, *Environmental Science and Technology,* **21**: 1194–1201.

McKone, T.E. (1993) Linking a PBPK Model for Chloroform with Measured Breath Concentrations in Showers: Implications for Dermal Exposure Models, *Journal of Exposure Analysis and Environmental Epidemiology,* **3**(3): 339–365.

McKone, T.E. and Howd, R.A. (1992) Estimating Dermal Uptake of Nonionic Organic Chemicals from Water and Soil, Part 1: Unified Fugacity-Based Models for Risk Assessments, *Risk Analysis,* **12**: 543–557.

Morris, R.D., Audet, A.M, Angelillo, I.F., Chalmers, T.C., and Mosteller, F. (1992) Chlorination By-Products and Cancer: A Meta-Analysis, *American Journal of Public Health,* **82**: 955–963.

Mountcastle, V.B. (1980) *Medical Physiology: Volume Two,* 14th ed., C.V. Mosby, St. Louis, MO, 1101–1102.

NCI (1976) *Report on Carcinogenesis Bioassay of Chloroform,* Report No. NTIS PB-264018, National Cancer Institute, Springfield, VA.

NTP (1985) *Toxicology and Carcinogenesis Studies of Chlorodibromomethane in F344/N Rats and B6C3F1 Mice (Gavage Studies),* U.S. Department of Health and Human Services, Technical Report Series No. 282, National Toxicology Program, Research Triangle Park, NC.

NTP (1987) *Toxicity and Carcinogenesis Studies of Bromodichloromethane (CAS No. 75-27-4) in F344/N Rats and B6C3F1 Mice (Gavage Studies),* U.S. Department of Health and Human Services, Technical Report Series No. 321, National Toxicology Program, Research Triangle Park, NC.

NTP (1989a) *Toxicology and Carcinogenesis Studies of Tribromomethane (Bromoform) in F344/N Rats and B6C3F1 Mice (Gavage Studies),* U.S. Department of Health and Human Services, Technical Report Series No. 350, National Toxicology Program, Research Triangle Park, NC.

NTP (1989b) *Bromoform: Reproduction and Fertility Assessment in Swiss CD-1 Mice When Administered by Gavage,* Report No. NTP-89-068, National Institute of Environmental Health Sciences, National Toxicology Program, Research Triangle Park, NC.

Patel, H., Ten Berg, W., and Cronin, M.T.D., (2002) Quantitative Structure–Activity Relationships (QSARs) for the Prediction of Skin Permeation of Exogenous Chemicals, *Chemosphere,* **48**: 603–613.

Pellizzari, E.D., Perritt, K., Hartwell, T.D., Michael, L.C., Whitmore, R., Handy, R.W., Smith, D., and Zelon, H. (1987a) *Total Exposure Assessment Methodology (TEAM) Study: Vol. II: Elizabeth and Bayonne, New Jersey, Devils Lake, North Dakota, and Greensboro, North Carolina,* U.S. Environmental Protection Agency, Washington, DC.

Pellizzari, E.D., Perritt, K., Hartwell, T.D., Michael, L.C., Whitmore, R., Handy, R.W., Smith, D., and Zelon, H. (1987b) *Total Exposure Assessment Methodology (TEAM) Study: Vol. III: Selected Communities in Northern and Southern California,* U.S. Environmental Protection Agency, Washington, DC.

Potts, R.O. and Guy, R.H. (1992) Predicting Skin Permeability, *Pharmacological Research,* **9**: 663–669.

Prah, J., Ashley, D., Blount, B., Case, M., Leavens, T., Pleil, J., and Cardinali, F. (2004) Dermal, Oral, and Inhalation Pharmacokinetics of Methyl Tertiary Butyl Ether (MTBE) in Human Volunteers, *Toxicological Sciences,* **77**: 195–205.

Rook, J.J. (1976) Haloforms in Drinking Water, *Journal of the American Water Works Association,* **68**: 168–172.

Rook, J.J. (1977) Chlorination Reactions of Fulvic Acids in Natural Waters, *Environmental Science and Technology,* **11**: 478–482.

Scheuplein, R.J. and Blank, I.H. (1971) Permeability of Skin, *Physiological Review,* **51**: 702–747.

Schnoor, J.L., Nitzschke, J.L., Lucas, R.D., and Veenstra, J.N. (1979) Trihalomethane Yields as a Function of Precursor Molecular Weight, *Environmental Science and Technology,* **13**: 1134–1138.

Shatkin, J.A. and Brown, H.S. (1991) Pharmacokinetics of the Dermal Route of Exposure to Volatile Organic Chemicals in Water: A Computer Simulation Model, *Environmental Research,* **56**: 90–108.

Squillace, P.J., Zogorski, J.S., Wilber, W.G., and Price, C.V. (1996) Preliminary Assessment of the Occurrence and Possible Sources of MTBE in Groundwater in the United States, 1993–1994, *Environmental Science and Technology,* **30**(5): 1721–1730.

Stewart, R.D. and Dodd, H.C. (1964) Absorption of Carbon Tetrachloride, Trichloroethylene, Tetrachloroethylene, Methylene Chloride, and 1,1,1-Trichloroethane through Human Skin, *Journal of the American Industrial Hygiene Association,* **25**: 439–446.

Symons, J.M., Bellar, T.A., Carsell, J.K., Demarco, J., Kropp, K.L., Robeck, G.G., Seeger, D.R., Slocum, C.J., Smith, B.L., and Stevens, A.A. (1975) National Organics Reconnaissance Survey for Halogenated Organics, *Journal of the American Water Works Association,* **67**: 708–729.

USEPA (1979) National Interim Primary Drinking Water Regulations, Control of THMs in Drinking Water, Final Rule, Office of Drinking Water, *Federal Register,* **44**(231): 68624–68707.

USEPA (1984) *Health Assessment Document for Chloroform, Part 1,* Report No. EPA 600/8-84-004a, U.S. Environmental Protection Agency, Washington, DC.

USEPA (1992) *Dermal Exposure Assessment: Principles and Applications,* Report No. EPA/600/8-91/011b, U.S. Environmental Protection Agency, Washington, DC.

Vena, J.E., Graham, S., Freudenheim, J., Marshall, J., Zielezny, M., Swanson, M., and Sufrin, G. (1993) Drinking Water, Fluid Intake and Bladder Cancer in Western New York, *Archives of Environmental Health,* **48**: 191–198.

Wallace, L.A. (1997) Human Exposure and Body Burden for Chloroform and Other Trihalomethanes, *Critical Reviews in Environmental Science and Technology,* **27**: 113–194.

Wallace, L.A., Pellizzari, E.D., Hartwell, T.D., Davis, V., Michael, L.C., and Whitmore, R.W. (1989) The Influence of Personal Activities on Exposure to Volatile Organic Compounds, *Environmental Research,* **50**: 37–55.

Wallace, L.A., Pellizzari, E., Hartwell, T., Sparacino, C., Sheldon, L., and Zelon, H. (1985) Personal Exposures, Indoor-Outdoor Relationships and Breath Levels of Toxic Air Pollutants Measured for 355 Persons in New Jersey, *Atmospheric Environment,* **19**: 1651–1661.

Weisel, C.P. and Chen, W.J. (1994) Exposure to Chlorination By-Products from Hot Water Uses, *Risk Analysis,* **14**: 101–106.

Weisel, C.P. and Shepard, T.A. (1994) Chloroform Exposure and the Body Burden Associated with Swimming in Chlorinated Pools, in *Water Contamination and Health: Integration of Exposure Assessment, Toxicology, and Risk Assessment,* Wang, R.G.M., Ed., Marcel Dekker, New York, NY, 135–147.

Weisel, C.P., Jo, W.K., and Lioy, P.J. (1992) Utilization of Breath Analysis for Exposure and Dose Estimates of Chloroform, *Journal of Exposure Analysis and Environmental Epidemiology,* Supp. 1: 55–70.

Wester, R.C. and Maibach, H.I. (1989) Human Skin Binding and Absorption of Contaminants from Ground and Surface Water during Swimming and Bathing, *Journal of the American College of Toxicology,* **8**: 853–860.

Westrick, J.J., Mello, J.W., and Thomas, R.F. (1983) *The Ground Water Supply Survey Summary of Volatile Organic Contaminant Occurrence Data,* U.S. Environmental Protection Agency, Technical Support Division, Office of Drinking Water and Office of Water, Cincinnati, OH.

Part IV

Ingestion

13 Ingestion Exposure

P. Barry Ryan
Emory University

CONTENTS

13.1 Synopsis..303
13.2 Introduction..304
13.3 Background...304
    13.3.1 Basic Physiology 304
    13.3.2 Contrasts with Other Routes 304
    13.3.3 Active vs. Passive Transport 305
13.4 Dietary Ingestion 306
    13.4.1 Water Ingestion...306
        13.4.1.1 Pesticides...306
        13.4.1.2 Metals 307
        13.4.1.3 VOCs 307
    13.4.2 Food Ingestion 308
        13.4.2.1 Metal Contamination..308
        13.4.2.2 Organics..310
        13.4.2.3 Ingestion of Foods Contaminated after Processing......................311
13.5 Ingestion of Nondietary Items 311
    13.5.1 Hand-to-Mouth 311
    13.5.2 Other Soil Ingestion 312
    13.5.3 Inhalation/Swallowing 312
    13.5.4 Intake through Inadvertent Ingestion 312
13.6 Summary and Further Directions for Research 312
13.7 Questions for Review 313
References 313

13.1 SYNOPSIS

Ingestion, both intentional and inadvertent, represents a major route of exposure to environmental contaminations. Ingestion of drinking water and foodstuffs containing environmental contaminants can result in significant exposure to many compounds with known adverse health effects. The gastrointestinal system offers an important protective mechanism to preclude absorption of harmful contaminants but, by its essential processes, affords other mechanisms that allow passage of possibly harmful materials into the body. In this chapter we will assess this route of exposure. We begin with a brief discussion of the physiology of the gastrointestinal system contrasting it with the other barriers to absorption: the lung epithelium and the skin. This is followed by a discussion of the primary pathways of ingestion exposure: water ingestion, food ingestion, and ingestion of nondietary items. At each stage, we will call upon current scientific work to address specifics of these issues. Since the mechanisms for ingestion and absorption of inorganic and organic compounds are different, we will investigate each separately. Further, we will use typical values of dietary and nondietary intake to calculate the likely exposure and intake of specific chemical compounds in water, food, and in nondietary items. These calculations, done for both children and adults, give insight into the levels of contaminants taken into the body through this route.

13.2 INTRODUCTION

An important route of exposure is ingestion of contaminated materials. The boundary active in ingestion exposure is the gastrointestinal epithelium. It is made up of multiple cell layers consisting of proteins, lipids, and water that act as a barrier to the outside world. However, this barrier must be permeable, for if it were not, no nutrients could be absorbed into the body and we would not survive. The permeability of this barrier has evolved over time to ensure proper transfer of materials across it to provide adequate nutrients. Unfortunately, during historical times, a significant change has occurred in the compounds present in the environment. The mechanisms in place to afford the absorption of essential nutrients from food also afford absorption of unwanted contaminants from the environment into the body with the potential for adverse health outcomes.

In this chapter we will assess this route of exposure. We begin with a brief discussion of the physiology of the gastrointestinal system, contrasting it with the other barriers to absorption: the lung epithelium and the skin. This is followed by a discussion of the primary pathways of ingestion exposure: water ingestion, food ingestion, and ingestion of nondietary items. At each stage, we will call upon current scientific work to address specifics of these issues.

13.3 BACKGROUND

13.3.1 BASIC PHYSIOLOGY

In order for an environmental contaminant to have an effect on any biological system, it must first pass the boundary of the system and thus be absorbed. For ingestion-based exposure, the primary mechanism is absorption through the gastrointestinal tract. The gastrointestinal tract may be viewed as a long tube commencing with the mouth and ending with the anus (Greene and Moran 1994; Hunt and Groff 1990; Rozman and Klaassen 1996). Topologically, the entire tract is outside the body, so that contaminants that are to have a toxic influence, other than those causing direct damage to the gastrointestinal tract itself (e.g., strong acids or bases) must be transported across the boundary. While this is typical of exposures through all routes, absorption across this boundary differs in that the basic purpose of the gastrointestinal tract is to extract nutrients from food and afford the absorption of such into the body. Thus this boundary is required to be permeable.

13.3.2 CONTRASTS WITH OTHER ROUTES

The epithelial layer of the lung is quite thin, only one or two cell layers thick, making passage across it relatively easy for many molecules as well as some macroscopic materials including small particles. The skin, which is designed to keep out contaminants, is quite thick with numerous cell layers of different construction. The gastrointestinal tract lies intermediate in thickness and varies in composition from the mouth, through the esophagus, stomach, and small and large intestines. Each component of the gastrointestinal system has a specific structure that optimizes the extraction of certain nutrients and water from ingested materials. To effect this extraction, certain digestive compounds are produced that break down the complex food matrix and afford the transfer. Structures within the system, particularly in the stomach, small intestines, and large intestines, facilitate this extraction and trans-barrier transport.

13.3.3   ACTIVE VS. PASSIVE TRANSPORT

Transport across the gastrointestinal tract can occur by two mechanisms. The first of these is *simple diffusion*. Nutrients, and other materials, that are present in higher concentrations on one side of a membrane that is permeable naturally diffuse through the membrane. The lining of the gastrointestinal tract is no exception to this; the gastrointestinal epithelium may be looked upon as a permeable membrane with ingested material on one side and the body on the other. If a given material is in higher concentration on the "outside" it may diffuse across the permeable barrier into the body and, thereby, be absorbed. The mechanisms for this process are varied. One important mechanism is the dissolution of organic materials into the organic-rich lipid layers in the membrane and the consequent migration through such layers. The lipid-based membranes are perfused with blood capillaries that afford a mechanism for removal of high-concentration nutrients. Organic contaminants, such as pesticides, ingested organic solvents, and other such materials, may be transported across the gastrointestinal epithelium in this passive fashion.

Inorganic materials, especially those that are ionic in nature, are not easily passed through the gastrointestinal epithelium since this medium is composed mostly of organic components. The concentration gradient supplies insufficient impetus to get the compound from one side of the membrane to the other; too much energy is required to "dissolve" the inorganic compound in the organic matrix despite the presence of the concentration difference. Under such conditions, *facilitated diffusion* may occur. In this process, a carrier material binds the molecule and facilitates its transport. The process is still passive in that it requires no direct energy input but does require the presence of the carrier species, typically a protein or other macromolecule. The molecule to be transported binds to the protein reversibly and is transported by the random motion of the macromolecule through the membrane where it is later released. Because the concentration of the compound is higher on the "outside," the random release of the compound affects the concentration "inside" more markedly.

Certain nutrients and other materials may be transported across the interface despite being present in lower concentrations in the gastrointestinal tract. Further, macroscopic materials (e.g., particles) that would diffuse only very slowly, may also be transported. However, such processes do not occur spontaneously and require energy to occur. This is referred to as *active transport*. Usually, this process occurs by attachment of the selected molecule to a large macromolecule, such as a protein, that is ideally suited to transport the molecule in question (Greene and Moran 1994; Rozman and Klaassen 1996) in a process analogous to facilitated diffusion described above. However, in this case, energy must be put into the system to, for example, change the structure of the protein allowing the transfer to occur.

Alternative strategies for transport may be employed as well. These include engulfing the molecule or a particle (a process called pinocytosis) by a component of the cell wall. Energy is required to close off the cell wall, forming a sac. The sac is transported across the membrane then released into the bloodstream on the "inside" (Hunt and Groff 1990). This mechanism is used to transport large molecules (e.g., dietary lipids and proteins) in an intact fashion into the body. The mechanism is similar to the sequestration of lipids in breast milk production (Lawrence 1994).

The combination of active and passive transport across the gut epithelium accounts for not only the transport of water and nutrients from ingested food, but also other materials contained within the ingested material that are not considered healthful. Both organic and inorganic components are transported by these mechanisms. Organic contaminants are transported primarily by diffusion, while inorganic components are transported primarily by facilitated diffusion. Larger, macroscopic particles are transported by pinocytosis.

13.4 DIETARY INGESTION

Dietary ingestion is the principal pathway associated with the ingestion route. Each day the typical adult ingests about 1 1/2 liters of water with the majority drinking between 0.9 and 2.1 L (USEPA 1997). The U.S. Environmental Protection Agency (USEPA 1997) suggests using 2 liters per day when modeling environmental exposures through drinking water. In one study, adults consumed somewhat less than one kilogram of food per day, with most falling within the 0.5–1.25 kg/day range (Scanlon et al. 1999; MacIntosh et al. 2001; MacIntosh, Kabiru, and Ryan 2001; Ryan, Scanlon, and MacIntosh 2001). We will use this number (1 kg) to estimate dietary intake of contaminants.

In addition to the naturally occurring waterborne compounds and nutrients in food, there is the potential to ingest contaminants found in these media. Such contaminants, even those found in very minute quantities, can have significant health impacts. We now continue our discussion by looking at the "intentional" exposures that occur through ingestion of water and food as part of our daily diet. Later we will discuss unintentional or inadvertent ingestion of contaminants that is associated with activities not normally considered as part of dietary ingestion.

13.4.1 Water Ingestion

Water sources range from individual wells to large-scale municipal water systems. In municipal systems, water purification usually takes place that is designed to eliminate many contaminants, most notably biological contamination including bacteria, viruses, and parasites. However, even in such systems, as well as untreated water supplies such as individual wells, many environmental contaminants remain in the water delivered to the tap. The treatment of water may also create new contaminants. One example is that trihalomethanes can be introduced into drinking water supplies through the interaction of chlorine-containing compounds and organic matter in the water. Further, contamination that occurs in the delivery system (e.g., lead in feeder lines and soldered piping) can contribute to the overall burden on contaminants consumed by individuals.

13.4.1.1 Pesticides

Most pesticides are organic in nature and do not dissolve well in water. This is particularly true of the persistent organochlorine pesticides such as DDT, chlordane, and toxaphene, which were used extensively in the U.S., from the 1930s through the 1970s. Though very persistent in the environments, these compounds are essentially insoluble in water and are measured in only very minute quantities in water supplies. The same can be said for the more modern pesticides, the organophosphate pesticides (e.g., chlorpyrifos, malathion, diazinon) and pyrethrin/pyrethroid pesticides (e.g., permethrin, cypermethrin, deltamethrin). Although these compounds degrade fairly rapidly in the environment, they display little solubility in water, and thus water ingestion does not contribute significantly to the overall body burden.

An exception to the general trends noted above occurs for the herbicide atrazine. Atrazine is a triazine herbicide used to control weed growth in farm crops and as a roadside herbicide to restrict the growth of weeds in such environments. Atrazine does not biodegrade very rapidly and is resistant to breakdown by sunlight. It is relatively soluble in water and can often be found in measurable quantities in water supplies. USEPA has placed controls upon the concentrations of atrazine allowed in water supplies. The Maximum Contaminant Limit (MCL) for atrazine has been set at 3 parts per billion (ppb) (USEPA 2004a). Further, if the concentration of this compound is found to be greater that 1 ppb, the municipal water control group must set up a monitoring program to evaluate atrazine concentrations. If the MCL is exceeded, measures must be taken to reduce the concentration of this compound and to inform the public of its presence.

Atrazine concentrations have a seasonal profile. The use of the herbicide, especially in agricultural areas, tends to peak in the spring when it is used for pre-emergent weed control. This use is reflected in an increase in measured atrazine concentrations in water supplies (MacIntosh, Hammerstrom, and Ryan 1999) in the early summer. An additional peak occurs in drinking water supplies in the late summer and early fall, possibly due to the residential use of this compound, again for weed control.

Typical atrazine concentrations found in municipal water supplies tend to be below the 1 ppb cutoff. Using this value and the expected ingestion intake of 2 liters of water per day, the average adult is likely to take in less than 1 μg of atrazine per day. Assuming a body weight of 70 kg, this corresponds to an average daily dose of less than 0.05 μg of atrazine/kg body weight/day.

13.4.1.2 Metals

Environmental metal contamination usually consists of inorganic compounds with a wide range of solubilities. Calcium chloride, which is often applied to surface streets in cold climates to improve roadway conditions, is very soluble and leaches very quickly into both surface and surficial aquifer water supplies. On the other hand, lead sulfide (galena) is nearly insoluble and may not be found in measurable quantities in water supplies even if the material is present in the local environment in large amounts.

Concentrations of many metals (e.g., lead and arsenic) are regulated by the USEPA in a manner similar to the way pesticides are regulated. The MCLs for lead and arsenic are 15 ppb and 10 ppb, respectively (USEPA 2004b). The USEPA has established a goal of zero concentration for each of these metals in the future. The concentration of lead and arsenic in water does display some seasonal variability (Ryan, Scanlon, and MacIntosh 2001).

Typical lead concentrations found in municipal water supplies tend to be near the 15 ppb cutoff. Using this value and the expected ingestion intake of 2 liters of water per day, the average adult is likely to take in less than 30 μg of lead per day. Assuming a body weight of 70 kg, this corresponds to an average daily dose of less than 0.5 μg of lead/kg body weight/day. Arsenic concentrations in drinking water supplies tend to be highly variable and depend on local geological formations. Most municipal water supplies have very low concentrations of arsenic. However, some locations, including certain locations in Utah and New Mexico, have concentrations at or even exceeding the 10 ppb MCL. Using this value, and the expected intake of 2 liters of water per day, one would ingest approximately 20 μg of arsenic per day. Assuming a body weight of 70 kg, this corresponds to an average daily dose of less than 0.3 μg of arsenic/kg body weight/day.

13.4.1.3 VOCs

Volatile organic compounds (VOCs) are low-molecular-weight organic compounds that are typically found in air. Although the normal route of exposure for these compounds, especially more polar chemical species, is through the inhalation route, some of these compounds have significant solubilities in water. Drinking water systems purified using chlorine often contain trace levels of chloroform and related compounds, due to the interaction of the purifying agent and organic matter initially present in the water supply. Ingestion of drinking water can result in exposure to these compounds. Other volatile organic compounds with significant solubilities include some solvents used in industry, such as trichloroethane (TCE) used as a degreasing agent and tetrachloroethylene (PERC), used in dry cleaning. Both can contaminate groundwater sources through spillage and can be a major exposure concern.

The USEPA has attempted to control such contamination through implementation of MCLs for these compounds as well. The USEPA allows 100 ppb of chloroform (measured as trihalomethanes), 200 ppb of TCE, but only 5 ppb for PERC in drinking water. Exposure to these compounds has been associated with various diseases of the liver and kidneys. The varying values for the MCLs reflect the potency of the effect in animals. Again, assuming 2 liters per day intake at the MCL, adults may be expected to take in approximately 200 μg of chloroform, 400 μg of TCE, and 10 μg of PERC per day. Most water supplies contain substantially less than the MCL for each of these compounds. However, some industrially contaminated supplies do contain more. Adults taking in these amounts have a body-weight corrected intake of about 3 μg of chloroform/kg body weight/day, about 6 μg of TCE/kg body weight/day, and about 0.2 μg of PERC/kg body weight/day.

13.4.2 FOOD INGESTION

The ingestion of contaminated food represents the most common pathway for dietary ingestion. Due to the varying chemical nature of foods, most contaminants can be found in some food materials. Metal contamination in soil used for agriculture may result in increased concentration in crops grown on the land. Organic compounds, particularly persistent organics such as the organochlorine pesticides, are often found in the fatty tissues of animals. Ingestion of these crops or tissues may lead to exposure to the contaminants contained in them. Let us consider some specific examples.

13.4.2.1 Metal Contamination

The contamination of foodstuffs by metals is a worldwide problem stemming partially from the use of these materials in industrialized societies. A significant amount of research is being done in this area in an attempt to characterize the magnitude of the problem, the sources of the contamination, and the potential outcomes through investigation of human uptake.

The international interest in metal contamination of the diet is evident through the diversity of research being done throughout the world. Cadmium exposures have been investigated by researchers in Japan (Ikeda et al. 2004). Researchers in France have investigated multiple metal exposures experienced by individuals dining in French restaurants (Noel, Leblanc, and Guerin 2003). Researchers in Kuwait have examined multiple food items for several metals using a market basket approach in which samples are gathered from markets and exposures inferred from dietary questionnaires (Husain et al. 2003). Large-scale investigations in the United Kingdom (Ysart et al. 1999, 2000) and in the United States (Pennington 2001; Pennington and Hernandez 2002) have attempted to characterize the total dietary input for certain metals. Also in the United States, a series of investigations designed to assess exposures to various pollutants in various media have added to our knowledge of diet-based metal exposures (Moschandreas et al. 2002; Ryan, Scanlon, and MacIntosh 2001; Scanlon et al. 1999). Smaller-scale investigations have been undertaken to understand the impact of diet lead content on children (Melnyk et al. 2000). Significant work is being done in the developing world as well (Alam, Snow, and Tanaka 2003). Exposures can be measured even in non-industrialized remote regions such as Baffin Island in Canada (Chan et al. 1995) suggesting the ubiquitous nature of dietary exposures to metals.

The commonness of the contamination of foodstuffs raises the question of the source of such contamination. Metals are, themselves, widely dispersed in the environment in both industrialized and nonindustrialized society. The growth of plant crops may be viewed as extracting nutrients from the soil in which the plant grows, coupling that with gathering energy from sunlight to afford the synthesis of large biomolecules in the plants themselves. Consumption of the plant material results in an energy source for animal species higher on the food chain. In addition to nutrients drawn from the soils, plants may also take up contaminants in the soil that have similar chemical properties. Consumption of such plant material may result in exposure to these contaminating species.

Evaluation of the uptake of contaminant species by plants has been a significant area of research as well. This research has been carried out in both laboratory settings, in which relatively high concentrations of metal compounds are introduced to the soil and uptake measured, and in natural settings, in which the source of the metal contamination may be viewed as "natural," although human activities are often the cause. Examples of such investigations abound. Laboratory investigations looking at plant uptake often focus on representative species under conditions of particular interest. An example is a study of metal uptake by dill and peppermint plants using a high-copper compost as the source (Zheljazkov and Warman 2004). These researchers were interested in the effects of using compost composed, in part, of industrial sludge and the consequent uptake of metals by plants. Their concern focuses on the use of such materials on crops and the potential for consequent increase in exposure experienced by individuals consuming those crops. Other laboratory-based investigations have been designed to look at the uptake of other food species, including non-crop species such as *Daphnia magna* or "water flea," a small crustacean that is an important species on the food chain intermediate between plants and higher organisms such as fish (Yu and Wang 2002), and midges (Malchow, Knight, and Maier 1995). These simpler biological systems are used as models for higher systems. Studies have also been done to determine the uptake of certain animal species in contact with contaminated sediments (Wang 2002).

While much information is gathered from laboratory investigations, measurement of uptake of metals in the natural setting offers insight into exposures likely experienced through the food chain. One study, done in Bangladesh, a country suffering from severe problems associated with large amounts of arsenic in drinking water, examined the uptake of various metals including arsenic, cadmium, lead, copper, and zinc by plants in typical agricultural soils (Alam, Snow, and Tanaka 2003). Their work suggested relatively low uptake of arsenic from such soils but did find elevated levels of other metals, suggesting differential uptake of the various metal species in food crops. Another investigation studied the uptake of grasses planted in a revegetation effort in an area contaminated with mine tailings containing arsenic (Milton and Johnson 1999). These investigators further examined the movement of arsenic through the food chain concentrating on, first, plant uptake, and then consumption of arsenic-containing plant material by various other biological species. They found evidence for bioaccumulation, or increase in concentration in species higher on the food chain, of arsenic. While commonly accepted in organic contamination (see below), bioaccumulation is less common in metal contamination unless the species is organometallic (existing in a combination of metal and organic components) in nature. Another investigation examined the bioaccumulation of selenium through the food chain from invertebrates consumed by birds to elevated concentration in their eggs (Fairbrother et al. 1999). Consumption of eggs by higher species, including humans, may result in increased exposures. Mercury exposure through dietary consumption of fish (Harris and Bodaly 1998) is of concern and has received attention even in the popular press. Other studies have investigated metal concentrations in tilapia (Suhendrayatna et al. 2001) and in the food chain starting with benthic organisms and continuing to sport fish often consumed by humans (Besser et al. 2001).

The presence of metals throughout the food web suggests that exposure to metals through the dietary pathway is likely to be very important in understanding the total exposure experienced by an individual. An additional important consideration is the nature of the metal found in the foodstuff. A key element is the bioavailability of the metal. If the metal is in an insoluble form, for example, lead as lead sulfide, exposure through dietary ingestion may offer little risk since the compound will not be absorbed across the gastrointestinal epithelium. Alternatively, an organic form, such as methyl mercury, may increase the amount absorbed substantially. Determination of the bioavailable fraction of metals in a food sample is often approached using a model of the human digestive system (Ruby et al. 1996). By examining the fraction of material available for absorption, rather than simply the amount present in the food itself, one can achieve a better understanding of the effect of ingestion exposure on health.

Typical metal concentrations in food are 10–50 ppb by weight for arsenic, cadmium, chromium, and lead (Ryan, Scanlon, and MacIntosh 2001; Scanlon et al. 1999). Assuming an adult dietary intake of 1 kilogram of food per day, expected intake for metals would be about 30 μg of metal/day or, for an adult, about 0.5 μg of metal/kg of body weight/day.

13.4.2.2 Organics

Organic compounds represent another large class of potential contaminants in food. Like metals, such contaminants are ubiquitous. Unlike metals, however, many organic compounds, most notably pesticides, are applied either directly to food crops or in surrounding soils. Certain organic compounds (e.g, polychlorinated biphenyls [PCBs], polychlorinated diphenyl ethers [PCDEs], and related compounds like dioxins) are associated with industrial use and industrial waste by-products.

Organic compounds display much more bioaccumulation than do most metal compounds. This can be seen most readily by drawing upon the chemistry adage "like dissolves like." Invertebrates in contact with soil and sediment absorb organic material through their outer layers, due to the organic nature of these materials. Once inside the body, organic contaminants readily dissolve in lipid layers. Since most metabolic processes are water-based, once sequestered in the fatty, lipid-rich components of the body, organic contamination tends to remain stable throughout the lifetime of the animal. Predator species higher on the food chain consume the animal, thereby taking in the organic contamination sequestered in these fat layers. In a similar manner to the lower animal, these contaminants become sequestered in *their* fatty deposits. As we move higher on the food chain, we note higher concentration of these chemical species in fat layers. Human beings, being at the top of the food chain, consume animals that have themselves consumed countless lower animals, thereby increasing their organic contaminant body burden.

The bioaccumulation process is important from the point of view of ingestion exposure because many organic compounds tend to persist in the environment, that is, they are not readily decomposed by chemical, biochemical, or photochemical degradation processes. Organochlorine pesticides like DDT (often called persistent organochlorine pesticides or POPs) have environmental lifetimes measured in decades. DDT itself, although banned for almost all use in the United States since the 1970s, is measurable in nearly every environmental sample taken. Food is no exception. Soil in agricultural areas often contains significant DDT and degradation products such as DDE. Grazing animals, such as cows and beef cattle, take in a large amount of soil while grazing. The POPs contained in the soil are then distributed throughout the fatty tissues of the animal's body. Later, when the meat or milk is consumed, one is exposed to these organic compounds.

Ingestion of feed animals from agricultural areas is not the only pathway for exposure to POPs. POPs are distributed throughout the food chain by the mechanisms described above. As one moves up the food chain, the concentrations of these compounds in the animal tissues becomes higher. Fatty tissues have especially high concentrations. Notable are concentrations of POPs and related compounds in marine fish and mammals (Brunstrom and Halldin 2000; Chan et al. 1997; Hobbs et al. 2003; Stapleton and Baker 2003; Storelli et al. 2003; Voorspoels, Covaci, and Schepens 2003) Human ingestion of animal tissues containing significant amounts of POPs and other organic compounds including PCBs, dioxins, and related compounds, have been noted around the world in various field studies. (See, for example, Binelli and Provini 2004; Brock et al. 1998; Falco et al. 2004; Lobscheid, Maddalena, and McKone 2004; Muntean et al. 2003; Patandin et al. 1999; Vongbuddhapitak et al. 2002; Whitmore et al. 1994.) McKone and co-workers (Lobscheid, Maddalena, and McKone 2004) have developed models of total intake of pesticides and related compounds and have estimated that the intake of POPs and PCBs are 1,000 times greater through the ingestion route than the inhalation route due to the high concentrations of these compounds in the diet (compared to air) and their relatively low volatility.

Several research groups have measured concentrations of certain organic compounds, in particular POPs, in the diet. Examples include (Ballantine and Simoneaux 1991; Darnerud et al. 2001; Kannan et al. 1992; Kipcic, Vukusic, and Sebecic 2002; MacIntosh et al. 2001; MacIntosh, Kabiru, and Ryan 2001; Nakata et al. 2002), but these only scratch the surface of this rich research field. There is substantial interest in this area with hundreds of research groups producing thousands of papers on the subject. Despite these extensive investigations, there is still much to be learned regarding the exposure to these ubiquitous contaminants in our diet.

13.4.2.3 Ingestion of Foods Contaminated after Processing

A recent concern raised by exposure assessors is the ingestion of compounds leached from food packaging materials. Phthalates, organic compounds used as plasticizers in many manufactured products, have been studied extensively in many industrialized countries throughout the world (Balafas, Shaw, and Whitfield 1999; Gruber, Wolz, and Piringer 1998; Petersen and Breindahl 2000; Saito et al. 2002; Tsumura et al. 2002). Phthalates belong to a large class of organic compounds referred to as "endocrine disruptors." They have been implicated in reproductive anomalies in animals and in humans. Phthalates leach into food from packing materials, including plastic and foam cartons, plastic wrap, and similar materials. Since they are organic in nature, they tend to leach most in fatty foods including meat, margarine, and butter. Total phthalate concentrations can reach 10 µg phthalate/kg of food; however, not all foods are wrapped in phthalate-containing materials, and overall expected exposure levels have not been determined.

13.5 INGESTION OF NONDIETARY ITEMS

Ingestion of food is a necessary activity for life. There is some risk of exposure to environmental contaminants through this process; however, one must eat in order to survive. Environmental scientists in collaboration with the food industry and government regulators are working diligently to reduce the hazards associated with the food supply. But there are other ingestion-related exposures not commonly considered. All individuals, but especially small children, ingest soil and other materials inadvertently. Such materials, if they contain contaminants, are another source of exposure and the inadvertent ingestion an additional pathway to this exposure. We will spend the next few paragraphs addressing these issues.

Specific activities involved in nondietary, ingestion-related exposures have been discussed from a modeling point of view (Zartarian et al. 2000). Their focus has been to develop methods of quantifying the transfer of contaminated material from surfaces and modeling such effectively. Hubal and co-workers (Hubal et al. 2000) present an interesting review of activities focusing on the needs exposure assessors see in understanding nondietary ingestion in children. They describe the need for high-quality data on temporal and spatial variability in contaminant levels in indoor environments, pesticide use in areas frequented by children, and detailed activity patterns for children and infants. By understanding these related aspects, we can begin to focus on the likely impact of nondietary exposure to this sensitive subpopulation that is likely to experience a larger fraction of their total exposure through this pathway

13.5.1 Hand-to-Mouth

The transfer of contaminated materials from surfaces, including carpeting, furniture, countertops, etc. to the hand, and consequently, to the mouth, may be a major source of ingestion exposure in adults but especially in children (Hubal et al. 2000). To obtain a better understanding of this process, numerous researchers have done field studies in which they have ascertained the amount of a specific compound, often pesticides or combustion-related polynuclear aromatic hydrocarbons, found on surfaces, then attempted to quantify the amount transferred to the hand. Exposure once on the hand can come from mouthing activities, an ingestion route, or through direct dermal absorption, a route not discussed in this chapter. Nishioka and co-workers (Nishioka et al. 2001) measured the concentrations of the herbicide 2,4-D on surfaces in a daycare center and attempted to estimate exposures experienced through the hand-to-mouth pathway. Their estimates suggested that nondietary ingestion at this daycare center would be approximately 1–10 µg/day, equivalent or greater than dietary ingestion, estimated to be about 1.3 µg/day in a typical diet. For a child, with a small body weight, often modeled at 20 kg, the hand-to-mouth, body-weight scaled dose may be as large as 0.5 µg 2,4-D/kg of body weight /day.

13.5.2 OTHER SOIL INGESTION

Many researchers have estimated that a child may ingest approximately 100 mg of soil per day from activities ranging from the hand-to-mouth activities described above, to actual direct ingestion in crawling infants (Wilson, Chuang, and Lyu 2001). Any contamination, be it metal contamination, organic, or even biological contamination, is ingested with such soil resulting in an important pathway of exposure. Even adults ingest a certain amount of soil and dust each day, although the amount is believed to be substantially less (USEPA 1997).

13.5.3 INHALATION/SWALLOWING

Yet another ingestion-related pathway of exposure involves the inhalation of contaminated dust followed by clearance and swallowing. The inhalation route of exposure is typically important for gases. However, particles from dust or soil can be inhaled as well. Larger particles are removed from the air stream in the nasal-pharyngeal region of the respiratory system. Smaller particles can penetrate more deeply into the lung. In the former case, particles are typically swallowed where they act similar to inadvertently ingested soil particles. For the smaller sizes, particles may be engulfed by pulmonary macrophages, or otherwise removed and then brought up from the lung by mucociliary clearance. Mucus is then swallowed at the back of the throat and the material enters the gastrointestinal system. Numerous studies have investigated the inhalation component of the ingestion pathway (Butte and Heinzow 2002; Chuang et al. 1999; Nishioka et al. 2001; Wilson, Chuang, and Lyu 2001) through measurement of dust resuspension, concentrations of various materials, most notably pesticides and polynuclear aromatic hydrocarbons, and modeling of exposure through inhalation rates (USEPA 1997). In many residences, the resuspension of dust and subsequent inhalation and ingestion can be the dominant pathway of exposure. Shalat and co-workers (Shalat et al. 2003) went a step further and measured biological markers of exposure, in this case a metabolite of the compound of interest found in the urine, to infer the magnitude of such exposures in children.

13.5.4 INTAKE THROUGH INADVERTENT INGESTION

Using the standard number of 100 mg of soil intake for a child, and an expected concentration of 100 ppm of lead in soil, a child might be expected to take in about 10 µg of lead each day from inadvertent soil ingestion, a value similar in magnitude to that taken in with the normal diet. Hence, a significant portion of the risk associated with metal exposure may come from inadvertent exposure. The intake of other metals would be on a similar scale. Using a 20 kg body weight for a young child, body-weight scaled intake would center on 0.5 µg of lead/kg body weight/day.

13.6 SUMMARY AND FURTHER DIRECTIONS FOR RESEARCH

We close this section by noting that, despite the large number of studies completed in the area of ingestion-related exposure, there is still much to be learned. Very few data have been collected that address how exposure through ingestion varies with time. Most studies perform a single data collection and assume that this represents exposures experienced over, for example, a year. However, there are likely to be strong seasonal variations in exposures similar to that observed for atrazine in drinking water due to variability in diet, the source of the food in the diet, and variability in pesticide use over the seasons. Our understanding of contaminant concentrations in the food supply comes from analysis of very few samples — often only a small handful of samples of a particular food item are fully characterized for their contaminant content. This, in turn, gives us a poor characterization of the exposures experienced by the populations as a whole. It may be that two seemingly identical apples were treated with vastly different amounts of pesticide and, therefore, contain pesticide residues differing by a thousandfold. Yet the data available to us to model exposure experienced via this route are limited, and we would likely assume identical exposures to individuals consuming them.

There are many avenues of investigation open to the curious exposure assessor. Better data can be gathered. Better methods of analysis can be implemented. Larger-scale investigations can be undertaken. We encourage the reader to examine these ideas and, perhaps, develop new and intriguing methods to afford more accurate assessment of exposure through this important route.

13.7 QUESTIONS FOR REVIEW

1. Describe some important differences between the epithelial barrier between the gastrointestinal system and the lung and skin epithelial barriers.
2. Assume the concentration of arsenic in drinking water is 3 ppb and in the food portion of the diet is 20 ppb. Assume that an individual ingests 2 L of water per day and consumes 1 kg of food. What is the expected intake of arsenic from these two sources? Report your answer in μg per day and in body weight scaled units (μg of arsenic/per kg of body weight/per day), assuming your own body weight.
3. Describe some sources of metals in the environment that can result in human exposure.
4. Draw a schematic diagram indicating the pathway that an inorganic contaminant may follow in the environment commencing with the source of the pollution and ending up in the circulating bloodstream of an individual.
5. Organic contaminants often follow completely different pathways than inorganic contaminants. Why is this? Draw a similar schematic indicating a pathway followed by a typical organic contaminant in moving through the environment from a source to the circulating bloodstream of an individual.
6. Why does one consider the exposure experienced through inhalation of particles of soil followed by swallowing and ingestion-related exposure rather than an inhalation-related exposure?
7. Describe in words how one would develop a model of ingestion-related exposure. Consider sources, transport, uptake, and bioavailability.
8. You have been charged with developing an exposure assessment study designed to assess the exposure to polychlorinated biphenyls (PCBs) through the ingestion route. PCBs are organic compounds, formerly used as cooling oils, that persist for a long time in the environment. Consider where you would likely find them in the environment and where they would appear in highest concentrations in the food chain. What questions would you ask of potential participants? What samples would you take in the environment? Would you take biological samples from individuals? What foods would you consider important? Sketch out a sampling strategy addressing these issues.

REFERENCES

Alam, M.G.M., Snow, E.T., and Tanaka, A. (2003) Arsenic and Heavy Metal Contamination of Vegetables Grown in Samta Village, Bangladesh, *Science of the Total Environment,* **308**(1–3): 83–96.

Balafas, D., Shaw, K.J., and Whitfield, F.B. (1999) Phthalate and Adipate Esters in Australian Packaging Materials, *Food Chemistry,* **65**(3): 279–287.

Ballantine, L.G. and Simoneaux, B.J. (1991) Pesticide Metabolites in Food, *ACS Symposium Series,* **446**: 96–104.

Besser, J.M., Brumbaugh, W.G., May, T.W., Church, S.E., and Kimball, B.A. (2001) Bioavailability of Metals in Stream Food Webs and Hazards to Brook Trout (*Salvelinus fontinalis*) in the Upper Animas River Watershed, Colorado, *Archives of Environmental Contamination and Toxicology,* **40**(1): 48–59.

Binelli, A. and Provini, A. (2004) Risk for Human Health of Some POPs Due to Fish from Lake Iseo, *Ecotoxicology and Environmental Safety,* **58**(1): 139–145.

Brock, J.W., Melnyk, L.J., Caudill, S.P., Needham, L.L., and Bond, A.E. (1998) Serum Levels of Several Organochlorine Pesticides in Farmers Correspond with Dietary Exposure and Local Use History, *Toxicology and Industrial Health,* **14**(1-2): 275–289.

Brunstrom, B. and Halldin, K. (2000) Ecotoxicological Risk Assessment of Environmental Pollutants in the Arctic, *Toxicology Letters,* **112**: 111–118.

Butte, W. and Heinzow, B. (2002) Pollutants in House Dust as Indicators of Indoor Contamination, *Reviews of Environmental Contamination & Toxicology,* **175**: 1–46.

Chan, H.M., Berti, P.R., Receveur, O., and Kuhnlein, H.V. (1997) Evaluation of the Population Distribution of Dietary Contaminant Exposure in an Arctic Population Using Monte Carlo Statistics, *Environmental Health Perspectives,* **105**(3): 316–321.

Chan, H.M., Kim, C., Khoday, K., Receveur, O., and Kuhnlein, H.V. (1995) Assessment of Dietary Exposure to Trace-Metals in Baffin Inuit Food, *Environmental Health Perspectives,* **103**(7-8): 740–746.

Chuang, J.C., Callahan, P.J., Lyu, C.W., and Wilson, N.K. (1999) Polycyclic Aromatic Hydrocarbon Exposures of Children in Low-Income Families, *Journal of Exposure Analysis and Environmental Epidemiology,* **9**(2): 85–98.

Darnerud, P.O., Eriksen, G.S., Johannesson, T., Larsen, P.B., and Viluksela, M. (2001) Polybrominated Diphenyl Ethers: Occurrence, Dietary Exposure, and Toxicology, *Environmental Health Perspectives,* **109**: 49–68.

Fairbrother, A., Brix, K.V., Toll, J.E., McKay, S., and Adams, W.J. (1999) Egg Selenium Concentrations as Predictors of Avian Toxicity, *Human and Ecological Risk Assessment,* **5**(6): 1229–1253.

Falco, G., Bocio, A., Llobet, J.M., Domingo, J.L., Casas, C., and Teixido, A. (2004) Dietary Intake of Hexachlorobenzene in Catalonia, Spain, *Science of the Total Environment,* **322**(1-3): 63–70.

Greene, H. and Moran, J. (1994) The Gastrointestinal Tract: Regulator of Nutrient Absorption, in *Modern Nutrition in Health and Disease,* Shils, M., Olson, J., and Shike, M., Eds., Lea & Febiger, Philadelphia, PA.

Gruber, L., Wolz, G., and Piringer, O. (1998) Analysis of Phthalates in Baby Food, *Deutsche Lebensmittel-Rundschau,* **94**(6): 177–179.

Harris, R.C. and Bodaly, R.A. (1998) Temperature, Growth and Dietary Effects on Fish Mercury Dynamics in Two Ontario Lakes, *Biogeochemistry,* **40**(2-3): 175–187.

Hobbs, K.E., Muir, D.C.G., Michaud, R., Beland, P., Letcher, R.J., and Norstrom, R.J. (2003) PCBs and Organochlorine Pesticides in Blubber Biopsies from Free-Ranging St. Lawrence River Estuary Beluga Whales (Delphinapterus leucas), 1994–1998, *Environmental Pollution,* **122**(2): 291–302.

Hubal, E.A., Sheldon, L.S., Zufall, M.J., Burke, J.M., and Thomas, K.W. (2000) The Challenge of Assessing Children's Residential Exposure to Pesticides, *Journal of Exposure Analysis and Environmental Epidemiology,* **10**(6 Pt 2): 638–649.

Hunt, S. and Groff, J. (1990) *Advanced Nutrition and Human Metabolism,* West Publishing, St. Paul, MN.

Husain, A., Sawaya, W., Al-Sayegh, A., Al-Amiri, H., Al-Sager, J., Al-Sharrah, T., Al-Kandari, R., and Al-Foudari, M. (2003) Screening Level Assessment of Risks Associated with Dietary Exposure to Selected Heavy Metals, Polycyclic Aromatic Hydrocarbons, and Radionuclides in Kuwait, *Human and Ecological Risk Assessment,* **9**(4): 1075–1087.

Ikeda, M., Ezaki, T., Tsukahara, T., and Moriguchi, J. (2004) Dietary Cadmium Intake in Polluted and Non-Polluted Areas in Japan in the Past and in the Present, *International Archives of Occupational and Environmental Health,* **77**(4): 227–234.

Kannan, K., Tanabe, S., Ramesh, A., Subramanian, A., and Tatsukawa, R. (1992) Persistent Organochlorine Residues in Foodstuffs from India and Their Implications on Human Dietary Exposure, *Journal of Agricultural and Food Chemistry,* **40**(3): 518–524.

Kipcic, D., Vukusic, J., and Sebecic, B. (2002) Monitoring of Chlorinated Hydrocarbon Pollution of Meat and Fish in Croatia, *Food Technology and Biotechnology,* **40**(1): 39–47.

Lawrence, R.A. (1994) *Breastfeeding: A Guide for the Medical Profession,* Mosby, St. Louis, MO.

Lobscheid, A.B., Maddalena, R.L., and McKone, T.E. (2004) Contribution of Locally Grown Foods in Cumulative Exposure Assessments, *Journal of Exposure Analysis and Environmental Epidemiology,* **14**(1): 60–73.

MacIntosh, D.L., Hammerstrom, K., and Ryan, P.B. (1999) Longitudinal Exposure to Selected Pesticides in Drinking Water, *Human and Ecological Risk Assessment,* **5**(3): 575–588.

MacIntosh, D.L., Kabiru, C., Echols, S.L., and Ryan, P.B. (2001) Dietary Exposure to Chlorpyrifos and Levels of 3,5,6-trichloro-2-pyridinol in Urine, *Journal of Exposure Analysis and Environmental Epidemiology,* **11**(4): 279–285.

MacIntosh, D.L., Kabiru, C.W., and Ryan, P.B. (2001) Longitudinal Investigation of Dietary Exposure to Selected Pesticides, *Environmental Health Perspectives,* **109**(2): 145–150.

Malchow, D.E., Knight, A.W., and Maier, K.J. (1995) Bioaccumulation and Toxicity of Selenium in Chironomus-Decorus Larvae Fed a Diet of Seleniferous Selenastrum-Capricornutum, *Archives of Environmental Contamination and Toxicology,* **29**(1): 104–109.

Melnyk, L.J.O., Berry, M.R., Sheldon, L.S., Freeman, N.C.G., Pellizzari, E.D., and Kinman, R.N. (2000) Dietary Exposure of Children in Lead-Laden Environments, *Journal of Exposure Analysis and Environmental Epidemiology,* **10**(6): 723–731.

Milton, A. and Johnson, M. (1999) Arsenic in the Food Chains of a Revegetated Metalliferous Mine Tailings Pond, *Chemosphere,* **39**(5): 765–779.

Moschandreas, D.J., Karuchit, S., O'Rourke, M.K., Lo, D., Lebowitz, M.D., and Robertson, G. (2002) Exposure Apportionment: Ranking Food Items by Their Contribution to Dietary Exposure, *Journal of Exposure Analysis and Environmental Epidemiology,* **12**(4): 233–243.

Muntean, N., Jermini, M., Small, I., Falzon, D., Furst, P., Migliorati, G., Scortichini, G., Forti, A.F., Anklam, E., von Holst, C., Niyazmatov, B., Bahkridinov, S., Aertgeerts, R., Bertollini, R., Tirado, C., and Kolb, A. (2003) Assessment of Dietary Exposure to Some Persistent Organic Pollutants in the Republic of Karakalpakstan of Uzbekistan, *Environmental Health Perspectives,* **111**(10): 1306–1311.

Nakata, H., Kawazoe, M., Arizono, K., Abe, S., Kitano, T., Shimada, H., Li, W., and Ding, X. (2002) Organochlorine Pesticides and Polychlorinated Biphenyl Residues in Foodstuffs and Human Tissues from China: Status of Contamination, Historical Trend, and Human Dietary Exposure, *Archives of Environmental Contamination and Toxicology,* **43**(4): 473–480.

Nishioka, M.G., Lewis, R.G., Brinkman, M.C., Burkholder, H.M., Hines, C.E., and Menkedick, J.R. (2001) Distribution of 2,4-D in Air and on Surfaces Inside Residences after Lawn Applications: Comparing Exposure Estimates from Various Media for Young Children, *Environmental Health Perspectives,* **109**(11): 1185–1191.

Noel, L., Leblanc, J.C., and Guerin, T. (2003) Determination of Several Elements in Duplicate Meals from Catering Establishments Using Closed Vessel Microwave Digestion with Inductively Coupled Plasma Mass Spectrometry Detection: Estimation of Daily Dietary Intake, *Food Additives and Contaminants,* **20**(1): 44–56.

Patandin, S., Dagnelie, P.C., Mulder, P.G.H., de Coul, E.O., van der Veen, J.E., Weisglas-Kuperus, N., and Sauer, P.J.J. (1999) Dietary Exposure to Polychlorinated Biphenyls and Dioxins from Infancy until Adulthood: A Comparison Between Breast-Feeding, Toddler, and Longterm Exposure, *Environmental Health Perspectives,* **107**(1): 45–51.

Pennington, J.A.T. (2001) Use of the Core Food Model to Estimate Mineral Intakes, Part 1. Selection of U.S. Core Foods, *Journal of Food Composition and Analysis,* **14**(3): 295–300.

Pennington, J.A.T. and Hernandez, T.B. (2002) Core Foods of the U.S. Food Supply, *Food Additives and Contaminants,* **19**(3): 246–271.

Petersen, J.H. and Breindahl, T. (2000) Plasticizers in Total Diet Samples, Baby Food and Infant Formulae, *Food Additives and Contaminants,* **17**(2): 133–141.

Rozman, K. and Klaassen, C. (1996) Adsorption, Distribution, and Excretion of Toxicants, in *Casarett & Doull's Toxicology. The Basic Science of Poisons,* Klaassen, C., Ed., McGraw-Hill, New York, NY.

Ruby, M., Davis, A., Schoof, R., Eberle, S., and Sellstone, C. (1996) *Environmental Science and Technology,* **30**: 422–430.

Ryan, P.B., Scanlon, K.A., and MacIntosh, D.L. (2001) Analysis of Dietary Intake of Selected Metals in the NHEXAS-Maryland Investigation, *Environmental Health Perspectives,* **109**(2): 121–128.

Saito, I., Ueno, E., Oshima, H., and Matsumoto, H. (2002) Levels of Phthalates and Adipates in Processed Foods and Migration of Di-Isononyl Adipate from Polyvinyl Chloride Film into Foods, *Journal of the Food Hygienic Society of Japan,* **43**(3): 185–189.

Scanlon, K.A., MacIntosh, D.L., Hammerstrom, K.A., and Ryan, P.B. (1999) A Longitudinal Investigation of Solid-Food Based Dietary Exposure to Selected Elements, *Journal of Exposure Analysis and Environmental Epidemiology,* **9**(5): 485–493.

Shalat, S.L., Donnelly, K.C., Freeman, N.C., Calvin, J.A., Ramesh, S., Jimenez, M., Black, K., Coutinho, C., Needham, L.L., Barr, D.B., and Ramirez, J. (2003) Nondietary Ingestion of Pesticides by Children in an Agricultural Community on the U.S./Mexico Border: Preliminary Results, *Journal of Exposure Analysis and Environmental Epidemiology,* **13**(1): 42–50.

Stapleton, H.M. and Baker, J.E. (2003) Comparing Polybrominated Diphenyl Ether and Polychlorinated Biphenyl Bioaccumulation in a Food Web in Grand Traverse Bay, Lake Michigan, *Archives of Environmental Contamination and Toxicology,* **45**(2): 227–234.

Storelli, M.M., Giacominelli-Stuffler, R., Storelli, A., and Marcotrigiano, G.O. (2003) Polychlorinated Biphenyls in Seafood: Contamination Levels and Human Dietary Exposure, *Food Chemistry,* **82**(3): 491–496.

Suhendrayatna, Ohki, A., Nakajima, T., and Maeda, S. (2001) Metabolism and Organ Distribution of Arsenic in the Freshwater Fish Tilapia Mossambica, *Applied Organometallic Chemistry,* **15**(6): 566–571.

Tsumura, Y., Ishimitsu, S., Kaihara, A., Yoshii, K., and Tonogai, Y. (2002) Phthalates, Adipates, Citrate and Some of the Other Plasticizers Detected in Japanese Retail Foods: A Survey, *Journal of Health Science,* **48**(6): 493–502.

USEPA (1997) *Exposure Factors Handbook*, Report No. EPA/600/P-95/002Fa, U.S. Environmental Protection Agency, Washington, DC.

USEPA (2004a) Consumer Factsheet on: ATRAZINE, http://www.epa.gov/safewater/contaminants/dw_contamfs/atrazine.html, last update: April 20, 2004, (accessed August 24, 2004).

USEPA (2004b) List of Drinking Water Contaminants & MCLs, http://www.epa.gov/safewater/mcl.html#1, last update: May 26, 2004, (accessed August 25, 2004).

Vongbuddhapitak, A., Atisook, K., Thoophom, G., Sungwaranond, B., Lertreungdej, Y., Suntudrob, J., and Kaewklapanyacharoen, L. (2002) Dietary Exposure of Thais to Pesticides during 1989–1996, *Journal of AOAC International,* **85**(1): 134–140.

Voorspoels, S., Covaci, A., and Schepens, P. (2003) Polybrominated Diphenyl Ethers in Marine Species from the Belgian North Sea and the Western Scheidt Estuary: Levels, Profiles, and Distribution, *Environmental Science & Technology,* **37**(19): 4348–4357.

Wang, W.X. (2002) Cd and Se Aqueous Uptake and Exposure of Green Mussels *Perna viridis*: Influences of Seston Quantity, *Marine Ecology-Progress Series,* **226**: 211–221.

Whitmore, R.W., Immerman, F.W., Camann, D.E., Bond, A.E., Lewis, R.G., and Schaum, J.L. (1994) Non-Occupational Exposures to Pesticides for Residents of 2 U.S. Cities, *Archives of Environmental Contamination and Toxicology,* **26**(1): 47–59.

Wilson, N.K., Chuang, J.C., and Lyu, C. (2001) Levels of Persistent Organic Pollutants in Several Child Day Care Centers, *Journal of Exposure Analysis and Environmental Epidemiology,* **11**(6): 449–458.

Ysart, G., Miller, P., Crews, H., Robb, P., Baxter, M., De L'Argy, C., Lofthouse, S., Sargent, C., and Harrison, N. (1999) Dietary Exposure Estimates of 30 Elements from the U.K. Total Diet Study, *Food Additives and Contaminants,* **16**(9): 391–403.

Ysart, G., Miller, P., Croasdale, M., Crews, H., Robb, P., Baxter, M., de L'Argy, C., and Harrison, N. (2000) 1997 U.K. Total Diet Study — Dietary Exposures to Aluminium, Arsenic, Cadmium, Chromium, Copper, Lead, Mercury, Nickel, Selenium, Tin and Zinc, *Food Additives and Contaminants,* **17**(9): 775–786.

Yu, R.Q. and Wang, W.X. (2002) Kinetic Uptake of Bioavailable Cadmium, Selenium, and Zinc by *Daphnia magna, Environmental Toxicology and Chemistry,* **21**(11): 2348–2355.

Zartarian, V.G., Özkaynak, H., Burke, J.M., Zufall, M.J., Rigas, M.L., and Furtaw, E.J. Jr. (2000) A Modeling Framework for Estimating Children's Residential Exposure and Dose to Chlorpyrifos via Dermal Residue Contact and Nondietary Ingestion, *Environmental Health Perspectives,* **108**(6): 505–514.

Zheljazkov, V.D. and Warman, P.R. (2004) Application of High-Cu compost to Dill and Peppermint, *Journal of Agricultural and Food Chemistry,* **52**(9): 2615–2622.

Part V

Multimedia Exposure

14 Exposure to Pollutants from House Dust

John W. Roberts
Engineering-Plus, Inc.

Wayne R. Ott
Stanford University

CONTENTS

14.1 Synopsis..319
14.2 Characteristics and Measurement of House Dust..320
14.3 Major Pollutants in House Dust...326
    14.3.1 Lead: Sources, Pathways, Trends, and Effects 326
    14.3.2 Reducing Lead Exposure from House Dust 330
    14.3.3 Pesticides, Polycyclic Aromatic Hydrocarbons (PAHs), Phthalates, and
        Other Toxic Pollutants..332
14.4 Control of House Dust 337
    14.4.1 Cleaning...337
    14.4.2 Why Vacuum Cleaners Do and Do Not Work......................................338
14.5 Master Home Environmentalist™ Program..339
14.6 Health Care Cost Savings 341
14.7 Making a Plan to Reduce Your Personal Exposure...342
14.8 Questions for Review 342
14.9 Acknowledgments 343
References..343

14.1 SYNOPSIS

House dust presents a special problem: it is not routinely measured; it has been found to contain a large number of toxic pollutants; and it presents potential health problems for infants and children who have small body mass and developing organs, spend large amounts of time near floors and carpets, and engage in hand-to-mouth activities. Specially designed vacuum cleaners, the HVS3 and HVS4, have been developed to collect large samples of house dust from carpets and bare floors, as well as dust from outdoor surfaces such as streets, sidewalks, and lawns. These standardized vacuums have been used since 1990 to collect house dust samples as part of numerous studies of American homes. The results demonstrate that concentrations of many toxic pollutants (lead, pesticides, and polycyclic aromatic hydrocarbons) in homes often exceed the soil "screening levels" established for Superfund sites. There has been a dramatic reduction in the lead in gasoline and children's blood since 1990. However, one in three children under 6 years of age still lives in a house with a lead-based paint hazard. Homes built before 1940 present the most risk for toddlers.

A major cause of the high pollutant concentrations in house dust is the preferential "track-in" of small dust particles from outdoor soil containing exterior paint of older homes, pesticides from lawns and gardens, air pollution fallout, allergens, and vehicular pollutants found on streets and sidewalks. Deep dust in ordinary carpets acts as a sink and source of these pollutants and is the major source of pollutants in surface dust. Human activity in a room causes carpet dust to be resuspended. The present expenditure of public funds at most Superfund sites is not efficient in reducing total exposure. The exposures from indoor dust and air are usually much higher than those due to many Superfund sites. Fortunately, pollutant concentrations and loadings from house dust can be reduced in most homes by relatively simple steps, such as removing shoes before entering the home, using a commercial-grade door mat, reducing carpets indoors, selecting carpets and floor surfaces that are easy to clean, dusting, and vacuuming frequently with a vacuum cleaner equipped with an agitator brush and a dirt finder. Many environmental pollutants reach people through only one carrier medium. For example, we are exposed to carbon monoxide (CO) only through the air we breathe. For single-medium pollutants, a multimedia exposure analysis is not appropriate. For others, a multimedia analysis may be necessary to determine the concentrations of each pollutant in each medium, which may include house dust, ambient air, drinking water, and food. Pollutants for which multimedia exposure analyses are appropriate include lead (Pb) and polycyclic aromatic hydrocarbons (PAHs).

14.2 CHARACTERISTICS AND MEASUREMENT OF HOUSE DUST

It should come as no surprise to anyone who has cleaned the interior of a home that a layer of particles called "house dust" accumulates rapidly on surfaces, such as floors, shelves, and window-sills. We have all observed layers of dust that have settled on tables, chairs, sofas, light fixtures, bookshelves, floors, cupboards, figurines, and on other surfaces in a home. Even after just a few weeks, the layer of dust becomes thick enough to be visible to the eye and is easily picked up on fingers touching horizontal surfaces. Anyone who has operated a vacuum cleaner can remember emptying clumps of gray-brown matter from the collection bag after an hour or two of vacuuming. Indeed, removal of house dust is such a common household chore that the vacuum cleaner, a motorized air movement filtering system equipped with a moving agitator brush, was invented to help residents collect house dust. Feather dusters, invented to dislodge the dust from household objects and surfaces, simply spread dust around and are of dubious value. Area rugs have been used for thousands of years and can be cleaned by hanging them on a wire and beating them or washing them with water. Wall-to-wall carpets were not popular before the vacuum cleaner was invented, in part because they could not be cleaned adequately.

How does one go about measuring quantitatively the exposures of children and adults to pollutants in house dust? In 1987, engineers John Roberts and Mike Ruby, working under a contract with the U.S. Environmental Protection Agency (USEPA), developed the High Volume Surface Sampler (HVS2), a special-purpose vacuum cleaner designed to collect house dust from surfaces in a standardized manner for subsequent chemical analysis for exposure assessment purposes (Roberts et al. 1991a). The HVS2 has a known and reproducible dust removal rate on various types of test surfaces and relatively constant efficiency at different loadings of surface dust. The HVS2 weighed 55 lbs, was expensive ($8,000), and difficult to operate. This first test model soon evolved into the HVS3, a 24-lb vacuum cleaner costing around $3,000 that could be used to collect a large, representative sample of house dust from indoor sources such as carpets, rugs, and bare floors, and dust from outdoor surfaces such as streets, sidewalks, lawns, and bare, packed dirt (Roberts et al. 1991b) (Figure 14.1). The HVS3 vacuum is recognized in American Standard of Testing Materials (ASTM) method 5438-00 and is now available commercially from CS3, Inc. (Sandpoint, Idaho).

Due to its high airflow volume (17–20 cubic feet per minute [cfm] or 8.1–9.5 l/s), the HVS3 dust sampler can collect a 10-minute sample that is large enough (2–200 grams, [g]) to permit subsequent laboratory chemical analyses and bioassays. This specialized vacuum cleaner system

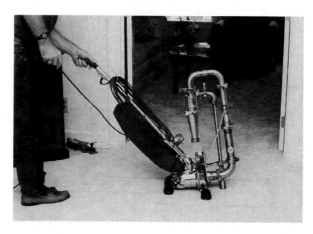

FIGURE 14.1 Special purpose vacuum, the HVS3, designed to collect a large quantity of dust in a standardized manner for laboratory analysis of the sample's chemical composition and pollutant concentration. (Courtesy of Dr. Robert Lewis, USEPA; see also Chapter 15 and Figure 15.4.)

maintains uniform sampling conditions by measuring and controlling the airflow and pressure drop across a 5-inch (14-cm) nozzle. A 3-inch (7.5-cm) cyclone collects more than 99% of the house dust that enters without any reduction in airflow. The cyclone "catch cup" can be used to transport the sample for laboratory analysis. The dust that is collected in the cyclone cup is usually sieved through a 100-mesh sieve so that only dust below 150 microns in diameter is analyzed. This cut point was selected because these particles stick to skin and hands better and present more risk. Approximately 140 of these specialized vacuum cleaners were in use throughout the world in 2005, and a number of scientific papers using the HVS3 have reported results for lead, pesticides, and other pollutants in house dust. These results can be compared with each other, since the operators of the HVS3 are all using the same standardized method. The HVS4, developed in 2001, is a 17-pound simplified version of the HVS3 that is easier to carry, operate, and purchase.

House dust, because of its proximity to pets, children, and adults, has considerable potential for exposing the occupants of a home to toxic and hazardous pollutants. Pollutants in dust increase the potential exposure of all people and pets, especially infants and toddlers who crawl and mouth their hands and other objects. Dogs and cats that clean their fur with their tongue will also ingest pollutants in house dust and soil that they contact. These animals may have high health risks from Pb in an old house that is being painted or remodeled. The average infant's daily dust ingestion rate, estimated to be 100 milligrams per day (mg/day), is more than two times that of adults. Eleven percent of toddlers may exhibit pica behavior, eating nonfood items, and may consume up to 10 g of soil and dust per day (Calabrese and Stanek 1991; Mahaffrey and Annest 1985). Potential risks to small children, when compared to adults, are further increased due to their smaller size, higher ratio of surface area to body weight, and the stage of development of their organs, nervous systems, and immune systems (Woodruff et al. 2003; Roberts et al. 1992). Figure 14.2 illustrates the role of house dust as a carrier medium for pollutants reaching a baby, compared with the quantity of food eaten, air breathed (6.3 m³/day), and water drunk (1 liter/day). House dust gets into indoor air, food, and water. However these routes of exposure to house dust are not important when compared with the 100 mg that the average infant ingests (Calabrese and Stanek 1991). A baby's dust intake from indoor air is estimated to be 0.1 mg/day. A general finding from research conducted thus far is that house dust often contains a great variety of pollutants that should be of concern for the health of residents (Roberts et al. 1992; Camann and Buckley 1994; Rudel et al. 2003; Maertens, Bailey, and White 2004). Pollutants found in house dust include lead, cadmium, chromium, mercury, arsenic, other toxic metals, polycyclic aromatic hydrocarbons (PAHs), polychlorinated biphenyls (PCBs), dichloro diphenyl trichloroethane (DDT), phthalates, fire retardants, and other persistent

EXPOSURE PATHWAYS FOR CHILDREN

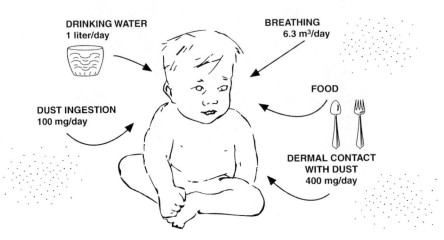

FIGURE 14.2 Estimated quantity of dust per day reaching a small child from dust ingestion, drinking water, breathing, food ingestion, and dermal contact.

pesticides. House dust is a major pathway for children and adults for the flame retardants polybrominated diphenylethers (PBDEs) (Stapleton et al. 2005; Jones-Otazo et al. 2005). House dust also contains pollutants that come from window cleaners, laundry detergents, spot removers, plastics, electronics, and carpeting. Concentrations of these pollutants may exceed USEPA health-based standards (Rudel et al. 2003).

Most of the pollutants in the ambient air may be deposited with particles on the roof and siding of houses and be washed down to the foundation soil and walkways. Measurements (Adgate et al. 1998) show that carrying the pollutant indoors on the shoes, or "track-in," is one of the largest sources of house dust in most U.S. homes (Figure 14.3). There appears to be a preferential track-in of small particles because they stick to shoes better (Roberts et al. 1996). The small particles have a higher surface area (and toxicity) per unit mass. This is the best theory to explain why the concentration of toxic pollutants found in house dust are higher than those found in soil around the house and can equal or exceed street dust concentrations found in front of the house (Roberts et al. 1992). Most parents do not allow their infants to play in street dusts because of the known toxic pollutants in vehicle exhaust, motor oil, and asphalt. However, increased awareness of the pollutants in house dust may cause parents to put a clean sheet down before they allow a baby to play on a carpet and also to learn effective methods of cleaning to protect themselves and their children.

House dust on carpets is easily resuspended into the air of a room where occupants breathe it as airborne particulate matter. In an experiment by one of the authors (Ott) in a Palo Alto, CA, home, two occupants purposely "stomped" on the carpet, plush shag, for 18 minutes after eating dinner and relaxing (Figure 14.4). The mass concentration of particulate matter of 5 micrometers or smaller (PM_5) measured at a height of 30 inches above the floor rapidly increased in the room as soon as the stomping began. The doors and windows were closed as usual, and the particle concentrations did not return to normal levels until about 1 1/2 hours later.

Most people may be unaware of the ease by which particles are stirred up in their homes, or resuspended, after just a small amount of physical indoor activity. Because of their small size, resuspended fine particles are not easily seen by the eye, even though they may surround us indoors. Indeed, one might conclude that we live in a "sea of particles" in our homes. The "particle loading" is often higher inside than outside and in some homes is much higher than in others, depending on the activities of the occupants, the presence of pets, and the steps taken to reduce house dust.

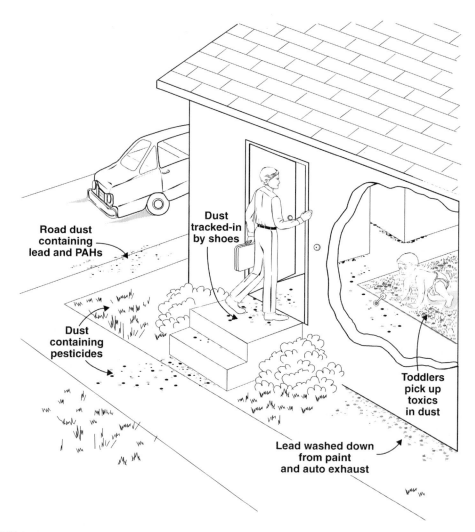

FIGURE 14.3 Dust tracked in from outdoors contains lead, pesticides, and other pollutants. The dust collects on floors and surfaces and becomes embedded deep in the weave of carpets. It becomes resuspended when people walk on floors, exercise, or breathe particles. It can be ingested directly by toddlers engaging in hand-to-mouth activities. Using a doormat, removing shoes, and frequently cleaning with an agitator-equipped vacuum can reduce indoor concentrations of these pollutants.

Overall, five important steps can help reduce the levels of house dust:

- Wiping shoes twice on a commercial grade door mat before entering the home
- Removing shoes while in the home
- Frequent and thorough vacuuming of carpets and furniture
- Removing deep dust from carpets with a vacuum with a power head and an embedded dirt finder
- Cleaning all surfaces inside the home on a regular basis

A commercial-grade doormat can be obtained by making a special order at a large hardware store. The "twister" mat often found in front of department stores is an example of a high-quality mat.

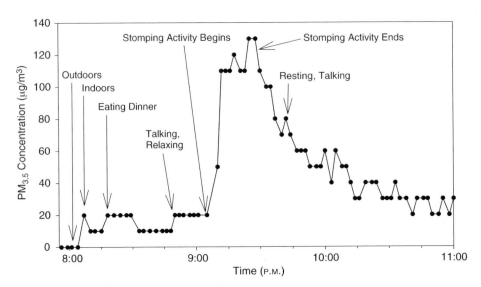

FIGURE 14.4 Particle mass concentration measured indoors before, during, and after 18 minutes of stomping activity by two people on an old shag carpet in an efficiency apartment in Palo Alto, CA, with the doors and windows closed. The piezobalance fine particle monitor was located at a height of 30 in. (0.76 m) and within 4 ft (1.2 m) of the two stomping persons, who rested before and after the stomping activity.

Persons with allergies, asthma, and heart problems have special health risks due to resuspended particles indoors. Books such as *Allergy-Free Living* (Howarth and Reid 2000) list dozens of steps a resident can take to decrease exposure to the major interior allergens and irritants by creating a healthy, allergy-free home and lifestyle. Asthma is a leading cause of chronic disease and absenteeism among schoolchildren in the United States, and an increasing incidence of asthma in the United States suggests that efforts to reduce fine particles (that include dust mites, mold, and cat dander) indoors could produce health benefits for both children and adults (Krieger et al. 2005).

Although it is possible to reduce resuspension of carpet dust by frequent vacuuming and other practical steps, many people do not realize that an extremely thorough vacuuming is needed to remove embedded dust particles. Even after extensive vacuuming, large amounts of dust remain deeply embedded in the fabric of the carpet. Some new vacuum cleaners are equipped with an embedded dirt finder, an electronic sensor system with red and green signal lights that the operator can easily see while vacuuming. The light is red when the vacuum is picking up particles, but it turns green when the noise made by the particle stream hitting a sensor plate drops below a fixed level. Removing the deep dust from an old carpet for the first time may require vacuuming some areas of the carpet for surprisingly long time periods, sometimes up to 45 minutes per square meter (min/m^2) of carpet, to get a green light on the dirt finder. In two studies of 11 and 10 older carpets in the Seattle area, the median time to remove the deep dust was 8 and 20 min/m^2, respectively (Roberts et al. 1999; Roberts, Glass, and Mickelson 2004). Once the "deep dust" is removed, it is relatively easy to keep it out by frequent vacuuming and use of a commercial grade doormat. A 6-meter long high-quality doormat reduced dirt track-in at an elementary school from 12 to 2 milligrams (mg) per person (Lehen 1983), a reduction of 83%.

One of the authors (Roberts, Glass, and Mickelson 2004) developed a "3-spot test," a method to estimate how much deep dust resides in a carpet and the time required to remove the dust. The method uses a vacuum cleaner equipped with an embedded dust finder, such as the upright Hoover Vacuum Model U 6445-900. The 3-spot test consists of the following procedure:

- Choose a spot in the center of the carpet.
- Start the vacuum and a stopwatch at the same time. Hold the vacuum in one place until the light turns green.
- Then move the vacuum quickly without stopping it to a second spot 3 feet away and hold it in one place until the light turns green. Repeat this procedure for a third spot. The three spots should form a triangle with equal sides.
- Note the total time required to clean all three spots.

When the time required for the dust detector light to turn green is less than 11 seconds for all 3 spots, the carpet is considered relatively clean with 6 seconds being an achievable goal. Roberts, Glass, and Mickelson (2004) found that the surface dust (g/m^2), deep dust (g/m^2), and dust collection rate (g/min) tended to drop rapidly at first and then much more slowly during vacuuming. An HVS4 was used to measure the surface dust. For this study the starting surface dust loading was 0.7–21.1 g/m^2, which decreased by 85–99% when the deep dust was removed. Vacuuming times ranging from 2.3–95 min/m^2 were required to remove the deep dust.

Until the deep dust is removed, it will tend to rise to the surface of the carpet and become surface dust. Activity in the room brings up deep dust. Lead concentrations from two sequential samples of surface dust on a carpet in a remodeled house increased from 180 parts per million (ppm) to 59,800 ppm (Roberts, Glass, and Mickelson 2004). The surface lead loading increased by a factor of 60 on the second sample. We theorized that a layer of lead paint from remodeling activity was uncovered as the deep dust was removed. Measuring the lead surface loading, in units of micrograms per square meter ($\mu g/m^2$), is one of the best ways to predict the amount of lead in the blood of an infant who crawls on such a carpet in an old house (Davies et al. 1990). Estimating deep dust with the 3-spot test and measuring the lead in the deep dust collected in the 3-spot test and a small area inside the 3-spot triangle, may also predict lead in the blood of an infant. Surface dust shows much more variation than deep dust. Since deep dust is the major source of the pollutants that appear in surface dust, removal of deep dust tends to reduce the risk from surface dust by more than 90% (Roberts, Glass, and Mickelson 2004).

There may be an analogy between vacuuming the deep dust from a carpet in a home and digging at an archeological site; both efforts uncover the historical record of past events. Every individual who enters a home interacts with its dust history. They take dust from a room when they leave and make a unique contribution to the dust with skin scales, clothing fibers, dust falling off clothes, and tracked-in dust. The DNA in skin scales and unique clothing fibers can theoretically be used to identify past room occupants (National Institute of Justice 2002). They take away a "fingerprint" of the dust from the room that collects on their clothes.

Replacing house carpets with bare wood floors will eliminate the tendency for carpets and backing to act as a reservoir for pollutants. Bare floors are easiest to clean. Flat and level loop carpets are also easy to clean, with carpet cleaning difficulties increasing for short plush carpets, deep plush, and shag carpets, respectively. Spilled liquids may initiate mold growth in a carpet that is not dried in 24 hours. Interface (www.interfaceinc.com) as well as Collins and Aikeman (www.powerbond.com) make carpets with low volatile organic compound (VOC) emissions and waterproof backing that are around 50% easier to clean. The cost of installing a Powerbond carpet in a Seattle apartment in 2005 was $3.50/ft².

While there may be health advantages for bare floors, the homeowner must consider the costs and benefits of different types of floor coverings. Ceramic, solid hardwood, laminated hardwood, and linoleum floors are long lasting but more expensive. Vinyl tile floors are lower in cost but have VOC emissions. Carpets reduce noise, are softer to walk on, and may reduce falls. Bare floors in classrooms may require architectural and surface material changes to reduce noise.

14.3 MAJOR POLLUTANTS IN HOUSE DUST

14.3.1 LEAD: SOURCES, PATHWAYS, TRENDS, AND EFFECTS

Lead is unique among the toxic heavy metals because of its abundance in the Earth's crust. Because of its easy isolation and low melting point, lead was among the first metals to be used by humans thousands of years ago. The environmental significance of lead is a result both of its utility and its abundance. World production of lead, about 4 million tons per year, is larger than the commercial production of any other toxic heavy metal (USEPA 1986).

Lead is present in food, water, air, soil, dustfall, paint, and other materials with which the general population comes into contact. Each of these is a potential pathway for human exposure to lead through inhalation or ingestion. The actual lead content in each environmental medium may vary by several orders of magnitude. Individual exposure is further complicated by different activity patterns and differences in indoor and outdoor microenvironments.

Centuries of mining, smelting, and usage have made the natural background concentration of lead difficult to determine. Geochemical data indicate that the concentrations of lead in most surface soils in the United States range from 10–30 ppm or $\mu g/g$ (USEPA 1986). Unlike gaseous pollutants, where 1 ppm denotes a part-per-million by volume, trace metals like lead usually are reported in mass ppm units, and 1 ppm by mass is equivalent to 1 microgram of lead per gram of soil, or 1 $\mu g/g$.

Trace amounts of lead occur naturally in air and water as a result of wind and rain erosion, and in air as a result of volcanic dusts, forest fires, sea salt, and the decay of radon. Natural background concentrations of airborne lead have been estimated as approximately 0.0006 $\mu g/m^3$ or 5×10^{-7} $\mu g/g$. Estimated natural fresh water and ocean water lead concentrations are about 0.5 $\mu g/L$ of water (5×10^{-4} $\mu g/g$ water) and 0.05 $\mu g/L$ (5×10^{-5} $\mu g/g$ water), respectively (USEPA 1986).

As a consequence of the diverse uses of lead in products, including its extensive history of use in gasoline and house paint in the United States, the present concentrations of lead in air, soil, and water are higher than these estimated background levels. Typical average nonurban lead concentrations in air are about 0.01 $\mu g/m^3$ (USEPA 2004). Prior to restrictions on the lead content in gasoline in the United States, typical ambient lead concentrations in U.S. cities averaged about 1.5 $\mu g/m^3$ (Woodruff et al. 2003).

Concentrations of lead in most urban water supplies are well below 10 $\mu g/L$ standard for water (0.01 $\mu g/g$ water) (Woodruff et al. 2003). However, values above 50 $\mu g/L$ have been reported in some locations on occasion. Suspended solids contain the major fraction of lead in river waters. Concentrations of lead in tap water may be considerably higher than those in municipal supplies. Lead values as high as 2,000 $\mu g/L$ have been reported for homes with lead pipes and lead-lined storage tanks. Efforts to reduce the effect of lead solder joints in copper pipes and fittings and greater use of plastic pipe has reduced the levels of lead in residential tap water. Running the tap a few minutes before using tap water can help reduce lead concentrations in cases where lead is added by a home's plumbing system.

The contribution of food to human exposure to lead is highly variable and not well quantified. Estimates of the daily intake of lead from food vary from about 100–350 $\mu g/day$. Historically, foods stored in lead-soldered cans or stored or served in imported glazed pottery were identified as having high lead content (ATSDR 1989). In the United States there have been efforts to reduce the contact between lead-containing containers and foods.

Use of lead as an antiknock additive in gasoline historically accounted for a major share of U.S. lead production. Consequently, motor vehicles constituted the major source of atmospheric lead emissions. As a result of legislation that limited the lead content of gasoline, the production and use of alkyl lead additives decreased in the United States beginning around 1978. Beginning at this same time, concentrations of lead in ambient air have shown a similar downward trend. In 1977 ambient air lead concentrations at 78 sites in the United States averaged 1.5 $\mu g/m^3$. By 1984 ambient air lead levels in the United States had dropped to an average of 0.3 $\mu g/m^3$. In 1995 ambient

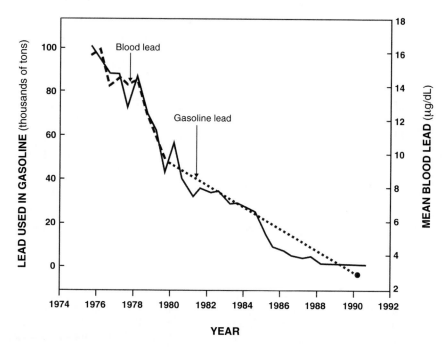

FIGURE 14.5 Measurements of blood levels in Americans based on the National Health and Nutrition Examination Survey (NHANES) from NHANES-II (1976–1980) and NHANES-III (1988–1991), compared with the amount of lead used in U.S. gasoline. (Personal communication with David M. Mannino, Centers for Disease Control, Atlanta, GA, 2002.)

air lead concentrations in the United States had continued to decline, reaching an average of about 0.1 $\mu g/gm^3$, which is close to the background concentrations measured years earlier in nonurban areas (USEPA 1986; Woodruff et al. 2003).

Lead happens to be one of the few environmental pollutants for which long-term data are available on an important biomarker of dose, the concentrations in the blood of U.S. citizens. Elevated blood lead levels are a predictor of adverse health effects, including anemia and brain damage. The U.S. National Center for Health Statistics periodically conducts the National Health and Nutrition Examination Survey (NHANES). In this survey, as many as 22,000 respondents are statistically selected and given a physical examination that includes a detailed questionnaire about their current health, diet, and activities. Blood drawn from a subset of the NHANES respondents is analyzed by the U.S. Centers for Disease Control (CDC) to determine blood lead concentrations. The results indicate a downward trend for blood lead concentrations of Americans over the 14-year period from 1976–1990 (Figure 14.5). This downward trend is continuing to the present time and roughly correlates with a similar downward trend in the lead content of gasoline (Woodruff et al. 2003). Measurements of average blood lead levels for the U.S. population based on NHANES-II, NHANES-III, and follow-up national surveys indicate that average blood lead levels for the U.S. population have continued to drop through 1999, but at a lower rate (Figure 14.6).

The median blood lead concentration of children 5 years old and younger dropped from 15 micrograms per deciliter ($\mu g/dL$) for the 1976–1980 period, to 2.2 $\mu g/dL$ for the 1999–2000 period. However, 10% of children during 1999–2000 still had blood lead concentrations above 4.8 $\mu g/dL$, a level which is still of considerable concern (Woodruff et al. 2003). The data presented in this chapter suggest that over 50% of adults 25–35 years of age in 2005 have had their intellectual potential reduced by exposure to lead in childhood.

Despite this downward trend in blood lead levels nationwide, many children and adults still live in older homes that contain risks from lead-based paints. The possibility that elevated blood

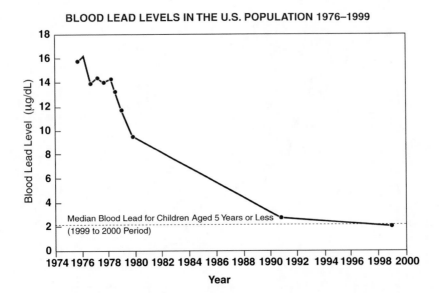

FIGURE 14.6 Measurements of blood lead levels in Americans based on the National Health and Nutrition Examination Survey (NHANES) from NHANES-II (1976–1980), NHANES-III (1988–1991), and NHANES for 1999 and beyond. (Personal communication with Larry Needham, Centers for Disease Control and Prevention, Atlanta, GA, 2004.)

lead concentrations is due to indoor contaminants continues to warrant scrutiny. A U.S Department of Housing and Urban Development (HUD) (Clickner et al. 2001) national survey of lead from a representative sample of U.S. housing found that one in three children under the age of 6 live in a house with a lead-based paint (LBP) hazard. In this study, the term "LBP hazard" referred to deteriorating paint, which indicated problems of lead-contaminated dust indoors or lead-contaminated soil from lead-based paint outdoors. A significant LBP hazard is defined by the HUD safe housing rule on lead as follows:

Lead-contaminated dust — Dust on floors (bare or carpeted) with greater than or equal to 40 $\mu g/ft^2$ lead, dust on windowsills greater than or equal to 250 $\mu g/ft^2$ lead as measured with wipe method

LBP or deteriorated LBP — Deterioration larger than 20 square feet (ft^2) exterior and 2 ft^2 interior or damage to 10% of the total surface area of interior small surface area components (window sills, baseboard, trim). LBP is defined as paint or other surface coatings (varnish, lacquer, or wallpaper over paint) that contains lead equal to or greater than 1.0 mg/cm^2

Bare, lead-contaminated soil — More than 9 ft^2 of bare soil with a concentration of equal to or greater than 2,000 ppm lead, or 400 ppm for soil frequented by child under the age of 6 years

Of the 16.4 million homes with children under the age of 6, 5.7 million (34%) have LBP hazards. About 17 million houses have interior dust hazards, 14 million houses have deteriorated LBP, and 6 million houses have soil lead hazards. Some 39% of homes with interior LBP exceed the LBP hazard rule. Some 6% of the 67 million houses without interior LBP exceed the same standard. Lead may be tracked in from the outside or tracked home from the job. Hobbies may also contaminate house dust with lead. Only 2% of the 87 million homes without deteriorated LBP have lead in the soil above 2000 ppm. Clean homes have fewer LBP hazards. Some 15% of all housing showed no evidence of cleaning (Clickner et al. 2001).

There are various physiological effects of lead that occur at the subcellular, cellular, and organ system level. A number of significant effects of lead on the blood have been observed in lead poisoning. These effects are prominent in clinical lead poisoning but are still present to a lower degree in persons with a lower level of lead exposure. Although a level of health concern for blood lead has been set at 10 μg/dL, it is believed that there is no true lower threshold at which adverse effects are nonexistent (Canfield et al. 2003). Anemia is a clinical feature of lead poisoning. In children, symptoms of anemia occur at blood lead levels of 40 μg/dL and in adults at levels of 50 μg/dL. The effects of lead on the nervous system range from acute intoxication behavior and fatal brain damage to more subtle changes associated with lower levels of exposure, such as reduced coordination and reduced intelligence (USEPA 1986, 1990; ATSDR 1989).

The most extreme effects of lead poisoning, severe irreversible brain damage, chronic enceph-alopathy (brain damage) symptoms, or death, have been known to occur at surprisingly low levels of blood lead. However, for most adults such damage does not occur until blood lead levels substantially exceed 120 μg/dL, although some evidence suggests that acute encephalopathy (brain damage) and death may occur at blood lead levels slightly below 100 μg/dL (ATSDR 1989). For children, the effective blood levels for producing encephalopathy or death are lower than for adults, with such effects being seen somewhat more often starting at blood lead concentrations of approx-imately 100 μg/dL. Studies also suggest that severe neural damage can exist without obvious symptoms, and children with high blood lead concentrations (greater than 80–100 μg/dL) are permanently impaired cognitively (USEPA 1986, 1990; ATSDR 1989).

The blood lead concentrations at which neurobehavioral deficits occur in children who exhibit no other symptoms appear to start at about 50–60 μg/dL. Mounting concerns about the adverse health effects of lead on children have given rise to a protective concentration set at 10 μg/dL, but there are concerns that subtle neurological and cellular effects such as lowered IQ could occur in some children below this level. Canfield et al. (2003) measured the blood lead concentrations of 172 children every 6 months for 5 years starting at age 6 months. The intelligence (IQ) was measured at 3 and 5 years. The IQ was inversely and significantly associated with lead in blood showing a decline of 4.6 points for every 10 μg/dL increase in blood lead concentrations. The IQ decline for children whose blood lead concentrations remained below 10 μg/dL was greater. IQ declined 7.4 points as average blood lead concentrations increased from 1 to 10 μg/dL.

Two pollutants for which multimedia exposure assessments are important are lead (Pb) and polycyclic aromatic hydrocarbons (PAHs). Multimedia scientific studies, although limited, have sought to quantify the contributions of outdoor air, indoor air, house dust, food, and drinking water to population exposures for these two multimedia pollutants.

The most effective exposure analysis for some pollutants in house dust is one that examines all relevant environmental carrier media by which the pollutant reaches a particular target, such as a human being. Ideally, this multimedia exposure analysis determines how often individuals in a population are exposed to concentrations that exceed critical thresholds, and the contribution of individual sources to the total exposure. A frequency distribution provides information on the percent of individuals exposed to different concentration levels for an averaging time of interest, such as 24 hours. For example, the percent of people exposed to outdoor air quality concentrations greater than the 24-hour air quality standard is useful for planning health-based initiatives.

Surprisingly, few multimedia exposure analyses have been carried out for several reasons. Significant concentrations of the pollutant of interest found only in one medium negate the need to study other media. The pollutant of interest may only travel through one carrier medium. Data on a pollutant in a specific carrier media are usually not available. Exposure levels themselves are imprecisely known, or it is unclear how the exposure actually occurs. Thus, there are few multimedia exposure analyses available in the literature.

Multimedia exposure analyses can be extremely important for setting regulatory policies. Determining the amount of exposure attributable to each carrier medium can help provide a decision maker with guidance about reducing total exposure in a manner that protects public health effectively and is least costly to society. Fortunately, the absence of complete multimedia exposure analyses for many agents is not as limiting as it might seem. Many pollutants, such as carbon monoxide (CO), reach human beings through only one carrier medium, inhaled air. It is not necessary to consider house dust, food, or drinking water when evaluating contact with CO. Similarly, many toxic volatile organic compounds (VOCs) present in the air we breathe are also found in food and drinking water at negligible concentrations. For these pollutants, a reasonably accurate exposure analysis may be done on indoor and outdoor air.

14.3.2 REDUCING LEAD EXPOSURE FROM HOUSE DUST

Adgate et al. (1998) used a chemical mass balance approach to find the major contributors of lead to indoor house dust obtained from 64 homes in Jersey City, NJ. To apply the chemical mass balance approach, the authors looked at the relative concentrations of lead and 16 other elements to help identify the "chemical profiles" of individual sources. For both coarse particles (up to 60 micrometers) and PM_{10} (particulate matter with aerodynamic diameters 10 microns or less) outdoor air was responsible for approximately two-thirds of the lead found in house dust. Suspended crustal materials from soil and deposited airborne particulate matter comprised the outdoor air sources. Interior lead-based paint sources contributed the remaining third. Thus, about one-third of the indoor lead was associated with old paint on surfaces, while two-thirds of the lead was from outdoor origin, including tracked-in yard soils and street dusts.

Lioy et al. (1998) studied the effectiveness of cleaning protocols in homes of children found to have moderate lead poisoning (blood lead levels of 10–20 µg/dL). Samples were collected in the homes at least twice, and in some cases three times, during the course of a yearlong randomized study in which half the homes used a heavy-cleaning protocol and the other half used a light-cleaning protocol. Both wipe sampling and a standard vacuum sampling were employed to evaluate the dust and Pb loading on bare and carpeted surfaces. When compared with the values seen on the first visit vacuum sampling, the results showed statistically significant decreases in lead and dust loading for the heavy-cleaning protocol. For the light-cleaning protocol, the sampling results found no significant reductions in dust and lead loading among any of the room surfaces sampled. In contrast, there were 75% and 50% reductions observed on the windowsills and on the bedroom floors of the homes in which the heavy-cleaning intervention took place.

The Superfund law requires a risk assessment when the concentrations of pollutants in residential outdoor soil exceed limits specified as the Preliminary Remediation Goals (PRGs). PRG levels for many toxic pollutants are listed in USEPA tables as the upper concentration limits acceptable for protecting people, including crawling children, from the health risks associated with outdoor soil (Smucker 1995). These levels are used by risk assessors to help decide the level of remediation at these sites. Surprisingly, concentrations of pollutants found in the dust of ordinary homes often exceed these Superfund PRG remediation limits. When these laws initially were written, the initial framers may not have been aware of the high residential indoor pollutant concentrations in American homes. For example, the PRG level for lead is 400 mg of lead per kg of material (that is, 400 µg/g or 400 ppm by mass).

Over a 7-year period from 1993–2000, more than 150 Master Home Environmentalist volunteers in Seattle agreed to collect house dust samples in their homes as part of their training and for a study conducted by one of the authors (Roberts). They used upright vacuums, most of which were equipped with dirt finders, to collect the samples on a section of carpet in their homes. The samples subsequently were sieved and analyzed by x-ray fluorescence mass spectrometry for lead content. The results showed that the percent of Seattle homes built before 1940 with indoor house dust lead concentrations above the 400 ppm Superfund PRG screening limit was 51% as shown in Figure 14.7. The horizontal axis of this logarithmic-probability graph lists the proportion of the houses below the stated concentration level given on the vertical axis. By comparison, only 4% of the homes built after 1940 had indoor house dust lead concentrations above 400 ppm by mass. This

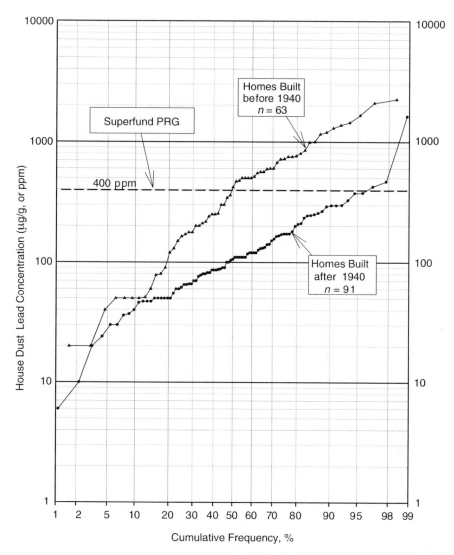

FIGURE 14.7 Distributions of lead concentrations measured in the house dust of 154 Seattle homes, comparing homes built before 1940 with those built after 1940. The Superfund Primary Remediation Goal of 400 ppm by mass (μg/g) also is shown (horizontal dashed line).

dramatic difference in indoor dust concentrations with the age of a home is evident in the difference in mean lead concentrations for these two groups of homes; the mean lead concentration for the 63 older homes was 519 ppm, vs. 146 ppm for the homes built after 1940.

The lead content of paint began to drop after 1940, and strict limits on the lead content of paint were imposed in 1978. Much of this lead found in house dust can be attributed to the soil around the homes. Lead in yard soil is attributed to a home's exterior paint and to road dust, but cannot be attributed to Superfund sites (Adgate et al. 1998). These results show that many American homes have interior lead concentrations on floors and carpets that are higher than the soil lead levels required to trigger a risk assessment under the Superfund Act. Remediation goals may be lowered below the PRG for individual chemicals when there are multiple pollutants at or near the PRG levels as found in house dust (Roberts 1998).

Children probably visit most Superfund sites rarely, if ever. They are much more likely to be exposed to lead while playing on floors and surfaces in their own homes, particularly if their home was built before 1940. This inconsistency between widespread exposures to high concentrations of several toxic pollutants in one's own home, compared with rare and unlikely exposure from most Superfund sites, has not been fully addressed or balanced in existing environmental laws. The spending of public funds on Superfund site cleanup has not been efficient in reducing the total exposure of children and adults but may have other benefits in terms of protecting groundwater and wildlife. The shift of a larger share of the cost of Superfund cleanups from corporations, who caused the pollution, to the public increases the need for evaluation of the related costs and benefits (Seelye 2002). The laws governing Superfund cleanups are not appropriate for controlling exposure in the home. A different approach is needed, which will be discussed later in this chapter.

The decrease in the blood lead levels of the U.S. population over more than a decade (Figure 14.5) is attributable primarily to the removal of lead from gasoline and, to a lesser extent, to the removal of lead from paints (USEPA 1986, 1990, 2004). However, many older homes still contain lead-based paint. Remodeling activities such as sanding and wall preparation can release large amounts of lead, in the form of dust, into the home. A certified contractor should perform these remodeling activities in older homes with great care. It is advisable that persons, especially pregnant women and small children, should not live in old homes during remodeling. It is probable that do-it-yourself home owners will continue to remodel old houses when these vulnerable people are present. If they take training, follow safety instructions, and take measures required to control lead dust, then they may be able to protect themselves and their families. HEPA vacuums should be used to clean up the dust at the end of each day. Working clothes should be changed before entering the clean part of the house. Plastic sheets should be installed and high quality doormats should be used to isolate remodeling areas from the rest of the house. Lead tests should be done to verify that the area where people live is safe.

Although lead has been removed from gasoline, it is still present as fine dust near roadways and streets, and the track-in of lead on shoes continues to be a source of concern (Roberts et al. 1996). One of the authors (JR) has noted a 50% drop in the average concentration of lead in house dust in homes in Seattle over the 1980–1998 period. The lead in the soil did not go away but became diluted or covered by more recent fallout with a lower lead content. Using doormats, removing shoes indoors, and frequent thorough vacuuming and cleaning of carpets, windowsills, and window wells are steps that can easily be taken to reduce the levels of lead and other pollutants inside the home. These steps may reduce the blood lead levels of small children living in the home.

14.3.3 PESTICIDES, POLYCYCLIC AROMATIC HYDROCARBONS (PAHs), PHTHALATES, AND OTHER TOXIC POLLUTANTS

A breast cancer study by Rudel et al. (2003) analyzed 89 organic compounds in air and house dust from 120 homes on Cape Cod. Phthalates (found in plasticizers and emulsifiers), o-phenylphenol (disinfectants), 4-nonyphenol (detergent metabolite), and 4-tert-butylphenol (adhesives) were the most abundant pollutants found in indoor air at concentrations in the 50–1,500 nanograms per cubic meter (ng/m^3) range. The most abundant pesticide detected in dust was permethrins with the banned pesticides heptachlor, chlordane, methoxychlor, and DDT frequently detected among the 27 pesticides discovered in dust. The concentrations detected exceeded PRG health-based guidelines established by the USEPA for 15 compounds. The study also detected 28 pollutants classed as endocrine disruptors and for which no guidelines are available. Existing guidelines designed to protect toddlers from exposure to individual pollutants via soil ingestion do not consider endocrine health effects.

PAHs, sometimes called polynuclear aromatic (PNA) compounds, are organic chemicals with one or more benzene rings. PAHs are formed from combustion processes, especially combustion where the oxygen supply is limited. Typical sources of PAHs are motor vehicles, diesel engines, fireplaces, residential wood burning, incineration, forest fires, cigarette and cigar smoke, cooking, barbecues, and incense burning. It is expected that the PAH benzo(a)pyrene with its high molecular weight and low vapor pressure will partition more toward dust than air.

TABLE 14.1
Pollutant Concentrations in House Dust in 362 Midwest Homes

Pollutant	Percent Detected %	Median μg/g, ppm	Maximum μg/g, ppm	Superfund PRG μg/g, ppm
Pesticides				
Aldrin	5	0.21	0.8	0.029
Chlordane	17	0.46	17.0	1.6
Chlorpyrifos (Dursban™)	67	0.54	324	180
DDE	10	0.17	5.3	1.7
DDT	25	0.25	103	1.7
Dieldrin	13	0.25	139	0.03
Heptachlor	7	0.28	3.4	0.11
Polycyclic Aromatic Hydrocarbons (PAHs)				
Benz(a)anthracene	90	0.71	89	0.62
Benzo(b)fluoranthene	94	1.16	181	0.62
Benzo(k)fluoranthene	92	1.18	142	6.2
Benzo(a)pyrene	89	1.12	139	0.062
Chrysene	97	1.26	161	62
Dibenz(a,h)anthracene	69	0.36	142	0.062
Indeno(1,2,3-c,d)pyrene	84	0.72	310	0.62

(*Source*: Camann and Buckley 1994. With permission.)

Road dust and house dust contain probable human carcinogens, or B2[1] pollutants. One important B2 PAH, benzo(a)pyrene, was found in the house dust of 89% of 362 Midwest homes as shown in Table 14.1. The median concentration was 1.12 μg/g (1.12 ppm) and the maximum concentration was 139 ppm, 2,000 times higher than the outdoor residential soil limit of 0.06 ppm (Camann and Buckley 1994). These concentrations were from a National Cancer Institute case-controlled epidemiological study of children with recently diagnosed acute lymphocytic leukemia designed to assess exposure to a range of pesticides and PAHs. The median concentration of benzo(a)pyrene found in indoor house dust (1.12 ppm) exceeded the Superfund PRG limit (0.26 ppm) for industrial outdoor soils by a factor of four. An initial and a second sample found DDT concentrations of 100 ppm in one 90-year old Oriental rug. DDT has been used to protect rugs from insects.

In this same National Cancer Institute study, the pesticide dieldrin was found in 13% of the homes, with a median and maximum concentration of 0.25 and 139 ppm, respectively (Camann and Buckley 1994). By contrast, the dieldrin level required to trigger a risk assessment under the Superfund law is 0.03 ppm for residential outdoor soils and 0.15 ppm for industrial outdoor soils (Smucker 2004). The median dieldrin level found in 13% of 362 Midwest homes was eight and two times larger than the Superfund PRG limit for residential and industrial soils, respectively. An analysis of PAH and pesticide data collected in this 1994 study has yet to be published.

Camann et al. (2000) collected house dust samples from 1,040 homes in Long Island, NY; Cape Cod, MA; Detroit, MI; Seattle, WA; Los Angeles, CA; and Yuma County, AZ. The 90th percentile DDT concentrations ranged from 0.08 ppm in Yuma to 3 ppm in Cape Cod (the Superfund PRG for DDT is 1.3 ppm). Permethrin was the semi-volatile organic compound with the largest concentration in house dust. The permethrin concentration exceeded 30 ppm for 10% of the homes in Cape Cod and Los Angeles. Ten percent of the homes on Long Island contained house dust with

---

[1] The "B2" group is a collection of PAH compounds that have been classified, due to their toxicity, as likely or probable human carcinogens. Most of these compounds are of low volatility.

concentrations of benzo(a)pyrene and benzo(a)anthracene that exceeded 30 ppm. The sum of the median concentrations of the seven B2 PAHs in house dust was 6.51 µg/g in the Midwest study which increases the exposure and potential risk of cancer (Camann and Buckley 1994).

Mutagenic activity is used to screen chemicals for potential carcinogenic activity. House dust from 12 out of 29 Seattle homes was found to be mutagenic (that is, potentially capable of causing changes in DNA), and 20 out of 29 dust samples interfered with DNA repair (Roberts, Ruby, and Warren 1987).

Maertens, Bailey, and White (2004) of Health Canada suggest that the house dust may be a significant source of exposure for children to hazardous substances, including mutagenic material. They found that the mutagenic activity of house dust tended to be higher than that found in the soil at hazardous waste sites. Location was more important than smoking, but other factors appeared more important than either smoking or location. Elevated risk assessments of greater than 1.5–2.5 lifetime cancers per 10,000 children are associated with exposure to PAHs in house dust at or beyond the 95th percentile because of the enhanced susceptibility of infants. PAHs were responsible for around 25% of the mutagenic activity in house dust.

Chuang et al. (1999) studied the PAH exposures of children of low-income families to determine PAH concentrations in air, dust, soil, and food. In 24 homes these investigators found that the highest PAH concentrations, in decreasing magnitude, were in house dust, entryway dust, and pathway soil (Table 14.2). In this table, the footnote ("a") shows six compounds that belong to this higher-risk category, with totals for each of the two groups listed at the bottom of the table: (1) the "Sum of B2 PAH" and (2) the "Sum of target PAHs."

The Chuang et al. (1999) study of 24 homes measured total PAH concentrations in a variety of environmental media. Inhaled air was the main source of total PAHs for both children and adults (Figure 14.8). Food was a significant source of PAHs for both adults (38% of the total) and children (26% of the total). Nondietary ingestion played a small role in the intake of total PAHs for both adults and children. Children receive about one-quarter (24%) of their B2 PAH intake from nondietary ingestion. The higher uptake of B2 PAHs by children can be attributed to their activities in the home, playing on carpets, hand-to-mouth behavior, and high levels of physical activity indoors. For both children and adults, however, the major source of B2 PAHs was food. If the total intake of PAH is normalized by body weight, then the children's potential daily doses of PAHs were higher than those of adults in the same household. In summary, inhalation was found to be an important pathway for children's exposure to total PAHs, but dietary and nondietary ingestion pathways were more important for children's intake of the "B2" PAHs, compounds ranked as "probable human carcinogens" by the USEPA Integrated Risk System.

Based on the Camann and Buckley (1994) study, it has been estimated that each day the average urban crawling child will ingest 110 nanograms of benzo(a)pyrene (BaP), one of the two most toxic PAHs (Ott and Roberts 1998). This intake raises a child's risk of acquiring cancer, and the amount is sobering: it is equivalent to what a child would get from smoking 3 cigarettes. Companies who exposed their workers to this much BaP would be required to develop an action plan to reduce their exposure. BaP, a stable combustion by-product, is expected to be higher in soil downwind of large cities, industrial centers, and urban areas. The quantity of BaP in the *plow layer*, the top 23 cm of soil, in southeast England increased by a factor of over 15 in 100 years (1880–1980) while the total PAH burden increased by a factor of four. The flux rate for BaP was 0.36 mg/m²-yr (Jones et al. 1989). The change from coal to oil to natural gas for heating and power generation and the implementation of emission controls will slow the rate of increase. The other PAHs, measured as part of the study referenced above, are not as stable as BaP and did not increase as much. The concentration of BaP in surface soil in home yards where there is no cultivation and mixing of surface soils may increase at a faster rate. A BaP concentration in house dust may increase over time as ambient fallout accumulates. Soil is a major source of track-in. However, emission reductions, education of the public on methods of reducing track-in, and removal of deep dust in carpets may result in a reduction of human exposure to BaP from house dust.

TABLE 14.2
Summary of PAH Concentrations (µ/g, or ppm by Mass) in House Dust, Entryway Dust, and Pathway Soil of 24 Low-Income Homes

Compound	House Dust				Entryway Dust				Pathway Soil			
	Mean	Standard Deviation	Minimum	Maximum	Mean	Standard Deviation	Minimum	Maximum	Mean	Standard Deviation	Minimum	Maximum
Naphthalene	0.33	0.85	0.02	4.30	0.11	0.26	0.01	1.31	0.01	0.01	<0.01	0.04
Acenaphthylene	0.08	0.06	0.01	0.27	0.04	0.06	<0.001	0.27	0.01	0.01	<0.001	0.03
Acenaphthene	0.05	0.04	<0.001	0.18	0.04	0.03	0.01	0.12	0.01	0.02	<0.001	0.08
Fluorene	0.12	0.24	0.02	1.22	0.04	0.03	0.01	0.06	0.01	0.01	<0.001	0.04
Phenanthrene	0.44	0.40	0.13	2.15	0.29	0.29	0.04	1.32	0.08	0.09	0.01	0.36
Anthracene	0.12	0.15	0.01	0.75	0.10	0.12	<0.001	0.40	0.03	0.05	<0.001	0.18
Fluoranthene	0.52	0.37	0.09	1.89	0.37	0.35	0.02	1.44	0.12	0.15	0.01	0.57
Pyrene	0.43	0.33	0.06	1.65	0.28	0.28	0.02	1.04	0.12	0.16	0.01	0.60
Benz(a)anthracene[a]	0.22	0.17	0.04	0.69	0.15	0.14	0.01	0.52	0.06	0.09	<0.001	0.32
Chrysene[a]	0.39	0.47	0.05	2.41	0.20	0.19	0.02	0.74	0.02	0.02	<0.001	0.08
Cyclopenta[c,d]pyrene	0.08	0.05	0.02	0.22	0.05	0.03	<0.04	0.13	0.02	0.04	<0.001	0.20
Benzo(b and k)fluoranthenes[a]	0.55	0.32	0.17	1.34	0.36	0.28	0.04	0.08	0.12	0.15	0.01	0.47
Benzo(e)pyrene	0.26	0.17	0.05	0.75	0.20	0.15	0.02	0.54	0.05	0.06	<0.001	0.20
Benzo(a)pyrene[a]	0.23	0.15	0.07	0.63	0.15	0.13	0.01	0.41	0.06	0.01	<0.001	0.35
Indeno(1,2,3-c,d)pyrene[a]	0.23	0.18	0.05	0.70	0.16	0.13	0.01	0.42	0.05	0.07	<0.001	0.28
Dibenzol(a,h)anthracene[a]	0.10	0.09	0.02	0.41	0.05	0.05	0.01	0.15	0.02	0.03	<0.001	0.13
Benzo(g,h,i)perylene	0.25	0.16	0.08	0.61	0.17	0.13	0.01	0.37	0.05	0.06	<0.001	0.21
Coronene	0.13	0.11	0.04	0.50	0.10	0.11	0.01	0.51	0.03	0.04	<0.001	0.19
Sum of B2 PAHs	1.73	1.25	0.46	5.98	1.08	0.86	0.13	2.87	0.40	0.51	0.02	1.77
Sum of target PAHs	4.52	2.91	1.25	15.2	2.85	2.05	0.42	6.58	0.96	1.10	0.06	3.84

[a] Denotes B2 PAHs.

(*Source:* Chuang et al. 1999. With permission.)

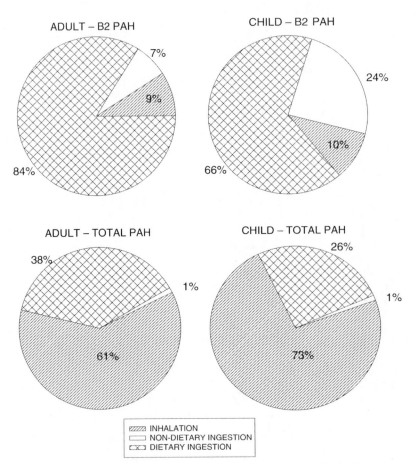

FIGURE 14.8 Estimated daily potential doses for B2 PAHs and total PAH. (From Chuang et al. 1999. With permission.)

It is not efficient to devote extensive financial and other resources for remediation projects on outdoor soils at remote Superfund sites located on private property that few people visit, while at the same time ignoring these same pollutants at higher concentrations in house dust which babies crawl in every day. The probability that a person comes into contact with the soil on a remote outdoor site is many times lower than the probability of coming into contact with the pollutant in one's own home. In addition, many tracked-in pollutants tend to accumulate in the home over time where they are protected from degradation by sunlight, moisture, and wind (Lewis et al. 1999). Indoor sources such as smoking and cooking may add more to these already high levels. As a result, the concentrations of pollutants inside homes are usually much higher than those in the soil immediately outside the home.

As with other policy-relevant findings in the emerging exposure sciences, the contradiction between the levels of concern at remote, outdoor Superfund sites and the higher concentrations of the same pollutants found routinely in ordinary homes has not been addressed. It is a major contradiction in health policy for environmental programs to place enormous emphasis on the low concentrations found in outdoor soil while placing little or no emphasis on the higher pollutant concentrations found inside the home where the most susceptible population (toddlers, the sick, and the elderly) may spend more than 90% of their time. It is unfortunate that a multimedia exposure assessment has not been conducted on the national level to help provide an enlightened policy for controlling indoor and outdoor pollutants in a manner that reduces exposure the most for each dollar spent. Health researchers and epidemiologists also need to conduct more research to determine if there is positive association between the high indoor pollutant concentrations found in American homes and adverse health outcomes, such as allergies, asthma, heart attacks, and cancer. The lack of adequate health research probably is a direct consequence of the lack of knowledge by the health research community of the high levels of exposure to toxic pollutants found in the American home. There is a need to search for efficient methods of measuring and reducing the total exposure of people to pollutants from indoor air and dust, outdoor air and soil, water, and food. It is difficult to plan for control of exposures that are not measured.

The original Superfund law was passed in response to concerns about pollutants dumped by Hooker Chemical Company at Love Canal. Virtually no scientific health studies or epidemiological research was done at the time to determine if the pollutant levels actually caused adverse health effects among the population surrounding the Love Canal site. The USEPA made some indoor and outdoor measurements at Love Canal and found high indoor VOC concentrations, which seemed surprising at the time. However, very few measurements from homes elsewhere in the U.S. were available for comparison to the concentrations measured in the homes near Love Canal. Some USEPA scientists, looking back on these data now, have concluded that the indoor VOC concentrations in the Love Canal homes were no higher than those found routinely in homes in isolated rural areas such as Devils Lake, ND, or in homes in industrial areas such as Bayonne–Elizabeth (Wallace 1987). Also, the indoor concentrations in the Love Canal homes showed a striking decrease when the occupants moved out, taking their belongings and furniture with them. It is likely that the high indoor concentrations of toxic pollutants observed in the homes at Love Canal were caused partly by the residents' own consumer products, furniture, and personal belongings, and were not caused alone by outdoor pollutants seeping indoors, as people assumed at the time.

With the possible exception of the Food Protection Act of 1996, none of the exposure data and findings presented in these chapters was available when major environmental laws were written. Very little was known then about the exposures people normally receive in their daily lives. Today, these environmental laws are still on the books and being implemented as originally written (see Chapter 21). Only one major environmental law, the Clean Air Act, has been revised since 1990. The findings from exposure assessment research confirm the need to conduct improved health research and to update the existing environmental statutes and policies. This is a mission that needs attention as we search for efficient ways to reduce suffering and minimize health costs. Reducing exposures in the home should be part of an integrated home health management plan that includes improved lifestyles, exercise, and diet. Perhaps talented outreach workers can empower families so they can make the necessary changes in the future (Krieger et al. 2005).

14.4 CONTROL OF HOUSE DUST

14.4.1 CLEANING

Cleaning reduces exposure to dust, mold, and bacteria. A yearlong study of the changes produced by effective cleaning in a large non-problem daycare center showed reductions of 52% in indoor air concentrations of dust and 40%, 88%, and 61% in colony-forming units/m^3 of bacteria, Gram-negative bacteria, and fungi, respectively. Monitoring was done for 5 months before and 7 months after effective cleaning method were introduced (Franke et al. 1997). Higher levels of cleaning are required in warm, moist, and damp climates to control the growth of dust mites, fungi (mold), and bacteria. The dust contains the food needed for these biological contaminants to grow.

Cleaning studies may be confounded, and more pollution may be left on the surface of a carpet than at the start if all the deep dust is not removed. Studies by Lewis (2002); Roberts et al. (1999); Roberts, Glass, and Mickelson (2004); Hilts, Hertzman, and Marion (1995); and Ewers et al. (1994) show that high loadings of lead, dust mite allergen, and other toxic pollutants may be brought to the surface where only part of the deep dust is removed. Ewers et al. (1994) found that high-efficiency filtration of 99.9% of particles less than 0.3 microns (HEPA) from vacuuming older carpets may increase the lead surface loading by a factor of 4 if the cleaning is not conducted for a sufficient time. Hilts, Hertzman, and Marion (1995) reporting the results of a study in British Columbia, Canada, noted that the surface lead loading increased in 3 out of 55 carpets of all ages following HEPA vacuuming.

Upholstery has also been found to contain high quantities of lead. However, it appears that it is easier to clean than carpets (Lewis 2002). Dust ground into carpets by foot traffic may make them harder to clean than upholstery. The use of a damp cloth is an effective way to remove dust from solid surfaces.

Shampooing carpets with a detergent may help to remove soot and stains, but has some risks from mold developing if the carpet and pad are not dry in 24 hours. A spilled glass of water or coffee on a carpet has the same potential. Lewis (2002) found that dry vacuuming removed 38% more lead than wet vacuuming. Removal of deep dust before commercial truck-mounted steam cleaning provides one option for restoring color to a carpet with less chance of leaving a damp carpet, dust, or increased surface lead.

Old carpets can be cleaned by carpet cleaning companies or individual families. Cleaning a carpet for 0.5–2 min/m^2 will not guarantee that less lead and dust will remain on the surface at the end than when the cleaning started. The Institute for Inspection Cleaning Restoration and Certification's (IICRC) 2002 S100 standard for cleaning has no objective requirement for determining if the carpet is clean at the end of the cleaning process. It suggests frequency of cleaning guidelines based on usage rather than on the dust remaining in the carpet. The S100 standard may help to reduce dust and related pollutants in most cases, but it is not adequate to fully protect small children and sensitive adults from pollutants that may remain in deep dust brought to the surface during cleaning. People with AIDS, asthma, allergies, cancer, heart disease, or lung disease can reduce their exposure to particles, allergens, lead, cadmium, pesticides, other carcinogens, bacteria, and many other pollutants by monitoring and controlling the deep and surface dust in carpets with a vacuum with an embedded dirt finder. Other methods of dust control mentioned in this chapter will also help. The three-spot test can be used to tell when the carpet is clean and estimate the time (min/m^2) required to remove deep dust (Roberts, Glass, and Mickelson 2004).

14.4.2 Why Vacuum Cleaners Do and Do Not Work

Not all vacuum cleaners and carpets are created equal. The combination of a canister vacuum without a power head and a shag or deep plush carpet can cause large amounts of dust (up to 400 g/m^2) and pollutants to collect in the carpet. Vacuums with a power brush are 2–6 times as effective as those without for cleaning carpets (Balough 1988). Deep plush or shag carpets are much more difficult to clean than flat carpets. An upright vacuum with a power-driven agitator should remove approximately 35–55% of the recently applied and embedded dust from a plush carpet, and 70–80% from a flat or level loop carpet (Balough 1988). These results were obtained by performance testing of the vacuum following the standard ASTM F 608-89 test. In contrast, the canister cleaner without a power-driven brush removes only 10% of such dust from a plush carpet, and 40–50% from flat carpets (Balough 1988). Tests of low pile carpets show they are easier to clean (Lewis 2002). New carpets are also easier to clean than old carpets (Lewis 2002). However, 10–25 times as much dust, lead, and pesticides may remain in an old rug in a home than is removed by normal vacuuming (Roberts et al. 1999; Fortune, Blanchard, and Ellenson 2000). The ASTM F 608-89 test, which is useful for comparing the efficiencies of new vacuums, does not apply to old carpets. There is a wide difference between the test dust used in the standard method and the highly variable and the deeper embedded dust in old carpets.

Canister cleaners work well on bare floors. However, trying to clean area rugs with a vacuum without a power head can leave behind large amounts of dust, lead, and other pollutants. Area rugs on bare floors act as dust magnets by removing the particles collected on shoes much better than a bare floor.

Vacuums fail to work effectively because of the following reasons.

- The bag is full. A full bag reduces airflow and dust pickup. Bags should be changed when they are half full.
- The belt is worn or broken.
- The brush is worn.
- Leaks reduce airflow through the nozzle.
- They are not used enough or not used effectively.
- They have no agitator.

The performance of vacuums tends to degrade with age if they are not maintained. Families should take their vacuum to a qualified technician when they are not working properly. More vacuuming is required near entrance doors and in high traffic areas. A vacuum with an embedded dirt detector tells the user where the dirt is and when the carpet is clean (Roberts et al. 1999; Roberts, Glass, and Mickelson 2004). Vacuuming in parallel swaths followed by vacuuming at 90 degrees to the original swaths increases effectiveness (Tucker 2003).

14.5 MASTER HOME ENVIRONMENTALIST™ PROGRAM

Many scientists throughout the world are researching human exposure to environmental pollutants, including scientists at the USEPA National Exposure Research Laboratory in Research Triangle Park, NC. Public and private organizations also have become involved in exposure reduction activities. For example, the American Lung Association of Washington has developed the Master Home Environmentalist™ (MHE) Program founded in 1992 by one of the authors (JR). The MHE Program is designed to help people reduce their exposure to indoor pollutants using assessments of the home undertaken by trained volunteers. Dickey (2001) has edited a training manual for the Master Home Environmentalist™ program that provides detailed information on the steps people can take to assess and reduce their exposure to pollutants in their daily lives and in their homes.

In this program, volunteers receive 35 hours of training by professionals in key areas of indoor pollution, communication, and outreach. The American Lung Association provides this training at no cost to the volunteer. The topic areas include toxicology, lead, dust, mold, moisture, biological contaminants, household chemicals, ventilation, tobacco cessation, asthma, allergies, behavior change, and multiple chemical sensitivity. The volunteers are trained to work with residents to complete a detailed Home Environmental Assessment List (HEAL™) to identify pollutants and address residents' concerns about the air and dust in their homes. The family and the volunteer negotiate the top three priorities for change that the family is willing to undertake in the next 6 weeks. The volunteers focus on low-cost or no-cost solutions and may recommend working with a doctor, the local health department, or an environmental professional if high health risks or high cost solutions are involved. Volunteers follow up with a phone call after 2 weeks.

The Master Home Environmentalist™ is an example of an outreach program designed to help members of the general public reduce their personal exposure to environmental pollutants. The program has spread from the Seattle office of the American Lung Association of Washington to Tacoma, Yakima, many counties in Washington State, Portland, OR, Boise, ID, New York, NY, Washington, DC, Fresno, CA, and Tulsa, OK. Both the American Lung Association and the USEPA would like to see this program extended nationwide. Over 700 volunteers have been trained as of this writing. Hundreds of thousands of volunteers are available to help their families, friends, and neighbors have healthier homes and lives. However, a budget is needed to pay for the recruiting, training, and coordination of these volunteers. Paid indigenous community health workers (CHW) have also been used to reduce home exposures. The CHWs were successful in empowering low-income families with asthmatic children to reduce exposure to dust mites, dust, and other asthma triggers in the home (Takaro et al. 2004).

Information is essential but often not sufficient to cause a family to change behavior in the home environment that will result in reduced indoor personal exposure. Encouragement and support from a doctor, nurse, counselor, outreach worker, friend, or volunteer can help the individual make progress through the stages that lead to permanent changes. General information about the home environment is of interest, but assessing the personal exposures in the home has a personal impact. Laws and regulation have reduced emissions and exposure to pollutants in outside air. Regulations have effectively reduced lead in paint and emissions from automobiles and building materials. These regulations have not eliminated personal exposures in the home that require changes in cleaning methods, ventilation, or use of home chemicals. Such changes must be voluntary. Trained volunteers and CHWs can reach where laws and regulations cannot. They can provide motivational interviews that meet the needs and specific exposures unique to each family discovered during the HEAL™. While everyone will profit from a healthier home, those individuals and families who suffer from asthma and allergies, or who are concerned about the indoor risks to babies and sensitive individuals, will be most willing to seek help in assessing and controlling indoor exposures. Sensitive individuals include those with cancer, heart disease, AIDS, poor immune systems, or who live or work in structures with sick building health problems.

Leung et al. (1997) and Primomo (2003) have evaluated the effectiveness of the MHE program in Seattle and Tacoma, respectively. Leung found changes in cleaning behavior after a home assessment by a MHE. Primomo found 60 of 64 families (94%) made at least one change in cleaning behavior. Also, 34 out of 39 families (87%) with an asthmatic member reported an improvement in the condition after a HEAL™ evaluation and changes to cleaning behavior. The HEAL™ evaluation by a community volunteer can be an effective way to increase knowledge about the home environment, asthma, and allergy triggers. Families who evaluate their home situations with a HEAL tend to engage in behavioral changes and perceive improvement in asthma and allergy symptoms.

Takaro et al. (2004) evaluated the effect of environmental intervention by CHWs to reduce exposure to asthma triggers in the homes of 274 low-income children with asthma, as part of the Seattle-King County Healthy Home Project I. The families were given allergy-control bedding covers, vacuum cleaners with a dirt finder, high quality doormats, and cleaning kits. CHWs made 5–9 visits over a year focusing on asthma trigger reduction. The 3-spot test (Roberts, Glass, and Mickelson 2004) was used to evaluate the effectiveness of carpet dust control and was performed before and after vacuuming. In 74 homes with data available, the average time to get three green lights on the dirt finder equipped vacuums dropped 61%, from 59 to 23 seconds, and the median time dropped 23%, from 15 to 11.5 seconds. Ten seconds to get three green lights is considered a clean carpet and correlates with less than 10 g/m^2 of deep dust (Roberts, Glass, and Mickelson 2004). This test was attractive to the CHWs because it could be done in less than a minute and both the CHW and the caregiver could see results immediately.

Giving the family a vacuum with a dirt finder appeared to motivate and empower the person who did the cleaning. Twenty-five percent of the families in the Seattle study did not have a vacuum at the beginning of the study. It is estimated that fewer than half of the families in this study had an effective vacuum. The average low-income family in this study increased its ability to remove the deep dust, dust mites, mold, and animal dander from carpets and overcame one of the major handicaps when compared to higher income groups. The average surface dust loading from the Seattle study dropped from 2.64 g/m^2 to 0.96 g/m^2 (Takaro et al. 2004) This compares favorably with the 1.4 g/m^2 average loading measured in 362 homes in the Midwest in the childhood leukemia study (Camann and Buckley 1994). The dust loading in both studies was measured with the HVS3. There is still room for improvement. Surface loading of 0.1 g/m^2 can be achieved when the deep dust is removed (Roberts et al. 1999; Roberts, Glass, and Mickelson 2004). Low-income families with children with asthma may be further handicapped by excessive mold and moisture, low or high temperatures, older and deteriorated housing, family stress, and poor access to health care.

14.6 HEALTH CARE COST SAVINGS

The high cost of health care is a serious problem in all countries of the world. An important goal of healthcare reform is to reduce the cost of health care. However, primary prevention programs have a greater potential to reduce disease, injury, disability, pain, premature death, as well as the enormous health care costs (WSDH 1994). Prevention is an attractive option to the consumer. Most families would rather manage exposure rather than disease after they are fully informed. Reducing exposure to toxic substances reduces risk of disease, suffering, and related costs.

Takaro et al. (2004) and Krieger et al. (2005) documented the reductions in exposures to asthma triggers, the severity of asthma, and health care costs as part of the Healthy Homes Project I in Seattle. The occurrence of symptoms during a 2-week period dropped from 4.5 to 0.9 (80%) at night, symptom days dropped from 8.0 to 3.2 (60%), and days with activity limitations dropped from 5.6 to 1.5 (73%). The proportion of participants using urgent health care services (unscheduled doctor visits and the emergency room) dropped from 23.4 to 8.4%. There was also a reduction in the use of medicines. The frequency of missed school dropped from 31.1 to 12.2%. The percent of parents who missed work dropped from 13.1 to 12.2%. There was a significant improvement in the quality of life of the parents who had fewer interruptions of sleep and work, as well as less worry.

The estimated annual medical cost savings was $1,205–$2,004 per child per year. This project reduced the total exposure of the children to lead, pesticides, phthalates, carcinogens, VOCs, and other pollutants found in dust and indoor air as well as asthma triggers. The phthalates found in house dust have been associated with asthma (Rudel et al. 2003; Bornehag et al. 2004). Additional research would be required to sort out the synergistic effect of such reductions. A control group received only allergy-control mattress and pillow covers plus one visit. The annual cost of the intervention in this group was $222 with a saving of $1,110–$1,890. An annual return on investment of 400–800% is very attractive. Similar savings would continue every year with very little additional cost.

The National Cooperative Inner-City Asthma Study (NCICAS) of 1,033 low-income children with asthma was done in Seattle and seven other cities (Sullivan et al. 2002). The children were given allergy-control pillow and mattress covers and the family provided education and social support in controlling asthma triggers in the home. NCICAS also reported a reduction in asthma severity. This study found that it was cost effective to spend an average of $245 per year per child to achieve a $9.20 savings per symptom-free day gained. The sicker the child the greater the health cost savings. Fewer visits to the doctor, hospital, or emergency room and less use of expensive medicines may be a large source of savings. The simple solution of pillow and mattress covers was the most cost effective way to reduce symptoms and costs. Managing exposure to asthma triggers rather than the associated disease could result in important social and financial benefits.

As more scientific knowledge becomes available from research now underway on exposure assessment methods, measurements, and models, it is likely that the public will demand more information and technical assistance on effective ways to reduce its everyday exposure to environmental pollutants. It is obvious that government as well as public and private groups have an important role to play in disseminating these research findings and other useful information to the public about how to reduce health risks by reducing exposures.

Some studies suggest that exposure to higher levels of indoor endotoxin in house dust in early life may protect against allergen sensitization that may lead to allergies and asthma (Gereda et al. 2000). Some parents may assume that less cleaning and more house dust is good for their children. Families who are short of time may be looking for reasons to do less cleaning. However, a study of 4,000 school-aged children in southern California found asthma diagnosis before age 5 was associated with exposure during the first year of life to wood or oil smoke, cockroaches, herbicides, pesticides, farm dust, or farm animals (Salam et al. 2003). Most house dust qualifies as hazardous waste (Roberts 1998). In the Cape Cod study 15 chemicals exceeded government health-based guidelines (Rudel et al. 2003). House dust is the largest source of lead for most crawling children and may also be a large source of benzo(a)pyrene, pesticides, and other carcinogens. Cancer rates for children have increased 26% from 1975–1998 (Woodruff et al. 2003). It is prudent to reduce exposure to the toxic material in house dust by cleaning while searching for a safe method for protecting small children against sensitization.

14.7 MAKING A PLAN TO REDUCE YOUR PERSONAL EXPOSURE

Families are faced with exposure to many pollutants in air, dust, soil, food, and water that can exceed health-based guidelines set by government agencies. It is fortunate that the highest exposures for the most vulnerable members of the family (infants, the ill, and the elderly) occur where the family has the most control — in the home. They need a summary of the best information on their personal exposures and action options so they can make strategic decisions to manage their total exposure, risks, and health. A questionnaire called the HEAL™, mentioned previously, designed by the American Lung Association's Master Home Environmentalist Program (MHEP)™, is one of the best tools available to empower your family. The HEAL evaluates potential concerns in your home and suggests priorities for action to reduce exposures. Families living in the Puget Sound region can call 206-441-5100 or those living near a MHEP program can ask for a trained volunteer to come to the home to do a HEAL free of charge. If you don't have a MHEP nearby, you can ask for a Do-it-Yourself HEAL on the Internet and complete it. The URL for the HEAL is at: www.alaw.org/air_quality/master_home_environmentalist/free_home_assessment.html. It is prudent to hire an environmental professional and talk to your doctor or health department as appropriate if there are high costs or health risks involved.

14.8 QUESTIONS FOR REVIEW

1. Why are relatively few examples of multimedia exposure analyses found in the published literature?
2. Identify pollutants other than lead and PAHs for which multimedia exposure analyses would be appropriate. These are pollutants for which exposure occurs through more than one environmental carrier medium.
3. Why is it important to identify the geophysical background concentration of a pollutant, even though identifying true "background" levels may be difficult?
4. Discuss the differences in the frequency distributions of lead concentrations in house dust for Seattle homes built before and after 1940.
5. All the information is available in Figure 14.7 to make histograms, for each of the two groups of Seattle homes, of the lead concentrations measured in the house dust. Use the sample sizes listed on the figure and estimate the lead quartiles visually from the graph. Sketch a histogram of the pre-1940 homes and a second histogram of the homes built after 1940. Divide the counts in each class interval of each histogram by the total number of homes and sketch a frequency distribution of the house dust lead concentrations for each group of homes. Plot these two frequency distributions showing the percentage of homes in each lead class interval on the same graph (or on graphs above one another) so the distributions of lead house dust concentrations in the two groups of homes can be easily compared on a linear scale.
6. Based on several of the studies cited in this chapter, what steps can you describe to clean a home thoroughly to reduce the exposure of its occupants to lead? What other steps might be taken that are not related to house dust?
7. Discuss the characteristics of the B2 PAHs that separate them from the other PAHs.
8. What is the most important carrier medium by which the B2 PAHs reach children and adults?

9. Even though some pesticides have been banned, people still can be easily exposed to them. How does exposure occur, and what are some of these common pesticides?
10. Using available literature sources, list examples of other B2 environmental pollutants in addition to those discussed in this chapter.

14.9 ACKNOWLEDGMENTS

The authors are grateful to Peter G. Hummer, environmental consultant, for his thoughtful suggestions and editing. Philip Dickey, staff scientist of the Washington Toxics Coalition, also made useful editorial comments.

REFERENCES

Adgate, J.L., Willis, R.D., Buckley, T.J., Chow, J.C., Watson, J.G., Rhodes, G.G., and Lioy, P.J. (1998) Chemical Mass Balance Source Apportionment of Lead in House Dust, *Environmental Science and Technology*, 32(1): 108–114.

ATSDR (1989) *Toxicological Profile for Lead*, Agency for Toxic Substances and Disease Registry, U.S. Department of Health and Human Services, Atlanta, GA.

Balough, J. (1988) Personal Communication, Director of Laboratories, The Hoover Company, North Canton, OH.

Bornehag, C.G., Sundell, J., Weschler, C.J., Sigsgaard, T., Lundgren, B., Hasselgren, M., and Engman, L.H. (2004) The Association between Asthma and Allergenic Symptoms in Children and Phthalates in House Dust: A Nested Case-Control Study, *Environmental Health Perspectives*, 112: 1393–1397.

Calabrese, E.J. and Stanek. E. (1991) A Guide to Interpreting Soil Ingestion Studies: II. Quantitative Evidence of Soil Ingestion, *Regulatory Toxicology and Pharmacology*, 13: 278–292.

Camann, D.E. and Buckley, J.D. (1994) Carpet Dust: An Indicator of Exposure at Home to Pesticides, PAHs, and Tobacco Smoke, Paper No. 141, presented at the 6th Conference of the International Society of Environmental Epidemiology and 4th Conference of the International Society for Exposure Analysis, Research Triangle Park, NC.

Camann, D., Colt, J., Teitelbaum, S., Rudel, R., Hart, R., and Gammon, M. (2000) Pesticide and PAH Distributions in House Dust from Seven Areas of the USA, Paper 570, presented at SETAC 2000, Society of Environmental Toxicology and Chemistry, Nashville, TN, November 16, 2000.

Canfield, R.L., Henderson, C.R., Cory-Slecta, D.A., Cox, C.R., Jusko, T.A., and Lanphear, B.P. (2003) Intellectual Impairment in Children with Blood Lead Concentrations below 10 µg per Deciliter, *New England Journal of Medicine*, 348(16): 1517–1526.

Chuang, J.C., Callahan, P.J., Lyu, C.W., and Wilson, N.K. (1999) Polycyclic Aromatic Hydrocarbon Exposures of Children in Low-Income Families, *Journal of Exposure Analysis and Environmental Epidemiology*, 9(2): 85–98.

Clickner, R.P., Marker, D., Viet, S.M., Rogers, J., Broene, P. (2001) *National Survey of Lead and Allergen in Housing, Vol. I: Analysis of Lead Hazards,* Westat, Inc., U.S. Dept. of Housing and Urban Renewal, Washington, DC.

Davies, D.J.A., Thorton, J., Watt, J.M., Culbard, E., Harvey, P.G., Sherlock, J.C., Smart, G.A., Thomas, J.F.A., and Quinn, M.J. (1990) Lead Intake and Blood Lead in Two-Year Old U.K. Urban Children, *Science of the Total Environment*, 90: 13–29.

Dickey, P., Ed. (2001) *Master Home Environmentalist Training Manual*, American Lung Association of Washington, Seattle, WA.

Ewers, L., Clark, S., Menrath, W., Succop, P., and Bornschien, R. (1994) Clean-Up of Lead in Household Carpets and Floor Dust, *American Industrial Hygiene Association Journal*, 55(7): 650–657.

Fortune, C.R., Blanchard, F.T., and Ellenson, W.D. (2000) *Analysis of Aged In-Home Carpeting to Determine the Distribution of Pesticide Residues between Dust, Carpet, and Pad Compartments*, Report No. EPA/600/R-00/030, U.S. Environmental Protection Agency, National Exposure Research Laboratory, Research Triangle Park, NC.

Franke, D.L., Cole, E.C., Leese, K.E., Foarde, K.K., and Berry, M.A. (1997) Cleaning for Improved Air Quality: An Initial Assessment of Effectiveness, *Indoor Air*, **7**: 41–54.

Gereda, J.E., Leving, D.Y.M., Thatayatlkom, A., Streib, J.E., Price, M.R., Klinnert, M.D., and Liv, A.H. (2000) Relation between House Dust Endotoxin Exposure, Type 1 T-cell Development, and Allergen Sensitization in Infants at High-Risk of Asthma, *The Lancet*, **355**: 1680–1683.

Hilts, S.R., Hertzman, C., and Marion, S.A. (1995) A Controlled Trial of the Effect of HEPA Vacuuming on Childhood Lead Exposure, *Canadian Journal of Public Health*, **86**(5): 345–350.

Howarth, P. and Reid, A. (2000) *Allergy-Free Living: How to Create a Healthy, Allergy-Free Home and Lifestyle*, Octopus Publishing Group Limited, London, U.K.

Jones, K.C., Strafford, J.A., Waterhouse, K.S., Furlong, E.T., Giger, W., Hites, R.A., Schaffner, C., and Johnson, A.E. (1989) Increases in the Polynuclear Aromatic Hydrocarbon Content of an Agricultural Soil over the Last Century, *Environmental Science and Technology*, **23**(1): 95–101.

Jones-Otazo, H.A., Clark, J.P., Diamond, M.L., Archbold, J.A., Ferguson, G., Harner, T., Richardson, G.M., Ryan, J.J., and Wilford, B. (2005) Is House Dust the Missing Exposure Path for PBDEs? An Analysis of Urban Fate and Human Exposure to PBDEs, *Environmental Science and Technology*, **38**(14): 5121–5130.

Krieger, J., Takaro, T.K., Song, L., and Weaver, M. (2005) The Seattle-King County Healthy Homes Project: A Randomized, Controlled Trial of a Community Health Worker Intervention to Decrease Exposure to Indoor Asthma Triggers, *American Journal of Public Health*, **95**: 652–659.

Lehen, A. (1983) *Concerning the Effectiveness of Clean-Off Zones*, TNO, Deutsches, eppich-Forschungsinstitut, Delft, Holland.

Leung, R., Koenig, J.Q., Simcox, N., van Belle, G., Fenske, R., Gilbert, S.A. (1997) Behavioral Changes Following Participation in a Home Health Promotion Program in King County, Washington, *Environmental Health Perspectives*, **105**(10): 1132–1135.

Lewis, R.D. (2002) *The Removal of Lead Contaminated House Dust from Carpets and Upholstery*, U.S. Department of Housing and Urban Renewal, Office of Healthy Homes and Lead Hazard Control, Washington, DC.

Lewis, R.G., Fortune, C.R., Willis, R.D., Camann, D.E., and Antley, J.T. (1999) Distribution of Pesticides and Polycyclic Aromatic Hydrocarbons In House Dust as a Function of Particle Size, *Environmental Health Perspectives*, **107**(9): 721–727.

Lioy, P.J., Yiin, L., Adgate, J., Weisel, C., and Rhoads, G.G. (1998) The Effectiveness of a Home Cleaning Intervention Strategy in Reducing Potential Dust and Lead Exposures, *Journal of Exposure Analysis and Environmental Epidemiology*, **8**(1): 17–35.

Maertens, R.M., Bailey, J., and White, P.A. (2004) The Mutagenic Hazards of Settled House Dust: A Review, *Mutation Research*, **567**: 401–425.

Mahaffrey, K.R. and Annest, J.L. (1985) Association of Blood Level, Hand to Mouth Activity, and Pica among Children, *Federation Proceedings*, **44**: 72.

National Institute of Justice (2002) *Understanding DNA Evidence: A Guide for Victim Service Providers*, Department of Justice, Washington, DC, www.ncjrs.org/pdffiles/nij/bc000657.pdf (accessed May 2, 2004).

Ott, W.R. and Roberts, J.W. (1998) Everyday Exposure to Toxic Pollutants, *Scientific American*, 278(2): 86–91, http://www.wannabee.org.au/archives/papers_research/IAQ/SciAm%20Indoor%20Air%20article/SciAM%20-%20Everyday%20Pollutants.pdf.

Primomo, J. (2003) Effectiveness Evaluation of the American Lung Association's Master Home Environmentalist Program: A Community-Based Outreach and Education Program, American Thoracic Society Annual Conference, Seattle, WA, May, 2003.

Roberts, J.W. (1998) Reducing Health Risks from Dust in the Home, in *Master Home Environmentalist Training Manual*, Dickey, P., Ed., Chapter 6, American Lung Association of Washington, Seattle, WA.

Roberts, J.W., Budd, W.T., Camann, D.E., Fortmann, R.C., Sheldon, L.C., and Lewis, R.G. (1991a) A Small High Volume Surface Sampler (HVS3) for Pesticides and Other Toxic Substances in House Dust, Paper No. 150.2, *Proceedings of the Annual Meeting of the Air and Waste Management Association*, Pittsburgh, PA.

Roberts, J.W., Budd, W.T., Ruby, M.G., Bond, A.E., Lewis, R.G., Wiener, R.G., and Camann, D.E. (1991b) Development and Field Testing of a High Volume Sampler for Pesticides and Toxics in Dust, *Journal of Exposure Analysis and Environmental Epidemiology*, **1**(1): 143–155.

Roberts, J.W., Budd, W.T., Ruby, M.G., Camann, D.E., Fortmann, R.C., Lewis, R.G., Wallace, L.A., and Spittler, T.M. (1992) Human Exposure to Pollutants in the Floor Dust of Homes and Offices, *Journal of Exposure Analysis and Environmental Epidemiology,* **2**(supp. 1): 127–146.

Roberts, J.W., Clifford, W.S., Glass, G., and Hummer, P.G. (1999) Reducing Dust, Lead, Bacteria, and Fungi in Carpets by Vacuuming, *Archives of Environmental Contamination and Toxicol*ogy, **36**: 477–484.

Roberts, J.W., Crutcher III, E.R., Crutcher IV, E.R., Glass, G., and Spittler, T. (1996) Quantitative Analysis of Road and Carpet Dust on Shoes, in *Measurement of Toxic and Related Air Pollutants,* Air and Waste Management Association, Pittsburgh, PA, 829–835.

Roberts, J.W., Glass, G., and Mickelson L. (2004) A Pilot Study of the Measurement and Control of Deep Dust, Surface Dust, and Lead in 10 Old Carpets Using the 3-Spot Test While Vacuuming, *Archives of Environmental Contamination and Toxicology,* **48**(1): 16–23.

Roberts, J.W., Ruby, M.G., and Warren, G.R. (1987) Mutagenic Activity in House Dust, in *Short-Term Bioassays in the Analysis of Complex Mixtures,* Sandu, S.S., DeMarini, D.M., Mass, M.J., Moore, M.M., and Mumfod, J.L., Eds., Plenum Press, New York, NY, 355–367.

Rudel, R.A., Camann, D.E., Spengler, J.D., Korn, L.R., and Brody, J.G. (2003) Phthalates, Alkylphenols, Polybrominated Diphenyl Ethers, and Other Endocrine-Disrupting Compounds in Indoor Air and Dust, *Environmental Science and Technology,* **37**: 4543–4553.

Salam, M.S., Li, Y.F., Langholtz, B., and Gilliland, F.D. (2003) Early Life Risk Factors for Asthma: Findings from the Children's Health Study, *Environmental Health Perspectives,* **112**(6): 760–765.

Seelye, K.Q. (2002) President Shifting Superfund Costs to Public, *San Francisco Chronicle,* February 24, 2002, A9.

Smucker, S.J. (2004) *Region IX Preliminary Remediation Goals (PRGs)* memorandum from the Technical Support Section, U.S. Environmental Protection Agency, Region 9, San Francisco, CA.

Stapleton, H.M., Dodder, M.G., Offenberg, J.H., Schantz, M.M., and Wise, S.A. (2005) Polybrominated Diphenyl Ethers in House Dust and Clothes Dryer Lint, *Environmental Science and Technology,* **38**(4): 925–931.

Sullivan, S.D., Weiss, K.B., Lynn, H., Mitchell, H., Gergen, P.J., and Evans, R. (2002) The Cost-Effectiveness of Inner-City Asthma Intervention for Children, *Journal of Allergy and Clinical Immunology,* **110**(4): 576–581.

Takaro, T.K., Krieger, J., Song, L., and Beaudet, N. (2004) Effect of Environmental Interventions to Reduce Exposure to Asthma Triggers in Homes of Low-Income Children in Seattle, *Journal of Exposure Analysis and Environmental Epidemiology,* **14**(supp. 1): S133–S143.

Tucker, D. (2003) Personal communication, Director of Laboratories, The Hoover Company, North Canton, OH.

USEPA (1986) *Air Quality Criteria for Lead,* Report No. EPA/60018-83/028AF, U.S. Environmental Protection Agency, Washington, DC, www.epa.gov/oar/aqtrends.html/toc.html, (accessed January 11, 2003).

USEPA (1990) *Air Quality Criteria for Lead: Supplement to the 1986 Addendum,* Report No. EPA/6000/8-89/049F, U.S. Environmental Protection Agency, Washington, DC.

USEPA (2004) More Details on Lead-Based on Data through 2002, Washington, DC, www.epa.gov/air-trends/lead2.html, (accessed May 4, 2002).

Wallace, L.A. (1987) *The Total Exposure Assessment Methodology (TEAM) Study: Summary and Analysis, Vol. 1,* U.S. Environmental Protection Agency, Washington, DC.

Woodruff, T.J., Axlerad, D.A., Kyle, A.D., Nweke, O., and Miller, G.G. (2003) *America's Children and the Environment: Measures of Contaminants, Body Burdens, and Illnesses,* 2nd ed., Report No. EPA 240-R-03-001, U.S. Environmental Protection Agency, Office of Children's Health Protection, Washington, DC, www.epa.gov/oar/aqtrends.html/toc.html, (accessed January 11, 2004).

WSDH (1994) *Public Health Improvement Plan,* Washington State Department of Health, Olympia, WA.

15 Exposure to Pesticides

Robert G. Lewis
U.S. Environmental Protection Agency (ret.)

CONTENTS

15.1 Synopsis...347
15.2 Introduction...348
15.3 Pesticide Regulation 348
15.4 Residential and Commercial Building Use...349
15.5 Air Monitoring Methods 350
15.6 House Dust Sampling Methods 355
15.7 Contact-Dislodgeable Residue Monitoring Methods...357
15.8 Handwipe Methods...361
15.9 Occurrence, Sources, Fate, and Transport in the Indoor Environment.............363
15.10 Exposure Risks and Health Effects...368
References 370

15.1 SYNOPSIS

There are at least 600 different pesticides in use and 45,000 to 50,000 pesticide formulations that may include one or more pesticides as active ingredients (a.i.). Pesticides consist of a wide variety of chemical compounds ranging from inorganic substances such as elemental sulfur and chromated copper arsenate, volatile organic compounds (VOCs) such as methyl bromide and paradichlorobenzene, semivolatile organic compounds (SVOCs) such as diazinon and chlorpyrifos, and nonvolatile organic compounds (NVOCs) such as 2,4-D and permethrin. In the United States, the Environmental Protection Agency (USEPA) is responsible for registering new pesticides and reviewing existing pesticides for re-registration to avoid unreasonable risks to human health and the environment. Potential human risks include acute (short-term) reactions, such as toxic poisoning or skin and eye irritation, as well as possible chronic (long-term) effects such as cancer, birth defects, or reproductive system disorders. The Food Quality Protection Act (FQPA) sets tolerances for all pesticides residues in food based on a "reasonable certainty" that they will do "no harm" to human health, but it also requires the USEPA to consider all routes of exposure when setting these tolerances.

Conventional pesticides made up 18% of the total pesticide usage in the United States in 1999, far behind that for chlorine and hypochlorites at 52%. Wood preservatives accounted for 16% of the quantity used, while specialty biocides and commodity pesticides amounted to 7% each. Of the total of 414 million kg of active ingredients used in conventional pesticide formulations, home and garden use accounted for 36 million kg, compared to 320 million kg for agricultural use and 57 million kg for other nonagricultural uses. Approximately 74% of all U.S. households reported using pesticides in 1999. About 56% said they used insecticides and disinfectants inside the home, and over 38% reported applying herbicides to their lawns and gardens. The major home and garden use of pesticide a.i. by individuals consisted of herbicides (68%), insecticides and miticides (18%), and fungicides (12%). An additional 14–16 million kg of paradichlorobenzene and 1–2 million kg of naphthalene were used in moth repellents, insecticides, germicides, and room deodorizers.

The major exposure of the general population of the United States to pesticides occurs in the home. The most commonly used pesticides are disinfectants. Pesticides used indoors can vaporize from treated surfaces, such as carpets and baseboards, can be resuspended into air attached to particles, and be tracked indoors where they accumulate in house dust. Typical pesticide concentrations in indoor air and house dust are 10–100 times higher than those found in outdoor air or surface soil. Many pesticides are SVOCs and vaporize when applied to indoor surfaces, and there may be significant temporal and spatial variations of pesticide concentrations in a home. Some pesticides, though banned, are found at appreciable concentrations in house dust.

Dermal exposures may occur when homeowners apply pesticides around the home or when residents come into contact with contaminated surfaces. Infants and toddlers constitute the population of greatest concern for incidental dermal exposure as they are more apt to have intimate contact with floors, turf, and other residential surfaces — and generally wear less clothing indoors. Very young children also frequently engage in mouthing of their hands, which may result in ingestion of dermal residues.

With the recent banning of popular organophosphate pesticides, the current trend is toward the use of pyrethroids and other pesticides that have very low vapor pressures. Hence, exposure to pesticides in house dust and through dermal contact with contaminated surfaces has become a major focus for research in a field that respiratory exposure concerns once dominated.

15.2 INTRODUCTION

A pesticide is defined as any substance used for controlling, repelling, or killing a pest (e.g., insect, weed, fungus, rodent). Pesticides cover a very large and varied range of substances and are subclassified according to their modes of action into many classes, including insecticides, acracides, herbicides, fungicides, rodenticides, avicides, larvicides, repellents, plant growth regulators, germicides (disinfectants), and other types of biocides. Pesticides are typically applied in formulations (which may include one or more active ingredients in solvents or on powdered substrates, along with other substances designed to enhance the effectiveness of the active ingredients. Thus the 600 or so active ingredients in use may make up 45,000–50,000 different formulations on the market (Baker and Wilkinson 1990). The USEPA classifies pesticides as conventional (insecticides/miticides, herbicides/plant growth regulators, fungicides, nematicides/fumigants, other), chlorine/hypochlorites (disinfectants, water purifiers), wood preservatives (creosote, pentachlorophenol, and chromated copper arsenate), specialty biocides (disinfectants and sanitizers), and other (sulfuric acid, insect repellents, zinc sulfate, moth control chemicals) (Kiely, Donaldson, and Grube 2004). The scope of this chapter is limited to conventional pesticides used in and around homes and offices.

15.3 PESTICIDE REGULATION

The marketing, use, and disposal of pesticides are regulated by the USEPA, principally under the Federal Insecticide, Fungicide and Rodenticide Act (FIFRA) (7 U.S.C. 136, *et seq.*) and the Food Quality Protection Act (FQPA) of 1996 (P.L. 104-170). The Agency is responsible for registering new pesticides and reviewing existing pesticides for re-registration to ensure that they will not present unreasonable risks to human health or the environment. FIFRA requires the USEPA to take into account economic, social, and environmental costs and benefits in making decisions. Registration and regulatory decisions are based on evaluation of data provided by the registrants from tests that may be specified by the agency. These required tests include studies to show whether a pesticide has the potential to cause adverse effects to individuals using pesticide formulations (applicators) and to persons who may be exposed post-application. Potential human risks include acute (short-term) reactions such as toxic poisoning or skin and eye irritation, as well as possible chronic (long-term) effects such as cancer, birth defects, or reproductive system disorders. Before pesticides can be registered for residential or institutional use, the USEPA requires that studies be performed to determine post-application dissipation rates and dislodgeable, or transferable, residues.

The Food Quality Protection Act not only sets tolerances for all pesticide residues in food based on a "reasonable certainty" that they will do "no harm" to human health, but it also requires the USEPA to consider all routes of exposure (i.e., "aggregate" exposure) when setting tolerances. When setting a food tolerance level, the USEPA must aggregate exposure information from all potential sources, including pesticide residues in specific foods of concern and those in other foods for which tolerances have already been set, residues in drinking water, and residues from other nondietary, nonoccupational uses of the pesticide (i.e., residential and other indoor/outdoor uses). FQPA further mandates that potential risks to infants and small children be specifically addressed. In order to assure "that there is a reasonable certainty that no harm will result to infants and children from aggregate exposure to the pesticide's chemical residues," FQPA calls for a "tenfold margin of safety for pesticide residues and other sources of exposure" to be applied to estimating risks to children, taking into account "potential pre- and post-natal toxicity." It allows the USEPA to use a different margin of safety "only if, on the basis of reliable data, such margin will be safe for infants and children."

An example of the USEPA regulatory process that relates to permissible indoor pesticide uses is the action taken in the year 2000 on chlorpyrifos (e.g., Dursban®). The insecticide had been one of the most heavily used pesticides for control of fleas and crawling insects indoors and grubs in residential lawns. It also was the primary termiticide that replaced chlordane after its discontinuation in 1988. Because of concern over its toxicity and the high potential for exposure from residential use, the USEPA reached an agreement with the manufacturer, DowElanco, in June 2000 to cancel and phase out nearly all indoor and outdoor residential uses of chlorpyrifos. This action eliminated the pesticide from products for indoor crack and crevice treatment, "broadcast"[1] flea control, total release foggers, post-construction termite treatment, lawn insect control, and pet care (shampoos, dips, and sprays). Remaining uses are limited to certified professional and agricultural applicators. The agreement also restricted the uses of chlorpyrifos on certain foods that pose the greatest dietary exposure risks to children. Similar restrictions were placed on diazinon later in the year 2000.

15.4 RESIDENTIAL AND COMMERCIAL BUILDING USE

Conventional pesticides are used both indoors and outdoors in homes, office buildings, schools, hospitals, nursing facilities, and other public institutions. A wide variety of pesticide products are available "off the shelf" for use by the homemaker. However, in recent years they are most readily available in ready-to-apply formulations rather than concentrates that require the user to dilute them before use. Preparations include those for control of flies, roaches, ants, spiders, and moths within the home; flea and tick sprays and shampoos for pets; insecticides for use on house plants and home gardens; and herbicides, insecticides and fungicides for lawn treatment. Most homemakers use disinfectants routinely as kitchen and bathroom cleaners, room deodorizers, or laundry aids. Many homeowners and landlords utilize professional pest control services for routine indoor treatments or lawn care. In many parts of the United States, pre- or post-construction treatment for termite protection is essential. Excluding disinfectants and insect repellents, the most common indoor uses are for control of cockroaches and ants (crack and crevice treatment, baits), flies (sprays, pest strips), fleas (broadcast sprays and foggers), and rodents (baits). Outdoor uses in addition to lawn and garden care include perimeter and crawl space treatments for termites and crickets.

The general populous of the United States, as well as that of most other countries, receives the majority of its exposure to pesticides inside the home. Studies have shown that about 90% of all

---

[1] Broadcast application refers to spreading of the pesticide over a wide area.

U.S. households use pesticides (USEPA 1979; Savage et al. 1981; Godish 1990; Whitmore et al. 1993). A national survey conducted by the USEPA during 1976–1977, revealed that more than 90% of U.S. homeowners used pesticides, with 84% using them inside the house, 21% in the garden, and 29% on the lawn (USEPA 1979). The survey found that over 90% of the households used disinfectants (antimicrobials), 36% used moth repellents, and 26% were treated with termiticides. The USEPA-sponsored National Home and Garden Pesticide Use Survey in 1990 found that 82% of the 66.8 million U.S. households used pesticides and that about 20% of them (ca. 16 million households) were commercially treated for indoor pests such as cockroaches, ants, or fleas (Whitmore et al. 1993). It further showed that some 18 million U.S. households use pesticides on their lawns, 8 million in the garden, and 14 million on ornamental plants. About 15% of U.S. residences with private lawns employ commercial lawn care companies that apply pesticides. A survey of 238 households in Missouri in 1989–1990 revealed that 98% of all families used pesticides at least one time per year, and 75% used them more than five times per year (Davis, Brownson, and Garcia 1992). The Missouri survey also determined that 70% of the respondents used household pesticides during the first 6 months of a child occupant's life.

Direct purchases of conventional pesticides for home and garden use accounted for 19% of the more than $11 billion spent on pesticide products in the United States in 2000 and 2001 (Kiely, Donaldson, and Grube 2004). Other nonagricultural commercial sales made up 14% of the total. Most of the latter was intended for commercial home, office, and institutional application. Home and garden use in 2001 consumed over 46 million kg of pesticidal active ingredients (a.i.), compared to 306 million kg for agricultural use and 50 million kg for other nonagricultural commercial use. On the basis of quantity of a.i., herbicides made up 71% of the total home and garden use by individuals, insecticides and miticides 17%, and fungicides 16%. Not included in these figures are 27 million kg of other pesticides including 1,4-dichlorobenzene (paradichlorobenzene) (moth repellent, insecticide, germicide, and deodorant), naphthalene (moth repellent), and N, N-diethyl-m-toluamide (DEET) (insect repellent). The most commonly used pesticides by homeowners during 2001 are shown in Table 15.1. This does not include pesticides applied to private residences by professional applicators, the most common of which in 1999 were 2,4-D, glyphosate, copper sulfate, pendimethalin, chlorothlonil chlorpyrifos, diuron, MSMA triclopyr and malathion.

Misuse of pesticides by homemakers and commercial applicators occurs all too frequently. In California during 1983–1986, nearly 300 cases of illness or injury reported to physicians were attributed to three popular household insecticides: chlorpyrifos, dichlorvos (DDVP), and propoxur (Edmiston 1987). In 1991, the American Association of Poison Control Centers received 78,177 calls regarding pesticide poisonings at 73 centers across the United States (Litovitz et al. 1991). Seventy percent of these involved insecticides. Pesticide poisoning calls ranked seventh in frequency (behind cleansers, analgesics, cosmetics, plants, cough and cold medications, and bites). The removal of many acutely toxic pesticides and concentrated formulations from the home market and improvements in packaging safety has reduced the number of poisoning incidents in recent years. However, of greater concern are potential chronic health effects that may derive from long-term exposures to pesticides in indoor environments.

15.5 AIR MONITORING METHODS

Air sampling can be classified as instantaneous (grab), real-time (or continuous), or integrative (over a period of exposure). Except for a few reactive pesticides present in air at relatively high concentrations (e.g., occupational levels), integrative sampling is necessary in order to obtain a sufficient quantity of the pesticide for laboratory analysis. Pesticide air sampling typically involves the collection of pesticides from air onto a solid sorbent or a combination trap consisting of a particle filter backed up by a sorbent trap. Solvent extraction and chemical analysis by gas chromatography or high performance liquid chromatography are most commonly employed.

TABLE 15.1

Approximate Annual Quantities of Conventional Pesticides Consumed by the Homeowner Market in the United States during 2001

Rank	Pesticide	Type[a]	Quantity Used, 10^6 kg/yr
1	2,4-D (in e.g., Weed-B-Gone®)	H	4–5
	[2,4-dichlorophenoxyacetic acid and salts]		
2	Glyphosate (e.g., Roundup®)	H	2–4
	[isopropylamine salt of N-(phosphonomethyl)glycine]		
3	Pendimethalin (e.g., Prowl®)	H	1–3
	[N-(1-ethylpropyl)-2,6-dinitro-3,4-xylidine]		
4	Diazinon	I	2–3
	[$0,0$-diethyl-O-{6-methyl-2-(1-methylethyl)-4-pyrimidinyl} phosphorothioate]		
5	MCPP (Mecoprop, in, e.g., Weed-B-Gone®)	H	1–2
	[2-(4-chloro-2-methylphenoxy) propionic acid		
6	Carbaryl (e.g., Sevin®)	I	1–2
	[Sevin® — 1-naphthylmethylcarbamate]		
7	Dicamba (e.g., Banvel®)	H	1–2
	[3,6-dichloro-2-methoxybenzoic acid; 3,6-dichloro-o-anisic acid and salts]		
8	Malathion	I	0.5–1.4
	[diethyl(dimethoxythiophosphorylthio)succinate]		
9	DCPA (e.g., Dacthal®)	F	0.5–1.4
	[dimethyl-2,3,5,6-tetrachlorobenzen-1,4-dicarboxylic acid]		
10	Benefin (Benfluralin)	H	0.5–1.4
	[N-Butyl-N-ethyl-a,a,a-trifluoro-2,6-dinitro-p-toluidine]		

[a] H = herbicide, I = insecticide.

Air sampling media that have been shown to be efficient for collection of conventional pesticides are polyurethane foam (Bidleman and Olney 1974; Orgill, Sehemel, and Petersen 1976; Lewis, Brown, and Jackson 1977; Lewis and MacLeod 1982; Billings and Bidleman 1980; Wright and Leidy 1982; Lewis, Fortmann, and Camann 1994); Chromosorb 102 (Thomas and Seiber 1974; Hill and Arnold 1979); Amberite® XAD-2 (Farewell, Bowes, and Adams 1977; Johnson, Yu, and Montgomery 1977; Lewis and Jackson 1982; Billings and Bidleman 1983; Williams et al. 1987; Leidy and Wright 1991; Wright, Leidy, and Dupree 1993; Lu and Fenske 1998); Amberlite XAD-4 (Woodrow and Seiber 1978; Jenkins, Curtis, and Cooper 1993); Tenax®-GC or TA (Billings and Bidleman 1980, 1983; Lewis and Jackson 1982; Lewis and MacLeod 1982; Roinestad, Louis, and Rosen 1993); Poropak®-R (Lewis and Jackson 1982), and Florisil® (Yule, Cole, and Hoffman 1971; Lewis and Jackson 1982). These sorbents appear to be about equally efficient for trapping most pesticides. Polyurethane foam (PUF) has enjoyed the most widespread popularity because it is more convenient to use and has much less resistance to airflow than the granular sorbents. However, a few of the more volatile pesticides may not collect efficiently on PUF.

Samples may be collected over 24-hour periods or for shorter periods of exposure time, depending upon the design of the study and the sensitivity of the method. When the usual gas or liquid chromatographic analysis procedures are used, air volumes of 0.01–1 m^3 are sufficient for occupational exposure levels (i.e., 0.1–10 mg/m^3) and 1–10 m^3 for nonoccupational exposures (i.e., 0.01–10 μg/m^3).

Methods for several pesticides at occupational levels in air are given in the *NIOSH Manual of Analytical Methods* (Eller and Cassinelli 1994). The NIOSH methods for organochlorine and organophosphate utilize small traps with a particle filter backed up by two Amberlite® XAD-2 resin

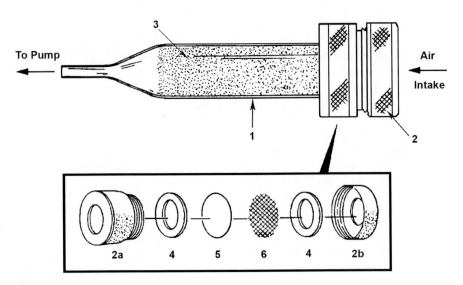

FIGURE 15.1 Simple air sampling cartridge with open-face particle filter. 1: Glass sorbent cartridge; 2: Particle filter holder; 2a: Filter holder, front element; 2b: Filter holder, rear element. 3: Sorbent (e.g., PUF); 4: PTFE filter gaskets; 5: Particle filter; 6: Filter support screen, stainless steel, 50% open area.

beds. They are designed to be used with personal sampling pumps at 0.2–1 L/min for a maximum sample volume of 60–240 L. Detection limits are in the 5–600 ng/m³ range.

There are two ASTM International methods designed primarily for determining airborne pesticides at nonoccupational levels. ASTM Standard D 4861 describes a sampling method and recommended analytical procedures for a broad spectrum of pesticides at concentrations in the 0.001–50 µg/m³ range (ASTM 2005a) and D 4947 is a specific method for chlordane and heptachlor (ASTM 2005b), which may still be found in the air inside homes built before 1978. D 4861 is based on USEPA Compendium Method TO-10A, and is the method used in many large surveys conducted by the Agency (USEPA 1999a). The sampling device employed by both ASTM methods consists of a 22-mm × 76-mm polyurethane foam (PUF) cylinder (plug), which has been used with and without a particle filter attached to the inlet. The PUF cartridge with or without an open-face particle filter (see Figure 15.1) is commercially available from several vendors (e.g., Supelco Model Orbo 1000®; SKC Cat. No. 226-124). A size-selective inlet for this method has been designed and used in several recent USEPA indoor air studies. It is an integral system incorporating either a 2.5 µm or 10 µm inlet based on a design by Marple et al. (1987) and can be used at flow rates up to 4 L/min for up to 24 hours (Camann et al. 1994). The glass sampling cartridge and particle filter are contained in a rugged high-density polypropylene case, which is highly resistant to breakage and tampering. The sampler, shown in Figure 15.2, is commercially available (URG Model 2000). The USEPA and ASTM methods are designed to be used with portable air sampling pumps capable of pulling about 4 L/min of air through the collector for a total sample volume not to exceed 5–6 m³. Depending on the analytical finish, the minimum detection limits of the ASTM methods range from 1 ng/m³ to 100 ng/m³. The World Health Organization has published a method that is essentially similar to D 4861 (Lewis 1993). Either sampler is suitable for both area sampling and personal exposure monitoring. For the latter purpose, they are usually worn by the study subject in the breathing zone with the inlets pointing downward (see Figure 15.3).

Most of the large studies employing TO-10A or ASTM D 4861 (e.g., the Non-Occupational Exposure Study, NOPES) have not used a particle filter; however, one is recommended if pesticides associated with respirable particulate matter are likely to be present. The backup PUF trap should always be used behind the particle filter, even for collection of nonvolatile pesticides (e.g., when sampling for airborne acid herbicides indoors). As much as 20% of airborne 2,4-D, applied as the

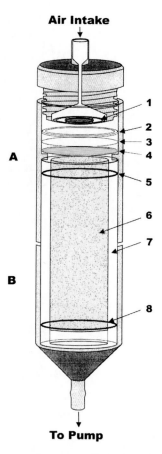

Air Intake

A

B

To Pump

FIGURE 15.2 Air sampling assembly with size-selective inlet, particle filter, and glass sorbent cartridge. Parts A and B are separable sections of shock-resistant case. Internal parts: 1: Impactor for size-selective inlet; 2: PTFE O-ring; 3: Particle filter; 4: Stainless steel filter support screen; 5 and 8: Rubber O-ring seals; 6: PUF or granular sorbent; 7: Glass sorbent cartridge.

trimethylamine salt, has been detected on the backup PUF plug, presumably due to hydrolysis to the semivolatile free acid (USEPA 1999b).

Except for herbicide salts, some pyrethroids, and a few other nonvolatile compounds, most pesticides will either be present in air primarily in the vapor phase or will volatilize from airborne particulate matter readily after collection on a filter (Lewis and Gordon 1996). Solid sorbent beds will collect most particulate-associated pesticides along with vapors; however, recent evidence suggests that some penetration of fine particulate matter (0.1 to 1μm) may occur with PUF and Florisil (Kogan et al. 1993). Fine particles were not found to penetrate XAD-2 beds, presumably due to their retention by static charge. It may be good practice, therefore, to use a particle filter in front of the sorbent bed. In this case, the filter and sorbent bed should be extracted together for analysis to provide for better detection and prevent misinterpretation of the analytical results with respect to original phase distributions. It should be noted that although very small particles have been shown to be poorly retained by the PUF plug, simultaneous, collocated sampling of residential indoor air with and without a quartz fiber particle filter showed no significant measurement differences even when sweeping and vacuuming activities took place in the same room (Camann, Harding, and Lewis 1990).

Air samples should be taken within homes or other buildings in the best locations for estimation of human exposure (e.g., family rooms, bedrooms, office spaces). Occupant activity logs may be

FIGURE 15.3 Open-face air sampler (left) and sampler with size-selective inlet (right) in use for personal exposure monitoring. (Courtesy of persons in the photograph.)

required in order to obtain accurate estimates of human exposure. The sampler may be conveniently positioned on a table, desk, or countertop, during which time it may be operated by means of a power converter/charger. For monitoring periods longer than 8 hours, the latter procedure will usually be necessary due to limited battery life and to cover the sleep period. Air intakes (inlets) should be positioned 1–2 m above the floor or ground and oriented downward or horizontally to prevent contamination by nonrespirable dustfall. If two or more samplers are to be used for collocated sampling, intakes should be at least 30 cm apart for low-volume samplers (1–5 L/min) and 1–2 m apart for high-volume samplers (up to 1,000 L/min).

Indoor residential sampling can be restricted because of available space or by homeowner objections. Equipment noise can also be an issue, depending on the size of the space being monitored, the acoustics of the area, and the presence of occupants. Noise from sampling equipment used in residences, schools, offices, and other relatively noise-free areas should be limited to 35 db (1 sones) at 8,000 Hz (ASTM 2003b). Many battery-operated portable pumps designed for personal respiratory exposure monitoring are quiet enough for this purpose, although additional acoustic insulation may be required for use in bedrooms and family rooms. Nonindustrial workplace monitoring is often more flexible to space and noise restrictions. Security of sampling equipment should be considered in the plan. Typically, samplers that cannot either easily be tampered with or changed by the homeowner or office worker, are preferable to those with exposed sampling elements or controls (e.g., the possibility of electrical power disruption or contamination by onlookers or passersby should be considered in the sampling plan for any effort).

15.6 HOUSE DUST SAMPLING METHODS

House dust is the major reservoir of pesticide residues that may be accessible for human exposure in the home environment (Lewis, Fortmann, and Camann 1994). Since infants and toddlers have high levels of intimate contact with floors and other dust-containing objects and engage in frequent mouthing activities, nondietary ingestion of house dust may constitute a major exposure pathway for them, especially to low- and nonvolatile pesticides. Although analysis of pesticides in dust is more complicated than that for air samples, knowledge of the pesticide content of house dust can provide a good indication of the overall contamination of the home environment (including the air) and afford useful estimates of relative exposure risks to inhabitants (Lewis et al. 1995; Lioy, Freeman, and Millette 2002).

Various methods have been used to collect dust from floors and upholstery (Que Hee et al. 1985; USEPA 1989; Roberts et al. 1991; Farfel et al. 1994; Ness 1994; Lanphear et al. 1995; USEPA 1995). The approach most commonly employed by industrial hygienists is based on drawing dust by means of a personal air sampling pump operating at 2–3 L/min onto a particle filter held in a plastic cassette. The filter cassette is held close to the surface being sampled. The method is sometimes referred to as the Dust Vacuum Method or DVM (Que Hee et al. 1985). A modification of this method developed by Midwest Research Institute (MRI) was the "Blue Nozzle" sampler, which utilized a 5-cm × 10-cm sampling nozzle and a 110 V rotary vane pump to draw larger quantities of dust through the same type of filter cassette (Constant and Bauer 1992; USEPA 1995). Its sampling efficiency was reported to be only 44–59% for dust sampled from bare concrete, linoleum, and wood floors. Another MRI design pulled dust through a rigid 2.5-cm i.d. plastic pipe into a cyclone and deposited it onto a filter in a cassette holder at the bottom of the cyclone (Dewalt et al. 1995). A handheld vacuum cleaner was used as the vacuum source.

Such vacuum sampling methods typically do not collect adequate quantities of dust for pesticide residue analysis and are not amenable to use on large surfaces such as floors. Consequently, the USEPA designed the HVS3 cyclone vacuum sampler (Figure 15.4), which is capable of collecting

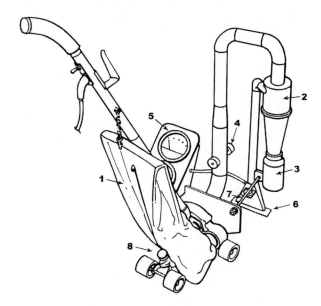

FIGURE 15.4 HVS3 cyclone vacuum sampler (CS$_3$, Bend, OR). 1: Commercial vacuum cleaner; 2: Cyclone with 5-µm cut-point; 3: Sample catch bottle; 4: Flow control valve; 5: Vacuum gauges; 6: 10-cm suction nozzle; 7: Nozzle level adjustment screw; 8: Platform level adjustment knob.

enough dust for pesticide residue analysis, at a constant sampling rate, and in a highly reproducible manner (Roberts et al. 1991, USEPA 1995). The sampler consists of a 1-cm × 12.4-cm flat sampling nozzle, particle collection cyclone, glass or PTFE catch bottle, and flow control system mounted on a standard upright vacuum cleaner to provide suction. The cyclone has a nominal cut point of 5 μm at a flow velocity of 40 cm/sec. In efficiency tests conducted according to ASTM F 608 (ASTM 2003a), the sampler has been shown to collect 67–69% of test dust from plush and level loop carpet and to trap 99% of the vacuumed dust in the cyclone catch bottle. Additional tests with spiked test dust have demonstrated 97+% recoveries of several common pesticides (Roberts et al. 1991). The method has been evaluated for efficiency at collecting and retaining floor dust and the pesticides associated with it. It has been subjected to round-robin testing and is the basis of ASTM International Standard D 5438 for collection of dust from carpets and bare floors (ASTM 2005c). The method has been used in many large and small USEPA studies (e.g., Lewis, Fortmann, and Camann 1994; Nishioka et al. 1999; Lewis et al. 2001; Wilson et al. 2004).

The amount of dust that can be collected by the HVS3 will vary greatly according to the dust loadings on the floor. Vacuuming of a 1 m^2 area of carpet typically collects 0.5–10 g of dust. The collected dust is retained in the catch bottle, which is capped and kept chilled or frozen until analyzed. The standardized methodology calls for the collected dust to be passed through a sieve to exclude particles larger than 150 μm prior to extraction and analysis. This cut-point is used because larger particles adhere poorly to the skin and present less of an exposure potential than smaller particles. The sieved sample normally ranges from 10–60% of the bulk dust sample (avg. ca. 50%). These amounts of dust will permit the quantitative measurement of 0.05 μg/g or less of most pesticides.

Home vacuum cleaners have also been used to collect floor dust for pesticide residue analysis (Starr et al. 1974; Davies, Edmundson, and Raffonelli 1975; Roinestad, Louis, and Rosen 1993; Colt et al. 1998; Rudel et al. 2003); however, standard (unlined) vacuum cleaner bags do not retain fine particles well. As much as 25–35% of the dust in the 2–4 μm size range may be lost during collection by penetration of the vacuum bags (IBR 1995). Additional losses of fine particles may occur due to adherence to the walls of the vacuum bags. The collection efficiency of particles smaller than 5 μm is also low for the HVS3, since the cyclone inlet cuts off at that point. Side-by-side comparisons of the HVS3 and a conventional upright vacuum cleaner in university dormitory rooms revealed the HVS3 to be more efficient for particles smaller than 20 μm (Willis 1995). It also showed that both types of vacuum devices collected particles down to at least 0.2 μm in diameter. Since concentrations of pesticides on house dust increase rapidly on particles smaller than 25–50 μm in diameter (Lewis et al. 1999), analytical results for dust collected with household vacuum cleaners may be lower than those obtained with the HVS3. However, no significant differences in the concentrations of pesticides were found by the National Cancer Institute (NCI) in house dust collected with the HVS3 from 15 homes and that collected from the homeowners' vacuum cleaner bags (Colt et al. 1998). Another study of nine daycare centers yielded higher results for pesticides and polycyclic aromatic hydrocarbons (PAHs) in dust collected in standard vacuum cleaner bags in most cases (USEPA 1999c). In the NCI study, the HVS3 sample was collected from carpets throughout the house, while the USEPA collected the HVS3 sample from a single room on one day. In both cases the bag sample was taken from the home or facility vacuum cleaner and represented dust collected over an unknown period of time and from multiple locations within the building. Consequently, concentration differences in the two types of samples reported by the USEPA may have reflected a lack of both spatial and temporal homogeneity of the dust.

15.7 CONTACT-DISLODGEABLE RESIDUE MONITORING METHODS

Pesticide residues are deposited onto indoor and outdoor residential surfaces after application of a pesticide formulation or by transfer from treated areas to nontargeted surfaces (e.g., transfer from lawn to carpet). Dislodgeable residues on exposed outdoor surfaces, such as turf, are typically short lived. Indoor surface residues, however, may persist for long periods of time since they are largely shielded from environmental degradation and dissipation in the indoor environment. Human contact with contaminated surfaces may dislodge a portion of these residues, resulting in their transfer to the skin or clothing, where they may be absorbed through the skin or ingested through mouthing. As in the case of house dust exposure, infants and toddlers constitute the population of greatest concern for incidental dermal exposure as they are more apt to have intimate contact with floors, turf, yard soil, and other residential surfaces and generally wear less clothing indoors (Fenske et al. 1990; Lewis, Fortmann, and Camann 1994; Zartarian and Leckie 1998). Very young children also frequently engage in mouthing of their hands, which may result in ingestion of dermal residues.

The term *dislodgeable* (or *transferable*) *residue* is defined as "that part of the residue of a chemical deposited on a solid surface which may be transferred by direct contact to human skin or clothing" (ASTM 2004). It is generally estimated by means of mechanical devices, although bare-skinned or clothed human subjects are sometimes used. Methods for determining dislodgeable residue transfer have included bare hand presses (Lewis, Fortmann, and Camann 1994), gloved hand presses (Roberts and Camann 1989), choreographed whole body dermal contact (Vaccaro 1993), whole body garments combined with an aerobic exercise routine (Ross et al. 1990), gauze wipes (Geno et al. 1996; Lu and Fenske 1998), and various sampling media that are pressed, dragged, or rolled across the surface at known contact pressures and rates or times (Hsu et al. 1990; Ross et al. 1991; Vaccaro 1993; Lioy, Wainman, and Weisel 1993; Edwards and Lioy 1999).

Three of the most popular devices used in recent years to estimate skin-transferable pesticide residues from carpets and floors are the polyurethane foam (PUF) roller (Hsu et al. 1990; Lewis, Fortmann, and Camann 1994), the drag sled (Vaccaro and Cranston 1990), and the California roller (Ross et al. 1991). All three methods have been rigorously evaluated and subjected to round-robin testing (USEPA 1997a) for inclusion in the FIFRA Subdivision K guidelines (USEPA 2000a).

The PUF roller (Figure 15.5) was designed to measure dislodgeable residues that may be transferred to a small child's skin from contact with floor surfaces. It is the basis of the ASTM Standard Practice D 6333 (ASTM 2004). Dislodgeable pesticide residues are collected by transfer to an annular ring of medium density (0.029 g/cm^3) open-cell, polyether-type polyurethane foam (8.9 cm o.d. × 8 cm wide), which is rolled across the floor at a constant speed and applied pressure. The PUF sampling ring is slipped over a cylindrical metal axle that functions as the front wheel of the PUF roller apparatus. The apparatus is typically constructed of aluminum and consists of a frame with two permanent rear wheels and the detachable axle cylinder on the front. Weights are attached to the roller frame to apply the desired downward force on the PUF roller ring (sampling pressure). A total weight of 3.9 kg provides a sampling pressure of 8,000 Pa, corresponding approximately to that of a 9-kg child crawling (6,900 Pa) or walking (8,600 Pa). A handle is connected at the rear of the roller frame to push or pull the device across the floor surface with the aid of a template or similar measuring device to identify the area to be sampled. The axle cylinder is fitted with a clean PUF ring, and the roller is pushed at a constant rate of approximately 10 cm/s over a distance of 1.0 m and then immediately pulled in the reverse direction back over the same sampling area at the same rate of speed, ending at the original starting position. The total surface area sampled is 800 cm^2. At the conclusion of the traverse, the PUF ring is removed from the detached axle cylinder and placed in a sealed container for transport to the laboratory for analysis.

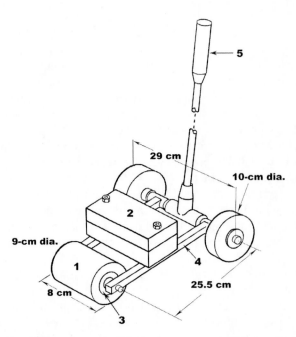

FIGURE 15.5 PUF roller with snap-in foam sampling ring and replaceable weights. 1: PUF ring (8.9 cm o.d. × 8 cm wide × 2.3 cm thick) on roller; 2: Weights (adjustable, 3.9 kg provide roller pressure of 8,000 Pa.); 3: Roller-axle (metal cylinder, spring-loaded, 4 cm dia × 8.5 cm long); 4: Frame (tiltable for roller loading and unloading); 5: Adjustable handle.

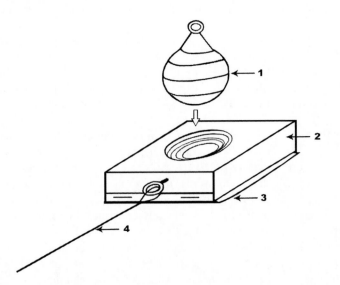

FIGURE 15.6 Drag sled (Dow Chemical Co., Midland, MI). 1: Weight (3.6 kg); 2: Sled (7.6 cm² × 2 to 4 cm high, typically made of wood covered with aluminum foil); 3: Collection medium (undyed denim cloth, 8 cm × 10 cm); 4: Drag line (e.g., fishing line).

The drag sled (Figure 15.6) is a simple device constructed of a 7.6-cm × 7.6-cm × 1.9-cm thick block of wood or other material, which is used to hold down a 10-cm × 10-cm piece of denim cloth (sampling medium) as it is dragged across the floor. The bottom of the block is covered with solvent-rinsed aluminum foil and the denim patch is placed over the foil. Staples or pushpins are

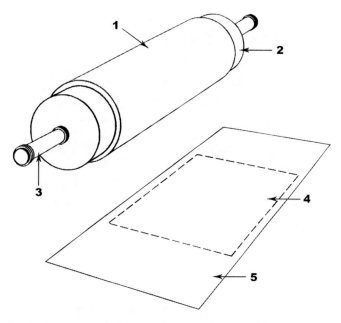

FIGURE 15.7 The California roller (California Department of Pesticide Regulation, Sacramento, CA). 1: Roller (PVC pipe, 13 cm dia. × 63 cm long); 2: Foam cover, 51 cm long and 1 cm thick); 3: Handles; 4: Collection medium (percale bed sheeting, 43 cm × 43 cm, 50% combed cotton and 50% polyester, 180 thread count); 5: Plastic sheet (over collection medium).

typically used to secure the sampling medium to the block. A 3.6-kg weight is centered on top of the block to provide a downward pressure of 4,500 Pa. The sled is pulled at 8 cm/s to 12 cm/s along a 1.2-m path by means of an attached cord (e.g., a 60-cm to 90-cm nylon fishing line), sampling 925 cm² of floor area.

The California roller (Figure 15.7) is basically a large, weighted "rolling pin" that is used to press a piece of polyester-cotton percale bedsheet onto the floor to collect residues. The roller consists of a large (13-cm o.d. × 63-cm long) cylinder constructed from polyvinyl chloride (PVC) pipe, covered with a foam cushion (1-cm thick × 51-cm long), and fitted with end caps and handles. The roller is weighted with 11.4 kg of steel shot ballast placed inside the cylinder to provide a total weight of 14.5 kg. The applied pressure is approximately 2,300 Pa. The sampling medium is a 43-cm × 43-cm square piece of bedsheet made from 50% combed cotton and 50% polyester, 180 thread count. When sampling, the bedsheet is placed on the floor surface, covered with a plastic sheet, and the roller is rolled over it ten times in each direction (20 passes). The California is very heavy and difficult to maneuver. It is also hard for the operator to use it without applying force that will translate to the roller and thereby increase the pressure applied to the collection sheet. As part of its evaluation of the device, the USEPA designed a sled (see Figure 15.8) for the device to circumvent the latter problem (USEPA 1997a).

The mode of sampling is different for each method. The PUF roller picks up residues by rolling contact of the pliable foam sampling medium with the floor surface, the drag sled operates through a wiping motion with a thick cotton fabric, and the California roller presses a thin cloth sheet against the surface. The characteristics of the three methods and their comparison with adult human hand presses are summarized in Table 15.2. All three mechanical methods have been subjected to comparative performance evaluation for collection of formulated pesticide residues from carpets and vinyl flooring. The performance of each method has also been compared to human hand presses (USEPA 1996d, 2000b). No biases have been observed with respect to the direction of traverse when the samplers were used on plush or level-loop nylon carpets. However, for carpets that do

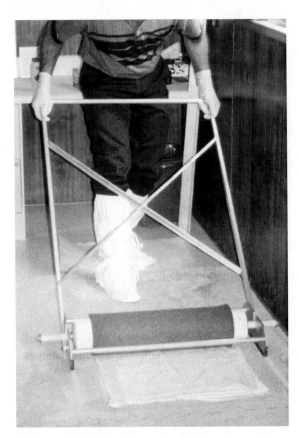

FIGURE 15.8 California roller in USEPA custom sled designed to eliminate influence of user on applied pressure.

not have uniform and level surfaces, some directional bias may be encountered, especially in the case of the drag sled.

The three dislodgeable residue samplers were also evaluated for use on turf grass (USEPA 1997b). The drag sled could not be used because it tended to capsize when pulled through grass. The PUF roller and California roller performed better, but problems were encountered with grass clippings and other debris that adhered to the sampling media and had to be mechanically removed prior to extraction and analysis of the sampling media. The problem was particularly serious for the PUF roller. The California roller also collected small amounts of clippings and debris, but these could be easily removed by the laboratory analyst. Both the PUF roller and a modified version of the California roller have been used to determine turf dislodgeable pesticide residues (Nishioka et al. 1996; Fuller et al. 2001).

Surface wipes are appropriate for determining dislodgeable residues from bare floors and other hard residential surfaces (e.g., table and countertops, windowsills, cabinets, appliances, dinnerware, and children's toys). However, residential wipe sampling techniques for pesticides have not yet been standardized and are unlikely to accurately reflect the transferability of surface residues to skin. Typically, cotton gauze or filter materials, dry or wetted with solvents, held in the hand are used to wipe a defined area (McArthur 1992). Surface areas of 100–1,000 cm^2 are normally defined for wiping by marking with tape or using a template (Ness 1994).

The U.S. Occupational Safety and Health Administration (OSHA) recommends glass-fiber filter material (air sampling filters), dry or wetted with 2-propanol, for determining surface residues of pesticides (OSHA 1999). A USEPA method uses two 10-cm × 10-cm surgical gauze sponges made

TABLE 15.2
Characteristics of the Three Principal Dislodgeable Residue Methods and Hand Press

Property	PUF Roller	Drag Sled	California Roller	Human Hand Press
Contact motion	Roll	Drag	Press	Press
Sampling medium (material)	Polyurethane foam cylinder (open-cell, polyether type, 0.029 g/cm³) 8.9-cm o.d. × 7.6-cm long	Denim weave cloth (predominantly cotton) 10-cm × 10-cm	Percale bed sheet (50% cotton, 50% polyester) 43-cm × 43-cm	Skin on palm of hand
Total area of exposed medium	212 cm²	57.8 cm²	1,849 cm²	64 cm²
Pressure exerted through sampling medium	8,000 Pa	5,900 Pa	2,300 Pa	6,900 Pa
Sampled carpet area	760 cm²	930 cm²	1,849 cm²	640 cm² (10 presses)
Number of passes over sampled carpet area	2	1	20	1
Sampling speed over carpet	10 cm/s	10 cm/s	20 cm/s	1 press/s

of 6-ply cotton, each wetted with 10 mL of 2-propanol, to collect pesticide residues from residential human hands and indoor surfaces (Geno et al. 1996; Lewis et al. 2001). For indoor surfaces, the sampled area is wiped with a straight-line hand motion with each of the two gauze dressings, frequently turning them to provide fresh wiping faces. The first gauze dressing is used to wipe in one direction; the second for wiping in a direction orthogonal to the first. The same technique has been used with the acetonitrile/phosphate buffer previously mentioned as the wetting agent to collect lawn herbicide residues deposited on windowsills and 29-cm × 29-cm Formica® sheets placed on tabletops inside homes (Nishioka et al. 2001).

Comparison of various wipe materials have found them to be comparable for hard, smooth surfaces but not for rough surfaces (Chavalnitikul and Levin 1984). Wet wipe sampling is generally not recommended for carpet, upholstery and other fabric-covered or soft surfaces because the solvent may be absorbed into the surface being sampled. Wipes of soft surfaces also are less likely than wipes of hard surfaces to reflect the dermal exposure potential (Ness 1994).

Human hand presses have also been used to determine the transfer of pesticides from surfaces to skin (Lewis, Fortmann, and Camann 1994; USEPA 2000b). However, they are generally not practical monitoring tools for determining transferable pesticide residues. The nature of human skin varies between individuals, as do the size and geometry of the hand; it is difficult to determine and control the pressure applied; the hand can only be pressed infrequently (e.g., once or twice per day) on treated surfaces for both safety and health considerations and because solvents used to recover the pesticide residue may affect the nature of the skin; and pesticide residues may be absorbed into the skin to the extent that they are not fully recoverable.

15.8 HANDWIPE METHODS

Hand rinses or washes have long been employed for collecting hand residues from agricultural workers and pesticide applicators (e.g., Durham and Wolfe 1962; Fenske et al. 1998). Various aqueous surfactant solutions, 1% aqueous sodium bicarbonate, or alcohols (either 2-propanol or ethanol in concentrated or dilute solutions) have been most widely used for this purpose. In the "bag wash" method, the wash solution is placed in a plastic bag (e.g., a 1.9-L sealable freezer bag) that is slipped over the hand and secured around the wrist, then shaken vigorously to assure good contact with the skin (Davis 1984). Another popular approach is to rinse the hand with solvent delivered by a laboratory wash bottle and collect the rinsate through a funnel into a bottle (Atallah, Cahill, and Whitacre 1982; Lewis, Fortmann, and Camann 1994). However, the use of hand rinses on small children has not been common practice. Lewis, Fortmann, and Camann (1994) reported problems in administering 2-propanol hand washes on young children due mostly to the reaction of the child to the sensation of coldness as the solvent evaporated from the skin.

When alcohols are used for hand washes or rinses, especially in concentrated form, evaporative losses may be encountered, particularly with the open hand rinse or basin approach. Alcohols may also extract plasticizers or other additives from plastic bags and should always be evaluated for analytical interferences before use. The use of aqueous solutions may be limited by analytical constraints, especially for the determination of low-level residues. Hydrolysis of some pesticides (especially carbamates and organophosphates) may also occur in aqueous solution.

Hand wipes for the determination of pesticide residues have seen limited use until recently. Geno et al. (1996) developed a handwipe technique utilizing two 10-cm × 10-cm surgical gauze sponges (see above). The technique calls for each of the two sponges to be wetted with 10 mL of pesticide grade 2-propanol and used sequentially to thoroughly wipe the entire surface of the hand, being careful to thoroughly wipe each digit and in between. Immediately following sampling, the sponges are placed in a solvent-rinsed, oven-dried wide-mouthed glass jar with a PTFE-lined lid and an additional 50 mL of 2-propanol is added to the container before sealing with PTFE tape and packing on dry ice for transport to the laboratory. While children are capable of performing the wipes of their own hands under the supervision of a field sampling technician, the wiping procedure should be performed by a parent or the technician if the child is under 2 years of age.

Hand wipes are much easier to perform and have been shown to be as effective or better than hand washes at recovering pesticide residues. Laboratory studies with adult volunteers have demonstrated that the 2-propanol wipe method quantitatively removes pesticides that were freshly applied to the skin (USEPA 1996d). Adult skin was spiked by rubbing and pressing the hand on aluminum foil onto which measured quantities of emulsifiable concentrate (EC) formulations of chlorpyrifos and pyrethrins had been deposited and allowed to dry. Immediately after the transfer (ca. 30 s), before the pesticides could be absorbed through the epidermis, the hand was wiped and the foil rinsed to recover all residues. Quantitative (ca. 100%) recoveries of residues transferred to the skin at levels ranging from 2 $\mu g/m^2$ to 250 $\mu g/m^2$ were obtained (Geno et al. 1996, USEPA 1996d). The recovery efficiency of the wipe was significantly higher than that reported for alcohol rinses. For example, Fenske and Lu (1994) reported a 43% recovery of chlorpyrifos residues from human hands with 10% 2-propanol in water immediately after exposure to dried formulations.

Similar studies with aqueous surfactants, human saliva, and a saliva surrogate as wetting agents for hand wipes have yielded recoveries of about 50% of freshly applied pesticides from the hand (Majumdar et al. 1995; USEPA 2000c). Two saliva surrogates were evaluated: 1.3% dioctyl sulfosuccinate sodium salt (DSS) in deionized water and a saliva substitute used to treat salivatory deficiency (Olsson and Axéll 1991). Each subject used his or her own saliva, collecting 50 mL in a PTFE bottle. Immediately after each hand press, the exposed hand was wiped sequentially with two gauze sponges, each wetted with 5 mL of saliva, the saliva substitute, DSS, or 2-propanol, and the sponges were then placed in containers with 50 mL of methanol pending analysis. The results of replicate analyses, corrected for analytical recovery, showed 40–60% recoveries with saliva and the surrogates compared to 90–100% with 2-propanol. Handwipes with saliva surrogates may more accurately estimate the potential dislodgeability of pesticide skin residues by mouthing action than alcohol wipes. However, problems with storage stability and analytical difficulties associated with water limit the usefulness of these wipes for routine exposure monitoring (Majumdar et al. 1995).

Gauze wetting with 2-propanol using the Geno et al. (1996) method have been successfully used on small children in several U.S. Environmental Protection Agency (USEPA) studies (e.g., Lewis et al. 2001; Morgan, Stout, and Wilson 2001; Wilson et al. 2004). More studies are needed, however, to better estimate the exposures of small children to residential pesticides. Better methodologies need to be developed and applied to more accurately determine surface-to-skin and skin-to-mouth transfer efficiencies, pesticide bioavailability from ingested dust, and the relationship of child activity patterns to residential exposures. Such studies are essential before reliable exposure assessments can be made.

15.9 OCCURRENCE, SOURCES, FATE, AND TRANSPORT IN THE INDOOR ENVIRONMENT

Pesticides applied indoors vaporize from treated surfaces (e.g. carpets and baseboards) and can resuspended into air on particles. Those applied to the foundation or perimeter of a building may penetrate into interior spaces, resulting in measurable indoor air levels. Pesticides used on the lawn or in the garden may be tracked indoors, where they can accumulate in house dust that may be resuspended into air. Pesticides may also be added to carpets, paints, and other furnishings and building materials at the time of manufacture. Consequently, the presence of pesticide residues in indoor air does not necessarily indicate their use within or on the premises.

The most commonly used household pesticides are disinfectants. Pine oil and phenols are widely used disinfectants that contribute to indoor air pollution. α-Pinene, a component of pine oil, has been recognized as a ubiquitous indoor air constituent for several years. It is a volatile organic compound (vapor pressure 6×10^{-1} kPa at $25°C$) and not amenable to the sampling methods previously discussed. The average concentration of α-pinene in 539 U.S. households has been reported to be 2.4 $\mu g/m^3$ (Wallace 1987). O-phenylphenol and 2-benzyl-4-chlorophenol (BCP) are the two of the most popular phenolic antimicrobials. In 1984, Reinert (1984) estimated an annual use of 680,000 kg of O-phenylphenol and 5.3 million kg of BCP in U.S. homes. In California, total sales in 1986 amounted to 146,600 kg of O-phenylphenol as the free acid and 102,576 kg as the sodium and potassium salts (CDPR 1988). By comparison, 88,563 kg of free BCP was sold, along with only 5,331 kg of the salts. O-phenylphenol was detected in the indoor air of 70–90% of residences monitored in Florida and Massachusetts during 1986–1987 at 0.03 to 1 $\mu g/m^3$ (Whitmore et al. 1994). In a small New Jersey study in 1991, it was found in 43% of the homes monitored at levels ranging from 0.04–0.06 $\mu g/m^3$ (Roinestad, Louis, and Rosen 1993). The use of O-phenylphenol in household disinfectants has declined in the United States in recent years.

Although its use in the United States has been very limited in recent years, pentachlorophenol (PCP) is still a widespread environmental pollutant and may be a major component of indoor air. Relatively little indoor air data on PCP exist due to the analytical difficulties in measuring it. It was, however, detected in over 80% of the air samples collected inside homes studied in North Carolina and was present at a mean concentration of 0.05 $\mu g/m^3$ (Lewis, Fortmann, and Camann 1994). PCP has also been reported at relatively high air levels inside office buildings (Levin and Hahn 1984). PCP has been used as a wood preservative, principally on porches and patio decking. It has also been used as a fungicide to protect leather goods, some paints, and other materials that may be used inside the home. In Europe, where use of PCP in indoor paints and stains has been common in the past, indoor exposures via the air route are considered potentially serious (e.g., Maroni et al. 1987). In the United States, it is ubiquitously present in household dust (Lewis et al. 1999).

Insecticides that have been commonly used indoors are pyrethroids (natural and synthetic), chlorpyrifos, diazinon, propoxur, dichlorvos, malathion, and piperonyl butoxide (and synergist). Two of the most widely used broad-spectrum insecticides have been permethrin and chlorpyrifos. Chlorpyrifos has been one of the most frequently detected insecticides in indoor air and remains so despite its banning from residential use in 2000.

The most heavily used household pesticides are acid herbicides. They are readily detected in house dust (generally at much higher concentrations than in surface soils), but rarely reported in indoor air. Phenoxyacetic acid herbicides (e.g., 2,4-D, dicamba, and mecoprop) and glyphosate are the most commonly used as post-emergence herbicides. Pre-emergence herbicides, such as triazines (e.g., atrazine, simazine) are used in lesser quantities.

Pesticides may be periodically introduced into indoor air by direct application (e.g., insect sprays and bombs; disinfectant sprays; room deodorizers). In addition, there are often sources that continually emit vapors into the living space (e.g., continuous evaporation of residues from crack and crevice treatments; emissions from pest control strips or other devices). Whether used inside the home or office, or outside on the lawn or garden, pesticides accumulate on indoor surfaces, especially in carpet dust, and also in upholstery and in or on children's toys (Lewis, Fortmann, and Camann 1994; Simcox et al. 1995; Nishioka et al. 1996; Gurunathan et al. 1998). Insecticides and disinfectants applied indoors may persist for extended periods, protected from sunlight, rain, temperature extremes, and most microbial action. Pesticides applied to the lawn or used in the garden may be tracked indoors by humans and pets (which are very efficient transporters), where they can persist for months or years, as opposed to perhaps days outside on the grass (Nishioka et al. 1996; Morgan, Stout, and Wilson 2001). Termiticides, the efficacy of which depends on persistence, continually migrate from the building foundation into the living space by both the air route and track in of foundation soil (Wright, Leidy, and Dupree 1994). Typically, pesticide concentrations in indoor air and house dust are 10–100 times those found in outdoor air and surface soil (Lewis et al. 1988; Lewis, Fortmann, and Camann 1994; Whitmore et al. 1994).

Most household pesticides are used during spring and summer months when pest infestations are greatest. Warmer weather may also increase volatilization of termiticides from the soil beneath the house resulting in increased contributions to living spaces. Consequently, air concentrations should be highest during these times. For example, in a USEPA study in Jacksonville, FL, the mean indoor air concentrations for the 23 detected pesticides were 0.11 $\mu g/m^3$ in summer, 0.05 $\mu g/m^3$ in spring, and 0.03 $\mu g/m^3$ in winter (Whitmore et al. 1994). The mean air concentrations of the four predominant pesticides (propoxur, diazinon, chlorpyrifos, and chlordane) found in Jacksonville were 0.32–053 $\mu g/m^3$ in summer, 0.11–0.25 $\mu g/m^3$ in spring, and 0.08–0.22 $\mu g/m^3$ in winter. Since insect populations are generally larger in more moderate climates, pesticide usage and corresponding exposure levels should be greater in the southern states. The aforementioned USEPA study found total indoor levels of pesticides to be 6–8 times higher in Jacksonville, FL, than in Springfield, MA, during the same seasons.

Many pesticides are semivolatile (vapor pressures between 10^{-2} kPa and 10^{-8} kPa) and tend to vaporize when applied to indoor surfaces (e.g., carpets and baseboards). The rate of volatilization will depend on the vapor pressure of the compound, the formulation (solvent, surfactants, microencapsulation, etc.), the ambient and surface temperatures, indoor air movement (ventilation), the type of surface treated, and the lapsed time after application. The vapor pressure of the pure pesticide is frequently known and may be of value for assessing the relative importance of potential respiratory exposures. Good compilations of vapor pressure data and other physical properties of pesticides may be found in Tomlin (2003), Howard (1991), and Wauchope et al. (1992).

All organophosphate (OP) insecticides are semivolatile and will vaporize from surfaces after applications. The most volatile OP is dichlorvos (vapor pressure = 7×10^{-3} kPa at 25°C). Dichlorvos has been a common household insecticide for many years, particularly in slow-release insecticide strips and flea collars for cats and dogs. Several studies showed that air concentrations of 100 $\mu g/m^3$ or greater were not unusual in rooms in which DDVP pest strips were deployed (Leary et al. 1974; Lewis and Lee 1976). In 1988, the USEPA estimated that lifetime exposures to dichlorvos vapors from pest strips may result in a 10^{-2} cancer risk (Consumers Union 1988). While the pesticide is still registered for use in pest strips, foggers, and flea collars, use in the United States has declined dramatically since 1988. The use of OP pesticides in and around residences has also been declining in favor of pyrethroid insecticides. Both chlorpyrifos (2.7×10^{-6} kPa at 25°C) and diazinon (1.2×10^{-5} kPa at 25°C) were banned from residential indoor use by the USEPA in 2000.

The majority of pyrethroid insecticides have low volatilities. The heavily used synthetic pyrethroid permethrin is classified as nonvolatile on the basis of its vapor pressure (1.3×10^{-9} kPa at 20°C) and is rarely found in indoor air. However, it was the major pesticide residue recently found in house dust (Lewis et al. 1999). Cypermethrin and cyfluthrin are two other low-volatility pyrethroids commonly used for indoor flea and cockroach control.

Most acid herbicides used on lawns (e.g., 2,4-D and glyphosate) are applied in the form of amine salts that possess very extremely low vapor pressures; hence, they may be found in air only at low concentrations and are mostly associated with suspended particulate matter. For example, 2,4-D was detected in the air inside of 64 of 82 homes commercially treated with the herbicide (Yeary and Leonard 1993). The time-weighted average concentration over 7 hours on the day of application was determined to be 34 ng/m³ in 6 homes with the highest measured air levels. 2,4-D has also been measured in residential indoor air for several days after application to lawns at concentrations of 1–15 ng/m³ (USEPA 1999b; Nishioka et al. 2001). Indoor air concentrations of 2,4-D increased from nondetectable before lawn treatment to 0.2–10 ng/m³ (10 µm inlet) after homeowner application, with about 65% of the total particulate 2,4-D associated with inhalable particles (<10 µg/m³). 2,4-D associated with <1 µm and smaller particles made up 25–30% of the total mass. Triazine herbicides, which are used for pre-emergent control of weeds, are more volatile than acid herbicide salts, but are still largely classified as nonvolatile. There have been very few reports of their presence in indoor air. Atrazine (3.7×10^{-8} kPa at 20°C) has been reported at concentrations of 3–12 ng/m³ in indoor air of farm residences and homes near agricultural operations (Camann et al. 1993; Mukerjee et al., 1997).

The influence of the volatility of a pesticide on measurable indoor air levels is evident by comparing semivolatile chlorpyrifos with nonvolatile permethrin. Room air concentrations of these two insecticides (monitored at 12.5 cm and 75 cm above the floor) were comparable (means 30 µg/m³ and 42 µg/m³, respectively) 0–2 hours after broadcast spraying, but the air levels of permethrin declined so rapidly that nothing (<1 µg/m³) could be detected in air after 8 hours, while chlorpyrifos levels remained high at 31 µg/m³ (Koehler and Moye 1995b). Immediately after spraying, permethrin was most likely present in air as an aerosol, which underwent little or no volatilization after deposition. Unfortunately, it is not possible to accurately predict rates of volatilization or expected air concentrations based on vapor pressures. Even when ambient conditions, substrates, and formulations are similar, emission rates for pesticides will depend on other factors such as the concentration and molecular structure of the active ingredient. Jackson and Lewis (1981) compared emission rates from three kinds of pest control strips in the same room under constant conditions of temperature (21 ± 1°C) and humidity (50 ± 20%) and found that room air concentrations over a period of 30 days were much higher for diazinon than for chlorpyrifos, but similar to those for propoxur. On day 2 room air levels were 0.76 µg/m³ for diazinon, 0.14 µg/m³ for chlorpyrifos, and 0.79 µg/m³ for propoxur. After 30 days, the air concentrations were 1.21, 0.16, and 0.70 µg/m³, respectively. The vapor pressure of diazinon is nearly 10 times higher than that of chlorpyrifos and nearly 100 times lower than that of propoxur (4×10^{-7} kPa at 20°C).

The rate of volatilization of microencapsulated pesticides is much slower than that of emulsifiable concentrates (Jackson and Lewis 1979; Koehler and Patterson 1991). Indoor air levels of chlorpyrifos applied as a microencapsulated formulation were measured at 3.1 µg/m³ 0–2 hours after broadcast spraying and 5.2 µg/m³ after 24 hours compared to 30 µg/m³ and 15 µg/m³, respectively, for the emulsifiable concentrate (Koehler and Moye 1995a). After 48 hours, levels were still about double for the emulsifiable concentrate application (8.5 µg/m³ vs. 4.0 µg/m³).

There may be significant temporal and spatial variations in the concentrations of pesticides in indoor air, especially if monitoring is performed after an indoor application. Air concentrations typically drop rapidly for about 3 days after application as the pesticide is absorbed into furnishings or dissipates to the outdoor air. However, concentrations of the more volatile pesticides may still be 20–30% of those on the day of application after as much as 21 days (Leidy, Wright, and Dupree 1993; Lewis, Fortmann, and Camann 1994). Dissipation rates will also depend on the method of application. Aerosol sprayers have been shown by some to result in higher post-application air levels than compressed air sprayers (Leidy, Wright, and Dupree 1993; Koehler and Moye 1995a). Likewise, broadcast spraying has been shown to result in higher air concentrations than crack and crevice treatments (Fenske and Black 1989). However, Lu and Fenske (1998) recently reported finding little differences in air concentrations for chlorpyrifos applied by pressurized broadcast spraying and aerosol (fogger) release. One day after treatments, chlorpyrifos levels averaged 23–64 $\mu g/m^3$ for broadcast application and 43–48 $\mu g/m^3$ for the fogger. Air concentrations of chlorpyrifos declined more rapidly after aerosol release than they did after broadcast spraying, declining on the second day by 75% and 50%, respectively. Seven days later, air concentrations were similar (aerosol: 0.4–8.6 $\mu g/m^3$; broadcast: 0.9–8 $\mu g/m^3$).

When there has not been a recent application, pesticide levels in the air of most rooms vary little with vertical distance from the floor (Leidy, Wright, and Dupree 1993; Lewis, Fortmann, and Camann 1994). However, after an application, air concentrations near the floor (in the breathing zone of small children) may be several times higher than those in the adult breathing zone (Fenske et al. 1990; Lewis, Fortmann, and Camann 1994; Lu and Fenske 1998). This is particularly true for broadcast and aerosol spray applications, where concentrations of chlorpyrifos have been found to be more than 4 times higher at 25 cm above the floor than at 100 cm 5–7 hours after application (Fenske et al. 1990). Lu and Fenske (1998) observed concentration gradients for chlorpyrifos when samples were taken at 25, 100, and 175 cm above the floor after broadcast and aerosol treatments of carpeted floors, but the gradients had largely disappeared after 3–5 days. Conversely, Lewis, Fortmann, and Camann (1994) found air concentrations of chlorpyrifos to be twice as high at 12 cm as at 75 cm 2 days after crack and crevice treatment and 1.5–2 times as high at 8 and 15 days post-application. Koehler and Moye (1995a) reported concentration gradients after broadcast application of chlorpyrifos up to 175 cm. Diazinon (1×10^{-5} kPa at 20°C) levels were found to be higher at 1.2 m than a ceiling level 21 days after crack and crevice treatment but to have equalized after 35 days (Leidy, Wright, and Dupree 1993). Although the synthetic pyrethroids d-phenothrin and d-tetramethrin have relatively low vapor pressures (1.6×10^{-7} and 9.4×10^{-7} kPa, respectively), they were found in room air immediately after crack and crevice treatment at 752 $\mu g/m^3$ and 1,040 $\mu g/m^3$, respectively, but declined rapidly to 2–3 $\mu g/m^3$. No differences in concentrations were observed at heights of 25 cm and 120 cm above the floor after the air levels dropped to several $\mu g/m^3$ (Matoba, Takimoto, and Kato 1998).

The only large-scale monitoring survey of pesticides in residential indoor air was the Non-Occupational Pesticides Exposure Study (NOPES) conducted by the USEPA during 1985–1990 (Whitmore et al. 1994). This Congressionally mandated study was carried out in two cities, Jacksonville, FL, and Springfield/Chicopee, MA, which were selected to represent high and low household pesticide usage areas, respectively. In each city, households were selected by statistical sampling of population census districts and stratified prior to monitoring into low-, medium-, and high-use categories. Approximately 175 homes in Jacksonville and 85 homes in Springfield (representing ca. 300,000 and 140,000 residents, respectively) were chosen for study of air, dermal, drinking water, and dietary exposures to 33 of the most commonly used household pesticides. Exposure measurements were made during the spring, summer, and winter seasons.

In Jacksonville, where three seasons of monitoring were conducted, 25–27 of the targeted pesticides were detected in indoor air. Total mean air concentrations were 2.3 $\mu g/m^3$ in summer, 1.2 $\mu g/m^3$ in spring, and 0.81 $\mu g/m^3$ in winter. Outdoors (on porches and patios), 9–16 were detected at total mean concentrations of 0.11, 0.03, and 0.06 $\mu g/m^3$ for summer, spring, and winter, respectively. Summertime monitoring was not performed in Springfield/Chicopee. During spring and winter, 20 and 16 pesticides were found indoors at total mean air concentrations of 0.40 and 0.10 $\mu g/m^3$, respectively. Ten and six were found outdoors at 0.03 and 0.01 $\mu g/m^3$, respectively. Seven of the most prevalent pesticides shown in Table 15.3 and Table 15.4 made up more than 90% of

TABLE 15.3
Most Frequently Detected Pesticides in Residential Indoor Air in Jacksonville, Florida, and Their Mean Concentrations (Non-Occupational Pesticide Exposure Study, 1986–1987)

Pesticide	Summer % of Households	Summer Mean Conc., ng/m³	Spring % of Households	Spring Mean Conc., ng/m³	Winter % of Households	Winter Mean Conc., ng/m³
Chlorpyrifos	100	370	88	205	29	120
Propoxur	85	528	93	222	95	162
o-Phenylphenol	85	96.0	84	70.4	79	59.0
Diazinon	83	421	83	109	83	85.7
Dieldrin	79	14.7	37	8.3	62	7.2
Chlordane	61	324	54	245	94	220
Heptachlor	58	163	71	155	92	72.2
Lindane (γ-HCH)	34	20.2	47	13.4	68	6.0
Dichlorvos	33	134	14	86.2	10	24.5
Malathion	27	20.8	32	14.9	17	20.4
α-Hexachlorocyclohexane	25	1.2	23	1.2	22	1.1
Aldrin	21	31.3	19	6.8	31	6.9
Bendiocarb	23	85.7	20	5.5	20	3.4
Carbaryl	17	68.1	1	0.4	0	0

TABLE 15.4
Most Frequently Detected Pesticides in Residential Indoor Air in Springfield/Chicopee, Massachusetts, and Their Mean Concentrations (Non-Occupational Pesticide Exposure Study, 1987–1988)

Pesticide	Spring % of Households	Spring Mean Conc., ng/m³	Winter % of Households	Winter Mean Conc., ng/m³
o-Phenylphenol	90	44.5	72	22.8
Chlordane	50	199	83	34.8
Heptachlor	50	31.3	70	3.6
Propoxur	49	26.7	38	17.0
Chlorpyrifos	29	9.8	30	5.1
Diazinon	16	48.4	10	2.5
Dieldrin	12	1.0	34	4.2
Lindane (γ-HCH)	10	0.5	21	9.5
Dichlorvos	2	4.3	1	1.5
Malathion	2	5.0	0	0

the total mass of all pesticides monitored. These were the indoor insecticides chlorpyrifos, propoxur, diazinon and dichlorvos; the termiticides chlordane and heptachlor (both discontinued in 1988); and the disinfectant O-phenylphenol.

15.10 EXPOSURE RISKS AND HEALTH EFFECTS

Pesticide residues found inside the home contribute significantly to the overall exposure of the general population. Of the several possible routes of exposure to household pesticides, the air route is one of the most important. The average resident of the United States spends the large majority of time indoors, mostly inside his or her own domicile. Full-time homemakers and small children may spend over 21 hours/day inside the home, with another 2.5 hours inside other buildings (stores, offices, etc.) or in transit vehicles (Robinson 1977; Shoaf 1991; USEPA 1996a,b,c). Even when employed, urban dwellers spend about the same amounts of time indoors, 63% of it at home and 28% at work. For many persons, at least 90% of our daily cumulative exposure to pesticides occurs at home. Results from simultaneous 24-hour indoor air and personal exposure monitoring in NOPES, showed that 85% of the total daily adult exposure to airborne pesticides was from breathing air inside the home (Whitmore et al. 1994).

While pesticide residues on residential surfaces represent an exposure risk to all occupants, small children are potentially the most vulnerable. Infants and toddlers spend more time indoors and generally wear less clothing than adults. They have higher body surface-to-weight ratios than adults, have more intimate skin contact with the floor, and may frequently have wet, sticky hands that are several times as efficient at dislodging surface residues than dry skin (Lewis 2005). They engage in much more mouthing activities (hands, toys, furniture) than adults. It has been estimated that children under the age of 5 ingest 2.5 times more soil and dust from around the home than adults, yet they possess only about 20% of the body weight (Hawley 1985). Consequently, pesticide residues that are on surfaces or that are transferred to the skin may be ingested, resulting in exposures that may rival or exceed respiratory exposures.

Whereas occupational inhalation exposure guidelines have been established for many pesticides, the United States has no current guidelines for non-occupational indoor air exposures. Workplace exposure limits are established by National Institute for Occupational Safety and Health (NIOSH), OSHA, and the American Conference of Governmental Industrial Hygienists, Inc. (ACGIH). Recommended time-weighted average (TWA) inhalation exposure limits are typically in the 0.5–5 mg/m^3 range for 8–10-hour exposures, far above the levels normally observed in indoor air in residences, public buildings, or commercial buildings outside of the pesticide control industry. The National Research Council and USEPA did issue interim guidelines for indoor air exposures to five termiticides in 1982 (NRC 1982). They were 1 $\mu g/m^3$ for aldrin and dieldrin, 2 $\mu g/m^3$ for heptachlor, 5 $\mu g/m^3$ for chlordane, and 10 $\mu g/m^3$ for chlorpyrifos. The four cyclodiene termiticides have been discontinued since 1989 and post-construction use of chlorpyrifos was phased out in 2000.

Guidelines for estimating nondietary oral exposure of pesticides by small children are not as well established as those for inhalation exposure. However, the USEPA assumes that the average soil and dust ingestion rate for small children is 100 mg per day (USEPA 1996c). Therefore, if the child's weight (in kg) and the pesticide content of house dust or yard soil (in $\mu g/g$) in areas contacted by the child are known, an estimate of exposure may be obtained in $\mu g/kg/day$. This exposure may then be compared to the USEPA reference dose (*RfD*) for the pesticide of concern. Thus, a risk assessment can be made in the same manner given below for inhalation exposure.

A number of pesticides have been classified by USEPA as known (Group A), probable (Group B), or possible (Group C) human carcinogens. Inorganic arsenic and chromium compound, which are used as wood preservatives, are the only Group A pesticides that may have residential applications. Since these are nonvolatile compounds, the potential for air route exposure to them is low. Several pesticides used for residential and institutional indoor pest control are classified as B2 carcinogens (probable human carcinogens based on animal data): *O*-phenylphenol, fenoxycarb, propoxur, and lindane. The former termiticides aldrin, chlordane, dieldrin, heptachlor, and heptachlor epoxide; the wood preservative pentachlorophenol (canceled in 1986); and perchloroethylene (a VOC, canceled in 1992) are also classified B2. Group C pesticides of concern include atrazine, benomyl, bromacil, dinoseb (canceled in 1992), ethalfluralin, *N*-octylbicycloheptene dicarboximide (MGK-264), and paradichlorobenzene (a VOC). The fungicide chlorothalonil has been under review and is likely to be classified as a probable or possible human carcinogen. It is no longer labeled for use on home lawns.

For suspect human carcinogens, the USEPA uses probability-based risk assessment. An acceptable nonoccupational risk for the general public is considered to be 1×10^{-6} or a 1 in a million chance above the background probability (currently 1:13) that an individual will develop cancer over his or her normal lifetime. The estimated excess lifetime cancer risks from inhalation exposure as determined by NOPES were highest for the now discontinued cyclodiene termiticides. In Jacksonville, FL, they ranged from 1×10^{-6} for heptachlor epoxide to 2×10^{-4} for heptachlor. In Springfield/Chicopee the range was from 5×10^{-7} for aldrin to 4×10^{-5} for heptachlor. For other pesticides, the lifetime excess cancer risks for Jacksonville ranged from 6×10^{-9} for 2,4-D (measured as the esters) to 2×10^{-6} for γ-hexachlorocyclohexane; for Springfield they ranged from 2×10^{-9} for chlorothalonil to 2×10^{-7} for 2,4-D. Noncancer risk assessments were not performed in NOPES.

The USEPA has published general guidelines on exposure assessment, which are applicable to indoor air inhalation exposures (USEPA 1996c). Specific guidelines for post-application residential pesticide exposure assessment have been recently published by the agency (USEPA 2000a). These guidelines cover sampling protocols, monitoring methods, analytical methods, and data presentation, but do not mandate the use of specific methods or protocols. They also cover the fundamentals of exposure and risk assessments, human activity patterns, and modeling. Under the pesticides registration process, the USEPA is generally interested in data on exposure to total airborne pesticide residues, as opposed to inspirable or respirable fractions, because of the potential for absorption of pesticides through the mouth and gastrointestinal tract, as well as the lung. The absorbed, or internal, dose received via the air route is assumed to be the same as the potential dose; i.e., dose is calculated on the basis of 100% absorption of the pesticide that potentially enters the body through breathing, assuming no filtration of particles by the nose. The average daily inhalation dose (ADD_I) in µg of pesticide absorbed per kg of body weight per day is, therefore, determined to be:

$$ADD_I = \frac{CR_I t_E}{w t_A} \tag{15.1}$$

where
ADD_I = average daily inhalation dose (µg/kg-day)
C = measured air concentration of period of exposure (µg/m^3)
R_I = inhalation rate (m^3/day)
t_E = duration of exposure (days)
t_A = averaging time (days)
w = body weight (kg)

For noncarcinogenic effects, t_A is taken to be equal to t_E. For carcinogenic or chronic health effects, t_A is 70 years (25,550 days), or the average human life span. When conducting a risk assessment, the calculated ADD_I may compared to the *RfD* for the specific pesticide of interest, which may be obtained from the USEPA *Integrated Risk Information System* or IRIS database (Dwyer 1998; USEPA 2003) or other sources.

The *RfD* is defined by the USEPA (2003) as "An estimate (with uncertainty spanning perhaps an order of magnitude) of a daily exposure to the human population (including sensitive subgroups) that is likely to be without appreciable risk of deleterious effects during a lifetime." It is determined from Equation 15.2:

$$RfD = \frac{NOAEL}{U_F M_F} \tag{15.2}$$

where
   $NOAEL$ = the "no observable adverse effects level" (mg/kg/day)
   U_F = composite uncertainty factors
   M_F = situation-specific modification factor

A U_F of 10 is used to account for variations in sensitivity among human subpopulations. An additional 10- to 100-fold uncertainty factor is invoked when animal toxicity data is used for extrapolation to human health effects (which is most often the case). If the extrapolation is from valid long-term animal studies, the additional U_F is 10; when extrapolating from less than chronic animal toxicity data and there are no useful long-term human data studies available, an additional 10-fold factor is applied. If the RfD is from a lowest observable adverse effects level ($LOAEL$), instead of a $NOAEL$, still another 10-fold uncertainty factor is applied. Finally, FQPA calls for another 10-fold uncertainty factor for children (*vide supra*). The M_F may vary from a default value of 1 up to 10 depending on the weaknesses or uncertainties in the scientific data used.

IRIS uses a default inhalation rate of 20 m³/day for respiratory exposure, which is close to the estimated average daily inhalation rates published by the International Commission on Radiological Protection for adults (ICRP 1981); i.e., 22.8 m³/day for adult males and 21.1 m³/day for adult females. However, for a child of 10 years of age, ICRP estimates 14.8 m³/day; an infant (age 1 year), 3.8 m³/day; and a newborn, 0.8 m³/day. Use of the IRIS methodology for determining reference doses for inhalation exposure risk assessment is described in detail by the USEPA (USEPA 1996c). Shoaf (1991) and Swartout et al. (1998) also provide good overall discussions on the use of reference dose in risk assessment.

Meaningful exposure and risk assessments require knowledge of human activity patterns as well as environmental concentrations. The amount of time spent indoors, the distribution of time spent within various rooms, and the levels of physical exertion must be known to accurately determine the extent of respiratory exposure in a given indoor environment. When small children are of concern, air measurements made in their breathing zones (10–75 cm above the floor) should also be considered. Computer-based models for prediction of potential residential inhalation exposures that take into account human activity patterns have been developed. One such program is the USEPA's *THERdbASE* (Pandian, Bradford, and Behar 1990; USEPA 1998). However, it has not been applied to semivolatile pesticides. As previously stated, risk assessments required to establish the safe use of pesticides must be based on the potential total, or aggregate, exposure from all sources. Multimedia, multipathway models for human exposure via the air, water, and soil ingestion routes have been published (e.g., McKone and Daniels 1991), but there have been few, if any, reported studies in which such models have been field tested for pesticide exposures. While models may be used to predict human exposure or health risks associated with pesticides, the accuracies of such models can only improve as better measurement data become available.

REFERENCES

ASTM (2003a) F 608-03, Standard Test Method for Evaluation of Carpet-Embedded Dirt Removal Effectiveness of Household Vacuum Cleaners, *Annual Book of ASTM Standards*, 11.03, ASTM International, West Conshohoken, PA, http://www.astm.org.

ASTM (2003b) D 6345-98 (rev. 2003), Standard Guide for Selection of Methods for Active, Integrative Sampling of Volatile Organic Compounds in Air, *Annual Book of ASTM Standards*, 11.03, ASTM International, West Conshohoken, PA, http://www.astm.org.

ASTM (2004) D 6333-98 (rev 2003) Standard Practice for Collection of Dislodgeable Residues from Floors, *Annual Book of ASTM Standards*, 11.03, ASTM International, West Conshohoken, PA, http://www.astm.org.

ASTM (2005a) D 4861-05, Standard Practice for Sampling and Selection of Analytical Techniques for Pesticides and Polychlorinated Biphenyls in Air, *Annual Book of ASTM Standards*, 11.03, ASTM International, West Conshohoken, PA, http://www.astm.org.

ASTM (2005b) D 4947-05, Standard Test Method for Chlordane and Heptachlor Residues in Indoor Air, *Annual Book of ASTM Standards*, 11.03, ASTM International, West Conshohoken, PA, http://www.astm.org.

ASTM (2005c) D 5438-05, Standard Practice for Collection of Floor Dust for Chemical Analysis, *Annual Book of ASTM Standards*, 11.03, ASTM International, West Conshohoken, PA, http://www.astm.org.

Atallah, Y.H., Cahill, W.P., and Whitacre, D.M. (1982) Exposure of Pesticide Applicators and Support Personnel to O-ethyl-O-(4-Nitrophenyl)phenyl Phosphonothioate (EPN), *Archives of Environmental Contamination and Toxicology*, 11: 219–225.

Baker, S.R. and Wilkinson, C.F., Eds. (1990) *The Effects of Pesticides on Human Health*, Princeton Scientific Publishing, Princeton, NJ, 5–33.

Bidleman, T.F. and Olney, C.E. (1974) High Volume Collection of Atmospheric Polychlorinated Biphenyls, *Bulletin of Environmental Contamination and Toxicology*, 11: 442–450.

Billings, W.N. and Bidleman, T.F. (1980) Field Comparisons of Polyurethane Foam and Tenax-GC Resin for High-Volume Sampling of Chlorinated Hydrocarbons, *Environmental Science and Technology*, 14: 679–683.

Billings, W.N. and Bidleman, T.F. (1983) High-Volume of Chlorinated Hydrocarbons in Urban Air Using Three Solid Sorbents, *Atmospheric Environment*, 17: 383–391.

Camann, D.E., Geno, P.W., Harding, H.J., Giardino, N.J., and Bond, A.E. (1993) Measurements to Assess Exposure of the Farmer and Family to Agricultural Pesticides, in *Measurement of Toxic & Related Air Pollutants: Proceedings of the 1993 U.S, EPA/A&WMA International Symposium*, Publication VIP-34, Air & Waste Management Association, Pittsburgh, PA, 712–717.

Camann, D.E., Harding, H.J., and Lewis, R.G. (1990) Trapping of Particle-Associated Pesticides in Indoor Air by Polyurethane Foam and Exploration of Soil Track-in as a Pesticide Source, in *Indoor Air '90*, Vol. 2, Walkinshaw, D.S., Ed., Canada Mortgage and Housing Corporation, Ottawa, Canada, 621–626.

Camann, D.E., Harding, H.J., Stone, C.L., and Lewis, R.G. (1994) Comparison of PM2.5 and Open-Face Inlets for Sampling Aerosolized Pesticides on Filtered Polyurethane Foam, in *Measurement of Toxic & Related Air Pollutants: Proceedings of the 1994 U.S, EPA/A&WMA International Symposium*, Publication VIP-39, Air & Waste Management Association, Pittsburgh, PA, 838–843.

CDPR (1988) *1987 Pesticide Residue Annual Reports*, California Department of Pesticide Regulation, Sacramento, CA.

Chavalnitikul, C. and Levin, L. (1984) A Laboratory Evaluation of Wipe Testing Based on Lead Oxide Surface Contamination, *American Industrial Hygiene Association Journal*, 45: 311–317.

Colt, J.S., Zahm, S.H., Camann, D.E., and Hartge, P. (1998) Comparison of Pesticides and Other Compounds in Carpet Dust Samples Collected from Used Vacuum Cleaner Bags and from a High-volume Surface Sampler, *Environmental Health Perspectives*, 106: 721–724.

Constant, P.C. and Bauer, K.M. (1992) Engineering Study to Explore Improvements in Vacuum Dust Collection, Final Report to the U.S. EPA under Contract 68-DO-0137, Midwest Research Institute, Kansas City, MO.

Consumers Union (1988) Warning: Pesticide Increases Cancer Risk, *Consumer Reports*, 53: 286.

Davis, J.E. (1984), Procedures for Dermal and Inhalation Studies to Assess Exposure to Pesticides, in *Determination and Assessment of Pesticide Exposure*, Siewierski, M., Ed., Elsevier, New York, NY, 123–131.

Davies, J.E., Edmundson, W.E.F., and Raffonelli, A. (1975) The Role of House Dust in Human DDT Pollution, *American Journal of Public Health*, 65: 53–57.

Davis, J.R., Brownson, R.C., and Garcia, G. (1992) Family Pesticide Use in the Home, Garden, Orchard, and Yard, *Archives of Environmental Contamination and Toxicology*, 22: 260–266.

Dewalt, G., Constant, P., Buxton, B.E., Rust, S.W., Lim, B.S., and Schwemberger, J.G. (1995) Sampling and Analysis of Lead in Dust and Soil for the Comprehensive Abatement Performance Study (CAPS), in *Lead in Paint, Soil and Dust: Health Risks, Exposure Studies, Measurement Methods, and Quality Assurance,* Beard, M.E. and Iske, A.S.D., Eds., ASTM International, West Conshohoken, PA, 227–248.

Durham, W.F. and Wolfe, H.R. (1962) Measurement of the Exposure of Workers to Pesticides, *Bulletin of WHO,* **26**: 75–91.

Dwyer, S.D. (1998) *EPA's Integrated Risk Information System on CD ROM,* Government Institutes, Inc., Rockville, MD.

Edmiston, S. (1987) *Human Illnesses/Injuries Reported by Physicians in California Involving Indoor Exposure to Pesticides Containing Chlorpyrifos, DDVP, and/or Propoxur 1983-1986,* Division of Pest Management Report No. HS-1431, California Department of Food and Agriculture, Sacramento, CA.

Edwards, R.D. and Lioy, P.J. (1999) The EL Sampler: A Press Sampler for the Quantitative Estimation of Dermal Exposure to Pesticides in Housedust, *Journal of Exposure Analysis and Environmental Epidemiology,* **9**: 521–529.

Eller, P. and Cassinelli, M., Eds. (1994) *NIOSH Manual of Analytical Methods,* 4th ed., Publication No. 94-113, National Institute for Occupational Safety and Health, U.S. Department of Health and Human Services, 3rd Supplement 2003-154, Schlecht, P.C. and O'Connor, P.F., Eds., Cincinnati, OH, http://www.cdc.gov/niosh/nmam.

Farfel, M.R., Lees, P.S.J., Lim, B., and Rohde, C.A. (1994) Comparison of Two Cyclone-Based Devices for the Evaluation of Lead-Containing Residential Dusts, *Applied Occupational and Environmental Hygiene,* **9**: 212–217.

Farewell, S.O., Bowes, F.W., and Adams, D.F. (1977) Evaluation of XAD-2 as a Collection Medium for 2,4-D Herbicides in Air, *Journal of Environmental Science and Health,* **B12**: 71–83.

Fenske, R.A. and Black, K.G. (1989) Dermal and Respiratory Exposures to Pesticides in and Around Residences, in *Measurement of Toxic & Related Air Pollutants: Proceedings of the 1993 US EPA/A&WMA International Symposium,* Publication VIP-34, Air & Waste Management Association, Pittsburgh, PA, 853–858.

Fenske, R.A., Black, K.G., Elkner, K.P., Lee, C.L., Methner, M.M., and Soto, R. (1990) Potential Exposure and Health Risks of Infants Following Indoor Residential Pesticide Applications, *American Journal of Public Health,* **80**: 689–693.

Fenske, R.A. and Lu, C. (1994) Determination of Handwash Removal Efficiency: Incomplete Removal of the Pesticide, Chlorpyrifos, from Skin by Standard Handwash Techniques, *American Industrial Hygiene Association Journal,* **55**: 425–432.

Fenske, R.A., Schulter, C., Lu, C., and Allen, E. (1998) Incomplete Removal of the Pesticide Captan from Skin by Standard Handwash Exposure Assessment Procedures, *Bulletin of Environmental Contamination and Toxicology,* **61**: 194–201.

Fuller, R., Klonne, D., Rosenheck, L., Eberhart D., Worgan, J., and Ross, J. (2001) Modified California Roller for Measuring Transferable Residues on Treated Turfgrass, *Bulletin of Environmental Contamination and Toxicology,* **67**: 787–794.

Geno, P.W., Camann, D.E., Harding, H.E., Villalobos, K., and Lewis, R.G. (1996) A Handwipe Sampling and Analysis Procedure for the Measurement of Dermal Contact to Pesticides, *Archives of Environmental Contamination and Toxicology,* **30**: 132–138.

Godish, T. (1990) *Air Quality*: Lewis Publishers, Chelsea, MI, 293–337.

Gurunathan, S., Robson, M., Freeman, N., Buckley, B., Roy, A., Meyer, R., and Bukowski, J. (1998) Accumulation of Chlorpyrifos on Residential Surfaces and Toys Accessible to Children, *Environmental Health Perspectives,* **106**: 9–16.

Hawley, J.K. (1985) Assessment of Health Risk from Exposure to Contaminated Soil, *Risk Analysis,* **5**: 289–302.

Hill, R.H. and Arnold, J.E. (1979) A Personal Sampler for Pesticides, *Archives of Environmental Contamination and Toxicology,* **8**: 621–628.

Howard, P.H. (1991) *Handbook of Environmental and Exposure Data for Organic Chemicals,* Vol. III (*Pesticides*), Lewis Publishers, Boca Raton, FL.

Hsu, J.P., Camann, D.E., Schattenberg, H., III, Wheeler, B., Villalobos, K., Kyle, M., Quarderer, S., and Lewis, R.G. (1990) New Dermal Exposure Sampling Technique, in *Total Exposure Assessment Methodology: A New Horizon*, Publication VIP-16, Air & Waste Management Association Publication, Pittsburgh, PA, 489–497.

IBR (1995) *Aerosol Retention Efficiency, Vacuum Bags*, Report No, 2866A to Hoover Co., InterBasic Resources, Inc., Grass Lake, MI.

ICRP (1981) *Report of the Task Group on Reference Man,* International Commission on Radiological Protection, Pergamon Press, New York, NY.

Jackson, M.D. and Lewis, R.G. (1979) Volatilization of Methyl Parathion from Fields Treated with Microencapsulated and Emulsifiable Concentrate Formulations, *Bulletin of Environmental Contamination and Toxicology,* **21**: 202–205.

Jackson, M.D. and Lewis, R.G. (1981) Insecticide Concentrations in Air after Application of Pest Control Strips, *Bulletin of Environmental Contamination and Toxicology,* **27**: 122–125.

Jenkins, J.J., Curtis, A.S., and Cooper, R.J. (1993) Two Small-Plot Techniques for Measuring Airborne and Dislodgeable Residues of Pendimethalin Following Application to Turfgrass, in *Pesticides in Urban Environments*, ACS Symposium Series 522, Racke, K.D. and Leslie, A.R., Eds., American Chemical Society, Washington, DC, 228–242.

Johnson, E.R., Yu, T.C., and Montgomery, M.L. (1977) Trapping and Analysis of Atmospheric Residues of 2,4-D, *Bulletin of Environmental Contamination and Toxicology,* **17**: 369–372.

Kiely, T., Donaldson, D., and Grube, A. (2004) *Pesticide Industry Sales and Usage, 2000 and 2001 Market Estimates*, Report No. 733-R-02-001, Office of Prevention, Pesticides and Toxic Substances, U.S. Environmental Protection Agency, Washington, DC, http://www.epa.gov/oppbead1/pestsales/99pestsales/market estimates1999.pdf.

Koehler, P.G. and Moye, H.A. (1995a) Chlorpyrifos Formulation Effect on Airborne Residues Following Broadcast Application for Cat Flea (Siphonaptera: Pulicidae) Control, *Journal of Economics and Entomology,* **88**: 918–923.

Koehler, P.G. and Moye, H.A. (1995b) Airborne Insecticide Residues after Broadcast Application for Cat Flea (Siphonaptera: Pulicidae) Control, *Journal of Economics and Entomology,* **88**: 1684–1689.

Koehler, P.G. and Patterson, R.S. (1991) Residual Effectiveness of Chlorpyrifos and Diazinon Formulations for German Cockroach (Orthoptera: Blattellidae) on Panels Placed in Commercial Food Preparation Areas, *Journal of Entomological Science,* **26**: 59–63.

Kogan, V., Kuhlman, M.R., Coutant, R.W., and Lewis, R.G. (1993) Aerosol Filtration by Sorbent Beds, *Journal of the Air & Waste Management Association,* **43**: 1367–1373.

Lanphear, B.P., Emond, M., Jacobs, D.E., Weitzman, M., Tanner, M., Winter, N.L., Yaki, B., and Eberly, S. (1995) A Side-by-Side Comparison of Dust Collection Methods for Sampling Lead-Contaminated House Dust, *Environment Research,* **68**: 114–123.

Leary, J.S., Keane, W.T., Cleve Fontenot, M.S., Feichtmeir, E.F., Schultz, D., Koos, B.A., Hirsch, L., Lavor, E.M., Roon, C.C., and Hine, C.H. (1974) Safety Evaluation in the Home of Polyvinyl Chloride Resin Strips Containing Dichlorvos (DDVP), *Archives of Environmental Health,* **29**: 308–314.

Leidy, R.B. and Wright, C.G. (1991) Trapping Efficiency of Selected Adsorbents for Various Airborne Pesticides, *Journal of Environmental Science and Health,* **B26**: 367–382.

Leidy, R.B., Wright, C.G., and Dupree Jr., H.E. (1993) Exposure Levels to Indoor Pesticides, in *Pesticides in Urban Environments*, ACS Symposium Series 522, Racke, K.D. and Leslie, A.R., Eds., American Chemical Society, Washington, DC, 283–296.

Levin, H. and Hahn, J. (1984) Pentachlorophenol in Indoor Air, in *Proceedings of the 3rd International Conference on Indoor Air Quality and Climate*, Berglund, B., Lindvall, T., and Sundell, J., Eds., Swedish Council for Building Research, Stockholm, Sweden, **5**: 123–128.

Lewis, R.G. (1993) Determination of Pesticides and Polychlorinated Biphenyls in Indoor Air by Gas Chromatography, Method 24, in *Environmental Carcinogens: Methods for Analysis and Exposure Measurement*, Vol. 12, Seifert, B., van de Wiel, H., Dodet, B., and O'Neill, I., Eds., International Agency for Research on Cancer, Lyon, France, 353–376.

Lewis, R.G. (2005) Residential Post-Application Pesticide Exposure Monitoring, in *Occupational and Residential Exposure Assessment for Pesticides,* Franklin, C.A. and Worgan, W.P., Eds., John Wiley & Sons, Ltd., Sussex, U.K.

Lewis, R.G. and Gordon, S.M. (1996) Sampling of Organic Chemicals in Air, in *Principles of Environmental Sampling*, 2nd ed., Keith, L.H., Ed., ACS Professional Reference Book, American Chemical Society, Washington, DC, 401–470.

Lewis, R.G. and Jackson, M.D. (1982) Modification and Evaluation of a High-Volume Air Sampler for Pesticides and Other Semivolatile Industrial Organic Chemicals, *Analytical Chemistry,* **54**: 592–594.

Lewis, R.G. and Lee Jr., R.E. (1976) Air Pollution from Pesticides: Sources, Occurrence, and Dispersion, in *Air Pollution from Pesticides and Agricultural Processes*, Lee, R.E., Jr., Ed., CRC Press, Boca Raton, FL, 5–50.

Lewis, R.G. and MacLeod, K.E. (1982) A Portable Sampler for Pesticides and Semi-Volatile Industrial Organic Chemicals, *Analytical Chemistry,* **54**: 310–315.

Lewis, R.G., Bond, A.E., Johnson, D.E., and Hsu, J.P. (1988) Measurement of Atmospheric Concentrations of Common Household Pesticides: A Pilot Study, *Environmental Monitoring and Assessment,* **10**: 59–73.

Lewis, R.G., Brown, A.R., and Jackson, M.D. (1977) Evaluation of Polyurethane Foam for High-Volume Air Sampling of Ambient Levels of Airborne Pesticides, Polychlorinated Biphenyls, and Polychlorinated Naphthalenes, *Analytical Chemistry,* **49**: 1668–1672.

Lewis, R.G., Camann, D.E., Chuang, J.C., Roberts, J.W., and Ruby, M.G. (1995) Measuring and Reducing the Pollutants in House Dust, *American Journal of Public Health,* **85**: 1168.

Lewis, R.G., Fortmann, R.C., and Camann, D.E. (1994) Evaluation of Methods for the Monitoring of the Potential Exposure of Small Children to Pesticides in the Residential Environment, *Archives of Environmental Contamination and Toxicology,* **26**: 37–46.

Lewis, R.G., Fortune, C.R., Blanchard, F.T., and Camann, D.E. (2001) Movement and Deposition of Two Organophosphorus Pesticides within Residences after Interior and Exterior Applications, *Journal of the Air & Waste Management Association,* **51**: 339–351.

Lewis, R.G., Fortune, C.R., Willis, R.D., Camann, D.E., and Antley, J.T. (1999) Distribution of Pesticides and Polycyclic Aromatic Hydrocarbons in House Dust as a Function of Particle Size, *Environmental Health Perspectives,* **107**: 721–726.

Lioy, P.J., Freeman, N.C.G., and Millette, J.R. (2002) Dust: A Metric for Use in Residential and Building Exposure Assessment and Source Characterization, *Environmental Health Perspectives,* **110**: 969–983.

Lioy, P.J., Wainman, T., and Weisel, C.P. (1993) A Wipe Sampler of the Quantitative Measurement of Dust on Smooth Surfaces: Laboratory Performance Studies, *Journal of Exposure Analysis and Environmental Epidemiology,* **3**: 315–330.

Litovitz, T.L., Bailey, K.M., Schmitz, B.F., Holm, K.C., and Klein-Schwartz, W. (1991) 1990 Annual Report of the American Association of Poison Control Centers, National Data Collection System, *American Journal of Emergency Medicine,* **9**: 461–509.

Lu, C. and Fenske, R.A. (1998) Air and Surface Chlorpyrifos Residues Following Residential Broadcast and Aerosol Pesticide Applications, *Environmental Science and Technology,* **32**: 1386–1390.

Majumdar, T.K., Camann, D.E., Ellenson, W.D., and Lewis, R.G. (1995) Determination of Pesticides and Their Stabilities in Salivary Matrices, in *Proceedings of the EPA/A&WMA International Conference on Measurement of Toxic and Related Air Pollutants*, Publication VIP-50, Air & Waste Management Association, Pittsburgh, PA, 569–571.

Maroni, M., Knoppel, H., Schlitt, H., and Righetti, S. (1987) Occupational and Environmental Exposure to Pentachlorophenol, Report EUR 10795EN, Commission of the European Communities, Ispra, Italy.

Marple, V.A., Rubow, K.L., Turner, W., and Spengler, J.D. (1987) Low Flow Rate Sharp-Cut Impactors for Indoor Air Sampling: Design and Calibration, *Journal of the Air Pollution Control Association,* **37**: 1303–1307.

Matoba, Y., Takimoto, Y., and Kato, T. (1998) Indoor Behavior and Risk Assessment Following Residual Spraying of *d*-Phenothrin and *d*-Tetramethrin, *American Industrial Hygiene Association Journal,* **59**: 191–199.

McArthur, B. (1992) Dermal Measurement and Wipe Sampling Methods: A Review, *Applied Occupational and Environmental Hygiene,* **7**: 599–606.

McKone, T.E. and Daniels, J.I. (1991) Estimating Human Exposure through Multiple Pathways from Air, Water, and Soil, *Regulatory Toxicology and Pharmacology,* **13**: 36–61.

Morgan, M.K., Stout, D.M., II, and Wilson, N.K. (2001) Feasibility Study of the Potential for Human Exposure to Pet-Borne Diazinon Residues Following Lawn Applications, *Bulletin of Environmental Contamination and Toxicology,* **66**: 295–300.

Mukerjee, S., Ellenson, W.D., Lewis, R.G., Stevens, R.K., Somerville, M.C., Shadwick, D.S., and Willis, R.D. (1997) Soil Characterizations Conducted in the Lower Rio Grande Valley of Texas, III, Residential Microenvironmental Measurements with Applications for Regional and Temporal-Based Exposure Assessments, *Environment International,* **23**: 657–673.

National Research Council (NRC) (1982) *An Assessment of the Health Risks of Seven Pesticides Used for Termite Control,* National Research Council, Committee on Toxicology, National Academy Press, Washington, DC.

Ness, S.A. (1994) *Surface and Dermal Monitoring for Toxic Exposures,* Van Nostrand Reinhold, New York, NY.

Nishioka, M.G., Burkholder, H.M., Brinkman, M.C., Gordon, S.M., and Lewis, R.G. (1996) Measuring Transport of Lawn-Applied Herbicide Acids from Turf to Home: Correlation of Dislodgeable 2,4-D Turf Residues with Carpet Dust and Carpet Surface Residues, *Environmental Science and Technology,* **30**: 3313–3320.

Nishioka, M.G., Burkholder, H.M., Brinkman, M.C., and Lewis, R.G. (1999) Distribution of 2,4-Dichlorophenoxyacetic Acid in Floor Dust throughout Homes Following Homeowner and Commercial Lawn Applications: Quantitative Effects on Children, Pets, and Shoes, *Environmental Science and Technology,* **33**: 1359–1365.

Nishioka, M.G., Lewis, R.G., Brinkman, M.C., Burkholder, H.M., and Hines, C. (2001) Distribution of 2,4-D in Air and on Surfaces Following Lawn Applications: Comparing Exposure Estimates for Young Children from Various Media, *Environmental Health Perspectives,* **109**: 1185–1191.

Olsson, H. and Axéll, T. (1991) Objective and Subjective Efficacy of Saliva Substitutes Containing Mucin and Carboxymethylcellulose, *Scandinavian Journal of Dental Research,* **99**: 316–319.

Orgill, M.M., Sehemel, G.A., and Petersen, M. (1976) Some Initial Measurements of Airborne DDT over Pacific Northwest Forests, *Atmospheric Environment,* **10**: 827–834.

OSHA (1999) Sampling for Surface Contamination, in *OSHA Technical Manual,* Section II, Chapter 2, Occupational Safety and Health Administration, U.S. Department of Labor, Washington, DC.

Pandian, M.D., Bradford, J., and Behar, J.V. (1990) Therdbase: Total Human Exposure Relational Database, in *Total Exposure Assessment Methodology: A New Horizon, Proceedings of the EPA/A&WMA Specialty Conference,* Publication VIP-16, Air & Waste Management Association, Pittsburgh, PA, 204–209.

Que Hee, S.S., Peace, B., Clark, C.S., Boyle, J.R., Bornshein, R.L., and Hammond, P.B. (1985) Evolution of Efficient Methods to Sample Lead Sources, Such as House Dust and Hand Dust, in the Homes of Children, *Environment Research,* **38**: 77–95.

Reinert, J.C. (1984) Pesticides in the Indoor Environment, in *Proceedings of the 3rd International Conference on Indoor Air Quality and Climate,* Stockholm, Sweden, 223–238.

Roinestad, K.S., Louis, J.B., and Rosen, J.D. (1993) Determination of Pesticides in Indoor Air and Dust, *Journal of the Association of Official Analytical Chemists International,* **76**: 1121–1126.

Roberts, J.W. and Camann, D.E. (1989) Pilot Study of a Cotton Glove Press Test for Assessing Exposure to Pesticides in House Dust, *Bulletin of Environmental Contamination and Toxicology,* **43**: 717–724.

Roberts, J.W., Budd, W.T., Ruby, M.G., Stamper, J.R., Camann, D.E., Sheldon, L.S., and Lewis, R.G. (1991) A High Volume Small Surface Sampler (HVS3) for Pesticides and Other Toxics in House Dust, in *Proceedings of the 84th National Meeting of the Air & Waste Management Association,* Paper No. 91-150.2, Vancouver, BC.

Robinson, J.P. (1977) *How Americans Use Time: A Social Psychological Analysis of Everyday Behavior,* Praeger Publishers, New York, NY.

Roinestad, K.S., Louis, J.B., and Rosen, J.D. (1993) Determination of Pesticides in Indoor Air and Dust, *Journal of the Association of Official Analytical Chemists International,* **76**: 1121–1126.

Ross, J., Thongsinthusak, T., Fong, H.R., Margetich, S., and Krieger, R.I. (1990) Measuring Potential Dermal Transfer of Surface Pesticide Residue Generated from Indoor Fogger Use: An Interim Report, *Chemosphere,* **20**: 349–360.

Ross, J., Thongsinthusak, T., Fong, H.R., Margetich, S., and Krieger, R.I. (1991) Measured Potential Dermal Transfer of Surface Pesticide Residue Generated from Indoor Fogger Use: Using the CDFA Roller Method, *Chemosphere,* **22**: 975–984.

Rudel, R.A., Camann, D.E., Spengler, J.D., Korn, L.R., and Brody, J.G. (2003) Phthalates, Alkylphenols, Pesticides, Polybrominated Diphenyl Ethers, and Other Endocrine-Disrupting Compounds in Indoor Air and Dust, *Environmental Science and Technology,* **37**: 4543–4553.

Savage, E.P., Keefe, T.J., Wheeler, H.W., Mounce, L.M., Helwic, L., Applehaus, F., Goes, E., Goes, T., Mihlan, G., Rench, J., and Taylor, D.K. (1981) Household Pesticide Usage in the United States, *Archives of Environmental Health,* **36**: 304–309.

Shoaf, C.R. (1991) Current Assessment Practices for Noncancer End Points, *Environmental Health Perspectives,* **95**: 111–119.

Simcox, N.J., Fenske, R.A., Wolz, S.A., Lee, I.-C., and Kalman, D.A. (1995) Pesticides in Household Dust and Soil, *Environmental Health Perspectives,* **103**: 1126–1134.

Starr, H.G., Jr., Aldrich, F.D., MacDougall, W.D., III, and Mounce, L.M. (1974) Contribution of Household Dust to the Human Exposure to Pesticides, *Pesticides Monitoring Journal,* **8**: 209–212.

Swartout, J.C., Price, P.S., Dourson, M.L., Carlson-Lynch, H.L., and Keenan, R.E. (1998) A Probabilistic Framework for the Reference Dose (Probabilistic RfD), *Risk Analysis,* **18**: 271–282.

Thomas, T.C. and Seiber, J.N. (1974) Chromosorb 102: Efficient Trapping of Pesticides from Air, *Bulletin of Environmental Contamination and Toxicology,* **12**: 17–26.

Tomlin, C.D.S., Ed. (2003) *The Pesticide Manual,* 13th ed., British Crop Protection Council, Berkshire, U.K.

USEPA (1979) *National Household Pesticide Usage Survey, 1976-1977,* Report No. 540/9-80-002, U.S. Environmental Protection Agency, Office of Prevention, Pesticides and Toxic Substances, Washington, DC.

USEPA (1989) *Development of a High Volume Surface Sampler for Pesticides in Floor Dust,* Report No. 600/4-88/036, U.S. Environmental Protection Agency, Office of Research and Development, Research Triangle Park, NC.

USEPA (1995) *Sampling House Dust for Lead: Basic Concepts and Literature Review,* Report No. 600-R-95-007, U.S. Environmental Protection Agency, Office of Research and Development, Research Triangle Park, NC.

USEPA (1996a) *Analysis of the National Human Activity Pattern Survey (NHAPS) Respondents from a Standpoint of Exposure Assessment,* Report No. 600/R-96/074, U.S. Environmental Protection Agency, Office of Research and Development, Washington, DC.

USEPA (1996b) *Descriptive Statistics Tables from a Detailed Analysis of the National Human Activity Pattern Survey (NHAPS) Data,* Report No. 600/R-96/148, U.S. Environmental Protection Agency, Office of Research and Development, Washington, DC.

USEPA (1996c) *Exposure Factors Handbook,* Report No. 600/P-95/002, U.S. Environmental Protection Agency, Office of Research and Development, Washington, DC, http://www.epa.gov/ordntrnt/ORD/WebPubs/exposure/.

USEPA (1996d) *Comparison of Methods to Determine Dislodgeable Residue Transfer from Floors,* Report No. EPA/600/R-96/089, U.S. Environmental Protection Agency, National Exposure Research Laboratory, Research Triangle Park, NC.

USEPA (1997a) *Round-Robin Testing of Methods for Collecting Dislodgeable Residues from Carpets,* Report No. EPA/600/R-97/119, U.S. Environmental Protection Agency, National Exposure Research Laboratory, Research Triangle Park, NC.

USEPA (1997b) *Evaluation of Dislodgeable Residue Collection Methods for Pesticide Applications on Turf,* Report No. 600/R-97/107, U.S. Environmental Protection Agency, National Exposure Research Laboratory, Research Triangle Park, NC.

USEPA (1998) *THERdbASE Exposure Assessment Software User Manual,* Version 1.2, U.S. Environmental Protection Agency, Research Triangle Park, NC.

USEPA (1999a) *Compendium of Methods for the Determination of Toxic Organic Chemicals in Ambient Air,* Report No. 600/R-96/010b, U.S. Environmental Protection Agency, Office of Research and Development, Cincinnati, OH, http://www.epa.gov/ttn/amtic/airtox.html.

USEPA (1999b) *Transport of Lawn-Applied 2,4-D from Turf to Home: Assessing the Relative Importance of Transport Mechanisms and Exposure Pathways,* Report No. 600/R-99/040, U.S. Environmental Protection Agency, National Exposure Research Laboratory, Research Triangle Park, NC.

USEPA (1999c) *Evaluation and Application of Methods for Estimating Children's Exposure to Persistent Organic Pollutants in Multiple Media,* Report No. EPA 600-R-98-164, U.S. Environmental Protection Agency, National Exposure Research Laboratory, Research Triangle Park, NC.

USEPA (2000a) *Post Application Exposure Guidelines: Series 875 - Group B*, U.S. Environmental Protection Agency, Office of Prevention, Pesticides, and Toxic Substances, Washington, DC.

USEPA (2000b) *Dermal Transfer Efficiency of Pesticides from New and Used Cut-Pile Carpet to Dry and Wetted Palms*, Report No. EPA/600/R-00/028, U.S. Environmental Protection Agency, National Exposure Research Laboratory, Research Triangle Park, NC.

USEPA (2000c) *Evaluation of Saliva and Artificial Salivary Fluids for Removal of Pesticide Residues from Human Skin*, Report No. EPA/600/R-00/041, U.S. Environmental Protection Agency, National Exposure Research Laboratory, Research Triangle Park, NC.

USEPA (2003) Integrated Risk Information System, U.S. Environmental Protection Agency, Washington, DC, http://www.epa.gov/iris.

Vaccaro, J.R. (1993) Risks Associated with Exposure to Chlorpyrifos and Chlorpyrifos Formulation Components, in *Pesticides in Urban Environments*, ACS Symposium Series 522, Racke, K.D. and Leslie, A.R., Eds., American Chemical Society, Washington, DC, 297–306.

Vaccaro, J.R. and Cranston, R.J. (1990) *Evaluation of Dislodgeable Residues and Absorbed Doses of Chlorpyrifos Following Indoor Broadcast Applications of Chlorpyrifos-Based Emulsifiable Concentrate*, Dow Chemical Company, Midland, MI.

Wallace, L.A. (1987) *The Total Exposure Assessment Methodology (TEAM) Study*, Report No. EPA/600/6-87/002, U.S. Environmental Protection Agency, Washington, DC.

Wauchope, R.D., Buttler, T.M., Hoensby, A.G., Augustijn-Becker, P.W.M., and Burt, J.P. (1992) The SCR/ARS/CWS Pesticide Properties Database for Environmental Decision-Making, in *Reviews of Environmental Contamination and Toxicology*, **123**: 1–69.

Whitmore, R.W., Immerman, F.W., Camann, D.E., Bond, A.E., Lewis, R.G., and Schaum, J.L. (1994) Non-Occupational Exposure to Pesticides for Residents of Two U.S. Cities, *Archives of Environmental Contamination and Toxicology*, **26**: 47–59.

Whitmore, R.W., Kelly, J. E., Reading, P.L., Brandt, E., and Harris, T. (1993) National Home and Garden Use Survey, in *Pesticides in Urban Environments*, ACS Symposium Series 522, Racke, K.D. and Leslie, A.R., Eds., American Chemical Society, Washington, DC, 18–36.

Williams, D.T., Shewchuck, C., Lebel, G.L., and Muir, N. (1987) Diazinon Levels in Indoor Air after Periodic Application for Insect Control, *American Industrial Hygiene Association*, **48**: 780–785.

Willis, R.D. (1995) *SEM Characterization of House Dusts Collected by Conventional Vacuum and the HVS3 Sampler*, Project Report, U.S. EPA Contract 68-D5-0049, ManTech Environmental Technology, Inc., Research Triangle Park, NC.

Wilson, N.K., Chuang, J.C., Iachan, R., Lyu, C., Gordon, S.M., Morgan, M.K., Özkaynak, H., and Sheldon, L.S. (2004) Design and Sampling Methodology for a Large Study of Preschool Children's Aggregate Exposures to Persistent Organic Pollutants in Their Everyday Environments, *Journal of Exposure Analysis and Environmental Epidemiology*, **14**: 260–274.

Woodrow, J.M. and Seiber, J.N. (1978) Portable device with XAD-4 Resin Trap for Sampling Airborne Residues of Some Organophosphorus Pesticides, *Analytical Chemistry*, **50**: 1229–1231.

Wright, C.G. and Leidy, R.B. (1982) Chlordane and Heptachlor in the Ambient Air of Houses Treated for Termites, *Bulletin of Environmental Contamination and Toxicology*, **28**: 617–623.

Wright, C.G., Leidy, R.B., and Dupree, H.E. (1993) Cypermethrin in the Ambient Air and on Surfaces in Rooms, Treated for Cockroaches, *Bulletin of Environmental Contamination and Toxicology*, **51**: 356–360.

Wright, C.G., Leidy, R.B., and Dupree Jr., H.E. (1994) Chlorpyrifos in the Air and Soil of Houses Eight Years after Its Application for Termite Control, *Bulletin of Environmental Contamination and Toxicology*, **52**: 131–134.

Yeary, R.A. and Leonard, J.A. (1993) Measurement of Pesticides in Air during Application to Lawns, Trees, and Shrubs in Urban Environments, in *Pesticides in Urban Environments*, ACS Symposium Series 522, Racke, K.D. and Leslie, A.R., Eds., American Chemical Society, Washington, DC, 275–281.

Yule, W.M., Cole, A.F.W., and Hoffman, I. (1971) A Survey of Atmospheric Contamination Following Spraying with Fenitrothion, *Bulletin of Environmental Contamination and Toxicology*, **6**: 289–296.

Zartarian, V.G. and Leckie, J.O. (1998) Dermal Exposure: The Missing Link, *Environmental Science and Technology*, **32**: 134A–137A.

16 Exposure to Dioxin and Dioxin-Like Compounds[1]

Daniel J. Stralka
U.S. Environmental Protection Agency

Harold A. Ball
U.S. Environmental Protection Agency

CONTENTS

16.1 Synopsis...379
16.2 Dioxin Toxicity..380
16.3 Toxicity Factors and Equivalence 382
16.4 Sources, Emissions, and Environmental Fate..384
16.5 Media and Food Levels...387
16.6 Sources and Pathways to Human Exposure 389
16.7 Summary and Future Directions 391
16.8 Questions for Review 391
References 392

16.1 SYNOPSIS

Dioxin and dioxin-like compounds (DLC) are a family of natural and human-made chemicals that are ubiquitous and biologically persistent. They are associated with a broad spectrum of adverse biological effects, both cancer and non-cancer. Dioxin entered the public lexicon as a result of a number of high-profile news stories over the past several decades. Although never intentionally produced, dioxins were later found to be significant chemical by-products in the synthesis of a range of chemical products. For example, dioxin was a common contaminant in products produced from chlorophenol, including Agent Orange, a chemical defoliant used in Vietnam, and a bactericide used for disinfection. Inappropriate disposal of waste from the manufacture of hexachlorophene led to significant exposure of residents of Times Beach, MO, to dioxin. The area was subsequently cleaned up by the U.S. Environmental Protection Agency (USEPA) under Superfund legislation. However, the major source of environmental dioxin release today is as a by-product of almost every combustion process. Dioxins then move through the environment where they bioconcentrate in animals and fish, which become a source of low-level exposure to the population. There are a number of PBTs, or persistent, bioaccumulative and toxic compounds, that are receiving international attention. The focus of this chapter is the occurrence and fate of dioxin in the environment, dioxin toxicity and current exposure to the population, and strategies to manage the general population risks associated with dioxin.

---

[1] The contents of this chapter do not necessarily reflect the views or policies of the U.S. Environmental Protection Agency.

379

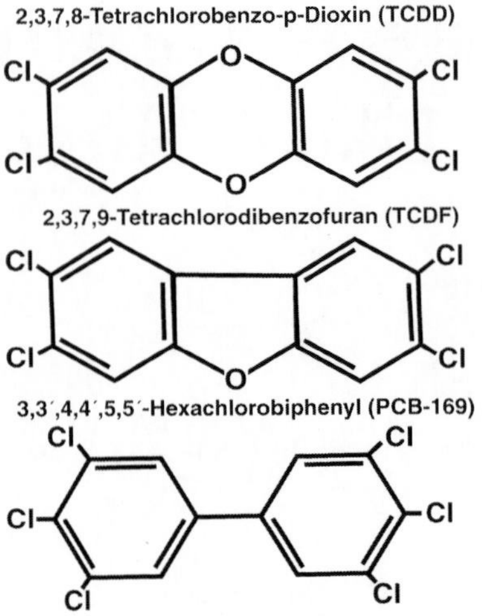

2,3,7,8-Tetrachlorobenzo-p-Dioxin (TCDD)

2,3,7,9-Tetrachlorodibenzofuran (TCDF)

3,3′,4,4′,5,5′-Hexachlorobiphenyl (PCB-169)

FIGURE 16.1 Example structures of polychlorinated dibenzo-*p*-dioxins, dibenzofurans and biphenyls.

16.2 DIOXIN TOXICITY

Dioxins are a class of compounds that have a wide range of toxic effects at very low doses. This group of compounds is defined by a similar physical structure, their affinity for an extracellular membrane protein, and is thought to express some of its effects through a common mechanism of action. This group of about 30 or so active compounds is defined by its binding to an extracellular receptor called the aryl hydrocarbon (Ah) receptor (Whitlock 1993). All of these compounds have a planar configuration, 2 or 3 rings and their binding is potentiated by being halogenated in the lateral positions (Figure 16.1). Congeners are specific compounds within each family of compounds that differ by the degree and extent of halogenation. These physical properties also make them resistant to enzymatic action and fat soluble, which leads to persistence and bioaccumulation or biomagnification in the environment. A model for this class of compounds is 2,3,7,8-tetrachlorod-ibenzo-*p*-dioxin (2,3,7,8-TCDD), which is the most studied of this group. While all these compounds express similar toxic responses, they differ in the doses necessary to elicit the same level of response (USEPA 2000b). This attribute of variable binding affinity for the Ah receptor is used to construct a relative ranking of all the compounds in this class. In the environment, exposure is usually to a complex mixture of these active dioxins, depending on the source of the exposure (USEPA 2000a). Almost any environmental soil and water sample will have trace amounts of dioxins when advanced analytical procedures are employed.

There is a wide range of possible toxic outcomes from exposure to dioxins (USEPA 2000b). These effects are usually delayed from exposure. The delay supports the receptor-mediated response model. Frank effects at high doses are gonadal and lymphoid tissue atrophy, wasting syndrome, and death. Lower doses can be expressed in humans as chloracne, a severe skin disease with acne-like lesions. Altered pigmentation, hyperplasia, and hyperkeratosis have also been reported. Carcinogenesis has been evaluated by both the International Agency for Research on Cancer (IARC 1997) and the USEPA (2000c), which rank 2,3,7,8-TCDD as a human carcinogen. Further studies evaluating the carcinogenic mechanism of action demonstrate that 2,3,7,8-TCDD is itself a weak mutagen but a very potent cancer promoter (USEPA 2000b). This is further evidence that TCDD is not genotoxic but is operating through a receptor-mediated response.

Other effects that are seen at still lower doses can range from biochemical effects like oxidative stress, enzyme induction, changes in hormone and growth factors, immunosuppression, and altered glucose tolerance, ultimately leading to diabetes. Organism life stage at exposure is also critical to the type and extent of effect expressed. For example, *in utero* exposures can lead to congenital malformations and subsequent developmental effects whereas similar exposures to an adult organism may or may not show effects.

There is a wide variation between species for any one specific effect, if expressed in the particular species at all, but the overall type and magnitude of effects is similar across species. There is also difference in magnitude of any response within species that can range as much as 10,000 times. Thus the ultimate disease outcome may be different within any population (USEPA 2000b).

For humans, the World Health Organization (WHO) has derived a tolerable daily intake (TDI), a daily average intake that would not present adverse effects, as 1–4 pg/kg-day (JECFA 2001) reduced from 10 pg/kg-day in 1998. This is based on no adverse effect levels (NOEAL) in animals with a safety factor of 100 (10× for extrapolating from animals to humans and 10× for response sensitivity within the human population). The USEPA in its recent dioxin exposure reassessment (USEPA 2000a) has estimated that our primary source of dioxin exposure is from food (meat, dairy, and fish) and at levels of about 1 pg/kg-day. In order to prioritize risks and the uncertainties of the various inputs into both the toxicity evaluation and the exposure assumptions, it is useful to look at the margin of exposure (MOE). The MOE is the simple ratio of a safe dose divided by the actual dose to a compound of interest. Generally, the higher the MOE the less the concern, and MOEs greater than 100 are not of great concern because there is a 100-fold margin of safety. For dioxins, the MOE is less than 10 suggesting that there may be effects being expressed in the human population. However, these effects may not be adverse in that there could be adaptability to some of the initial effects, which would be interpreted as variability within the human population. Regardless, the low MOE suggests that exposure should be minimized to increase the margin of safety.

The initial steps that define the receptor response are the agonist binding to the Ah receptor on the cell surface, being internalized, moving into the nucleus and binding to DNA and regulating gene expression (Birnbaum 1994) (Figure 16.2). Multiple genes may be affected, both up- and down-regulated. This multiple gene response is a similar cascade of events initiated by hormones with relatively small concentrations having an effect that is greatly multiplied through gene regulation. One protein in particular that seems to be the most sensitive indicator of dioxin exposure is the expression of mixed function oxidases called cytochrome P-450 in the endoplasmic reticulum. In particular, isoforms of cytochrome P-450 are expressed (1A1 in lung and skin, and 1A2 in liver). These enzymes play a critical role in a number of cellular processes from hormone synthesis to cellular homeostasis. A function of these enzymes is to insert an active oxygen species into a typically lipid soluble compound. This oxygen molecule, once inserted into the compound, can act as a point of attachment for the detoxifying enzyme systems to chemically add compounds that increase water solubility, thus enhancing partitioning in the body and ultimately the body's ability to clear it. Unfortunately, the process of inserting the active oxygen species into the lipophilic compound can also result in a more reactive species that can then react with other cellular components and possibly lead to changes that ultimately could be expressed as cancer.

The variability seen within a species and between species may be due to the effects later in the chain of events caused by the receptor binding or due to the structural affinity of the Ah receptor. Whatever the cause of this variability, the binding affinity for the Ah receptor is proportional to the level of response. It is this attribute that has been used to extrapolate the wealth of information specific to 2,3,7,8-TCDD to the other active dioxin compounds. In balancing all the different endpoints or effects measured in various species, several groups have derived numerous schemes for describing the potency of the individual dioxins. The most recent consensus group sponsored by the WHO has derived a set of toxicity equivalent factors (TEFs) for mammals, birds, and fish (Van den Berg et al. 1998).

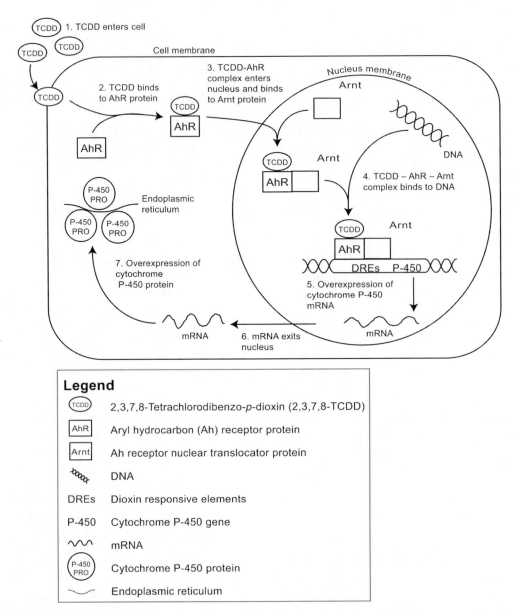

FIGURE 16.2 Mechanism of dioxin uptake into the cell. (From USEPA 2002b, modified by Willa AuYeung.)

16.3   TOXICITY FACTORS AND EQUIVALENCE

Polychlorinated dibenzo-*p*-dioxins (CDDs) are a family of tricyclic aromatic compounds consisting of chlorinated benzene rings joined by an oxygenated ring. CDDs along with some polychlorinated dibenzofurans (CDFs) and certain polychlorinated biphenyls (PCBs) make up a group of chemicals that are termed dioxin-like compounds (DLC). There are a total of 75 different CDD congeners, 135 CDF congeners, and 209 different PCB congeners. The subsets of this class of compounds that are considered "dioxin-like" are those congeners that are characterized by similar structure, physical-chemical properties, and toxic response. The dioxin-like CDDs and CDFs are characterized by chlorine substitution at the 2,3,7,8 positions on the benzene rings. Some of the PCBs have

TABLE 16.1
Toxic Equivalency Factors (TEF-WHO$_{98}$) for Dioxins and Dioxin-Like Compounds

Dioxin (D) Congener	TEF	Furan (F) Congener	TEF	Dioxin-Like PCB (P)	TEF
2,3,7,8-TCDD	1.0	2,3,7,8-TCDF	0.1	3,3',4,4'-TCB (77)	0.0001
1,2,3,7,8-PeCDD	1.0	1,2,3,7,8-PeCDF	0.05	3,4,4',5-TCB (81)	0.0001
1,2,3,4,7,8-HxCDD	0.1	2,3,4,7,8-PeCDF	0.5	2,3,3',4,4'-PeCB (105)	0.0001
1,2,3,6,7,8-HxCDD	0.1	1,2,3,4,7,8-HxCDF	0.1	2,3,4,4',5-PeCB (114)	0.0005
1,2,3,7,8,9-HxCDD	0.1	1,2,3,6,7,8-HxCDF	0.1	2,3',4,4',5 -PeCB (118)	0.0001
1,2,3,4,6,7,8-HpCDD	0.01	1,2,3,7,8,9-HxCDF	0.1	2',3,4,4',5-PeCB (123)	0.0001
1,2,3,4,6,7,8,9-OCDD	0.0001	2,3,4,6,7,8-HxCDF	0.1	3,3',4,4',5-PeCB (126)	0.1
		1,2,3,4,6,7,8-HpCDF	0.01	2,3,3',4,4',5-HxCB (156)	0.0005
		1,2,3,4,7,8,9-HpCDF	0.01	2,3,3',4,4',5'-HxCB (157)	0.0005
		1,2,3,4,6,7,8,9-OCDF	0.0001	2,3',4,4',5,5'-HxCB (167)	0.00001
				3,3',4,4',5,5'-HxCB (169)	0.01
				2,3,3',4,4',5,5'-HpCB (189)	0.0001

CDD — chlorodibenzo-*p*-dioxin; CDF — chlorodibenzo-*p*-furan; CB — chlorobiphenyl; T — tetra; Pe — penta; Hx — hexa; Hp — hepta; O — octa.

Source: Van den Berg et al. 1998.

dioxin-like character when chlorine is substituted at four or more of the lateral positions and no more than one of the ortho positions (USEPA 2000c).

In the environment, the dioxin-like compounds are typically found as a mixture of congeners. Consequently, an approach was developed in 1989 to estimate the risks associated with exposure to mixtures of CDDs and CDFs (USEPA 1989). Here, toxic equivalency factors (TEFs) were determined for the various congeners based upon relative toxicity when compared to the well-studied 2,3,7,8-tetrachlorodibenzo-*p*-dioxin (TCDD), the most toxic member of the group. TEFs reflect the differing potencies of compounds that all initiate a similar cascade of events, compared to TCDD, which is assigned a TEF of 1.0. Adopted internationally, this approach is not exact but is thought to have an uncertainty within a factor of 10. The current WHO consensus dioxin TEFs are presented in Table 16.1 (Van den Berg et al. 1998).

In a complex mixture, the Toxic Equivalency concentration (TEQ) is determined by multiplying the concentration of each congener in the mixture by its corresponding TEF, and summing all the products to give a single 2,3,7,8-TCDD equivalent as follows:

$$\text{Total Toxic Equivalency (TEQ)} = \Sigma C_i TEF_i$$

where C_i equals the concentration of the individual congener (i) in the complex mixture.

The accepted nomenclature for this TEQ scheme is TEQ$_{DFP}$-WHO$_{98}$, where TEQ represents the toxic equivalency of the mixture as 2,3,7,8-TCDD. The subscripts DFP indicate that dioxins (D), furans (F), and dioxin-like PCBs (P) are included in the scheme. The subscript 98 following WHO displays the year changes made to the TEF scheme were published.

The currently accepted TEFs presented in Table 16.1 include 7 dioxin congeners, 10 furan congeners, and 12 dioxin-like PCB congeners. However, in human tissue samples and food products, only five of these congeners, TCDD, 1,2,3,7,8-PCDD, 1,2,3,6,7,8-HxCDD, 2,3,4,7,8-PeCDF, and PCB 126, account for over 70% of the total TEQ (USEPA 2000b).

16.4 SOURCES, EMISSIONS, AND ENVIRONMENTAL FATE

Dioxin and dioxin-like compounds are inadvertently formed by natural processes (including forest fires) and a number of human activities. Dioxin can be a product of industrial processes in such industries as paper, metal smelting, and chemical manufacturing. PCBs are no longer manufactured in the United States but they were once widely used and are present in the environment. The major sources of dioxin formation today, however, are combustion related.

Several mechanisms have been proposed to explain the appearance of DLC in combustion emissions (Lustenhouwer, Olie, and Hutzinger 1980) including: (1) the DLC can be a contaminant in the material being burned but is not destroyed; (2) the DLC can be a breakdown product of larger, complex organic molecules reacting in the presence of chlorine and heat; and (3) de novo synthesis of the DLC from unrelated precursor molecules involving "heterogeneous, surface-catalyzed reactions between carbonaceous particulates and an organic or inorganic chlorine donor" (USEPA 2001b). Such surface-catalyzed reactions are thought to be the dominant mechanism for DLC formation. These reactions typically occur as combustion gases are cooling and take place in a temperature range between 200°C and 400°C (Kilgroe et al. 1990). The reaction is promoted by the presence of molecular chlorine, which chlorinates DLC precursors through substitution reactions. Chloride ions from the fuel or from atmospheric sources can participate if condensed to chlorine through the Deacon reaction catalyzed by copper (Griffin 1986; Gullett, Bruce, and Beach 1990). Copper also acts to catalyze the condensation reactions of chlorinated aromatic rings to form the DLC backbone molecular structure (Gullett et al. 1992). If present, sulfur acts as an inhibitor to the reaction, apparently by reacting with and depleting the chlorine present and by poisoning the copper catalyst (Griffen 1986; Raghunathan and Gullet 1996).

Recently, the USEPA updated its inventory of sources of DLC release to the environment in the United States (USEPA 2001a). Here, the most reliable data on source emission rate and dioxin concentration was used to estimate total dioxin releases for the years 1987 and 1995. A summary of the emissions data is presented in Table 16.2. Due to data limitations in the study, only the dioxin and furans were considered and TEQs were determined using the TEQ_{DF}-WHO_{98} scheme. The year 1987 was selected because prior to that time little CDD/CDF emissions data were available, and it was a time period before there was widespread installation of controls to limit CDD/CDF emissions. The year 1995 was selected as the most recent year for which reliable activity-level data were available for many source categories and also a year prior to which numerous regulatory and non-regulatory efforts to reduce formation and release of dioxin-like compounds had been implemented. In 1987 all known human source activity (for which reliable estimates could be made) in the United States contributed 13,998 grams TEQ_{DF}-WHO_{98} to the environment. By 1995, these same sources contributed 3,253 grams TEQ_{DF}-WHO_{98} to the environment, a reduction of approximately 80%. Since 1995, the USEPA has adopted a number of regulations that should reduce the DLC emissions from various sources, including: municipal waste combustors, medical waste incinerators, hazardous waste incinerators, cement kilns burning hazardous waste, and pulp and paper facilities using chlorine bleached processes (USEPA 1998, 2002). Although only recently identified as a significant source, uncontrolled backyard burning of household trash is and will continue to be a significant contributor to the national DLC budget (Lemieux et al. 2000).

In 1995, the environmental releases of DLC were 96% to the atmosphere, about 3% to land, and about 1% to water (Figure 16.3). A schematic of the transport of DLC through the environment is presented in Figure 16.4 (USEPA 2000a). As discussed above, DLC are primarily released to the atmosphere from combustion sources and transported. DLC are then deposited and adsorbed into plant matter, or adsorbed onto soils. Where plant matter is used as feed for farm animals, the animals consume the feed and take up associated DLC. DLC in soils are washed into sediments and bioaccumulate through the food chain to fish. In animals and fish, dioxins tend to accumulate in fatty tissues. Dioxin has very low solubility in groundwater and is preferentially associated with soils and sediments.

TABLE 16.2
Inventory of Sources of Dioxin-Like Compounds in the United States (TEQ$_{DF}$-WHO$_{98}$) 1987 and 1995

Inventory Source	1987 Emissions gTEQ$_{DF}$/yr	1995 Emissions gTEQ$_{DF}$/yr	Percent Reduction 1987–1995
Municipal Solid Waste Incineration, air	8877.0	1250.0	86
Backyard Refuse Barrel Burning, air	604.0	628.0	−4
Medical Waste Incineration, air	2590.0	488.0	81
Secondary Copper Smelting, air	983.0	271.0	72
Cement Kilns (hazardous waste burning), air	117.8	156.1	−33
Sewage Sludge (land applied), land	76.6	76.6	0
Residential Wood Burning, air	89.6	62.8	30
Coal-Fired Utilities, air	50.8	60.1	−18
Diesel Trucks, air	27.8	33.5	−21
Secondary Aluminum Smelting, air	16.3	29.1	−79
2,4-D, land	33.4	28.9	13
Iron Ore Sintering, air	32.7	28.0	14
Industrial Wood Burning, air	26.4	27.6	−5
Bleached Pulp and Paper Mills, water	356.0	19.5	95
Cement Kiln (nonhazardous waste burning), air	13.7	17.8	−30
Sewage Sludge Incineration, air	6.1	14.8	−143
Ethylene Dichloride/Vinyl Chloride, air	NA	11.2	NA
Oil-Fired Utilities, air	17.8	10.7	40
Crematoria, air	5.5	9.1	−65
Unleaded Gas, air	3.6	5.6	−56
Hazardous Waste Incineration, air	5.0	5.8	−16
Lightweight Ag Kilns (hazardous waste), air	2.4	3.3	−38
Commercially Marketed Sewage Sludge, land	2.6	2.6	0
Kraft Recovery Boilers, air	2.0	2.3	−15
Petroleum Refining Catalyst Regeneration, air	2.24	2.21	1
Leaded Gasoline, air	37.5	2.0	95
Secondary Lead Smelting, air	1.29	1.72	−33
Bleached Pulp and Paper Mill Sludge, land	14.1	1.4	90
Cigarette Smoke, air	1.0	0.8	20
Ethylene Dichloride/Vinyl Chloride, land	NA	0.73	NA
Primary Copper, air	0.5	0.5	0
Ethylene Dichloride/Vinyl Chloride, water	NA	0.43	NA
Boilers and Industrial Furnaces, air	0.78	0.39	50
Tire Combustion, air	0.11	0.11	0
Drum Reclamation, air	0.08	0.08	0
Carbon Reactivation Furnace, air	0.08	0.06	25
Totals	13,998	3,253	77

NA = not available

Source: USEPA 2001a.

DLC compounds are extremely stable in the environment. The only environmentally significant transformation processes for these congeners are believed to be atmospheric photooxidation and photolysis of nonsorbed species in the gaseous phase or at the soil or water-air interface (USEPA 2000a). Consequently, in media where photodegradation is not possible, the ultimate sink for DLC

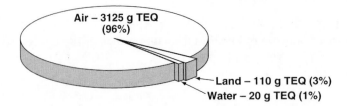

FIGURE 16.3 Environmental releases of dioxins and furans in the U.S. 1995 (TEQDF-WHO98). (From USEPA 2001a.)

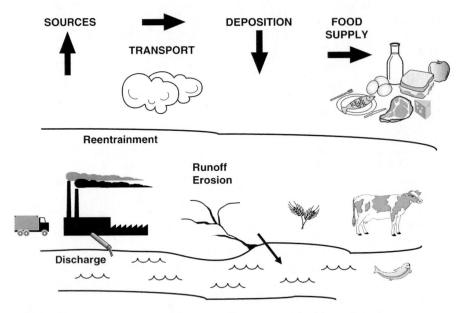

FIGURE 16.4 Pathways for entry of dioxin-like compounds into the terrestrial and aquatic food chains. (From USEPA 2000a.)

TABLE 16.3
Mean Background Concentrations of Dioxin-Like Compounds in the U.S. (TEQ$_{DF}$-WHO$_{98}$)

Environmental Media	Background Concentration (TEQ$_{DF}$-WHO$_{98}$)	Environmental Screening Level
Urban Soil (ppt, pg/g)	9.3	3.9
Rural Soil (ppt, pg/g)	2.7	3.9
Urban Air (pg/m³)	0.12	0.045
Rural Air (pg/m³)	0.013	0.045
Sediment (ppt, pg/g)	5.3	5.8*
Water (ppq, pg/L)	0.00056	0.45

* Upper effects threshold for freshwater sediments.

Source: Buchman 1999; USEPA, 2003, 2004b.

is deep soil or sediments. In 2000, the USEPA reported estimates for background levels of DLC in environmental media based on data from a variety of studies conducted at different locations in North America (USEPA 2000a). These estimates are reported in Table 16.3. For this estimate, the USEPA utilized available data from locations described as "background" and not from areas impacted by local sources of contamination. Due to the limited available data, the USEPA indicated that these data cannot be considered to be definitive national means. Nonetheless, the "environmental media concentrations found in the United States were consistent across the various studies, and were consistent with similar studies in Europe....The limited data on dioxin-like PCBs in environmental media are summarized in the document, but were not deemed adequate for estimating background levels. Because of the limited number of locations examined, however, it is not known if these ranges adequately capture the full national variability, if significant regional variability exists making national means of limited utility, or if elevated levels above this range could still be the result of background contamination processes" (USEPA 2000a). Table 16.3 also presents environmental screening levels for air, water, and soil (USEPA 2004b) that are used by the USEPA to screen contaminated sites for potential risk from direct contact to the individual media and, in the case of sediments, by the National Oceanic and Atmospheric Administration (Buchman 1999) to identify potential impacts of contaminated sites on coastal resources and habitats. As is clear from Table 16.3, the background concentration of DLC in air, soil, and sediments is within an order of magnitude of current environmental screening levels.

However, in 2000, the USEPA found that environmental levels of DLC appear to be declining. The USEPA indicated that "Concentrations of CDD/CDFs in the environment were consistently low for centuries until the 1930s. Then, concentrations rose steadily until about the 1960s, at which point concentrations began to drop. Evidence suggests that the drop in concentrations is continuing to the present" (USEPA 2000a). This finding is based on several lines of evidence, including sampling of sediment cores in North America, and a review of trends in environmental loading. Further monitoring of environmental levels will be needed to reduce the uncertainty in these projections and confirm this trend.

16.5 MEDIA AND FOOD LEVELS

DLCs are transported in the atmosphere and deposited onto vegetation and soils. Since DLCs are quite persistent, they bioconcentrate in both the terrestrial and aquatic food chains. As discussed below, the primary route of population exposure to DLCs in the environment is through the consumption of food with small concentrations of this contaminant.

In the past 20 years, significant effort has gone into determining the concentration of DLC in the food supply of both the United States and Europe. Schecter et al. (1997) reported data on the concentrations of TEQ_{DFP} in common food groups obtained from supermarkets in the United States in 1995. Their findings are shown in Figure 16.5. It is interesting to note that the dioxin-like PCBs contribute a significant portion of the total TEQ in much of the food tested. In 2000, the USEPA summarized the data available for concentration of DLCs in North American food (USEPA 2000a). Using current data on food consumption rates, and data on typical inhalation, water consumption, and soil exposure rates, combined with the data on DLC concentration in food and environmental media, the USEPA calculated DLC intake rates for a typical adult in the United States (USEPA 2000c). This data is presented in Table 16.4. In summary, an individual with a typical diet in the United States would ingest approximately 0.9 pg TEQ_{DFP}-WHO_{98}/kg-d. The primary source (96%) of individual exposure to DLC is through food consumption, with 3% from inhalation and 1% from soil (Figure 16.6). The USEPA's review of available congener-specific data indicated that 65% of the total TEQ was from the dioxin and furans, while 35% was from the dioxin-like PCBs.

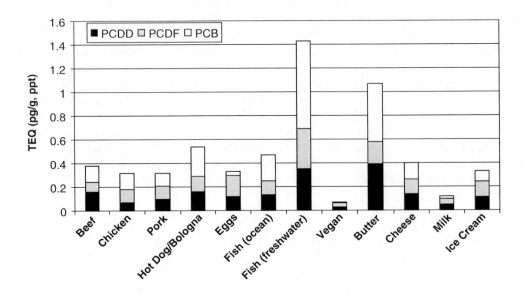

FIGURE 16.5 DLC levels in North American foods in 1995. (From Shecter et al. 1997. With permission.)

TABLE 16.4
Typical Adult Intake of DLC in the U.S. (TEQ$_{DFP}$-WHO$_{98}$)

Source	Concentration (TEQ)	Contact Rate	Intake (pg TEQ/kg-d)	Intake (%)
Beef	0.264 pg/g	0.67 g/kg-d	0.19	20
Fish/shellfish — freshwater	2.2 pg/g	5.9 g/d	0.184	20
Dairy	0.178 pg/g	55 g/d	0.14	15
Other meats	0.221 pg/g	0.35 g/kg-d	0.076	8
Milk	0.027 pg/g	175 g/d	0.067	7
Fish/shellfish — marine	0.51 pg/g	9.6 g/d	0.07	7
Pork	0.292 pg/g	0.22 g/kg-d	0.065	7
Poultry	0.094 pg/g	0.5 g/kg-d	0.047	5
Eggs	0.181 pg/g	0.24 g/kg-d	0.043	5
Inhalation	0.12 pg/m^3	13.3 m^3/d	0.023	2
Vegetable fat	0.093 pg/g	17 g/d	0.023	2
Soil ingestion	11.6 pg/g	50 mg/d	0.0082	0.9
Soil dermal contact	11.6 pg/g	12 g/d	0.002	0.2
Water	0.0005 pg/L	1.4 L/d	1.1E-5	0.001
Total			0.94	

Source: USEPA, 2003.

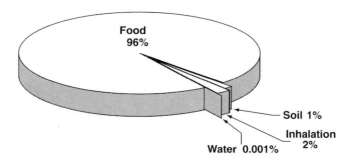

FIGURE 16.6 Sources of average adult DLC exposure. (From USEPA 2000b.)

16.6 SOURCES AND PATHWAYS TO HUMAN EXPOSURE

In order to address how and to what extent people are exposed to dioxins, let us first look at environmental concentrations. As shown in Table 16.3, concentrations in soil, water, and air of dioxins are low compared to current environmental screening levels. Therefore there should be little exposure from direct contact to these media. However, dioxins are resistant to breakdown, are fat soluble or lipophilic, and they have relatively long half-lives in biological organisms (estimated as 7–13 years in humans). These physical properties combine to bioaccumulate dioxins through the food chain.

Tracking dioxins in the environment suggests that dioxins can be transported long distances in the air and are found in most places around the globe (USEPA 2000a). Once released into the air the main routes of deposition are from air-to-plants and air-to-water/sediment (Figure 16.4). Even though these media concentrations are relatively low, dioxins' persistence allows them to be concentrated into fatty tissue and be passed up the food chain. The environmental mass balance can be viewed as different fluxes between compartments made up of various environmental media with partitioning being favored into more lipophilic compartments, such as biota. It is estimated that more than 95% of human exposure is from diet, primarily dairy and meat products (USEPA 2000b) (Figure 16.5 and Figure 16.6). These estimates are for the general population, but there are certain segments of the population that may be of concern for higher exposure. Those that are in close proximity to point sources and also produce their food near those sources could be at increased exposure. Also certain dietary practices could increase exposure, such as consumption of a high percentage of animal fats, for example, by northern Native Peoples. Subsistence fishers are of particular concern due to the varying levels of dioxins in different fish species — and their increased total fish consumption. Nursing infants are also a population of concern due to their higher consumption of fat in breast milk relative to their body weight during a developmentally important growth stage. The American Academy of Pediatrics promotes breastfeeding for the first 6 months and continuation up to 1 year (AAP 1997). The advantages for breastfeeding for both the child and mother far outweigh any deleterious effects that may come from dioxins; however, mothers in highly exposed populations should discuss their situation with their primary healthcare provider. Since the mid 1980s, the WHO has carried out an international monitoring program of DLC in human milk that has demonstrated, on average, a 40% decrease in overall TEQ from 1993–2001 (WHO 1989, 1996; Malisch and van Leeuwen 2003).

Similar to the example of mass balance in environmental media, a chemical partitioning into the body can be thought of as being divided among compartments representative of the major tissues, with an associated dynamic flux between these different tissue compartments. (For an example of the compartmental approach, see Chapter 4 by Ferro and Hildemann.) This is a pharmacodynamic model that is quite useful in assessing the overall equilibrium concentrations of highly lipophilic compounds such as DLC. Compounds with short residence times in the body are easily evaluated by looking at intakes since the time course of the associated effective concentration

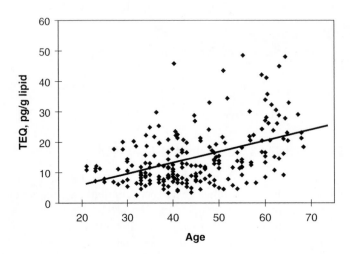

FIGURE 16.7 DLC serum levels with age. (From ATSDR 1999.)

is relatively short. However, with compounds with long half-lives, a rather short exposure can lead to a significant body concentration over a long period. This is the concept of a body burden that is recommended for evaluating dioxin exposure (USEPA 2000c). Since the primary source of dioxin exposure is from food, types and total amount consumed change over a lifetime, and dioxin exposure can be quite variable. But since the half-life is long the body burden should reflect exposure over the long term. Figure 16.7 illustrates the body burden increase seen with age. This change is reflective of the overall temporal trend of dioxins measured in sediment cores. Analysis of sediment cores shows an increase in dioxin levels from the 1920s to a peak extending from the 1950s to the 1970s, followed by a decrease (Cleverly et al. 1996; Czuczwa, Niessen, and Hites, 1985). Thus, serum levels in the older population who were exposed to higher doses earlier in life reflect higher body burdens than younger individuals who were exposed to lower doses. Therefore, over time there does appear to be a decrease in body burden (Figure 16.7).

Food monitoring studies jointly conducted by the USEPA and the U.S. Department of Agriculture (USDA) in 1996 had one curious outcome. As part of an overall assessment of sources of dioxins in the food supply, various foods were analyzed, such as beef, pork, poultry, eggs, and dairy products. Chickens from farms in the southern states were elevated relative to other parts of the country (Ferrario et al. 1997). Further investigation of the source of the dioxins revealed contamination in a minor component in the commercial feed, ball clay, added as an anticaking agent. The source of the clay was a natural, mined product that has no known human source of contamination (Ferrario, Bryne, and Cleverly 2000). Even though the different lines of evidence demonstrate that human-made dioxins may be the largest contributor to dioxin exposure in the environment, natural sources do exist and can be significant contributors under the right circumstances. With the removal of ball clay from the feed preparation, dioxin levels decreased to the national average. Research efforts by the USEPA's Dioxin Exposure Initiative (USEPA 2004a) have begun to evaluate the quality of animal feeds.

The USEPA maintains that the U.S. food supply is safe and national trends in the overall exposure and body burden of dioxins are decreasing with the further control of sources. Nonetheless, the margin of exposure is small (USEPA 2000c). There are additional measures individuals can do to reduce their exposure, such as following the National Dietary Guidelines (USDA 2000), which focus on reducing overall fat intake. Today, adults in the United States consume 34% of their overall caloric intake from fats. Since the major source of dioxins in our diet is associated with animal fats in their various forms, reducing the overall intake would reduce the exposure. Currently the guidelines recommend no more than 30% of daily average caloric intake from fat. While the elimination of all fat would not be a healthful option, simple things like trimming visible fat from meat, not consuming the skin on poultry and fish, selecting preparation methods that allow the fat to drip away, and replacing whole milk with skim milk could easily reduce average fat consumption to these levels.

16.7 SUMMARY AND FUTURE DIRECTIONS

As we have seen in this chapter, DLC are a family of natural and human-made compounds that are potent human toxicants. DLC have a broad spectrum of both cancer and non-cancer adverse biological health effects. DLC are produced primarily through combustion processes and are transported through the atmosphere where a fraction is subject to photocatalytic degradation processes. The remaining DLC are deposited in terrestrial and aquatic environments where they are persistent and tend to bioconcentrate in animals and fish. The primary route of population exposure to DLC is typically through consumption of food with low-level DLC contamination. There is evidence to support the hypothesis that there is a downward trend in dioxin emissions, environmental levels, and human body burden. Continued long-term monitoring is required to confirm this hypothesis.

Over the past several decades, there has been an increased focus on the occurrence and exposure to DLC in the environment both in the United States and among the international community. Current efforts are directed at: (1) reducing DLC emissions through better control of combustion technology where possible; (2) improving the detection limits and lowering the cost of available analytical methods; (3) collecting additional data on DLC in the environment and in foods to increase our confidence in current exposure models; and (4) understanding developing methods to interrupt the cycle of DLC through the food supply. These efforts constitute the current direction in the overall development of a strategy to reduce risk associated with environmental dioxin exposure.

16.8 QUESTIONS FOR REVIEW

1. Discuss the ways that environmental dioxin contamination could be an environmental justice issue. Potential sensitive populations to discuss include subsistence fishers, those who raise domestic chickens, cultures with special dietary practices, etc.
2. What could be done to minimize individual exposure to animal fats?
3. What changes in food production practices could reduce the cycling of dioxin in the national food supply?
4. Given that environmental dioxin levels appear to be on the decline, what additional governmental control strategies are appropriate? Should the government focus on collecting data on dioxin levels in environmental media and the food supply to confirm the effectiveness of current control strategies?
5. The USEPA estimates the upper-bound risk to the general population from current dioxin exposure may exceed a 1 in 1000 increased chance of experiencing cancer. Given that the current national cancer rate is 1 in 3, is dioxin a significant problem?
6. Should backyard barrel burning of household trash be banned?

REFERENCES

American Academy of Pediatrics (AAP) (1997) Breast-Feeding and the Use of Human Milk, *Pediatrics*, **100**(6): 1035–1039.

ATSDR (1999) Health Consultation (Exposure Investigation) Calcasieu Estuary (aka Mossville) Lake Charles, Calcasieu Parish, Louisiana, Cerclis No, LA0002368173, prepared by Exposure Investigation and Consultation Branch, Division of Health Assessment and Consultation, Agency for Toxic Substances and Disease Registry, Atlanta, GA, November 19, 1999.

Birnbaum, L. (1994) Evidence for the Role of the Ah Receptor in Response to Dioxin, in *Receptor-Mediated Biological Process: Implications for Evaluating Carcinogenesis, Progress in Clinical and Biological Research*, Spitzer, H.L., Slaga, T.J., Greenlee, W.F., McClain, M., Eds., Vol. 387, Wiley-Liss, Inc., New York, NY, 139–154.

Buchman, M.F. (1999) NOAA Screening Quick Reference Tables, NOAA HAZMAT Report 99-1, Coastal Protection and Restoration Division, National Oceanic and Atmospheric Administration, Seattle, WA.

Cleverly, D., Monetti, M., Phillips, L., Cramer, P., Heit, M., McCarthy, S., O'Rourke, K., Stanley, J., and Winters, D. (1996) A Time-Trends Study of the Occurrences and Levels of CDDs, CDFs and Dioxin-Like PCBs in Sediment Cores from 11 Geographically Distributed Lakes in the United States, *Organohalogen Compounds*, **28**: 77–82.

Czuczwa, J.M., Niessen, F., and Hites, R.A. (1985) Historical Record of Polychlorinated Dibenzo-p-Dioxins in Swiss Lake Sediments, *Chemosphere*, **14**: 1175–1179.

Ferrario, J., Byrne, C., Lorber, M., Saunders, P., Leese, W., Dupuy, A., Winters, D., Cleverly, D., Schaum, J., Pinsky, P., Deyrup, C., Ellis, R., and Walcott, J. (1997) A Statistical Survey of Dioxin-Like Compounds in United States Poultry Fat, *Organohalogen Compounds*, **32**: 245–251.

Ferrario, J. B., Bryne, C.J., and Cleverly, D.H. (2000) 2,3,7,8-Dibenzo-*p*-Dioxins in Mined Clay Products from the United States: Evidence for Possible Natural Origin, *Environmental Science and Technology*, **34**: 4524–4532.

Griffin, R.D. (1986) A New Theory of Dioxin Formation in Municipal Solid Waste Combustion, *Chemosphere*, **15**: 1987–1990.

Gullett, B.K., Bruce, K.R., and Beach, L.O. (1990) Formation of Chlorinated Organics during Solid Waste Combustion, *Waste Management & Research*, **8**: 203–214.

Gullett, B.K., Bruce, K.R., Beach, L.O., and Drago, A.M. (1992) Mechanistic Steps in the Production of PCDD and PCDF During Waste Combustion, *Chemosphere*, **25**: 1387–1392.

IARC (1997) *Polychlorinated Dibenzo-para-Dioxins and Polychlorinated Dibenzofurans,* Monographs on the Evaluation of Carcinogenic Risks to Humans, 69, International Agency for Research on Cancer, WHO Press, Geneva, Switzerland.

JECFA (2001) *Polychlorinated Dibenzodioxins (PCDDs), Polychlorinated Dibenzofurans (PCDF) and Coplanar Polychlorinated Biphenyls (PCBs)*, Report No. TRS 909-JECFA 57, Joint Food and Agriculture Organization of the United Nations and World Health Organization Expert Committee on Food Additives.

Kilgroe, J.D., Nelson, P.L., Schindler, P.J., and Lanier, W.S. (1990) Combustion Control of Organic Emissions from Municipal Waste Combustors, *Combustion, Science and Technology*, **74**: 223–244.

Lemieux, P.M., Lutes, C.C., Abbott, J.A., and Aldous, K.M. (2000) Emissions of Polychlorinated Dibenzo-p-dioxins and Polychlorinated Dibenzofurans from the Open Burning of Household Waste in Barrels, *Environmental Science and Technology*, **34**(3): 377–384.

Lustenhouwer, J.W.A., Olie, K., and Hutzinger, O. (1980) Chlorinated Dibenzo-p-Dioxins and Related Compounds in Incinerator Effluents, *Chemosphere*, **9**: 501–522.

Malisch, R. and van Leeuwen, F.X.R. (2003) Results of the WHO-Coordinated Exposure Study on the Levels of PCBs, PCDDs and PCDFs in Human Milk, Dioxin 2003, 23rd International Symposium on Halogenated Environmental Organic Pollutants and POPs, August 24–29, Boston, MA, **64**: 140–143.

Raghunathan, K. and Gullett, B.K. (1996) Role of Sulfur in Reducing PCDD and PCDF Formation, *Environmental Science and Technology*, **30**: 1827–1834.

Schecter, A., Cramer, P., Boggess, K., Stanley, J., and Olsen, J.R. (1997), Levels of Dioxins, Dibenzofurans, PCB and DDE Congeners in Pooled Food Samples Collected in 1995 at Supermarkets across the United States, *Chemosphere*, **34**(5–7): 1437–1447.

USDA (2000) Dietary Guidelines for Americans, Home and Garden Bulletin 232.

USEPA (1989) Interim Procedures for Estimating Risks Associated with Exposures to Mixtures of Chlorinated Dibenzo-*p*-Dioxins and -Dibenzofurans (CDDs and CDFs), Report No. EPA/625/3-89.016, U.S. Environmental Protection Agency, Risk Assessment Forum, Washington, DC.

USEPA (1998) 40 CFR Parts 63, 261, and 430, National Emission Standards for Hazardous Air Pollutants for Source Category: Pulp and Paper Production; Effluent Limitations Guidelines, Pretreatment Standards, and New Source Performance Standards: Pulp, Paper, and Paperboard Category, April 15, 1998, *Federal Register,* **63**(72): 18503–18751.

USEPA (2000a) Exposure and Human Health Reassessment of 2,3,7,8-Tetrachlorodibenzo-*p*-Dioxin (TCDD) and Related Compounds, Part I: Estimating Exposure to Dioxin-Like Compounds, Vol 3: Properties, Environmental Levels, and Background Exposures, Draft, Report No. EPA/600/P-00/001Bc, U.S. Environmental Protection Agency, Washington, DC.

USEPA (2000b) Exposure and Human Health Reassessment of 2,3,7,8-Tetrachlorodibenzo-*p*-Dioxin (TCDD) and Related Compounds, Part II: Health Assessment for 2,3,7,8-Tetrachlorodibenzo-*p*-dioxin (TCDD) and Related Compounds, Draft, Report No. EPA/600/P-00/001Be, U.S. Environmental Protection Agency, Washington, DC.

USEPA (2000c) Exposure and Human Health Reassessment of 2,3,7,8-Tetrachlorodibenzo-*p*-Dioxin (TCDD) and Related Compounds, Part III: Integrated Summary and Risk Characterization for 2,3,7,8-Tetrachlorodibenzo-*p*-Dioxin (TCDD) and Related Compounds, Draft, Report No. EPA/600/P-00/001Bg, U.S. Environmental Protection Agency, Washington, DC.

USEPA (2001a) Database of Sources of Environmental Releases of Dioxin-Like Compounds in the United States, Report No. EPA/600/C-01/012, March 2001, U.S. Environmental Protection Agency, Washington, DC.

USEPA (2001b) Risk Burn Guidance for Hazardous Waste Combustion Facilities, Report No. EPA530-R-01-001, July 2001, U.S. Environmental Protection Agency, Washington, DC.

USEPA (2002) 40 CFR Parts 63, 264, 265, 266, 270, and 271, NESHAP: Interim Standards for Hazardous Air Pollutants for Hazardous Waste Combustors (Interim Standards Rule); Final Rule, February 13, 2002, *Federal Register,* **67**(30): 6792–6818.

USEPA (2003) Exposure and Human Health Reassessment of 2,3,7,8-Tetrachlorodibenzo-*p*-Dioxin (TCDD) and Related Compounds, Part III: Integrated Summary and Risk Characterization for 2,3,7,8-Tetrachlorodibenzo-*p*-Dioxin (TCDD) and Related Compounds [NAS Review Draft], Washington, DC, http://www.epa.gov/ncea/pdfs/dioxin/nas-review/. Accessed April 19, 2006.

USEPA (2004a) Dioxin Exposure Initiative Ongoing Projects, U.S. Environmental Protection Agency, Washington, DC, http://cfpub.epa.gov/ncea/cfm/recordisplay.cfm?deid=54925. Accessed 19 April 2006.

USEPA (2004b) Preliminary Remediation Goals, U.S. Environmental Protection Agency, Region 9, San Francisco, CA, http://www.epa.gov/region09/waste/sfund/prg/index.html. Accessed April 19, 2006.

Van den Berg, M., Birnbaum, L., Bosveld, A.T.C., Brunstrom, B., Cook, P., Feeley, M., Giesy, J.P., Hanberg, A., Hasegawa, R., Kennedy, S.W., Kubiak, T., Larsen, J.C., van Leeuwen, F.X.R., Liem, A.K.D., Nolt, C., Peterson, R.E., Poellinger, L., Safe, S., Schrenk, D., Tillitt, D., Tysklind, M., Younes, M., Waern, F., and Zacherewski, T. (1998) Toxic Equivalency Factors (TEFs) for PCBs, PCDDs, PCDFs for Humans and Wildlife, *Environmental Health Perspectives,* **106**(12): 775–792.

Whitlock, J.P., Jr. (1993) Mechanistic Aspects of Dioxin Action, *Chemical Research in Toxicology,* 6(6): 754–763.

WHO (1989) Levels of PCBs, PCDDs and PCDFs in Breast Milk, Environmental Health Series No. 34, World Health Organization, Regional Office for Europe, Copenhagen, Denmark.

WHO (1996) Levels of PCBs, PCDDs and PCDFs in Breast Milk, Second Round of WHO-Coordinated Exposure Study, Environmental Health in Europe No. 3, World Health Organization, Regional Office for Europe, Copenhagen, Denmark.

17 Biomarkers of Exposure

Lance A. Wallace
U.S. Environmental Protection Agency (ret.)

CONTENTS

17.1 Synopsis...395
17.2 Introduction...395
17.3 Volatile Organic Compounds 396
17.4 Semivolatile Organic Compounds..400
17.5 Metals 401
17.6 Summary..402
17.7 Questions for Review 403
References 404

17.1 SYNOPSIS

Biomarkers of exposure are chemicals found in the body providing evidence of environmental exposure to that chemical or to a precursor chemical. Biomarkers have been utilized in occupational studies for more than a century, and in environmental studies more recently. Sometimes a biomarker is the most telling evidence of the effectiveness of an environmental regulation or societal behavioral change, as was shown by the decline of lead in blood following its removal from gasoline and the decline of cotinine (a tobacco derivative) in blood of children coinciding with the decline of cigarette smoking in this country. Some principles governing the use of biomarkers are described —they must be specific to the chemical of interest, quantitatively related to its level in the environmental medium, and be amenable to precise analytical measurements. Other factors affecting selection of biomarkers include the willingness of persons to provide exhaled breath, blood, urine, bone, fat, saliva, or other samples typically employed for biomarker identification. In some cases, biomarkers have provided important information that was not available by normal methods of measurement, as in the discovery that mainstream smoke (the smoke inhaled by the smoker, which is not measurable by air quality monitors) provides the dominant source of benzene and other aromatic compounds for active smokers. Biomarkers for three major pollutant groups are discussed.

17.2 INTRODUCTION

Biomarkers have been an important component of exposure studies for many years. A biomarker is a chemical compound found in or excreted from body tissues. It might be the *parent compound*, the compound as it existed in the environment before being absorbed by the body. Or it may be a *metabolite*, a new compound created in the body from the parent compound. In the latter case, to be useful as a biomarker, the metabolite must be traceable to the parent compound in some quantitative fashion.

We must distinguish here between *biomarkers of exposure* and *biomarkers of effects*. Biomarkers of exposure are used to identify that exposure has occurred, and in some cases can also identify the *route of exposure* (e.g., inhalation or ingestion). We shall see later that biomarkers can sometimes identify unsuspected *sources of exposure* as well. Biomarkers of effect, on the other hand, are (generally) changes within the body in response to an exposure that can be linked to later health effects (for example, an alteration in DNA due to exposure to a mutagen). The dividing line is not always clear —some biomarkers can be considered as both types. This chapter deals mainly with biomarkers of exposure.

What are the general principles that make a biomarker of exposure useful? First, we must distinguish among the possible purposes for exploring biomarkers of exposure. One purpose is to link the biomarker with exposure through a given route. In that case, the biomarker will not provide a quantitative estimate of exposure through a given pathway if it enters the body via two or more routes of exposure. For example, chloroform can be inhaled, ingested in drinking water, or absorbed through the skin while bathing — therefore a measure of chloroform in the body cannot be related unambiguously to any of these routes of exposure in a quantitative manner. Similarly, a biomarker can not provide a quantitative estimate of exposure if it can be created by more than one chemical. For example, benzene can be metabolized into phenol, but exposure to phenol itself may add to the total amount in the body; thus a measure of phenol alone in the body is not sufficient to determine benzene exposure. This is a major problem in using many metabolites as biomarkers; the problem is averted by using the parent compound as the biomarker.

Finally, if it is desired to use the biomarker as a quantitative reflection of exposure, the relation between the level of the biomarker and the level of exposure must be determined. For the case of exhaled breath, it is necessary first to determine the fraction of the parent chemical that is exhaled in breath under equilibrium conditions. This usually requires chamber studies in which human volunteers breathe the chemical at a known and constant concentration for a lengthy period of time.

Almost every type of human tissue has been investigated for containing possible useful biomarkers of exposure. Breath, blood, bone, teeth, fat, hair, saliva, fingernails, urine, and feces are among the most common biomarker tissues. Some "dual" biomarkers, capable of providing information about both mother and child, include the placenta, meconium (early excretion from newborns), and cord blood.

Similarly, almost every major environmental pollutant has been investigated using biomarkers. This includes volatile organic compounds (VOCs) and semivolatile organic compounds (SVOCs). The latter group includes pesticides, polychlorinated biphenyls (PCBs), polyaromatic hydrocarbons (PAHs), furans, dioxins, and cotinine (a metabolite of nicotine). Metals such as lead, cadmium, and arsenic are also studied in blood, bone, hair, teeth, and nails. We shall treat each major group of pollutants in turn.

17.3 VOLATILE ORGANIC COMPOUNDS

VOCs, which include extremely high-volume products such as benzene, toluene, styrene, and tetrachloroethylene, have a very long history of serving as parent compound biomarkers of exposure, beginning with occupational studies in the early twentieth century. In a series of studies carried out for Dow Chemical, Stewart and colleagues immersed the thumbs of workers in the pure liquid VOC to determine dermal absorption characteristics (Stewart and Dodd 1964). Blood levels of the parent compounds were measured as biomarkers of exposure.

Later, exhaled breath was explored as a second medium for measuring VOCs. Breath has several advantages over blood — it is a simpler (gaseous as opposed to a liquid of great complexity) medium to sample and analyze, and more people are willing to have their breath sampled than to have their blood taken. A pioneer in analyzing breath for VOCs was Boguslaw Krotoszynski, who studied the breath of a number of nurses in Chicago for many target VOCs (Krotoszynski, Bruneau, and O'Neill 1979). Other early studies were carried out of benzene in breath in Scandinavia (Berlin et al. 1980).

More recently, environmental studies have identified the most common VOCs and have studied many of them for use as biomarkers. The largest and one of the earliest of these studies was the Total Exposure Assessment Methodology (TEAM) Study, carried out by the U.S. Environmental Protection Agency (USEPA) in 1979–1987. This series of studies targeted 32 VOCs. Personal exposures and outdoor concentrations were measured over two consecutive 12-hour periods (day and night) for about 800 persons in 8 cities (Lioy, Wallace, and Pellizzari 1991; Wallace et al. 1984, 1985, 1988, 1991). Since all participants were chosen according to strict probabilistic design, the 800 participants actually represented about 800,000 residents of the cities. Each person provided a breath sample and the parent compound was quantified. This provided an unequalled opportunity not only of identifying the most common VOCs in the air but also the corresponding levels in our bodies. (See Chapter 7 for a more complete discussion of the TEAM study.)

The TEAM study showed that concentrations of nearly all the VOCs were two to five times higher indoors than outdoors (Wallace 1993). This meant that the main sources of exposure were indoors, and included consumer products and building materials (Wallace et al. 1987a). The breath samples confirmed these findings, since the breath concentrations often exceeded the outdoor concentrations, an impossibility if outdoor levels were the sole source of exposure. The higher indoor concentrations also translated into higher risks for the five human or suspected human carcinogens included in the TEAM studies (benzene, chloroform, *para*-dichlorobenzene, trichloroethylene, tetrachloroethylene) (Wallace 1991).

Although a relationship could be seen between the previous 12-hour integrated exposure and the resulting breath concentration, since the temporal profile of the exposure was unknown, the TEAM study results could not be used to establish a quantitative relationship between breath levels and exposure. Therefore, a series of chamber studies was initiated to measure breath levels under controlled (steady) exposures (Gordon et al 1988; Pellizzari, Wallace, and Gordon 1992; Wallace, Pellizzari, and Gordon 1993; Wallace et al. 1997). These studies were able to establish the crucial parameters that allow breath concentrations to be used to calculate previous exposures. These parameters include the fraction f of the VOC that is exhaled under steady-state conditions and the residence times τ_i of the VOC in the various body compartments, where i indexes the compartment type. For example, tetrachloroethylene, which is hardly metabolized at all, is breathed out at about the same concentration that it is breathed in (f is close to 1) whereas xylenes are metabolized freely (f is about 0.1). The compartments of interest are the blood itself, the organs that receive large amounts of blood, the muscles, and the fat. The residence times in these four compartments were very similar for most of the VOCs tested, very roughly about 3 minutes for the blood, 30 minutes for the organs, 3 hours for the muscles, and 3 days for the fat (Wallace, Pellizzari, and Gordon 1993; Wallace et al. 1997). These estimates, although rough, were sufficiently useful to allow for designing future chamber studies in the most efficient manner.

The TEAM Study findings using breath as the biomarker were confirmed a decade later by studies using blood as the biomarker. About 800 persons (the same number as in the TEAM studies) selected by a nationwide probabilistic design, provided blood samples that were analyzed for a similar set of target VOCs (Ashley et al. 1996).

This points up a tremendous advantage of biomarkers: simple surveys of biomarkers in the human body such as the TEAM and Centers for Disease Control (CDC) studies can help narrow down the list of all chemicals in production to those that are truly important in exposure. For example, there are thousands of VOCs. However, the TEAM and CDC studies both identified the same 12 or so VOCs that were prevalent in the exhaled breath (and therefore blood) of the participants. These "dirty dozen" are listed in Table 17.1. (Left off the list are some *endogenous* VOCs [created in the body] such as acetone and isoprene.)

Including a biomarker in studies of exposure has several advantages. For one, it provides the ability to recognize unmeasured pathways. For example, in the TEAM studies, smokers were found to exhale 10 times as much benzene as nonsmokers (Figure 17.1). (The personal monitors showed a factor of only 1.5 between smokers and nonsmokers.) This allowed the identification of smoking

TABLE 17.1
The Most Prevalent VOCs in Our Breath and
Blood and Their Major Sources

Compound	Major Sources
Limonene	Scented products
α-Pinene	Scented products
para-Dichlorobenzene	Air fresheners, moth crystals
Chloroform	Chlorinated water
Carbon tetrachloride	Global background
Tetrachloroethylene	Dry-cleaned clothes
Benzene	Cigarettes, gasoline
Toluene	Paints, adhesives, cigarettes, gasoline
Xylenes	Paints, adhesives, cigarettes, gasoline
Ethylbenzene	Paints, adhesives, cigarettes, gasoline
Styrene	Paints, adhesives, cigarettes
Methylene chloride	Paint remover

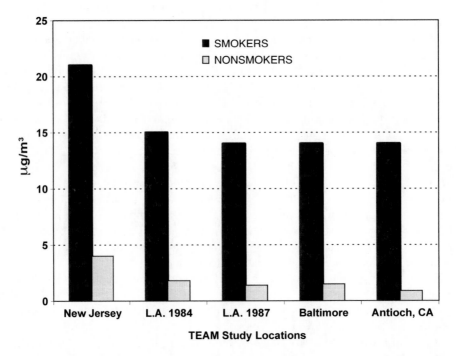

FIGURE 17.1 Benzene in exhaled breath of smokers and nonsmokers in several TEAM study locations. About 700 persons are represented.

as the single largest source of benzene exposure for smokers (Wallace et al. 1987b; Wallace 1990). Without the breath measurements, this finding would not have been made.

Breath measurement methods continue to be improved. Although earlier methods involved breathing into a 20-L Tedlar™ bag, a later improvement used a 1-m tube into which persons could breath normally (Pellizzari, Zweidinger, and Sheldon 1993; Raymer et al. 1990). The first part of the breath (which includes the unwanted "dead space," air that did not undergo exchange with the blood at the alveoli) is wasted into the air through the end of the tube, but the last (desired) alveolar part of their breath is sampled continuously by a *critical orifice* into either a Tenax cartridge or an evacuated canister (Thomas, Pellizzari, and Cooper 1991). Later a small 1-L canister was used to sample directly from the person's lungs using a disposable plastic tube (Pleil and Lindstrom 1995). The person wastes the first part of his or her breath into the air and then clamps the lips around the tube and opens the valve on the evacuated canister, which pulls the last liter of alveolar air from the lungs. A continuous breath sampler was developed by Battelle under contract to the USEPA (Gordon, Kenny, and Kelly 1992). This sampler is particularly well suited to study dermal absorption, since it is possible to isolate the person from inhalation exposure by having him or her breathe pure air while being exposed to VOCs in liquid next to the skin. The observed sudden increase in breath concentrations after the VOC diffuses into the blood is a direct measure of the time taken to cross the stratum corneum. This was the method used to study dermal absorption of both chloroform and methyl tert-butyl ether (MTBE) as described in Chapter 12. The continuous breath sampler also allows a large number of measurements to be made during the decay period following exposure, which is helpful in determining residence times in tissues. Because of the expense of analyzing Tenax cartridges or canisters, they are not cost-effective when many measurements are required over a short time.

Biomarkers of chloroform exposure are largely limited to breath and blood, since it is such a volatile chemical that it does not stay long in the body. Chloroform is unusual among the VOCs in that three routes of exposure are routinely available: inhalation, ingestion of drinking water, and dermal absorption (Aggazzotti et al. 1993; Lindstrom, Pleil, and Berkoff 1997; Wallace 1997). Therefore neither the breath nor blood biomarkers are particularly useful as quantitative estimates of exposure through any single route. However, work has been done to relate breath levels to exposure through inhalation separately and dermal absorption separately. In the latter series of studies, subjects sat in a bathtub or stood in a shower and breathed pure air through a delivery system, while exhaling into a continuous breath analyzer (Gordon et al. 1992, 1998; Corley, Gordon, and Wallace 2000). This approach was useful in identifying for the first time the exact time (6–9 minutes) required for chloroform to diffuse through the stratum corneum (the skin's protective layer) and reach the bloodstream. Without using breath as the biomarker, it would have been very difficult to determine this time. Establishing the diffusion time is important, because exposures for less than this time would not yield much of an uptake. These studies also established the extraordinary effect of water temperature in affecting dermal exposure, with hot baths at 40°C providing 30 times the chloroform uptake of lukewarm baths at 30°C (Gordon et al. 1998).

Persons living near dry cleaning shops can be exposed to emissions of tetrachloroethylene (Verberk and Scheffers 1980). In "vertical" cities such as Manhattan, many families live directly above dry cleaning shops on the ground floor. A series of studies carried out by Judith Schreiber of the New York State Department of Health showed large indoor air exposures to residents above dry cleaning shops, and measurements in mothers' milk (which contains about 3% fat) confirmed that the chemical was entering their bodies in substantial amounts (Schreiber 1992; Schreiber et al. 2002). This disturbing finding led to more stringent laws on tetrachloroethylene emissions from dry cleaner shops located in residential buildings in New York state.

A VOC of recent interest is methyl *tert*-butyl ether (MTBE). MTBE found wide use in the 1990s as a highly oxygenated additive to gasoline, helping to reduce pollutants in the exhaust by promoting more complete combustion. However, MTBE was later found to be one of the most prevalent VOCs in groundwater. Also, some persons reported minor symptoms from the introduction of MTBE into their local supplies of gasoline. Ultimately, the EPA banned MTBE from use. However, its continued existence in groundwater raises the question of whether it might enter our bodies through dermal absorption while bathing or showering. Therefore a study was undertaken of MTBE absorption during bathing using the same approach as the previous study using chloroform (Gordon 2003b). Results indicated that MTBE diffuses much more slowly than chloroform through the stratum corneum and were therefore reassuring to regulators and the public. Other studies have used breath, blood, or both as biomarkers to determine exposure to MTBE through inhalation

(Buckley et al. 1997; Cain et al. 1996; Gordon 2003a; Johanson, Nihlen, and Lof 1995; Lindstrom and Pleil 1996) or through inhalation, ingestion, and dermal absorption (Prah et al. 2003).

Another VOC of considerable recent interest, due to the discovery of strong carcinogenic potential, is 1,3-butadiene, a product of combustion and found in both auto exhaust and cigarette smoke (Löfroth et al. 1989). Measurements of 1,3-butadiene in air have been few because of its reactivity with standard sorbents. The major urinary metabolite is dihydroxy-butyl-mercapturic acid, and that has been a target of several biomarker studies (Urban et al. 2003). Henderson et al. (2001) hypothesized that the carcinogenic effect may be associated with diepoxide in blood, which is high in mice (a species strongly affected by 1,3-butadiene) but low in rats (not strongly affected).

Another VOC of interest is formaldehyde. Formaldehyde is ubiquitous because of its use in pressed wood and particleboard, major components of cabinets, stairs, etc. in homes. Recently it has been classified as a human carcinogen (IARC 2005). Because of its high reactivity, it is not suitable for breath analysis; however, one metabolite in urine has recently been used to estimate formaldehyde uptake (Bono et al. 2005).

Recently, it has been discovered that certain heating processes in food production produce high levels of acrylamide in french fries, potato chips, and crisp breads (Swedish National Food Agency 2002). Since this chemical has carcinogenic and mutagenic properties, considerable research has ensued, some of it employing biomarkers such as hemoglobin adducts in blood (Dybing et al. 2005).

17.4 SEMIVOLATILE ORGANIC COMPOUNDS (SVOCS)

Some pollutants are *lipid soluble* and end up in fat. For example, many of the chlorinated pesticides, such as DDT, chlordane, and aldrin, have been found in fat. *Polychlorinated biphenyl compounds* (PCBs), a class of over 200 chemicals widely used in electrical equipment such as transformers and fluorescent light fixtures, before being banned, may be found in fat as well. *Furans* and *dioxins*, other classes with hundreds of chemicals, also migrate to fat tissues (Safe 1990). Accumulation in fat continues over a lifetime.

Although it is difficult to collect adipose tissue for sampling, fat can be easily collected from human breast milk. A study in 1981–1982 of DDT, PCBs, and a widely used pesticide, β-hexachlorocyclohexane, occurring in mothers' milk in 10 countries was carried out by the World Health Organization (WHO) in 1981–1982 (WHO 1983). This study using biomarkers was immediately successful in identifying approximate levels of exposure in these countries, whereas previous environmental monitoring was somewhat scattered and unable to provide a clear picture of human exposure.

Other studies of nursing mothers have identified increased exposure to PCBs and DDT (Rogan et al. 1986). Since these amounts have accumulated over the mother's lifetime, the first born that is breastfed receives higher amounts of these chemicals than later-born children. (Nonetheless, the documented advantages of breastfeeding are considered to outweigh the small increased risk from this exposure.)

Blood levels of these pesticides and PCBs are also indicative of exposure, although for a shorter time period. A recent analytical technique developed at the EPA allows a broad screening method for four organophosphate pesticides including diazinon and chlorpyrifos, 16 organochlorines including aldrin and dieldrin, DDT and its metabolite DDE, lindane and pentachlorophenol, five pyrethroid pesticides, and nine PCBs in blood (Liu and Pleil 2001).

An example of the use of a metabolite as a biomarker is cotinine, the main metabolite of nicotine. Cotinine levels in blood, urine, or saliva may be used to estimate exposure to environmental tobacco smoke (ETS). Sensitive cotinine measurements have been used to show that a very large majority of children are exposed to some tobacco smoke (CDC 2004, 2005). As smoking levels have declined, cotinine levels have declined as well (Figure 17.2).

A biomarker for an important class of pesticides, the pyrethrins (used mainly indoors as insecticides), is chrysanthemumdicarboxylic acid in urine (Elflein et al. 2003). However, for most

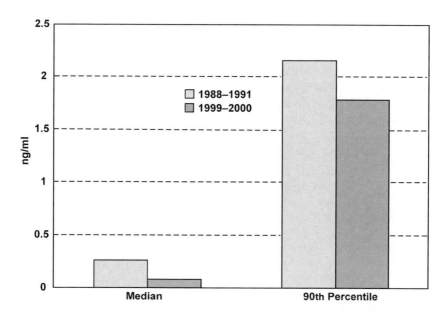

FIGURE 17.2 Decline in serum cotinine levels in children (3–11 years). (Data from CDC 2004.)

pesticides, biomonitoring studies have focused on cytogenetic (genetic changes within cells) end-points, namely, chromosomal aberrations (CA), micronuclei (MN) frequency and sister-chromatid exchanges (SCE) (Bolognesi 2003).

A "product" related to birth is meconium, the initial fecal material of the newborn. Meconium was investigated as a possible matrix for biomarkers of exposure to pesticides (Hong, Gunter, and Randow 2002). Three of sixty newborns in the former East Germany had measurable levels of DDE (the main metabolite of DDT) in their meconium, although the pesticide had been banned 25 years before. The authors concluded that this was evidence of the long-term global health problem associated with DDT.

Another study of DDT exposure used stored serum samples from pregnant mothers who had their children between 1959 and 1966 (Longnecker et al. 2001). Out of 2,380 children, 361 were born preterm, and 221 were small for gestational age. The median maternal DDE concentration was 25 mg/L, several times higher than the current U.S. concentration. The odds ratios of these outcomes increased with increasing concentration of serum DDE. Their findings strongly suggest that DDT use increases preterm births, which is a major contributor to infant mortality.

A biomarker for exposure to polyaromatic hydrocarbons (PAH) is 1-hydroxypyrene, a major metabolite found in urine (Jongeneelen et al. 1988; Kim et al. 1998). However, as with many metabolites, variations in enzyme activity across people and nonspecificity (one metabolite may be produced by many precursors) limit the ability to provide quantitative estimates of previous exposure.

17.5 METALS

Human tissues such as hair and fingernails are often exploited for biomarkers for metals such as arsenic (Mandal, Ogra, and Suzuki 2003). Although metals may not have a long half-life in blood, they last much longer in these tissues due to forming complexes with the fibrous proteins (keratins) found in abundance in nails and hair. The nails and hair give a picture of longer-term exposure (weeks to months) compared with measurements in blood or urine (days to weeks). A problem with using hair is that airborne deposits of the metal can coat the outside, leading to interferences with determining the concentration in the hair from previous internal dose. Another difficulty is that often the oxygenated form of the metal determines its toxicity. For example, hexavalent chromium (chromium VI) is much more toxic than its more common trivalent form (chromium III). Therefore chemical extraction and analysis must be very sophisticated in order to separate different valence states of the same metal.

In twentieth-century environmental studies, probably the most widely known biomarker is lead in blood. For many years, lead in blood has been used for determining the success of environmental actions taken to reduce lead in the environment, including, most notably, its removal from gasoline (an almost complete reduction of 99.8% between 1976 and 1990), but also from use as solder in food and soft-drink cans (a drop from 47% of all cans using lead solder in 1980 to none in 1991) and in interior paints (Pirkle et al. 1998). Blood lead levels are used to characterize both exposures and as input to health effect studies. (A main effect of lead appears to be reducing intelligence as measured by IQ tests.)

Methylmercury is another pollutant that *bioaccumulates*, particularly in "top predator" fish such as tuna and swordfish. The major source of mercury in the environment is coal-burning power plants. Mercury vapor falls out onto soil and into streams and rivers, ultimately finding a final sink in the ocean. Bacteria transform some of it into methylmercury, a far more toxic form. Pregnant women can pass the methylmercury through the placenta to their fetuses, where it affects brain development (Davidson et al. 1998). Since the infant's brain cannot be studied directly, a "stand-in" biomarker, methylmercury levels in the mother's hair, is used to estimate exposure and dose (Cernichiari et al. 1995). The most recent large-scale study of methylmercury in U.S. adults shows that all women of childbearing age (16–49) were below the level (56 mg/L) of mercury in blood that has been associated with neurodevelopmental effects; however, 5.7% of the women were within a factor of 10 of this level, putting them above the recommended margin of safety (CDC 2005).

Proteomics, the study of all proteins produced by a particular genome, has advanced in recent years. Four proteins produced in blood by exposure to arsenic and cadmium were recently studied in two Chinese populations with high exposure to one or the other of these toxic metals, and the researchers succeeded in identifying an interaction causing increased health effects over the amount produced by exposure to each metal alone (Nordberg et al. 2005). Protein adducts in blood can also be produced by other pollutants such as benzene (Rappaport et al. 2005).

17.6 SUMMARY

The most common biomarkers and biomarker tissues for the major pollutant groups are shown in Table 17.2. Three important recent nationwide studies have been carried out that employ biomonitoring methods. NHANES, the National Health and Nutrition Examination Survey, measures a

TABLE 17.2
Pollutant Groups and Their Commonly Deployed Biomarker Tissues

Pollutant Group	Parent Compound	Metabolite
VOCs	Breath, blood	Blood, urine
Tetrachoroethylene	Breath, blood, mothers' milk	None
SVOCs (pesticides, PCBs, PAHs, dioxins/furans)	Blood, fat, mothers' milk	Blood, urine
Metals	Blood, bone, hair, cord blood, placenta, feces	
Carbon monoxide	Breath, blood	Blood (carboxyhemoglobin)
Environmental tobacco smoke (ETS)	Breath, blood (2,5-dimethylfuran)	Saliva, blood (cotinine)

large number of blood and tissue biomarkers in a sample of some 16,000 U.S. residents. Lead in blood and cotinine in saliva are two of the biomarkers that were measured in the 1996–2000 NHANES study. The most recent NHANES study included 148 chemicals, of which 38 were measured for the first time in U.S. residents (CDC 2005). Among the important new results were the findings that children's exposure to cigarette smoke, as measured by cotinine levels in saliva, have been reduced by 68%; blood lead levels continue to decline for all age groups; and mercury levels are low in women of childbearing age. NHEXAS, the National Human Exposure Assessment Survey, a follow-up to the TEAM studies, took place in the USEPA's region V (including Michigan, Minnesota, and Illinois, among other states), Arizona, and Baltimore. NHEXAS included measurements of VOCs, metals, and pesticides in air, water, soil, and blood and urine (Clayton et al. 1999). Finally, a nationwide study in Germany has included measurements of metals, VOCs, pentachlorophenol, and other compounds in air, water, soil, house dust, and blood and urine (Seifert et al. 2000). These large-scale and expensive studies are an indication that biomonitoring is now a mainstream technique in environmental and health studies.

17.7 QUESTIONS FOR REVIEW

1. Why study chemicals in the body to determine exposure if the exposure can be measured directly by monitoring air or drinking water?
2. How does measuring biomarkers complement environmental measurements in exposure studies?
3. State the difference between biomarkers of exposure and biomarkers of effect. Can you think of some biomarkers that might be both?
4. What biomarker measurement was used to determine that benzene was a main constituent of tobacco smoke exposure?
5. What biomarker measurement was used to confirm that removing lead from gasoline led to reduced exposure to lead?
6. What biomarker was used to confirm that reduction in cigarette smoking led to less exposure for children?
7. What advantages does breath have compared to blood as a choice for measuring biomarkers of VOC exposure?
8. Name 6 of the 12 most common VOCs in our breath and blood. What are the main sources of exposure for each?
9. For most VOCs, breath and blood are preferred media for biomarkers. However, for tetrachloroethylene, mothers' milk was the most important way to measure exposure to dry cleaning emissions for persons living above dry cleaning shops. What feature of tetrachloroethylene made it more suitable than other VOCs for utilizing mothers' milk as a medium? [Answer: lack of metabolism so that the parent compound can collect in fat. 2nd Answer (advanced knowledge required): lower volatility compared to other VOCs means more of it can reach the fat before being exhaled.]
10. You are designing a chamber study to determine the residence time of two VOCs (tetrachloroethylene and *ortho*-xylene) in blood. Your limit of detection (LOD) in breath is 1 $\mu g/m^3$ for each VOC. What concentration of each will you need to supply to the chamber to be sure of exceeding the LOD? Hint: consider *f* for each chemical. [Answer: Since tetrachloroethylene is hardly metabolized ($f = 1$), breath levels after sufficiently long exposure will be close to the exposure level, so slightly more than 1 $\mu g/m^3$ would be sufficient to get the breath level above the LOD; but xylenes are efficiently metabolized ($f = 0.1$) so at least 10 $\mu g/m^3$ would be required to be able to measure the breath level.]

REFERENCES

Aggazzotti, G., Fantuzzi, G., Righi, E., Tartoni, P., Cassinadri, T., and Predieri, G. (1993) Chloroform in Alveolar Air of Individuals Attending Indoor Swimming Pools, *Archives of Environmental Health*, **48**: 250–254.

Ashley, D.L., Bonin, M.A., Cardinali, F.L., McCraw, J.M., and Wooten, J.V. (1996) Measurement of Volatile Organic Compounds in Human Blood, *Environmental Health Perspectives*, **104**(supp. 5): 871–877.

Berlin, M., Gage, J.C., Gulberg, B., Holm, S., Knutsson, P. and Tunek, A. (1980) Breath Concentration as an Index of the Health Risk from Benzene, *Scandinavian Journal of Work Environment and Health*, **6**: 104.

Bolognesi, C. (2003) Genotoxicity of Pesticides: A Review of Human Biomonitoring Studies, *Mutation Research*, **543**: 251–272.

Bono, R., Vincenti, M., Schiliro, T., Scursatone, E., Pignata, C., and Gilli, G. (2005) N-Methylenvaline in a Group of Subjects Occupationally Exposed to Formaldehyde, in *Toxicology Letters*, http://dx.doi.org/10.1016/j.toxlet.2005.07.016 (accessed September 8, 2005).

Buckley, T.J., Prah, J.D., Ashley, D., Wallace, L.A., and Zweidinger, R.A. (1997) Body Burden Measurements and Models to Assess Inhalation Exposure to Methyl Tertiary Butyl Ether (MTBE), *Journal of the Air & Waste Management Association*, **47**(7): 739–752.

Cain, W., Leaderer, B., Ginsberg, G., Andrews, L., Cometto-Muniz, J., Gent, J., Buck, M., Berglund, L., Mohseinin, V., Monahan, E., and Kjaergaard, S. (1996) Acute Exposure to Low-Level Methyl Tertiary Butyl Ether (MTBE): Human Reactions and Pharmacokinetic Response, *Inhalation Toxicology*, **8**: 21–48.

CDC (2004) *Second National Report on Human Exposure to Environmental Chemicals*, NCEH Pub # 02-0716, Centers for Disease Control, National Center for Environmental Health, Division of Laboratory Sciences, Atlanta, GA.

CDC (2005) *Third National Report on Human Exposure to Environmental Chemicals*, Centers for Disease Control, National Center for Environmental Health, Division of Laboratory Sciences, Atlanta, GA, http://www.cdc.gov/exposurereport/3rd/ (accessed April 23, 2006).

Cernichiari, E., Toribara, T.Y., Liang, L., Marsh, D.O., Berlin, M., Myers, G.J., Cox, C., Shamlaye, C.F., Choisy, O., Davidson, P.W., and Clarkson, T.W. (1995) The Biological Monitoring of Methylmercury in the Seychelles Study, *Neurotoxicology*, **16**(4): 613–628.

Clayton, C.A., Pellizzari, E.D., Whitmore, R.W., Perritt, R.L., and Quackenboss, J.J. (1999) National Human Exposure Assessment Survey (NHEXAS): Distributions and Associations of Lead, Arsenic, and Volatile Organic Compounds in EPA Region 5, *Journal of Exposure Analysis and Environmental Epidemiology*, **9**(5): 381–392.

Corley, R.A., Gordon, S.M., and Wallace, L.A. (2000) Physiologically Based Pharmacokinetic Modeling of the Temperature-Dependent Dermal Absorption of Chloroform by Humans Following Bath Water Exposures, *Toxicological Sciences*, **53**: 13–23.

Davidson, P.W., Myers, G.J., Cox, C., Axtell, C., Shamlaye, C., Sloane-Reeves, J., Cernichiari, E., Needham, L., Choi, A., Wang, Y., Berlin, M., and Clarkson, T.W. (1998) Effects of Prenatal and Postnatal Methylmercury Exposure from Fish Consumption at 66 Months of Age: The Seychelles Child Development Study, *Journal of the American Medical Association*, **280**(8): 701–707.

Dybing, E., Farmer, P.B., Andersen, M., Fennell, T.R., Lalljie, S.P.D., Müller, D.J.G., Olin, S., Petersen, B.J., Schlatter, J., Scholz, G., Scimeca, J.A., Slimani, N., Törnqvist, M., Tuijtelaars, S., and Verger, P. (2005) Human Exposure and Internal Dose Assessments of Acrylamide in Food, *Food and Chemical Toxicology*, **43**: 365–410.

Elflein, L., Berger-Preiss, E., Preiss, A., Elend, M., Levsen, K., and Wunsch, G. (2003) Human Biomonitoring of Pyrethrum and Pyrethroid Insecticides Used Indoors: Determination of the Metabolites *E-Cis/Trans-*Chrysanthemumdicarboxylic Acid in Human Urine by Gas Chromatography-Mass Spectrometry with Negative Chemical Ionization, *Journal of Chromatography B*, **795**: 195–207.

Gordon, S. (2003a) *Inhalation Exposure to Methyl* tert-*Butyl Ether (MTBE) Using Continuous Breath Analysis*, Final Report, EPA Contract # 68-D-99-011, U.S. Environmental Protection Agency, National Exposure Research Laboratory, Research Triangle Park, NC.

Gordon, S. (2003b) *Human Exposure to Methyl* tert-*Butyl Ether (MTBE) while Bathing with Contaminated Water*, Final Report, EPA Contract # 68-D-99-011, U.S. Environmental Protection Agency, National Exposure Research Laboratory, Research Triangle Park, NC.

Gordon, S., Wallace, L.A., Pellizzari, E.D., and O'Neill, H. (1988) Breath Measurements in a Clean-Air Chamber to Determine 'Wash-Out' Times for Volatile Organic Compounds at Environmental Concentrations, *Atmospheric Environment*, **22**: 2165–2170.

Gordon, S.M., Kenny, D.V., and Kelly, T.J. (1992) Continuous Real-Time Breath Analysis for the Measurement of Half-Lives of Expired Volatile Organic Compounds, *Journal of Exposure Analysis and Environmental Epidemiology*, Supplement 1: 41–54.

Gordon, S.M., Wallace, L.A., Callahan, P.J., Kenny, D.V., and Brinkman, M.C. (1998) Effect of Water Temperature on Dermal Exposure to Chloroform, *Environmental Health Perspectives*, **106**(6): 337–345.

Henderson, R.F., Bechtold, W.E., Thornton-Manning, J., and Dahl, A.R. (2001) Urinary Butadiene Diepoxide: A Potential Biomarker of Blood Diepoxide, *Toxicology*, **160**: 81–86.

Hong, Z., Gunter, M., and Randow, F.F.E. (2002) Meconium: A Matrix Reflecting Potential Fetal Exposure to Organochlorine Pesticides and Its Metabolites, *Ecotoxicology and Environmental Safety*, **51**: 60–64.

IARC (2005) *Monographs on the Evaluation of Carcinogenic Risks to Humans, Vol. 88: Formaldehyde, 2-butoxyethanol, and 1-tert-butoxy-2-propanol*, International Agency for Research on Cancer, Lyon, France.

Johanson, G., Nihlen, A., and Lof, A. (1995) Toxicokinetics and Acute Effects of MTBE and ETBE in Male Volunteers, *Toxicology Letters*, **82/83**: 713–718.

Jongeneelen, F.J., Anzion, R.B., Scheepers, P.T., Bos, R.P., Henderson, P.T., Nijenhuis, E.H., Veenstra, S.J., Brouns, R.M., and Winkes, A. (1988) 1-Hydroxypyrene in Urine as a Biological Indicator of Exposure to Polycyclic Aromatic Hydrocarbons in Several Work Environments, *Annals of Occupational Hygiene*, **32**: 35–43.

Kim, J.H., Stansbury, K.H., Walker, N.J., Trush, M.A., Strickland, P.T., and Sutter, T.R. (1998) Metabolism of Benzo[A]Pyrene and Benzo[A]Pyrene-7,8-Diol by Human Cytochrome P450 1B1, *Carcinogenesis*, **19**: 1847–1853.

Krotoszynski, B.K., Bruneau, G.M., and O'Neill, H.J. (1979) Measurement of Chemical Inhalation Exposure in Urban Populations in the Presence of Endogenous Effluents, *Journal of Analytical Toxicology*, **3**: 225–234.

Lindstrom, A.B. and Pleil, J.D. (1996) Alveolar Breath Sampling and Analysis to Assess Exposures to Methyl Tertiary Butyl Ether (MTBE) during Motor Vehicle Refueling, *Journal of the Air & Waste Management Association*, **46**: 676–682.

Lindstrom, A.B., Pleil, J.D., and Berkoff, D.C. (1997) Alveolar Breath Sampling and Analysis to Assess Trihalomethane Exposures during Competitive Swimming Training, *Environmental Health Perspectives*, **105**: 636–642.

Lioy, P.J., Wallace, L.A., and Pellizzari, E.D. (1991) Indoor/Outdoor and Personal Monitor and Breath Analysis Relationships for Selected Volatile Organic Compounds Measured at Three Homes During New Jersey TEAM — 1987, *Journal of Exposure Analysis and Environmental Epidemiology*, **1**(1): 45–61.

Liu, S. and Pleil, J. (2001) Human Blood and Environmental Media Screening Method for Pesticides and Polychlorinated Biphenyl Compounds Using Liquid Extraction and Gas Chromatography — Mass Spectrometry Analysis, *Journal of Chromatography B*, **769**: 155–167.

Löfroth, G., Burton, R.L., Forehand, L., Hammond, S.K., Seila, R., Zweidinger, R.B., and Lewtas, J. (1989) Characterization of Environmental Tobacco Smoke, *Environmental Science and Technology*, **23**: 610–614.

Longnecker, M.P., Klebano, M.A., Zhou, H., and Brock, J.W. (2001) Association between Maternal Serum Concentration of the DDT Metabolite DDE and Preterm and Small-for-Gestational-Age Babies at Birth, *Lancet*, **358**: 110–114.

Mandal, B.K., Ogra, Y., and Suzuki, K. (2003) Speciation of Arsenic in Human Nail and Hair from Arsenic-Affected Area by HPLC-Inductively Coupled Argon Plasma Mass Spectrometry, *Toxicology and Applied Pharmacology*, **189**: 73–83.

Nordberg, G.F., Jin, T., Hong, F., Zhang, A., Buchet, J.P., and Bernard, A. (2005) Biomarkers of Cadmium and Arsenic Interactions, *Toxicology and Applied Pharmacology*, **206**: 191–197.

Pellizzari, E.D., Wallace, L.A., and Gordon, S.M. (1992) Elimination Kinetics of Volatile Organics in Humans Using Breath Measurements, *Journal of Exposure Analysis and Environmental Epidemiology*, **2**(3): 341–356.

Pellizzari, E.D., Zweidinger, R.A., and Sheldon, L.S. (1993) Breath Sampling, in *Environmental Carcinogens: Methods of Analysis and Exposure Measurement — Indoor Air*, Vol. 12, Seifert, B., Van de Wiel, H.J., Dodet, B., and O'Neill, I.K., Eds., IARC Publication No. 109, World Health Organization, Lyon, France, 228–236.

Pirkle, J.L., Kaufmann, R.B., Brody, D.J., Hickman, T., Gunter, E.W., and Paschal, D.C. (1998) Exposure of the U.S. Population to Lead, 1991–1994, *Environmental Health Perspectives*, **106** (11): 745–750.

Pleil, J.D. and Lindstrom, A.B. (1995) Collection of a Single Alveolar Exhaled Breath for Volatile Organic Compounds Analysis, *American Journal of Industrial Medicine*, **27**: 109–121.

Prah, J., Ashley, D., Blount, B., Case, M., Leavens, T., Pleil, J., and Cardinali, F. (2003) Dermal, Oral, and Inhalation Pharmacokinetics of Methyl Tertiary Butyl Ether (MTBE) in Human Volunteers, *Toxicological Sciences*, **77**(2): 195–205.

Rappaport, S.M., Waidyanatha, S., Yeowell-O'Connell, K., Rothman, N., Smith, M.T., Zhang, L., Qu, Q., Shore, R., Li, G., and Yin, S. (2005) Protein Adducts as Biomarkers of Human Benzene Metabolism, *Chemico-Biological Interactions*, **153–154**: 103–109.

Raymer, J.H., Thomas, K.W., Cooper, S.D., Whitaker, D.A., and Pellizzari, E.D. (1990) A Device for Sampling Human Alveolar Breath for the Measurement of Expired Volatile Organic Compounds, *Journal of Analytical Toxicology*, **14**: 337–344.

Rogan, W.J., Gladen, B.C., McKinney, J.D., Carreras, N., Hardy, P., Thullen, J., Tingelstad, M., and Tully, M. (1986) Poly-Chlorinated Biphenyls (PCBs) and Dichlorodiphenyl Dichloroethane (DDE) in Human Milk, *American Journal of Public Health*, **76**: 172–177.

Safe, S. (1990) Determination of 2,3,7,8-TCDD Toxic Equivalent Factors (Teqs), Support for the Use of the in Vitro ALIII Induction Assay, *Chemosphere*, **28**: 791–802.

Schreiber, J. (1992) An Exposure and Risk Assessment Regarding the Presence of Tetrachloroethene in Human Breast Milk, *Journal of Exposure Analysis and Environmental Epidemiology*, **2**(supp. 2): 15–26.

Schreiber, J.S., Hudnell, H.K., Geller, A.M., House, D.E., Aldous, K.M., Force, M.S., Langguth, K., Prohonic, E.J., and Parker, J.C. (2002) Apartment Residents' and Day Care Workers' Exposures to Tetrachloroethylene and Deficits in Visual Contrast Sensitivity, *Environmental Health Perspectives*, **110**(7): 655–664.

Seifert, B., Becker, K., Hoffman, K., Krause, C., and Schulz, C. (2000) The German Environmental Survey 1990/92 (GerESII): A Representative Population Study, *Journal of Exposure Analysis and Environmental Epidemiology*, **10**(2): 115–125.

Stewart, R.D. and Dodd, H.C. (1964) Absorption of Carbon Tetrachloride, Trichloroethylene, Tetrachloroethylene, Methylene Chloride, and 1,1,1-Trichloroethane through Human Skin, *Journal of the American Industrial Hygiene Association*, **25**: 439–446.

Swedish National Food Agency (2002) Press release, April 24, 2002, http://www.slv.se/engdefault.asp (accessed on September 23, 2005).

Thomas, K.W., Pellizzari, E.D., and Cooper, S.D. (1991) A Canister-Based Method for Collection and GC/MS Analysis of Volatile Organic Compounds in Human Breath, *Journal of Analytical Toxicology*, **15**: 54–59.

Urban, M., Gilch, G., Schepers, G., van Miert, E., and Scherer, G. (2003) Determination of the Major Mercapturic Acids of 1,3-Butadiene in Human and Rat Urine Using Liquid Chromatography with Tandem Mass Spectrometry, *Journal of Chromatography B*, **796**: 131–140.

Verberk, M.M. and T.M.L. Scheffers (1980) Tetrachloroethylene in Exhaled Air of Residents Near Dry Cleaning Shops, *Environment Research*, **21**: 432–437.

Wallace, L.A. (1990) Major Sources of Exposure to Benzene and Other Volatile Organic Compounds, *Risk Analysis*, **10**: 59–64.

Wallace, L.A. (1991) Comparison of Risks from Outdoor and Indoor Exposure to Toxic Chemicals, *Environmental Health Perspectives*, **95**: 7–13.

Wallace, L.A. (1993) A Decade of Studies of Human Exposure: What Have We Learned? *Risk Analysis*, **13**: 135–139.

Wallace, L.A. (1997) Human Exposure and Body Burden for Chloroform and Other Trihalomethanes, *Critical Reviews in Environmental Science and Technology*, **27**: 113–194.

Wallace, L.A., Nelson, W.C., Pellizzari, E.D., and Raymer, J.H. (1997) A Four-Compartment Model Relating Breath Concentrations to Low-Level Chemical Exposures: Application to a Chamber Study of Five Subjects Exposed to Nine VOCs, *Journal of Exposure Analysis and Environmental Epidemiology*, **7**(2): 141–163.

Wallace, L.A., Nelson, W.C., Ziegenfus, R., and Pellizzari, E.D. (1991) The Los Angeles TEAM Study: Personal Exposures, Indoor-Outdoor Air Concentrations, and Breath Concentrations of 25 Volatile Organic Compounds, *Journal of Exposure Analysis and Environmental Epidemiology*, **1**(2): 37–72.

Wallace, L.A., Pellizzari, E.D., and Gordon, S. (1993) A Linear Model Relating Breath Concentrations to Environmental Exposures: Application to a Chamber Study of Four Volunteers Exposed to Volatile Organic Chemicals, *Journal of Exposure Analysis and Environmental Epidemiology*, **3**(1): 75–102.

Wallace, L.A., Pellizzari, E., Hartwell, T., Perritt, K., and Ziegenfus, R. (1987b) Exposures to Benzene and Other Volatile Organic Compounds from Active and Passive Smoking, *Archives of Environmental Health*, **42**: 272–279.

Wallace, L.A., Pellizzari, E., Leaderer, B., Hartwell, T., Perritt, R., Zelon, H., and Sheldon, L. (1987a) Emissions of Volatile Organic Compounds from Building Materials and Consumer Products, *Atmospheric Environment*, **21**: 385–393.

Wallace, L.A., Pellizzari, E.D., Hartwell, T., Rosenzweig R., Erickson, M., Sparacino, C., and Zelon, H. (1984) Personal Exposure to Volatile Organic Compounds: I. Direct Measurement in Breathing-Zone Air, Drinking Water, Food, and Exhaled Breath, *Environmental Research*, **35**: 293–319.

Wallace, L.A., Pellizzari, E., Hartwell, T., Sparacino, C., Sheldon, L., and Zelon, H. (1985) Personal Exposures, Indoor-Outdoor Relationships and Breath Levels of Toxic Air Pollutants Measured for 355 Persons in New Jersey, *Atmospheric Environment*, **19**: 1651–1661.

Wallace, L.A., Pellizzari, E.D., Hartwell, T.D., Whitmore, R., Perritt, R., and Sheldon, L. (1988) The California TEAM Study: Breath Concentrations and Personal Exposures to 26 Volatile Compounds in Air and Drinking Water of 188 Residents of Los Angeles, Antioch, and Pittsburgh, CA, *Atmospheric Environment*, **22**: 2141–2163.

WHO (1983) *Assessment of Human Exposure to Selected Organochlorine Compounds through Biological Monitoring*, Slorach, S. and Vaz, R., Eds., World Health Organization, Swedish National Food Administration, Uppsala, Sweden.

Part VI

Models

18 Mathematical Modeling of Indoor Air Quality

Wayne R. Ott
Stanford University

CONTENTS

18.1 Synopsis 411
18.2 Introduction 412
18.3 Derivation of the Mass Balance Equation 412
18.4 Response of Compartment to Step and Rectangular Source Time Functions 417
18.5 Assumption of Uniform Mixing 421
18.6 Measuring the Ventilatory Air Change Rate 424
18.7 Calculating Indoor Source Strengths 428
18.8 Peak Estimation Approach 431
18.9 Incorporating Particle Deposition Rates 432
18.10 Indoor Particle Sources 434
18.11 Questions for Review 442
18.12 Acknowledgments 443
References 443

18.1 SYNOPSIS

Indoor settings are important for analyzing human exposure to pollutants, because people spend so much time indoors. Although the most accurate way to determine pollutant concentrations in indoor air is by measurement, mathematical models capable of predicting indoor concentrations have a variety of uses and can complement indoor measurements. Indoor air quality models, once validated, offer a practical methodology for understanding and predicting indoor concentrations for use in exposure analysis and for assessing health risks. This chapter derives the basic equations used to model indoor air quality in a well-mixed chamber, which is relevant to a room, a house, or a motor vehicle under certain conditions. There are three important cases: (a) a source emitting inside the indoor setting, (b) a pollutant infiltrating from outdoor ambient air into the indoor setting, and (c) a combination of the two cases by the principle of superposition. This chapter discusses a source emitting pollutants within a microenvironment, and Chapter 8 discusses a pollutant infiltrating from outdoors. Analytical solutions for the single-compartment model have a variety of practical applications, and these solutions always depend on the source emission rate as a function of time. The validity of the assumption of a well-mixed microenvironment is supported by experiments after the source ends its emissions. Practical formulas are useful for estimating the source strength from experimental data, and indoor modeling requires information on parameters such as the air change rate, deposition rate, and physical volume of the indoor setting. Indoor source emission rates are useful for generalizing the findings from one indoor setting to another, and indoor models allow the exposure analyst to ask "what if" questions about changes in indoor air quality and human exposure.

18.2 INTRODUCTION

As discussed in other chapters, indoor air quality is extremely important to exposure analysis for pollutants in the air carrier medium, because humans have a high probability of being indoors when they are exposed. The most accurate and reliable way to determine indoor air quality is by making direct measurements of pollutant concentrations in indoor settings. However, such measurements sometimes require skilled professionals and complex, expensive laboratory equipment. A complementary approach is to develop a mathematical model for predicting indoor pollutant concentrations that can be validated by measurements. Validation is important to show that the model is sufficiently accurate for its intended application. A basic mathematical model has been evaluated in indoor settings and found to be sufficiently accurate for many applications. This mathematical model is designed to predict pollutant concentrations as a function of time using parameters that are specific to a particular microenvironment, such as the volume, air exchange rate, deposition rate, and emission rate of the indoor source as a function of time.

Indoor air quality models usually are based theoretically on the "mass balance" equation, which follows from a classical law of physics that matter can be neither created nor destroyed at the macro-atomic scale. An additional assumption is that a pollutant emitted into the indoor air of an ordinary microenvironment, such as a room or a motor vehicle, mixes quickly, causing its concentration to be approximately spatially uniform at any instant of time. These two assumptions — conservation of mass and uniform mixing — allow a system of equations to be derived theoretically that can be used to predict the indoor concentration. Although one can easily imagine situations that do not meet these conditions exactly, the model predicts the interior pollutant concentrations in many real microenvironments with surprisingly high accuracy. The model also helps provide insight into the basic factors that affect indoor air quality and the basic processes involved.

This chapter derives the basic form of the mass balance equation and discusses several specific solutions that are useful for predicting air pollutant concentrations in indoor microenvironments that can be treated as a single mixing compartment. The resulting analytical solutions usually can be applied using a hand calculator, and it is not necessary to solve the differential equations using computerized step-by-step approximations, or Runge-Kutta methods. The analytical solutions that we derive can predict the indoor concentration as a function of time — that is, the "concentration time series" — and it is relatively straightforward to calculate the predicted average concentration or the expected maximum concentration in a room from the model. It often is useful to compare the predicted concentrations from these models with actual measured concentrations in real indoor settings, and good agreement usually is obtained. The quality of the predictions often depends on the accuracy of the parameters, and therefore parameter estimation is an important step in the indoor modeling process.

18.3 DERIVATION OF THE MASS BALANCE EQUATION

The basic linear differential equation — the *mass balance equation* — commonly used to model indoor air quality in a well-mixed compartment, has a great variety of other applications in physics, engineering, chemistry, and biology. Though it may be familiar from other scientific disciplines, it is instructive to explore the theory and assumptions that underlie the model when it is applied to indoor air quality problems. This section derives the mass balance equation for a single compartment to help provide an "intuitive" feeling for its principles, limitations, and practical applications in everyday settings.

Solutions to the mass balance model have a long history of application to indoor air quality problems (Bridge and Corn 1972; Jones 1974; Repace and Lowrey 1980; Dockery and Spengler 1981; Traynor, Anthon, and Hollowell 1982; Wadden and Scheff 1983; Nazaroff and Cass 1989; Switzer and Ott 1992). Comparison of predictions using the indoor mass balance model with concentration measurements indoors generally have shown good agreement in a variety of microenvironments, such as the parlor of a large San Francisco home (Klepeis, Ott, and Repace 1999), a small single-story home (Ott, Klepeis, and Switzer 2003), the passenger compartment of a motor vehicle (Ott, Langan, and Switzer 1992), a Menlo Park sports tavern (Ott, Switzer, and Robinson 1996), and two airport smoking lounges (Klepeis, Ott, and Switzer 1996). Because the mass balance model predicts the indoor air pollutant concentration time series from which the maximum concentration and time-averaged concentration are easily obtained, it is an important tool for exposure analysis.

There are three basic cases to which this model usually is applied: (a) predicting the concentrations indoors for sources emitting pollutants that are located indoors, and (b) predicting the concentrations indoors for ambient concentrations that are infiltrating into an indoor setting from outdoors, and (c) a combination of the two. The solutions from the first two cases can be combined by linear superposition to give the third case. This chapter discusses the first case — a source emitting pollutants within a microenvironment — and Chapter 8 discusses particulate matter infiltrating indoors from outdoors.

Consider an indoor point source emitting pollutant at a rate $g(t)$, where the emission rate $g(t)$ may be any function of time t. Consider a well-mixed compartment or chamber of volume v containing this single source at a point, and assume $x(t)$ is the concentration inside the chamber, regardless of where that sampling point is located (Figure 18.1). We shall return to examine this "well-mixed" assumption later, but for now assume that mixing occurs so rapidly throughout the chamber due to random internal air movements that the concentration is approximately the same everywhere at any instant of time t. Assume also that (1) the chamber does not leak, and (2) that the pollutant is relatively nonreactive chemically so that there are no "sinks" that would consume the pollutant inside the compartment. As we shall see, it is possible later to modify this assumption of no indoor sinks and to include a parameter for the deposition and removal of particulates on the interior chamber surfaces. Assume for convenience that the source emitting pollutants inside the chamber is located at a particular point in the chamber and therefore is a *point source*. Assume

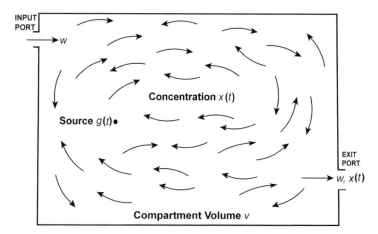

FIGURE 18.1 Simplified schematic of a single chamber or room with fresh air entering the compartment through an input port at flow rate w, mixing in the chamber, and then flowing out of the chamber through the exit port at the same flow rate w. A source inside the chamber emits the pollutant at the emission rate $g(t)$ as a function of time t.

also that the initial interior concentration is zero, or $x(t) = 0$ and time $t = 0$. Finally, assume that fresh air free of the pollutant enters the chamber through an *input port* at a constant rate w and is incompressible and at constant temperature so that it leaves the chamber through an *exit port* at the same flow rate w. For example, a pump sending fresh air into the chamber at a constant flow rate of w cubic meters per hour would cause the air to leave the chamber at the same constant flow rate of w cubic meters per hour. The idealized point source is assumed to add only mass flow (e.g., grams per minute) to the chamber; it does not alter the amount of air that flows into or out of the chamber.

To apply mass balance modeling concepts to an indoor source in this chapter, we first consider $G_{source}(T)$, the total quantity of pollutant mass emitted inside the chamber between the initial time period $t = 0$ and some final time period of interest $t = T$:

$$G_{source}(T) = \int_0^T g(t)dt \tag{18.1}$$

where

  $G_{source}(T)$ = total mass contributed by the source over time T (e.g., mg)
  $g(t)$ = emission flow rate as a function of time t (e.g., mg/min)

If fresh air is flowing through the chamber at the constant flow rate w with no internal sinks to consume the pollutant, then $Q_{lost}(T)$, the amount of pollutant mass leaving the chamber through the exit port from time $t = 0$ to time $t = T$, is the integral of the product of the flow rate w and the concentration $x(t)$ leaving the chamber:

$$\text{Total Mass Lost} = Q_{lost}(T) = \int_0^T wx(t)dt \tag{18.2}$$

where

  $Q_{lost}(T)$ = total mass lost from the chamber over time T (e.g., mg)
  $x(t)$ = concentration of pollutant in the air exiting the chamber (e.g., mg/m³)
  w = flow rate of air exiting the chamber (e.g., m³/min)

Finally, because the concentration $x(t)$ is assumed to be the same everywhere inside the chamber of volume v in m³, the total mass contained inside the chamber at time T is given by the product of the chamber volume v and the concentration $x(T)$:

$$\text{Total Mass inside the Chamber at Time } T = vx(T) \tag{18.3}$$

By the mass balance concept, the total amount of pollutant emitted in the chamber has to be fully accounted for, since mass can be neither created nor destroyed on this macro-atomic scale. Thus, from time $t = 0$ to time $t = T$ the total mass of pollutant generated inside the chamber (Equation 18.1) minus the amount of mass that has departed from the chamber (Equation 18.2) must equal the amount of mass remaining inside the chamber (Equation 18.3):

$$\begin{pmatrix} \text{Mass Generated} \\ \text{between Time} \\ t = 0 \text{ and } t = T \end{pmatrix} - \begin{pmatrix} \text{Mass Lost} \\ \text{between Time} \\ t = 0 \text{ and } t = T \end{pmatrix} = \begin{pmatrix} \text{Mass inside} \\ \text{Chamber} \\ \text{at Time } t = T \end{pmatrix} \tag{18.4}$$

The mass balance in Equation 18.4 is expressed mathematically as

$$Q_{source}(T) - Q_{lost}(T) = vx(T)$$ (18.4a)

Substituting Equation 18.1, Equation 18.2, and Equation 18.3 into the general mass balance Equation 18.4a with assumed initial conditions inside the chamber of $x(t) = 0$ for $t = 0$, we obtain the following version of Equation 18.4a expressed as integrals over time T:

$$\int_0^T g(t)dt - \int_0^T wx(t)dt = vx(T)$$ (18.5)

where

$g(t)$	= source emission rate (M/T)
w	= volumetric airflow rate (L^3/T)
v	= volume (L^3)
$x(t)$	= pollutant concentration (M/L^3)
t	= time (T)
T	= total time period of interest (T)

These ideas are based on conservation of mass within a well-mixed compartment. The difference between the amount of pollutant generated and the amount lost is the basis for deriving practical analytical solutions for calculating indoor air concentrations under a variety of circumstances. Pollutant loss can occur in many ways, such as loss through ventilation by air flowing out of the compartment, chemical reactions of the pollutant with other constituents in the compartment, or, in the case of particles, by settling, coagulation, and deposition on surfaces. For pollutants that are relatively nonreactive chemically — such as carbon monoxide (CO), sulfur hexafluoride (SF_6), and benzene — pollutant removal occurs primarily through ventilation from air moving from the compartment to the external surroundings, or interior-to-exterior airflow.

Figure 18.1 shows an idealized chamber with the fresh air entering through an input port and the air and pollutant leaving through an exit port, but even the most tightly sealed home has many cracks and small openings through which air can flow and infiltrate. Without this inadvertent air exchange between indoor and outdoor air, the occupants of a tightly closed home would suffocate. Homes and offices also have ventilation systems that exchange indoor air with the air from outside by moving it mechanically through ducts. The overall airflow of a room or compartment, including the effect of the ventilation system and the airflow through the cracks and gaps around doors and windows, is described by the ventilatory "air change rate" of the compartment, which is measured quantitatively as the number of air changes per unit time. One "air change" is an amount of fresh air entering (or leaving) the compartment that is equal to the volume v of the entire compartment.

Equation 18.5 can be written as a linear differential equation by differentiating each term. Differentiating the integral in the first term simply removes the integral sign and leaves just $g(t)$; differentiating the second term and substituting $T = t$ removes the integral sign and leaves the product of the volumetric flow rate and the concentration $wx(t)$. Finally, differentiating the term on the right-hand side of the equal sign gives the term $v\dfrac{dx(t)}{dt}$, and Equation 18.5 is written in differential notation as

$$g(t) - wx(t) = v\frac{dx(t)}{dt}$$ (18.6)

It is convenient to write Equation 18.6 with all the terms relating to the pollutant variable of interest $x(t)$ on the left-hand side of the equal sign, and with the source emission variable term $g(t)$ on the right-hand side:

$$v\frac{dx(t)}{dt} + wx(t) = g(t) \tag{18.7}$$

Dividing by the airflow rate w,

$$\frac{v}{w}\frac{dx(t)}{dt} + x(t) = \frac{g(t)}{w} \tag{18.8}$$

In solving linear differential equations of this kind, $g(t)$ often is considered the "input" or "forcing function" to the system, and $x(t)$ is considered the "output" or variable of interest that the analyst seeks to understand or predict.

Setting $g(t) = 0$ in Equation 18.8 gives the following homogeneous form of this first-order linear differential equation:

$$\frac{v}{w}\frac{dx(t)}{dt} + x(t) = 0 \tag{18.9}$$

For convenience, we define the parameter $\phi = w/v$, which, for the case of a pollutant without internal sinks, is the air change rate of the system. In some published papers, the air change rate ϕ parameter is denoted as "a" and is called the "air exchange rate" or the "ventilation rate":

$$\frac{1}{\phi}\frac{dx(t)}{dt} + x(t) = 0 \tag{18.10}$$

An equally relevant parameter is the *residence time* τ, which is the reciprocal of the air change rate; that is, $\tau = 1/\phi$. Since ϕ for a house is the number of house volumes of air per unit time that are being exchanged, its reciprocal τ is the amount of time required for an amount of air equal to one house volume to be exchanged. The homogeneous forms given by Equation 18.9 and Equation 18.10 describe the "natural response" of the system — the system's natural reaction to a momentary source impulse of infinitesimal time duration. The natural response is similar to the tendency that a pendulum has to swing at just one rate, provided that one gives it an initial push to start it swinging.

Equation 18.10 is solved in a straightforward manner in calculus by dividing both terms by $x(t)$ and then integrating:

$$\int \frac{dx(t)}{x(t)} = -\int \phi dt$$

$$\log_e x(t) + C = -\phi t \tag{18.11}$$

Here C is a constant of integration, and taking the anti-logarithm of Equation 18.11:

$$x(t) = e^{-C}e^{-\phi t} = Ae^{-\phi t} \tag{18.12}$$

In Equation 18.12, the parameter $A = e^{-C}$ is a constant of integration, and its value can be determined from the initial and final conditions of the system. It can be demonstrated that Equation 18.12 is a solution to Equation 18.10 by substituting Equation 18.12 into Equation 18.10 and showing that the two resulting terms on the left side of the equal sign, $-Ae^{-\phi t}$ plus $Ae^{-\phi t}$, add to zero.

We would like to consider the solution to Equation 18.8 for a source with an emission rate g_{source} that is *constantly emitting* and does not vary with time:

$$\frac{1}{\phi} \frac{dx(t)}{dt} + x(t) = \frac{g_{source}}{w} \qquad (18.13)$$

From our physical understanding, the initial concentration inside the chamber or room must be $x(0) = 0$. Thus, we can try $x(t) = Ae^{-\phi t} + B$ as a solution, where A and B are unknown constants. Substituting $t = 0$ into this expression for the initial conditions, $x(0) = A + B = 0$, which implies that $A = -B$. Thus, a solution is of the form $x(t) = -Ae^{-\phi t} + A = A(1 - e^{-\phi t})$, and we need to find the value of the constant A. By looking at Equation 18.13, we see that if g_{source} continues to emit for a very long time period, then there will be a "final condition" in which the concentration is stable and does not change, and thus $dx/dt = 0$. Substituting $dx/dt = 0$ into Equation 18.13 gives the solution $x(t) = (g_{source})/w$ for this final condition, and therefore $A = (g_{source})/w$. Therefore, the resulting solution for a constantly emitting source that starts at $t = 0$ is given by the solution $x(t) = (g_{source}/w)$ $(1 - e^{-\phi t})$.

18.4   RESPONSE OF COMPARTMENT TO STEP AND RECTANGULAR SOURCE TIME FUNCTIONS

For the special case discussed above in which the source begins to emit at a constant rate $g(t) = g_{source}$ at time $t = 0$ and continues its emission rate uniformly for $t > 0$, the solution to Equation 18.8, where $\phi = w/v$, is of the following form based on the above discussion:

$$x(t) = \frac{g_{source}}{w}(1 - e^{-\phi t}) \quad \text{for } t \geq 0 \qquad (18.14)$$

The source $g(t)$ in this case is called a "step" time function, because it rises abruptly from the origin and then remains constant at an emission rate g_{source} indefinitely, resembling a single step from a staircase (heavy dashed line in Figure 18.2).

The solution for the indoor concentration as a function of time given by Equation 18.14 in response to a step source function is a concave-downward curve with its initial slope given by $x'(0) = (g_{sourec})(\phi/w) = (g_{source})/v$ at the origin, and the slope gradually decreases toward zero as time increases. As the time t becomes increasingly large without limit, the chamber approaches the asymptotic concentration $(g_{source})/w$ (horizontal dashed line in Figure 18.2). This asymptotic concentration can be viewed as an "equilibrium concentration" — the concentration at which the chamber reaches a balance between the rate at which the pollutant is internally generated by the source and the rate at which it is removed. The step time function also is called the Heaviside function, named after an engineer who first described it mathematically (see Blanchard, Devaney, and Hall 1998).

A pilot light of a gas stove in a kitchen that emits carbon monoxide (CO) continuously at a fixed rate after its installation behaves as a Heaviside or step source time function indoors. However, not all indoor sources abruptly begin emitting and have a fixed emission rate. A more common indoor source is one that abruptly begins emitting and then stops emitting after a finite time period. If the source emits at a relatively constant rate while it is on, it is described as a "rectangular" time

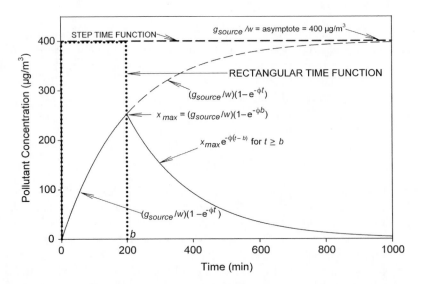

FIGURE 18.2 Solution to the mass balance equation (solid line) for a rectangular source time function (dotted lines) with asymptote $(g_{source}/w) = 400$ and time duration $b = 200$ min. A step source time function starting at time $t = 0$ also is shown (long dashed lines). These results would apply to a 500 m³ house with a ventilation rate of 0.3 air changes per hour and an SF₆ source emission rate of $g_{source} = 1.0$ mg/min.

function (dotted lines in Figure 18.2). Examples of rectangular indoor sources are the cigarette, the cigar, and the incense stick. This source time series can be viewed as a combination of two Heaviside functions: a positive-going step time function that begins at time $t = 0$ combined with a negative-going, or upside down, step time function of equal magnitude that occurs when the source ends at time $t = b$.

Example. The equations shown for the step indoor source and the rectangular source in Figure 18.2 are general and are illustrated by an example. Consider a large home with its doors and windows closed. Suppose the home's interior volume is 500 m³ and it behaves as a single compartment with an air change rate of 0.3 hr⁻¹ (ach). The air change rate (ach) is the rate at which an amount of air equal to the house's volume is exchanged with outdoor air. Suppose the home contains a relatively nonreactive pollutant such as CO that is emitted indoors for 200 minutes at a constant rate of 1 mg/min:

$$g_{source} = 1 \text{ mg/min}$$

$$b = 200 \text{ min}$$

$$\phi = 0.3 \text{ hr}^{-1}$$

$$v = 500 \text{ m}^3$$

Using these parameter values and recalling from the discussion of Equation 18.9 and Equation 18.10 that the airflow rate $w = \phi v$, the asymptotic indoor concentration will be given by:

$$\frac{g_{source}}{w} = \frac{g_{source}}{\phi v} = \frac{1 \text{ mg/min}}{\left(0.3 \dfrac{\text{air changes}}{\text{hour}}\right)\left(\dfrac{1 \text{ hour}}{60 \text{ min}}\right)\left(\dfrac{500 \text{ m}^3}{\text{air change}}\right)} = 0.4 \text{ mg/m}^3 \quad (18.15)$$

Typically the concentrations in the air of indoor settings are not reported as mg/m^3 but as μg/m^3, so the asymptotic concentration calculated for this house of 0.4 mg/m^3 would be reported as 400 μg/m^3. If the source emitting at 1 mg/min stops its emission for the time period ≥ 200 min, then the maximum concentration predicted by the model will occur at time $t = b = 200$ min and is given by applying Equation 18.14:

$$x_{max} = \frac{g_{source}}{w}(1 - e^{\phi b}) = \frac{g_{source}}{\phi v}(1 - e^{\phi b})$$

$$= \frac{1 \text{ mg/min}}{\left(0.3 \dfrac{\text{air change}}{\text{hour}}\right)\left(\dfrac{1 \text{ hour}}{60 \text{ min}}\right)\left(\dfrac{500 \text{ m}^3}{\text{air change}}\right)}(1 - e^{-(0.3)(1/60)(200)})$$

(18.16)

$$= (0.4 \text{ mg/min})(1 - e^{-1.0}) = (0.4)(0.632)$$

$$= 0.253 \text{ mg/m}^3 = 253 \text{ μg/m}^3$$

The resulting solution for this example is plotted in Figure 18.2, in which the maximum concentration reached at time $t = 200$ min is 254 μg/m^3 for a rectangular source with an emission rate of 1 mg/min and an asymptotic concentration of 400 μg/m^3. After the source cutoff time at $t = 200$ min, the predicted concentration suddenly begins to decrease with time, following an exponential decay curve of the same form as Equation 18.12:

$$x(t) = x_{max}e^{-\phi(t-b)} = 254e^{-(0.005)(t-200)} \quad \text{for } t \geq 200 \text{ min} \quad (18.17)$$

One can verify this equation by substituting $t = 200$, which gives $x(200) = 254e^0 = 254$ μg/m^3, the predicted maximum concentration in the compartment. Equation 18.16 uses the air change rate expressed in minutes, and $\phi = (0.3$ air changes/hour$)$ $(1$ hour/60 min$) = 0.005$ min^{-1}, or air changes per min (acm).

The abrupt change in slope of $x(t)$ at $t = 200$ in Figure 18.2 is caused by the end of the source emission, and the result gives two continuous exponential functions that are joined together at the stopping time $t = b$. The value of the concentration $x(200)$ will be the same for both curves at the discontinuity at $t = 200$, but the slopes of the curves will differ at the discontinuity. Because two continuous curves are joined at this point, the solution to the single-compartment mass balance equation for a rectangular source consists of two *piecewise continuous* functions that are joined together at the source stopping time $t = b$.

It is instructive to examine the concentration time series predicted by the model in this 500 m^3 home as the ventilation rate increases from $\phi = 0.3$ ach to $\phi = 4$ ach (Figure 18.3). A ventilation rate of 0.3 ach is reasonable for a home with all the doors and windows closed, but air change rates as high as 4 ach could occur with several windows wide open to "air out" the house (Howard-Reed, Wallace, and Ott 2002). For a rectangular source time function emitting at 1 mg/min, the maximum concentration in the home decreases from 157.4 μg/m^3 at a ventilation rate of 0.3 ach to 29.96 μg/m^3 at 4 ach. The shapes of the concentration curves are different for the different ventilation rates. For the lowest air change rate, the indoor concentration increases almost linearly

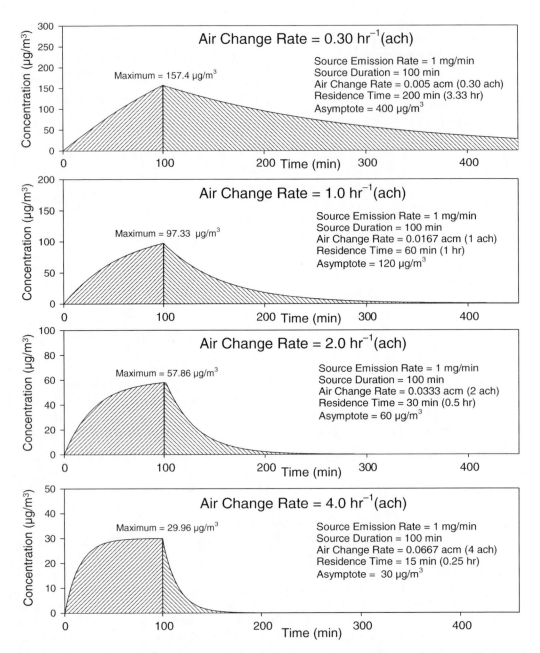

FIGURE 18.3 Indoor concentration predicted by the mass balance model for a rectangular source that begins emitting at 1 mg/min at $t = 0$ and ends at $t = 100$ min, with a mixing volume of 500 m³ and ventilation rates ranging from 0.3 ach (top) to 4 ach (bottom). These results are applicable to pollutants such as CO that do not have reactive or deposition tendencies.

while the source is emitting, and then the concentration decreases very gradually with time after the source ends at $t = 100$ min, remaining at a significant value of 27 µg/m³ at 450 min, or 5.8 hours after the source's emission rate has dropped to zero. A 500 m³ home with $\phi = 0.3$ ach is the same as the example shown in Figure 18.2, except that the rectangular source in Figure 18.3 emits for 100 min instead of 200 min. The curvature of concentration versus time for both the "source-on" and "source-off" periods becomes greater as the ventilation rate increases. For the high air

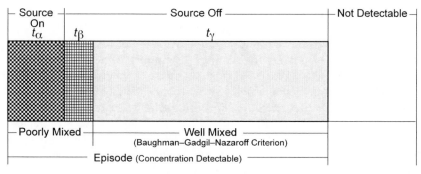

FIGURE 18.4 Schematic representation of three different concentration time periods for a pollutant emitted into a room. The concentration is below the limit of detection and therefore not measurable in the "Not Detectable" region after time t_γ. (From Mage and Ott 1996. With permission.)

change rate of 4 ach, there is a sharp curvature for each of the two graphs, and the concentration plots resemble the rectangular shape of the source emission time function. The change in shape of the concentration curves with an increased air change rate is due to the larger value of ϕ in the exponents of Equation 18.14 and Equation 18.17. The reduced maximum concentration with an increased air change rate is due primarily to the value of $w = \phi v$ in the denominator of Equation 18.17.

18.5 ASSUMPTION OF UNIFORM MIXING

A critical assumption of the indoor mass balance model is that the air is sufficiently well mixed so that the concentration in the compartment is approximately spatially uniform at any instant of time. A source that emits a pollutant for a finite time period inside a chamber, an automobile, or a room of a home typically can be characterized by several important stages in time. First, there is the source-on time period — the time during which the pollutant is emitting — which, in this figure, lasts from $t = 0$ until $t = t_\alpha$. Examples are smoking a cigarette, broiling a hamburger, and using a spray can indoors. The notation b in the above example for the cigarette is the same as the more general notation of t_α for the source-on, or "α-period." Mage and Ott (1996) proposed three different mixing time periods for a source emitting in a chamber or a room (Figure 18.4). Typically a cigarette is represented by a rectangular time function that lasts about 10 minutes, but not all sources have uniform emission rates. Some sources, like freshly painted surfaces, have very long source-on periods as they "outgas" volatile organic compounds as the material cures and hardens. Chlordane emitted from the soil beneath a home's foundation is another example of a source with a long source-on time. Other sources, like burning the popcorn in a microwave or cooking a hamburger in a kitchen frying pan, have relatively brief source-on, or α-periods.

During the time in which the source is emitting (the α-period), the concentration field across a room will be nonuniform, since there will be relatively high concentrations very close to the source. For example, the concentrations a few centimeters from the tip of a burning cigarette are higher than elsewhere in the room, and therefore it is not possible for the room to be spatially uniform over the entire mixing volume during the α-period (Figure 18.4). The α-period is a poorly mixed stage in which concentrations vary spatially across the room. Furtaw et al. (1996) reported that the concentrations measured in a chamber very close to the source (less than 1 m) were two to three times higher than elsewhere in the room during the source-on period. Their chamber measurements close to the source revealed that small "concentration eddies" occurred as a random process, and the average of all these random eddies was higher than the average calculated by the mass balance equation, although individual eddies may have lasted for very brief time periods.

McBride et al. (1999) studied a house by releasing tracer pollutants — sulfur hexafluoride gas, CO, and particulate matter — in the living room of a two-story house and verified a "proximity effect" near the source. For CO, as many as 40 points were monitored while the tracer gas was released at a constant rate from a single point in the room — the coffee table where a cigarette might be smoked. Many high concentration eddies near the point source were observed to last for brief time periods; these eddies are called "microplumes." Microplumes are high-concentration pulses of short duration that typically increase in frequency and amplitude close to a source that is actively emitting. Typically, there is considerable microplume activity at distances less than 0.5 m from the source, with occasional microplumes at 3 m from the source, and relatively few microplumes at more than 5 m. Microplume activity at 0.5–1 m from the source can be sufficient to elevate the average concentration to more than 3 times the hourly average predicted by the mass balance equation, but this gradient appears only during the source emission, or α-period. A stationary indoor monitor located 5 m from the source in the same room will read a concentration that agrees closely with the concentration predicted by the mass balance equation, despite the proximity effect near the center of the room where the source is located.

At time $t > t_\alpha$ in this conceptual model, most indoor sources stop emitting (that is, the cigarette ends or the toast stops burning), and then the source-off time period begins. In Figure 18.4, the source-off period is divided into two periods, a transition time period called the β-period lasting for time duration t_β and a well-mixed time period called the γ-period. Ideally, the γ-period would meet a criterion originally proposed by Baughman, Gadgil, and Nazaroff (1994), in which the standard deviation of the concentrations at all the points in the room is less than 10% of the mean concentration. This criterion for a well-mixed room is equivalent to requiring the ratio of the standard deviation of all the concentrations at all the points in the room to the mean, or the *spatial coefficient of variation*, to be less than 0.1. In an experiment releasing sulfur hexafluoride tracer gas with 41 sampling points in a 31 m³ room, they found that the room was poorly mixed for only about 5–10% of the residence time early in the release period. This finding is consistent with other experimental results showing that the β-period generally is small compared with the γ-period in most indoor settings. The duration of the α-period, of course, depends on the emission activity of the source itself. A cigarette typically lasts about 10 minutes, which is small compared with a γ-period lasting for 3 hours or more in a home with a typical ventilation rate.

An unusual experiment was conducted in a sports tavern in Menlo Park, CA, to determine whether a real microenvironment is well mixed when a pollutant is emitted from a point source indoors (Ott, Switzer, and Robinson 1996). The Oasis Beer Garden serves beer, pizza, hamburgers, and light foods. When crowded, this sports tavern can hold 80–100 persons, and its floor area is 210 m². Using a tape to measure its interior dimensions, its physical volume was calculated as 548 m³. After obtaining permission from the management, the investigators placed monitoring instruments at three interior locations when the tavern was not occupied by customers. They set up CO monitoring instruments at three widely spaced tables: a central table, a southwest corner booth, and a northwest corner booth. At each table, they placed a Langan L15 CO Personal Exposure Measurerer™ equipped with a DataBear battery-powered data logger (Langan Products, Inc., San Francisco, CA) that measures and records CO concentrations automatically at 10-second intervals. The two corner booths were approximately 12 m apart, and the central table was approximately 7 m from each corner booth. At the central table, a machine was used to smoke four cigars rapidly, two at a time, to provide a strong source of CO emissions.

The experiment was conducted during nonbusiness hours, and no other sources of CO were present. The manager of the tavern tried to set the operation of the ventilation system fans, doors, and windows to their normal positions. At 8:47:30 A.M., two investigators mechanically smoked two pairs of cigars at the central table. When the first pair of cigars ended, two more cigars were ignited, and the total source-on period lasted 5.5 minutes, making $t_\alpha = 5.5$ min (see Figure 18.5).

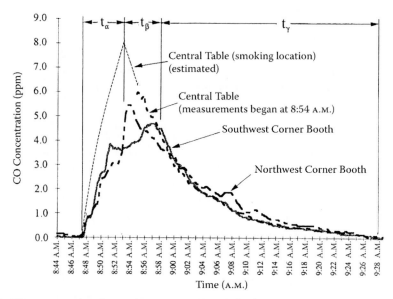

FIGURE 18.5 CO concentration time series measured in a 548 m³ tavern at three locations after investigators smoked four cigars at a central area. The α-period is t_α = 5.5 min; the β-period is t_β = 5 min; and the γ-period is t_γ = 30 min. The northwest corner booth and southwest corner booths were approximately 12 m from each other; the central table sampling location was 7 m from the northwest corner booth and 6.7 m from the southwest corner booth. The air change rate was 7.5 air changes per hour. (From Mage and Ott 1996. With permission.)

The cigars were extinguished by plunging them into a glass of water at 8:53 A.M. Thereafter the CO concentrations at the widely spaced locations decayed. Because measurements were made at only three locations, calculation of the spatial coefficient of variation described above would not be meaningful, but the time at which the β-period ended and the γ-period began was estimated from the graph. The concentrations at the two far corner booths did not become similar until 8:58 A.M. Thus, Figure 18.5 suggests that the poorly mixed, source-off transition β-period lasted for t_β = 5 min, while the well-mixed period during the concentration decay lasted for t_γ = 30 min. Although the time durations of the α- and β-periods were 14% and 12% of the total 40.5-minute indoor episode, respectively, the important issue regarding uniform mixing is the differences in the average concentration measured at each location during these periods (Table 18.1).

During the α-period, the CO average at the two distant corner booths (2.25 and 1.90 ppm) differed by 16%, whereas the estimated CO at the central table where the source was located (4.43 ppm) was about twice as high. During the β-period, the two corners differed, on the average, by only 0.43 ppm, and during the γ-period they differed by 0.13 ppm. When all three periods are combined, the average CO concentration in the southwest corner was 1.65 ppm, and the average CO concentration in the northwest corner was 1.76 ppm, a difference of 6.7%.

Although data were not available from the central table for the full episode, the missing values were estimated by the Mage and Ott (1996) by extrapolating the exponential concentration curve back to the point at which the source was extinguished and by using the mass balance model to predict an exponential rise up from the origin to the same point.

The averages for the β- and γ-periods at the central table are close to the averages for the two corner booths. A patron who is present during the full smoking episode would experience almost the same average exposure of 1.7 ppm CO from the cigars regardless of whether the person were located in either of the two corner booths, while another person sitting at the central table where the cigars were smoked would experience an exposure of 2.1 ppm, about 25% higher than the average concentration at either booth. Although the two corner booths are 12 m apart, their concentration differences are not great even during the α- and β-periods. Thus, a person sitting at

---

TABLE 18.1
Average CO Concentrations (ppm) Measured at
Three Locations after Four Cigars Were Smoked
in a 548 m³ Sports Tavern[a]

Averaging Period	Southwest Corner Booth	Northwest Corner Booth	Central Table
α	2.25	1.90	4.43[a]
β	4.16	4.59	5.39[a]
γ	1.14	1.27	1.12
Overall Episode[b]	1.65	1.76	2.10[a]

[a] A portion of the interval over which the average was computed was not measured during this period and was estimated by the authors.

[b] The average of the concentration at all three locations is 1.84 ppm.

Source: Mage and Ott (1996).

---

a corner booth would experience an average CO exposure from the cigars that is only 20% less than the exposure of a person sitting 7 m away at the central table where the cigars were smoked. Over the entire 40.5-minute episode, the three average concentrations are almost within $\pm$ 10% of the overall average of 1.84 ppm, and the 10% figure has been adopted by the American Society for Testing and Materials (ASTM) as a criterion for a uniformly mixed concentration (ASTM Rule E 741), which is essentially the same as the Baughman–Gadjil–Nazaroff criterion discussed above (Mage and Ott 1996).

Since the time of this field study, the tavern no longer permits smoking inside. Soon after smoking was prohibited in this tavern by a local city ordinance, the State of California prohibited smoking indoors in many public locations, such as restaurants, taverns, stores, casinos, and public buildings.

18.6 MEASURING THE VENTILATORY AIR CHANGE RATE

A variety of methods have been used to measure the air exchange rate of an indoor setting, such as an office or a home, and these methods generally rely on the solutions to the mass balance model. A typical approach for determining the ventilatory air change rate ϕ is to release a quantity of tracer gas, such as sulfur hexafluoride (SF_6), and to measure the rate of decay of the indoor concentration with time. SF_6 is a clear, odorless, relatively nonreactive gas that is available in pressurized cylinders, and few other sources of SF_6 are found naturally indoors or outdoors, so nearly all the SF_6 concentrations measured in an indoor setting can be attributed to the release of this tracer gas. Once the source emission of the tracer gas stops and concentrations in the indoor microenvironment become reasonably spatially uniform, the decay of the concentration of this gas will be influenced entirely by the ventilatory air change rate and will follow the exponential solution of the mass balance model given by Equation 18.12.

Taking the natural logarithm of the exponential decay solution to the mass balance equation in Equation 18.12 produces an equation expressing a linear relationship between $\log_e x(t)$ and time t:

$$\log_e x(t) = -\phi t + \log_e A \qquad (18.18)$$

To apply this equation to experimental measurements, the analyst plots the logarithm of the measured SF_6 concentrations vs. time and examines the rate of decay. Equation 18.18 is in the same form as the standard equation for a straight line with slope m and intercept B:

$$y(t) = mt + B \qquad (18.19)$$

where

$y(t)$	$= \log_e x(t)$
m	$= -\phi$
t	$= \text{time}$
B	$= \log_e A \ (or \ A = \text{constant} = e^B)$

Usually, the measurements are a number of individual data points, and performing a linear regression analysis of the natural logarithm of the concentration vs. time gives a least-squares or best-fit estimate of the slope of the resulting straight line. The negative value of this slope is the ventilatory air change rate $\phi = a$. Another approach is to plot the concentration measurements on semi-log graph paper vs. time. However, the widely available linear regression statistical analysis programs give a better fit of the model to the observed data than trying to fit a line by eye, and the linear regression computer programs also provide a statistical measure of how well the line fits the data: the coefficient of determination R^2.

An important step in the curve-fitting process is to subtract any background concentration that may be present from the observed concentration *before* taking the logarithms. Overlooking the background concentration can cause serious error in the estimation of the air change rate, because it will change the apparent slope of the semi-log plot of the observations. Therefore, if a nonzero background concentration c is present, a more accurate formula for determining the air change rate using the mass balance equation requires subtraction of c from the logarithm of the concentration:

$$\log_e \left\{ x(t) - c \right\} = -\phi t + \log_e A \qquad (18.20)$$

For tracer gases like SF_6, which have very low ambient background concentrations and virtually no ordinary indoor sources, the background concentration c may turn out to be a relatively small number, but the estimated background concentration still should be subtracted from the observed SF_6 concentration before taking the logarithm.

Example. In the living room of a California home, 99.99% pure SF_6 was released from a pressurized test cylinder at a constant flow rate of 200 cc/min for 47 minutes. The tracer gas cylinder was turned on at 4:05 P.M., which corresponds to 0.0833 hours after 4:00 P.M., and it was turned off at 4:52 P.M., or 0.8667 hours after 4:00 P.M. (Figure 18.6). A sensitive real-time infrared absorption SF_6 air monitoring instrument (Brüel & Kjaer Model 1302 Multigas Monitor) made concentration measurements approximately every minute, and the mixing volume of the home was determined earlier to be $v = 460$ m^3. All the external doors and windows were closed during this experiment, which sometimes is called "Ventilation State 0" for a home.

The SF_6 concentration in the house rose sharply soon after the gas was turned on at $t = 0.0833$ hr, and it reached a maximum value of about 14 ppm (Figure 18.6a), although it did not exhibit a single sharp peak as the theoretical solution for a rectangular source in Figure 18.2 predicted. As a first step in the analysis, the SF_6 concentration was plotted on a linear graph in Figure 18.6a as a function of time, and the background SF_6 concentration of about 0.01 ppm was determined by noting the low levels observed before the experimental gas release and after the end of the decay period 16 hours later. The decay portion of the curve in Figure 18.6 is similar to the exponential decay curves discussed earlier.

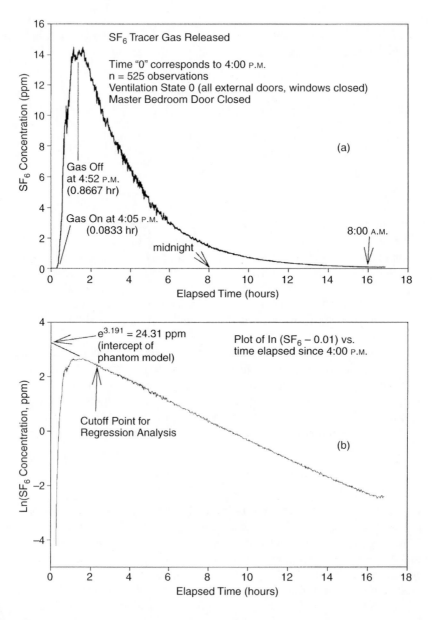

FIGURE 18.6 Plots showing (a) SF_6 concentration and (b) the natural logarithm of the SF_6 concentration in a 4-bedroom, 460 m³ California home after SF_6 tracer gas was emitted at 200 cc/min for 47 minutes, beginning at 4:05 P.M. The elapsed time begins at 4:00 P.M., and a 0.01 background concentration has been subtracted.

Next the analyst evaluated how well the decay curve follows an exponential function by setting the vertical axis to the natural logarithm and by determining how straight the line is for this semi-log plot. Figure 18.6a and Figure 18.6b both were plotted using Sigma-Plot (Systat Software, Richmond, CA), and the semi-log plot in Figure 18.6b showed that the decay portion of the curve was, to a large degree, linear. After examining both graphs in Figure 18.6, the analyst decided to focus on the portion of the decay curve that begins at $t = 2.26$ hr. Therefore, it was necessary to cut and paste the SF_6 data from one column of the spreadsheet to another column and then to subtract the background level of 0.01 ppm from each concentration reading in the adjusted column. Last, a linear regression analysis of $\log_e\{x(t) - 0.01\}$ vs. time t in Figure 18.7a produced a slope

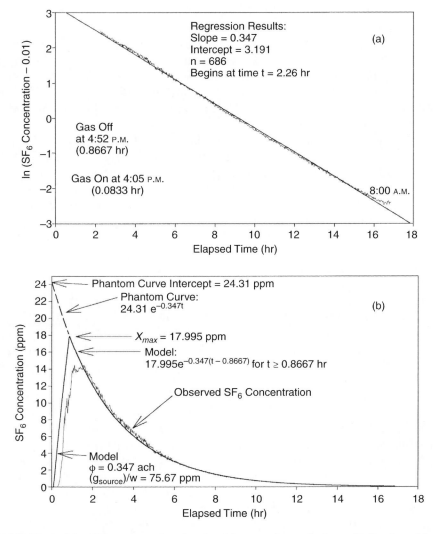

FIGURE 18.7 Plots of the SF_6 concentration showing (a) regression analysis results for the stable portion of the decay curve (beginning at 2.26 hours), and (b) times-series solution to mass balance model based on the parameters obtained from the regression analysis.

of 0.347 hr^{-1} for $n = 686$ observations. Extending the straight line from the regression analysis to the origin gave $\log_e A = 3.191$, so $A = e^{3.291} = 24.31$ ppm.

These parameter values yield the following equation for the natural logarithm of the adjusted concentration as a function of time:

$$\log_e \left\{ x(t) - 0.01 \right\} = -0.347t + \log_e 24.31 \qquad (18.21)$$

Taking anti-logarithms yields the following exponential solution:

$$x(t) = 24.31e^{-0.347t} + 0.01 \qquad (18.22)$$

Because the straight line in Figure 18.7a was, in effect, extended back to the vertical axis and origin, the portion of the exponential curve between $t = 0$ and $t = 2.26$ hr does not follow through any measured observation points but instead uses the model to extrapolate from the exponentially decaying part of the curve after $t > 2.26$ hr. As a result, this portion of the curve is called a "phantom curve" because it is based on a solution to the mass balance model and not on the observations themselves in this region. Choosing the time at $t = 0.8667$ hr as the cutoff point for the source emission, we can use Equation 18.22 to calculate the expected maximum concentration as $x_{max} = 24.32e^{-(0.347)(0.8667)} + 0.01 = 18.01$ ppm.

Careful examination of the model-predicted concentration in Figure 18.7b reveals a time delay between the exponential rise and the measured indoor SF_6 concentration. To study this time delay, Ott, McBride, and Switzer (2002) placed two integrating nephelometers, very sensitive real-time particle monitors employing light-scattering techniques, 5 m on either side of an incense point source to measure this time delay more accurately than has been done before. The two nephelometers made particle measurements at 2-second intervals to determine the first arrival time of the first observable concentration (the "first hit" time). With calm indoor air (< 0.2 m/s), the first microplumes of pollutant reached the monitors at the outer edges of the room more rapidly than was predicted for molecular diffusion alone, and the "first hit" times at 5 m ranged from 1.7 to 17 minutes, averaging 4.9 minutes (or approximately 0.017 m/s).

Ott, McBride, and Switzer (2002) conclude that the pollutant is transported by indoor microplumes that, in turn, are caused by natural turbulence in the room, even without the operation of any indoor fan or ventilation system. The air exchange rate of the home was 0.2–0.3 ach, corresponding to an air residence time of 3.3–5 hours. The average travel time was approximately 5 minutes for the pollutant to move 5 m across the room in either direction, which helps explain why a room becomes uniformly mixed rapidly, thereby meeting the well-mixed conditions of the mass balance equation.

A source proximity effect is important for indoor sources that emit pollutants continuously — such as gas pilot lights, carpet glue, and certain consumer products — where personal exposures close to the source are likely to be higher than indoor air levels, and parts of the microenvironment can be perpetually in an α-period. A source proximity effect also may occur when a person prepares meals and is close to cooking sources or when a person engages in painting or other hobby activities and is very close to a source emitting a pollutant. The source proximity effect is one reason that a significant difference can occur between personal exposure to indoor pollutants and the concentrations measured by a stationary indoor monitor or predicted by an indoor air quality model that assumes spatially uniform concentrations.

In the early days of indoor air quality studies, it was often assumed that winds outside a home had an important effect on the air change rate of a home, but a study of more than 300 air change rate measurements in two well-insulated homes found that external winds had relatively little effect on indoor ventilation rates. The simple opening of a window, however, had a large effect on a home's air change rate (Figure 18.8). Opening the single window of a large home by just 12.7 cm (5 in.) caused the home's ventilation rate to more than double, increasing from 0.34 to 0.84 ach, as shown by the abrupt change in slope of the SF_6 concentration decay curve (Howard-Reed, Wallace, and Ott 2002).

18.7 CALCULATING INDOOR SOURCE STRENGTHS

The basic model derived earlier in Equation 18.8 and the concept of mass balance for a single compartment is useful for many different kinds of indoor applications:

$$\frac{1}{\phi} \frac{dx(t)}{dt} + x(t) = \frac{g(t)}{w} \tag{18.23}$$

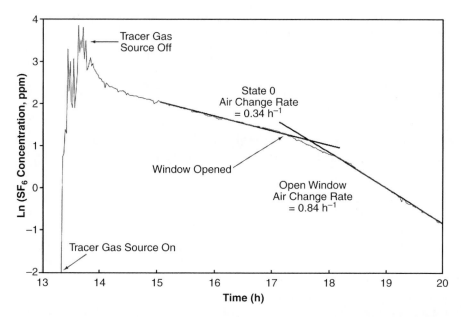

FIGURE 18.8 Effect on SF$_6$ concentration decay of opening a window in a large home. When all the windows and external doors were closed ("Ventilation State 0"), the air change rate (negative slope of curve) was 0.34 hr^{-1} (ach). When a single window was opened only 12.7 cm (5 in), there was an abrupt increase in slope, and the air change rate increased from 0.34 to 0.84 ach. (From Howard-Reed, Wallace, and Ott 2002.)

In some exposure applications, the analyst is interested not just in the times series solutions or the maximum concentration but also in the average concentration over some averaging time T, since health effects often are associated with the concentration averaged over some time period T. Although time-weighted averages can be obtained from the time series solutions, they also can be obtained directly from the basic differential equations themselves. The time-weighted averages are useful for exposure analysis and for indoor air quality studies, but most books on differential equations do not deal with the average value of the variables in the differential equations.

Suppose we divide every term in Equation 18.23 by the averaging time T and then integrate each term, assuming the initial conditions of $x(t) = 0$ at time $t = 0$:

$$\frac{1}{\phi T} \int_0^T dx(t) + \frac{1}{T} \int_0^T x(t)dt = \frac{1}{wT} \int_0^T g(t)dt \tag{18.24}$$

The two terms in Equation 18.24 that contain $x(t)$ and $g(t)$ integrated over from the time $t = 0$ to $t = T$ and divided by T give the average values:

$$\overline{x(T)} = \frac{1}{T} \int_0^T x(t)dt \tag{18.25}$$

$$\overline{g(T)} = \frac{1}{T} \int_0^T g(t)dt \tag{18.26}$$

These equations show that the mean, or average value, of any time-varying function is given by the area under the curve divided by the averaging time T. Substituting Equation 18.25 and Equation 18.26 into Equation 18.24 and performing the integration gives a useful result:

$$\frac{x(T)}{T\phi} + \overline{x(T)} = \frac{1}{w}\overline{g(T)}$$ (18.27)

Solving Equation 18.27 for the average of the emission over time T and substituting $\phi = 1/\tau$ to express the residence time τ gives an expression for the mean concentration over time T:

$$\overline{x(T)} = \frac{\overline{g(T)}}{w} - \frac{\tau}{T}x(T)$$ (18.28)

Equation 18.28 is exact, and it is completely general, because it makes no assumptions about the source emission rate $g(t)$ as a function of time, except for the initial conditions of $x(t) = 0$ at $t = 0$. Thus, it applies to any indoor emission rate time series and to any concentration time series solution of the mass balance equation; it is valid for a rectangular source function, a triangular source time function, or any other type of source time function. We see that the ratio of the residence time τ to the averaging time T plays an important role in Equation 18.28, because this ratio becomes smaller as the averaging time becomes larger, causing the last term on the right side to become smaller. This subtracted term $(\tau/T)x(T)$ on the right side of Equation 18.28 sometimes has been called a "trend adjustment" term, because it adjusts for the portion of the curve that is missing from the integration process since it was cut off for times greater than T. I have sometimes referred to Equation 18.28 as the "golden rule of averages," because of its importance and its many practical uses. A similar relationship based on average values can be derived for any linear differential equation.

For an exponentially decaying function or any other concentration that decreases with time, $x(T)$ becomes smaller as the averaging time T becomes larger, and T in the denominator also causes the right-hand term on the right side of Equation 18.28 to approach zero as T increases. Thus, for a large averaging time T compared with the air residence time τ, the average concentration in Equation 18.28 is approximately equal to the average of the emission rate divided by w:

$$\overline{x(T)} \cong \frac{\overline{g(T)}}{w}$$ (18.29)

Although Equation 18.29 is not exact, it gives an approximation of the relationship between the average concentration and average source that is often of practical use in indoor air experiments and exposure models. It has been used to develop a table of design values for ventilating indoor locations with smoking activity (Ott 1999).

It also is possible to use Equation 18.28 to obtain a useful exact expression for the mean source emission rate as a function of the mean concentration and the air change rate ϕ and mixing volume v:

$$\overline{g(T)} = \phi v\overline{x(T)} + \frac{vx(T)}{T}$$ (18.30)

The integral in Equation 18.26 by itself gives the area under the curve of the source emission rate $g(t)$, which is equal to the total emissions G released by the source and is called the "source strength" to distinguish it from the source emission rate $g(t)$. The source strength G is expressed in mass units such as milligrams (mg), while the source emission rate $g(t)$ is expressed as mg/min:

$$G = Q_{source}(T) = T\,\overline{g(T)} = \int_0^T g(t)dt \tag{18.31}$$

In most modeling applications, G refers to a source that has reached a zero emission rate just prior to time T, even though the form of the time series $g(t)$ is not specified, and therefore this equation is general. Substituting $\overline{g(T)} = G/T$ from Equation 18.31 into Equation 18.27 and noting that $w = \phi v$ gives the following exact expression for computing the total emissions G:

$$G = T\phi v\overline{x(T)} + vx(T) \tag{18.32}$$

Notice that Equation 18.32 allows the analyst to calculate the source strength G using only the physical parameters of a well-mixed room and the average concentration measured in that room or compartment, as well as the instantaneous concentration at the very end of the averaging time T. Because computing the average requires the area under the concentration curve of the observed data, this approach has been called the "area-under-the-curve" method for determining the source strength.

Finally, for the special case of a rectangular source time function emitting at the constant rate $g(t) = g_{source}$ for the time period from $t = 0$ to $t = b$, the source strength G will be the product of the fixed emission rate and the source duration:

$$G = g_{source}b \tag{18.33}$$

Substituting G from Equation 18.33 into Equation 18.32 and solving for g_{source} gives the following practical equation for computing the source emission rate in an indoor experiment:

$$g_{source} = \frac{Tv}{b}\left\{\phi\overline{x(T)} + \frac{x(T)}{T}\right\} \tag{18.34}$$

Brauer et al. (2000) used this equation to compute the source emission rate for the concentration measured in a chamber from a cigarette and for cooking sources.

18.8 PEAK ESTIMATION APPROACH

The area-under-the-curve approach for estimating the source strength described above has the advantage that it is exact and that all the parameters can be obtained in an experiment. However, a simpler method has been suggested for determining the source that requires primarily the "peak" concentration estimated from the experiment. To find the peak concentration for a "single pulse" source of very short time duration b, we first rewrite Equation 18.14 using the residence time $\tau = 1/\phi$ instead of the air change rate:

$$x(t) = \frac{g_{source}}{w}(1 - e^{-t/\tau}) \quad for\ t \geq 0 \tag{18.35}$$

For the case of a rectangular source of extremely short time duration $t = b$, we substitute the following Taylor series expansion of $e^{-b/\tau}$:

$$x(b) = \frac{g_{source}}{w}\left\{1 - 1 + \frac{b}{\tau} - \left(\frac{b}{\tau}\right)^2 + \left(\frac{b}{\tau}\right)^3 - \ldots\right\} \tag{18.36}$$

Assuming that the ratio b/τ is small because of the small time duration b, we can approximate Equation 18.36 by eliminating all terms of power 2 and above due to their relatively small contribution:

$$x(b) = x_{max} \cong \frac{g_{source}}{w}\frac{b}{\tau} \tag{18.37}$$

Since the product $(g_{source})b$ for a rectangular source has the physical interpretation in Equation 18.33 as the "total quantity of mass released" and $v = w\tau$, this result also can be written as

$$x(b) = x_{max} \cong \frac{\text{total mass released G}}{v} \tag{18.38}$$

This result has a practical interpretation. It shows that the maximum concentration x_{max} is related approximately to the total mass released divided by the volume of the compartment, provided that the source duration b is small compared to the residence time τ.

Substituting $w = \phi v = v/\tau$ into Equation 18.37 for a rectangular source in a well-mixed room with emission rate g_{source} and duration b, where $b \ll \tau$, we obtain an important relationship that provides the basis for using the peak concentration in a well-mixed room to calculate the source emission rate:

$$g_{source} \cong \frac{vx_{max}}{b} \tag{18.39}$$

In summary, to obtain the source emission rate for a short-duration source, we multiply the volume and the maximum concentration together and then divide by the source duration b. Since there may be a problem determining the correct value of x_{max} from actual measurements because the maximum occurs during the transition from the α-period to the β-period, when the room is not well-mixed, the "phantom curve" method can be used. Here, the smooth decay curve from the γ-period is extended backward to estimate the likely maximum concentration at time $t = b$ in a manner similar to the tavern example shown in Figure 18.5 and the ventilatory air change rate example in Figure 18.6. Thus, the analyst uses the mass balance model to extrapolate the measured curve from the well-mixed γ-period to the more poorly mixed α- and β-periods, predicting the concentrations likely to have occurred had the room been well-mixed over the entire time period. Extending the exponential solutions of the mass balance model in this manner to time periods when the mixing assumption is not perfectly met by the phantom curve method provides a powerful and easy-to-use method for estimating the source strength of indoor pollutant sources.

18.9 INCORPORATING PARTICLE DEPOSITION RATES

The indoor model developed thus far does not include a pollutant removal process, or a "sink." Particles tend to "plate out" or deposit on indoor surfaces, in addition to their loss due to ventilation, so the indoor model must include a sink removal term for particles. The removal process by deposition of particles on surfaces is similar to the death rates of biological populations. With other factors held constant, the rate at which members of a population die (number of deaths per unit time) is proportional to the population present at time t. It is easy to understand why more people are expected to die per day of natural causes in a large city such as Chicago than in a small city such as Bakersfield, CA, simply because the overall population is greater in the larger city. If there are no births or new arrivals, then the population will decline, and the death rate will be proportional to the population present at any time. We assume that the deposition rate of particles, like the death rate of biological populations, is linearly related to the number of particles present at any time. Therefore, the rate at which the particles are deposited on surfaces is assumed to be proportional to the concentration of particles present in the compartment at time t.

We can incorporate this assumption about particle deposition into the basic derivation of the mass balance equation by changing $Q_{lost}(T)$ in Equation 18.4a on page 415 to include a new term for the product of the concentration $x(t)$ and a deposition parameter u, in which u represents the deposition rate of particles in a manner analogous to the airflow rate w:

$$Q_{lost}(T) = \int_0^T wx(t)dt + \int_0^T ux(t) \tag{18.40}$$

Incorporating this new parameter u, which can be described as the volume of particles per unit time deposited, into our earlier equation for pollutant loss (Equation 18.4a) and following the same steps that were followed before to derive the general mass balance differential equation (Equation 18.8), we obtain:

$$\left(\frac{v}{w+u}\right)\frac{dx(t)}{dt} + x(t) = \frac{g(t)}{w+u} \tag{18.41}$$

Considering the parameters inside the parentheses in the first term, it is possible to define a new parameter ϕ_p that represents just the particle decay rate and is analogous to the ventilatory air exchange rate $\phi = a$ used earlier in Equation 18.8 to model a nonreactive air pollutant that has no indoor sinks:

$$\phi_p = \frac{w+u}{v} = \frac{w}{v} + \frac{u}{v} = a + k \tag{18.42}$$

where

ϕ_p = particle decay rate (hr^{-1} or min^{-1})
a = ventilatory air change rate (hr^{-1} or min^{-1})
k = particle deposition rate (hr^{-1} or min^{-1})

Although the overall particle decay rate ϕ_p is analogous to the ventilatory air change rate $\phi = a$ for a pollutant without sinks and is expressed in the similar time units, the decay rate ϕ_p should not be expressed as "air changes per hour (ach)," because it is not a true flow rate. As can be seen from Equation 18.42, the particle decay rate ϕ_p is always equal to or greater than the ventilation rate, or $\phi_p \geq a$, because the deposition process causes the particles to disappear more rapidly in a room than if the ventilation process were acting alone. Introducing the new parameter ϕ_p into the mass balance equation causes few problems in the analysis, because the earlier solutions behave much as they did before, except that now ϕ_p becomes the important exponential parameter. With this new parameter, there also is associated a new effective residence time for particles, or $\tau_P = 1/\phi_P$.

Example. Consider a chamber experiment with a cigarette in which the chamber volume is $v = 56.6$ m^3, the averaging time period is $T = 51$ min, the cigarette burn time is $b = 14.2$ min. If the decay rate is $\phi_p = 0.019$ min^{-1}, or 1.14 hr^{-1}, and the measured average concentration in the room over the 51-minute time period is $x(51) = 266$ µg/m^3, then Equation 18.34 can be used to compute the particle source emission rate with ϕ_p substituted in place of ϕ:

$$g_{source} = \frac{51 \text{ min}}{14 \text{ min}} (56.6 \text{ m}^3) \left[(0.019 \text{ min}^{-1})(266 \text{ µg}/\text{m}^3) + \frac{156 \text{ µg}/\text{m}^3}{51 \text{ min}} \right] \quad (18.43)$$

$$= 1{,}670 \text{ µg}/\text{m}^3 \cong 1.7 \text{ mg/min}$$

This example is taken from a paper by Brauer et al. (2000), and the source emission rate of 1.7 mg/min for a 14.2-minute emission time period, or a total emissions of $G = (1.7 \text{ mg/min})(14.2 \text{ min}) = 24$ mg approximately, is within the range of normal source strengths (7 to 23 mg per cigarette, depending on the brand; see Figure 9.5 on page 214) for cigarettes reported in the scientific literature. Using this source strength to calculate the indoor particle concentrations with a typical particle decay rate and the typical volume of a home or a bedroom will result in surprisingly high indoor fine particle concentrations, which are similar to the relatively high indoor concentration levels from smoking that have been measured in real homes. It is common for the indoor air of a home with a smoker to exceed the federal outdoor National Ambient Air Quality Standards for fine particulate matter indoors, which has health implications for the other residents, including children living with the smoker.

18.10 INDOOR PARTICLE SOURCES

Previous sections of this chapter discussed the theoretical basis for applying the mass balance model to computing the indoor particle concentration time series from a known source emission rate, source duration, physical volume, ventilation rate, and deposition rate. It is important to consider how well any mathematical model works in practical applications in real settings. Fortunately, the mass balance model has proved itself to be surprisingly accurate and useful for a great many indoor applications and is an important tool for exposure analysis. This section presents several examples of its application. Predicting indoor particle concentrations from a cigarette in a model that includes a deposition term is relatively straightforward.

Consider a single cigarette smoked in a room with a constant fine particle source emission rate of $g_{source} = g_{cig} = 1.4$ mg/min. Suppose the cigarette is smoked in a 41 m^3 room with a particle decay rate of $\phi_P = 0.6$ hr^{-1} (0.01 min^{-1}). Substituting these parameter values into the solution for a rectangular source time function, Equation 18.14, with $\phi = \phi_P$ and $w = \phi_P v$ gives the following solution for the cigarette's source-on time period:

$$x(t) = \frac{g_{cig}}{w}(1 - e^{-\phi_P t}) \qquad \text{for } t \geq 0$$

$$= \frac{(1.4 \text{ mg/min})(1000 \text{ µg/mg})}{(0.01 \text{ changes/min})(41 \text{ m}^3/\text{change})}(1 - e^{-0.01t}) \qquad (18.44)$$

$$= 3{,}414.6(1 - e^{-0.01t}) \qquad \text{for } 0 \leq t \leq b$$

Equation 18.44 can be used to compute the maximum concentration x_{max} at time $t = b$:

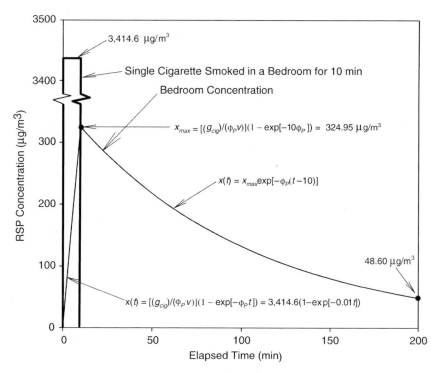

FIGURE 18.9 Example showing fine particle mass concentration predicted by the mass balance model for a cigarette smoked in a 41 m³ room for 10 minutes with a decay rate of $\phi_p = 0.6$ hr⁻¹ ($\phi_p = 0.01$ min⁻¹) and a source emission rate of $g_{cig} = 1.4$ mg/min. The vertical axis has been broken to show the top of the rectangular function representing the cigarette, which would cause an asymptote of 3,414.6 µg/m³ if the cigarette continued indefinitely instead of ending after 10 minutes.

$$x_{\max} = 3,414.6\,(1 - e^{-0.01t}) = 324.95\,\mu g/m^3 \tag{18.45}$$

Finally, substituting these values of the parameters x_{max}, $\phi = \phi_p$, and b into Equation 18.17 gives a concentration time function for the cigarette's source-off period:

$$x(t) = x_{\max}e^{-\phi_P(t - b)} = 324.95\,e^{-0.01(t - 10)} \quad \text{for} \quad t \geq 10 \tag{18.46}$$

The concentration time series for the cigarette predicted by the model in Equation 18.44 and Equation 18.46 over 200 minutes shows that the theoretical concentration rises very rapidly during the smoking period, then declines slowly over the next 3 hours or more (Figure 18.9). A sharp concentration rise, followed by a long decay period, shows the typical indoor air quality behavior when a cigarette is smoked indoors.

One way to compute the average value of the concentration time series in Figure 18.9 for any averaging time $T > 10$ min is by using the basic form of Equation 18.28. Suppose we wish to compute the average concentration predicted by the model in the room over the period $T = 200$ min. Using this method, we first must compute the mean of the source emission rate:

$$\overline{g(t)} = \frac{g_{cig}b}{T} = \frac{(1.4 \text{ mg/min})(10 \text{ min})}{200 \text{ min}} = 0.07 \text{ mg/min} \tag{18.47}$$

Noting that $x(T) = x(200) = 48.6\,\mu g/m^3$ and $\tau_P = 1/\phi_P = 1/0.01 = 100$ min, the mean concentration for the entire time period from $t = 0$ min to $t = 200$ min is computed from Equation 18.28 on page 430 in a single step:

$$\overline{x(T)} = \frac{\overline{g(T)}}{w} - \frac{\tau_P}{T}x(T)$$

$$= \frac{(0.07\ \text{mg/min})(1,000\,\mu g/mg)}{(0.01\ \text{min}^{-1})(41\ \text{m}^3)} - \frac{100\ \text{min}}{200\ \text{min}}(48.6\ \mu g/m^3) \qquad (18.48)$$

$$= 146.2\ \mu g/m^3$$

Using the approach in Equation 18.48 (the golden rule of averages) generally is easier than forming the individual exponential expressions for the average of each of the two piecewise continuous exponential functions and combining them (see Equation 18.49 to Equation 18.51 in Questions 2 and 3 and also Equation 18.52 in Question 4). Both approaches give the same value for the average concentration.

The above analysis describes a single cigarette smoked in a bedroom, but most real smokers smoke a sequence of cigarettes, one after another, or often a pack (20 cigarettes) per day. A computer model, called the Sequential Cigarette Exposure Model (SCEM), has been developed based on the rectangular solutions to predict the concentration time series for any cigarette activity pattern (Ott, Langan, and Switzer 1992). Figure 18.10 shows an example comparing polycyclic aromatic hydro-carbon (PPAH) concentrations measured in a 460 m³ home with PPAH concentrations predicted by the SCEM, which is written in QuickBASIC and runs on a PC. In this example, a real-time monitor that can measure PPAH concentrations at 1-minute time intervals (EcoChem, Inc., Murrieta, CA) was placed in an upstairs bedroom of the house. All exterior doors and windows were closed, and six Marlboro Regular cigarettes were smoked downstairs in the living room approximately 35 minutes apart. The concentration time series generated by SCEM actually consisted of 12 piecewise continuous exponential time functions, and the measured PPAH concentration in the upstairs bedroom showed reasonably good agreement with the predicted concentration time series. There was evidence of a time delay like the delay phenomenon described earlier, and there also was evidence that each cigarette smoked downstairs affected the concentration measured upstairs and throughout the house in rooms with doors open. In Figure 18.10, the distance between the downstairs source and the upstairs monitor was more than 13 m, or 39 ft.

In another indoor air experiment, an English muffin was toasted too long in a kitchen toaster, and the fine particulate mass concentration was measured using a piezoelectric microbalance monitor (Figure 18.11; TSI Model 8510 piezobalance). The piezobalance measures the $PM_{3.5}$ mass concentration reading every 2 minutes, which also is called respirable suspended particles (RSP), a size range that is of interest in cigarette smoke exposure studies. In this experiment, the exponential curve predicted by the mass balance model happened to coincide with the observed maximum concentration (see "coincident point"). A "char index" developed to describe cooking activity characterized this toast as "90% blackened" and is described in a larger study (Zartarian, Ott, and Wallace 1998). The source strength was calculated from these data using the peak estimation approach in Equation 18.38, giving $G = x_{max}v = (300\ \mu g/m^3)(460\ m^3) = 138$ mg.

In another experiment, Orville Redenbacher's butter popcorn was cooked in a microwave oven for 5 minutes instead of the usual time recommended on the label (halting cooking when the popping sounds occur at an interval of about 2–3 seconds). The 5-minute cooking time was estimated to be about 1-minute longer than the recommended cooking time, and a large quantity of smoke emerged when the microwave oven door was opened. Overcooking of popcorn by a minute or two often occurs by accident, and portions of the popcorn were still edible, although

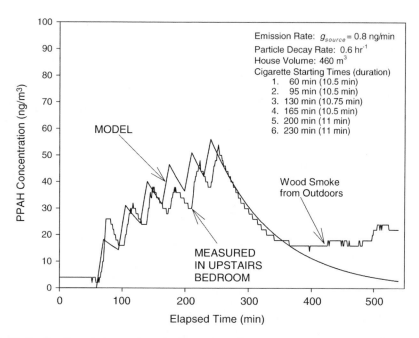

FIGURE 18.10 Predicted and observed particulate polycyclic aromatic hydrocarbon (PPAH) concentration measured in a 2-story house in which six Marlboro Regular Filter cigarettes were smoked approximately 35 minutes apart. The six cigarettes were smoked downstairs in the living room, and the PPAH concentrations wee measured in an upstairs bedroom at approximately a 13 m (or 39 ft) distance from the source. Near the elapsed time of $t = 350$ min (approximately 7:30 P.M.), fireplace activity in the surrounding neighborhood caused the background concentration to rise abruptly.

about 20% was charred too much to be eaten. Linear regression of the 77 measured fine particle (RSP, or PM$_{3.5}$) concentrations during the decay period yielded a particle decay rate of $\phi_P = 1.056$ hr^{-1} and produced a reasonably good agreement with a straight line fit by linear regression methods (Figure 18.12).

When the exponential solution to the mass balance model based on this particle decay rate is plotted alongside the concentration measurements, the exponential decay model fits the popcorn measurements quite well (Figure 18.13). This figure illustrates the phantom curve method for extending the exponential model to the "coincident" point, the time $t = t_s$ when the door of the microwave oven was opened and the smoke suddenly poured out. Because this emission time is so brief, the emission rate $g(t)$ cannot be characterized over a finite time period, and the emissions can be viewed as a near-instantaneous release of the pollutant. The good fit of the model to the data does not begin until time $t = t_e$, where we assume that the air of the home becomes sufficiently well mixed for the exponential solution to apply correctly.

In another experiment, a hamburger patty (ground round) was cooked in a frying pan for lunch. Salt and pepper were sprinkled on the patty to give it flavor, and a spatula was used to press the cooking burger firmly down against the pan to help insure that it was well done. The resulting hamburger patty surface was slightly charred and singed in a few places, but the well-done burger was edible and used to make a tasty hamburger sandwich. A total of eight 2-minute PM$_{3.5}$ piezobalance concentration measurements were made during the decay period, with about 20 minutes skipped between each measurement (Figure 18.14, top), and a semi-log plot of the data followed a reasonably straight line (Figure 18.14, bottom). Plotting the exponential decay solution to the mass balance equation with $\phi_P = 0.00889$ min^{-1} (0.533 hr^{-1}) from the regression analysis gave a fairly good fit of the model to the observed concentrations data on a linear graph (Figure 18.14, top).

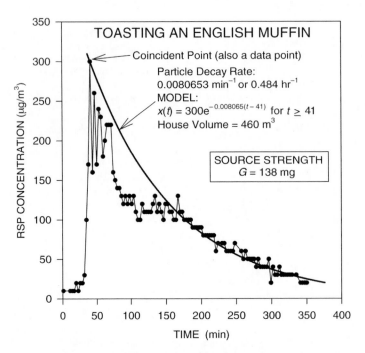

FIGURE 18.11 Fine particle mass concentration (PM$_{3.5}$) measured in the kitchen of a home in Redwood City, CA, after toasting an English muffin for too long (90% charred). The model parameters were estimated by the method of least squares using linear regression. Here, the phantom curve's coincident point happened to correspond to one data point.

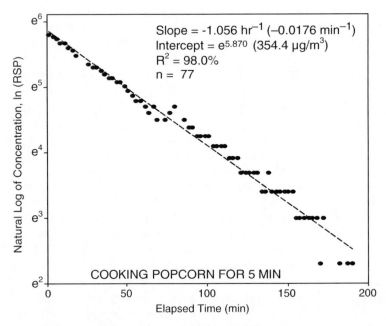

FIGURE 18.12 Semi-logarithmic plot of PM$_{3.5}$, or RSP, mass concentration time series for the decay curve generated by cooking Orville Redenbacher's butter popcorn for 5 minutes in a microwave oven, along with the straight line fit by linear least-squares regression.

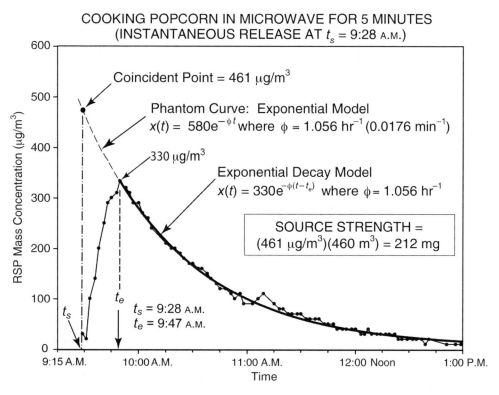

FIGURE 18.13 Comparison of exponential decay model fit by least-squares regression with RSP concentration for popcorn that was cooked for 5 minutes in a microwave oven (i.e., overcooked). This figure illustrates construction of an exponential phantom curve and calculation of the source strength for a source that has nearly an instantaneous emission release time. The instantaneous release occurred at t_s = 9:28 A.M. when the door of the microwave oven was opened, or 13 minutes after the graph began at 9:15 A.M. A smooth measured decay curve did not appear until t_e = 9:47 A.M. with its first data point at 330 μg/m^3, which was 32 minutes after the beginning of the graph. The particle decay rate of ϕ_p = 1.056 hr^{-1} that was estimated from the semi-log plot in Figure 18.12 allows us to write the equation $330 = x_o e^{-(1.056)(32/60)}$ for this data point, where x_o is the intercept at the origin. Solving this equation gives x_o = 580 μg/m^3 at the origin, producing a large portion of the phantom curve that is extrapolated from the measured decay curve. Finally, substituting the release time of t_s into this equation (13 minutes after the start of the graph), we obtain a peak concentration of 461 μg/m^3, which, based on a house volume of 460 m^3, gives the source strength for the burned popcorn of G = 212 mg.

In another experiment, a professional chef was invited to cook "blackened catfish," a Louisiana delicacy, in the same home in which the other experiments were performed. When the fish is cooked according to this recipe, the charred skin causes considerable smoke, and the chef found it necessary to use the kitchen hood fan. Linear regression of the natural logarithm of the indoor PM$_{3.5}$ concentrations vs. time for 79 data points gave a particle decay rate parameter of ϕ_P = 0.0165 min^{-1} (0.99 hr^{-1}) with R^2 = 99.2%. Measurement of the particle concentrations from cooking the blackened catfish dinner gave a similar times-series plot in both the kitchen and the living room (Figure 18.15). The similarity of concentrations in different rooms of the home in these studies helps support the important assumption of the mass balance model that this home behaves as a well-mixed compartment.

By using the peak estimation approach to determine the source strength by solving Equation 18.38 as $G = vx_{max}$, the fine particle source strengths measured for a variety of common indoor sources are summarized in Table 18.2. The results show that the occasional burning of toast or cooking of smoky foods such as blackened catfish can generate more particulate matter emissions

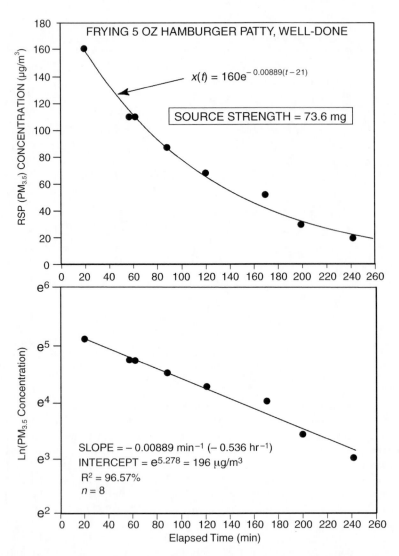

FIGURE 18.14 Measured RSP concentration after cooking a 5 oz. hamburger patty, well-done with $n = 8$ observations, and $R^2 = 96.6\%$ (top), along with the semi-logarithmic plot of the same data showing a relatively straight line (bottom).

indoors than an ordinary cigarette. However, the cigarette still is a more serious source of indoor particulate air pollution, because a smoker may smoke a pack (20 cigarettes) per day, while a family probably will cook foods that generate smoke less frequently than once per day. High indoor particle concentrations from cooking are likely to occur on a random basis, though in some homes elevated emissions may occur more frequently than is commonly assumed. More study is needed of cooking activity patterns, habits, and their effect on indoor pollutant concentrations. In contrast with cooking, because smoking is habitual, a person who routinely smokes as many as 20 cigarettes per day at home is likely to cause elevated indoor particle concentrations in the home almost all day, nearly every day.

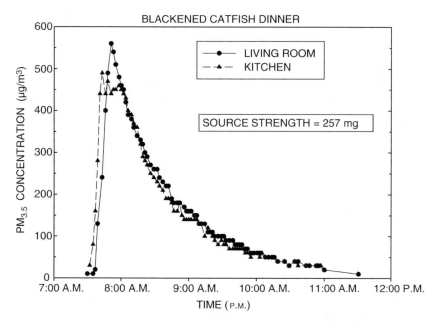

FIGURE 18.15 $PM_{3.5}$ mass concentrations measured in two rooms of a home after a blackened catfish dinner for two was prepared by a professional chef. Following this recipe produces considerable smoke, which caused the cook to turn on the kitchen hood ventilation fan. The similarity of the measurements in two rooms helps illustrate that this home behaves as a well-mixed compartment.

TABLE 18.2
Summary of $PM_{3.5}$ Source Strengths from Indoor Experiments

Indoor Source	Location	Source Duration (min)	Decay Rate ϕ_P (hr^{-1})	Total Emissions G (mg)
Frying hamburger, medium	Two-story house, 460 m³	–	0.46	69
Frying 5 oz hamburger, well-done	"	10	0.54	74
Frying 4.5 oz hamburger, well-done	"	8	0.63	78
Cooking two filet mignon steaks, rare	"	9	0.90	51
Blackened catfish, dinner for two	"	–	0.99	257
Burning popcorn	"	–	1.1	212
Accidental toast burning	"	–	–	178
Toasting English muffin, 90% char	"	–	0.48	138
Toasting wheat bread, 70% char	Bedroom, 41 m³	2.8	–	6
Toasting wheat bread, 85% char	"	3.1	0.77	11
Toasting wheat bread, 95% char	"	3.7	–	37
Marlboro filter cigarette	Two-story house, 460 m³	10	–	15
Kentucky research cigarette	"	7	–	37
Ashton cigar	"	10	–	32
Paul Garmirian cigar	Parlor, 97 m³	90	1.2	88
Marlboro filter cigarette	"	9	1.3	17

18.11 QUESTIONS FOR REVIEW

1. Discuss and describe the differences between the α-period, β-period, and the γ-period with regard to the mixing of a pollutant emitted in a room.

2. Derive the following equations for the average concentration as a function of time for a rectangular source input with emission rate g_{source} that begins at $t = 0$ and stops emitting at $t = b$:

$$\overline{x_A(T)} = \frac{g_{source}}{w}\left(1 - \frac{\tau}{T}\left[1 - e^{-T/\tau}\right]\right) \qquad \text{for} \quad 0 < T \le b \qquad (18.49)$$

$$\overline{x_B(T)} = \frac{\tau}{T-b}\frac{g_{source}}{w}\left(1 - e^{-b/\tau}\right)\left(1 - e^{-\frac{T-b}{\tau}}\right) \qquad \text{for} \quad b < T \qquad (18.50)$$

The first equation for $\overline{x_A(T)}$ is the average for the first piecewise continuous portion of the concentration curve from t = 0 to $t = b$ while the second equation for $\overline{x_B(T)}$ is the average for only the second continuous part of the curve for $b < T$.

3. Sketch the areas under the two curves of the piecewise continuous solution to the mass balance equation for which the two equations above apply. Obtain a general expression for the overall average for $b < T$ by substituting the equations above into the following expression for the overall mean concentration:

$$\overline{x(T)} = \frac{b}{T}\overline{x_A(b)} + \frac{(T-b)}{T}\overline{x_B(T)} \qquad (18.51)$$

4. The overall average indoor concentration $\overline{x(T)}$ for $b \le T$ is given by the following equation:

$$\overline{x(T)} = \frac{g_{source}}{w}\left(\frac{b}{T} - \frac{\tau}{T}\left[e^{-\frac{T-b}{\tau}} - e^{-T/\tau}\right]\right) \qquad \text{for} \quad b \le T \qquad (18.52)$$

Show that Equation 18.52 can be derived by substituting Equation 18.49 and Equation 18.50 into Equation 18.51. Discuss the differences between this mathematical approach for computing the average concentration and the "golden rule of averages" approach shown in Equation 18.28.

5. Suppose a person at home chooses to burn sage, a fiber material that is unscented and similar to incense. It is usually burned indoors for spiritual purposes. Suppose that sage is burned indoors for 30 minutes, releasing a total quantity of 300 mg of $PM_{2.5}$ particulate matter. The shape of the emission rate curve with time is not known, even though the area under the curve (total emissions) is known. If the decay rate in the well-mixed home is $\phi_P = 0.82$ hr^{-1} and the volume is $v = 460$ m^3, compute the average particle concentration after 1 hour using the golden rule of averages (Equation 18.28). After 3 hours, what is the average particle concentration? Starting at 1/2 hour after the 30 minutes end, plot a graph of the predicted concentration in the home for each successive hour. Make a similar graph of the predicted average indoor concentration in the home. Notice we can make

an exact prediction of the indoor concentrations after the time $t = b$, even though little is known about the original source emission rate $g(t)$ as a function of time.

18.12 ACKNOWLEDGMENTS

Grateful appreciation is given to the Flight Attendant Medical Research Institute (FAMRI) for partial funding of this chapter and its research.

REFERENCES

Baughman, A.V., Gadgil, A.J., and Nazaroff, W.W. (1994) Mixing of a Point Source Pollutant by Natural Flow Convection within a Room, *Indoor Air*, **4**: 114–122.

Blanchard, P., Devaney, R.L., and Hall, G.R. (1998) *Differential Equations*, Brooks/Cole Publishing Company, Pacific Grove, CA.

Brauer, M., Hirtle, R., Lang, B., and Ott, W. (2000) Assessment of Fine Aerosol Contributions from Environmental Tobacco Smoke and Cooking with a Portable Nelphelometer, *Journal of Exposure Analysis and Environmental Epidemiology*, **10**: 136–144.

Bridge, D.P. and Corn, M. (1972) Contribution to the Assessment of Exposure of Nonsmokers to Air Pollution from Cigarette and Cigar Smoke in Occupied Spaces, *Environmental Research*, **5**: 192–209.

Dockery, D.W. and Spengler, J.D. (1981) Indoor-Outdoor Relationships of Respirable Sulfates and Particles, *Atmospheric Environment*, **15**: 335–343.

Furtaw, E.J., Pandian, M.D., Nelson, D.R., and Behar, J. (1996) Modeling Indoor Air Concentrations Near Emission Sources in Imperfectly Mixed Rooms, *Journal of the Air and Waste Management Association*, **46**: 861–868.

Howard-Reed, C., Wallace, L.A., and Ott, W.R. (2002) The Effect of Opening Windows on Air Change Rates in Two Homes, *Journal of the Air and Waste Management Association*, **52**: 147–159.

Jones, R.M. (1974) Application of a Mathematical Model for the Buildup of Carbon Monoxide from Cigarette Smoking in Rooms and Houses, *American Society of Heating, Refrigeration, Air Conditioning Engineers (ASHRAE) Journal*, **16**: 49–53.

Klepeis, N.E., Ott, W.R., and Repace, J.L. (1999) The Effect of Cigar Smoking on Indoor Levels of Carbon Monoxide and Particles, *Journal of Exposure Analysis and Environmental Epidemiology*, **9**: 622–635.

Klepeis, N.E., Ott, W.R., and Switzer, P. (1996) A Multiple-Smoker Model for Predicting Indoor Air Quality in Public Lounges, *Environmental Science and Technology*, **30**(9): 2813–2820.

Mage, D.T. and Ott, W.R. (1996) Accounting for Nonuniform Mixing and Human Exposure in Indoor Environments, in *Characterizing Sources of Indoor Air Pollution and Related Sink Effects*, ASTM publication code number (PCN): 04-12870-17, Tichenor, B.A., Ed., American Society for Testing and Materials, West Conshohocken, PA, 263–278.

McBride, S.J., Ferro, A.R., Ott, W.R., Switzer, P., and Hildemann, L.M. (1999) Investigations of the Proximity Effect for Pollutants in the Indoor Environment, *Journal of Exposure Analysis and Environmental Epidemiology*, **9**: 602–621.

Nazaroff, W.W. and Cass, G.R. (1989) Mathematical Modeling of Indoor Aerosol Dynamics, *Environmental Science and Technology*, **23**: 157–166.

Ott, W.R. (1999) Mathematical Models for Predicting Indoor Air Quality from Smoking Activity, *Environmental Health Perspectives*, **107**(supp. 2): 375–381.

Ott, W.R., Klepeis, N., and Switzer, P. (2003) Analytical Solutions to Multicompartment Indoor Models with Application to Environmental Tobacco Smoke Measured in a Home, *Journal of the Air and Waste Management Association*, **53**: 918–936.

Ott, W.R., Langan, L., and Switzer, P. (1992) A Time Series Model for Cigarette Smoking Activity Patterns: Model Validation for Carbon Monoxide and Respirable Particles in a Chamber and an Automobile, *Journal of Exposure Analysis and Environmental Epidemiology*, **2**(supp. 2): 175–200.

Ott, W.R., McBride, S., and Switzer, P. (2002) Mixing Characteristics of a Continuously Emitting Point Source in a Room, in *Proceedings of the 9th International Conference on Indoor Air Quality and Climate*, Paper No. 4B4o6, Monterey, CA, June 30–July 5.

Ott, W.R., Switzer, P., and Robinson, J. (1996) Particle Concentrations Inside a Tavern Before and After Prohibition of Smoking: Evaluating the Performance of an Indoor Air Quality Model, *Journal of the Air & Waste Management Association*, **46**: 1120–1134.

Repace, J.L. and Lowrey, A.H. (1980) Indoor Air Pollution, Tobacco Smoke, and Public Health, *Science*, **208**: 464–472.

Switzer, P. and Ott, W. (1992) Derivation of an Indoor Air Averaging Time Model from the Mass Balance Equation for the Case of Independent Source Inputs and Fixed Air Exchange Rates, *Journal of Exposure Analysis and Environmental Epidemiology*, **2**(supp. 2): 113–135.

Traynor, G.W., Anthon, D.W., and Hollowell, C.D. (1982) Technique for Determining Emissions from a Gas-Fired Range, *Atmospheric Environment*, **16**: 2979–2988.

Wadden, R.A. and Scheff, P.A. (1983) *Indoor Air Pollution: Characterization, Prediction, and Control*, Wiley & Sons, New York, NY.

Zartarian, V.G., Ott, W.R., and Wallace L.A. (1998) Experiments to Measure the Effect of Cooking Activities on Indoor Air Quality in Homes, Paper No. 178 O, presented at the 8th Annual Conference of the International Society of Exposure Analysis, Boston, MA.

19 Modeling Human Exposure to Air Pollution

Neil E. Klepeis
Stanford University

CONTENTS

19.1   Synopsis 445
19.2   Introduction 445
19.3   Basic Formulas Used to Model Inhalation Exposure 446
19.4   An Illustrative Exposure Simulation 448
19.5   Human Activity Pattern Data 450
19.6   Practical Uses of Exposure Modeling 456
19.7   Review of Some Existing Inhalation Exposure Models 457
19.8   Advancing the Science of Exposure 460
        19.8.1 Models as Theory 460
        19.8.2 The Vanguard of Exposure Modeling 460
                19.8.2.1 Direct Evaluation of Model Predictions 461
                19.8.2.2 Understanding the Local Dispersion of Indoor and
                         Outdoor Pollutants 462
                19.8.2.3 Human Factors 463
19.9   Questions for Review 465
19.10   Acknowledgments 466
References 466

19.1 SYNOPSIS

This chapter is an introduction to the simulation of human exposure to air pollution. It includes a review of basic inhalation exposure models, in which air concentrations are matched with individual human activity patterns. Since people spend most of their time inside buildings, and the modeling of indoor pollutant concentrations is simpler than for outdoor pollutants, the emphasis is on indoor exposures. Separate sections are devoted to residential exposure to secondhand tobacco smoke and a recent representative survey of U.S. time–location patterns. Material is included on the advantages associated with the modeling of exposure as part of exposure assessment studies with respect to public health objectives. The final section discusses possible future directions in exposure modeling, including general approaches to model evaluation.

19.2 INTRODUCTION

Exposure to air pollution occurs whenever a human being breathes air in a location where there are trace amounts of one or more airborne toxins. To model exposure to airborne elements, one uses the conceptually simple approach of matching the locations that each exposed person visits with the time-averaged or dynamic air pollutant concentrations that are thought to exist in each visited location. Exposure models simulate exposures for either real or hypothetical individuals and populations.[1] Inhalation exposure models do not strictly take into account the inhaled dose of toxic airborne species, but only the presence of air pollutants near the breathing zone of a person.[2]

The modeling ideas introduced in this chapter apply equally well to indoor and outdoor sources of air pollution. However, people spend most of their time indoors, and it is generally easier to model indoor pollutant behavior from simple first principles. Therefore, the focus of this chapter is on exposure occurring inside buildings.

19.3 BASIC FORMULAS USED TO MODEL INHALATION EXPOSURE

An important concept to understand in this chapter is the canonical mathematical formalism used to describe human exposure. How do exposure modelers go about calculating exposure?

Two fundamental pieces of information are necessary to calculate exposure: (1) the whereabouts of the human beings who are being exposed; and (2) the concentration of pollutants in different locations. These two inputs are typically obtained simultaneously in the course of a single exposure study, or they may be drawn from two or more independent studies. In more sophisticated exposure models, they may be simulated using either deterministic or stochastic algorithms. Regardless of the complexity associated with specifying inputs for a given model, the same basic equation underlies all exposure models.

The mathematical formulation of exposure to air pollutants was first established by Fugas (1975), Duan (1982), and Ott (1982, 1984) and was dubbed the *indirect exposure assessment approach* in contrast to direct approaches in which exposure is measured using personal monitoring equipment. These early researchers introduced the concept of calculating exposure as the sum of the product of time spent by a person in different locations and the time-averaged air pollutant concentrations occurring in those locations. In this formulation, locations are termed microenvironments, and they are assumed to have homogeneous pollutant concentrations. The standard mathematical formula for integrated exposure is written as follows:

$$E_i = \sum_{j=1}^{m} C_{ij} T_{ij} \qquad (19.1)$$

where T_{ij} is the time spent in microenvironment j by person i with typical units of minutes, C_{ij} is the air pollutant concentration person i experiences in microenvironment j with typical units of micrograms per cubic meter [$\mu g/m^3$], E_i is the integrated exposure for person i [$\mu g/m^3$ min], and m is the number of different microenvironments.

The calculation amounts to a weighted sum of concentrations with the weights being equal to the time spent experiencing a given concentration. Each discrete time segment and its associated discrete concentration need not be sequential in time (i.e., there may be discontinuities in time and space), although Equation 19.1 is usually applied to contiguous time segments adding up to some convenient duration, such as a single day. Average personal exposure in concentration units of $\mu g/m^3$ is calculated by dividing E_i by the total time spent in all microenvironments.

---

[1] Simulation, in general, involves the artificial depiction of events with the intention of closely mimicking reality.
[2] You can find a general definition for exposure to all kinds of air pollution in Chapter 2 of this book.

The basis for the temporally and spatially discrete Equation 19.1, in which C_{ij} are supplied as average concentrations or concentrations that are constant during each corresponding time period T_{ij}, can be considered to arise theoretically from a fully continuous formulation:

$$E_i = \int_{t_2}^{t_1} C_i\left(t, x, y, z\right) dt \tag{19.2}$$

where $C_i(t, x, y, z)$ is the concentration occurring at a particular point occupied by the receptor i at time t and spatial coordinate (x, y, z), and t_1 and t_2 are the starting and ending times of a given exposure episode.

Time-dependent personal exposure profiles can be measured using real-time personal monitoring devices that are affixed to people as they move within and between all the locations that are a part of their daily routines. If discrete microenvironments are considered rather than fully continuous space, then the following semi-continuous formulation applies:

$$E_i = \sum_{j=1}^{m} \left(\int_{t_{j1}}^{t_{j2}} C_{ij}\left(t\right) dt \right) \tag{19.3}$$

where $C_{ij}(t)$ is the concentration experienced by the receptor in the discrete microenvironment j at a particular point in time t over the time interval defined by $[t_{j1}, t_{j2}]$, where t_{j1} is the starting time for the microenvironment and t_{j2} is the ending time.

Whereas in Equation 19.2 the exposure trajectory of the receptor is followed explicitly with no discontinuities, in Equation 19.3 there are no time discontinuities within any given microenvironment, but microenvironments need not correspond to contiguous time periods. With this formulation it is easy to see how arbitrary exposure profiles can be constructed by combining a variety of distinct microenvironment episodes—each with their own distinct concentration profile. The sum of integrals in Equation 19.3 can be written as a fully discrete sum of {average-concentration × elapsed-time} products (i.e., the form of Equation 19.1).

If the same microenvironment concentrations are used for every person, a simple population version of Equation 19.1 can be derived in terms of the total time spent by all receptors in each microenvironment:

$$\tilde{E} = \sum_{j=1}^{m} C_j \tilde{T}_j \tag{19.4}$$

where m is the number of microenvironments visited, C_j is the average pollutant concentration in microenvironment j assigned to every person i, $\tilde{E}$ is the integrated exposure over all members of the population, $\tilde{T}_j = \sum_{i=1}^{n} T_{ij}$ (i.e., the total time spent by all persons in microenvironment j) and n is the total number of people in the population being modeled. If each person spends the same total amount of time across all microenvironments, $T = T_i = \sum_{j=1}^{m} T_{ij}$, then the average personal

exposure for the population in units of concentration (e.g., $\mu g/m^3$) for the population is:

$$\bar{E}_c = \frac{1}{nT} \sum_{j=1}^{m} C_j \tilde{T}_j \qquad (19.5)$$

19.4 AN ILLUSTRATIVE EXPOSURE SIMULATION

To provide a concrete focal point for later discussions of exposure models, this section presents the application of a real simulation model to the case of residential secondhand tobacco smoke (SHS) exposure. This example should help to address what may be the most basic question for a newcomer to exposure modeling: What does the output of an actual exposure model look like?

The SHS exposure model we will be using treats multizonal pollutant and human location dynamics by incorporating dynamic pollutant emissions and household dispersion and the complex spatial trajectories of smoking and nonsmoking household members. In keeping with the fundamental exposure formulation presented above, the occurrence of an exposure event depends on the concurrence in time and space of pollutant concentrations and a human being.

Our model incorporates a dynamic mass-balance indoor air quality (IAQ) model that accounts for (1) airborne particle emissions from smoking activity in any room at any moment in time, (2) outdoor air exchange rates, (3) transport of particles between rooms, (4) particle removal via outdoor air exchange, (5) and particle loss through surface deposition.[3] The central assumption of the indoor air model is instantaneous mixing of airborne particles within each room. While the model includes consideration of natural leakage ventilation through building cracks and airflow across interior doorways, it does not consider airflow across open windows or changes in airflow due to the operation of a central air handling system.

The input parameter values for the model have been selected so that they fall approximately in the middle range of values reported in the scientific literature. The hypothetical house, whose layout is pictured in Figure 19.1, has five zones on a single level with a total volume of 220 m³. In this house, the hallway mediates airflow between each of the three main rooms, and the bathroom is connected only to the bedroom. The whole-house leakage air exchange rate is 0.5 ach, and airflow rates through open and closed doors are assumed to be 100 and 1 m³/h, respectively. The size-integrated deposition rate for SHS particles, which adhere irreversibly to household surfaces, is 0.1 ach. The duration of each cigarette smoked in the house is assumed to be 10 minutes, with each cigarette having 10 milligrams of total particle emissions.

Although the above physical input parameter values are held fixed, pollutant emissions and house airflow characteristics can change over time due to the behavior of household occupants who may smoke cigarettes in different rooms and close doors of rooms they occupy. To supply realistic movement patterns for people in the house, a pair of time–location profiles, corresponding to a smoker and a nonsmoker, was randomly sampled from an empirical activity pattern diary data set (these data are described in Section 19.5). The occupants are assumed to be spouses who sleep in one bedroom.

In this simulation, the smoker consumes 15 cigarettes in the main rooms of the house between about 7:00 A.M. and 8:00 P.M. The SHS particle concentration time profiles in each room of the

---

[3] The IAQ model is defined by a set of n coupled differential equations, one corresponding to each room. The differential equations are solved numerically using a Runge–Kutta algorithm to obtain dynamic airborne particle concentrations in each room of the house.

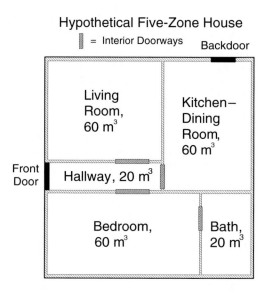

FIGURE 19.1 Floor plan for a hypothetical five-zone house, which provides the environment for an illustrative simulation of secondhand tobacco smoke room and personal exposure. The house has three main rooms of equal size plus a master bathroom and a hallway. The main rooms are interconnected via doorways to the centrally located hallway. See Figure 19.2 and Figure 19.3 for the simulation results.

house resulting from these cigarettes are presented in Figure 19.2 for the case when doors are generally left open in the house, except during time spent in the bathroom or sleeping in the bedroom (the door-open case). Figure 19.3 shows the case for when the smoking room door is closed during smoking episodes in which the nonsmoker and smoker occupy separate rooms (the door-closed case). In addition to room concentrations, each figure also shows the time–location patterns and exposure profiles of the smoker and nonsmoker house occupants and the smoker's active smoking profile.

For the door-open case, the 24-hour average SHS particle concentrations are highest in the living room and kitchen-dining room (69 and 49 μg/m^3, respectively), where most of the cigarettes are smoked. The SHS exposure of the smoker (not including his or her direct exposure from smoking the cigarettes) is comparable to the 24-hour concentrations in the rooms with the most smoking (57 μg/m^3). In contrast, the nonsmoker spends part of the time either out of the house or in rooms away from active smoking, so his or her 24-hour SHS particle exposure is significantly lower than that of the smoker (38 μg/m^3). For different nonsmoker time–location patterns where a person might spend either more or less time in the same room as the smoker, exposure can approach or exceed that of the smoker, or perhaps be much lower.

For the door-closed case, where doors to rooms are closed when the active smoker is alone in a room where he or she smokes, the 24-hour average living room concentration is much higher than before (91 μg/m^3), whereas all of the other rooms have lower average concentrations. This situation arises because the living room is the location where the smoker spends most of his or her time alone. The smoker's exposure increases dramatically to 81 μg/m^3 in the closed-door case due to the significant amount of time he or she spends in a smoke-filled room with practically no air exchange with other parts of the house. The nonsmoker experiences elevated peak levels close to 400 μg/m^3 upon entering the smoke-filled living room in the closed-door case vs. only about 200 μg/m^3 when the doors were left open.

These simulation results illustrate how the zonal character of a house can result in quite different SHS concentration in different rooms, as well as significant differences in 24-hour exposures for different household occupants. Taking the simulation approach a few steps further, it would be possible to explore how changes in multiple door and window positions, central air handling, and

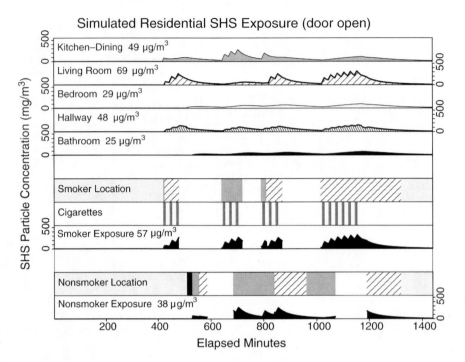

FIGURE 19.2 Simulated 24-hour time profiles for room particle concentrations [µg/m³] (top panels), selected occupant-specific behavior patterns, and occupant exposure [µg/m³] (middle and bottom panels) for the case when doors are left open in the house, except when occupants are sleeping or in the bathroom. Each profile starts and ends at midnight. Occupant-specific activity profiles are included for the cigarette and location behavior of a single smoker–nonsmoker pair. The 24-hour average room and exposure are included in the appropriate panels. The simulated exposure profile for each person is positioned below each group of behavior profiles. The grayscale shading and hatch patterns that have been used to draw each room concentration match the fill patterns used in the location profiles. White space in the activity profiles corresponds to "absent from house" and "inactive" conditions for location and cigarette profiles, respectively. Filled segments correspond to the opposite condition.

active filtration can affect residential SHS exposure. Using time-diaries of household occupants sampled from a real population, one can estimate frequency distributions of exposure for typical time–location patterns.

19.5 HUMAN ACTIVITY PATTERN DATA

The strong influence of human activity patterns on exposure is evident from Equation 19.1 and the results of the example exposure simulation presented above, where the movement of house occupants between different rooms has a sizable impact on 24-hour average exposures. Human activity data are routinely collected as part of individual exposure assessment studies. Several large-scale human activity pattern databases are also available for populations in North America.

The most detailed and representative human activity and location study conducted for the U.S. population is the National Human Activity Pattern Survey (NHAPS), which was sponsored by the U.S. Environmental Protection Agency (USEPA) and carried out in the early-to-mid 1990s (Klepeis et al. 2001). Both NHAPS, and the subsequent Canadian Human Activity Pattern Survey (CHAPS) (Leech et al. 1996), were patterned after a set of studies conducted in California (Jenkins et al.

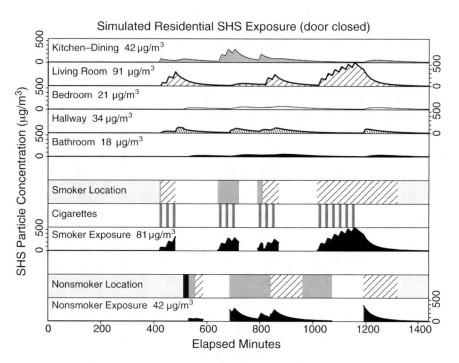

FIGURE 19.3 Simulated 24-hour time profiles for room particle concentrations [μg/m³] (top panels), selected occupant-specific behavior patterns, and occupant exposures [μg/m³] (middle and bottom panels) for the case when doors are closed in smoking rooms during smoking episodes when the smoker and nonsmoker are in separate rooms (i.e., the door is left open during smoking episodes only when the smoker and nonsmoker are in the same room). See Figure 19.2 and its caption for more information on the plot and for simulation results when the smoker's door is always left open during smoking episodes. Notice how the concentrations in the living room, during times when the active smoker is alone, are much higher when the doors are closed. Consequently, when the nonsmoker enters the living room soon after smoking has stopped, he or she receives a higher exposure than if the door had been open for the entire smoking episode.

1992; Wiley et al. 1991a,b). The USEPA's consolidated human activity database (CHAD) contains data from many recent human activity surveys, including NHAPS (McCurdy et al. 2000).[4]

The NHAPS respondents comprise a representative cross-section of 24-hour daily activity patterns in the contiguous United States.[5] The 9,386 NHAPS respondents, who were interviewed by telephone, gave a minute-by-minute diary account of their previous day's activities, including the places they visited and the presence of a smoker in each location.[6] Detailed information was provided on the rooms that each respondent visited while in residences, whether their own or one they were visiting. Since NHAPS contains the precise sequence and duration of human locations for a large sample of people, with room-specific categories for time spent at home, it presents a rich resource for use in understanding the frequency distribution of exposures to a variety of pollutants for which a single 24-hour period is an appropriate time scale, e.g., for secondhand smoke exposure in the residential indoor environment.

---

[4] The NHAPS data are also available at the ExposureScience.Org website, http://exposurescience.org, along with other exposure-related materials, including research articles and modeling software.

[5] Note that NHAPS is biased because it undersamples people who are homeless, on vacation, or without telephones, and excludes those who are institutionalized or in the military.

[6] The time reported in the presence of a smoker may be a biased predictor of actual secondhand tobacco smoke exposure, because of complications surrounding awareness of smokers, smoke persistence, and proximity to smokers.

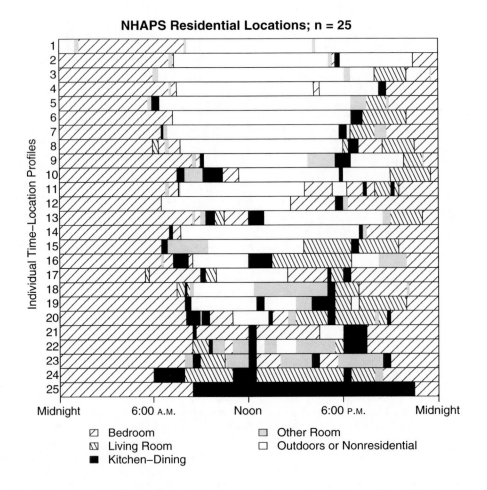

FIGURE 19.4 Residential time–location profiles for a random sample of 25 out of the 9,386 NHAPS respondents living in detached houses in the contiguous United States. White space indicates time that was spent outside of the home or away from home. The timelines are sorted from bottom to top by the amount of time spent at home.

Figure 19.4 illustrates the character of the NHAPS time–location data using plots of stacked timelines across different residential locations. The plot shows 25 randomly sampled NHAPS respondent diaries, each represented by a horizontal strip with different patterns and shades designating the different rooms the respondent was reported to visit. The four residential locations depicted in this figure are a reduced but exhaustive set derived from the 15 total residential locations that were coded for each NHAPS respondent. Figure 19.5 contains a plot of the time–location profiles for 25 randomly sampled participants from the USEPA's PTEAM study conducted in Riverside, CA (Özkaynak et al. 1993, 1996). This study was an exposure monitoring study, which was not focused on the gathering of time-activity patterns, but which provides another example of empirical activity pattern data. The most striking feature of the time–location plots in Figure 19.4 and Figure 19.5 is the overwhelming amount of time spent at home over a 24-hour time block. Even the portion of each sample that spent the least amount of time at home still spent the bulk of the 12-hour period between 8 P.M. and 8 A.M. at home.

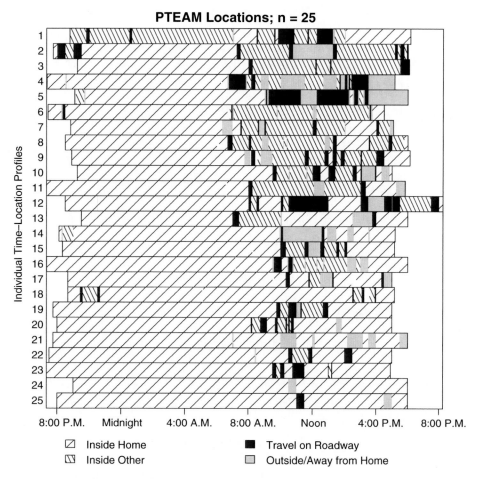

FIGURE 19.5 Time–location profiles for a random sample of 25 of the 178 participants in the USEPA's PTEAM study conducted in Riverside, CA. White space before and after each profile indicates time not accounted for in the study. The timelines are sorted from bottom to top by the amount of time spent at home.

Aggregate statistics for comprehensive time spent by NHAPS respondents in six locations over the 24-hour day are given in Table 19.1. These include the overall average time spent in each location taken across all of the NHAPS respondents, the overall average percentage of time spent in each location, the percentage of respondents that reported being in each location (i.e., the doers), and the average time spent by the doers in each location. More analysis of the NHAPS diary, disaggregated by demographic and health variables, is available from Klepeis et al. (2001), Klepeis, Tsang, and Behar (1996), and Tsang and Klepeis (1996). The results presented here indicate that over 90% of time is spent indoors or in a vehicle and that the home is undeniably the location where one spends the bulk of one's life. All but a very small percentage of sampled Americans spent time in their own home on the day just before they were interviewed, being at home for an average time of more than 16 hours, or two thirds of the day.

A conspicuous feature of the time spent in different rooms of detached homes by NHAPS respondents, as evident from the per-room statistics presented in Table 19.2, is that almost 98% of interviewed Americans spend time in the bedroom for more than 9 hours, on average, which is 58% of the time spent, on average, in any location in or around the house. Taken together, the kitchen, living room, and bedroom account for over 85% of the total time spent at home, with 5%

TABLE 19.1
Overall Weighted Statistics for Time Spent by NHAPS Respondents in Six Different Group Locations over a 24-Hour Period[a]

Location	Average Time (min)	Average Time %[b]	Doer %	Doer Average Time (min)
In a Residence[c]	990	68.7	99.4	996
Office-Factory	78	5.4	20.0	388
Bar-Restaurant	27	1.8	23.7	112
Other Indoor	158	11.0	59.1	267
In a Vehicle	79	5.5	83.2	95
Outdoors	109	7.6	59.3	184

[a] Means and percentages have been calculated using sample weights.

[b] This overall average percentage time spent was calculated by dividing the mean number of minutes spent by NHAPS respondents in each location by the total time spent on the diary day (i.e., 24 hour = 1,440 min).

[c] The "In a Residence" category includes time spent in one's own home or in another person's home.

TABLE 19.2
Overall Statistics for Time Spent by NHAPS Respondents Living in Detached Homes in Different Rooms of Their Residences over a 24-Hour Period[a]

Location	Average Time (min)	Average Time %[b]	Doer %	Doer Average Time (min)
Kitchen	75.3	7.2	77.2	97.6
Living, Family, Den	199.5	19.3	81.4	245.2
Dining Room	13.8	1.3	19.5	70.6
Bathroom	24.5	2.7	70.9	34.5
Bedroom	547.4	58.0	97.6	560.6
Study, Office	9.8	0.9	4.3	227.1
Garage	3.2	0.3	2.7	117.2
Basement	5.2	0.5	3.7	141.4
Utility, Laundry	3.9	0.4	5.3	72.7
Pool, Spa	1.0	0.1	1.0	98.4
Yard, Outdoors	40.2	3.6	28.7	140.1
Room to Room[c]	54.6	5.0	40.6	134.5
In and Out of House	6.3	0.6	6.6	94.5
Other, Verified	1.9	0.2	1.5	129.1
Refused to Answer	0.3	0.0	0.3	131.4

[a] All statistics are unweighted.

[b] The overall average percentage time spent was calculated by averaging the individual percentages of time spent in each residential location, which are taken over the total time spent by each individual in all residential locations. This total time spent in residential locations varied from individual to individual.

[c] The room-to-room location was likely a fallback for respondents who were unsure where they were, or who visited many rooms over a short time period.

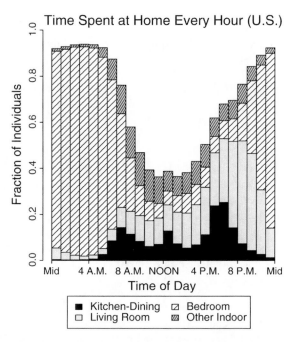

FIGURE 19.6 Stacked bar chart showing the overall fraction of NHAPS respondents living in detached houses who spent time in various locations in their homes during each hour of the day.

taken up with time reported as moving from room to room, which may have been a fallback category for some respondents, and less than 5% for any other house location.

Figure 19.6 presents the fraction of NHAPS respondents that spent the bulk of each hour of the day in different rooms of their detached houses, focusing on the most predominant rooms (i.e., kitchen or dining room, living room, and bedroom). From this figure, it is apparent that the largest fraction of individuals are in the bedroom until about 9 A.M. and after 11 P.M., as might be expected. During the middle of the day, and especially between 6 P.M. and 10 P.M., more Americans are in the kitchen and living room than in any other rooms of the house, although about 40–60% of Americans are away from home between the hours of 9 A.M. and 6 P.M.

Although NHAPS offers a rich and representative human activity pattern data set, the data are somewhat limited for use in understanding exposures occurring in complex environments. The interaction of individuals in a house environment cannot be fully characterized by independent activity profiles from unassociated individuals, such as those collected as part of NHAPS. In addition, the NHAPS time-diary data do not contain information on activities that are likely to affect pollutant emission or removal in a given location, such as the operation of appliances, the smoking of cigarettes, filtration practices, or flow-related activities involving windows, doors, or mechanical air handling. Nevertheless, NHAPS, and similar databases, can be used to explore frequency distributions of residential exposure occurring in multiple-person households by superimposing hypothetical or separately observed window, door, ventilation, and source-related activity patterns onto time–location patterns, and by matching individual time–location diaries for persons in a hypothetical household based on selected temporal or demographic characteristics, such as age, gender, or day of the week. This general approach for a single pair of matched NHAPS respondents was used in the example simulations presented above in Section 19.4.

19.6 PRACTICAL USES OF EXPOSURE MODELING

Who uses exposure models? Are they really helpful to professionals in the health and environment fields? To help shed light on these questions, consider the following:

1. You are an academic researcher involved in a large European health study where you must estimate the exposure of persons in different European cities to airborne particulate matter using only projected concentrations in different fixed locations based on a relatively small number of measurements and data on human travel habits between homes and work or school.
2. You are a scientist working with the USEPA and you need to estimate the exposure of Americans to airborne toxic metals as part of a risk assessment that will determine whether or not a product can be marketed, although unfortunately you do not have the budget for a multimillion dollar personal monitoring survey.
3. You are a graduate student in epidemiology studying respiratory disease in rural Indian villages, but you only have enough resources to measure average particle concentrations in selected locations and to gather crude activity diaries from the village residents.

In these three situations, direct information on exposure is lacking, and, therefore, the characterization of potential exposures for each group of affected people requires one to synthesize available information on airborne pollutant concentrations and human behavior patterns. By using an exposure model, the investigator in each of these cases can quantify the exposure distribution of study subjects and examine the likely influence of each location and other exposure factors. Conversely, without making use of an exposure model, only broad inferences could be made about potential exposures.

Although exposure models do not contain any data on the probability of ill health, and they do not assess the acquired dose of particular chemical species, they are still useful to health researchers, practitioners, and the general public. The rest of this section discusses specific areas in public health where inhalation exposure modeling can be useful. A summary is given in Table 19.3.

TABLE 19.3
Public Health Uses of Inhalation Exposure Models

Area	How Exposure Model Results Are Used
Epidemiology	As epidemiologists try to establish links between exposure to toxic pollutants and specific disease outcomes, they are assisted in the construction of questionnaires and diaries by accurate and reliable information on how exposure occurs and which exposure variables are most important.
Education	The results of exposure models can be used to educate the general public on how much exposure to toxic air pollutants it may receive in a variety of everyday situations.
Intervention	Efforts by health practitioners to intervene in unhealthy situations where persons are being exposed to toxic agents benefit from data on effective exposure reduction measures. This information can be used in ongoing dialogs with family members to facilitate the empowerment of individuals and to accentuate their involvement in reducing exposure.
Risk Assessment	When estimating the health risk of populations who are exposed to specific kinds of toxic air pollutants, the exposure for the affected population must be estimated before being combined with toxicological data. Models are an inexpensive and flexible way to provide the exposure data for a wide variety of situations.
Air Quality Guidelines	The modeling of exposure to air pollutants has a large role in establishing guidelines for acceptable indoor and outdoor levels of pollution that rely on the estimation of health risk associated with air pollutants for different likely scenarios.

One of the most important uses of exposure modeling in environmental health is the identification and exploration of physically effective means to mitigate exposure to toxic species. Once a link has been established between typical exposure levels and disease, models can be used to establish situations where unhealthful conditions might arise. Exposure modeling results can be used to make informative brochures or reference documents for the public and health researchers alike. Apart from the effectiveness of specific physical measures, successful interventions also depend on changes in human behavior patterns. The knowledge imparted by the modeling of exposure lends itself to critical discussions between family members or coworkers that evaluate specific mitigation strategies, which may be especially practical or attractive to particular households or workplaces.

Links between acute and chronic adverse health effects and exposure to toxic airborne pollutants are established by environmental epidemiologists and toxicologists. Epidemiological studies typically rely on questionnaires or diaries that could be revised and expanded in light of sophisticated model-based information on how exposure occurs in homes and other locations. The same kind of exposure information, combined with data on a compound's toxicity, can be used in health risk assessments, which estimate the probability of ill health resulting from typical uses of common products.

Using the results of risk assessment, governmental agencies, such as the USEPA and the California Air Resources Board (CARB), work to establish standards for levels of ambient air pollution, which are designed to protect the health of persons in the United States, particularly those who live in cities suffering from motor-vehicle-induced smog. Unfortunately, these ambient air quality standards were not designed to be applicable to the range and intensity of the toxic constituents in indoor air pollution or air pollutant emissions from short-lived local outdoor sources. Beyond the technical difficulties of characterizing indoor exposure patterns, the enactment of explicit indoor air quality standards is fairly problematic from a policy perspective, likely because of issues related to jurisdiction and enforcement.

However, building standards for indoor ventilation have already been established by the American Society of Heating, Refrigerating, and Air Conditioning Engineers (ASHRAE). These standards can be used as a basis for designing healthy homes. In addition, indoor air concentration guidelines have already been successfully established for the case of lung cancer risk due to radon gas. These guidelines were created using estimates of risk based on established health and exposure data.[7] In the future, formal concentration, building, and product use guidelines might be set by the USEPA, or some other regulatory agency, for other specific types of indoor air pollution, such as secondhand smoke, through the use of exposure simulation. By applying the machinery of a sophisticated exposure model, the likelihood of exceeding a particular indoor air quality concentration could be associated with specific building conditions and human behavior patterns.

19.7 REVIEW OF SOME EXISTING INHALATION EXPOSURE MODELS

At this point in the chapter, we have seen that exposure models can be useful to scientists studying the interaction between human health and the human environment. But what are some examples of real models that currently exist? In this section, we introduce some of the more well-developed exposure models for airborne contaminants that have appeared in the scientific literature or that are currently undergoing active use and refinement. All of these exposure models apply the basic exposure formula given in Equation 19.1 in which airborne pollutant concentrations and receptor movements are superimposed, revealing patterns in human exposure.

[7] See http://www.epa.gov/radon/risk_assessment.html and http://www.epa.gov/radon/pubs/ for more information on residential radon guidelines.

Although all of the models use a common formula, the inputs required by each different model for assigning or simulating pollutant concentrations and human activity patterns may be quite different. Two elements that most current models have in common are (1) they derive outdoor concentrations from raw empirical levels of ambient air pollution measured at fixed sites and (2) they use time–location profiles obtained from empirical human activity pattern data. Raw activity patterns are sometimes manipulated to artificially generate long-term (multiday) time profiles or profiles for multiple associated persons.

When treating indoor exposure, some models draw from empirical distributions of indoor concentrations, while others use a single or multizone indoor air quality model to predict indoor levels, as for the example presented in Section 19.4. Indoor air quality models are capable of simulating time-varying or time-averaged room concentrations using specified indoor source emission rates and the physical characteristics of a building, such as the airflow rates between rooms, the rate of air exchange with the outdoors, room volumes, and rates of chemical or physical transformation. These models generalize knowledge of the physical and chemical behavior of pollutants to arbitrary buildings and environmental conditions, which, while adding another level of overall model complexity, can be convenient when exploring the determinants of exposure.

The many inhalation exposure models currently under development can be crudely divided into two camps, the "exploratory" and the "regulatory," according to their primary intended purpose. Table 19.4 lists a number of existing inhalation exposure models, categorizing them by their general status in either camp. Seigneur et al. (2002) and Price et al. (2003) both present fairly in-depth descriptions of many of these listed models, as well as others. The entries in the table reflect significant efforts by a regulatory agency or efforts that have an associated article in the referee-scientific literature, or both. As a whole, they are reasonably reflective of the current state of inhalation exposure modeling, including efforts by government, academia, and industry. Most of the models listed are, or have been, made available in distributable (executable) form.

The first camp is comprised of models that are exploratory and limited in scope, focusing on a particular domain of exposure scenarios. Their purpose is primarily in developing methods or approaches, establishing mechanisms of exposures, empirically testing model assumptions, and exploring model predictions as part of a formal sensitivity or uncertainty analysis. For this camp, the prediction of exposures for arbitrary populations is less a priority than is understanding how exposure occurs in a given setting. The essence of the exploratory approach is to conduct carefully controlled computer-based simulation experiments to isolate the effects of a small number of key variates on the outcome variate of interest. Some of the earliest examples of exploratory models are those by Sparks, Tichenor, and White (1993), Sparks (1991), Koontz and Nagda (1991), and Wilkes et al. (1992), which track the behavior of household occupants and follow pollutant concentrations between rooms, incorporating detailed physical mechanisms of emissions and pollutant dynamics.

The second camp of models has a much broader scope and is intended to support regulatory mandates, such as the estimation of population health risk. Those who use these models are generally interested in applying them to large groups of people, and therefore they may incorporate sophisticated sampling techniques (e.g., Monte Carlo or Latin Hypercube sampling), and stratification of model inputs and outputs according to geographic or demographic characteristics. They describe multiple sources of pollution and a range of different settings where exposure can occur.

In recent years, there has been an emphasis on the regulatory type of model with organizations such as the USEPA investing considerable resources in its NEM, HAPEM, SHEDS, and APEX series of models (McCurdy 1995; Rosenbaum 2002; Burke, Zufall, and Özkaynak 2001; Richmond et al. 2002). Because inhalation is likely the most important exposure route for many toxic chemicals and is mechanistically one of the simplest routes of exposure, and since extensive air quality regulations are already concerned with air quality (e.g., the U.S. Clean Air Act), these inhalation exposure models are among the most well-developed, especially by in-house or contracted researchers at regulatory agencies (e.g., USEPA or CARB). They tend to be statistically based, sampling

TABLE 19.4
Examples of Some Existing Regulatory and Exploratory Inhalation Exposure Models

Reference	Acronym	Class[a]	Developer[b]	Full Name or Description
Ott et al. (1988); Ott (1984)	SHAPE	Expl	EPA	Simulation of Human Activity Patterns and Exposure
McKone (1987)	–	Expl	LLNL	Residential inhalation exposure model for volatile compounds in tap water
Traynor, Aceti, and Apte (1989)	–	Expl	LBNL	A "macromodel" for indoor exposure to combustion products
Sparks (1988, 1991); Sparks et al. (1993)	RISK	Expl	USEPA	Descendant of EXPOSURE and INDOOR models; simulates multizone indoor air concentrations, individual exposure, and risk
Koontz and Nagda (1991)	MCCEM	Expl	–	Multichamber Chemical Exposure Model
Wilkes et al. (1992, 1996, 2002)	MAVRIQ/TEM	Expl	Carnegie-Mellon	Model for Analysis of Volatiles and Residential Indoor Air Quality/Total Exposure Model
McCurdy (1995)	NEM-pNEM	Reg	USEPA	(Probabilistic) National Exposure Model; criteria pollutants
Macintosh et al. (1995)	BEADS	Reg	Harvard	Benzene Exposure and Absorbed Dose Simulation
Koontz, Evans, and Wilkes (1998); Koontz and Niang (1998); Rosenbaum et al. (2002)	CPIEM	Reg	CARB	California Population Indoor Exposure Model
Burke, Zufall, and Özkaynak (2001)	SHEDS-PM	Reg	USEPA	Stochastic Human Exposure Dose Simulation–Particulate Matter
Rosenbaum (2002)	HAPEM	Reg	USEPA	Hazardous Air Pollutant Exposure Model; mobile source air toxics
Dols and Walton (2002)	CONTAM	Expl	NIST	Multizone simulation of airflows, contaminant concentrations, and personal exposure.
Richmond et al. (2002)	APEX/TRIM Expo	Reg	USEPA	Air Pollutants Exposure Model and Total Risk Integrated Methodology Exposure Event Module; criteria and hazardous air pollutants
Kruize et al. (2003); Hänninen et al. (2003)	EXPOLIS	Reg	EXPOLIS	European population particle exposure model
de Bruin et al. (2004)	EXPOLIS	Reg	EXPOLIS	European population carbon monoxide exposure model
Briggs et al. (2003)	–	Expl	Northampton	Residential Radon exposure model

[a] Regl: Regulatory models used for development or enforcement of government regulations or for related risk assessments. These models are typically applied to large populations and require extensive data inputs that are representative of the population being modeled; Expl: Exploratory models used for intensive scientific study of particular exposure scenarios. These models typically treat an individual or narrowly defined cohort of people and have facilities for a detailed treatment of residences or some other specific microenvironment.

[b] USEPA: U.S. Environmental Protection Agency, Washington, DC; NIST: National Institute of Standards and Technology, Gaithersburg, MD; CARB: California Air Resources Board, Sacramento, CA; EXPOLIS: European Exposure Assessment Project; LLNL: Lawrence Livermore National Laboratory, Livermore, CA; LBNL: Lawrence Berkeley National Laboratory, Berkeley, CA; Northampton: Contributed by academic researchers in Northampton, U.K. Information and downloads for the APEX, TRIM, HAPEM, and HEM regulatory models for criteria pollutants and air toxics can be accessed from the USEPA Web site at the following URL: http://www.epa.gov/ttn/fera/.

from empirical or parameterized distributions of observed air concentrations and aggregate times spent in broad location categories (e.g., home, outdoors, or automobile).

There exists a massive database of ambient air quality data to support regulatory and other predictive population exposure models, as mandated under regulations such as the U.S. Clean Air Act. There is also a growing database of personal inhalation exposure monitoring data from studies such as EXPOLIS, NHEXAS, TEAM, PTEAM (Koistinen et al. 2001; Sexton, Kleffman, and Callahan 1995; Pellizzari et al. 1995; Wallace 1987; Özkaynak et al. 1996), and others, and a large database of microenvironmental inputs, including indoor air quality model parameters, to support the scope of regulatory modeling efforts. The American Chemistry Council has funded two recent in-depth reviews of datasets and reports having relevance to exposure modeling (Koontz and Cox 2002; Boyce and Garry 2002). The USEPA's "Exposure Factors Handbook" and "Child-Specific Exposure Factors Handbook" are two fairly comprehensive resources of appropriate inputs for predictive exposure models (USEPA 1997, 2002).[8] An online European Exposure Factors Source-book, called Expofacts, provides access to electronic datasets containing exposure-related information for many different European countries.[9]

19.8 ADVANCING THE SCIENCE OF EXPOSURE

19.8.1 MODELS AS THEORY

Exposure models exist because they are of practical value in estimating the health impact of particular products or behavior patterns. But more fundamentally, the development and application of models form the basis for advancement in exposure theory.

Any given empirical survey of human exposure can only address a limited domain of possible exposures and scenarios over a restricted period of time. On the other hand, models, such as the residential exposure model described in Section 19.4, are well equipped to describe the complex interaction between elements of exposure, including environmental characteristics, human behavior, and pollutant dynamics, and their evolution in time. Exposure models generalize experimental findings across a range of complicated and arbitrary scenarios and time scales, encapsulating the current state of scientific knowledge related to a particular environmental health problem. They consolidate a wide range of submodels, survey data, and expert opinions into an adaptable quantitative framework, which can be used to explore relationships between various exposure factors (e.g., as part of a formal sensitivity analysis).

Because they can predict exposures for arbitrary situations and human populations, models facilitate the generation of testable hypotheses concerning the mechanisms by which exposure occurs and, therefore, fulfill a need of the utmost importance in any field of science. Whether conceptual or quantitative, models provide direction for future studies, and therefore the driving force for scientific advancement. In this way, the development and application of exposure models lie at the heart of exposure science. Once model predictions are compared to empirical data, the model assumptions can be revised and theoretical mechanisms of exposure can be updated, thereby completing the cycle of scientific inquiry.

19.8.2 THE VANGUARD OF EXPOSURE MODELING

As evidenced by the material presented in this chapter, exposure modeling is already in a fairly advanced state of development. However, as with any scientific endeavor, there are many remaining frontiers and areas of uncertainty that need further investigation. The following subsections contain discussion of three topic areas that are at the leading edge of exposure modeling research and, therefore, of exposure research in general. These areas are (1) the direct evaluation of predictive

---

[8] See http://www.epa.gov/ncea/pdfs/efh/front.pdf
[9] See http://www.epa.ktl.fi/expofacts for more information on Expofacts.

TABLE 19.5
Future Directions in Exposure Modeling Research

Area	General Problem Description
Direct Model Evaluation	There is currently a need for more direct evaluation of exposure predictions against the results of empirical personal exposure surveys. Ideally, these new surveys would collect data expressly for the purpose of testing the performance of exposure models, including relevant and complete data on houses, airflow parameters, and human activity patterns.
Indoor Air Modeling	Because indoor air quality modeling is central to many inhalation exposure models, efforts should be made to test and parameterize these models in a variety of situations that are relevant to specific sources, human behavior, and environmental conditions.
Mixing and Proximity Effects	The nonuniform mixing (dispersion and dilution) of air pollutants in both indoor and outdoor settings for times when a proximate source is active can lead to elevated exposures for persons spending time near the active source. This effect needs more study. Dispersion of hazardous chemical and biological agents, perhaps from intentional and malevolent releases, is an especially pressing area of study.
Longitudinal Activity Patterns	There is currently a dearth of multiple day human activity pattern data. Most activity data are limited to a single 24-hour period. It is currently unclear how much human location and activity for a given person may change with time.
Multiperson Household Activity Patterns	Most available human activity data are limited to a single person per household. Since the interaction between persons in a household is likely to impact exposure, surveys of multiple persons in a population of homes should be conducted.
Detailed Activity Categories	Activity pattern studies are sometimes designed under limited budget circumstances or for use in a large variety of modeling analyses. For best use in characterizing specific types of exposure, activity pattern surveys should use focused location and activity categories, such as information on source proximity, room size and type, window and door position, and use of household pollutant sources.
Time-Series Analysis	Exposure models must be applied across a variety of time scales. More work needs to be done in understanding how concentrations and human activities, and therefore human exposures, vary in time. How the distribution of exposure changes as a function of averaging time and the correlation of exposure in time, i.e., its autocorrelation, are two issues that deserve attention.
Social Ecologies	Exposure models currently treat the ecology of a household or other exposure environment in a fairly distracted way, focusing more on pollutant levels and cross-sectional time–location patterns. For the purpose of identifying both technologically and socially effective means to reduce exposure, modelers should consider the complex, nonlinear social dynamics of persons having different roles and demographics in each modeled environment.

exposure models, (2) the improved characterization of the dispersion of indoor and outdoor pollutant concentrations, and (3) the inclusion of more detailed human social factors into exposure model design. Some specific topics associated with these areas are also summarized in Table 19.5.

19.8.2.1   Direct Evaluation of Model Predictions

While models are very useful for exploring the effect of different variables on human exposure to air pollution, it is important to have knowledge about their limitations in predicting real exposure for individual cases or for populations. The evaluation of exposure models can be conducted on several levels. For example, one could proceed by validating the different component elements of an exposure model. How accurate are its predictions of pollutant concentrations or human time–location patterns? Or, more directly, the exposure metrics produced by models could be compared with empirical surveys of exposure that make use of personal monitoring devices. A careful comparison of the distribution of simulated and observed exposures, taking into account specific housing characteristics and occupant behavior patterns, allows for an evaluation of the general performance of the model, as well as calibration of the model input parameters and the interpretation of features in the empirical exposure frequency distribution.

Currently, there have not been very many attempts to compare the output of population exposure models with the results of personal exposure surveys. One problem is that few large-scale exposure surveys exist. Another problem is that surveys, because of limited time and fiscal budgets, tend to measure less detailed information than is typically used as input for exposure models, making it difficult to gain insight into the causes of discrepancies between theory and experiment. Nevertheless, it is still possible to gauge the overall accuracy of exposure models using available survey data.

As part of the USEPA's PTEAM study (Özkaynak et al. 1993, 1996), which was a representative study of personal particle exposures for persons residing in a city in California in the early 1990s, and other monitoring studies (e.g., Williams et al. 2003a,b; Wallace et al. 2003; Liu et al. 2003; Allen et al. 2003), some of the necessary variables to facilitate a comparative analysis were recorded, e.g., house air exchange rates, time spent by individuals at home, smoking activity, cooking activity, cleaning activity, and approximate house size and room types. Unfortunately, specific information on the timing of sources either in or out of the house was not collected. In spite of these deficiencies, a systematic comparison between the results of a dynamic exposure simulation model and the results of intensive monitoring studies would be desirable.

Additional exposure surveys for the purpose of validating an exposure model should be conducted. A large validation study with a more complete set of variables, similar to or exceeding the level of the PTEAM effort, is expected to be very expensive and time-consuming. A more manageable approach might involve using carefully scripted location and activity profiles for a small number of houses, where the level of information detail could be expanded, including the use of real-time, or nearly real-time, monitoring of room concentrations, personal exposures, personal activities, house configuration, and environmental characteristics. An improvement over most time-activity diaries that are administered to study participants would be to include greater resolution in time and space on the locations and activities of subjects in their homes, including the rooms that were visited, the positions of doors and windows, as well as the use of combustible products and other sources of air pollution. With a systematic analysis across these study factor combinations, an airborne exposure simulation model could be thoroughly tested across a variety of important scenarios.

19.8.2.2 Understanding the Local Dispersion of Indoor and Outdoor Pollutants

The PTEAM study and other investigations have shown that concentrations measured in the personal breathing zone of subjects can be significantly elevated relative to concentrations measured at fixed points in the rooms of a house. This phenomenon has come to be called the "personal cloud" effect and is thought to arise from various causes, such as the compartmental nature of homes, the resuspension of dust from carpeting or clothing, or the effect of close proximity between a source and the individual being exposed. This last possibility, the "proximity effect," has been studied in recent research. The proximity effect is generally related to the issue of airborne pollutant mixing.

Although some recent efforts, such as those by Ribot et al. (2002), involve modeling the distribution of indoor pollutants in single rooms using computational fluid dynamics (CFD), the central assumption of many zonal indoor air quality models is that of uniform mixing of pollutants in individual rooms. Under this assumption, any emitted pollutant is instantaneously mixed throughout the zone of release. In reality, it takes a finite amount of time for emissions to mix within a room so that the average exposure one receives while immediately next to an active pollutant source may be larger than the average exposure at a more distant location, such as on the other side of the room.

It may be possible that the average microenvironmental concentration over a sufficiently long period is not much different than the theoretical well-mixed case. Based on a number of published studies that have evaluated the performance of multizone indoor air models or investigated the phenomena of indoor mixing and source proximity (Table 19.6), the general behavior of indoor models seems accurate, although under some airflow conditions or when the human receptor and source are in close proximity, the assumption of uniform mixing may break down. This possibility deserves more attention. A careful investigation into the proximity effect for emissions from an assortment of household products, especially one that characterizes the distribution of exposures as a function of distance and averaging time, is warranted.

The modeling of concentrations and exposures near active outdoor pollutant sources is expected to be even more complex than for indoor sources. Unlike for indoor settings, the persistence of local pollutant emissions in outdoor settings is very short. Therefore, there is no buildup or homogenization of emissions as there is indoors, and the forces of mixing and dispersion, driven from wind and turbulent air currents, are of key importance in determining local concentrations. Proximate exposure to local outdoor sources of air pollution is an emerging area of study with respect to common pollutants, such as tobacco smoke, but also with respect to releases of hazardous chemical and biological agents, perhaps due to deliberate destructive intent.

19.8.2.3 Human Factors

The nature of human activities makes up half of the exposure equation (Equation 19.1). However, this critical aspect of exposure modeling has thus far received relatively little attention as compared to that given to the measurement and modeling of environmental concentrations. For example, there is currently insufficient information available on human activity patterns on the household level. Two large-scale activity pattern studies conducted across the United States and in California have produced timelines of human movement for broad locations outside of their home and between specific rooms of their houses (see Section 19.5). Although several recent exposure assessment studies have incorporated multiday activities in their design (e.g., Liu et al. 2003; Wallace et al. 2003; Wallace and Williams 2005), no studies have appeared for large populations that have collected multiday human activity pattern data simultaneously for two or more members of a household.

To fully understand how exposure to residential air pollutants, such as secondhand smoke, occurs, it is important to consider dependencies among members of a household, and possible changes in activity patterns from day to day, perhaps in response to particular exposure-relevant initiatives or changing source behaviors. The amount of time that occupants spend together and in which rooms, and what activities are performed together or apart, coupled with the time-dependent nature of particular pollutant generating patterns (e.g., smoking, cooking, cleaning, and door, window, and centralized air handling configurations), will all affect exposures. Careful consideration of relative movements for occupants of different ages and relationships (e.g., child and caregiver), would allow for a better understanding of how different demographic groups are exposed.

In addition to a lack of longitudinal data for multiple household members, much of the activity pattern data collected to date consists of fairly crude location and activity categories. Future exposure studies should be focused on specific types of exposure and measure detailed information on exposure-related human activities events as much as possible. The use of detailed microenvironment and behavior categories results in a record of the micro-level behavior of human beings, which can play a critical role in how exposure occurs and help to identify new and better strategies for reducing or eliminating exposure. The collection of detailed information may be prohibitive for large studies, although the advent of sophisticated electronic monitoring equipment may facilitate data gathering and management. Modern information technology, including microsensors and remote digital loggers, holds the promise of the more efficient collection of real-time activity patterns and other exposure-related data. The use of geographic information systems (GIS), which

TABLE 19.6
Studies Evaluating Models of Residential Multizone Transport of Indoor Air Pollutants, Single-Zone Mixing, and Source-Proximity Effects

Study	Source	Method	Conclusions/Results
De Gids and Phaff (1988)	Tracer gas	Real-time CO monitoring in a house	Good agreement between measured and modeled CO
Sparks et al. (1991)	Moth cakes, kerosene heater, dry-cleaned clothes, aerosol spray, applied wet products	VOC and particle samples in a house	Multizone model does a good job of predicting indoor pollutant concentrations
Miller, Leiserson, and Nazaroff (1997); Miller and Nazaroff (2001)	Cigarettes and tracer gas	Real-time tracer and particle monitoring in a house	Good agreement between two-zone model and measurements
Ott, Klepeis, and Switzer (2003)	Cigarettes	Real-time particle and CO monitoring in a house	Good agreement between measurements and two-zone model parameterized from same experiment; error surface shows relative insensitivity to flow parameters
Baughman, Gadgil, and Nazaroff (1994)	Tracer gas	Grab sampling of SF_6 at 41 points in a chamber	Mixing times range from 7–15 min under natural convection in which heat was added from solar radiation or an electrical heater
Drescher et al. (1995)	Tracer gas	Real-time monitoring of CO at 9 points in a chamber	Mixing times range from 2–15 min for forced convection
Mage and Ott (1996)	Cigarettes and tracer gas	Real-time particle and CO monitoring in a tavern and house	Use of a uniformly mixed assumption to determine average exposures is generally valid for an intermittent source if the source-off well-mixed time period is large compared to the source-on plus mixing time periods
Klepeis (1999)	Cigarettes and tracer gas	Real-time particle and CO monitoring in a house, tavern, and smoking lounge	Mixing of air pollutant in medium-to-large rooms is fairly rapid in real locations under typical conditions on the order of 12–15 min before average concentrations at separated points are within 10% of the room mean
Furtaw et al. (1996)	Tracer gas	Real-time SF_6 monitoring in a chamber	Average concentration at a distance of 0.4 m from the source was double the theoretical well-mixed concentration for typical flow rates
McBride et al. (1999)	Tracer gas and incense stick	Real-time CO and particle monitoring	Proximity to active particle sources of 1 m resulting in mean concentrations averaging three times higher than those at a fixed distant location; proximate CO concentrations were also much higher than distant ones during source-on periods

Abbreviations: CO, carbon monoxide; VOC, volatile organic compound; SF_6, sulphur hexafluoride.

can track health, demographic, and environmental data could be used to manage and facilitate the analysis of exposure variation for households across neighborhoods, cities, regions, and nations.

As mentioned earlier in this chapter, the identification of effective mitigation strategies for exposure is an important application of exposure modeling. However, while strategies for reducing exposure may at times appear to be straightforward from a technological or logistical perspective, there may be sizable hurdles to overcome in terms of house roles, personalities, habits, and scheduling. For public health intervention projects focusing on households, as with urban planning or peace and nation-building efforts that have significantly larger scope, a roadmap for improvement or recovery handed down "from above" is not enough, and is likely to fail. The people who live in the households (or cities) must desire the change. Ideally, they understand the process, and are fully informed and involved, as it progresses.

For example, for the case of residential secondhand smoke (SHS) exposure, it is likely not enough that SHS has known adverse health effects and that strategies such as isolating smokers, opening windows, or using filtration devices may be effective means of removing indoor air pollutants. When this knowledge is fed into the ecology of a smoking household by healthcare workers, the media, or other elements of society, it may or may not have any lasting and beneficial effect. Nonlinear effects related to social pressure on smokers, interpersonal roles, personal empowerment, and feelings of involvement in evaluating, identifying, or implementing effective means of decreasing SHS exposure are likely to play important roles in the eventual reduction or elimination of exposure.

In the next phase in exposure science, physical models of pollutant dynamics might be fused with informed social models of human dynamics. New developments in quantitative social science, including techniques of agent-based modeling (Epstein and Axtell 1996), show promise as a means of understanding the complex, changing relationships in human ecologies. Some exposure modelers are beginning to recognize that life-stage and life-role variables must be included in studies of human activity in order to sufficiently explain and understand the variation in human behavior that impacts exposure (Graham and McCurdy 2004). Exposure science could benefit by drawing from fields such as cognitive psychology, sociology, and geography, which are focused on human behavior and the interactions among individuals and between individuals and the environment.

The enhanced use of social variables in theoretical descriptions of exposure would allow exposure assessors to study how factors such as roles, empowerment, knowledge, perception, and beliefs contribute to a particular exposure landscape, and could help facilitate the identification of both physically and socially practical means for reducing or eliminating dangerous exposures.

19.9 QUESTIONS FOR REVIEW

1. What are four specific uses of exposure models in public health?
2. Make a diary of the locations you visit over the course of a full day, including the times you enter and leave each location. Calculate the percentage of time spent in each type of location. In which location did you spend the most amount of time? The least? How do you think these percentages might change from day to day? Are these percentages likely to be different for other members of your family?
3. When designing a household activity pattern survey for the purpose of gathering data that will be used in modeling human exposure, describe four important design considerations.
4. Given the table of average airborne PM_{10} concentrations for different locations provided below, calculate your 24-hour average exposure to airborne particles based on the activity patterns you recorded in Question 2. What percentage of total exposure was accounted for by the microenvironmental concentrations? Now add a 35 $\mu g/m^3$ "personal cloud"

or "proximity effect," to time spent in one of the locations, such as might result from cigarette smoking. How does this change the contribution of that microenvironment to your total exposure?

Location	Average Airborne PM_{10} Concentration ($\mu g/m^3$)
Home	90
Office–Factory	40
Bar–Restaurant	200
Other Indoor	20
In a Vehicle	45
Outdoors	35

5. Derive the time-averaged canonical exposure formula (Equation 19.1) from the semi-continuous dynamic formula (Equation 19.3).
6. Derive the population version of the canonical exposure formula (Equation 19.4).
7. What are two major assumptions that are typically made when modeling indoor exposure to airborne pollutants?
8. Discuss at least three strategies for reducing exposure to secondhand tobacco smoke in the home. Which strategies do you think would be the least and the most effective? Describe specific simulation experiments you might use to explore the effectiveness of different mitigation strategies.

19.10 ACKNOWLEDGMENTS

Grateful appreciation is given to the Flight Attendant Medical Research Institute (FAMRI) for partial funding of this research.

REFERENCES

Allen, R., Larson, T., Sheppard, L., Wallace, L.A., and Liu, L.-J.S. (2003) Use of Real-Time Light Scattering Data to Estimate the Contribution of Infiltrated and Indoor-Generated Particles to Indoor Air, *Environmental Science and Technology*, **37**: 3484–3492.

Baughman, A.V., Gadgil, A.J., and Nazaroff, W.W. (1994) Mixing of a Point Source Pollutant by Natural Convection Flow within a Room, *Indoor Air*, **4**(2): 114–122.

Boyce, C.P. and Garry, M.R. (2002) Review of Information Resources to Support Human Exposure Assessment Models, *Human and Ecological Risk Assessment*, **8**(6): 1445–1487.

Briggs, D.J., Denman, A.R., Gulliver, J., Marley, R.F., Kennedy, C.A., Philips, P.S., Field, K., and Crockett, R.M. (2003) Time Activity Modeling of Domestic Exposures to Radon, *Journal of Environmental Management*, **67**(2): 107–120.

Burke, J.M., Zufall, M.J., and Özkaynak, H. (2001) A Population Exposure Model for Particulate Matter: Case Study Results for $PM_{2.5}$ in Philadelphia, PA, *Journal of Exposure Analysis and Environmental Epidemiology*, **11**(6): 470–489.

de Bruin, Y.B., Hänninen, O., Carrer, P., Maroni, M., Kephalopoulos, S., di Marco, G.S., and Jantunen, M. (2004) Simulation of Working Population Exposures to Carbon Monoxide Using EXPOLIS-Milan Microenvironment Concentration and Time-Activity Data, *Journal of Exposure Analysis and Environmental Epidemiology*, **14**(2): 154–163.

De Gids, W.F. and Phaff, J.C. (1988) Recirculation of Air in Dwellings: Differences in Concentrations between Rooms in Dwellings Due to the Ventilation System, Paper 17, in *Proceedings from the 9th AIVC Conference*, Gent, Belgium, 301–310.

Dols, W. and Walton, G.N. (2002) CONTAMW 2.0 User's Manual: Multizone Airflow and Contaminant Transport Analysis Software, National Institute of Standards, Building and Fire Research Laboratory, Gaithersburg, MD.

Drescher, A.C., Lobascio, C., Gadgil, A.J., and Nazaroff, W.W. (1995) Mixing of a Point-Source Indoor Pollutant by Forced-Convection, *Indoor Air*, **5**(3): 204–214.

Duan, N. (1982) Microenvironmental Types: A Model for Human Exposure to Air Pollution, *Environment International*, **8**: 305–309.

Epstein, J.M. and Axtell, R. (1996) *Growing Artificial Societies: Social Science from the Bottom Up*, Brookings Institution Press and MIT Press, Washington, DC and Cambridge, MA.

Fugas, M. (1975) Assessment of Total Exposure to Air Pollution, in *Proceedings of the International Conference on Environmental Sensing and Assessment*, Vol. 2 of IEEE #75-CH1004-1ICESA, Las Vegas, NV, Paper No. 38-5.

Furtaw, E.J., Pandian, M.D., Nelson, D.R., and Behar, J.V. (1996) Modeling Indoor Air Concentrations Near Emission Sources in Imperfectly Mixed Rooms, *Journal of the Air and Waste Management Association*, **46**(9): 861–868.

Graham, S.E. and McCurdy, T. (2004) Developing Meaningful Cohorts for Human Exposure Models, *Journal of Exposure Analysis and Environmental Epidemiology*, **14**(1): 23–43.

Hänninen, O., Kruize, H., Lebret, E., and Jantunen, M. (2003) EXPOLIS Simulation Model: PM2:5 Application and Comparison with Measurements in Helsinki, *Journal of Exposure Analysis and Environmental Epidemiology*, **13**(1): 74–85.

Jenkins, P., Phillips, T., Mulberg, E., and Hui, S. (1992) Activity Patterns of Californians: Use of and Proximity to Indoor Pollutant Sources, *Atmospheric Environment*, **26A**(12): 2141–2148.

Klepeis, N.E. (1999) Validity of the Uniform Mixing Assumption: Determining Human Exposure to Environmental Tobacco Smoke, *Environmental Health Perspectives*, **107**(supp. 3): 357–363.

Klepeis, N.E., Nelson, W.C., Ott, W.R., Robinson, J.P., Tsang, A.M., Switzer, P., Behar, J.V., Hern, S.C., and Engelmann, W.H. (2001) The National Human Activity Pattern Survey (NHAPS): A Resource for Assessing Exposure to Environmental Pollutants, *Journal of Exposure Analysis and Environmental Epidemiology*, **11**(3): 231–252.

Klepeis, N.E., Tsang, A.M., and Behar, J.V. (1996) Analysis of the National Human Activity Pattern Survey (NHAPS) Responses from a Standpoint of Exposure Assessment, Report No. EPA/600/R-96/074, U.S. Environmental Protection Agency, Washington, DC.

Koistinen, K.J., Hänninen, O., Rotko, T., Edwards, R.D., Moschandreas, D., and Jantunen, M.J. (2001) Behavioral and Environmental Determinants of Personal Exposures to PM2:5 in EXPOLIS- Helsinki, Finland, *Atmospheric Environment*, **35**(14): 2473–2481.

Koontz, M.D. and Cox, S.S. (2002) Literature Review on Microenvironmental Modeling, in *Indoor Air – Proceedings of the 9th International Conference on Indoor Air Quality and Climate*, Levin, H., Ed., Monterey, CA, 304–309.

Koontz, M.D. and Nagda, N.L. (1991) A Multichamber Model for Assessing Consumer Inhalation Exposure, *Indoor Air*, **1**(4): 593–605.

Koontz, M.D. and Niang, L.L. (1998) California Population Indoor Exposure Model (CPIEM), Version 1.4F, A933-157 User's Guide, Geomet Technologies, prepared for California Air Resources Board, Sacramento, CA.

Koontz, M.D., Evans, W.C., and Wilkes, C.R. (1998) Development of a Model for Assessing Indoor Exposure to Air Pollutants, A933-157, Geomet Technologies, Inc. prepared for California Air Resources Board, Sacramento, CA.

Kruize, H., Hänninen, O., Breugelmans, O., Lebret, E., and Jantunen, M. (2003) Description and Demonstration of the EXPOLIS Simulation Model: Two Examples of Modeling Population Exposure to Particulate Matter, *Journal of Exposure Analysis and Environmental Epidemiology*, **13**(2): 87–99.

Leech, J.A., Wilby, K., McMullen, E., and Laporte, K. (1996) The Canadian Human Activity Pattern Survey: A Report of Methods and Population Surveyed, *Chronic Diseases in Canada*, **17**(3-4): 118–123.

Liu, L.-J.S., Box, M., Kalman, D., Kaufman, J., Koenig, J., Larson, T., Lumley, T., Sheppard, L., and Wallace, L.A. (2003) Exposure Assessment of Particulate Matter for Susceptible Populations in Seattle, *Environmental Health Perspectives*, **111**: 909–918.

Macintosh, D.L., Xue, J.P., Özkaynak, H., Spengler, J.D., and Ryan, P.B. (1995) A Population Based Exposure Model for Benzene, *Journal of Exposure Analysis and Environmental Epidemiology*, **5**(3): 375–403.

Mage, D.T. and Ott, W.R. (1996) The Correction for Nonuniform Mixing in Indoor Environments, in *Characterizing Indoor Air Pollution and Related Sink Effects*, Tichenor, B.A., Ed., American Society for Testing and Materials, West Conshohocken, PA, 263–278.

McBride, S.J., Ferro, A.R., Ott, W.R., Switzer, P., and Hildemann, L.M. (1999) Investigations of the Proximity Effect for Pollutants in the Indoor Environment, *Journal of Exposure Analysis and Environmental Epidemiology*, **9**(6): 602–621.

McCurdy, T. (1995) Estimating Human Exposure to Selected Motor Vehicle Pollutants Using the NEM Series of Models: Lessons to be Learned, *Journal of Exposure Analysis and Environmental Epidemiology*, **5**(4): 533–550.

McCurdy, T., Glen, G., Smith, L., and Lakkadi, Y. (2000) The National Exposure Research Laboratory's Consolidated Human Activity Database, *Journal of Exposure Analysis and Environmental Epidemiology*, **10**(6): 566–578.

McKone, T.E. (1987) Human Exposure to Volatile Organic Compounds in Household Tap Water: the Indoor Inhalation Pathway, *Environmental Science and Technology*, **21**(12): 1194–1201.

Miller, S.L. and Nazaroff, W.W. (2001) Environmental Tobacco Smoke Particles in Multizone Indoor Environments, *Atmospheric Environment*, **35**(12): 2053–2067.

Miller, S.L., Leiserson, K., and Nazaroff, W.W. (1997) Nonlinear Least-Squares Minimization Applied to Tracer Gas Decay for Determining Air Flow Rates in a Two-Zone Building, *Indoor Air*, **7**(1): 64–75.

Ott, W. (1982) Concepts of Human Exposure to Air Pollution, *Environment International*, **7**: 179–196.

Ott, W.R. (1984) Exposure Estimates Based on Computer Generated Activity Patterns, *Journal of Toxicology: Clinical Toxicology*, **21**(1-2): 97–128.

Ott, W.R., Klepeis, N.E., and Switzer, P. (2003) Analytical Solutions to Compartmental Indoor Air Quality Models with Application to Environmental Tobacco Smoke Concentrations Measured in a House, *Journal of the Air and Waste Management Association*, **53**(8): 918–936.

Ott, W.R., Thomas, J., Mage, D.T., and Wallace, L.A. (1988) Validation of the Simulation of Human Activity and Pollutant Exposure (SHAPE) Model Using Paired Days from the Denver, CO, Carbon Monoxide Field Study, *Atmospheric Environment*, **22**(10): 2101–2113.

Özkaynak, H., Xue, J., Spengler, J., Wallace, L., Pellizzari, E., and Jenkins, P. (1996) Personal Exposure to Airborne Particles and Metals—Results from the Particle TEAM Study in Riverside, California, *Journal of Exposure Analysis and Environmental Epidemiology*, **6**(1): 57–78.

Özkaynak, H., Xue, J.,Weker, R., Butler, D., and Spengler, J. (1993) The Particle Team (PTEAM) Study: Analysis of the Data; Volume III, Contract Number 68-02-4544, U.S. Environmental Protection Agency, Research Triangle Park, NC.

Pellizzari, E., Lioy, P., Quackenboss, J., Whitmore, R., Clayton, A., Freeman, N., Waldman, J., Thomas, K., Rodes, C., and Wilcosky, T. (1995) Population-Based Exposure Measurements in EPA Region 5—A Phase I Field Study in Support of the National Human Exposure Assessment Survey, *Journal of Exposure Analysis and Environmental Epidemiology*, **5**(3): 327–358.

Price, P. S., Koontz, M., Wilkes, C., Ryan, B., Macintosh, D., and Georgopoulos, P. (2003) Construction of a Comprehensive Chemical Exposure Framework Using Person Oriented Modeling, The Lifeline Group, Developed for the American Chemistry Council.

Ribot, B., Chen, Q. Y., Huang, J. M., and Rivoalen, H. (2002) Numerical Simulations of Indoor Air Quality in a French House: Study of the Distribution of Pollutants in Each Room, Modeling Contaminant Exposure, in *Indoor Air—Proceedings of the 9th International Conference on Indoor Air Quality and Climate*, Levin, H., Ed., Monterey, CA, 524–529.

Richmond, H.M., Palma, T., Langstaff, J., McCurdy, T., Glen, G., and Smith, L. (2002) Further Refinements and Testing of APEX(3.0): EPA'S Population Exposure Model for Criteria and Air Toxic Inhalation Exposures, presented at the ISEA/ISEE Joint Conference, Vancouver, BC, August 11–15.

Rosenbaum, A. (2002) The HAPEM4 User's Guide: Hazardous Air Pollutant Exposure Model, Version 4, ICF Consulting, prepared for the U.S. Environmental Protection Agency, Office of Air Quality, Planning, and Standards, Research Triangle Park, NC.

Rosenbaum, A., Cohen, J., Kavoosi, F., Lum, S., and Jenkins, P. (2002) The California Population Indoor Exposure Model, Version 2: A User-Friendly Assessment Tool for Population Exposure to Air Pollutants, presented at the ISEA/ISEE Joint Conference, Vancouver, BC, August 11–15.

Seigneur, C., Pun, B., Lohman, K., and Wu, S. Y. (2002) *Air Toxics Modeling*, Atmospheric and Environmental Research, Inc., San Ramon, CA, prepared for the Coordinating Research Council, Inc. and the U.S. Department of Energy, Office of Heavy Vehicle Technologies, National Renewable Energy Laboratory, Golden, CO.

Sexton, K., Kleffman, D.E., and Callahan, M.A. (1995) An Introduction to the National Human Exposure Assessment Survey (NHEXAS) and Related Phase I Field Studies, *Journal of Exposure Analysis and Environmental Epidemiology*, **5**(3): 229–232.

Sparks, L.E. (1988) *Indoor Air Model, Version 1.0*, Report No. EPA 600/8-88-097a, U.S. Environmental Protection Agency, Research Triangle Park, NC.

Sparks, L.E. (1991) *EXPOSURE Version 2: A Computer Model for Analyzing the Effects of Indoor Air Pollutant Sources on Individual Exposure*, Report No. EPA-600/8-91-013, U.S. Environmental Protection Agency, Research Triangle Park, NC.

Sparks, L.E., Tichenor, B.A., and White, J.B. (1993) Modeling Individual Exposure from Indoor Sources, in *Modeling of Indoor Air Quality and Exposure*, Nagda, N.L., Ed., No. ASTM STP-1205, American Society for Testing and Materials, Philadelphia, PA, 245–256.

Sparks, L.E., Tichenor, B.A., White, J.B., and Jackson, M.D. (1991) Comparison of Data from an IAQ Test House with Predictions of an IAQ Computer Model, *Indoor Air*, 1(4): 577–592.

Traynor, G.T., Aceti, J.C., and Apte, M.G. (1989) Macromodel for Assessing Residential Concentrations of Combustion-Generated Pollution: Model Development and Preliminary Predictions for CO, NO_2, and Respirable Suspended Particles, LBL-25211, Lawrence Berkeley National Laboratory, Berkeley, CA.

Tsang, A.M. and Klepeis, N.E. (1996) *Descriptive Statistics Tables from a Detailed Analysis of the National Human Activity Pattern Survey (NHAPS) Data*, Report No. EPA/600/R-96/148, U.S. Environmental Protection Agency, Washington, DC.

USEPA (1997) Exposure Factors Handbook, Report No. EPA/600/P-95/002F, U.S. Environmental Protection Agency, Office of Research and Development, Washington, DC.

USEPA (2002) Child-Specific Exposure Factors Handbook (Interim Report), Report No. EPA/600/P-00/002B, U.S. Environmental Protection Agency, Office of Research and Development, National Center for Environmental Assessment, Washington Office, Washington, DC.

Wallace, L.A. (1987) *The Total Exposure Assessment Methodology (TEAM) Study: Summary and Analysis: Volume I*, U.S. Environmental Protection Agency, Washington, DC.

Wallace, L.A. and Williams, R.W. (2005) Use of Personal-Indoor-Outdoor Sulfur Concentrations to Estimate the Infiltration Factor and Outdoor Exposure Factor for Individual Homes and Persons, *Environmental Science and Technology*, **39**(6): 1707–1714.

Wallace, L.A., Mitchell, H., O'Connor, G.T., Liu, L.-J., Neas, L., Lippmann, M., Kattan, M., Koenig, J., Stout, J., Vaughn, B.J., Wallace, D., Walter, M., and Adams, K. (2003) Particle Concentrations in Inner-City Homes of Children with Asthma: The Effect of Smoking, Cooking, and Outdoor Pollution, *Environmental Health Perspectives*, **111**: 1265–1272.

Wiley, J., Robinson, J.P., Cheng, Y., Piazza, T., Stork, L., and Pladsen, K. (1991a) Study of Children's Activity Patterns, Contract No. A733-149, California Air Resources Board, Sacramento, CA.

Wiley, J., Robinson, J.P., Piazza, T., Garrett, K., Cirksena, K., Cheng, Y., and Martin, G. (1991b) Activity Patterns of California Residents, California Air Resources Board, Contract No. A6-177-33, Sacramento, CA.

Wilkes, C.R., Blancato, J.N., Hern, S.C., Power, F.W., and Olin, S.S. (2002) Integrated Probabilistic and Deterministic Modeling Techniques in Estimating Exposures to Water-Borne Contaminants: Part 1, Exposure Modeling, in *Indoor Air—Proceedings of the 9th International Conference on Indoor Air Quality and Climate*, June 30–July 5, 2002, Levin, H., Ed., Monterey, CA, 256–261.

Wilkes, C.R., Small, M.J., Andelman, J.B., Giardino, N.J., and Marshall, J. (1992) Inhalation Exposure Model for Volatile Chemicals from Indoor Uses of Water, *Atmospheric Environment*, **26A**(12): 2227–2236.

Wilkes, C.R., Small, M.J., Davidson, C.I., and Andelman, J.B. (1996) Modeling the Effects of Water Usage and Co-Behavior on Inhalation Exposures to Contaminants Volatilized from Household Water, *Journal of Exposure Analysis and Environmental Epidemiology*, **6**(4): 393–412.

Williams, R., Suggs, J., Rea, A., Sheldon, L., Rodes, C., and Thornburg, J. (2003a) The Research Triangle Park Particulate Matter Panel Study: Modeling Ambient Source Contribution to Personal and Residential PM Mass Concentrations, *Atmospheric Environment*, **37**(38): 5365–5378.

Williams, R.W., Suggs, J., Rea, A., Leovic, K., Vette, A., Croghan, C., Sheldon, L., Rodes, C., Thornburg, J., and Ejire, A. (2003b) The Research Triangle Park Particulate Matter Panel Study: PM Mass Concentration Relationships, *Atmospheric Environment*, **37**(38): 5349–5363.

20 Models of Exposure to Pesticides

Robert A. Canales
Harvard University

James O. Leckie
Stanford University

CONTENTS

20.1 Synopsis...471
20.2 Introduction...472
20.3 Activity Data 472
20.4 Routes of Exposure 473
    20.4.1 Dermal Exposure 473
    20.4.2 Nondietary Ingestion Exposure 475
    20.4.3 Dietary Ingestion Exposure 475
20.5 Exposure Models...475
    20.5.1 The CalTOX Approach...476
    20.5.2 Calendex™ 476
    20.5.3 LifeLine™ 477
    20.5.4 The Cumulative and Aggregate Risk Evaluation System (CARES)....477
    20.5.5 The Stochastic Human Exposure and Dose Simulation (SHEDS) 478
20.6 Issues in Modeling Exposure 478
    20.6.1 Combining Data for Model Inputs..478
    20.6.2 Temporal and Spatial Variability..479
    20.6.3 Co-Occurrence of Environmental Contaminants 480
    20.6.4 Estimating Dose from Exposure 480
20.7 Questions for Further Review 481
References 481

20.1 SYNOPSIS

This chapter is devoted to the exploration of models of exposure to pesticides, their complexities, and their emergence after the passing of the Food Quality Protection Act of 1996. A discussion begins with a basic component of exposure models — human activity patterns — and establishes a foundation for understanding the many models that have been proposed for estimating exposure. The quantification of dietary ingestion, nondietary ingestion, and dermal exposure is surveyed. Several exposure models are explored in terms of inputs, modeling objectives, and methods. The chapter ends with a number of other issues important in modeling exposure to pesticides. These issues include temporal and spatial variability, dose estimates, the co-occurrence of contaminants, and the use of literature data for model inputs.

20.2 INTRODUCTION

Interest in measuring and modeling exposure to pesticides through all important pathways has increased since the development of the Food Quality Protection Act (FQPA) of 1996. The impetus for the Act was a report by the National Research Council (NRC) entitled *Pesticides in the Diets of Infants and Children* (NRC 1993). The report examined policies of government agencies in regulating pesticide residues in foods. The influence of the NRC publication is evident in the Act's resolution that there should be a consistent, health-based standard for all pesticides used on foods, pesticide regulation should address the vulnerability of potentially sensitive groups (e.g., children), and older pesticides should be reassessed and perhaps replaced with safer substitutes. The report also recognized the importance of accounting for exposures from all non-occupational sources and from all routes of exposure (i.e., "aggregate exposure"), and the effects of exposure to multiple pesticides with "common mechanisms of toxicity" (i.e., "cumulative exposure") (USEPA 1997).

20.3 ACTIVITY DATA

Several types of aggregate exposure models exist and they may be characterized in a number of ways. For instance, models may be categorized by their purpose (e.g., screening analysis, heuristic exploration, regulatory compliance), the form of necessary input data (e.g., probability distributions, point estimates), or their output (e.g., lifetime risk, short-term exposure, population estimates). A crucial distinction unique to human exposure models is the type and detail of activity data.

The U.S. Environmental Protection Agency (USEPA) recognizes that models for assessing exposure can be structured around different forms of activity pattern data (Cohen Hubal et al. 1998). The different approaches dictate the level of detail, the level of accuracy, the inherent uncertainty, and the computational structure of an exposure model. The form of the output and the adequacy of the exposure and dose estimates for risk determinations are also affected by the qualitative nature of the activity pattern data. The common forms of activity pattern data include macro-, meso-, and micro-level activities (Figure 20.1).

Macro-level activities describe human activities by general location and general activity. Normally collected via diaries or interviews, examples of this form of activity include playing outdoors, reading indoors, or working in an office. This information is typically used to estimate inhalation

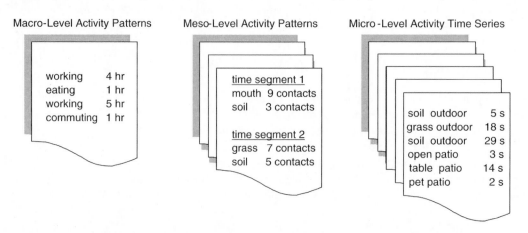

FIGURE 20.1 Forms of activity patterns used in modeling exposure.

exposures, but it has been used in assessing children's residential exposure to pesticides (Cohen Hubal et al. 1998). The quantification of exposure from the use of macro-level activities, however, inherently lumps several details, such as dermal contacts and objects inserted into the mouth, that may be important in estimating nondietary ingestion and dermal exposure (Cohen Hubal et al. 1998; Michelson and Reed 1975).

Using meso-level activities, exposure can be modeled with more specific data on microenvironments and contact behavior. Meso-level activity data track some characteristic of activity over a certain time frame (Figure 20.1). As an example, an individual may complete 9 hand-to-mouth contacts and 3 contacts with soil some time within a 30-minute period, and 7 contacts with grass and 5 contacts with soil some time in the following 30 minutes. These data may be further reduced to an hourly or daily basis. Data collected in this manner may provide an opportunity to preserve the sequence of activities, at least at some level, and characterize the variability of the data (Reed, et al. 1999). It is obvious that data requirements for the meso-level activity approach are more extensive when compared to the collection of macro-level activities (Cohen Hubal et al. 1998). Furthermore, since there is little regard to order within the arbitrary time segment, this activity form may not adequately describe sequential contacts necessary for including certain phenomena when estimating dermal and nondietary ingestion exposure.

A more detailed activity pattern form can account for sequential micro-level contact events and microenvironments, and allow for the estimation and inclusion of other time-varying factors (Figure 20.1). In estimating dermal exposure this information is an improvement over other activity forms since detailed data on sequential contacts are necessary for appropriately handling removal mechanisms and loading from multiple media. This more specific representation of activities is also useful for understanding the fundamental mechanisms and behaviors influencing exposure. Data collection and modeling tasks associated with micro-level activity, however, are more extensive when compared to other forms of activity data. Collecting such data may be infeasible for large populations.

20.4 ROUTES OF EXPOSURE

Aggregate exposure models use activity data to assign environmental concentrations and related transfer parameters to route-specific equations. Separate equations then exist for inhalation, nondietary ingestion, dietary ingestion, and dermal exposure. Since inhalation exposure is treated exhaustively in several chapters in this book, including Chapter 4 and Chapter 19, it will not be discussed here. This chapter will focus on dermal exposure and dietary and nondietary ingestion.

20.4.1 DERMAL EXPOSURE

The dermal route is perhaps the most complex of the more important exposure routes and the most difficult to model (Zartarian and Leckie 1998). A number of slightly different equations have been developed by different groups to estimate dermal exposure. All the equations have in common a term for the concentration of the substance in the medium (e.g., soil, water) and a term for the "transfer factor" governing how the substance moves from the medium to the skin. Some include a term for the area of the skin involved. Many include a term for the frequency of contact with the medium, using activity pattern data to describe such frequencies (Moschandreas et al. 2001). Other models (Zartarian et al. 2000; Price, Young, and Chaisson 2001) consider dermal exposure via dislodgeable residues, which may be some combination of liquid residues, house dust, and soil.

The OP Case Study Group Non-Dietary Subcommittee (1999) and Price, Young, and Chaisson (2001) utilize a version of what the USEPA's National Exposure Research Laboratory (NERL) terms the "macroactivity approach." The general version of this approach uses the following equation:

$$E_{der} = ED \times TC_{der} \times C_{surf} \tag{20.1}$$

where

E_{der} = dermal exposure resulting from the completion of an activity (mg)
ED = duration of activity (hr)
TC_{der} = dermal transfer coefficient (cm^2/hr)
C_{surf} = dislodgeable contaminant loading on surface (mg/cm^2)

The key to this approach is the transfer factor or transfer coefficient, which provides a lumped measure of contaminant transfer within a microenvironment or after some general activity.

Other models, such as the one by Moschandreas et al. (2001), use rates of dermal contact to incorporate activities into dermal exposure estimates. A unitless term representing the proportion of contaminant on a surface transferred to the skin is utilized rather than a lumped transfer coefficient. This unitless term, or transfer efficiency, may be a function of the type of contact event, surface, or environmental media. NERL's general equation for such a model is:

$$E_{der} = C_{surf} \times TF \times SA \times EV \tag{20.2}$$

where

E_{der} = dermal exposure associated with a given event (mg/day)
C_{surf} = dislodgeable contaminant loading on surface (mg/cm^2)
TF = fraction available for transfer from surface to skin (unitless)
SA = surface area contacted (cm^2/event)
EV = event frequency (events/day)

Models have also been developed that consider sequential dermal contacts (Zartarian 1996; Canales 2004). While these models are relatively more complicated, they tend to account for phenomena such as removal from the skin (e.g., hand-to-mouth contacts, negative transfers, washing) and accumulation due to a number of media (e.g., liquids, air, residues, soil, house dust). The use of sequences of dermal contacts also permits the development of detailed exposure time profiles and greater insight into mechanisms and behaviors that result in dermal exposure (Zartarian 1996). The general equation below is applied to sequential contacts and the concentration, surface area, and exposure factor terms change as a function of the contact or the type of medium.

$$DE = C_{medium} \times EF_{medium} \times \left(\frac{A_{medium}}{A_{skin}} \right) \tag{20.3}$$

where

DE = spatially averaged dermal exposure after contact event [M/L^2]
C_{medium} = concentration in medium [M/L^3], [M/L^2], or [M/M]
EF_{medium} = exposure factor for particular medium [L], [–], or [M/L^2]
(A_{medium}/A_{skin}) = fraction of skin area covered with medium [–]

Total dermal exposure is then the sum of exposure from each contact.

20.4.2 NONDIETARY INGESTION EXPOSURE

Nondietary ingestion exposure, a generally overlooked route of ingestion exposure, occurs when an agent on surfaces (e.g., toys, food, fingers) or in nondietary media (e.g., soil, house dust) is ingested via hand-to-mouth or object-to-mouth contact, rather than through the ingestion of food. Several models created to meet the needs of the FQPA of 1996 consider the route, but few agree on methodologies. Pang et al. (2002), for instance, account for daily incidental ingestion (ng/day) of carpet dust and soil, using the number of carpet contacts per day multiplied by an ingestion rate per contact. Pang et al. also consider an absorption factor, so that their approach actually results in a daily *dose*, an amount crossing the gut surface, rather than an exposure. Cohen Hubal et al. (2000), however, consider contaminants on hands and on objects, and use surface concentrations multiplied by transfer fractions and contact frequency in quantifying nondietary ingestion exposure.

20.4.3 DIETARY INGESTION EXPOSURE

Dietary ingestion exposure is a function of food consumption patterns and the concentration of chemical in the food (Cohen Hubal et al. 2000). The product of these two factors is calculated for each item of food and summed over all food consumed over some time period. Typically the result is a daily ingestion exposure in units of micrograms per day.

While this approach is very simple, complications arise when considering the variability in chemical concentrations in food, person-to-person differences in consumption patterns, and variations in dietary profiles across age, gender, ethnic groups, and geographic regions. The primary issue when estimating dietary exposure then is not the exposure equation but rather the source of the inputs to the equation.

If the study population is small, or if adequate resources are available relative to the size of the study, 24-hour duplicate diet sampling is often employed. This method involves collecting a duplicate portion of all food and beverages consumed by an individual over a 24-hour period. Upon analysis, both the concentration of the contaminant in the food and the total mass of food are known, and an estimate of the daily ingestion exposure can be calculated. If an estimate of ingestion exposure is required over longer time periods or for a larger population, this daily information can then be extrapolated via modeling.

In studies examining a relatively large population, databases of food and beverage consumption are often utilized. One such database consists of the combined data from the Continuing Survey of Food Intakes by Individuals (CSFII) and the National Health and Nutrition Examination Survey (NHANES). The NHANES consists of a multistage sample representative of the United States. Both surveys recorded consumption data for a single day per person for close to 10,000 people. Several groups were oversampled, such as adolescents, minorities, and pregnant women, to allow for more precise estimates. The CSFII was conducted by the U.S. Department of Agriculture (USDA) as several 1-year surveys to provide 1-day and 3-day dietary intake data. The data have been mined to produce reports specifying eating patterns by sex, age, race, and geographic region. The integrated database is titled *What We Eat in America — National Health and Nutrition Examination Survey* (Dwyer et al. 2001).

20.5 EXPOSURE MODELS

Prior to the 1990s there were few models that specifically addressed aggregate and cumulative exposure assessment. Of those existing models that explored aggregate exposure, the majority had divergent overall goals with exposure as an intermediate step, or they examined exposure with an inadequate level of detail.

The Food Quality Protection Act's new requirements have led to an evolution in exposure assessment methodologies. Several aggregate models are reviewed below. While others exist, these models were chosen for review based on their range of methods and the availability of documentation (i.e., either in peer-reviewed journals or as workshop summaries). As these models are utilized and evaluated they may change in equation form, structure, or level of detail. In time some may be deemed more useful than others and may become the predominant models used in assessing exposure to pesticides and other contaminants. This review then represents a snapshot of these aggregate models during this evolution. Model inputs, overall objectives, general methods, implementation, and special features are discussed.

20.5.1 THE CalTOX APPROACH

Developed at the University of California at Berkeley and the Lawrence Berkeley National Laboratory, the multimedia fate and transport fugacity-based model, CalTOX, has been extended to explore exposure and dose by investigating pathways that contribute significantly to risk. This approach calculates chemical concentrations in exposure media and estimates the dose of a single chemical through multiple routes. Pathways and routes considered include dietary ingestion of produce, meats, tap water, and mother's milk; nondietary ingestion of soil and surface water while swimming; inhalation of vapors and particles; and dermal contact with soil and water. Additional extensions of CalTOX address uncertainty, variability, and sensitivity analysis (Bennett, Kastenberg, and McKone 1999).

Inputs for the fate and transport portion of CalTOX include bioconcentration factors, landscape characteristics, partitioning coefficients, and other chemical-specific factors. Toxicological variables such as cancer potency factors and lethal dose values are utilized for an assessment of risk. Human uptake estimates require statistically averaged physiological exposure factors (e.g., breathing rate, surface area, food intake rate, soil adherence to skin) and macro-level activity patterns (e.g., time spent sleeping, time spent outdoors) (Bennett, Kastenberg, and McKone 1999).

The CalTOX approach first creates characteristics of individuals. With these hypothetical individuals and the predicted environmental media concentrations, a distribution of potential dose is estimated. This methodology incorporates nested Monte Carlo simulation and results in CalTOX's unique ability to assess the variability and uncertainty of outputs. Another unique feature of CalTOX is its accessibility and ease of use. Portions of the model are freely available and the majority of calculations require only Microsoft Excel® software. The interface is user-friendly and there are several available documents describing the model's construct, variables, and use (Bennett, Kastenberg, and McKone 1999).

One could argue that attempting to estimate exposure and dose from an uncertain fate and transport model propagates error, whereas using known exposure media concentrations could help to eliminate this error. Since CalTOX is a fate and transport model, a single discharge or source term must be defined. For certain chemicals or scenarios where the source is unknown or multiple sources are present, CalTOX may need to be modified. Another issue is that the model uses averaged macro- and meso-level activity patterns and therefore may not be equipped to handle the details of dermal and nondietary ingestion exposure. The model is also better suited to estimate exposures over a year or an entire lifetime. CalTOX is then limited in its ability to assess exposures for acute or intermediate scenarios (Fryer et al. 2004).

20.5.2 CALENDEX™

The Novigen Calendex™ System performs aggregate and cumulative estimates of consumer and occupational exposures to chemicals. In assessing human exposure to and risks from pesticides used in residential settings exposure levels are calculated for each day and for each individual in the target population. The system relies upon the U.S. Department of Agriculture's Continuing Survey of Food Intakes by Individuals (CSFII) to provide both dietary consumption data and demographic variables of the population. In calculating an individual's dose, the model also uses exposure factors (e.g., body weight, breathing rate, activity patterns), use pattern information of pesticides, and environmental concentration data prior to, during, and after pesticide applications (Petersen et al. 2000). This calendar-based probabilistic model then links information on the frequency of pesticide use with the probability of an exposure occurring and calculated exposures for different routes to estimate a daily aggregate dose (Novigen Sciences, Inc. 1998). The model also permits the inclusion of temporal aspects by modeling changes in exposure and dose due to changes in environmental concentrations over time. A macro-activity approach is taken when estimating post-application dermal dose rates. For nondietary ingestion exposures, meso-level activities are used in the form of hand-to-mouth events per hour.

Calendex™ has been criticized for its lack of transparency, extrapolations of short-term data from the CSFII to simulate longitudinal exposures, and lack of tracking mechanisms to analyze contributions to the model output (Kendall et al. 2000). Additionally, the model relies on the CSFII data for both consumption and demographic data. This survey, however, is known to focus on lower socioeconomic classes that are not representative of the entire U.S. population. In addition to the lack of transparency, significant professional judgment is necessary to apply the model (Fryer et al. 2004).

20.5.3 LifeLine™

The LifeLine approach, a collaborative effort between the Hampshire Research Institute, TAS-ENVIRON, and ChemRisk, aims to characterize inter-individual variations in pesticide doses and aggregate exposures received from multiple sources. Databases used in the LifeLine approach include the National Home and Garden Pesticide Use Survey, the *Exposure Factors Handbook*, U.S. birth records, mobility surveys, and the USDA's Food Consumption Survey. Other inputs include sample data of pesticide residues in the home, tap water, and food (Muir et al. 1998).

LifeLine constructs characteristics (e.g., age, gender, socioeconomic status, body weight, diet, location of home, frequency of pesticide use) of individuals, and simulates each day in these hypothetical individuals' lives (i.e., from birth to 85 years) using probabilistic rules. Macro-level activity patterns and transitions between activities are assigned probabilistically, while meso-level activity patterns are assigned on a yearly basis dependent upon the individual's characteristics. These characteristics and patterns then define the likelihood of exposure to a source and are used in conjunction with pesticide concentration data to determine a total dose. Distributions of dose can be viewed as a function of year or season (Muir et al. 1998).

Notable features in the LifeLine approach include the ability to enter inputs as deterministic values or probability distributions, the use of transition rules for an individual's mobility and pesticide use, and the consideration of a number of indoor and outdoor microenvironments when estimating residential exposure. The developers are also committed to making LifeLine user-friendly and available for public and professional use.

Since this scheme was created to estimate chronic and lifetime exposures, it may not be well suited to estimate short-term exposure events (i.e., less than a day). Furthermore, since LifeLine was expressly created for the FQPA, it may not be easily adapted to consider other chemicals besides pesticides and nonresidential exposures (Fryer et al. 2004). According to a Federal Insecticide, Fungicide and Rodenticide Act (FIFRA) advisory panel, the model may also be more complex and more difficult to use than necessary (Kendall et al. 2000).

20.5.4 The Cumulative and Aggregate Risk Evaluation System (CARES)

CARES, funded by CropLife America, represents a collaborative effort between EXP Corporation, Novigen Sciences, Alceon Corporation, Cambridge Environmental, Inc., Summit Research Services, the University of Georgia, and Sielkin & Associates Consulting, Inc. CARES aims to estimate cumulative and aggregate risk from dietary, drinking water, and residential exposures to pesticides (Baugher et al. 1999).

Model inputs include data from the U.S. Census, Public Use Microdata Samples (PUMS), the Continuing Survey of Food Intakes by Individuals (CSFII), peer-reviewed literature, product attributes, and the *Exposure Factors Handbook*. The model uses census and survey data to create a reference population, and constructs daily profiles of individuals. These daily profiles are then accumulated to estimate short-term, intermediate, or chronic exposures. Additional features include the ability to accept deterministic or probabilistic inputs, the propagation of population variability, and a module to identify important factors contributing to exposure (Baugher et al. 1999).

After a review of the CARES model conducted by the FIFRA advisory panel, several criticisms emerged. For example, the system focuses heavily on dietary exposures and realistically representing population demographics, but uses unrealistic activity patterns extracted from Jazzercise® scenarios and lacks data for residential exposures. The model construct is fairly inflexible and its calculations are not transparent. Additionally sensitivity analysis is difficult to conduct and the framework does not allow for the assessment of variability between replications of individuals (Roberts et al. 2002). Similar to the LifeLine model, the application of CARES may not be easily adapted beyond residential exposures to pesticides (Fryer et al. 2004).

20.5.5 THE STOCHASTIC HUMAN EXPOSURE AND DOSE SIMULATION (SHEDS)

The SHEDS models were developed by the USEPA's National Exposure Research Laboratory. The models aim to improve risk assessments by considering variability in exposure and dose calculations for at-risk populations and by helping to prioritize measurement needs (International Life Sciences Institute [ILSI] 2001). All versions of SHEDS are physically based, probabilistic models designed to estimate exposure beyond the screening level.

The primary version to support the FQPA is SHEDS-Pesticides. Several case studies were explored in developing SHEDS-Pesticides. With each study a stand-alone model was constructed. Residential-SHEDS focuses on children's dermal and nondietary ingestion exposure and dose to the pesticide chlorpyrifos applied in and around the home (Zartarian et al. 2000). SHEDS-PM models daily inhalation exposures to particulate matter from ambient and indoor sources (Burke, Zufall, and Özkaynak 2001). Children's aggregate exposures to arsenic and chromium from treated play sets and decks are explored in the SHEDS-Wood model. The SHEDS-Air Toxics model aims to model aggregate exposures to urban hazardous air pollutants (HAPS) such as benzene, formaldehyde, and metals from industrial sources (USEPA 2003).

SHEDS-Pesticides evaluates ingestion, inhalation, and dermal exposure and dose specifically for children aged 1–6 years (Fryer et al. 2004). The underlying engine of SHEDS uses Monte Carlo methods and the USEPA's Consolidated Human Activity Database (CHAD). The macro-level activity diaries contained in CHAD are utilized in the inhalation models and are supplemented with consumption and contact data for ingestion and dermal exposure estimates (Burke, Zufall, and Özkaynak 2001; Zartarian et al. 2000).

As with other models, some believe the SHEDS models are only applicable for a limited range of scenarios and chemicals. Any customization of the models requires the user to have relatively sophisticated data as inputs. Additionally the models may not properly evaluate longitudinal exposure trends (Fryer et al. 2004).

20.6 ISSUES IN MODELING EXPOSURE

20.6.1 COMBINING DATA FOR MODEL INPUTS

As needed for the FQPA, the aim of many modeling projects is to predict population exposures for policy development. Unfortunately measuring parameters relevant for a large population may be difficult and expensive. It may be necessary to combine or extrapolate data from smaller studies. Several issues may arise, however, in using these smaller studies.

For example, in the developing field of exposure assessment, measurement techniques to collect relevant environmental media and contaminants often differ from study to study given that few standard techniques have been established. In measuring contaminants in house dust, for instance, sampling devices include vacuum cleaners, moistened wipes, hand presses, rollers, dust fall plates, and dust settling mats. The amount of dust captured by the various methods most likely differs even though each technique may be trying to capture the same true value. Nonetheless, a lack of data often persuades the modeler to combine data from different sources — perhaps sources collecting data using different measurement techniques. The effects of combining such datasets on the resulting estimates have not been adequately studied.

Combining time activity data from various sources also poses problems. Activity data are typically collected as categorical data of microenvironments or objects contacted. From study to study, however, these categories may differ, making it difficult to group discrete activities. Adding to the difficulty, activities may be reported as frequency data or as a time series. Data may also be collected at the micro-level for dermal contacts and nondietary mouth insertions or at the macro-level for use in inhalation exposure assessment.

The Consolidated Human Activity Database (CHAD) is a collection of activities from existing human activity pattern studies. In order to combine the data, however, information in some studies was modified according to a predefined format (e.g., a limited number of microenvironments). Given the potential loss of information, the original raw activity data are also provided if more detail is needed.

A separate concern is the extrapolation of data. In particular, there is a need for human activity data for use in modeling longitudinal exposures. These estimates require human activity pattern data spanning several years. Collecting longitudinal activities, however, requires extensive resources, time, and subject participation. Existing activity pattern data, therefore, typically span only a few hours to a few days (e.g., via diaries, recall, videography techniques). Ideally existing short-term data could be used to extrapolate to longer periods of time. Unfortunately methods for extrapolation are not well defined. Partially this is because short-term activities have not been studied to assess inter- and intra-variability, distinct changes in activities over long periods are unknown, and there is little knowledge as to how well the existing activities represent specific populations or age groups. Past methods to simulate longitudinal activities from short-term data have included the use of probabilistic models, random sampling of activities, and repetition of activities representing particular days of the week, seasons, or age groups. Whether or not the activities resulting from these methods realistically represent longitudinal activities is still unknown.

20.6.2 TEMPORAL AND SPATIAL VARIABILITY

There are also temporal characteristics to consider. In measuring contaminant concentrations in air, for example, data sets with measurements taken at different sampling periods may be difficult to combine. While you can always average, say, 1-hour measurements, over a day to conform to other daily measurements, there is typically a loss of detail. This detail in hourly variability may be important to evaluate extreme exposures or to explore contaminant sources. Similarly, attention must be paid to specific designs and goals of individual studies. The study design may be aiming to collect values that represent a continuous time series, seasonal variation, or annual averages. Values from studies with different goals therefore may not be representative for a particular project/exposure assessment.

Similar to temporal characteristics, spatial characteristics must be examined. The first issue is one of scale. Are measurements meant to capture the effects of global, regional or local sources? Once again using the example of concentrations in air, measurements may represent personal exposure, microenvironmental concentrations, or ambient concentrations. Each of these different scales is typically used in a unique way in inhalation exposure models, and combining raw data from different scales could result in unrealistic estimates.

Another issue is one of how well the data represent the exposure scenarios of interest. The National Human Exposure Assessment Survey (NHEXAS), for example, aimed at collecting measurements of multiple chemicals to identify predictors of exposure in multiple geographic locations (Callahan et al. 1995). Sites included the upper Midwest, Arizona, and Maryland. But are data from these sites useful for other locations, or are they representative of a larger population? Can indoor surface concentrations be used to assess dermal exposures in South Texas? Can microenvironmental air concentrations from all sites be combined to estimate inhalation exposures across the entire United States?

20.6.3 Co-Occurrence of Environmental Contaminants

Exposure modeling efforts typically focus on single contaminants. Although dealing with one chemical at a time makes aggregate exposure assessment more feasible, the reality is that few data exist on the co-occurrence of chemicals and their interactions.

A particular case of interest is that of estimating exposure from the co-occurrence of environmental concentrations of pesticides. Such considerations are required by the FQPA. In the absence of quantitative data from sampling media, modelers consider information regarding pesticide properties, transport, and timing of applications. Co-occurrence may then be determined via modeling techniques where probabilities are assigned to the coincidental occurrence of pesticides or where fate and transport models are applied to a region to examine the resulting pesticide spatial distributions. More research should be aimed at collecting the concentrations of multiple contaminants per sample, understanding the phenomena of co-occurrence, and in modeling such phenomena for use in human exposure models.

20.6.4 Estimating Dose from Exposure

Modeling methods for assessing exposure are, for the most part, mathematically simple. Often the most elaborate portions of exposure models entail data storage and management. Other pieces of the human health risk model are historically more complex. For example, models of the fate of contaminants in microenvironments and the transport between microenvironments often involve a system of differential equations and perhaps a number of uncertain parameters. Mechanistic models that specifically estimate dose from exposure can be very complicated as well. Although the focus of the chapter has been exposure modeling, methods for estimating dose are also of concern, since it is the dose that is important in assessing risk and health effects. Furthermore several models claiming to explore pesticide exposure result in a final estimate representing intake or dose.

There are a number of complicated models for estimating dose from dermal exposure. A portion of these are box models with separate compartments representing the layers of skin (i.e., stratum corneum, viable epidermis, papillary dermis). These models are typically based in mass transfer theory, and include chemical-specific properties and parameters, and transfer under steady- or unsteady-state conditions. The majority of dermal exposure models account for dermal dose by simply assuming a fraction of the skin surface concentration is absorbed into the skin.

But how different are dose estimates when using a complex model vs. a simple absorption fraction? Is there a need to combine the most detailed exposure models with the most detailed dose models? To complicate matters, the differences in modeling approaches may be hidden due to the variability within populations and the uncertainties in parameters. Further research needs to explore these issues.

20.7   QUESTIONS FOR FURTHER REVIEW

1. Techniques have been developed for measuring certain pesticides in the air, in soil, in water, and on surfaces. Some may say then that it is not necessary to create models for pesticide exposure, since measurement techniques are available. Do you agree or disagree with this statement? Provide reasoning behind your argument.

2. While models exist to estimate exposure for each route separately, there may be value in creating an aggregate model that considers correlations between model inputs. What might be some important correlations in environmental concentrations? How might dermal and nondietary ingestion exposure be correlated?

3. Although the importance of aggregate exposure has been discussed, assessing exposure through multiple routes may not always be crucial depending on the contaminant of interest. In general, what factors require an aggregate exposure assessment? Name four compounds for which an aggregate assessment may be justified.

4. You would like to estimate the dermal exposure of compound X using Equation 20.3. Notice that in defining the concentration and exposure factor terms, a number of different dimensions are given. You know that compound X possibly exists in the gas phase, in soils, and as a residue on surfaces. If you would like the final spatially averaged dermal exposure estimate to be in units of $\mu g/cm^2$, what are the units of the environmental concentrations and corresponding exposure factors? Be specific in describing the units (e.g., μg of X/cm^2 of skin, μg of X/cm^3 of air).

5. A researcher is conducting a study to determine the lead exposure to preschool age children through the dietary route. She selects a number of children and uses the duplicate diet technique to estimate the mass of lead in their food and the mass of their food for several days. She also collects biological samples from the same children. When comparing her calculated exposure to biomarker data, the biological data seem to show a greater mass of lead when compared to estimates based on the data collected from the duplicate diets. Assuming there is no intake via the dermal and inhalation routes, how might you account for this discrepancy? What measurements would you collect to test your theory?

REFERENCES

Baugher, D.G., Bray, L.D., Breckenridge, C.B., Burmaster, D.E., Crouch, E.A.C., Farrier, D.S., MacIntosh, D.L., Mellon, J.E., Sielken, R.L., and Stevens, J.T. (1999) *Cumulative & Aggregate Risk Evaluation System (CARES), Conceptual Model*, January 8.

Bennett, D.H., Kastenberg, W.E., and McKone, T.E. (1999) A Multimedia, Multiple Pathway Risk Assessment of Atrazine: The Impact of Age Differentiated Exposure Including Joint Uncertainty and Variability, *Reliability Engineering and Systems Safety*, **63**: 185–198.

Burke, J.M., Zufall, M.J., and Özkaynak, H. (2001) A Population Exposure Model for Particulate Matter: Case Study Results for $PM_{2.5}$ in Philadelphia, PA, *The Journal of Exposure Analysis and Environmental Epidemiology*, **11**(6): 470–489.

Callahan, M.A., Clickner, R.P., Whitmore, R.W., Kalton, G., and Sexton, K. (1995) Overview of Important Design Issues for a National Human Exposure Assessment Survey, *The Journal of Exposure Analysis and Environmental Epidemiology*, **5**(3): 257–282.

Canales, R.A. (2004) The Cumulative and Aggregate Simulation of Exposure Framework, Ph.D. diss., Stanford University, Stanford, CA.

Cohen Hubal, E.A.C., Sheldon, L.S., Burke, J.M., McCurdy, T.R., Berry, M.R., Rigas, M.L., Zartarian, V.G., and Freeman, C.G. (2000) Children's Exposure Assessment: A Review of Factors Influencing Children's Exposure, and the Data Available to Characterize and Assess That Exposure, *Environmental Health Perspectives*, **108**(6): 475–486.

Cohen Hubal, E.C., Thomas, K., Quackenboss, J., Furtaw, E., and Sheldon, L. (1998) *Dermal and Non-Dietary Ingestion Exposure Workshop*, Report No. EPA 600/R-99/039, U.S. Environmental Protection Agency, Research Triangle Park, NC.

Dwyer, J., Ellwood, K., Moshfegh, A.J., and Johnson, C.L. (2001) Integration of the Continued Survey of Food Intakes by Individuals and the National Health and Nutrition Examination Survey, *Journal of the American Dietetic Association*, **101**(10): 1142–1143.

Fryer, M.E., Collins, C.D., Colvile, R.N., Ferrier, H., and Nieuwenhuijsen, M.J. (2004) *Evaluation of Currently Used Exposure Models to Define a Human Exposure Model for Use in Chemical Risk Assessment in the U.K.*, Imperial College, London, U.K., April.

ILSI (2001), *Aggregate Exposure Assessment: Model Evaluation and Refinement Workshop Report*, International Life Sciences Institute, Washington, DC.

Kendall, R.J., Matsumura, F., Needleman, H., Ferson, S., Hattis, D., Heeringa, S., MacDonald, P., McKone, T., Reed, N.R., and Odiott, O. (2000) *FIFRA Scientific Advisory Panel Meeting, a Set of Scientific Issues Being Considered by the Environmental Protection Agency Regarding: Residential Exposure Models-REx, Calendex™ Model, Aggregate and Cumulative Assessments Using LifeLine™ — A Case Study Using Three Hypothetical Pesticides*, September 27–29, http://www.epa.gov/scipoly/sap/2000/september/modelsreport.pdf.

Michelson, W.M. and Reed, P. (1975) The Time Budget — Behavioral Research in Environmental Design, in *Behavioral Research Methods in Environmental Design*, Michelson, W.M., Ed., Hutchinson and Ross, Stroudsburg, PA.

Moschandreas, D.J., Ari, H., Daruchit, S., Dim, Y., Lebowitz, M.D., O'Rourke, M.K., Gordon, S., and Robertson, G. (2001) Exposure to Pesticides by Medium and Route: The 90th Percentile and Related Uncertainties, *The Journal of Environmental Engineering*, **127**(9): 857–864.

Muir, W.R., Young, J.S., Benes, C., Chaisson, C.F., Waylett, D.K., Hawley, M.E., Sandusky, C.B., Sert, Y., DeGraff, E., Price, P.S., Keenan, R.E., Rothrock, J.A., Bonnevie, N.L., and McCrodden-Hamblin, J.I. (1998) A Case Study and Presentation of Relevant Issues on Aggregate Exposure, in *Aggregate Exposure Workshop Report to EPA/ILSI Workshop*, Olin, S.S., Ed., ILSI Press, Washington, DC.

Novigen Sciences, Inc. (1998) The Novigen Calendex™ System, Calendex™ Brochure, http://www.exponent.com//practices/foodchemical/calendexbro2.pdf.

NRC (1993) *Pesticides in the Diets of Infants and Children*, National Research Council, National Academies Press, Washington, DC.

OP Case Study Group, Non-Dietary Subcommittee (1999) *REx, Residential Exposure Assessment, Generic Methods, Case Study: Lawn Care Products*, December 15.

Pang, Y., MacIntosh, D.L., Camann, D.E., and Ryan, P.B. (2002) Analysis of Aggregate Exposure to Chlorpyrifos in the NHEXAS-Maryland Investigation, *Environmental Health Perspectives*, **110**(3): 235–240.

Petersen, B.J., Youngren, S.H., Walls, C.L., Barraj, L.M., and Petersen, S.R. (2000) *Calendex™: Calendar-Based Dietary & Non-Dietary Aggregate and Cumulative Exposure Software System*, Novigen Sciences, Inc.

Price, P.S., Young, J.S., and Chaisson, C.F. (2001) Assessing Aggregate and Cumulative Pesticide Risks Using a Probabilistic Model, *Annals of Occupational Hygiene*, **45**(1001): S131–S142.

Reed, K.J., Jimenez, M., Freeman, N.C.G., and Lioy, P.J. (1999) Quantification of Children's Hand and Mouthing Activities through a Videotaping Methodology, *The Journal of Exposure Analysis and Environmental Epidemiology*, **9**(5): 513–520.

Roberts, S.M., Odiott, O., Thrall, M.A., Adgate, J.L., Durkin, P., Engel, B., Freeman, N., Hattis, D., Heeringa, S., MacDonald, P., Portier, K., Potter, T.L., Reed, N.R., Zeise, L. (2002) *FIFRA Scientific Advisory Panel Meeting, a Set of Scientific Issues Being Considered by the Environmental Protection Agency Regarding: Cumulative and Aggregate Risk Evaluation System (CARES) Model Review*, April 30–May 1.

USDA (2004) *Food and Nutrient Database for Dietary Studies, 1.0*, Agricultural Research Service, Food Surveys Research Group, U.S. Department of Agriculture, Beltsville, MD.

USEPA (1997) *1996 Food Quality Protection Act, Implementation Plan, Prevention, Pesticides and Toxic Substances*, U.S. Environmental Protection Agency, Washington, DC.

USEPA (2003) *Stochastic Human Exposure and Dose Simulation Model for Air Toxics, Significant Research Findings: National Exposure Research Laboratory Research Abstract*, http://www.epa.gov/nerl/research/2003/g1-5.pdf (accessed November 12, 2005).

Zartarian, V.G. (1996) A Physical-Stochastic Model for Understanding Dermal Exposure to Chemicals, Ph.D. diss., Stanford University, Stanford, CA.

Zartarian, V.G., and Leckie, J.O. (1998) Dermal Exposure: The Missing Link, *Environmental Science & Technology*, **3**(3): 134A–137A.

Zartarian, V.G., Özkaynak, H., Burke, J.M., Zufall, M.J., Rigas, M.L., and Furtaw, E.J., Jr. (2000) A Modeling Framework for Estimating Children's Residential Exposure and Dose to Chlorpyrifos via Dermal Residue Contact and Nondietary Ingestion, *Environmental Health Perspectives*, **108**(6): 505–514.

Part VII

Policy

21 Environmental Laws and Exposure Analysis

Anne C. Steinemann
University of Washington

Nancy J. Walsh
Emory University

CONTENTS

21.1 Synopsis 487
21.2 Introduction 488
21.3 Clean Air Act (CAA) 488
21.4 Comprehensive Environmental Response, Compensation, and Liability Act
      (CERCLA) and Superfund Amendments and Reauthorization Act (SARA)..................490
21.5 Consumer Product Safety Act (CPSA)..492
21.6 Federal Food, Drug, and Cosmetic Act (FFDCA) and Food Quality
      Protection Act (FQPA)..493
21.7 Federal Insecticide, Fungicide, and Rodenticide Act (FIFRA) 495
21.8 Occupational Safety and Health Act (OSH Act) 496
21.9 Resource Conservation and Recovery Act (RCRA) 498
21.10 Toxic Substances Control Act (TSCA) 499
21.11 Conclusions 501
21.12 Questions for Review..502
21.13 Acknowledgments 502
References 512

21.1 SYNOPSIS

This chapter analyzes how major federal laws address human exposure. The analysis provides a surprising conclusion. Our environmental regulations, designed to protect human health, offer scant protection against major sources of pollutant exposure that endanger human health. The largest of these sources — common consumer products and building materials — are virtually untouched by existing laws. One reason is that our regulatory approach focuses on outdoor emissions and effluents, rather than on exposures, even though exposures are how pollutants reach humans and affect health. Another reason is that no federal agency or law specifically regulates indoor environments, where most of our exposure currently occurs. For example, our primary exposure to many "hazardous air pollutants" (HAPs) occurs indoors. Yet existing regulations focus on HAPs outdoors, essentially ignoring the high levels of HAPs found indoors. Moreover, the laws contain exclusions and loopholes that enable significant exposures to occur. For instance, everyday household products can be exempt from testing and disclosure of their toxic chemical constituents. Finally, the laws generally have not incorporated advances in the science of exposure analysis, such as use of personal exposure monitors to obtain exposure data, which would make the laws more effective.

21.2 INTRODUCTION

Earlier chapters examined different types of exposures and how they occur. A natural question is *why* these exposures occur, given that we have numerous regulations that aim to protect human health. This chapter investigates that question by conducting an in-depth analysis of 10 federal laws that have relationships to human exposure. For each law, we look at three main issues: What is the law and its goals? How does the letter of the law address human exposure? How effective is the law in actually reducing exposures? We also provide a content analysis of the laws (Table 21.1), which shows that "exposure" was mentioned 311 times total in these laws. Despite its frequent mention on paper, exposure is often treated only superficially in practice. For instance, many laws rely on vague estimates of exposure rather than actual exposure measurements. This chapter uncovers the gaps between the intent and implementation of the laws, and highlights the important yet unfulfilled role of exposure science in the laws. We conclude with a proposed Human Exposure Reduction Act (Table 21.2), designed to address current deficiencies and provide a more effective framework for environmental regulation and public health protection.

21.3 CLEAN AIR ACT (CAA)[1]

Prior to the Clean Air Act of 1970, the nation's air pollution laws contained many useful provisions, but they were coupled with cumbersome procedures that made controlling air pollution remarkably slow. The prior air pollution laws also relied primarily on state and regional actions to control air pollution, with scant federal enforcement. The 1970 CAA revamped air pollution controls in the United States, increased enforcement powers, and transferred regulatory authority for air pollution from the Department of Health, Education and Welfare to the newly created U.S. Environmental Protection Agency (USEPA).

The Clean Air Act authorizes the USEPA Administrator to establish nationally uniform air quality standards, called National Ambient Air Quality Standards (NAAQS), intended to protect public health and the environment. The NAAQS exist only for a class of pollutants called the "criteria" air pollutants: carbon monoxide (CO), ozone (O_3), sulfur dioxide (SO_2), nitrogen dioxide (NO_2), particulate matter ($PM_{2.5}$ and PM_{10}), and lead (Pb) (see Chapter 4 on Inhalation Exposure).

To establish a NAAQS, Section 108 of the Clean Air Act requires a judgment by the Administrator that the air pollutant under consideration (1) "has an adverse effect on public health and welfare," (2) "results from numerous or diverse mobile or stationary sources," and (3) is one for which the Administrator "plans to issue air quality criteria."[2] Once these three conditions are met for a particular air pollutant, the USEPA publishes its intention to establish a NAAQS in a process called "listing," which starts a time clock, and the Administrator must issue, within 12 months after listing, an "air quality criteria document."[3] This document compiles available data on the public health and environmental effects of the pollutant at various levels in ambient air and provides the scientific basis for determining the NAAQS values. The USEPA periodically updates the air quality criteria documents, which exist for all pollutants with NAAQS, to include recent scientific findings.

The 1990 Clean Air Act Amendments included a list of 189 hazardous air pollutants (HAPs), or air toxics, with authority for the USEPA to revise the list based on new data. HAPs are defined as substances "which present, or may present, through inhalation or other routes of exposure, a

---

[1] 42 U.S.C. §§ 7401-7671q (2002).
[2] 42 U.S.C. § 7408(a)(1) (2002).
[3] 42 U.S.C. § 7408 (2002).

threat of adverse human health effects."[4] (See Table 21.3 for a current list of HAPs.) In contrast to the criteria air pollutants, the HAPs have no ambient standards, because HAPs are assumed to have no known safe levels of exposure, or health threshold. With no known health threshold, it has been difficult to adopt air quality standards for the HAPs, and attempts to create such standards before 1990 were largely unsuccessful. Instead, the 1990 amendments focused on mandatory reductions in emissions of HAPs using the maximum achievable control technology (MACT), rather than attaining ambient air standards. MACT standards for existing sources of pollution can be no less stringent than the average emission limitation achieved by the best performing 12% of existing sources in a similar source category or subcategory.[5]

The framers of the CAA intended to limit pollutant levels in air that people breathe by limiting pollutant levels in ambient air. Yet the original CAA does not define "ambient air,"[6] and the USEPA has limited its interpretation of ambient air to the regulation of outdoor air, or "air external to buildings."[7] Thus the law has focused on pollutant levels in outdoor air, virtually neglecting indoor air, even though human exposure to all but a few pollutants is higher indoors than outdoors (Wallace 1991). (See Chapter 1 and the Introduction.) Further, nowhere does the law mention the term "indoor air" (see Table 21.1).

Because of this limited interpretation, the USEPA does not currently exercise authority over indoor air pollution under the CAA. However, the CAA does not restrict the USEPA's authority to regulate indoor air, and the USEPA indeed regulates indoor air when it regulates ambient air, because outdoor air infiltrates indoors. Also, because the air quality criteria documents include exposure studies, the CAA considers, to some extent, the pollutant concentrations that people actually breathe, not ambient concentrations alone.

The USEPA has also used its authority under the National Emission Standards for Hazardous Air Pollutants (NESHAPS)[8] program of the CAA to ban indoor activities that affect emissions into the atmosphere (such as the spraying of asbestos insulation). Thus, the USEPA may have reduced exposure to indoor pollutants as an inadvertent result of controlling emissions to the atmosphere.

In 1998, standards were passed under the CAA to regulate consumer products if they contribute to at least 80% of the volatile organic compound (VOC) emissions outdoors in areas that violate the NAAQS for ozone.[9] However, these standards exempt some of the most significant sources of VOC exposures indoors, such as air fresheners, insecticides, adhesives, and moth-proofing products.[10] Many of these exempted products also contain one or more of the HAPs, and many HAPs have been found at higher levels indoors than outdoors, because of indoor sources (Table 21.3).

The CAA relies on a national network of outdoor air monitoring stations, but no similar network exists for measuring exposure. Indeed, nearly all the existing exposure data collected on the U.S. population come from large-scale research studies such as TEAM and NHEXAS (See Chapters 1, 3, 7, 13, and 15). These research studies were conducted in a single time period, and in only a few cities; their aim was to develop new exposure methods, rather than to measure exposure routinely. Thus, we lack long-term data on trends in exposure to these same pollutants. The other major set of relevant exposure data is the national biomonitoring studies conducted by the Centers for Disease Control (CDC, 2001, 2003, 2005) (see Chapter 17 on Biomarkers).

Overall, the Clean Air Act's regulation of criteria air pollutants has been effective in reducing ambient concentrations nationwide, and four of the six ambient air pollutants have shown a major decline throughout the United States over the last 20 years (USEPA 2005a). Nevertheless, because

---

[4] 42 U.S.C. § 7412(b)(2) (2002).

[5] 42 U.S.C. § 7412(d)(3) (2002).

[6] 42 U.S.C. § 7409 (2000).

[7] 40 C.F.R. § 50.1(e) (2002).

[8] 42 U.S.C. § 7412 (2000).

[9] Clean Air Act § 183(e), 42 U.S.C. §7511b(e): National Volatile Organic Compound Emission Standards for Consumer Products, Fed. Reg. 48819-48847 (1998), 40 C.F.R. §§ 59.201-59.214 (2003).

[10] 40 C.F.R. §§ 59.201(c)(1)-(7) (2003).

CAA does not specifically address indoor air, major sources of exposure to VOCs, particles, and pesticides have received negligible attention under the CAA.

21.4   COMPREHENSIVE ENVIRONMENTAL RESPONSE, COMPENSATION, AND LIABILITY ACT (CERCLA)[11] AND SUPERFUND AMENDMENTS AND REAUTHORIZATION ACT (SARA)[12]

Congress enacted CERCLA in 1980 to address releases of hazardous substances endangering public health. CERCLA established a "Superfund" from taxes on the chemical and petroleum industries to pay for cleanup of abandoned hazardous waste sites.[13] Under CERCLA, the USEPA must maintain a National Priorities List (NPL), which is "the list of national priorities among the known releases or threatened releases of hazardous substances, pollutants, or contaminants…intended primarily to guide the EPA in determining which sites warrant further investigation" (USEPA 2004). Once a site is on the NPL, the USEPA conducts a remedial investigation to determine the nature and extent of the contamination at the site, and a feasibility study to identify and evaluate cleanup strategies.

The USEPA responds to hazardous substances at Superfund sites through "removal" and "remedial" actions. Removal actions are generally short-term (less than 1 year) and low-cost (under $2 million), intended to address actual or potential releases of hazardous substances. Remedial actions are generally longer-term and more extensive, such as treating or containing contaminated soil, constructing underground walls to control the movement of groundwater, and incinerating hazardous wastes.

CERCLA set forth a new liability scheme, commonly referred to as the "polluter pays" program, for cleanup costs and other damages relating to releases of hazardous substances. Regardless of whether the USEPA or a private entity conducts a cleanup, CERCLA makes any owner or operator of contaminated property, or transporter or handler of a hazardous substance, a "potentially responsible party" with regard to the costs related to a release of such hazardous substance, essentially shifting the burden of proof to those entities to disprove their responsibility for any release.[14]

In 1986, Congress enacted SARA to reauthorize the Superfund tax and amend CERCLA and other statutes relating to hazardous substances. Among other changes, SARA established an emergency response and citizen right-to-know program involving state response authorities, encouraged greater citizen participation in cleanup decisions, and increased the funds for Superfund. In the first 5 years after CERCLA's enactment, the federal government collected $1.6 billion for cleaning up abandoned or uncontrolled hazardous waste sites. In 1986, SARA increased the size of the Superfund to $8.5 billion. In 1990, Congress reauthorized the Superfund program through 1994, adding $5.1 billion (USEPA 2005d). In 1995, the taxing authority of CERCLA/SARA expired, and has not been reauthorized by Congress.

CERCLA focuses on releases of hazardous substances into the environment. A "release" is defined broadly to include almost any discharge or leak into the environment.[15] Thus, a release, the trigger for CERCLA coverage, is not tied directly to actual human exposure to contaminants. Also, CERCLA/SARA allows the USEPA to rely upon calculations of risk derived from estimates of exposure in its *Exposure Factors Handbook* (USEPA 2005e), rather than actual exposure data.

---

[11] 42 U.S.C. §§ 9601-9675 (2000).

[12] Pub. L. 99-499, 100 Stat. 1613, codified in scattered sections of the Internal Revenue Code and in amendments to CERCLA at 42 U.S.C. §§ 9601-9675 (2000).

[13] Title II, § 221 of CERCLA, previously codified at 42 U.S.C. § 9631-9633, now codified at 26 U.S.C. § 9507 (2000); 42 U.S.C. § 9605 (2000).

[14] 42 U.S.C. § 9607(a) (2000).

[15] 42 U.S.C. § 9601(22) (2000).

CERCLA also created the Agency for Toxic Substances and Disease Registry (ATSDR) in the U.S. Department of Health and Human Services, and directed the administrators of the ATSDR and the USEPA to maintain registries and databases on toxic substances and their impact on human health. SARA broadened ATSDR's responsibilities with respect to environmental and public health by directing ATSDR and the USEPA to jointly prepare a list (ATSDR 2003), in order of priority, of at least initially 100 (and eventually over 250) hazardous substances that "pose the most significant potential threat to human health due to their known or suspected toxicity to humans and the potential for human exposure to such substances at facilities on the NPL."[16]

SARA also mandated ATSDR to perform a health assessment for each facility on the NPL and other contaminated sites. The ATSDR health assessments are intended to assist in determining whether actions should be taken to reduce human exposure to hazardous substances from a facility and whether additional information on human exposure and associated health risks is needed. Again, these assessments do not require actual measurements of human exposure.

Responding to increasing public concern about radon and other indoor air quality hazards, SARA also set forth the Radon Gas and Indoor Air Quality Research Act of 1986, which created a program within the USEPA to "(1) gather data and information on all aspects of indoor air quality in order to contribute to the understanding of health problems associated with the existence of air pollutants in the indoor environment, (2) coordinate federal, state, local, and private research and development efforts relating to the improvement of indoor air quality; and (3) assess appropriate federal government actions to mitigate the environmental and health risks associated with indoor air quality problems."[17]

This Act did not provide the USEPA with authority to promulgate standards for indoor air quality. Indeed, the Act explicitly states that "[n]othing in this title shall be construed to authorize the Administrator to carry out any regulatory program or any activity other than research, development, and related reporting, information dissemination, and coordination activities specified in this title."[18] In 1989, as part of its responsibilities under SARA, the USEPA established the Federal Interagency Committee on Indoor Air Quality (CIAQ) to coordinate the activities of the federal government on issues relating to indoor air quality (USEPA 2005c). The CIAQ meets quarterly, generally for half a day. Given the lack of directive for CIAQ to recommend regulations for indoor air quality, and the highly scientific nature of the topics on its meeting agendas, it is not surprising that the CIAQ's activities have not attracted significant attention from the public or created any impetus for regulation in this field.

CERCLA/SARA has been critiqued from several perspectives, and one of them is lack of attention to exposure and health effects. For instance, CERCLA/SARA does not contain any requirement to study the link between contamination at a site and human exposure to the contaminants. Also, both ATSDR and the USEPA, in their approach to risk assessment, rely upon the *Exposure Factors Handbook*, which is subject to inaccuracies. In addition, CERCLA/SARA does not require medical monitoring or exposure data to determine actual health impacts from contaminated sites. Another critique is that CERCLA has been inefficient, with resources consumed by litigation rather than directed to cleanup of sites. The USEPA has "delisted" from the NPL only 308 sites (with 1,239 sites remaining on the list) in nearly 25 years (USEPA 2005f).

For CERCLA/SARA to provide meaningful protection of public health, scientific advances in exposure analysis should be incorporated into the laws. Nevertheless, a problem remains: many of the same pollutants of concern at Superfund sites, where negligible human exposure occurs, are already present as indoor air pollutants in homes and workplaces, where significant human exposure occurs.

---

[16] Pub. L. No. 99-499, § 110, 42 U.S.C. § 9604(i)(2)(A) (2000).
[17] Pub. L. No. 99-499 § 403(a).
[18] Pub. L. No. 99-499 § 404.

21.5 CONSUMER PRODUCT SAFETY ACT (CPSA)[19]

Congress enacted the Consumer Product Safety Act (CPSA) in 1972 to address risks posed by consumer products. The CPSA is an "umbrella statute" that established the Consumer Product Safety Commission (CPSC), and provided the limits of its authority. When the CPSC finds an unreasonable risk of injury associated with a consumer product, it may develop a standard to reduce or eliminate the risk through notice and comment rulemaking.[20] If the CPSC determines that a consumer product poses an imminent danger, it can issue a mandatory recall of the product. It is this function of the CPSC with which consumers are probably most familiar.

The CPSA relies on voluntary consumer product safety standards.[21] The CPSA provides for mandatory reporting of (1) known failures of consumer products to meet applicable standards, (2) information suggesting a product defect that could create a substantial risk of injury, and (3) information suggesting an inherent unreasonable risk of serious injury or death. The CPSC may impose labeling requirements only if there is substantial evidence that a warning is "reasonably necessary" to prevent or reduce unreasonable risks of injury.[22]

Significantly, the CPSA does not require the listing of all ingredients in products. The CPSA requires the CPSC to maintain the confidentiality of trade secret or other confidential information (such as product formulation) provided to the CPSC.[23] The U.S. Supreme Court has concluded that the confidentiality provisions in the CPSA prohibit the CPSC from disclosing information deemed confidential under the CPSA, even in response to requests under the Freedom of Information Act.[24]

The CPSA includes a finding by Congress that "existing federal authority to protect consumers from exposure to consumer products presenting unreasonable risks of injury is inadequate."[25] The CPSA defines the term "risk of injury" to mean "a risk of death, personal injury, or serious or frequent illness."[26] The CPSA contends with the risk of cancer posed by consumer products, but it requires a Chronic Hazard Advisory Panel to determine that the product is a carcinogen before the CPSC can initiate any rulemaking procedures.[27] Also, Congress excluded from the CPSA's coverage many dangerous or potentially dangerous consumer products that are regulated by other statutes, some of which present significant potential for exposure to dangerous substances (including food, drugs, cosmetics, tobacco products, and pesticides).

In practice, the CPSA has regulated exposure by banning a few products that present exposure risks (such as lead-based paint products), specifying safety standards for a few exposure-related products (such as products using chlorofluorocarbons), and specifying label requirements for several chemicals and other dangerous substances (such as charcoal, fireworks, and art products).

However, the CPSA offers little protection to Americans from everyday exposures to hazardous chemicals in consumer products. An example is the prevalence of synthetic fragrances (toxic VOCs) found in air fresheners, laundry supplies, cleaners, and personal care products. Because of confidentiality provisions, a manufacturer need only list "fragrance" on the label, not the actual chemicals, even though more than 95% of chemicals used in fragrances are known toxics, sensitizers, and carcinogens (USHR 1986).

The CPSA has also had limited effect with regard to exposures because it, like many other agency-creating federal laws, requires the regulating agency to estimate the costs and benefits of proposed rules regulating or banning dangerous products.[28] Cost-benefit analyses pose particular

---

[19] Pub. L. No. 92-573, 86 Stat. 1207 (1972), codified at 15 U.S.C. §§ 2051-2084 (2002).

[20] 15 U.S.C. § 2056(a) (2002).

[21] 15 U.S.C. § 2056 (b)(1) (2002).

[22] 15 U.S.C. § 2056(a) (2002); *see also* 58 Fed. Reg. 8013, 8015 (1993).

[23] 15 U.S.C. § 2055 (2002).

[24] *GTE Sylvania, Inc. v. Consumer Product Safety Commission*, 447 U.S. 102, 100 S. Ct. 2051 (1980).

[25] 15 U.S.C. § 2051 (2002).

[26] 15 U.S.C. § 2052(a)(3) (2002).

[27] 15 U.S.C. § 2080 (2002).

[28] 15 U.S.C. § 2058(f) (2002).

difficulties when trying to identify and monetize a range of possible health effects related to exposure, and when discounting future and uncertain outcomes into present value amounts. In sum, the CPSA is a significant statute in some consumer product areas, but does not close many important gaps in protecting people from hazardous exposures.

21.6   FEDERAL FOOD, DRUG, AND COSMETIC ACT (FFDCA)[29] AND FOOD QUALITY PROTECTION ACT (FQPA)[30]

The Federal Food, Drug, and Cosmetic Act (FFDCA) of 1954 and the Food Quality Protection Act (FQPA) of 1996 are discussed together because a key focus of the FQPA was changing the FFDCA's framework for addressing pesticide residues in food. The breadth of these laws is far-reaching. Together they regulate the safety, effectiveness and labeling of drugs, cosmetics, and medical devices, while also dealing with food safety.

The federal government began regulating food and drug safety in 1906, but the FFDCA made several significant changes to weaker predecessor laws. Among these changes, the FFDCA extended federal control to cosmetics and therapeutic devices, required new drugs to undergo safety testing before marketing, and eliminated the requirement to prove intent to defraud in drug misbranding cases (USFDA 2005). To illustrate the extent to which these laws address exposure, this chapter will focus on two significant issues: pesticide residues in food, and cosmetic-drug distinctions.

Pesticide Residues in Food: In 1958, the Food Additives Amendment to the FFDCA was enacted, requiring manufacturers of new food additives (such as preservatives or colors) to establish the safety of the additives. The Delaney Clause, contained in Section 409, states that no additive can be considered safe if "it is found to induce cancer when ingested by man or animal, or if it is found, after tests which are appropriate for the evaluation of the safety of food additives, to induce cancer in man or animal."[31]

Pesticide residues in processed food products were considered food additives and, thus, subject to the zero-risk standard (i.e., no cancer risk allowed) of the Delaney Clause. Yet pesticide residues in raw foods were treated separately, under Section 408, and subject to a less restrictive standard of balancing risks and benefits. Also, under Section 402, "flow through" exemptions allowed pesticide residues on raw foods to remain in processed foods, at a tolerance specified for raw foods, notwithstanding the zero-risk language of the Delaney Clause. Thus, pesticide residues were regulated under different standards.

In 1987, a National Academy of Sciences (NAS) report concluded that the Delaney Clause, and its implementation by the Food and Drug Administration (FDA) and the USEPA, was an unworkable framework that could be creating higher risk of cancer by distinguishing between pesticide residues in raw and in processed foods (NRC 1987). The NAS explained that the Delaney Clause required the USEPA to prohibit new pesticides if they had any carcinogenic effect, even if they were considered safer overall than existing pesticides, illustrating the "Delaney Paradox" (NRC 1987).

In 1988, the USEPA promulgated a final rule, stating that it would permit carcinogenic pesticide residues in raw and processed foods if the risk of cancer was *de minimis*, which the USEPA defined as a 1 in 1 million risk of cancer over a lifetime. The courts struck down this interpretation as being inconsistent with the plain language of the Delaney Clause, which prohibited *any* carcinogenic risk in processed foods.[32] In 1993, the NAS issued another report on the Delaney Clause, which concluded that current studies underestimated risks to infants and children from pesticide residues,

---

[29] Pub. L. No. 75-717, 52 Stat. 1040, codified at 21 U.S.C. §§ 321-397 (2000).
[30] Pub. L. No. 104-170.
[31] 21 U.S.C. § 348(c)(3)(A) (2000).
[32] *Les v. Reilly*, 968 F.2d 985 (9th Cir. 1992).

and suggested adding a safety margin to risk assessment of pesticide residues in foods consumed by infants and children (NRC 1993).

These battles over the Delaney Clause and pesticide residues ultimately led to a compromise, reflected in the unanimous passage of the Food Quality Protection Act (FQPA) by Congress in 1996. The FQPA amended the FFDCA and the Federal Insecticide, Fungicide, and Rodenticide Act (FIFRA) to remove pesticide residues in food products from the scope of the Delaney Clause (i.e., pesticide residues were no longer considered "food additives"), and to establish a single health-based safety standard, "reasonable certainty of no harm," for regulation of risks to human health from pesticide residues in food.

Specifically, the FQPA provides that levels of pesticide residues in food products are acceptable if "there is a reasonable certainty that no harm will result from the aggregate exposure to the pesticide chemical residue."[33] The FQPA does not define "reasonable certainty," but a 1-in-1 million lifetime risk of cancer (which the USEPA had tried to implement through regulation previously) is the standard that Congress expected the USEPA Administrator to apply.[34] The FQPA also requires that a safety factor 10 times lower than for adults be applied to tolerances for infants and children, unless data show that children are not more susceptible to the health risks from pesticide residues.

The FQPA moves beyond the idea that cancer is the only health risk for regulatory purposes; the new law requires the USEPA to consider not only carcinogenicity but also estrogenic and other hormone disruptor effects from pesticide residues.[35] Also, the FQPA requires the USEPA to consider information on "cumulative effects" of pesticide residues on consumers, and aggregate exposure levels of consumers to pesticide residues in food and other media (including drinking water and home lawn care products) from non-occupational sources.[36] The USEPA must also consider the special consumption patterns of infants and children, and the cumulative effects of exposure to pesticide residues in infants and children. If the USEPA determines that total risk from all currently registered uses of a pesticide and other substances with a "common mechanism of toxicity" exceeds the safety standard, the USEPA must cancel one or more uses of the pesticide or reduce the tolerance levels for those uses, and the USEPA is prohibited from registering new uses of the pesticide (USDA 1997). Thus, in concept, the FQPA represents an important advance in considering total exposure.

Cosmetic–Drug Distinctions: According to the FFDCA, the term "cosmetic" means "(1) articles intended to be rubbed, poured, sprinkled, or sprayed on, introduced into, or otherwise applied to the human body or any part thereof for cleansing, beautifying, promoting attractiveness, or altering the appearance, and (2) articles intended for use as a component of any such articles; except that such term shall not include soap."[37]

This definition in the FFDCA is significant because (a) cosmetics, unlike drugs (but like most consumer products), are not subject to pre-market review, and (b) cosmetic manufacturers, unlike drug manufacturers, are not required to register with the FDA. Just as the Delaney Clause came to be perceived as unworkable, the cosmetic–drug distinction in the FFDCA has come under criticism for ignoring that many cosmetic products have drug-like effects. In addition, cosmetics often contain toxic chemicals that are regulated under other laws, but unregulated in consumer products. Further, the FDA cannot require companies to conduct safety studies of their cosmetics before marketing. For instance, only 11% of the 10,500 ingredients that the FDA documented in products have been assessed for safety by the cosmetic industry's review panel (EWG 2005).

The FFDCA drug testing and medical device provisions are some of the more science-based federal laws, relying upon years of studies with control groups. In contrast, cosmetic provisions

---

[33] 21 U.S.C. § 346a(b)(2)(A) (2000).
[34] H.R. Rep. No. 104-669, pt. 2 at 41 (1996).
[35] 21 U.S.C. § 346a (2000).
[36] 21 U.S.C. § 346a(b)(2)(D)(vi) (2000).
[37] 21 U.S.C. § 321(i) (2000).

contain an outdated framework that largely ignores risks and levels of exposures to hazardous substances in cosmetic products.

Although the FQPA has made the FFDCA's food safety components more exposure-oriented, there is no indication in the short history of the FQPA that it has reduced exposures to pesticides. When passed, it was estimated that the FQPA would significantly reduce pesticide use. However, total use of pesticides in the United States continues to grow (USEPA 2002a), and the FQPA's focus on pesticide effects on infants and children appears not to have affected USEPA considerations of tolerances (Cross 1997). A difficulty in implementing the FQPA is the EPA's lack of comprehensive exposure data, which makes it nearly impossible to accurately assess the impacts that the FQPA assumes can be assessed, such as consideration of cumulative exposure levels from pesticides and other toxic substances.

Since 1961, the FDA has conducted a "Total Diet Study" (also known as "Market Basket Study") to determine the levels of toxic chemicals, such as pesticide residues, and nutrients in approximately 280 core foods in the U.S. food supply. The market baskets are generally collected four times each year, once in each of four geographic regions of the United States each year. In the 2003 Total Diet Study, pesticide residues were found in 37.3% of the domestic samples, the most common being DDT (12%), malathion (7%), endosulfan (7%), dieldrin (6%), and chlorpyrifos-methyl (6%) (CFSAN 2005). Although both DDT and dieldrin were banned in the 1970s, their residues persist in our food supply. Moreover, many of the pesticides found in the Total Diet Study were also found in indoor air (see Chapter 15 on Pesticide Exposure), such as chlorpyrifos, dieldrin, malathion, and diazinon, although these pesticides did not make it on the list of HAPs (Table 21.2).

The FQPA required the USEPA to report to Congress within 4 years on the USEPA's progress in implementing the law. The USEPA report states that out of 612 pesticides eligible for re-registration between 1996 and 1999, 231 cases were voluntarily canceled, while the USEPA "canceled, deleted or declared not eligible for re-registration" only 21 pesticide products. The USEPA re-registered 189 pesticides, with the remainder subject to decision (USEPA 1999). These numbers indicate that the FQPA may have weeded out some of the most dangerous pesticides, but it does not demonstrate that existing pesticides or exposure levels are safe.

21.7   FEDERAL INSECTICIDE, FUNGICIDE, AND RODENTICIDE ACT (FIFRA)[38]

The Federal Insecticide, Fungicide, and Rodenticide Act (FIFRA) of 1947 has evolved from a law designed to protect farmers from damaged feed to a system of regulating pesticide use on food and in the environment. FIFRA defines "pesticide" broadly to include substances used to control mold and mildew in water or on stored grains, as well as fumigants, mothballs, rat poison, and other substances used to control pests. Under FIFRA, the USEPA is responsible for "registering," or licensing pesticide products for use in the United States.

Pesticide registration decisions are based on an assessment of the potential effects of a product on human health and the environment, when used according to label directions. For a pesticide to be registered, the applicant must submit data to the USEPA on its use and effects, but the EPA does not routinely measure exposure to pesticides or conduct its own health effect studies of pesticides. Also, the emphasis is on protecting pesticide applicators rather than members of the general public. If the USEPA registers a pesticide, it specifies the approved uses and conditions of use of the pesticide.[39]

The USEPA attempts to make a judgment whether a pesticide chemical residue is safe by considering whether there is "a reasonable certainty that no harm will result from aggregate exposure

---

[38] Pub. L. No. 80-104, and including the FIFRA amendments of 1988, Pub. L. No. 100-532 and amendments to FIFRA and the Food Quality Protection Act in Pub. L. No. 104-170. FIFRA is codified at 7 U.S.C. §§ 136-136y (2000).

[39] 7 U.S.C. § 136a(d)(1) (2000).

to the pesticide chemical residue, including all anticipated dietary exposures and all other exposures for which there is reliable information."[40] FIFRA provides that, with limited exceptions, "any pesticide chemical residue in or on a food shall be deemed unsafe" unless "a tolerance for such pesticide chemical residue in or on such food is in effect under this section and the quantity of the residue is within the limits of the tolerance."[41] Two of the factors the USEPA must use in establishing a tolerance for a pesticide are "available information concerning the cumulative effects of such residues and other substances that have a common mechanism of toxicity" and "available information concerning the aggregate exposure levels of consumers (and major identifiable subgroups of consumers) to the pesticide chemical residue and to other related substances, including dietary exposure…and exposure from other non-occupational sources."[42]

The modern version of FIFRA mentions exposure but does not require direct measurements of exposure, and often there are no data available on the actual exposure of the public. The Non-Occupational Pesticide Exposure Study (NOPES) of 1985–1986 was the first study to measure the public's everyday exposure to pesticides (See Chapter 15 on Pesticide Exposure). In addition, despite FIFRA's attention to exposure, total pesticide use continues to grow (USEPA 2002a).

Further, a significant weakness of FIFRA is its lack of clear regulation on inert ingredients. Although FIFRA requires manufacturers to disclose each active ingredient of a pesticide on the consumer label with the percentage of that active ingredient by weight, they need not specify the "inert ingredients" in the pesticides. Yet inert ingredients often predominate in pesticide products and can be more toxic than the active ingredients (USEPA 2005g). For example, a study of 85 consumer pesticide products found that 72% contained over 95% inert ingredients, and more than 200 of these "inerts" were classified as hazardous pollutants in other federal environmental statutes (NY 1996). In September 1997, the USEPA issued a regulation (Pesticide Regulation 97-6) encouraging, but not requiring, manufacturers of pesticides to stop using the term "inert ingredients" and to use the term "other ingredients" as a substitute, because the USEPA surveys showed that consumers erroneously believed "inert ingredients" were harmless. Thus, while FIFRA addresses hazardous exposure to pesticides, it does not address all hazardous ingredients in pesticide products or require exposure measurements.

21.8 OCCUPATIONAL SAFETY AND HEALTH ACT (OSH ACT)[43]

Congress enacted the Occupational Safety and Health Act (OSH Act) in 1970 to assure "safe and healthful working conditions" for private sector American workers.[44] The OSH Act contains a general duty provision requiring employers to provide working conditions "free from recognized hazards that are likely to cause death or serious physical harm" to employees.[45] In 1971, the Act established, under the Department of Labor, the Occupational Safety and Health Administration (OSHA), which administers the Act.

The OSH Act also authorized OSHA to adopt certain national consensus health and safety standards and established federal standards (called "interim standards" in the OSH Act) soon after the effective date of the OSH Act, without the notice-and-comment procedure applicable to most federal regulations.[46] In 1971, OSHA acted pursuant to this authority to promulgate a rule setting 425 permissible exposure limits (PELs) for air contaminants.[47] Most of the PELs were consensus standards that had been recommended in 1968 by the American Conference of Governmental

---

[40] 21 U.S.C. § 346a(b)(2)(A)(ii) (2000).

[41] 7 U.S.C. § 346a(a)(1) (2000).

[42] 21 U.S.C. § 346a(b)(2)(D)(v)(vi) (2000).

[43] 29 U.S.C. §§ 651 et seq. (1970).

[44] 29 U.S.C. § 651(b) (2000).

[45] 29 U.S.C. § 654(a) (2000).

[46] 29 U.S.C. § 655(a) (2000).

[47] 29 C.F.R. § 1910.1000 (2000).

Industrial Hygienists (ACGIH) and were applicable to many government contractors under the Walsh-Healey Act prior to the enactment of the OSH Act.[48]

The OSH Act created a notice-and-comment procedure whereby OSHA could amend, delete, or modify the interim standards with permanent health and safety standards "reasonably necessary or appropriate to provide safe or healthful employment and places of employment."[49] However, OSHA has been largely unsuccessful in its efforts to adopt new standards. Over 30 years after OSHA promulgated its "interim standards," they continue to constitute the bulk of OSHA's health standards relating to exposure to contaminants.

Air quality standards that have been adopted under the OSH Act are much less protective of health than the national ambient air quality standards adopted under the CAA. For example, the outdoor ambient NAAQS for carbon monoxide is a 9 parts per million (ppm) average for 8 hours, not to be exceeded in a community more than once per year, while the similar personal exposure standard adopted under the OSH Act is 50 ppm average for 8 hours (see Chapter 4 on inhalation exposure).

The OSH Act leaves several regulatory gaps with respect to exposure to pollutants. Most fundamentally, it only protects employees, and only in certain workplaces. In 2000, 41% of the U.S. population was employed (CQC 1994), and, based on national diary studies of the U.S. population's activities, Klepeis et al. (2001) found that Americans, on average, spent only 5.4% of their time in an office or factory. The OSH Act provides no coverage to persons outside the workplace, or to persons who work in their homes. Even in the workplace, the OSH Act does not cover all employers, and can exclude state and local government employees (unless the state has a plan approved by OSHA), private sector employees (if excluded by the state plan), and federal employees (except by executive order).

The courts have given OSHA little leeway in pursuing its statutory missions. In 1980, the U.S. Supreme Court reviewed OSHA's standard for benzene exposure and interpreted the OSH Act's requirements with respect to promulgation of permanent health and safety standards on toxic substances.[50] The Court held that OSHA must first demonstrate that a significant risk of material health impairment exists at the current levels of exposure to the toxic substance.[51] The Court implied that OSHA must quantify the risk by showing how many employees would become ill or die from the current exposure level. The Court further explained that OSHA must demonstrate that the proposed new standard prevents material impairment of health of employees to the extent feasible.[52] Thus, even if a proposed standard would improve worker health, it may be invalidated for not going far enough to prevent material impairment of health.

In 1989, after several unsuccessful attempts to adopt or revise rules on specific toxic substances, OSHA issued an Air Contaminants Standard that updated 212 of the 1971 interim standards and added 164 new health standards for toxic substances.[53] In *AFL-CIO v. OSHA*,[54] the U.S. Court of Appeals for the Eleventh Circuit vacated OSHA's order establishing the Air Contaminants Standard. The Court found that OSHA had conducted a "generic" analysis of the toxic substances and had not explained the significant risk posed by the existing level of exposure for each toxic substance and how the new standard protected workers from material impairments of health relating to each toxic substance.[55] The Court also held that OSHA had failed to explain the feasibility of the Air Contaminants Standard with respect to each toxic substance with regard to each industry affected.[56]

---

[48] 41 U.S.C. § 35 (2000); 53 Fed. Reg. 20960, 20966 (1988).

[49] 29 U.S.C. § 652(8) (2000).

[50] *Industrial Union Dept., AFL-CIO v. American Petroleum Inst.,* 448 U.S. 607 (1980) (plurality opinion).

[51] *Id.* at 614-615.

[52] *American Textile Mfrs. Inst., Inc. v. Donovan,* 452 U.S. 490, 512 (1981).

[53] 54 Fed. Reg. 2332 (1989).

[54] 965 F.2d 962 (1992).

[55] *Id.* at 975-79.

[56] *Id.* at 980-82.

OSHA acknowledged in the rulemaking process that it had taken a generic approach to estimate risks and feasibility posed by the toxic substances in the Air Contaminants Standard, explaining that "it would take decades to review currently used chemicals and OSHA would never be able to keep up with the many chemicals which will be newly introduced in the future."[57]

OSHA's efforts to establish exposure limits to toxic substances have generally not been successful because it is difficult for OSHA to develop the administrative record to demonstrate a significant risk of material health impairment. OSHA is also hamstrung by a small budget to use for scientific research (OSHA 2002), and weak penalty provisions in the OSH Act such that violations must result in an employee's death in order for the employer to be subject to criminal sanctions.[58] Given these constraints, OHSA relies upon self-monitoring and other regulatory schemes that are unlikely to be effective. For instance, the studies and data used to set permissible exposure limits are often generated by the employers who can have a conflict of interest.

The OSH Act established the National Institute for Occupational Safety and Health (NIOSH), under the Department of Human Health and Services, to research occupational health and safety conditions and provide information and training. Notably, NIOSH conducts a Health Hazard Evaluation (HHE) program that responds to requests by an employee, employee representative, or employer to investigate workplace health and safety issues. NIOSH has responded to hundreds of requests for HHE studies, which have been important in documenting indoor air quality problems such as VOCs from off-gassing materials, mold and fungal growth, improper ventilation systems, and pesticides.

In 1994, OSHA published a proposed comprehensive indoor air quality rule. According to OSHA, "The proposal would require employers to write and implement indoor air quality compliance plans that would include inspection and maintenance of current building ventilation systems to ensure they are functioning as designed…" (OSHA 2005). The proposed indoor air quality rule went through a notice-and-comment period, but OSHA never promulgated a final (enforceable) version due to changes in the U.S. Congress in 1994. The Indoor Air Quality rule remains an inactive item on OSHA's long-term agenda.

Overall, the OSH Act has improved worker safety standards in certain workplaces, but provides little protection outside those venues. The OSH Act provides no coverage for homes and other non-industrial environments, where many people work. In addition, OSHA has tended to focus on single hazards within industrial workplaces (such as large machinery), rather than multiple and often invisible hazards within typical office buildings (such as indoor air pollutants).

21.9 RESOURCE CONSERVATION AND RECOVERY ACT (RCRA)[59]

The Resource Conservation and Recovery Act (RCRA) of 1976 had two components: (1) "cradle-to-grave" regulation of hazardous waste, governing generation, transport, treatment, storage, and disposal; and (2) management of the growing volume of non-hazardous solid waste generated in the United States. The goals of RCRA include protecting human health and conserving energy. In 1984, Congress added regulation of underground storage tanks to RCRA's purview.

RCRA focuses on active and proposed waste sites, whereas CERCLA governs abandoned and inactive waste sites. RCRA defines "hazardous wastes" as wastes that may "cause, or significantly contribute to an increase in mortality or an increase in serious irreversible, or incapacitating reversible, illness" or "pose a substantial present or potential hazard to human health or the environment when improperly treated, stored, transported, or disposed of, or otherwise managed."[60]

---

[57] 53 Fed. Reg. at 20963.

[58] 29 U.S.C. § 666(e) (2000).

[59] Pub. L. No. 94-580, 90 Stat. 2806 (1976), codified at 42 U.S.C. §§ 6901-6992k (2002).

[60] 42 U.S.C. § 6903(5) (2002).

RCRA exempts household hazardous waste, such as pesticides, motor oil, and certain cleaners. Also, the federal courts have held that RCRA does not require the USEPA to list as hazardous all wastes that might satisfy the criteria for hazardous wastes, so its coverage is not comprehensive.[61]

RCRA's hazardous waste program requires applicants for landfill permits to provide information on "the potential for the public to be exposed to hazardous wastes or hazardous constituents through releases related to the [hazardous waste facility]" including "the potential magnitude and nature of the human exposure resulting from such releases."[62] RCRA also provides authority for the USEPA or state officials to require a health assessment of landfill activities by ATSDR whenever the landfill poses a substantial potential risk to human health, although this assessment is based usually on estimates instead of measurements.

RCRA does not limit the creation or use of hazardous wastes directly, although it makes it very expensive to transport and control hazardous wastes. Indeed, at least in its first few decades, RCRA created strong financial incentives to dispose of those wastes illegally. From a global and environmental justice perspective, RCRA has been criticized because it allows the export of hazardous wastes.

RCRA has significantly changed how most companies in the United States deal with hazardous wastes. The USEPA estimates that since RCRA was enacted, the annual generation of hazardous wastes in the United States has been reduced from 300 million tons to 40 million tons (USEPA 2002b). Although RCRA and its regulations provide a detailed scheme for addressing hazardous and other solid wastes, they have not eliminated, or regularly assessed, the exposure from such wastes.

21.10 TOXIC SUBSTANCES CONTROL ACT (TSCA)[63]

The Toxic Substances Control Act (TSCA) of 1976 authorizes the USEPA to secure information on all new and existing chemicals (or mixtures) sold in interstate commerce, and to control those chemicals that cause "unreasonable risk to public health or the environment."[64] Earlier laws did not require the screening of new chemicals or the control of existing substances until damage occurred.

In Title I of TSCA, Congress gave the USEPA authority to require by rule that chemical manufacturers test existing chemicals after the USEPA finds that (1) a chemical may present an unreasonable risk of injury to human health or the environment, or the chemical is produced in substantial quantities that could result in significant human or environmental exposure, (2) the available data to evaluate the chemical are not adequate, and (3) testing is necessary to develop such data.[65]

Title II (Asbestos Hazard Emergency Response) was added in 1986 to regulate asbestos abatement in schools. Title III (Indoor Radon Abatement) was added in 1988 and provides assistance to states in dealing with public health risks from radon. Title IV (Lead Exposure Reduction) was added in 1992 and provides assistance to states in reducing environmental lead contamination and lead exposure, especially in children. (See Chapter 14 on house dust exposure.)

Before manufacturing any new chemical or putting an existing chemical to a significant new use, the manufacturer must notify the USEPA. In deciding whether a chemical is "new," a manufacturer consults the TSCA Chemical Substance Inventory, which the USEPA maintains. If the substance is not on the Inventory, it is considered new. For an "existing" chemical, the Inventory can help to determine restrictions on its manufacture and use under TSCA. The Chemical Substance

---

[61] E.g., *Natural Res. Def. Council v. EPA*, 25 F.3d 1063 (D.C. Cir. 1994).

[62] 42 U.S.C. § 6939a (2002).

[63] Pub. L. No. 94-469, 90 Stat. 2003 (1976), codified at 15 U.S.C. §§ 2601-2692 (2002).

[64] 15 U.S.C. §2605(a) (2000).

[65] 15 U.S.C. § 2603(a) (2002).

Inventory currently contains over 82,000 chemicals (GAO 2005). Though the Chemical Substance Inventory is public information, significant portions of the Inventory are confidential.

If the USEPA determines that a chemical presents "an unreasonable risk of injury to health or the environment," then the USEPA may limit, or prohibit outright, the production of the chemical, or regulate its disposal, use, or marketing.[66] But TSCA does not define "unreasonable risk," and the USEPA has faced difficulties in proving this and other standards in order to take action. Further, the USEPA may impose limitations only "to the extent necessary to protect adequately against such risk using the least burdensome requirements."[67] For instance, if unreasonable risk could be managed by placing a warning label on the chemical, then the EPA could not ban or otherwise restrict use of that chemical (GAO 2005).

TSCA also contains several provisions that hinder the USEPA's ability to exercise such control. For instance, *Corrosion Proof Fittings v. Environmental Protection Agency*,[68] a landmark case decided by the Fifth Circuit Court of Appeals, shows how TSCA's restrictive language has hindered the USEPA's implementation of TSCA. In *Corrosion Proof Fittings,* the court vacated a rule the USEPA had issued under TSCA banning all use of asbestos in products. The court noted that the normal "arbitrary and capricious" standard of review for administrative agencies did not apply to rules promulgated by the USEPA pursuant to Section 2605(a) of TSCA. Congress had specified in TSCA a stricter "substantial evidence" standard of review for such rules. Applying that standard of review, the court concluded that the USEPA had not shown that human exposure to asbestos was substantial, or that less-restrictive limitations than a total ban on asbestos could prevent an unreasonable risk of injury to health or the environment. The court held that TSCA requires a balancing of costs and benefits in promulgating a new rule under Section 2605(a). The record in the *Corrosion Proof Fittings* case showed that the asbestos rule would cost the industry up to $74 million per life saved. Meanwhile, the USEPA could not show that adequate substitutes existed for many asbestos products. The court determined that, given such costs to industry, the USEPA could not show that asbestos presented such an unreasonable risk of injury to health or environment that a total ban was required. Given these strict legal standards, it has been very difficult and expensive for the USEPA to prove that a chemical in use presents an unreasonable risk to health.

In theory, TSCA offers an important advance in the assessment and control of chemicals, and gives the USEPA significant authority to reduce exposure. Yet TSCA's impact has been much narrower than the Act's words would suggest. Since the enactment of TSCA, the USEPA has promulgated rules under TSCA to place restrictions on only five existing chemicals/chemical classes and only four new chemicals (GAO 2005).

TSCA's effectiveness is also constrained by lack of data. There are few test data on short-term health effects and far fewer data on long-term chronic health effects. The USEPA does not have the funds to adequately test new products and does not require the producers to provide such data. Instead, the USEPA uses a method known as structural activity relationship analysis to compare new chemicals with chemicals of like molecular structure that have been tested to predict health effects. This method, while a useful screening tool, is subject to inaccuracies in predicting toxicity. The USEPA takes action on approximately 10% of the premanufacture notices (PMNs) submitted; only 2–3% of the total number of PMNs submitted undergo a detailed review by the USEPA, while the remaining 7–8% of PMNs are analyzed through the structural activity relationship analysis (USEPA 2005b).

The USEPA reviewed the risks of only 2% of the 62,000 chemicals already in use when the agency began to review new chemicals (GAO 1994). At the current rate of testing of existing chemicals, it could require hundreds of years to fully test chemicals in use. Once a chemical is approved and production begins, there is little monitoring of changes in production, use, and

---

[66] 15 U.S.C. § 2605(a) (2002).
[67] 15 U.S.C. § 2605(a) (2002).
[68] 947 F.2d 1201 (5th Cir. 1991).

exposure. Although the USEPA has taken action on about 3,500 new chemicals (out of about 32,000) submitted for review under TSCA, the "EPA's reviews of new chemicals provides limited assurance that health and environmental risks are identified before the chemicals enter commerce" (GAO 2005). Thus, tens of thousands of chemicals have not been evaluated for acute or chronic exposure; further, for those already evaluated, the process may have assessed risks inadequately.

Another factor limiting TSCA's effectiveness is the Act's confidentiality requirements. The USEPA must treat as confidential much of the information that manufacturers submit under TSCA, which also prevents the USEPA from sending this information to officials who have a responsibility to protect the public. For instance, about 95% of the PMNs for new chemicals submitted by chemical companies contain some information that is claimed as confidential (GAO 2005). Many of these claims to confidentiality may be unjustifiable, but the USEPA lacks the resources to challenge a significant portion of these claims.

Further, the USEPA lacks exposure data that it can use to justify regulation of chemicals in use. The USEPA can request test data from industry only when the USEPA can prove that the chemical may present an unreasonable risk of injury to health or the environment, or may lead to significant or substantial human exposure,[69] which the USEPA generally cannot prove without that additional data from industry.

As part of the implementation of TSCA, an Interagency Testing Committee (ITC) has been formed to make recommendations to the USEPA Administrator for testing existing chemicals. ITC knows which chemicals it would like to test but there is a lack of test data. When these data do exist, they are usually confidential, which greatly reduces their value in reducing human exposure (GAO 1994).

Finally, certain types and amounts of chemicals are excluded from TSCA's coverage. Pesticides, tobacco, and tobacco products, radioactive material, foods, food additives, drugs, and cosmetics are excluded from TSCA regulations.

In summary, TSCA has limited impact due to strict legal tests in the Act and USEPA's lack of resources and administrative support to use its authority under TSCA to review, measure, and control exposures to the majority of chemicals in commercial use. Without increases in exposure data, resources, administrative support, and changes in the laws, the potential effectiveness of TSCA may remain unfulfilled.

21.11 CONCLUSIONS

We have found that federal environmental regulations, which seek to protect human health, are missing major pollutant exposures that imperil human health. Exposures from sources indoors are currently far greater than exposures from sources outdoors. Yet these indoor environments, such as homes, offices, schools, and vehicles, are largely unregulated and unmonitored.

Fortunately, because many significant exposures are within our control, we can reduce health risks through relatively simple and cost-effective actions, such as using less toxic consumer products and building materials. Unfortunately, many people are unaware of the major sources of pollutant exposures, their health effects, and ways to reduce those exposures. Thus, a gap exists between regulation and risk, and between science and public awareness.

We can use the science of human exposure to bridge that gap by understanding what, where, and when pollutants come in contact with humans. We have successfully reduced outdoor pollutant levels, thanks to our environmental laws. But our regulatory lens needs to focus on *total* exposures to pollutants in order to reduce some major health risks that remain.

To this end, a group of scientists has proposed a Human Exposure Reduction Act (HERA), which is presented in Table 21.2. The HERA seeks to more efficiently and cost-effectively protect health by incorporating exposure into environmental laws.

---

[69] 15 U.S.C. § 2603(a) (2002).

21.12 QUESTIONS FOR REVIEW

1. Most of our exposure to pollutants occurs within indoor environments, such as homes, offices, and schools. Certain laws address some aspects of indoor air quality, yet no law provides comprehensive coverage.
   a. Select a law, and discuss how it covers an aspect of indoor air quality.
   b. Now discuss how the law does not cover indoor air exposures, even though it would seem within the scope of the law to do so.
   c. Finally, for this law, suggest revisions to address exposure more effectively.
2. Regulation often relies on a cost-benefit analysis as a prerequisite to action, such as banning a product or a hazardous chemical. What are some limitations of this cost-benefit approach? Respond with particular attention to benefits and costs related to exposure and human health. What would you suggest as an alternative regulatory approach or criterion?
3. In these laws, the health outcomes considered are usually cancer mortalities, even though exposures are linked to a range of other mortality and morbidity effects. Why do you think the laws focus on cancer as a basis for risk assessment and regulation? What other types of exposure-related health effects are overlooked?
4. Superfund (CERCLA/SARA) has been criticized for spending relatively large amounts of money to address sites that pose relatively little exposure risk to humans. Why do you think this has occurred? How might results from exposure studies be used to allocate resources?
5. You are working inside a state government building, which is being renovated. Employees are becoming sick due to exposures to materials and products used in the renovation. Which law(s) might apply in this case to reduce exposure, and how? Which law(s), if they were stronger, might have prevented this "sick building" incident from happening?
6. Most of the environmental laws are "source oriented" rather than "receptor oriented" in that they focus on emissions/effluents instead of exposures. Why do you think this is the case?
7. Fragrances (synthetic compounds) in products represent significant sources of human exposure to toxic VOCs, and are linked to a range of adverse health effects such as seizures, headaches, and breathing difficulties. Consumers often believe that if a product is sold in a store, then it must be "safe." Why, then, are these products allowed to be sold, and why do you think consumers continue to buy them, given the health risks?

21.13 ACKNOWLEDGMENTS

The authors thank Wayne Ott and Lance Wallace for their invaluable reviews and comments, John Roberts for his contributions to the TSCA section, Luiz Cavalcanti and Julie Horowitz for their research assistance, and Dan Ribeiro and Deborah Livingstone for their exceptional editing on this chapter and throughout the textbook.

TABLE 21.1
Content Analysis of Major Environmental Laws

Name of Law	"Exposure"	"Ambient Air"	"Indoor Air"	"Outdoor Air"
Clean Air Act (CAA)	29	175	0	0
Clean Air Act Amendments	10	13	0	0
Comprehensive Environmental Response, Compensation, and Liability Act (CERCLA)	41	2	0	0
Superfund Amendments and Reauthorization Act (SARA)	14	0	18	0
Consumer Product Safety Act (CPSA)	4	0	0	0
Federal Food, Drug, and Cosmetic Act (FFDCA)	35	0	0	0
Food Quality Protection Act (FQPA)	33	0	0	0
Federal Insecticide, Fungicide, and Rodenticide Act (FIFRA)	35	0	0	0
Occupational Safety and Health Act (OSH Act)	20	0	0	0
Resource Conservation and Recovery Act (RCRA)	27	2	0	0
Toxic Substances Control Act (TSCA)	63	2	0	0

TABLE 21.2
Human Exposure Reduction Act (HERA)

Intent:

The purpose of a proposed Human Exposure Reduction Act (HERA) is to assess and reduce human exposure to pollutants in all carrier media, and to more efficiently and effectively reduce health costs. A major objective is to reduce exposures to children because of their increased susceptibility to health effects caused by environmental pollutants.

Findings:

(1) Americans are regularly exposed to a range of pollutants that enter their bodies and can harm health

(2) Existing laws do not adequately address many of the major sources of pollutant exposures and health risks

(3) Pollutant exposures are not measured routinely among the American population in exposure field studies, even though these measurements provide critical information and can be made with high accuracy with today's science of exposure analysis

(4) The American public is generally unaware of their personal exposures to toxic pollutants through daily activities and would benefit from increased education on this issue

(5) A more effective regulatory approach would address the sources that contribute most to human exposures and health risks, without abandoning existing protection

Objectives:

(1) Identify and measure sources of human exposure to pollutants from all environmental media (air, water, soil, food, dust, dermal) in a single balanced approach

(2) Reduce pollutant exposure, with priorities based on relative contributions of each source to cumulative exposure

(3) Conduct regular monitoring and exposure studies of the general population, and quantify changes in human exposure to pollutants over time

(4) Assess and compare the exposure reduction effects of different environmental regulations and initiatives affecting all environmental media and exposure routes

(5) Require testing and labeling of consumer products, building materials, and other significant sources of pollutant exposures

(6) Perform independent studies of toxicity and health effects of chemicals, chemical mixtures, and other pollutants found in everyday life

(7) Conduct research to develop new exposure measurement methods, including personal monitoring systems and exposure models that will facilitate routine, less expensive exposure monitoring studies

(8) Conduct health studies that combine epidemiology with actual exposure measurements

(9) Support programs that provide training, information, and public outreach about pollutant exposures and reduction strategies in homes, workplaces, schools, and other environments

(10) Support academic institutions in their education of a new generation of human exposure scientists

(11) Support research to develop new technologies and products that reduce pollutant exposures, such as new building materials, new consumer product formulations, and easy-to-employ control systems

Principles:

(1) To effectively protect health and reduce health costs of the American public, we need to understand accurately the causes of human exposures to environmental pollutants, the ways in which these exposures can be altered, and the trends in population exposures over time

(2) To understand trends in population exposures over time, routine exposure monitoring programs are needed, much like the nationwide ambient air and water monitoring networks now operating, but instead focused on routinely measuring the status and changes in personal exposures

(3) Environmental epidemiology should include, wherever possible, as an integral part of its methodology, the direct measurement of physical, chemical, and biological indicators of actual exposure, in addition to surrogate indicators of exposure such as questionnaires

(4) Research should be conducted to improve our understanding of the individual variation in the susceptibility of different persons to environmental chemicals, as well as the nature, extent, and variability of the population exposure to these pollutants

TABLE 21.2 (CONTINUED)
Human Exposure Reduction Act (HERA)

(5) Regulations adopted under existing environmental laws should be evaluated comparatively across all media (air, water, soil, food, dust, dermal) for their effectiveness in reducing population exposure, especially the exposure of susceptible persons and children

(6) Accurate information should be available to the user of a consumer product about whether the product contains an appreciable concentration of certain listed chemicals, the specific concentrations of those listed chemicals, the likely exposure that might result from using the product, and the ways that the product should be used safely to reduce or eliminate these exposures

(7) Educational programs should be developed and experts should be trained in the emerging science of exposure analysis and exposure assessment, including outreach programs to demonstrate new methods for reducing exposure in the everyday lives of our citizens

(8) Government should require testing, labeling, and evaluation of the toxic pollutants emitted by consumer products and building materials, just as they require for food and drugs. At a minimum, the manufacturer should submit accurate and complete information about the toxic pollutants their products contain and the levels of exposure that might result from typical use

Source: Ott, W.R., Roberts, J.W., Steinemann, A.C., Repace, J., Gilbert, S.G., Moschandreas, D.J., and Corsi, R.L. (2002), and adapted in Moschandreas, D. (2003).

TABLE 21.3
List of Hazardous Air Pollutants (HAPs) Established in the Clean Air Act, as Amended in 1990

CAS Number	Chemical Name	Found in TEAM Studies[1]	Found in Household Products[2]	Found in Indoor Air in Other Studies
75070	Acetaldehyde			√4,5,6
60355	Acetamide			
75058	Acetonitrile			
98862	Acetophenone			
53963	2-Acetylaminofluorene			
107028	Acrolein			√4,7
79061	Acrylamide			
79107	Acrylic Acid			
107131	Acrylonitrile			√4,5,7
107051	Allyl Chloride	√		
92671	4-Aminobiphenyl			√4
65233	Aniline			√8
90040	o-Anisidine			
0	Antimony Compounds			√9,10
0	Arsenic Compounds (inorganic including arsine)			√4,5
1332214	Asbestos			
71432	Benzene (including benzene from gasoline)	√	√	√4,5,7,11,12,13, 14,15,16,17,18, 19,20
92875	Benzidine			
98077	Benzotrichloride			
100447	Benzyl chloride	√		
0	Beryllium Compounds			
92524	Biphenyl			
117817	Bis(2-ethylhexyl)phthalate (DEHP)	√		
542881	Bis(chloromethyl)ether	√		
75252	Bromoform	√		
106990	1,3-Butadiene	√		√5,7,8,15,17,18
0	Cadmium Compounds			√4,5,21
156627	Calcium cyanamide			
133062	Captan	√		
63252	Carbaryl	√		
75150	Carbon disulfide			√22
56235	Carbon tetrachloride	√	√	√7,18,19
463581	Carbonyl sulfide			
120809	Catechol			√4,5
133904	Chloramben			
57749	Chlordane	√		
7782505	Chlorine			
79118	Chloroacetic acid			
532274	2-Chloroacetophenone			
108907	Chlorobenzene	√		√7
510156	Chlorobenzilate			
67663	Chloroform	√	√	√12,13,14,17,18
107302	Chloromethyl methyl ether			
126998	Chloroprene	√		

TABLE 21.3 (CONTINUED)
List of Hazardous Air Pollutants (HAPs) Established in the Clean Air Act, as Amended in 1990

CAS Number	Chemical Name	Found in TEAM Studies[1]	Found in Household Products[2]	Found in Indoor Air in Other Studies
0	Chromium Compounds			√4,5,21
0	Cobalt Compounds			
0	Coke Oven Emissions			
1319773	Cresols/cresylic acid (isomers and mixture)			
95487	o-Cresol			√10
108394	m-Cresol			√10
106445	p-Cresol			√10
98828	Cumene			√16,17
0	Cyanide Compounds			
94757	2,4-D, salts and esters			
3547044	DDE	√		
334883	Diazomethane			
132649	Dibenzofurans			
96128	1,2-Dibromo-3-chloropropane			
84742	Dibutylphthalate	√		
106467	1,4-Dichlorobenzene(p)	√		√7,12,16,17,18,20
91941	3,3-Dichlorobenzidene			
111444	Dichloroethyl ether (Bis(2-chloroethyl)ether)			
542756	1,3-Dichloropropene			
62737	Dichlorvos	√		
111422	Diethanolamine			
121697	N,N,Diethyl aniline (N,N-Dimethylaniline)			
64675	Diemethyl Sulfate			
119904	3,3'-Dimethoxybenzidine			
60117	Dimethyl aminoazobenzene			
119937	3,3'-Dimethyl benzidine			
79447	Dimethyl carbamoyl chloride			
68122	Dimethyl formamide			
57147	1,1-Dimethyl hydrazine			√4,5
131113	Dimethyl phthalate	√		
77781	Dimethyl sulfate			
534521	4,6-Dinitro-o-cresol, and salts			
51285	2,4-Dinitrophenol			
121142	2,4-Dinitrotoluene			
123911	1,4-Dioxane(1,4-Diethyleneoxide)	√	√	√7
122667	1,2-Diphenylhydrazine			
106898	Epichlorohydrin (1-Chloro-2,3-epoxypropane)			
106887	1,2-Epoxybutane			
140885	Ethyl acrylate			
100414	Ethyl benzene	√	√	√11,16,17,18
51796	Ethyl carbamate (Urethane)			√4,5
75003	Ethyl chloride (Chloroethane)			
106934	Ethyl dibromide (Dibromoethane)	√		

TABLE 21.3 (CONTINUED)
List of Hazardous Air Pollutants (HAPs) Established in the Clean Air Act, as Amended in 1990

CAS Number	Chemical Name	Found in TEAM Studies[1]	Found in Household Products[2]	Found in Indoor Air in Other Studies
107062	Ethylene dichloride (1,2-Dichloroethane)	√	√	
107211	Ethylene glycol			√20
151564	Ethylene imine (Aziridine)			
75218	Ethylene oxide	√		
96457	Ethylene thiourea			
75343	Ethylidene dichloride (1,1,-Dichloroethane)			
0	Fine mineral fibers			
50000	Formaldehyde			√11,20,23
76448	Heptachlor			
118741	Hexachlorobenzene	√		
87683	Hexachorobutadiene			
77474	Hexachlorocyclopentadiene			
66721	Hexachloroethane			
822060	Hexamethylene-1,6-diisocyanate			
680319	Hexamethylphosphoramide			
110543	Hexane	√	√	
302012	Hydrazine			√4
7647010	Hydrochloric acid			
7664393	Hydrogen fluoride (hydrofluoric acid)			
123319	Hydroquinone			√9,10
78591	Isophorone			
0	Lead Compounds	√		√5,10,21
58899	Lindane (all isomers)	√		
108316	Maleic anhydride			
0	Manganese Compounds			√9,10
0	Mercury Compounds			√9,10
67561	Methanol			√8,11
72435	Methoxychlor			
74839	Methyl bromide (bromomethane)	√		
74873	Methyl chloride (chloromethane)	√	√	
71556	Methyl chloroform (1,1,1-Trichloroethane)	√	√	
78933	Methyl ethyl ketone (2-Butanone)	√	√	√20,23
60344	Methyl hydrazine			
74884	Methyl iodide (Iodomethane)			
108101	Methyl isobutyl ketone (Hexone)	√		
624839	Methyl isocyanate			
80626	Methyl methacrylate			
1634044	Methyl tert butyl ether			
101144	4,4′-Methylene bis(2-chloroaniline)			
75092	Methylene chloride (Dichloromethane)		√	
101688	Methylene diphenyl diisocyanate (MDI)			
101779	4,4-Methylenedianiline			
91203	Naphthalene	√		√8,18
0	Nickel Compounds			√4,5
98953	Nitrobenzene			

TABLE 21.3 (CONTINUED)
List of Hazardous Air Pollutants (HAPs) Established in the Clean Air Act, as Amended in 1990

CAS Number	Chemical Name	Found in TEAM Studies[1]	Found in Household Products[2]	Found in Indoor Air in Other Studies
92933	4-Nitrobiphenyl			
100027	4-Nitrophenol			
79469	2-Nitropropane			
684935	N-Nitroso-N-methylurea			
62759	N-Nitrosodimethylamine			√4,5
59892	N-Nitrosomorpholine			√4
56382	Parathion	√		
82688	Pentachloronitrobenzene (Quintobenzene)			
87865	Pentachlorophenol	√		
108952	Phenol	√		√9,20,24
106503	p-Phenylenediamine			
75445	Phosgene			
7803512	Phosphine			
7723140	Phosphorus			
85449	Phthalic anhydride			
1336363	Polychlorinated biphenyls (Aroclors)			√24
0	Polycylic organic matter	√		
1120714	1,3-Propane sultone			
57578	β-Propiolactone			
123386	Propionaldehyde			
114261	Propoxur (Baygon)	√		
78875	Propylene dichloride (1,2-Dichloropropane)	√		
75569	Propylene oxide	√	√	
75558	1,2-Propylenimine (2-methyl aziridine)			
91225	Quinoline			
106514	Quinone			
0	Radionuclides (including radon)			
0	Selenium Compounds			
100425	Styrene	√		√5, 8,11,12,16,17
96093	Styrene oxide	√		
1746016	2,3,7,8-Tetrachlorodibenzo-p-dioxin			
79345	1,1,2,2-Tetrachloroethane	√		
127184	Tetrachloroethylene (perchloroethylene)	√	√	
7550450	Titanium tetrachloride			
108883	Toluene		√	√8,11,12,15,16,23
95807	2,4-Toluene diamine			
584849	2,4-Toluene diisocyanate			
95534	o-Toluidine			
8001352	Toxaphene (chlorinated camphene)			
120821	1,2,4-Trichlorobenzene			√22
79005	1,1,2-Trichloroethane	√		
79016	Trichloroethylene	√	√	
95954	2,4,5-Trichlorophenol			
88062	2,4,6-Trichlorophenol			
121448	Triethylamine			

TABLE 21.3 (CONTINUED)
List of Hazardous Air Pollutants (HAPs) Established in the Clean Air Act, as Amended in 1990

CAS Number	Chemical Name	Found in TEAM Studies[1]	Found in Household Products[2]	Found in Indoor Air in Other Studies
1582098	Trifluralin			
540841	2,2,4-Trimethylpentane			
108054	Vinyl acetate			
593602	Vinyl bromide			
75014	Vinyl chloride			√4,5,14,25
75354	Vinylidene chloride (1,1-dichloroethylene)	√		
1330207	Xylenes (isomers and mixture)	√		
95476	o-Xylenes	√	√	√11,26
108383	m-Xylenes	√	√	√11,22,26,27
106423	p-Xylenes	√	√	√11,22,26,27

References:

[1] 42 U.S.C. § 7412(b)(2) (2002); hydrogen sulfide, caprolactam, and glycol ethers delisted from the original list of HAPs.

[2] Includes the TEAM studies of volatile organic compounds, pesticides (NOPES), and particulate matter (PTEAM). See Chapters 1, 3, 7, 8, and 15 in text.

[3] Sack, T.M., Steele, D.H., Hammerstrom, K., and Remmers J. (1992) A Survey of Household Products for Volatile Organic Compounds, *Atmospheric Environment*, 26A(6): 1063–1070.

[4] USDHHS (1989) Reducing the Health Consequences of Smoking, 25 Years of Progress, A Report of the Surgeon General, U.S. Department of Health and Human Services, Rockville, MD.

[5] Hoffmann, D. and Hoffmann, I. (1990) Cigars — Health Effects and Trends, in *Smoking and Tobacco Control Monograph*, Table 15, Carcinogens in tobacco smoke, NIH Publication No. 98-4302, February, National Cancer Institute, National Institutes of Health.

[6] Zhang, J., He, Q., and Lioy, P.J. (1994) Characteristics of Aldehydes: Concentrations, Sources and Exposures for Indoor and Outdoor Residential Microenvironments, *Environmental Science and Technology*, 28: 146–152.

[7] Sheldon, LS., Clayton, A., Jones, B., Keever, J., Perritt, R., Smith, D., Whitaker, D., and Whitmore, R. (1991) Indoor Pollutant Concentrations and Exposures, Final report, California Air Resources Board, Sacramento, CA.

[8] National Institute of Occupational Safety and Health (NIOSH) (1994) *Pocket Guide to Chemical Hazards*, Centers for Disease Control & Prevention, U.S. Department of Health and Human Services, June.

[9] Sax, N.I. (1984) *Dangerous Properties of Industrial Materials*, 6th ed., Van Nostrand Reinhold, New York, NY.

[10] USDHEW (1979) Smoking and Health, A Report of the Surgeon General, U.S. Department of Health, Education, and Welfare, Washington, DC.

[11] Brown, S.K. (2002) Volatile Organic Pollutants in New and Established Buildings in Melbourne, Australia, *Indoor Air*, 12(1): 55–63.

[12] Adgate, J.L., Bollenbeck, M., Eberly, L.E., Stroebel, C., Pellizzari, E.D., and Sexton, K. (2002) Residential VOC Concentrations in a Probability-Based Sample of Households with Children, Levin, H., Ed., in *Indoor Air, Proceedings of the 9th International Conference on Indoor Air Quality and Climate*, Santa Cruz, CA, 1: 203–208.

[13] Clayton, C.A., Pellizzari, E.D., Whitmore, R.W., Perritt, R.L., and Quackenboss, J.J. (1999) National Human Exposure Assessment Survey (NHEXAS): Distributions and Associations of Lead, Arsenic and Volatile Organic Compounds in EPA Region 5, *Journal of Exposure Analysis and Environmental Epidemiology*, 9: 381–392.

[14] Foster, S.J., Kurtz, J.P., and Woodland, A.K. (2002) Background Indoor Air Risks in Selected Residences in Denver, Colorado. Levin, H., Ed., in *Indoor Air, Proceedings of the 9th International Conference on Indoor Air Quality and Climate*, Santa Cruz, CA, 1: 932–937.

TABLE 21.3 (CONTINUED)
List of Hazardous Air Pollutants (HAPs) Established in the Clean Air Act, as Amended in 1990

[15] Gordon, S.M., Callahan, P.J., Nishioka, M.G., Brinkman, M.C., O'Rourke, M.K., and Lebowitz, M.D. (1999) Residential Environmental Measurements in the National Human Exposure Assessment Survey (NHEXAS) Pilot Study in Arizona: Preliminary Results for Pesticides and VOCs, *Journal of Exposure Analysis and Environmental Epidemiology*, 9: 456–470.

[16] Heavner, D.L., Morgan, W.T., and Ogden, M.W. (1995) Determination of Volatile Organic Compounds and ETS Apportionment in 49 Homes, *Environment International*, 21: 3–21.

[17] Heavner, D.L., Morgan, W.T., and Ogden, M.W. (1996) Determination of Volatile Organic Compounds and Respirable Suspended Particulate Matter in New Jersey and Pennsylvania Homes and Workplaces, *Environment International*, 22: 159–183.

[18] Van Winkle, M.R. and Scheff, P.A. (2001) Volatile Organic Compounds, Polycyclic Aromatic Hydrocarbons and Elements in the Air of Ten Urban Homes, *Indoor Air*, 11: 49–64.

[19] Mukerjee, S., Ellenson, W.D., Lewis, R.G., Stevens, R.K., Somerville, M.C., Shadwick, D.S., Willis, R.D. (1997) An environmental scoping study in the lower Rio Grande Valley of Texas, III, Residential Microenvironmental Monitoring for Air, House Dust, and Soil, *Environment International*, 23(5): 657–673.

[20] Otson, R., Fellin, P., and Tran, Q. (1994) VOCs in Representative Canadian Residences, *Atmospheric Environment*, 28: 3563–3569.

[21] Lioy, P.J., Freeman, N.J., and Millette, J.R. (2002) Dust: A Metric for Use in Residential and Building Exposure Assessment and Source Characterization, *Environmental Health Perspectives*, 110(10): 969–973.

[22] Girman, J.R., Hadwen, G.E., Burton, L.E., Womble, S.E., and McCarthy, J.F. (1999) Individual Volatile Organic Compound Prevalence and Concentrations in 56 Buildings of the Building Assessment Survey and Evaluation (BASE) study, in *Indoor Air 99, Proceedings of the 8th International Conference on Indoor Air Quality and Climate*, Vol. 2, Raw, G., Aizlewood, C., and Warren, P., Eds., Construction Research Communications Ltd., London, 460–465.

[23] Lindstrom, A.B., Proffitt, D., and Fortune, C.R. (1995) Effects of Modified Residential Construction on Indoor Air Quality, *Indoor Air*, 5: 258–269.

[24] Rudel, R.A., Camann, D.E., Spengler, J.D., Korn, L.R., and Brody, J.G. (2003) Phthalates, Alkylphenols, Polybrominated Diphenyl Ethers, and Other Endocrine-Disrupting Compounds in Indoor Air and Dust, *Environmental Science and Technology*, 37(20): 4543–4553.

[25] Kurtz, J.P. and Folkes, D.J. (2002) Background Concentrations of Selected Chlorinated Hydrocarbons in Residential Indoor Air, in *Indoor Air 2002, Proceedings of the 9th International Conference on Indoor Air Quality and Climate*, Vol. 1, Levin, H., Ed., Santa Cruz, CA, 920–925.

[26] Daisey, J.M., Hodgson, A.T., Fisk, W.J., Mendell, M.J., and Ten Brinke, J. (1994) Volatile Organic Compounds in Twelve California Office Buildings: Classes, Concentrations and Sources, *Atmospheric Environment*, 28: 3557–3562.

[27] Sheilds, H.C., Fleischer, D.M., and Weschler, C.J. (1996) Comparisons among VOCs Measured in Three Types of U.S. Commercial Buildings with Different Occupant Densities, *Indoor Air*, 6: 2–17.

REFERENCES

ATSDR (2003) Agency for Toxic Substances and Disease Registry, http://www.atsdr.cdc.gov/clist.html (accessed December 6, 2005).

CDC (2001) *First National Report on Human Exposure to Environmental Chemicals*, Centers for Disease Control and Prevention, U.S. Department of Health and Human Services, Atlanta, GA, http://www.cdc.gov/exposurereport/ (accessed December 6, 2005).

CDC (2003) *Second National Report on Human Exposure to Environmental Chemicals*, Centers for Disease Control and Prevention, U.S. Department of Health and Human Services, Atlanta, GA, http://www.cdc.gov/exposurereport/ (accessed December 6, 2005).

CDC (2005) *Third National Report on Human Exposure to Environmental Chemicals*, Centers for Disease Control and Prevention, U.S. Department of Health and Human Services, Atlanta, GA, http://www.cdc.gov/exposurereport/ (accessed December 6, 2005).

CFSAN (2005) Center for Food Safety and Applied Nutrition, http://www.cfsan.fda.gov/~dms/ pes03rep. html#summary (accessed December 6, 2005).

CQC (1994) Census Questionnaire Content, Bureau of the Census, http://www.census.gov/apsd/cqc/cqc20.pdf (accessed December 6, 2005).

Cross, F.B. (1997) The Consequences of Consensus: Dangerous Compromises of the Food Quality Protection Act, *Washington University Law Quarterly*, **75**: 1157.

EWG (2005) Environmental Working Group, http://www.ewg.org/reports/skindeep/index.php (accessed December 6, 2005).

GAO (1994) Toxic Substances Control Act: Legislative Changes Could Make the Act More Effective, Chapter Report, 09/26/94, GAO/RCED-94-103, U.S. General Accounting Office, Washington, DC, www.mapcruzin.com/scruztri/docs/gao94103.htm (accessed December 6, 2005).

GAO (2005) Chemical Regulation: Options Exist to Improve EPA's Ability to Assess Health Risks and Manage Its Chemical Review Program, GAO-05-458, U.S. General Accounting Office, Washington, DC, http://www.gao.gov/docsearch/abstract.php?rptno=GAO-05-458 (accessed December 6, 2005).

Klepeis, N.E., Nelson, W.C., Ott, W.R., Robinson, J.P., Tsang, A.M., Switzer, P., Behar, J.V., Hern, S.C., and Engelmann, W.H. (2001) The National Human Activity Pattern Survey (NHAPS): A Resource for Assessing Exposure to Environmental Pollutants, *Journal of Exposure Analysis and Environmental Epidemiology*, **11**: 231–252.

Moschandreas, D. (2003) The Whence, Wherefore and Whither of the New Scientific Discipline of Environmental Inquiry: Exposure Analysis, The 2002 Wesolowski Lecture, *Journal of Exposure Analysis and Environmental Epidemiology*, **13**(4): 247–255.

NRC (1987) *Regulating Pesticides in Food: the Delaney Paradox*, Committee on Scientific and Regulatory Issues Underlying Pesticide Use Patterns and Agricultural Innovation, National Research Council.

NRC (1993) *Pesticides in the Diets of Infants and Children*, Committee on Pesticides in the Diets of Infants and Children, National Research Council.

OSHA (2002) OSHA Trade News Release, http://www.osha.gov/pls/oshaweb/owadisp.showdocument?ptable=NEWS_RELEASES&pid=1182 (accessed December 6, 2005).

OSHA (2005) Indoor Air Quality in the Workplace, http://www.osha.gov/pls/oshaweb/owadisp.show_document?p_table=UNIFIED_AGENDA&p_id=5042 (accessed December 6, 2005).

Ott, W.R., Roberts J.W., Steinemann A.C., Repace J., Gilbert S.G., Moschandreas D.J., and Corsi R.L. (2005) The Proposed Human Exposure Reduction Act. (Full text available from lead author, and reprinted in Moschandreas, D., 2003.)

State of New York (1996) *The Secret Hazards of Pesticides: Inert Ingredients*, Office of the Attorney General, Environmental Protection Bureau, New York State, February, http://www.oag.state.ny.us/environment/inerts96.html (accessed December 6, 2005).

USEPA (1999) *Implementing FQPA: Progress Report,* U.S. Environmental Protection Agency, August.

USEPA (2002a) *Pesticide Industry Sales and Usage: 1998 and 1999 Market Estimates*, U.S. Environmental Protection Agency.

USEPA (2002b) 25 Years of RCRA: Building on Our Past to Protect Our Future, http://www.epa.gov/epaoswer/general/k02027.pdf (accessed December 6, 2005).

USEPA (2004) National Priorities List, http://www.epa.gov/superfund/sites/npl/ (accessed December 6, 2005).

USEPA (2005a) Air Trends, http://www.epa.gov/airtrends/ (accessed December 6, 2005).

USEPA (2005b) Chemical Categories Report, New Chemicals Program, http://www.epa.gov/oppt/newchems/pubs/chemcat.htm (accessed December 6, 2005).

USEPA (2005c) Interagency Committee on Indoor Air Quality, http://www.epa.gov/iaq/ciaq/index.html (accessed December 6, 2005).

USEPA (2005d) Key Dates in Superfund, http://www.epa.gov/superfund/action/law/keydates.htm (accessed December 6, 2005).

USEPA (2005e) National Center for Environmental Assessment, http://cfpub.epa.gov/ncea (accessed December 6, 2005).

USEPA (2005f) National Priorities List, http://www.epa.gov/superfund/sites/query/queryhtm/npldel.htm (accessed December 6, 2005).

USEPA (2005g) Pesticides: Regulating Pesticides, http://www.epa.gov/opprd001/inerts/lists.html (accessed December 6, 2005).

USFDA (2005) U.S. Food and Drug Administration, http://www.fda.gov/opacom/backgrounders/miles.html (accessed December 6, 2005).

USHR (1986) *Neurotoxins: At Home and the Workplace*, Report 99-827, Report by the Committee on Science and Technology, U.S. House of Representatives, September 16.

Wallace, L.A. (1991) Comparison of Risks from Outdoor and Indoor Exposure to Toxic Chemicals. *Environmental Health Perspectives*, **95**: 7–13.

Index

A

Absorbed dose, 45–47
  definition, 57
  dermal dose and, 54
  example, 50
  inhalation, 82
Absorption (uptake), 50, 53, 59, *See also* Dermal exposure
  definition, 57, 276
  direct measurement methods, 270–272
  enhancing, 256
  skin structure and, 260
  solute permeability and dermal dose, 262
  toxic metals (by plants), 308–309
  water temperature effects, 286, 287, 294
Absorption barrier, 35, 46, 50
  dermal exposure scenario, 53–54
  ingestion exposure scenario, 54, 55
Accumulated dose, 47
Accumulation mode particles, 183
Acetylation, defined, 276
Acrylamide, 400
Activated charcoal, 99, 160–161
Active dose, 35, 47
Active Smoking Count (ASC), 218–219
Active transport, 305
Activity patterns, *See* Human activity patterns
Acute exposure, 50, 55, 57
Adipose tissue, 260
  definition, 276
  SVOC contamination, 400
Administered dose, 35, 47
Aerosol pollutants, 10
Agency for Toxic Substances and Disease Registry
  (ATSDR), 491
Agent
  defined, 39
  examples, 49, 52, 55
Agent-based modeling, 465
Agent Orange, 379
Aggregate dose, 36
Aggregate exposure, 36, 256–257, *See also* Pesticides
Aggregate exposure models, 472–473, 476–478, *See also*
  Exposure models
AIDS education approach, 24
Air conditioning effects, 188–189
Air Contaminants Standard, 497–498
Air exchange rate, 188, 190, 223
  indoor air quality model, 415, 416, 418–421
  intake fraction and, 240, 243–244
  measuring, 424–428
  outdoor winds and, 428
  residence time and, 10, 416

ventilation design standard, 223, 224
Air filters, 190, 193
Air fresheners, 19, 101, 153, 156, 159
Air pollutants, *See specific pollutants*
Air Pollutants Exposure Model (APEX), 132
Air quality, indoor, *See* Indoor air quality
Air Quality Index (AQI), 216
Air sampling and analysis methods, 160–165, *See also*
  Personal exposure monitors
  active and passive modes, 160
  capture (whole-air) sampling, 162
  comparison of methods, 162–163
  MS-based techniques, 163–166
  multisorbent systems, 161
  olfactory analysis, 165
  pesticide monitoring, 350–354
  real-time techniques, 163–166
  solid-phase microextraction, 161–162
  sorbents, 99, 160–161
Alaskan air quality, 16, 155
Aldrin, 367, 368, 369, 400
Allergies, 324, 337
  allergen sensitization, 341
Alveolar respiration rate, 94
Alveoli, 84
Amberlite® XAD-2, 351–352
American Chemistry Council, 460
American Conference of Governmental Industrial
  Hygienists (ACGIH), 368
American Lung Association, 159, 339
American Society for Testing and Materials (ASTM)
  standards, 352, 356, 357, 424
American Society of Heating, Refrigerating, and Air
  Conditioning Engineers (ASHRAE)
  standards, 223, 224, 457
Amorphous, defined, 276
Anemia, lead poisoning and, 329
Angina, 117
Animal models of percutaneous absorption, 272
Antioch–Pittsburg, California, 150
APEX, 132, 458
Applied dose, 35, 44, 46
Arenas, 131
Arsenic, 257, 307, 309, 321, 368, 402
Asbestos, 82, 89, 91, 500
ASHRAE standards, 223, 224, 457
Asthma, 91, 324, 337
  trigger control programs, 339–341
ASTM standards, 352, 356, 357, 424
Atlanta, Georgia, 187
Atmospheric dispersion models, 5, 48
Atmospheric pressure chemical ionization (APCI), 163,
  166

515

Atmospheric sampling glow discharge ionization (ASGDI), 163, 166
Atrazine, 306–307, 364, 365, 368
At-risk subpopulations, *See also* Children and infants
    CO exposure, 117, 120
    dermal exposure, 257
    dioxin exposure risk, 389
    house dust exposure, 324
    particle exposure studies, 186–187
    pesticide exposures, 368
    VOC exposure, 154
Austin, Texas, 118
Automobile emissions, *See* Motor vehicle emissions
Automobile manufacturers, 120–121
Available dose, 47
Average exposure, 38
Avon, U.K., 155

B

Background levels
    CO, 50, 116
    definition, 57
    dioxins, 386–387
Baffin Island, Canada, 308
Bag wash method, 362
Bangladesh, 309
Basal cell population, 276
Bathroom deodorants, 19–21, 153, 155–156, 349, *See also para*-Dichlorobenzene
Bay Area Air Quality Management District (BAAQMD), 22–24, 123
Bayonne–Elizabeth, New Jersey, 150
Beaumont, Texas, 150
Bendiocarb, 367
Benomyl, 368
Benzene, 154, 396
    activities increasing exposures, 18
    biomarkers and, 396
    breath measurements, 396
    carcinogenicity, 154
    commuter exposure, 155
    health effects, 148
    impurities, 148
    indoor vs. outdoor exposures, 16
    regulating stationary (outdoor) vs. indoor sources, 22–24
    smoker vs. nonsmoker levels, 397–398
    source apportionment, 22
    sources, 10, 101
    TEAM studies findings, 22
    tobacco smoke, 152, 154
    vehicle emissions, 154
Benzo(a)anthracene, 334
Benzo(a)pyrene, 108, 332, 333, 334, 342
2-Benzyl-4-chlorophenol (BCP), 332
Bicycling, 130–131
Bioaccumulation, 309, 310, 400, 402
Bioavailability
    defined, 47, 57

of metals, 309
Bioavailable dose, 46, 82
Biogenic particulate matter, 91
Biologically effective dose, 35, 47
Biomarkers, 395–404
    CO exposure (carboxyhemoglobin), 50
    definition, 47, 57
    of exposure vs. of effects, 395–396
    meconium, 401
    metals, 401–402
    national surveys using biomonitoring, 402–403
    pesticides, 400–401
    routes of exposure and, 396
    semivolatile organic compounds, 400–401
    tobacco smoke exposure (nicotine and cotinine), 210, 400
    VOCs, 396–400
Blood analysis for VOCs, 166–167
Blue Nozzle sampler, 355
Body burden, 8, 57
    breath measurements, 9
    dioxin exposure evaluation, 390
Boston, Massachusetts, 121, 125, 129, 186, 187
Bounding estimate, 57
Brain development, 402
Breast milk contamination, 389, 399, 400
Breath analysis, 8–9, 165–166
    CO and carboxyhemoglobin, 116
    continuous analyzer, 292–295, 399
    dermal absorption measure, 292–295
    VOC levels, 17–19, 163–165, 292–295, 396–399
Bromacil, 368
Brownian diffusion, 89
Brownian motion, 183
Building materials, 152–154
1,3-Butadiene, 148, 152, 154, 157, 400
Butylated hydroxytoluene (BHT), 153
Butyldiglycolacetate, 153

C

Cadmium, 308, 309, 321, 402
Calcium chloride, 307
Calendex™, 476–477
California Air Resources Board, 159
California EPA, 202
California motor vehicle emissions standard, 136–137, 140
California Proposition 65, 16, 153
California roller, 267, 357, 359–361
California vehicle emissions standards, 115
CalTOX, 476
Canadian Human Activity Pattern Survey (CHAPS), 450
Cancer, 148, 159, *See also* Carcinogens
    indoor air pollution and, 334, 337, 342
    lifetime risk from pesticide exposure, 369
    particle exposure and, 183
Cape Cod, Massachusetts, 333
Capillaries, defined, 276
Capture sampling, 162
Carbaryl, 274, 367

Carbon composition of particles, 184
Carbon monoxide (CO), 82, 83, 113–141
    air quality standards, 119–120
    ambient trends, 114, 140
    at-risk subpopulations, 117, 120
    background level, 50, 116
    carboxyhemoglobin, 50, 92, 116, 117
    Clean Air Act Amendments, 119
    commercial microenvironments, 123–125
    commuter exposure surveys, 117, 127–129, 136–139
    co-occurring air pollutants, 114–115
    criteria air pollutants, 82, 488
    defective exhaust systems, 116
    direct approach to measuring, 125–129
    early exposure studies, 114, 117–119
    emissions standards and, 115–116, 140
    exposure models, 123, 131–132
    fixed-site monitors, 120
    health effects, 92, 116, 120, 134
    human activity patterns and, 132–133
    identifying microenvironments, 123
    indirect approach to measuring, 126, 129–130
    indoor cigar smoke dispersion experiment, 422–424
    international comparisons, 138–139
    mortality estimates, 134
    nationwide population exposure estimates, 121
    NRC report, 114
    occupational exposures, 117, 127, 130
    personal monitors for measuring, 103, 115, 125, 126,
        See also Personal exposure monitors
    policies affecting vehicle emissions, 133–136
        mortality effects, 134
        standards and commuter exposure, 136
        transportation investment and commuter exposure,
            134–136
    probability-based sampling design example, 72–74
    recreational exposures, 130–131
    residential exposures, 130
    resistance to mass transfer, 88
    respiratory uptake, 84
    scenario illustrating exposure definition framework,
        49–52
    SHAPE model, 12
    single-media analysis, 330
    source proximity effect, 422
    standards for mechanically ventilated buildings, 223
    street and sidewalk concentrations, 118–119
    superposition principle, 119
    TEAM studies, 48, 103, See also Total Exposure
        Assessment Methodology (TEAM) studies
    tobacco smoke, 225–226
    total exposure estimation, 122
    total human exposure approach, 6
    uptake by bloodstream, 92–93
Carbon tetrachloride, 18, 150, 154, 157
Carboxyhemoglobin, 50, 116
    associated health effects, 92
    NHANES data, 121
    smokers and, 117
Carcinogenic risk assessment, 154, 220, 226, 228, 369
Carcinogens, 148

chloroform and trihalomethanes, 287
Consumer Product Safety Act, 492
dioxins, 380
house dust, 333–334, 342
labeling requirements, 153
pesticides, 368–369
radon, 457
secondhand tobacco smoke, 153
tobacco smoke, 152, 204–207, 222
VOCs, 156
CARES, 477–478
Carpet, chemicals in, 153
Carpet cleaning, 337–338
    3-spot test, 324–325, 338
    low-income families and, 339–340
Carpet dust, See House dust
Catalytic converters, 114, 120, 138–139
Centers for Children's Environmental Health and Disease
        Prevention Research, 257
Centers for Disease Control (CDC)
    lead exposure studies, 327
    secondhand smoke research, 203
    VOC exposure studies, 397
CERCLA, 490–491
Chapel Hill, North Carolina, 150
Charcoal cloth surrogate skins, 265
Chemical protective clothing, 270
Chemical Substance Inventory, 499–500
Chicago, Illinois, 118, 152, 396
Children and infants
    brain development, 402
    breast milk contamination, 389, 399, 400
    daycare centers, 28, 257, 311, 337, 356
    dermal exposures, 257, 348, 357
    exposure prevention impact, 341
    house dust exposure, 312, 321, 341–342
    lead exposure, 308, 312, 327–329, 331–332, 341
    nondietary ingestion of substances, 311–312, 321, 334,
        355, 368
    PAH exposures, 334
    pesticide exposures, 349, 368
        estimation guidelines, 368
        exposure and risk assessment guidelines, 370
        hand wipes for assessing, 362–363
        regulation, 349
    secondhand smoke exposure, 212
Children's Health Act, 257
Chimneys, 185, 246
China, 154, 192
Chloracne, 380
Chlordane, 306, 332, 364, 367, 368, 400, 421
Chlorinated drinking water, 158, 287, 306, 307–308
Chloroform, 101, 276, 286
    activities increasing exposures, 18
    biomarkers and, 396
    breath measurements, 290, 292–295
    carcinogenicity, 148, 154, 287
    dermal absorption measures, 292–295, 399
    exposure biomarkers, 399
    indoor exposure sources, 10, 286
    inhalation exposure during showers, 158, 286, 292

inhalation exposure from swimming, 292
multiple carrier media, 7
outdoor levels, 154
skin permeability coefficient, 288
TEAM study findings in homes, 286
Tenax™ limitations, 102
U.S. vs. European residential exposures, 151
water contamination and volatilization, 151, 158, 287
water quality standard, 307–308
water temperature effects on dermal absorption, 286,
    287, 294
Chlorophenol, 379
Chlorothalonil, 350, 369
Chlorpyrifos, 306, 347, 349, 350, 362–368, 400, 495
Christie, Agatha, 12
Chrominated copper arsenate (CCR), 257, 347
Chromium, 309, 321, 368, 402
Chronic exposure, 50, 55, 57
Chronic obstructive pulmonary disease, 183, 186
Chrysanthemumdicarboxylic acid, 400
Cigar smoke, 213
    indoor dispersion experiment, 422–424
Cigarette brands, 214
Cigarette smoke, *See* Secondhand tobacco smoke
Cilia, 84
Clean Air Act (CAA) and amendments, 82, 115, 119, 133,
    139, 154, 214, 337, 460, 488–490
    benzene reductions, 154
    effect on CO exposure, 134
    effectiveness, 489–490
    hazardous air pollutants (HAPs), 488–489, 506–510
Coal-fired power plant emissions, 21, 154
Coal stoves, 193
Coarse particulate matter, 82, 91, 183, 184
Collagen, defined, 276
Commuter exposures, 121
    activity patterns, 14, *See also* Human activity patterns
    benzene, 155
    California TEAM study findings, 22
    CO exposure studies, 117–119, 121, 127–129
    El Camino Real surveys, 136–137
    emissions standards and, 134–136
    human activity patterns, 133
    intake fraction and in-vehicle exposures, 245–246
    secondhand tobacco smoke, 202, 225–226
    transportation investments and, 134–136
Comprehensive Environmental Response, Compensation,
    and Liability Act (CERCLA), 490–491
Computational fluid dynamics, 462
Concentration, defined, 39, 42–43, 57
Concentration time series, indoor air quality model,
    419–421
Congestion Management and Air Quality (CMAQ)
    improvement program, 134
Consolidated Human Activity Database (CHAD), 132, 451
Consumer Product Safety Act (CPSA), 492–493
Consumer Product Safety Commission (CPSC), 492
Contact boundary, 35, 40–41, 57
    examples, 49, 52, 54, 55
    inhalation exposure, 82
Contact volume, 41–42, 49, 57

dermal exposure scenario, 52, 53
    ingestion exposure scenario, 54, 55
Contact volume element, 42, 49
Contact volume thickness, 57
Continuing Survey of Food Intakes by Individuals (CSFII),
    475, 476
Continuous breath analyzer, 292–295, 399
Cooking-associated air pollution, 104, 182, 185–186
    char index, 434
    chimneys and, 185
    exposure reduction measures, 192–193
    household energy ladder, 185
    indoor air quality model application, 434–441
    personal monitor limitations, 109
    smoking-related particle exposure vs., 440
    source proximity effect, 428
Copper arsenate, chrominated (CCT), 257, 347, 384
Copper sulfate, 350
Corneocytes, 260–261, 276
*Corrosion Proof Fittings v. Environmental Protection
    Agency*, 500
Cosmetic-drug distinctions, 494–495
Cotinine, 202, 210, 400
    blood level trends, 395
    pharmacokinetics and dosimetry, 226–228
    secondhand smoke misclassification problem, 211
Criteria pollutants, 82, 114, 119–120, 488
Cross-sectional study designs, 186
Crustal material, 184
Cumulative and Aggregate Risk Evaluation System
    (CARES), 477–478
Cumulative dose, 36
Cumulative exposure, 36, 472
Cyclohexanone, 153
Cyfluthrin, 365
Cypermethrin, 306, 365
Cytochrome P-450, 381
Cytoskeletal, 276

D

2,4-D, 311, 347, 350, 352, 364, 365, 369
Data Quality Objectives (DQO), 70
Daycare centers, 28, 257, 311, 337, 356
Dayton, Ohio, 118
DDE, 400, 401
DDT
    blood screening method, 400
    dermal exposure scenario, 52–54
    food contamination, 310, 495
    in house dust, 321, 332
    meconium levels, 401
    in rugs, 333
    TEAM studies, 102
    water contamination, 306
Decane, 17
Decipols, 165
DEET, 350
Delaney Clause, 493–494
Delaware Air Quality Survey, 224–225

Delivered dose, 47
Deltamethrin, 306
Denver, Colorado, 12, 103, 118, 123, 126, 127, 128
Deposition mechanisms, 89–90
Dermal exposure, 255–278, 348, 473–474
    absorption measurement, 270–272, 292–295
    at-risk subpopulations, 257
    child and infant exposure, 257, 348, 357
    definitions and related terms, 52, 257–258
    dermal dose analysis, 258
    dermal exposure analysis, 258
    diffusion process, 53–54
    direct measurement methods, 265
        continuous breath measurement, 292–295, 399
        fluorescent tracer techniques, 266–267
        removal techniques, 266
        surface sampling techniques, 267–269
        surrogate skin techniques, 265–266
    dose and, 256–258, 262–263, 480
    factors affecting dermal dose, 262–263
    glossary of terms, 276–278
    health effects, 257
    indirect absorption measures, 292
    lag time, 276, 285–286, 289–290, 295
    macroactivity approach, 473–474
    mechanisms and pathways, 263–265
    micro-level activities, 473
    modeling, 473–474, 480, See also Exposure models
        two-compartment kinetic model, 288–289
    protective clothing efficacy, 270
    residue transfer coefficients, 269
    scenario illustrating exposure definition framework,
        52–54
    skin characteristics and function, 259–262
        permeability coefficients, 287–289
        skin surface properties, 263
    soil adherence, 269
    solute permeability and dermal dose, 262
    studies, 269–270, 273–274
    time profile, 52
    VOCs and activities in water, 285–296, See also Water
        immersion dermal exposures
Desmosome junctions, 260, 276
Desquamated, 276
Detroit, Michigan, 118, 333
Developing world, particulate pollution effects, 182
Diazinon, 306, 347, 349, 363–367, 400, 495
Dicamba, 364
1,4-Dichlorobenzene, See para-Dichlorobenzene
Dichlorvos, 350, 363, 364, 367
Dieldrin, 333, 367, 368, 400, 495
Dietary ingestion, 54, 306–311, See also Ingestion
        exposure; Water contaminants
    data collection, 475
    dioxins and dioxin-like compounds, 387–391
    metal contamination, 307–309
    PAHs, 334
    pesticide regulation, 493–494
    pesticides, 306–307, 310, 475, 495
    phthalates from packaging, 311
    VOCs, 307–308

Diffusion
    Brownian, 89
    definition, 276
    Fick's Law, 42
    respiratory tract mass transfer, 84, 89
    transport across GI tract, 305
Diffusion cells, 271–272
Diffusion charger, 107–108
Dinoseb, 368
Dioxin and dioxin-like compounds, 379–392, 400
    at-risk subpopulations, 389
    background level, 386–387
    dietary contamination, 387–391
    dioxin-like congeners, 381–382
    emissions and fate, 384–387
    media, 387
    multiple gene response, 381
    pathways to human exposure, 389–391
    reducing exposure, 390–391
    regulation, 384
    sources, 379, 384
    stability, 385
    structure and properties, 380
    toxicity, 380–381
    toxicity factors and equivalence, 382–383
Dioxin Exposure Initiative, 390
Direct approach to exposure analysis, 8
Disinfectants, 347–350, 363, 364
Dislodgeable residue, 357
Diuron, 350
Doormats, 21, 320, 323, 324
Dose, 38
    absorbed, See Absorbed dose
    "aggregate" and "cumulative," 36
    definitions and related terms, 35, 38–39, 45–47, 58
    dermal exposure and, 256–258
    estimating from exposure model, 480
    full risk model, 5
Dose rate, 50, 58
Dosimeters, 265
Drag sled, 267, 357, 358–359
Drinking water contaminants, See Water contaminants
Dry-cleaned clothes, 101, 156–157, 159
Dry cleaning establishments, 156–157, 399
Duct-mounted electrostatic precipitators, 190, 193
Dust, See House dust
Dust Vacuum Method (DVM), 355

E

Economic development and air quality, 138–139
Edwards and Lioy (EL) sampler, 267, 269
Effective dose, 47, 82
Elastin fibers, 276
El Camino Real commuter exposure surveys, 136–137
Electrostatic precipitators, 190, 193
Eliminated dose, 47, 59
Elizabeth, New Jersey, 189
Emergency Planning and Community Right-to-Know Act,
    25

Endocrine disruptors, 311, 332
Endosulfan, 495
England, 130
Environmental justice concerns, 246–247
Environmental laws and policies, 487–490, *See also* U.S.
            Environmental Protection Agency; *specific*
            laws
    air contaminant permissible exposure limits, 496–497
    Air Contaminants Standard, 497–498
    content analysis of major laws, 503
    cosmetic–drug distinctions, 494–495
    CO standards, 119–120
    cost-benefit analyses, 492–493
    dioxin regulation, 384
    effectiveness, 134–136, 238, 489–490
    effects on commuter CO exposures, 134–136
    emphasis on outdoor stationary sources, 22–24, 154,
            159, 332, 336
    exposure models and, 131–132, 457, 458–459, 477–478
    hazardous waste regulation, 498–499
    intake fraction and environmental justice, 246–247
    labeling requirements, 16, 153, 492
    leaded gasoline ban, 115–116, 326, 332, 395, 402
    multimedia exposure analysis, 329–330
    new and existing chemicals regulation, 499–500
    particulate exposure, 82, 214, 488
    pesticide regulation, 348–349, 368, 492–496
    prioritization using intake fraction, 244–245
    proposed Human Exposure Reduction Act, 501,
            504–505
    proposed indoor air quality rule, 498
    public education, 24–26
    review of specific laws, 488–512
    smoking bans, 24, 228
    smoking regulation issues, 202
    source-exposure relationships and, 5
    source-oriented vs. receptor-oriented approaches,
            12–13, 25
    vehicle emissions standards, 115
    workplace exposures, 368, 496–498
Environmental Protection Agency (EPA), *See* U.S.
            Environmental Protection Agency
Environmental tobacco smoke, *See* Secondhand tobacco
            smoke
Epidermal structure and function, 259–262
Ethalfluralin, 368
Ethanol breath analyzer, 8
Ethyl Corporation study, 105
Ethylene glycol, 153
European Exposure Factors Sourcebook, 460
European urban studies, 131, 151
EXPOLIS, 151, 460
Exposimeter, 25–26
Exposure agent
    defined, 39
    examples, 49, 52, 55
Exposure analysis
    definitions and related terms, *See* Exposure terminology
            and definitions
    direct approach, 8, 125–129
    full risk model, 4–6

indirect approach, 10–12, 126, 129–130, 446
measurement science, 27
multimedia approach, 329–330
overemphasis on traditional pollutant sources, 4–5
receptor-oriented approach, 3, 8, 12–14
source-oriented approaches, 3, 12–13, 25
total human exposure concept, 6–12, 256
Exposure assessment, defined, 47, 58
Exposure concentration
    defined, 43, 58
    examples, 49, 52, 54
Exposure dose, defined, 38
Exposure duration
    defined, 44, 58
    examples, 50, 55
Exposure event, 50, 55, 58
Exposure frequency, 44, 50, 55, 58
Exposure loading, 42, 43, 52, 58
Exposure mass, 35, 43, 54, 58
Exposure models, 131
    APEX, 132, 458
    Calendex™, 476–477
    CARES, 477–478
    combining data, 478–479
    continuous and semi-continuous formulations, 447
    co-occurring air pollutants, 480
    CO population exposure, 131–132
    database resources, 460
    definition, 58
    direct evaluation of, 461–462, 463
    dose estimation, 480
    "exploratory" and "regulatory" types, 458
    future development, 463–465
    gaseous pollutant uptake, 84–88
    human activity patterns, 10, 12, 27, 48, 370, 450–455,
            463, 472–473
    human factors in, 463–465
    indirect approach, 10–12, 446
    indoor air quality, 411–443, *See also* Indoor air quality
            modeling
        experiments, 422–424
        mass balance equation, 187–188, 412–417
        uniform mixing assumptions, 411, 413, 421–424
    inhalation exposures, 445–466, *See also* Inhalation
            exposure modeling
    Lifeline™, 477
    models as theory, 460
    NAAQS Exposure Model, 131–132
    NEMS, 131–132, 458
    pesticide exposures, 368, 471–481, *See also* Pesticide
            exposure models
    practical applications, 456–457
    quantitative structure-activity relationships, 288
    Random Component Superposition (RCS), 12
    review of existing models, 457–460
    secondhand tobacco smoke, 217–220
        active smoking count, 218–219
        habitual smoker model, 220
        Sequential Cigarette Exposure Model, 436
    SHAPE, 12, 131–132
    SHEDS, 12, 458, 478

source apportionment, 21–24
standard formula, 446
STREET, 137
temporal and spatial variability, 479–480
Exposure pathway, 39, 54, 58, 264
Exposure period, 44, 54, 58
Exposure point, 39, 58
Exposure route, 4, 39, 273–274, *See also* Dermal exposure;
      Ingestion exposure; Inhalation exposure
    biomarkers and, 396
    definition, 58
    examples, 52, 54
    multiple carrier media, 7
Exposure scenarios
    definition, 49
    illustrating exposure/dose definition framework, 49–55
Exposure surface, 40, 57
Exposure terminology and definitions, 33–60
    agent, 39
    concentration, 42–43
    contact boundary, 40–41
    contact volume, 41–42
    criteria for framework, 37
    dose and related terms, 35, 38, 45–47
    environmental health literature, 37–38
    exposure, 35, 39–40, 58
    glossary, 57–59
    illustrative examples, 49–55
        CO inhalation exposure, 49–52
        DDT dermal exposure, 52–54
        ingestion exposure, 54–55
    photography analogy, 56
    practical implications of exposure theory, 47–48
    professional standards, 36, 56
    radiobiological context, 37
    spatially related definitions, 44–45
    target, 39
    temporally related definitions, 35, 38, 44–45, 50
    theoretical framework, 34–35
    use of "aggregate" and "cumulative" terms, 36
Exposure time profile, 44, 52, 55, 59
Exposure trading, 159
Exposurist profession, 26–27
External dose, 46
Extrathoracic region, 84

F

Facilitated diffusion, 305
Fate and transport models, 5, 476
Federal Food, Drug, and Cosmetic Act (FFDCA), 493–495
Federal Insecticide, Fungicide and Rodenticide Act
    (FIFRA), 348, 495–496
Federal Interagency Committee on Indoor Air Quality
    (CIAQ), 491
Federal laws and policies, *See* Environmental laws and
    policies
Fenoxycarb, 368
Fiberglass, 89
Fick's Law, 42

Filters, 190, 193
Fine particulate matter ($PM_{2.5}$), 82, 91, 104, 183
    infiltration factor, 188–191
    NAS/USEPA research program, 186–187
    pesticide air sampling method, 353
    secondhand tobacco smoke, 212, 216, *See also*
        Respirable suspended particles; Secondhand
        tobacco smoke
    total particle exposure, 192
Fingernails, toxic metal levels, 401
Fire retardants, 321–322
Fish contamination, 309, 389, 402
Flight attendants, 228
Fluorescence microscopy, 276
Fluorescent tracer techniques, 266–267
Food ingestion, *See* Dietary ingestion
Food Protection Act, 337
Food Quality Protection Act (FQPA), 36, 256, 347, 348,
    349, 471, 472, 475, 493–495
Formaldehyde, 148, 153, 154, 157
France, 308
Fungicides, 348, 350, 369, *See also* Pesticides
Furans, 400

G

Gallup, George, 66
Garages, 124–127, 155
Garment samplers, 265–266, 276
Gas chromatography (GC), 350
    GC/FID, 99–100, 163
    GC/MS, 163
Gas exchange region of lung, 84
Gasoline additives
    lead, 115–116, 326, 332, 395, 402
    MMT, 105
    MTBE, 151, 399
Gasoline vapor, 155
Gas ranges and air quality, 127, 130, 133
    particulate combustion products, 183
    source proximity effect, 428
    step source time function, 417
Gastrointestinal physiology, 305
Geographic information systems (GIS), 463
Germany, 151, 153, 403
Glyphosate, 350, 364
Golden rule of averages, 430
Gravitational settling, 89
Grimm monitor, 107

H

Habitual Smoker Model, 220
Hair, toxic metal levels, 401
Hair follicles, 260, 262
Hand rinses, 269–270
Hand-to-mouth ingestion, 311, 475
Hand washes, 266, 362
Handwipe methods, 361–363

HAPEM, 458
Harvard Impactor, 104
Harvard 6-City Study, 104, 184, 223
Harvard University School of Public Health, 105
Hawaiian air quality, 129, 134–135
Hazardous air pollutants (HAPs), 488–489, 506–510
Hazardous waste
    house dust, 334, 341–342
    regulation, 498–499
    tobacco smoke pollutants, 203–207
Headspace SPME, 167
Health effects, *See also* Mortality estimates
    CO exposure, 92, 116, 120, 134
    cost savings for controlling indoor air pollution,
        341–342
    dermal exposures, 257
    dioxin toxicity, 380
    estimating using intake fraction, 245
    exposure model applications, 456–457
    inhaled airborne pollutants (table), 83
    lead, 321, 327, 329
    particle exposure, 91–92, 181, 182–184
        house dust, 324, 337, 341–342
    pesticide exposures, 257, 347, 348–349, 368–370
    pollution emissions-to-effects relationship, 239
    premature birth, 401
    secondhand tobacco smoke exposure, 203, 208, 229
    sick building syndrome, 148, 149
    unventilated indoor cooking, 186
    VOC exposure, 148–150, 159
    worker productivity impacts, 148, 150
Health Hazard Evaluation (HHE), 498
Healthy Homes Project I, 341
HEAL™, 339–340, 342
Heart disease, 117, 120, 183, 202, 324, 338, 340
Heaviside function, 417
Heavy metals, *See* Metals
HEPA filters, 193
Heptachlor, 332, 367, 368, 369
Herbicides, 10, *See also* Pesticides
    in house dust, 364
    nondietary ingestion, 311
    use patterns, 347, 350
    water contamination, 306–307
Hexachlorocyclohexane, 367, 369, 400
Hexachlorophene, 379
High performance liquid chromatography (HPLC), 350
High Volume Surface Sampler 2 (HVS2), 320
High Volume Surface Sampler 3 (HVS3), 319, 320–321,
        355–356
High Volume Surface Sampler 4 (HVS4), 321
Home Environmental Assessment List (HEAL™),
        339–340, 342
Honolulu, Hawaii, 129, 135–136
Horny layer, 259, 276
Hospitality industry, secondhand smoke in, 223–224, 229
House dust, 319–343, *See also* Particle exposure
    age of house and, 330–331
    allergen sensitization, 341
    carcinogens, 333–334, 342
    characteristics, 320

child and infant exposure, 319, 321, 355
    exposure estimation guidelines, 368
controlling, 323–325, 337–339
    bare floors and, 325
    cleaning, 337–339
    comparing cleaning protocols, 330
    doormats, 21, 320, 323, 324
    health care cost savings, 341–342
    Healthy Homes Project I, 341
    Home Environmental Assessment List, 339–340,
        342
    low-income families and, 339–340
    vacuum cleaners, 324–325
exposure estimation guidelines, 368
as hazardous waste, 334, 341–342
health effects, 321, 337, 341–342
inhalation/swallowing exposure pathway, 312
lead surface loading, 325
major pollutants in, 321–322, 326–327
multimedia exposure analysis, 329–330
PAHs, 321, 332–335, 356
pesticides in, 321–322, 332, 333, 342, 348, 363–364
pet health risks, 321
reducing lead exposure, 330–332
reducing personal exposure, 342
resuspended particles, 322
sample collection and measurement, 320–321, 355–356
sources, 322, 332
surface sampling, 268
toxicity mitigation activities, 21
track-in, 322, 332
vacuuming, 319–321, 355–356
Household energy ladder, 185
Houston, Texas, 188–189
Human activity patterns
    activities increasing VOC exposures, 17–21
    activity pattern data definition, 57
    Canadian Human Activity Pattern Survey (CHAPS),
        450
    CO exposure and, 132–133
    combining data from different sources, 479
    data extrapolation, 479
    exposure models, 10, 12, 27, 48, 370, 450–455, 462,
        472–473
    macro-level, 472–473
    meso-level, 473
    micro-level, 473
    National Human Activity Pattern Survey (NHAPS), 14,
        132–133, 450–455
    particle exposures and, 185
    pesticide exposure and risk assessment, 370
    social models, 465
    time budgets, 14, 15
    USEPA's database (CHAD), 132, 451
Human Exposure Reduction Act (HERA), 501, 504–505
HVS2, 320
HVS3, 319, 320–321, 355–356
HVS4, 319, 321
Hydrocortisone, 273
Hydrophilic, 276
1-Hydroxypyrene, 401

I

Impaction, 89
Incense, 151
Incinerators, 384, 385
Indianapolis, Indiana, 105
Indirect approach to exposure analysis, 10–12, 129–130, 446
Indoor air quality, 14–17, 185, 193, 212, 366–367, *See also* House dust; Particle exposure; *specific exposures*
  activities increasing VOC exposures, 17–21, *See also* Human activity patterns; Volatile organic compound (VOC) exposure
  air conditioning effects, 188–189
  air exchange rate and, *See* Air exchange rate
  at-risk subpopulations, 324
  bare floors and, 325
  chloroform, *See* Chloroform
  CO field studies, 123–125, 130, *See also* Carbon monoxide
  cooking and, *See* Cooking-associated air pollution
  estimated associated mortality, 181
  exposure reduction measures, 159, 192–193, *See also under* Particle exposures
  garages, 124–127, 155
  Home Environmental Assessment List, 339–340
  household energy (or fuel) ladder, 185
  infiltration factor, 188–191
  intake fraction difference between indoor and outdoor releases, 240
  Love Canal residences, 337
  major pollutants, 332–337
  multimedia exposure analysis, 329–330
  new buildings, 151
  OSHA and, 497–498
  outdoor pollutant concentration comparison, 16, 17, 154, 397
  outgassed tobacco pollutants, 228–229
  outside wind effects, 428
  particles of outdoor origin, 187–188
  particulate pollutants, 185–193, *See also* Particle exposure
  pesticide exposure guidelines and regulation, 368
  pesticide sources, fate, and transport, 363–367, *See also* Pesticides
  public risk awareness, 25, 158–159
  regulatory emphasis on outdoor sources vs., 22–24, 487–488
  Relative Source Apportionment of Exposure, 21–24
  "sea of particles," 322
  significance of, 10, 14
  source apportionment, 22–24
  source proximity effect, 422
  Superfund site air quality comparison, 319, 320, 331–332, 336
  TEAM studies, 9–10, 101, 104, 185, *See also* Total Exposure Assessment Methodology (TEAM) studies
  tobacco smoking and, *See* Secondhand tobacco smoke
  USEPA and, 489

  ventilation design for permitted smoking, 220
  ventilation standards, 223–224, 457
  VOC sources and levels, 17–21, 150–155, 397, *See also* Volatile organic compound (VOC) exposure
Indoor air quality modeling, 411–443, *See also* Exposure models; Inhalation exposure modeling
  concentration time series and change in ventilation rate, 419–421
  golden rule of averages, 430
  indoor-generated source strength, 188, 190
  mass balance equation, 187–188, 412–417
  particle deposition rates, 432–434
  particle source applications, 434–441
    cigarette smoke, 434–441, 448–450
    food cooking/burning experiments, 434–441
  peak estimation approach, 431–432
  phantom curve, 428, 437
  response to source time functions, 417–421
  Sequential Cigarette Exposure Model, 436
  source strength calculation, 428–431
  time-weighted averages, 429
  understanding pollutant dispersion, 462–463
  uniform mixing assumptions, 411, 413, 421–424, 462–463
    cigar smoke dispersion experiment, 422–424
  ventilation rate measurement, 424–428
Inertial impaction, 89
Infant brain development, 402
Infants, *See* Children and infants
Infiltration factor, 188–191
Ingestion exposure, 303–313, 321, *See also* Dietary ingestion; Water contaminants
  active vs. passive transport, 305
  bioaccumulation, 309, 310
  contrasts with other routes, 304
  data collection, 475
database resources, 475
  directions for research, 312–313
  dust inhalation/swallowing, 312
  GI physiology, 304
  nondietary materials, 311–312, 321, 355
    hand-to-mouth, 312, 475
    inadvertent intake, 312
    PAHs, 334
    pesticides, 311, 368, 475
    pica behavior, 59, 321
    soil, 312
  pesticides, 311, 368, 475
  scenario illustrating exposure definition framework, 54–56
Inhalable particulate matter (PM_{10}), 82, 104, 183, 212, 214
Inhalation dose, 50, 82
Inhalation exposure, 81–97, *See also* Carbon monoxide; Particle exposure; *specific pollutants*
  air contaminant permissible exposure limits, 496–497
  Air Contaminants Standard, 497–498
  breath or blood concentrations and dose correlation, 92
  CO uptake example, 92–96, *See also* Carbon monoxide
  dust inhalation/swallowing, 312
  hazardous air pollutants (HAPs), 488–489, 506–510
  health effects, 83, 91–92

intake dose, 82
intake fraction, *See* Intake fraction
mass transfer model, 84–88
MTBE, 399–400
PAHs and children, 334
particulate pollutants, 89–92, 183, *See also* Particle
    exposure
pesticide exposure and risk assessment guidelines, 370
pesticide exposure guidelines and regulation, 368
pesticides and cancer risk, 368–369
pollutants of concern, 82–83
respiratory uptake, 81, 83–84
volatilized chloroform from showers, 158, 292
Inhalation exposure modeling, 445–466, *See also* Exposure
    models; Indoor air quality modeling; Pesticide
    exposure models
    basic formulas, 446–448
    database resources, 460
    direct evaluation of predictions, 461–462, 463
    exposure mitigation applications, 457, 465
    future development, 463–465
    human activity patterns, 370, 450–455
    human factors, 463–465
    models as theory, 460
    pollutant dispersion, 462–463
    practical applications, 456–457
    regulatory models, 131–132, 458–459
    review of existing models, 457–460
    secondhand smoke exposure simulation, 448–450
Insecticides, 347, 350, *See also* Pesticides; *specific*
    chemicals
    indoor air pollution and, 363
    misuse, 350
    vapor pressure, 365
Instantaneous dose rate, 45
Instantaneous exposure, 6, 38, 44
Instantaneous point exposure, 39, 43, 52, 58
Instantaneous spatially integrated intake dose, 46
Institute for Inspection Cleaning Restoration and
    Certification (IICRC), 338
Intake, defined, 58
Intake dose
    definition, 35, 45–46, 58
    example, 50
    ingestion exposure scenario, 54
    inhalation exposure, 82
Intake fraction, 237–249
    air-exchange rate and, 240, 243–244
    defining, 237, 239
    difference between indoor and outdoor releases, 240
    emissions-to-effects relationship, 239–240
    environmental justice concerns, 246–247
    estimating health impacts, 245
    estimating using one-compartment model, 241–244
    factors influencing, 241
    on-road particulate matter sources, 245–246
    population, proximity and persistence, 240
    prioritizing emission reduction initiatives, 244–245
    self pollution, 238, 246
    studies, 242
    typical values, 240–241

Intelligence quotient (IQ), 329
Intensity, defined, 39, 58
Interception, 89
Intermodal Surface Transportation Efficiency Act (ISTEA),
    134
Internal dose, 35, 46
International Agency for Research on Cancer (IARC), 202
International Commission on Radiological Protection, 370
International commuter exposure comparisons, 138–139
International Programme on Chemical Safety (IPCS), 34,
    36, 56, 149
International Society for Exposure Analysis (ISEA), 34, 36,
    56
Interstitial fluid, 84
Interstitial space, 84
In vitro
    definition, 276
    dermal exposure measurement, 270–272
In vivo
    definition, 276
    dermal exposure measurement, 270, 272
Iontophoresis, 256, 276
Irish pubs, 226
Isolated perfused porcine skin flap (IPPSF), 272–274

J

Jacksonville, Florida, 102, 364, 366–367, 369
Japan, 308
Judgmental samples, 69–70

K

Kanawha Valley, West Virginia, 154
Keratin, 260–261, 276
Kuwait, 308

L

Labeling
    California Proposition 65 requirements, 16, 153
    Consumer Product Safety Act, 492
Lag time, 276, 285–286, 289–290, 295
Lamellar, 276
Landfill permits, 499
Landon, Alfred E., 66
Langan monitor, 103
Las Vegas, Nevada, 216
Lead, 257, 326–332, 402
    atmospheric emissions, 326
    blood level trends, 402
    child and infant exposure, 308, 312, 319, 327–329,
        331–332, 341
    criteria air pollutant, 82, 488
    food contamination, 309
    in gasoline, 115–116, 326, 332, 395, 402
    health effects, 327, 329
    in house dust, 21, 321
        carpet dust, 325

cleaning and vacuuming, 337–338
house age and, 330–331
pet health risks, 321
reducing exposure, 330–332, 341
track-in, 332
upholstery, 338
inhalation exposure, 82
multiple carrier media, 7
nondietary ingestion, 312
regulation, 499, *See also* Environmental laws and
policies; *specific applicable laws*
U.S. population blood levels, 332
water contamination, 307, 326
Lead-based paints, 327–328, 331–332, 402
Lead sulfide, 307, 309
Legislation, *See* Environmental laws and policies
Lifeline™, 477
Limonene, 101, 148, 151, 153
Lindane, 274, 367, 368, 400
Lioy-Wainman-Weisel (LWW) wipe sampler, 267, 270
Lipophilic
definition, 276
dermal exposure, 263, 272, 273
dioxin toxicity, 381
VOCs, 157, 167
Longifolene, 153
Long Island, New York, 333
Longitudinal study designs
commuter exposure surveys, 136
exposure modeling, 477, 478, 479
particle exposure, at-risk subpopulations, 186
Los Angeles, California, 105, 121, 123, 126, 127, 129, 150,
155, 187, 189, 238, 247, 333
Love Canal, 152, 337
Low-income families, 339–341
Lung anatomy, 84, 183, 304
Lung cancer, 183
Lycopene, 55
Lymphatic system, 276

M

Macro-level activities, 472–473
Malathion, 274, 306, 350, 363, 367, 495
Manganese (Mn), 54, 105
Margin of exposure (MOE), 381
Market basket studies, 55, 495
Marple personal exposure monitor, 104, 106
Mass balance equation for indoor air quality, 187–188,
412–417
Mass loading, 277
Mass spectrometry (MS) based breath analyzer, 163–166,
292
Mass transfer equation, 42, 87
resistor model, 87
Mass transfer modeling, gaseous pollutant uptake, 84–88
Master Home Environmentalist™ program, 330, 339–340
Mathematical modeling, *See* Exposure models; Indoor air
quality modeling

Maximum achievable control technology (MACT),
488–489
Meconium, 401
Mecoprop, 364
Medium, defined, 58
Medium intake rate, 50, 54, 58
Menlo Park, California, 422
Mercury, 309, 321, 402
Meso-level activities, 473
meta-Dichlorobenzene (*m*-DCB), 18
Metals, 257, 326, *See also specific metals*
bioaccumulation, 309, 402
bioavailability, 309
biomarkers, 401–402
food contamination, 308–309
in house dust, 321
water contamination, 307, 326
Methoxychlor, 332
Methyl bromide, 347
Methylcyclopentadienyl manganese tricarbonyl (MMT),
105
Methylene chloride, 102, 116, 148, 152
Methyl ethyl ketone, 152
Methylmercury, 309, 402
Methyl-*tert*-butyl ether (MTBE), 151, 295, 399
Mexico City, 138–139
Microdialysis, 272, 277
Microenvironments
activities increasing exposures, 17, *See also* Human
activity patterns; Indoor air quality
ambient exposure factor, 191
definition, 59, 123
exposure models and, 446, *See also* Exposure models
field surveys of commercial settings, 123–125
indirect approach to exposure analysis, 10
personal daily exposure to particles, 214–215
secondhand tobacco smoke concentrations, 212–215
total exposure estimation and, 123
uniform mixing assumptions, 411, 419–424
experiments, 422–424
Micro-level activities, 473
Microplumes, 422
MIE personal Data RAM™ (pDR), 106–107
Minnesota Children's Pesticide Exposure Study
(MNCPES), 269–270
Miticides, 350, *See also* Pesticides
MMT, 105
Modeling, *See* Exposure models; Indoor air quality
modeling; Inhalation exposure modeling
Monte Carlo simulation, 476
Mortality estimates
CO, 134
indoor pollution associated, 181
particle exposure, 184
secondhand tobacco smoke exposure, 202, 203, 229
Mosquito coils, 151
Moth repellents, 19, 101, 156, 348, 350
Motor vehicle emissions, 22, 119–120, *See also* Carbon
monoxide; Commuter exposures
at-risk subpopulations, 154
auto manufacturer perspectives, 120

early studies, 114, 117–119
effects of standards, 140
El Camino Real commuter exposure surveys, 136–137
intake fraction, *See* Intake fraction
international comparisons of commuter exposure,
        138–139
lead, 326
manganese (from MMT), 105
models for inventorying, 115
outdoor air quality and, 154
ozone formation, 117
policies affecting, 133–136
respirable suspended particle levels, 215
self-pollution, 238, 246
standards, 115
transportation system design and, 134–136
VOCs in, 154
Motor vehicle inspection and maintenance programs, 136
Moving average exposure, 50, 59
MTBE, 151, 295, 399
Mucus, 84
Multimedia exposure analysis, 329–330
Multiple carrier media, 7
Multiple chemical sensitivity, 148, 149
Multiple gene response, 381
Multisorbent systems, 102, 161
Multistage probability design, 8
Murder mystery analogy, 12–14

N

NAAQS Exposure Model (NEM), 131–132, 458
Nanoparticles, 183
Naphthalene, 348, 350
Nasal-pharyngeal region, 84, 92
Nashville, Tennessee, 186
National Academy of Sciences (NAS)
    Delaney Clause and, 493
    exposure assessment guidelines, 44
    exposure definitions, 56
    particle exposure research program, 186, 187, 194
    personal monitor development and, 99, 100, 109
    secondhand smoke research, 202
National air monitoring stations (NAMS), 120
National Air Pollution Control Administration (NAPCA),
        119
National Ambient Air Quality Standards (NAAQS), 82,
        114, 119–120, 139, 214, 488
    exposure model (NEM), 131–132, 458
    periodic review, 131
National Cancer Institute (NCI), 202
National Cooperative Inner-City Asthma Study (NCICAS),
        341
National Dietary Guidelines, 390–391
National Emission Standards for Hazardous Air Pollutants
        (NESHAPS), 489
National Health and Nutrition Examination Survey
        (NHANES), 121, 327, 402–403, 475
National Health and Nutrition Examination Survey III
        (NHANES III), 210, 227

National Home and Garden Pesticide Use Survey, 350
National Human Activity Pattern Survey (NHAPS), 14,
        132–133, 450–455
National Human Exposure Assessment Survey (NHEXAS),
        13, 67–69, 160, 403, 460, 480
National Institute for Occupational Safety and Health
        (NIOSH), 202, 368, 498
National Organics Reconnaissance Survey, 158
National Priorities List (NPL), 490
National Research Council (NRC)
    CO management report, 114
    pesticide exposure policy report, 472
National Toxicology Program (NTP), 202
Negative dose, defined, 47, 59
NEM, 131–132, 458
Net dose, 47
Netherlands, 131, 151
New buildings, 152
    sick building remediation, 152
    sick building syndrome, 148, 149
New Orleans, Louisiana, 158
New York City, New York, 187
New York State Energy Research and Development
        Authority, 223
Nicotine
    cotinine metabolite, 202, 210, 226–228, 400
    smoking pharmacokinetics and dosimetry, 226–228
    tobacco smoke atmospheric tracer, 202, 215–217
Nitrate composition of particulate matter, 184
Nitrogen dioxide (NO_2), 82, 105, 488
n-Octylbicycloheptene dicarboximide (MGK-264),
        368–369
Non-dispersive infrared (NDIR), 120
Nonmethane hydrocarbons (NMHC), 82, 100
Non-Occupational Pesticides Exposure Study (NOPES),
        366, 496
4-Nonyphenol, 332

O

Occlusion, 277
Occupational exposures
    CO, 117, 127, 130
    OSHA and, 496–498
    pesticide regulation, 368
    secondhand tobacco smoke, 210, 216, 223–224, 228,
        229
    smoking bans and, 24, 228
    VOC health effects, 148
Occupational Safety and Health Act (OSH Act), 496–498
Occupational Safety and Health Administration (OSHA),
        368, 496–498
    proposed indoor air quality rule, 498
n-Octylbicycloheptene dicarboximide (MGK-264),
        368–369
Olfactory analysis, 165
Oral mucosa, 260
Organic, defined, 277
OSHA, 368
Ozone, 88

criteria air pollutants, 488
early air pollution studies, 117
personal monitors for measuring, 105

P

Packaging materials, 311
Paints, 153
Palo Alto, California, 123–125
Papillary layer, 277
para-Dichlorobenzene (*p*-DCB), 101
    California ban, 159
    carcinogenicity, 148, 154, 369
    residential microenvironments, 19–21
    sources and uses, 10, 18–19, 153, 155–156, 348, 350
Parathion, 274
Parking garages, 124–127
Particle-bound polycyclic aromatic hydrocarbons (PPAH), 221–222, 224
Particle characteristics, 183–184
Particle deposition mechanisms, 89–90
Particle deposition rates, indoor air quality model, 432–434
Particle exposure, 83, 91, 181–195, *See also* House dust; Inhalation exposure
    airborne particle sizes, 82, 91, 104, 183, 488
    ambient exposure factor, 191–192
    average urban levels, 182
    cooking and, *See* Cooking-associated air pollution
    criteria air pollutants, 82, 488
    developing world problems, 182
    dust inhalation/swallowing, 312
    exposure reduction measures, 190, 192–193
    food cooking/burning experiments, 434–441, *See also* Cooking-associated air pollution
    Harvard 6-City Study, 184
    health effects, 83, 91–92
        causal agents, 184
        estimated associated mortality, 181, 182–183
    high-risk subpopulation studies, 186–187
    homes with vs. without smokers, 184
    human activity patterns and, 185
    indoor air quality model, 411–443, *See also* Indoor air quality modeling
        infiltration factor, 188–191
        mass balance equation, 187–188, 412–417
        particle source applications, 434–441
    modeling inhalation exposure, *See* Inhalation exposure modeling
    neighborhood pollution, 185
    non-tobacco smokes, 184
    particle composition, 184
    particle shape and, 91
    particles of outdoor origin, 187
    personal monitors for measuring, 100, 104–107
    regulation, 82, 214, 488, *See also* Environmental laws and policies
    residence time and, 91–92
    residential "sea of particles," 322
    resuspended dust, 322
    separating outdoor from indoor particles, 187–188
SHEDS-PM simulation, 12
TEAM studies, 48, 104, 184–185, 186
tobacco smoke, 214–215, *See also* Respirable suspended particles; Secondhand tobacco smoke
    cooking-related particle exposure vs., 440
    outgassed tobacco pollutants, 228–229
    power plant emissions vs., 21
total particle exposure, 192
uptake, 89–92
vacuum sampler efficiencies, 356
Particle monitors, 100, 104–107
Particle polycyclic aromatic hydrocarbons (PPAH), 221–222, 436
Particle size classifications, 82, 91, 104, 183
Particle Total Exposure Assessment Methodology (PTEAM) studies, 184–185, 186, 452, 460, 462
Pasadena, California, 117
Passive smoking, *See* Secondhand tobacco smoke
Passive transport, 305
Patch samplers, 265, 277
PCBs, 321, 381, 384, 387, 400, *See also* Dioxin and dioxin-like compounds
Pendimethalin, 350
Penetration coefficient, 188, 190
Pentachlorophenol (PCP), 332, 368, 400
Perchloroethylene (PERC), 10, 368
Percutaneous absorption measurement methods, 270–272, *See also* Dermal exposure
Permeability coefficients, 287–288
Permethrin, 306, 332, 333, 347, 363, 365
Permissible exposure limits (PELs), 496–497
Personal cloud effect, 27, 48, 104, 185, 462
Personal exposure monitors, 8, 48, 99–110, 163, *See also* Air sampling and analysis methods
    basic design, 100
    CO, 103, 125, 126
    definition, 59
    diffusion charger, 107–108
    Grimm monitor, 107
    Harvard University multipollutant monitor design, 105
    house dust sampling, 355
    inaccuracy related to exposure definitions, 47
    location on body, 48
    MIE pDR, 106–107
    multisorbent sampling cartridges, 102
    NHEXAS, 160
    noise, 354
    PAHs, 108–109
    particle exposure, 100, 104–107
    pesticides, 102
    Piezobalance, 107
    sorbents, 160–161
    TEAM studies, 8, 101, *See also* Total Exposure Assessment Methodology (TEAM) studies
    thermal desorption, 101, 160
    very volatile organics, 101–102
    VOCs, 99–102
Pesticide exposure models, 370, 471–481, *See also* Exposure models

Calendex™, 476–477
CalTOX, 476
CARES, 477–478
combining data, 478–479
co-occurring air pollutants, 480
dose estimation, 480
exposure routes, 473–475
human activity data, 472–473
Lifeline™, 477
SHEDS, 478
temporal and spatial variability, 479–480
Pesticides, 10, 83, 347–370, *See also specific pesticides*
absorption studies, 273–274
air monitoring methods, 350–354
at-risk subpopulations, 368
bioaccumulation, 310
breast milk contamination, 400
broadcast spraying, 366
carcinogens, 368–369
child and infant exposure, 349, 368
child exposure, 257
commercial building use, 349–350
contact-dislodgeable residue monitoring, 357–361
co-occurring pollutants, 480
definition, 348, 495
dietary ingestion, 310, 493–494
exposure biomarkers, 400–401
exposure estimation guidelines, 368
exposure risks, 368
handwipe methods, 361–363
health effects, 257, 347, 348–349, 368–370
in house dust, 321–322, 332, 333, 342, 348, 363–364
house dust sampling methods, 355–356
indoor sources, fate, and transport, 363–364
ingestion exposure, 475
misuse and poisonings, 350
nondietary, 311, 368, 475
persistence, 364, 495
personal monitors for measuring, 102
protective clothing efficacy, 270
regulation, 347, 348, 368, 492–496
effect on use trends, 495
in food, 493–494
inert ingredients, 496
registration, 347, 495–496
residential assessment guidelines, 369
residential use, 349–350
solute permeability, 262
subclassification, 348, 368
surface sampling, 357–361
USEPA's reference dose (*RfD*), 368, 369–370
use patterns, 347–348, 350
volatilization and air levels, 364–366
water contamination, 306–307
workplace exposure limits, 368
Pest strips, 364
Pets, 321
Phantom curve, 428, 437
Pharmacokinetic, defined, 277
Phenols, 153, 332, 396
Phenothrin, 366

Phenoxyethanol, 153
4-Phenylcyclohexene (4-PC), 153
o-Phenylphenol, 332, 363, 367, 368
Phillipsburg, New Jersey, 186
Phoenix, Arizona, 121
Phonophoresis, 256, 277
Photographic exposure, 56
Phthalates, 311, 321, 332
Pica behavior, 59, 321
Piezobalance, 107
Pilot light, 417, 428
α-Pinene, 148, 151, 153, 363
Pinocytosis, 305
Piperonyl butoxide, 363
Plant uptake of toxic metals, 308–309
$PM_{2.5}$, *See* Fine particulate matter; Particle exposure
$PM_{3.5}$, *See* Respirable suspended particles
PM_{10}, *See* Inhalable particulate matter
Policy issues, *See* Environmental laws and policies
Polychlorinated biphenyls (PCBs), 321, 381, 384, 387, 400
Polycyclic aromatic hydrocarbons (PAHs)
biomarker, 401
child and infant exposure, 334
combustion particles, 183
in house dust, 321, 332–335, 356
particle-bound PAH in tobacco smoke, 221–222, 436
personal monitors for measuring, 108–109
Polyurethane foam (PUF)
pesticide air sampling, 351
roller sampling method, 268, 357–358
Popcorn, 434–435
Population, defining, 65
Population exposure models, 131–132, *See also* Exposure models
Porcine, defined, 277
Positive matrix factorization (PMF), 190–191
Potential dose, 35, 47
Power plant emissions, 21
Preliminary Remediation Goals (PRG), 330, 333
Presidential election polls, 66
Preterm birth, 401
Priority lanes, 134–136
Probabilistic NEM for CO (pNEM/CO), 132
Probability-based sampling, 8, 65
data collection, 70
example, 72–74
National Human Exposure Assessment Survey (NHEXAS), 67–69
sample selection method, 66–67
sample size, 70
sampling frame, 67
sampling frame stratification, 71–72
scientific method, 70–71
statistical analysis, 72
supporting robust inferences, 70
TEAM studies, 101, 184
when required or not, 69–70
Progesterone, 273
Propoxur, 350, 363, 364, 365, 367, 368
Propylene glycol, 153
Protective clothing, 270

Proteomics, 402
Proton transfer reaction (PTR), 163, 166
Proximity effect, 422, 428
Public education, 24
    Master Home Environmentalist™ program, 339
Pulmonary region, 84, 92

Q

Quantitative structure-activity relationship (QSAR), 288

R

Radiobiological exposure, 37
Radon, 457, 499–500
Random Component Superposition (RCS), 12
Random sampling, 8, 67, 70–71
Real-time sampling and analysis techniques, 163–166, 292–295
Receptor-oriented approach, 3, 8, 12–14
Recreational exposures, 130–131
Rectangular time function, 417–418
Redwood City, California, 218–219
Reference dose (*RfD*), 368, 369–370
Relative Source Apportionment of Exposure (RSAE), 21–24
Research Triangle Park, North Carolina, 106, 150, 151
Residence time, 10, 91–92, 95, 416
Resistor model, 87
Resource Conservation and Recovery Act (RCRA), 498–499
Respirable suspended particles (RSPs, $PM_{3.5}$), 212, *See also* Secondhand tobacco smoke
    applications, 223–226
    atmospheric tracer, 202, 215–217
    food cooking/burning experiments, 434–441
    Habitual Smoker Model, 220
    other particulate sources and, 215
    personal daily secondhand smoke exposure, 214–215
    reduction due to regulation, 224
    smoking rates and emissions, 213–214
    time-varying concentrations, 221–222
Respiratory infections, 183
Respiratory tract regions, 84
Respiratory uptake, *See* Inhalation exposure
Reticular layer, defined, 277
RIOPA study, 188
Risk assessment
    aggregate and cumulative terms, 36
    aggregate pesticide exposure, 370
    dermal absorption measurement, 272
    exposure as "wasteland" of, 25
    exposure from foods for children or infants, 494
    exposure model applications, 47–48, 456–457, 478
    identifying target of risk, 4
    Superfund Act requirements, 330, 331, 333
    suspected carcinogens, 369
    USEPA guidelines, 369
    VOC carcinogens, 154, 220, 226, 228

Risk communication, 24–26
Riverside, California, 104, 184, 452
Roosevelt, Franklin D., 66
Rule of 1,000, 240

S

Safe Drinking Water Act (SDWA), 158, 287
Sample, defined, 66
Sample size, 70
Sampling, 184, *See also* Air sampling and analysis methods; Skin exposure sampling techniques
    judgmental samples, 69–70
    multistage probability design, 8
    presidential election polls, 66
    probability-based, 65–77, 184, *See also* Probability-based sampling
    random sampling, 8, 67, 70–71
    sample selection, 66–67
    scientific method, 70–71
Sampling frame, 67
    stratifying, 71–72
San Francisco Bay area, 22–24, 123
San Jose, California, 120, 123–124
Scandinavian air quality, 152
School buses, 238, 246
Scientific method, 70–71
"Sea of particles," 322
Seattle, Washington, 106, 187, 333, 334
Sebaceous glands, 260, 262, 277
Secondhand tobacco smoke, 83, 201–231, 430
    atmospheric tracers (nicotine and respirable suspended particles), 202, 215–217
    biomarkers (nicotine and cotinine), 210, 400
    child exposure, 212
    cigarette brands, 214
    cigarettes vs. cigars, 213
    cooking-related particle exposure vs., 440
    Delaware Air Quality Survey, 224–225
    dosimetry, 226–228
    exposure modeling, 434–436, 448–450, *See also* Exposure models
        active smoking count, 218–219
        habitual smoker model, 220
        indoor cigar smoke dispersion experiment, 422–424
        Sequential Cigarette Exposure Model, 436
        time-averaged models, 217–220
        time-varying concentrations, 221–222
    future issues, 228–229
    health effects, 208
    hospitality industry, 223–224
    in-vehicle exposure, 225–226
    mechanically ventilated buildings, 223
    microenvironmental air pollution, 212
        concentrations, 212–213
        daily exposure to particulate air pollution, 214–215
        smoking rates and emissions, 213–214
    misclassification problems, 210–212
    mortality estimates, 202, 203
    occupational exposures, 210, 216, 223–224, 228, 229

outgassing of deposited tars, 228–229
pollutants from, 152–154, 203–208
    CO, 225–226
        particle-bound PAH, 221–222, 436
        particle exposures vs. power plants, 21
    public policy and education outcomes, 24
    research initiatives, 202
    residential exposures, 223
    smoking bans, 24, 228
    smoking prevalence and trends, 209
Selected ion flow tube (SIFT), 163, 166
Selenium, 309
Self-pollution, 238, 246
Semivolatile organics (SVOCs), 347, *See also* Pesticides
    biomarkers, 400–401
    personal monitors for measuring, 102
Seoul, South Korea, 151
Sequential Cigarette Exposure Model (SCEM), 436
Settling velocity, 89
SHAPE model, 12, 131–132
SHEDS, 12, 458, 478
Showering or bathing, VOC dermal exposure and, 158,
            285–296, *See also* Water immersion dermal
            exposures
    direct absorption measure using continuous breath
            analyzer, 292–295, 399
    exposure reduction measures, 159
    indirect absorption measures, 292
    water temperature effects, 286, 287, 294
Sick building remediation, 152
Sick building syndrome (SBS), 148, 149, 152
Simazine, 364
Simulation of Human Activities and Exposures (SHAPE)
            model, 12, 131–132
Skin exposure sampling techniques, 265, *See also* Dermal
            exposure
    fluorescent tracer, 266–267
    surface sampling, 267–269
    surrogate skin, 265–266
    washing, rinsing, or wiping, 266
Skin loading, 44
Skin permeability coefficients, 287–289
Skin structure and function, 259–262, 304
Skin surface properties, dermal dose and, 263
Smoking, *See* Secondhand tobacco smoke
Smoking bans, 24, 228
Smoking regulation issues, 202
Social models, 465
Soil adherence to skin, 269
Soil composition of particulate matter, 184
Soil matrix, defined, 277
Soil particle, defined, 277
Solid-phase microextraction (SPME), 161–162, 167
Solvent effects on dermal dose, 262
Solvent extraction, 160, 350
Sorbents, 99, 160–161, 351
    multisorbent systems, 102, 161
    polyurethane foam, 102
    Tedlar®, 103
    Tenax™, 100–102
Source, defined, 59

Source apportionment of exposure, 21–24
Source-off periods, 422
Source-on periods, 421
Source-oriented approach, 3, 12–13, 25
Source proximity effect, 422, 428
South Korea, 151
Space heating, 130
Spacecraft, 152, 161
Spatial coefficient of variation, 422
Spatially averaged exposure, 42, 43
Spatially integrated exposure, 43
Spectroscopy, defined, 277
Sporting events, 131
Springfield, Massachusetts, 102, 364, 366–367, 369
Squamous, defined, 277
Stanford University exposure analysis curriculum, 27
Statistical sampling techniques, 8
Step time function, 417
Stochastic Human Exposure and Dose Simulation
            (SHEDS), 12, 458, 478
Stratification, 71–72
Stratified, defined, 277
Stratified random sampling, 8, 71
Stratum corneum, 53, 259, 277, 285
    function and structure, 260–261
    shedding and hydration, 261–262
    VOC diffusion, 399
Stratum germinativum, 259, 277
Stratum granulosum, 259, 277
Stratum lucidum, 259, 277
Stratum spinosum, 259, 277
STREET model, 137
Stressor, 39, 59
Styrene, 18, 396
Subchronic exposure, 50, 59
Sulfate composition of particulate matter, 184
Sulfur-based infiltration factor model, 189, 191
Sulfur dioxide (SO_2), 88
    criteria air pollutants, 82, 488
    personal monitor, 105
Sulfur hexafluoride (SF_6), 422, 424–428
Sulfur pesticides, 347
SUMMA polishing techniques, 162
Superfund Amendments and Reauthorization Act (SARA),
            337, 490–491
    risk assessment requirement, 330–331
Superfund Preliminary Remediation Goals (PRG), 330, 333
Superfund sites, 25, 490
    residential indoor air quality comparison, 319, 320,
            331–332, 336
Surface roughness, defined, 277
Surface sampling techniques, 267–269, 357–361
Surfactant
    dermal dose and, 262
    lung, 84
    skin wiping, 266
Surgeon General, 202
Surrogate skin techniques, 265–266
Swimming and dermal exposure, 286, *See also* Water
            immersion dermal exposures
    dermal absorption measures, 292

water temperature effects, 294

T

Target of exposure
    defined, 39, 59
    example, 49
TCDD, 380, 383
Tedlar®, 103, 120
Temperature effects on chloroform absorption, 286, 287,
        294
Tenax™, 100–102, 150, 161
Termiticides, 349, 350, 364, 368, 369
4-*Tert*-butylphenol, 332
2,3,7,8-Tetrachlorodibenzo-*p*-dioxin (TCDD), 380, 383
Tetrachloroethylene, 156–157, 396, 399
    absorption studies, 288
    breath measurements, 397
    carcinogenicity, 148, 154
    reducing exposure, 159
    sources and uses, 10, 18, 101
    water contamination, 307
Tetrahydrofuran, 153
Tetramethrin, 366
Texanol, 153
THERdbASE, 370
Thermal desorption, 101, 160
Thoracic region, 84
Three-spot test, 324–325, 338
Time-averaged exposure, 35, 44, 52, 55, 59
    environmental tobacco smoke, 217–220
    indoor air quality model, 429
Time-integrated exposure, 35, 38, 44, 50, 52, 55, 59
Time-integrated spatially integrated dose, 45
Time profile, 44, 52, 55, 59
Times Beach, Missouri, 379
Toasters, 183
Tobacco smokers, 202, *See also* Secondhand tobacco
        smoke
    breath benzene levels, 397–398
    carboxyhemoglobin levels, 117
    prevalence and trends, 209, 228
    rectangular source time function, 418
    smoking regulation, 24, 202, 228
Toilet deodorants, *See* Bathroom deodorants
Tollbooth workers, 154
Toluene, 151, 152, 153, 396
Tomato, 55
Toronto, Canada, 105
Total exposure
    defined, 59
    estimating for CO, 122–123
Total Exposure Assessment Methodology (TEAM) studies,
        8, 13, 148, 150, 156, 460
    indoor air quality findings, 9–10
    indoor chloroform findings, 286
    particle exposure studies (PTEAM), 48, 104, 184–186,
        452, 462
    particle monitors, 104
    personal exposure monitors, 48, 101

pesticide findings, 102
    probability-based sampling design, 101
Total human exposure approach, 6–12, 256
Total suspended particles, 104, 212
Toxaphene, 306
Toxic equivalency concentration (TEQ), dioxins, 383–388
Toxic equivalency factors (TEFs), 383
Toxic metals, *See* Metals
Toxic Organic (TO) Compendium Methods, 163, 164
Toxics Release Inventory, 25
Toxic Substances Control Act (TSCA), 499–501
Tracheobronchial region, 84, 92
Track-in pollution, 322, 332, 336
Transportation Conformity Rule (TCR), 134–136
Transportation investments, commuter exposure outcomes,
        134–136
1,1,1-Trichloroethane, 10, 17–19, 166, 307–308
Trichloroethylene, 18, 148, 288
Triclopyr, 350
n-Tridecane, 153
Trigeminal nerve, 149
Trihalomethanes (THMs), 158, 287, 306, 307–308, *See also*
        Chloroform
TXIB, 153

U

Ultrafine particles, 183
Uniform mixing assumptions, 411, 413, 421–424, 462–463
    experiments, 422–424
United Kingdom, 308
Upholstery cleaning, 338
Uptake (absorption), 50, 53, 57, 59, 82, 94, *See also* Dermal
        exposure; Inhalation exposure
    CO model, 92–93
    gaseous pollutants, 84–88
    particulate pollutants, 89–92
    respiratory, 83–84
    VOC model, 94–96
Uptake dose, 54
Urine analysis, 167
U.S. Clean Air Act Amendments, *See* Clean Air Act (CAA)
        and amendments
U.S. Environmental Protection Agency (USEPA), 488
    "aggregate" and "cumulative" term usage, 36
    aggregate exposure quantification mandate, 256–257
    Chemical Substance Inventory, 499–500
    CO exposure models, 132
    CO standards, 119–120
    consolidated human activity database, 451
    dermal exposure model, 289–290
    dioxin regulation, 384, 390
    exposure model resources, 460
    exposure models, 131–132, 458
    human activity data, 14, 132–133, 159
    indoor air quality and, 489
    NAAQS, *See* National Ambient Air Quality Standards
    new and existing chemicals regulation, 499–501
    particle classification system, 82, 104, 183
    particle exposure research program, 186–187

personal monitor development, 99, 100
pesticide air sampling method, 352
pesticide classification, 348, 368
pesticide reference dose, 368, 369–370
pesticide regulation, 347, 348–349, 495
residential pesticide assessment guidelines, 369
setting food tolerance levels, 349
surface sampling methods, 359–361
TEAM studies, *See* Total Exposure Assessment
        Methodology (TEAM) studies
THERdbASE, 370
TO Compendium methods, 163, 164
toxic metal regulation, 307
Transportation Conformity Rule, 134
vehicle emissions models, 115
VOC standards for water, 307–308
U.S. National Aeronautics and Space Agency (NASA),
        152

V

Vacuum cleaning, 337–338, 356
    efficacy, 338–339
    low-income families and, 339–340
    3-spot test, 324–325, 338
Vacuum sampling methods, 319–321, 355–356
Valdez, Alaska, 16, 155
Vapor pressure and volatilization of pesticides, 364–366
Vehicle, defined, 278
Ventilation rate, 243–244, *See also* Air exchange rate
    Habitual Smoker Model, 220
    indoor air quality model, 415, 416, 418
    measuring, 424–428
    outdoor winds and, 428
    predicted concentration time series, 418
Ventilation standards, 223, 224, 457
Very volatile organics (VVOCs), 101–102, 161
Vinyl chloride, 102, 148
Vinyl flooring, 153
Volatile organic compound (VOC) exposure, 83, 147–168,
        See also Benzene; Chloroform; *para-*
        Dichlorobenzene; Pesticides;
        Tetrachloroethylene; *other specific*
        compounds
    activities increasing exposures, 17–21
    air pollutants, 150–157, *See also specific pollutants*
    air sampling and measurement, 160–165
    biomarkers, 396–400
    body fluids, 165–167
    breast milk contamination, 399
    breath measurements, 163–165, 292–295, 396–399
    carcinogenicity, 154, 159
    Clean Air Act standards, 489
    common indoor sources, 10
    dermal exposure and activities in water, 285–296,
            399–400, *See also* Water immersion dermal
            exposures
    drinking water contaminants, 158, 399
    European urban studies, 151
    exposimeter, 25–26

exposure reduction measures, 153, 159
exposure while showering, *See* Showering or bathing,
        VOC dermal exposure and
health effects, 148–150, 159
indirect dermal absorption measures, 292
indoor sources and levels, 150–154
indoor vs. outdoor exposures, 154, 397
Love Canal residences, 337
most prevalent in breath and blood, 398
new buildings, 152
occupational exposures, 148, 154
olfactory analysis, 165
outdoor vs. indoor levels, 154
parameters affecting dermal absorption, 287–291
personal monitors for measuring, 99–102
pesticides, 347, *See also* Pesticides
policy implications, 159
public risk awareness, 153, 158–159
residence time and, 95
secondhand tobacco smoke, 152
sick building remediation, 152
sick building syndrome, 148, 149
single-media analysis, 330
skin permeability coefficients, 287–289
source apportionment, 22–24
sources, 101
synthetic fragrances, 492
TEAM studies, 101, 148, 150, 397, *See also* Total
        Exposure Assessment Methodology (TEAM)
        studies
uptake models, 94–96
USEPA standards for water, 307–308
USEPA TO Compendium methods, 163, 164
water contaminants, 307–308, *See also* Water
        contaminants

W

Washington, DC, 12, 72–74, 103, 118, 123, 126, 138–139,
        151
Water contaminants, 157, 158, 287, 303, 306–308, 326, 399
Water immersion dermal exposures, 158, 285–296
    absorption measure using continuous breath analyzer,
            292–295, 399
    chloroform sources, 286
    indirect dermal absorption measures, 292
    lag time, 289–290, 295
    parameters affecting absorption, 287–291
    uptake estimation, 53
    water temperature effects, 286, 287, 294
Wetting agents, 266
Whole-body samplers, 265–266
Winds and indoor air change rate, 428
Wood preservatives, 347, 368, *See also* Formaldehyde
Worker productivity impacts, 148, 150

X

Xenobiotics-metabolizing enzymes, 263

X-ray fluorescence analysis, 189
Xylenes, 151, 152, 397
    activities increasing exposures, 17, 18

Y

Yuma County, Arizona, 333